Radiation Therapy Dosimetry

A Practical Handbook

Radiation Therapy Dosimetry

A Practical Handbook

Edited By

Arash Darafsheh

CRC Press
Taylor & Francis Group
Boca Raton London New York

CRC Press is an imprint of the
Taylor & Francis Group, an **informa** business

First edition published 2021
by CRC Press
6000 Broken Sound Parkway NW, Suite 300, Boca Raton, FL 33487-2742

and by CRC Press
2 Park Square, Milton Park, Abingdon, Oxon, OX14 4RN

© 2021 Taylor & Francis Group, LLC

CRC Press is an imprint of Taylor & Francis Group, LLC

Library of Congress Cataloging-in-Publication Data

Names: Darafsheh, Arash, editor. Title: Radiation therapy dosimetry : a practical handbook / edited by Arash Darafsheh. Description: First edition. | Boca Raton : CRC Press, 2021. | Includes bibliographical references and index. Identifiers: LCCN 2020040810 (print) | LCCN 2020040811 (ebook) | ISBN 9781138543973 (hardback) | ISBN 9781351005388 (ebook) Subjects: LCSH: Radiotherapy. | Radiation dosimetry. Classification: LCC RM849 .R325 2021 (print) | LCC RM849 (ebook) | DDC 615.8/42--dc23 LC record available at https://lccn.loc.gov/2020040810LC ebook record available at https://lccn.loc.gov/2020040811

ISBN-13: 9781138543973 (hbk)
ISBN-13: 9780367686772 (pbk)
ISBN-13: 9781351005388 (ebk)

DOI: 10.1201/9781351005388

Typeset in Minion
by KnowledgeWorks Global Ltd.

Contents

Preface

This book is intended to cover the everyday use and underlying principles of radiation dosimeters used in radiation oncology clinics. It provides an up-to-date reference spanning the full range of current modalities with emphasis on practical know-how. The main audience is medical physicists, radiation oncology physics residents, and medical physics graduate students.

This book contains 30 chapters in four main parts. The chapters in Part I deal with the fundamentals of radiation dosimetry, basic physics of various dosimeters, practical dosimetry considerations, and Monte Carlo applications in medical physics. Part II covers the brachytherapy dosimetry. The chapters in Part III deal with external beam radiation therapy dosimetry; dosimetry of cutting edge techniques from radiosurgery to MR-guided systems to small fields and proton therapy as well as advanced range verification techniques in proton therapy are addressed in Part III. Part IV is on the dosimetry of imaging modalities in radiotherapy.

I sincerely thank all of the authors for their outstanding contributions to this book; their efforts have brought all of the chapters to a very high standard. I am grateful to my mentors, mentees, and colleagues in Washington University in St. Louis and University of Pennsylvania. I am also grateful to the publishing team for their assistance, guidance, and advice throughout the project.

Arash Darafsheh
Washington University in St. Louis, St. Louis, Missouri

About the Editor

Arash Darafsheh, Ph.D., is an associate professor of Radiation Oncology, a certified medical physicist by the American Board of Radiology (ABR), and the PI of the Optical Imaging and Dosimetry Lab at the Department of Radiation Oncology at the Washington University School of Medicine in St. Louis. He holds Ph.D. and M.Sc. in Optical Science and Engineering, and an M.Sc. in Radiation Medicine Engineering. His current research interests include optical methods in medical physics, detector development for radiotherapy, ultra-high dose rate FLASH radiotherapy, photodynamic therapy, and super-resolution microscopy. He has served as a mentor for many graduate students, postdoctoral research fellows, and clinical residents. He has published over 90 journal and conference papers, six book chapters, and one patent. He has been awarded research grants from the National Institutes of Health (NIH) and the American Association of Physicists in Medicine (AAPM). He is a member of AAPM and senior member of the Optical Society of America (OSA) and SPIE-the international society for optics and photonics. He has served as an associate editor for *Medical Physics* and as a reviewer for numerous scientific journals.

Contributors

John A. Antolak
Mayo Clinic
Rochester, Minnesota

Milad Baradaran-Ghahfarokhi
Washington University School of Medicine
St. Louis, Missouri
and
Department of Radiation Oncology
Vanderbilt University Medical Center
Nashville, Tennessee

Riccardo Dal Bello
German Cancer Research Center - DKFZ
Division of Biomedical Physics in Radiation Oncology
University of Heidelberg
Department of Physics and Astronomy
Heidelberg, Germany

Giordano Biasi
Centre for Medical Radiation Physics
School of Physics
University of Wollongong
Wollongong, NSW, Australia
Peter MacCallum Cancer Centre
Melbourne, VIC, Australia

Douglas Bollinger
Department of Radiation Oncology
University of Pennsylvania
Philadelphia, Pennsylvania

Sarah Boswell
Accuray Incorporated
Madison, Wisconsin

Raman Caleb
Department of Radiation Oncology
Washington University School of Medicine
St. Louis, Missouri

Xinyuan Chen
Department of Radiation Oncology
Washington University School of Medicine
St. Louis, Missouri

Wesley S. Culberson
Department of Medical Physics
School of Medicine and Public Health
University of Wisconsin – Madison
Madison, Wisconsin

Arash Darafsheh
Department of Radiation Oncology
Washington University School of Medicine
St. Louis, Missouri

Christopher L. Deufel
Radiation Oncology Department
Mayo Clinic
Rochester, Minnesota

Larry A. DeWerd
Department of Medical Physics
School of Medicine and Public Health
University of Wisconsin-Madison
Madison, Wisconsin

Sonja Dieterich
University of California Davis
Department of Radiation Oncology
Sacramento, California

Eric S. Diffenderfer
Department of Radiation Oncology
University of Pennsylvania
Philadelphia, Pennsylvania

George X. Ding
Department of Radiation Oncology
Vanderbilt University School of Medicine
Nashville, Tennessee

Nesrin Dogan
Department of Radiation Oncology
University of Miami
Miami, Florida

Dongsu Du
Department of Radiation Oncology
City of Hope Medical Center
Duarte, California

Thomas Dvergsten
Department of Radiation Oncology
Washington University School of Medicine
St. Louis, Missouri

Hermann Fuchs
Department of Radiation Oncology
Division of Medical Radiation Physics
Medical University Vienna
Wien, Austria

Christoph Fuerweger
European Cyberknife Center Munich
Munich, Germany

Dietmar Georg
Department of Radiation Oncology
Division of Medical Radiation Physics
Medical University Vienna
Wien, Austria

Carri Glide-Hurst
Department of Human Oncology
University of Wisconsin-Madison
Madison, Wisconsin

Rachael L. Hachadorian
Thayer School of Engineering
Dartmouth College
Hanover, New Hampshire

Nicholas Hardcastle
Centre for Medical Radiation Physics
University of Wollongong
Wollongong, NSW, Australia
and
Peter MacCallum Cancer Centre
Melbourne, VIC, Australia

Nazanin Hoshyar
New York-Presbyterian/Queens Department of
    Radiation Oncology
Flushing, New York

Nisy Elizabeth Ipe
Consultant, Shielding Design, Dosimetry and
    Radiation Protection
San Carlos, California

Perry B. Johnson
University of Florida Health Proton
    Therapy Institute
Jacksonville, Florida

Kevin C. Jones
Department of Radiation Oncology
Rush University Medical Center
Chicago, Illinois

Rao Khan
Department of Radiation Oncology
Washington University School of Medicine
St. Louis, Missouri

Michele M. Kim
Department of Radiation Oncology
University of Pennsylvania
Philadelphia, Pennsylvania

Nels C. Knutson
Department of Radiation Oncology
Washington University School of Medicine
St. Louis, Missouri

Tomas Kron
Department of Physical Sciences
Peter MacCallum Cancer Centre
Melbourne, VIC, Australia
and
Sir Peter MacCallum Depatment of Oncology
University of Melbourne
Parkville, VIC, Australia
and
Centre for Medical Radiation Physics
University of Wollongong
Wollongong, NSW, Australia

Stephen F. Kry
Department of Radiation Physics
The University of Texas MD Anderson Cancer Center
Houston, Texas

Joerg Lehmann
Department of Radiation Oncology
Calvary Mater Newcastle
Waratah, NSW, Australia
and
Institute of Medical Physics
University of Sydney
Sydney, NSW, Australia
and
School of Mathematical and Physical Sciences
University of Newcastle
Newcastle, NSW, Australia

Manju Liu
Department of Radiation Oncology
William Beaumont Hospital
Royal Oak, Michigan

Ruirui Liu
Department of Radiation Oncology
Washington University School of Medicine
St. Louis, Missouri

Peta Lonski
Peter MacCallum Cancer Centre
Melbourne, VIC, Australia

Jessica Lye
Australian Clinical Dosimetry Service
Australian Radiation Protection and Nuclear Safety
  Agency
Olivia Newton-John Cancer Wellness and Research
  Centre, Austin Health
Melbourne, Australia

C.-M. Charlie Ma
Department of Radiation Oncology
Fox Chase Cancer Center
Philadelphia, Pennsylvania

Paulo Magalhaes Martins
German Cancer Research Center - DKFZ
Division of Biomedical Physics in Radiation Oncology
Heidelberg, Germany
and
Instituto de Biofísica e Engenharia Biomédica - IBEB
Faculty of Sciences of the University of Lisbon
Lisbon, Portugal

Sina Mossahebi
Department of Radiation Oncology
University of Maryland School of Medicine
Baltimore, Maryland

Firas Mourtada
Radiation Oncology
Helen F. Graham Cancer Center and Research Institute
Christiana Care Health System
Newark, Delaware

Jennifer O'Daniel
Department of Radiation Oncology
Duke University School of Medicine
Durham, North Carolina

Katia Parodi
Department of Experimental Medical Physics
Ludwig-Maximilians-Universität München
Garching b. München, Germany

Brian W. Pogue
Thayer School of Engineering
Dartmouth College
Hanover, New Hampshire

Francisco J. Reynoso
Department of Radiation Oncology
Washington University School of Medicine
St. Louis, Missouri

Mark J. Rivard
Brown University
Rhode Island Hospital
Providence, Rhode Island

Anatoly B. Rosenfeld
Centre for Medical Radiation Physics
School of Physics
University of Wollongong
Wollongong, NSW, Australia

Joao Seco
German Cancer Research Center - DKFZ
Division of Biomedical Physics in Radiation
    Oncology
University of Heidelberg
Department of Physics and Astronomy
Heidelberg, Germany

Brayden Schott
Department of Radiation Oncology
Washington University School of Medicine
St. Louis, Missouri

Blake R. Smith
Department of Radiation Oncology
University of Iowa
Iowa City, Iowa

Karen Chin Snyder
Department of Radiation Oncology
Henry Ford Health System
Detroit, Michigan

Matthew T. Studenski
Department of Radiation Oncology
University of Miami
Miami, Florida

Baozhou Sun
Department of Radiation Oncology
Washington University School of Medicine
St. Louis, Missouri

Reza Taleei
Department of Radiation Oncology
Sidney Kimmel Medical College at Thomas Jefferson
    University
Philadelphia, Pennsylvania

Irwin I. Tendler
Thayer School of Engineering
Dartmouth College
Hanover, New Hampshire

Georg Weidlich
Zap Surgical
San Carlos, California

Ning Wen
Department of Radiation Oncology
Henry Ford Health System
Detroit, Michigan

Tianyu Zhao
Department of Radiation Oncology
Washington University School of Medicine
St. Louis, Missouri

I

Radiation Dosimeters and Dosimetry Techniques

CHAPTER 1

Fundamentals of Radiation Physics and Dosimetry

Blake R. Smith

University of Iowa
Iowa City, Iowa

Larry A. DeWerd

University of Wisconsin
Madison, Wisconsin

CONTENTS

1.1 ABSORBED DOSE

One of the most important quantities that concerns the practice of medical physics is dose. This quantity is prescribed by physicians to treat tumors, set by radiation safety officers as exposure limits to workers, and referenced by radiobiologists while performing cell irradiation studies. This quantity, however, is rather arbitrary and ill-defined by itself. Dose, D, is defined as a point quantity from the fundamental quantities of energy and mass as:

$$D = \frac{dE}{dm} \rightarrow \frac{\Delta E}{\Delta m}, \left[\frac{J}{kg} \right] \qquad (1.1)$$

In a theoretical sense, this point quantity refers to the energy deposited within an infinitesimal amount of mass (and thus volume). Realistically, energy deposition from atomic and subatomic events is discrete with respect to an infinitesimal volume for which the definition of dose is better represented as an average over a specified space leading to the adjacent expression in Equation 1.1. For example, a whole-body dose of 4 Gy has about a 50% chance of killing an adult human in 60 days [1]. On the other hand, patients undergoing radiation therapy often receive 40 Gy or more to their tumors, but in this context the dose is fairly localized with a large amount of energy deposited within a smaller volume.

It is also important to specify the medium that the dose is referred to. Dose to water versus dose to air imply subtle, but extremely important, differences in the amount of energy expended in the medium as well as how the energy was transferred. While different, both quantities are reported using the same units of Gray (Gy) which is defined as $1\,\text{Gy} = 1\,\text{J kg}^{-1} = 6.24 \times 10^9\,\text{MeV g}^{-1}$. The subject of absorbed dose differences and dependences is paramount to the understanding of detector response, which will be a subject of further discussion later on.

Calculating dose during discrete particle transport through a medium can be difficult where only a handful of scenarios exist that can be solved analytically. However, more complicated problems can be solved using Monte Carlo methods where the path a particle experiences is simulated discretely. Dose is only delivered to matter through charged particles. Uncharged particles, such as photons and neutrons, will traverse through a medium unimpeded until either an elastic or inelastic interaction occurs. During inelastic interactions, energy is released from the uncharged particles to the medium potentially transferring kinetic energy to charged particles. Those liberated charged particles expend their kinetic energy to the surrounding medium putting other charged particles into motion or producing uncharged particles and the process repeats. Solving these radiation transport problems requires an in-depth understanding of the types of interaction that can occur, the probability of their occurrence, and the kinematics following these interactions.

Let us consider the definition of dose a little more closely. Of concern is the transfer of energy to matter within a medium from charged particles. Following the definition of dose provided in Equation 1.1, dose to a fragment of matter within the medium can be determined from the kinetic energy loss, ΔE, that a particle experiences across a fragment of the medium with mass, m. If we model the kinetic energy loss of the particle in discrete, straight-line steps of length Δx, then we can relate the total path, s, that the particle travels within the fragment of matter to the kinetic energy lost by that particle and imparted to the medium. This is referred to as the stopping power of the medium, $\left(\frac{\Delta E}{\Delta x}\right)$, which is fundamentally related to the force acted on by the medium to slow the particle down. This loss of energy is then related to dose by

$$D = \frac{s}{m} \cdot \left(\frac{\Delta E}{\Delta x}\right) = \frac{s}{V} \cdot \left(\frac{\Delta E}{\rho \Delta x}\right) \tag{1.2}$$

$$\rightarrow D = \frac{1}{A} \cdot \left(\frac{\Delta E}{\rho \Delta x}\right) = \phi \cdot \left(\frac{\Delta E}{\rho \Delta x}\right) \text{ for multiparticle scenario} \tag{1.3}$$

where the normalization of the particle's stopping power to the density of the medium, ρ, is referred to as the mass stopping power. Dose is rarely defined or calculated from a single particle as a single particle's trajectory can vary substantially if the particle were to travel from the same initial conditions. The energy loss on this scale is largely stochastic in nature. The progression of Equation 1.3 illustrates how the calculation and definition of dose change from the single-particle scenario to a more familiar scenario of an incident fluence of charged particles, ϕ, upon an incident area, A, and volume, V, of the medium. The above example is referred to as the thin-film approximation, which subtly assumes no changes in the rate of energy loss of the particle as it traverses through the medium. While fine for our conceptual discussion of dose, more consideration is necessary to comprehensively describe the energy loss of charged particles. For further discussion, consider a general relation between dose and the transport of charged particles,

$$D = \int_0^{E_{\max}} \phi'(E) \left(\frac{dE}{\rho dx}\right)_c dE \tag{1.4}$$

For the calculation of dose, $\phi'(E)$ in Equation 1.4 is used for the differential fluence with respect to the energy of the charged particles and $\left(\frac{dE}{\rho dx}\right)_c$ is the portion of the stopping power responsible for collisional energy losses from the incident charged particle undergoing multiple Coulombic scattering with the surrounding orbital electrons in the medium. Another assumption necessary to allow us to calculate dose accurately is to limit our energy fluence spectrum from energetic, knock-on electrons, known as δ-rays, which are produced within our region of interest and deposit their energy elsewhere. This restriction is also referred to as charged particle equilibrium (CPE) and is necessary for the calculation of dose. Simply, CPE requires that the charged particle fluence which enters our region of interest also leaves the region. In a sense, CPE is a spatial assumption that the charged particle fluence remains constant throughout the region of interest and the point that dose is defined at. The problem becomes more complicated to determine the differential fluence spectrum at a point, requiring a more complete understanding of charged particle transport through matter.

1.2 CHARGED PARTICLE TRANSPORT

Elastic and inelastic collisions are the two fundamental types of interactions that occur in a two-body system consisting of an incident projectile and target particle. While the products and participants can vary widely among all atomic and subatomic particles, the general trends are consistent. Elastic collisions are characterized by a preservation of the total kinetic energy and momentum of the system before and after the collision. A corollary to this assumption is that the products after the collision must remain identical to the constituents before the collision. Inelastic collisions do not preserve the equality of products before and after the collision. This often leads to the loss of kinetic energy of the system to the form of intrinsic energy, which includes the production of daughter particles, excitation of atomic or subatomic particles, and the emission of bremsstrahlung. Inelastic interactions can be further categorized by their distance from the central atom from which they occur, known as their impact parameter, b. Soft collisions, when $b \gg$ atomic radius, result in small energy losses of the primary particle traversing through the medium. The majority of these energy losses are a result of the atomic excitation or ionization of valence electrons from the atoms within the medium. Hard collisions occur near the atom's atomic radius, $b \approx$ atomic radius, and result in the liberation of inner-shell orbiting electrons. Nuclear electric field interactions are the least common of the three and occur when the incident charged particle is within the nuclear radius of the atom, $b \ll$ atomic radius. The production of bremsstrahlung is classified as a nuclear electric field interaction.

The mass stopping power, initially utilized in Equation 1.2, can be divided into soft and hard collisional losses in addition to a radiative loss component

$$
\begin{aligned}
\frac{dE}{\rho dx} &= \left(\frac{dE}{\rho dx}\right)_c + \left(\frac{dE}{\rho dx}\right)_r \\
&= \left[\left(\frac{dE}{\rho dx}\right)_s + \left(\frac{dE}{\rho dx}\right)_h\right]_c + \left(\frac{dE}{\rho dx}\right)_r
\end{aligned}
\tag{1.5}
$$

where the subscripts s, h, and r correspond to the soft, hard, and radiative components of the stopping power, respectively. The mass stopping powers defined in Equation 1.5 are conventionally given in units of MeV cm^2 g^{-1}. The portion of radiative energy loss is carried away from the region of interest in the form of photons. Therefore, in our calculation of dose, we are primarily concerned with the contributions of soft and hard collisional stopping powers. Derived by Bethe [2], assuming that the kinetic energy is much greater than the electrons orbital energy of an atom (referred to as the Born approximation), the soft collisional stopping power of an incident charged particle with charge ze and the surrounding medium with an effective atomic number of Z is:

$$
\left(\frac{dE}{\rho dx}\right)_s = 2\pi r_0^2 m_e c^2 \left(\frac{N_A Z}{A}\right)\left(\frac{z^2}{\beta^2}\right)\left[\ln\left(\frac{2m_e c^2 \beta^2}{I^2(1-\beta^2)}H\right) - \beta^2\right]
\tag{1.6}
$$

where $\left(\frac{N_A Z}{A}\right)$ is effectively the number of electrons per unit mass in the medium defined from Avogadro's constant, N_A, and the mass number of the medium, A. The quantity $\left(\frac{z^2}{\beta^2}\right)$ describes the ratio of the atomic number of the incident charged particle, z, and the Lorentz factor, $\beta = v/c$, associated with the incident charged particle's velocity, v, relative to the speed of light, c. The product of the classical electron radius, mass of an electron, m_e, and Avogadro's constant is sometimes condensed as a single constant where $2\pi r_0^2 m_e c^2 N_A \approx 0.1535$ MeV cm^2 g^{-1}. The variable H is an arbitrary cutoff used to separate the evaluation of hard and soft collisions. The right portion of the equation, contained in square brackets, is an approximation of the change in kinetic energy experienced by the traversing particle from the force acted upon it from the electric field of nearby orbital electrons. This is classically integrated as a function of distance from the electrons, assuming they are populated uniformly from the traversing particle. An overview of the derivation of this approximation is beyond the scope of this discussion but can be found in most modern graduate physics textbooks [2–4]. An important term that arises within this component of Equation 1.6 is the mean excitation potential, I, which is the mean of all excitation and ionization potentials in the medium. A rule of thumb is that $I \sim 10\,Z$, except in cases for mediums with low atomic number. The component of the collisional stopping power resulting from hard collisions is given by

$$\left(\frac{dE}{\rho dx}\right)_h = 2\pi r_0^2 m_e c^2 N_A \frac{Z}{A}\frac{z^2}{\beta^2}\left(\ln\left(\frac{E_{max}}{H}\right)-\beta^2\right) \quad (1.7)$$

where E_{max} in the hard collision portion of the total mass collisional stopping power is the maximum energy that can be transferred in a head-on collision with an atomic electron. It can be approximated for a heavy charged particle ($m_{particle} \gg m_e$) traversing the medium as

$$E_{max} = 2m_e c^2\left(\frac{\beta^2}{1-\beta^2}\right) \quad (1.8)$$

Accounting for hard collisions, the mass collisional stopping power for a heavy charged particle is given by

$$\left(\frac{dE}{\rho dx}\right)_c = 2\pi r_0^2 m_e c^2 \left(\frac{N_A Z}{A}\right)\left(\frac{z^2}{\beta^2}\right)$$

$$\times\left[\left[\ln\left(\frac{2m_e c^2\beta^2}{I^2(1-\beta^2)}H\right)-\beta^2\right]\right.$$

$$\left.+\left[\ln\left(\frac{E_{max}}{H}\right)-\beta^2\right]\right],$$

and simplifying,

$$\left(\frac{dE}{\rho dx}\right)_c = 4\pi r_0^2 m_e c^2 \left(\frac{N_A Z}{A}\right)\left(\frac{z^2}{\beta^2}\right)\left[\ln\left(\frac{2m_e c^2\beta^2}{I(1-\beta^2)}\right)-\beta^2\right] \quad (1.9)$$

While not shown above, there are further corrections which have been derived for this expression of stopping power to account for relativistic scaling, shell correction to account for decreased interaction with K-shell electrons at lower particle velocities, and polarization effects of dense media.

The final component of the total stopping power is the mass radiative stopping power. According to Podgorsak [4], this expression for electrons or positrons is

$$\left(\frac{dE}{\rho dx}\right)_r = \frac{1}{137}r_0^2 Z^2 \frac{N_A}{A}\left(E_{K_0}+m_e c^2\right)B_{rad} \quad (1.10)$$

where B_{rad} is a slow-changing function slightly dependent on the atomic number, and E_{K_0} is the initial total kinetic energy of the light charged particle. For an initial kinetic energy of 1 MeV, 10 MeV, and 100 MeV, the corresponding values of B_{rad} are approximately 6, 10,

and 15, respectively. Note that the above expression is only valid for electrons and positrons. The generation of Bremsstrahlung from heavy charges particles is negligible as the intensity of the emitted Bremsstrahlung varies inversely with the square of the incident particle's mass. For instance, while a carbon ion would have a charge number 12 times that of an electron, the amount of Bremsstrahlung produced is still minuscule relative to an electron since the mass of a carbon nucleus is over 22,000 times larger than the electron's mass.

Our current treatment of stopping power has ignored the presence of δ-rays produced from hard collisions. If we have δ-ray CPE, then this is not an issue. However, since the condition of δ-ray CPE is not commonly satisfied, our treatment and calculation of dose to small volumes must account for the energy carried away from the region of interest. The restricted stopping power is the portion of the total collision stopping power which includes all of the soft collision energy losses but only hard collision losses resulting in delta rays of kinetic energy less than Δ. This assumes that any δ-rays produced below this threshold deposit their energy locally within the defined region of interest and are related to the unrestricted collisional stopping power as

$$\lim_{\Delta \to E_{K_0}}\left(\frac{dE}{\rho dx}\right)_\Delta = \left(\frac{dE}{\rho dx}\right)_c \quad (1.11)$$

Restructuring Equation 1.9 to accommodate a restricted stopping power prohibits expressing E_{max} using Equation 1.8 as it must be replaced with the energy cutoff, Δ

$$\left(\frac{dE}{\rho dx}\right)_\Delta = 2\pi r_0^2 m_e c^2 \left(\frac{N_A Z}{A}\right)\left(\frac{z^2}{\beta^2}\right)\left[\ln\left(\frac{2m_e c^2\beta^2\Delta}{I^2(1-\beta^2)}\right)-2\beta^2\right] \quad (1.12)$$

A direct application of stopping powers outside of determining dose is predicting range. An estimate of a particle's range, in terms of mass thickness ($\rho \times l$ in units of g cm^{-2}), in matter can be determined from

$$\mathcal{R}_{CSDA} = \int_0^{E_{K_0}}\left(\frac{dE}{\rho dx}\right)^{-1}dE \quad (1.13)$$

Equation 1.13 is referred to as the continuous slowing down approximation (CSDA) of a particle's range, which is the average length traveled by a charged particle when it slows down from its initial total kinetic energy to a final kinetic energy close to zero. This is often taken as a conservative estimate which overpredicts the range. For heavy charged particles, $\mathcal{R}_{\text{CSDA}}$ is a fairly indicative and close approximation. However, this metric fails to adequately predict the observed range for lighter ions, such as electrons and positrons, since the particle's path is assumed straight and $\mathcal{R}_{\text{CSDA}}$ does not take into account the influence of scatter on the particle's range.

1.3 PHOTON INTERACTIONS

Our treatment of dose to this point has focused upon the kinetic energy loss of charged particles traversing through a medium. However, the question remains on how these charged particles were initially set in motion. Some medical applications require the use of a beam of charged particles, namely, electrons or protons, although some heaver ion beams have been used albeit less frequently. For calculations of dose from these directly ionizing beams of charged particle radiation, our previous discussion of dose is sufficient. However, a large majority of radiation applications in medicine rely upon the use of indirectly ionizing radiation, namely, photons produced from Bremsstrahlung, called x-rays, or from the radioactive decay of nuclei, denoted as γ-rays. The calculation of dose from indirectly ionizing radiation is a two-step process: a photon must first interact within a medium and then impart kinetic energy to a charged particle placing it in motion. The kinetic energy is then transferred from the charged particle to the surrounding medium following the mechanics discussed in the previous section. For the purposes of this discussion, we will focus solely on photon–electron interactions as they are the most prominent interactions concerned within the practice of medical physics dosimetry.

There are four primary photon electron interactions that our discussion of dose is concerned with and they are Rayleigh, photoelectric, Compton, and pair and triplet production interactions. However, only the latter three emit charged particles after the interaction takes place and dominate within the ranges of photon energies encountered in the practice of medical physics. Each of these interactions contributes to a loss of incident photon fluence through a medium. The occurrence of an interaction taking place can be thought of as a dart hitting a binary dart board – either the dart hits the bull's eye or it misses. The probability that the bull's eye is hit

is related to the size of the dart and the size of the bull's eye relative to the dart board. This is analogous to a fluence of uncharged radiation, ψ (cm^{-2}), of particles incident upon a slab of matter composed of many potential atomic "targets" each with a little cross-sectional area, σ_a (cm^2). The number of targets within a unit volume of the matter can be estimated as $N_A \cdot \rho / A$ from the mass density, ρ, and mass number of the substance, A. The rate that the initial fluence reduces in depth due to the number of interactions that occur is

$$-\frac{d\psi}{dl} = \frac{\sigma_a \psi N_A \rho}{A} = \mu\psi \qquad (1.14)$$

If the target for the reaction is instead an electron bound (or loosely bound) to an atom, the atomic cross section is replaced with the combined cross section of all the electrons bound to the atom, $\sigma_e \cdot Z = \sigma_a$. The reduction of fluence, or attenuation, can be modeled as a single variable, μ, the total attenuation coefficient in units of cm^{-1} or the unit probability per depth an interaction occurs. This quantity is also expressed as the mass attenuation coefficient by normalizing μ to the density of the medium. This simple differential equation can be solved to obtain the uncollided fluence at a distance l in the medium

$$\psi(l) = \psi_0 e^{-\mu l} = \psi_0 e^{-\mu/\rho \cdot \Sigma} \qquad (1.15)$$

where ψ_0 is the initial fluence and Σ is used to define the mass thickness through a material. The total mass attenuation coefficient can be broken down into components for each of the three aforementioned interaction types

$$\frac{\mu}{\rho} = \frac{\tau}{\rho} + \frac{\sigma}{\rho} + \frac{\kappa}{\rho} \qquad (1.16)$$

where $\frac{\tau}{\rho}$, $\frac{\sigma}{\rho}$, and $\frac{\kappa}{\rho}$ represent the mass attenuation coefficients for photoelectric, Compton, and pair and triplet production, respectively.

The photoelectric effect absorption coefficient is dominant at low energies, around the rest mass of an electron. As such, this particular interaction is responsible for much of the contrast observed in radiological imaging systems. For this interaction to occur, the incident photon must coherently interact with the entire atom and impart enough energy to an orbital election to overcome its binding energy to the atom. The interaction results in a fluorescence photon in addition to at least one electron, possibly more if an Auger electron is emitted. The work of Heitler [5] approximated the K-shell component to the photoelectric effect absorption coefficient as

$$\tau_K = \frac{8\pi r_0^2}{3} 4\sqrt{2} \left(\frac{Z^5}{137^4} \right) \left(\frac{m_e c^2}{h\nu} \right)^{7/2} \qquad (1.17)$$

As indicated from Equation 1.17, the interaction cross section is largest at the binding energy of the k-shell, $E_B = \left(\frac{Z}{\left(n_{level}^{th}=1 \right) \cdot 137} \right)^2 \cdot \frac{m_e c^2}{2}$. This implies that the cross section spikes at the shell-binding energies and also suggests that the medium is transparent to its own fluorescence photons.

The Compton scattering effect occurs between an incident photon of energy, $h\nu$, and an unbound, stationary electron resulting in a scatter photon at angle θ_γ, and energy

$$h\nu' = \frac{h\nu}{1 + \frac{h\nu}{m_e c^2} \left(1 - \cos\left(\theta_\gamma \right) \right)} \qquad (1.18)$$

The scattered electron is then emitted with an energy of $E_e = h\nu - h\nu'$ at an angle of

$$\tan\left(\theta_e \right) = \frac{1}{\left(1 + \alpha_0 \right) \tan\left(\theta_\gamma / 2 \right)} \qquad (1.19)$$

where $\alpha_0 = \frac{h\nu}{m_e c^2}$. The interaction cross section for the Compton effect was initially derived by Klein and Nishina [6]

$$\sigma_e = 2\pi r_0^2 \left[\left(\frac{1+\alpha_0}{\alpha_0^2} \right) \left(\frac{2\left(1+\alpha_0 \right)}{1+2\alpha_0} - \frac{\ln\left(1+2\alpha_0 \right)}{\alpha_0} \right) \right.$$

$$\left. + \frac{\ln\left(1+2\alpha_0 \right)}{2\alpha_0} - \frac{1+3\alpha_0}{\left(1+2\alpha_0 \right)^2} \right] \qquad (1.20)$$

and

$$\sigma = \sigma_e \frac{N_A Z \rho}{A} \qquad (1.21)$$

For incident photon energies larger than 1 MeV, the assumption of an unbound electron is very good, especially since the scattered electron tends to receive most of the incident photon's energy. However, for lower incidents energies, a form factor function is employed to account for the binding energy of the electron. However, the impact that this correction factor has upon our estimate of dose is not very large since the photoelectric effect dominates at such low energies.

Pair production is a manifestation of Einstein's mass–energy relationship where a positron and an electron are produced from a photon interaction within an atom's electric field. In the event of triplet production, two electrons and a positron are produced. The minimum threshold photon energy for this interaction to occur is

$$h\nu \geq \begin{cases} 2m_e c^2 & \text{if pair production} \\ 4m_e c^2 & \text{if triplet production} \end{cases} \qquad (1.22)$$

Bethe and Heitler [7] initially derived the atomic cross section assuming that the interaction is far from the nucleus, thereby ignoring the effects of the nuclear electric field.

$$\kappa_{pair} = \frac{r_0^2 Z^2}{137} \bar{P} \qquad (1.23)$$

where $\bar{P} \approx \frac{28}{9} \ln\left(\frac{2h\nu}{m_e c^2} \right) - \frac{218}{27}$. The cross section for triplet production is similar to that of pair production but requires an extra term, C, to account for election exchange effects of the form

$$\kappa_{triplet} = \frac{\kappa_{pair}}{C \cdot Z} \qquad (1.24)$$

where C is a slowly changing value ranging between 1.6 for 5 MeV < $h\nu$ < 20 MeV and 1.1 for 20 MeV < $h\nu$ < 100 MeV. The total triplet and pair production cross section is

$$\kappa = \kappa_{pair} + \kappa_{triplet} = \frac{r_0^2}{137} Z \bar{P} \left(Z + \frac{1}{C} \right) \qquad (1.25)$$

1.4 QUANTITIES USED TO DESCRIBE IONIZING RADIATION

Up until this point, we have treated the interaction of indirectly ionizing radiation stochastically. The interaction cross sections recently discussed describe the probability of one particular event occurring and the consequential attenuation of the primary beam. While these quantities are useful for the understanding of radiation transport through matter, they are not so useful to practically calculate dose. In our previous discussion, the emphasis was placed primarily on attenuation of the primary beam. However, as we progress through our understanding of dose, it is practical to define nonstochastic quantities to predict the expectation value of energy loss in the indirectly ionizing beam of radiation. For the purposes of this discussion, we will discuss these quantities with respect to photons representing the energy fluence of the beam as $\Psi(E) = \psi(E) \cdot E$. However, analogs exist for other indirectly ionizing radiations such as neutron beams.

The first nonstochastic quantity that describes the interaction of indirectly ionizing radiation with matter is Terma, T, which is defined as the quantity of total energy transferred to matter. This quantity reflects the first stage of energy deposition from indirectly ionizing radiation with the surrounding medium and describes the rate of attenuation of the primary beam. Using our definitions of the interaction cross sections in the previous section, we can define Terma from the incident energy spectrum, up to the maximum energy, E_{max} mathematically as

$$T = \int_0^{E_{max}} \left(\frac{\mu}{\rho}\right)_E \cdot \Psi'(E) dE \rightarrow T \equiv \frac{\mu}{\rho} \Psi, \text{ if monoenergetic}$$
(1.26)

which has units of J kg^{-1}. Following an interaction between the beam of indirectly ionizing radiation and matter, a portion of energy from the primary beam may be transferred to charge particles in the medium. This quantity is referred to as kerma, K, and is defined formally as

$$K = \int_0^{E_{max}} \left(\frac{\mu_{tr}}{\rho}\right)_E \cdot \Psi'(E) dE \rightarrow K \equiv \frac{\mu_{tr}}{\rho} \Psi, \text{ if monoenergetic}$$
(1.27)

Once an interaction occurs, the amount of energy transferred to the medium's charged particles is stochastic in nature. The equations we used to specify the mass attenuation cross sections for the three types of photon interactions with matter, Equations 1.17, 1.20, and 1.25, can be adjusted to reflect the portion of energy loss from the primary beam and are referred to as the respective mass energy transfer coefficients. Analogous to the total attenuation coefficient, the mass energy transfer

coefficient is also defined by the constituents of each interaction.

$$\frac{\mu_{tr}}{\rho} = \frac{\tau_{tr}}{\rho} + \frac{\sigma_{tr}}{\rho} + \frac{\kappa_{tr}}{\rho}$$
(1.28)

Since the energy transfer coefficient is defined as the expected portion of energy initially transferred to charged particles, we can also define it as

$$\frac{\mu_{tr}}{\rho} = \frac{\mu}{\rho} \cdot \frac{\bar{E}_{tr}}{h\upsilon}$$
(1.29)

where $\bar{E}_{tr}$ is the average energy transferred to charged particles from an incident photon of energy, $h\upsilon$.

Let's consider photoelectric interactions. Upon an event's occurrence, we know that the incident photon's energy will be divided among an ionized L- or K-shell electron, a fluorescent photon, and potentially an Auger electron. Thus, the fraction of the energy imparted to the electrons is the difference between the incident photon energy and the portion lost due to fluorescence. The mass energy transfer coefficient can then be determined by

$$\frac{\tau_{tr}}{\rho} = \frac{\tau}{\rho}\left(\frac{h\upsilon - P_K Y_K h\upsilon_K - (1-P_K)P_L Y_L h\upsilon_L}{h\upsilon}\right)$$
(1.30)

where $P_K Y_K h\upsilon_K$ describes the expected fraction of energy loss from K shell fluorescence of photon energy $h\upsilon_K$ with a yield of Y_K occurring with a probability of P_K in addition to the energy losses should the interaction take place with an L-shell electron instead of a K-shell electron.

For the case of Compton interactions, the mean energy transferred to an electron can be determined from weighting the scattered electron energy following Equation 1.18 with the respective cross section value calculated using Equation 1.20 while normalizing to the integral of the cross section over all scattered electron energies.

$$\sigma_{e,\,tr} = \int_0^{E_{max}} \frac{d\sigma_e}{dE} \cdot \frac{E}{h\upsilon} dE$$
(1.31)

It follows that the Compton mass energy transfer coefficient is

$$\frac{\sigma_{tr}}{\rho} = \sigma_{e,tr} \frac{N_A Z}{A} = \frac{\bar{E}_{tr}}{h\upsilon} \cdot \frac{\sigma}{\rho}.$$
(1.32)

Pair and triplet production mass energy transfer coefficient is fairly straightforward as the energy initially transferred to the charged particles is the residual amount of energy left after the creation of the electron(s) and positron

$$\frac{\kappa_{tr}}{\rho} = \frac{\kappa}{\rho}\left(\frac{h\nu - 2m_e c^2}{h\nu}\right) \tag{1.33}$$

Kerma should not be confused with dose. While the two quantities are similar, even defined by the same unit of Gy, dose maintains an element of locality where energy must be deposited, whereas kerma simply states the initial kinetic energy transferred to charge particles at an interaction point. As the charge particle traverses through the medium, a portion of its transferred kinetic energy is expended by both collisional and radiative loss

$$K = K_c + K_r \tag{1.34}$$

The portion of the energy not radiated away is defined as the collision kerma, K_c,

$$K_c = \int_0^{E_{max}} \left(\frac{\mu_{en}}{\rho}\right)_E \Psi'(E)dE \to K_c \equiv \frac{\mu_{en}}{\rho}\Psi \text{ if monoenergetic} \tag{1.35}$$

The mass energy absorption coefficient, $\left(\frac{\mu_{en}}{\rho}\right)$, can be stated generally as

$$\frac{\mu_{en}}{\rho} = \frac{\tau_{en}}{\rho} + \frac{\sigma_{en}}{\rho} + \frac{\kappa_{en}}{\rho} = (1-g)\frac{\mu_{tr}}{\rho} \tag{1.36}$$

where the term $1-g$ denotes the fraction of the initial charged particle's kinetic energy that wasn't radiated away by photons. By the definition of collision kerma, we know the amount of energy transferred and kept by the charged particles from the incident uncharged radiation until they come to rest. Conceptually, if CPE exists within a defined volume, then we would know that any particle that left the specified volume with some kinetic energy would be replaced with another particle with the same kinetic energy that the first particle left with. Thus, in this special set of circumstances, we can state that dose is equivalent to the collision kerma. Much of the practice in modern dosimetry rests upon the application and understanding of these quantities. Rarely does CPE exist outside of well-controlled experiments. However, this is the basis from which modern dosimetry protocols rely on to provide accurate, traceable standards of radiation dose and kerma.

1.5 RADIATION DOSIMETRY

Dose is arguably one of the most important radiological metrics utilized in the treatment of cancer. It is also used to benchmark the constancy of the output from a medical linear accelerator. Intercomparisons of clinical radiation treatment outcomes reference delivered dose and dose rates. Radiation safety limits are set from studies that have reported in analogs of dose. Most pertinent to the practice of therapy physics is the measurement of absorbed dose to water from a medical linear accelerator. A direct measurement of dose to a medium from fundamental quantities, while possible, is very difficult and susceptible to large uncertainty unless great care is taken to minimize errors in setup and fully characterize the assumptions made within the measurement. If every clinic was responsible for this measurement, large discrepancies could arise between departments. Instead, it is safer and more practical for one entity to maintain the standard device, which measures a reference quantity. This standard can then be disseminated to several clinics using a precise instrument that can be easily calibrated with the standard. For example, an ionization chamber is both a practical and precise instrument that has been used to transfer the standard of absorbed dose to water measured from a calorimeter to a clinic.

Ionization chambers are the most common instrument to measure dose to water, which is the standard quantity used to calibrate the output of medical linear accelerators. However, depending on the primary quantity used for chamber calibration, the determination of dose to water by the physicist changes. Over the years, this primary quantity has changed historically from air kerma to absorbed dose to water from a ^{60}Co beam of radiation with a near-monoenergetic photon spectrum center around an average energy of 1.25 MeV. Ionization chambers are conceptually simple consisting of an air-filled cavity that collects the charge from ionizing radiation interacting with the gas molecules residing in the cavity. Further discussion of ionization chambers and their applications are discussed elsewhere [8], but for the context of this chapter, most portable chambers consist of a wall encapsulating a collection volume that is charged to establish a voltage potential with a collection electrode.

The current challenge at hand is to relate the measured signal in the chamber to the dose to water at the center of the chamber as if the chamber was not present to begin with. This exact problem has been the primary focus of several published reports. Today, the United States recognizes the task group report numbers 21 (TG-21) [9] and 51 (TG-51) [10] from the American Association of Physicists in Medicine (AAPM) as the primary protocols to determine the dose to water using an ionization chamber that was either calibrated to a known air kerma or absorbed dose to water, respectively.

1.5.1 Cavity Theory

The determination of dose within one medium as if it were occupied by a different material relies on the use of cavity theory, and the basis of much of the TG-21 and TG-51 formalisms is grounded in these applications. The foundations of cavity theory are based on Bragg–Gray theory [11]. Consider a plane-parallel field of photon radiation as shown in Figure 1.1.

If the spectrum of charged particles, ϕ, was known, the dose at a point, p, in the medium, w, can be determined from the spectrum's differential fluence, ϕ'

$$D_{\text{p in w}} = \int \phi'_w(E)\left(\frac{dE}{\rho dx}\right)_{E,w} dE \qquad (1.37)$$

If now the region surrounding point p was changed to air, the dose at p is then equivalent to

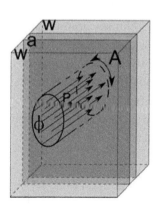

FIGURE 1.1 Simple slab geometry illustrating the principles of Bragg–Gray cavity theory. A uniform fluence of charged particles, ϕ, within an area, A, traverses a thin slab of material, a, sandwiched between material slabs composed of w. The dose at point, p, can be evaluated for both materials a and w assuming the Bragg–Gray conditions.

$$D_{\text{p in a}} = \int \phi'_a(E)\left(\frac{dE}{\rho dx}\right)_{E,a} dE \qquad (1.38)$$

Let's assume now the following:

1. The thickness, t, of the layer is much smaller than the range of charge particles traversing the layer. Therefore, no particles either start or stop in the layer.

2. The change in the layer's material from the surrounding medium does not substantially change the fluence at p.

These assumptions are commonly referred to as the Bragg–Gray conditions and imply that

$$\phi'_a(E) = \phi'_w(E) = \phi'(E) \qquad (1.39)$$

The final result provides a general relation that relates the dose in the two media

$$\frac{D_{\text{p in a}}}{D_{\text{p in w}}} = \frac{\int \phi'(E)\left(\frac{dE}{\rho dx}\right)_{E,a} dE}{\int \phi'(E)\left(\frac{dE}{\rho dx}\right)_{E,w} dE} \qquad (1.40)$$

and upon normalizing, the result with the total fluence in both the numerator and denominator

$$\frac{D_{\text{p in a}}}{D_{\text{p in w}}} = \frac{\int \phi'(E)\left(\frac{dE}{\rho dx}\right)_{E,a} dE / \int \phi'(E) dE}{\int \phi'(E)\left(\frac{dE}{\rho dx}\right)_{E,w} dE / \int \phi'(E) dE} \qquad (1.41)$$

reveals that the doses in the two media are related by the ratio of their mean stopping powers, $\bar{S}$, denoted by both a superscript and a subscript nomenclature

$$\frac{D_{\text{p in a}}}{D_{\text{p in w}}} = \frac{\left(d\bar{E}/\rho dx\right)_a}{\left(d\bar{E}/\rho dx\right)_w} = \frac{\bar{S}_a}{\bar{S}_w} = \bar{S}_w^a \qquad (1.42)$$

While great as a first-order approximation, Bragg–Gray theory falls short for large discrepancies in the atomic numbers and density between the two media. This suggests that the two Bragg–Gray assumptions were not well satisfied. The work of Spencer and Attix [3, 12] sought to improve upon the basis of Bragg–Gray theory by accounting for the generation and influence of δ-rays by including restricted stopping powers in the absorbed dose determination at a point

$$D = \int_{\Delta}^{E_0} \phi^{e,\delta} \cdot S(E,\Delta) dE \qquad (1.43)$$

where $\phi^{e,\delta}$ is the equilibrium fluence accounting for the existence of delta rays and the restricted stopping powers, $S(E,\Delta)$, are evaluated from an energy Δ, which is just large enough for the charged particles to traverse the cavity. Any particles with energy less than Δ are assumed to neither be able to enter the cavity nor transport energy. Using the Spencer–Attix cavity theory, the relationship between the doses in two media is therefore simply the ratio of the mean restricted mass stopping powers, $\bar{L}_\Delta$.

$$\frac{D_{\text{p in a}}}{D_{\text{p in w}}} = \frac{\bar{L}_{\Delta,a}}{\bar{L}_{\Delta,w}} = \bar{L}_{\Delta,w}^a \qquad (1.44)$$

Bragg–Gray and Spencer–Attix cavity theories build from assumptions on the charge particle fluence between contiguous media. The relationship of dose between materials reflects the change in energy loss from the charged particle field traversing through the medium. However, a similar relation can be made with respect to the uncharged radiation field from our definitions of collision kerma. Similar to CPE, radiation equilibrium (RE) is used to describe a spatial quality of point where the radiation fluence entering a specified volume is the same as the radiation fluence exiting the volume. Because the charged particle fluence is a direct consequence of the incident uncharged radiation field, it is conceptually simple to understand that if RE exists, then CPE exists. However, the inverse of this statement is not true. It follows from our earlier discussion of collision kerma that if RE exists within a volume of space, then the relationship in absorbed dose between two media is

$$\frac{D_{\text{p in a}}}{D_{\text{p in w}}} = \frac{K_{\text{c,p in a}}}{K_{\text{c,p in w}}} = \frac{(\bar{\mu}_{\text{en}}/\rho)_a}{(\bar{\mu}_{\text{en}}/\rho)_w} = \left(\frac{\bar{\mu}_{\text{en}}}{\rho}\right)_w^a \qquad (1.45)$$

The relation shown in Equation 1.45 reflects the uncharged radiation field under the conditions of RE

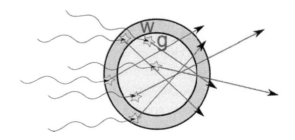

FIGURE 1.2 Burlin cavity theory discriminates charges produced within the wall (w) and cavity air (g) as they contribute to the total dose within the chamber. Considering each of these contributions separately bridges the gap between the assumptions made between very small chambers, which utilize the Bragg–Gray assumptions, and very large cavity volumes where the influence of the wall is negligible.

most commonly achieved with large-cavity chambers using the subscript–superscript ratio notation presented in Equation 1.42. A subtle but important note to take away is the difference between large-cavity and Bragg–Gray/Spencer–Attix theory: one relates the expected conditions of energy transfer with respect to the charged particle field verses the incident uncharged radiation field. The merger of these two theories can be extended to our chamber measurement as shown in Figure 1.2 since most chambers do not behave completely within the confines of one of these theories due to the presence of a wall.

This merger is known as Burlin cavity theory. If the volume of the cavity is very large in comparison to the wall thickness, then the fluence of electrons from the wall material will exponentially decrease as they are attenuated in the gas volume. However, the fluence of electrons generated within the gas volume will exponentially increase leading to the conditions governing large cavity theory. On the other hand, if the chamber volume is very small, the majority of electrons traversing the cavity originate from the wall material. The Burlin cavity relation can be expressed as

$$\frac{D_{\text{g}}}{D_{\text{w}}} = d \cdot \bar{S}_{\text{w}}^{\text{g}} + (1-d) \cdot \left(\frac{\bar{\mu}_{\text{en}}}{\rho}\right)_w^g \qquad (1.46)$$

where d is the portion of electrons generated from the wall that contribute to the equilibrium fluence in the cavity, and $1-d$ is the portion of electrons from the cavity that contribute to the equilibrium fluence in the cavity. These four theories build the foundation from which

TABLE 1.1   List of the Four Primary Theories Covered in this Chapter and Their Suitable Applications

Chamber Theory	Domain of Validity	Assumptions
Bragg–Gray	$\rho_g t_g \ll \rho_w t_w$	CPE, the majority of charges are produced in the chamber wall
Spencer–Attix	$\rho_g t_g \ll \rho_w t_w$	The majority of charges are produced in the chamber wall
Large Chamber	$\rho_g t_g \gg \rho_w t_w$	RE $\to$ CPE is dominated by the charges produced in the chamber cavity
Burlin	Any	CPE, charges produced in the cavity replaces those attenuated from the wall

Note: These theories share an implicit set of assumptions, which include a homogeneous wall, w, and cavity gas, g, material, uniform radiation field across the entire chamber, little-to-no attenuation of the incident radiation field throughout the chamber volume, and a charge particle equilibrium spectra that is shared between the wall and cavity air.

the current protocols in radiation dosimetry are derived and practiced and are summarized in Table 1.1.

1.5.2 Overview of TG-21

The AAPM TG-21 established a protocol to determine dose to a medium of water following an exposure calibration of a chamber in a ^{60}Co beam of radiation [9]. While a full discussion of chamber calibration and standards is outside the scope of this chapter, a brief discussion will be discussed to provide context of the problem to the reader. Chapter 2 will handle ionization chambers in detail.

Seeing how humans are considered to be equivalent to water, the quantity of absorbed dose to water is a desirable metric to reference. It is the concern of the medical physicist to accurately and precisely quantify how much dose is delivered in a variety of scenarios. However, it is good experimental practice to always maintain a well-established reference that can benchmark all other measurements. This reference is universally recognized as the dose to water given:

1. Measurements are performed with a reference-class chamber (i.e., a very stable chamber whose design mitigates any influence of charges produced outside the intended collection volume and wall) in a water tank with dimensions of at least 30 cm × 30 cm × 30 cm.

2. Depth of chamber is placed at the depth corresponding to the peak dose, as this is the only location along a depth profile that CPE is (nearly) established.

3. The source position relative to the surface of the water is 100 cm.

Furthermore, most chambers cannot accurately determine dose solely from the fundamental quantity of charge collected. They must be calibrated at a specified point with an instrument, called a standard, which can directly measure the intended primary quantity. For the TG-21 protocol, the intended quantity that chambers are calibrated in is the kerma in air from a ^{60}Co radiation field. At a specified point one meter from the source, the air kerma that has been measured by the physicist can be used to determine the dose to a point from the measured quantity of exposure. Exposure is defined as the charge released per unit mass from a beam of radiation.

$$K_c = X \cdot k \cdot \left(\frac{\overline{W}}{e}\right)_{gas} \tag{1.47}$$

describes the relationship for the measured exposure, X, in units of Roentgen (R) to collision kerma in units of Gy. The constant k converts from the unit of Roentgen to the charges produced per unit mass of air, $k = 2.58 \times 10^{-4}$ C kg^{-1} R^{-1}, and the quantity $\left(\frac{\overline{W}}{e}\right)_{gas}$, approximately 33.97 J C^{-1}, is the energy required to create a charge pair in dry air. The standard used for ^{60}Co energies at the US standards lab was a known-volume graphite chamber with a well-known collection volume. Based on the measured charge from ionized gas in the cavity, the exposure can be calculated from

$$X = \frac{Q}{k \cdot m_{chamber\ gas}} \left(\frac{\overline{L}}{\rho}\right)^{chamber\ wall}_{chamber\ gas} \left(\frac{\overline{\mu}_{en}}{\rho}\right)^{air}_{chamber\ wall} \beta^{-1}_{wall}\left(\Pi_i K_i^{-1}\right) \tag{1.48}$$

Radiation equilibrium is assumed within the chamber's collection volume and Spencer–Attix cavity theory allows us to relate the collision kerma from the air inside the cavity to the surrounding cavity wall. If we assume that the photon energy fluence is unperturbed by the chamber wall, and that the majority of charges contributing the wall kerma are from the wall than the surrounding air, large-chamber cavity theory can be applied to determine the kerma to the room air from the kerma in the wall and Spencer–Attix cavity theory can be used to determine the dose relationship between the wall and air inside the chamber cavity. The correction factor β accounts for the ratio of dose to collision kerma since true CPE is not fully established, but still very close (within 0.5%). The remaining correction factors, K_i^{-1}, are employed to account for electrometer calibration, chamber ion collection efficiency, water vapor content in the air, stem scatter, and the influence of wall thickness on the collision kerma. Since the air temperature and pressure in the room will influence the charge collected from the chamber, a standard lab corrects the raw charge reading to a standard temperature and pressure. The product of Equation 1.48 is the determination of exposure from air kerma in the room air at the center of the chamber from the measured ionized charges that are a result of the air kerma within the chamber.

The response, N_X, of any chamber can be related to the known radiation field by

$$N_X = \frac{X}{M} \tag{1.49}$$

where M is the measured charge normalized to standard temperature and pressure from the ionization chamber for a known delivered exposure, X. However, this only relates an arbitrary response to the radiation field established at a lab. What we need is a transferrable quantity unique to the chamber that reflects the dose to the chamber. Using Burlin cavity theory, we can work back from the exposure response of the chamber to the dose response of the chamber in air.

$$N_{\text{chamber gas}} =$$

$$N_X \frac{k\left(\dfrac{\bar{W}}{e}\right) A_{\text{ion}} A_{\text{wall}} \beta_{\text{wall}}}{d\left(\dfrac{\bar{L}}{\rho}\right)^{\text{chamber wall}}_{\text{chamber gas}} \left(\dfrac{\bar{\mu}_{\text{en}}}{\rho}\right)^{\text{air}}_{\text{chamber wall}} + (1-d)\left(\dfrac{\bar{L}}{\rho}\right)^{\text{chamber cap}}_{\text{chamber gas}} \left(\dfrac{\bar{\mu}_{\text{en}}}{\rho}\right)^{\text{air}}_{\text{chamber cap}}} \tag{1.50}$$

where d in this case is the ratio of charges that contribute to the charged particle fluence in the chamber's collection volume from the wall and the surrounding medium. The parameter d is equal to unity in the special case when the material composition of the chamber wall and the buildup cap are the same. The correction

factors A_{ion} and A_{wall} correct for the collection efficiency and attenuation and scatter in the wall and buildup cap, respectively. Note that these quantities are only well-defined for a ^{60}Co radiation field. The calculation of the dose to the gas volume in the chamber can simply be determined from

$$D_{\text{gas in chamber}} = M \cdot N_{\text{chamber gas}} \tag{1.51}$$

However, this is of little use if we wish to calculate dose to the water with our chamber. Working backward, the value of D_{water} can then be calculated from $D_{\text{gas in chamber}}$ from

$$D_{\text{medium}} = M_{\text{raw}} N_{\text{chamber gas}} \left(\frac{\bar{L}}{\rho}\right)^{\text{medium}}_{\text{chamber gas}} P_{\text{ion}} P_{\text{repl}} P_{\text{wall}} P_{\text{TP}} \tag{1.52}$$

where $N_{\text{chamber gas}}$ is now considered a chamber-specific calibration coefficient as defined in Equation 1.50, P_{ion} is the inverse of the ionization collection efficiency of the chamber, P_{repl} is the replacement correction for the perturbation in the fluence due to the placement of the chamber in the medium, and P_{wall} is the wall correction when the chamber wall and phantom composition are different, and the air density inside the chamber. Using Burlin cavity theory, P_{wall} can be analytically determined as

$$P_{\text{wall}} = \frac{\left[d\left(\bar{L}/\rho\right)^{\text{wall}}_{\text{chamber gas}} \left(\bar{\mu}_{\text{en}}/\rho\right)^{\text{medium}}_{\text{wall}} + (1-d)\left(\bar{L}/\rho\right)^{\text{medium}}_{\text{chamber gas}}\right]}{\left(\bar{L}/\rho\right)^{\text{medium}}_{\text{chamber gas}}} \tag{1.53}$$

Inserting the expression for P_{wall} from Equation 1.53 into Equation 1.52 to determine the absorbed dose to the medium is given as

$$D_{\text{medium}} = M_{\text{raw}} N_{\text{gas}} P_{\text{ion}} P_{\text{repl}} P_{\text{TP}} \left(d\left(\frac{\bar{L}}{\rho}\right)^{\text{wall}}_{\text{chamber gas}} \left(\frac{\bar{\mu}_{\text{en}}}{\rho}\right)^{\text{medium}}_{\text{wall}} \right.$$

$$\left. + (1-d)\left(\frac{\bar{L}}{\rho}\right)^{\text{medium}}_{\text{chamber gas}} \right) \tag{1.54}$$

Alternatively, this expression of D_{medium} can be derived directly from the expression of $D_{\text{gas in chamber}}$ using Burlin cavity theory. It should be noted that the stopping powers and mass absorption attenuation coefficients are now defined for the beam used at the clinic, no longer the ^{60}Co environment at the calibration lab. Further corrections must be made to M_{raw} at the clinic that are also referenced to the same conditions of standard temperature and pressure, collection efficiency, polarity effects, and electrometer response that were considered during the chamber's calibration. In addition to the ionization chamber, electrometers are also calibrated to traceable standards of resistance and capacitance and carry their own calibration coefficients that must be applied to the charge reading. Specifics about these calibration factors and how they are determined will be discussed in Chapter 2.

1.5.3 Overview of TG-51

The TG-21 report provides an analytical method to determine the dose to water with an ionization chamber from an exposure-based calibration in air. Conceptually, TG-21 is a process outlining the theory of how to relate the kerma at a specified point in air to the charges collected within an ionization chamber's gas cavity to the expected quantity of dose in a medium of water. While this report offers a fundamental insight into the physics behind modern dosimetry, it is an outdated protocol. There are several assumptions that were made, which limits its achievable accuracy in the determination of absorbed dose to water to around 3–4%. However, better accuracies are desired in the practice of modern radiotherapy dosimetry of around or less than 2%. For example, TG-21 does not provide specific corrections for the influence of the central electrode on the incident fluence, and there is a string disconnect between the beams of radiation used to calibrate the chamber response and the measurement of dose at the clinic. At a calibration lab, the chamber response is calibrated from a radiation field of 1.17 MeV and 1.33 MeV ^{60}Co γ-rays, whereas a medical linear accelerator at a clinic produces a Bremsstrahlung distribution of 6 MV or higher energy x-rays. Finally, TG-21 allows mediums other than water to perform the measurement in, such as water-mimicking plastic phantoms.

The current absorbed dose-to-water protocol that is practiced clinically is TG-51. This protocol is different than TG-21 as TG-51 is completely water based, water-mimicking plastics are not allowed. Ionization chambers are calibrated to an absorbed dose to water, not exposure, while also simplifying the absorbed dose-to-water

calculation by using Monte Carlo-derived correction factors [10]. These correction factors are simulated to determine chamber perturbation factors between beam quality and chamber components. For the context of this report, beam quality, Q, refers to a type of radiation beam in a sense that it uniquely describes the emitted spectrum. Following the TG-51 formalism, the dose to a medium of water for a beam quality, Q, and chamber charge measurement, M, is

$$D_w^Q = M \cdot N_{D,w}^Q \qquad (1.55)$$

where $N_{D,w}^Q$ is the absorbed dose to water calibration factor for an ionization chamber in the beam quality, Q. The beam quality expressed in Equation 1.55 is the radiation field of the Medical Physicist's medical linear accelerator. However, the calibration of the ionization chamber is still performed in a ^{60}Co beam quality, $Q = {}^{60}$Co providing an $N_{D,w}^{Q={}^{60}Co}$. To convert between the calibration beam quality and the user's beam quality, a k_Q factor is employed

$$N_{D,w}^Q = k_Q \cdot N_{D,w}^{{}^{60}Co} \qquad (1.56)$$

In parallel to the TG-21 protocol, a substantial amount of information and physics is tied into the single k_Q factor. Factors including the absorbed dose differences, fluence perturbation from the presence of the chamber, volume averaging of the chamber, and differences in radiation beam type have been calculated or measured for several clinical chambers [13–15]. The analytical approximations that were made using cavity theory are replaced with a Monte-Carlo-derived correction factor, k_Q, of the form

$$k_Q = \frac{N_{D,w}^Q}{N_{D,w}^{{}^{60}Co}} = \frac{D_w^Q/M_Q}{D_w^{{}^{60}Co}/M_{{}^{60}Co}} = \frac{D_w^Q}{D_w^{{}^{60}Co}} \cdot \frac{M_{{}^{60}Co}}{M_Q} \qquad (1.57)$$

Equation 1.57 is composed of four elements. It is commonly described in literature using the central form in Equation 1.57 but is perhaps better described pedagogically as the rightmost expression in Equation 1.57 using a series of ratios. The first ratio relates the dose to water at the point of measurement between both beam qualities. The second ratio describes the response of the chamber, M, which can be easily measured or simulated by tallying the dose to the sensitive collection volume of air in the chamber. Thus, each k_Q factor is unique to a chamber and beam. Large tables exist that have tabulated various k_Q factors for specific ionization chamber make and models used within a particular beam of radiation.

Beam quality is unique for each medical linear accelerator as inherent difference exists among different manufacturers of medical linear accelerators, the nominal manufacturing dimensions of components, and the room scatter conditions. Therefore, each linac's beam quality must be experimentally measured based on the applied peak voltage and exiting particle type (i.e., a photon or electron beam and the respective energy). The beam quality for a photon beam with any medical linear accelerator is defined at the percent depth dose of 10 cm in water, also referred to as $\%dd(10)_x$. Prior to measuring the percent depth dose of a photon beam, the relative depth placement of the chamber must be adjusted to implicitly correct the chamber's inherent perturbation of the charged particles produced in the water. This shift for a cylindrical chamber with radius r_{cav} is approximated as $0.6 \cdot r_{cav}$ upstream for photon beam qualities – thus either the chamber should be displaced downward before measurement or the depth dose curve shifted upward after the measurement.

The determination of a beam's k_Q factor using a specific ionization chamber from the measured $\%dd(10)_x$ will reflect the incident beam's average energy. The Monte Carlo work of Muir and Rogers [13, 15] has investigated the relationship between the absorbed dose response of several chambers among a multitude of medical linear accelerator brands and their producible beam energies. Since the beam's energy will have an influence on the chamber's response relative to its calibration conditions in a ^{60}Co beam, as shown in Equation 1.57, the physicist can correct this response using an analytical fit to the Monte Carlo data supplied in the TG-51 addendum [16] that has been normalized to the expected chamber's response in ^{60}Co.

Electron beams are also defined by a unique k_Q factor, but the determination of an electron beam's k_Q is slightly different than a photon beam k_Q factor. An electron beam quality factor is determined from

$$k_Q = P_{gr}^Q \cdot k_{R_{50}} \qquad (1.58)$$

where $k_{R_{50}}$ is a chamber-specific factor defined by the range that beam's distal portion of its depth dose profile falls to 50%. If a cylindrical chamber is used, then an additional correction term, P_{gr}^Q, is necessary to account for the nonuniform gradient across the chamber's collection volume. P_{gr}^Q is unity for plane-parallel chambers. $k_{R_{50}}$ is a product of two other factors considered uniquely for electron beams

$$k_{R_{50}} = k'_{R_{50}} \cdot k_{ecal} \qquad (1.59)$$

Conceptually, the factor k_{ecal} is a photon-to-electron conversion factor unique for each ionization chamber model. The factor of $k'_{R_{50}}$ then corrects the product of $k_{ecal} \cdot N_{D,w}^Q$ to a specific electron beam quality. The depth of where an electron beam falls to 50% requires two steps. The initial depth measured with an ionization chamber is referred to as the 50% depth ionization, I_{50}. This point is adjusted to account for gradient effects in the beam to achieve the depth dose at 50%, R_{50}. Considering these additional factors, the absorbed dose to water from an electron beam is expressed as

$$D_w^{Q_{electron}} = MP_{gr}^{Q_{electron}} k'_{R_{50}} k_{ecal} N_{D,w}^{^{60}Co} \qquad (1.60)$$

The TG-51 addendum [16] estimates that the uncertainty in the measurement of $\%dd(10)_x$ by a clinical physicist is about 2%. This, in turn, corresponds to a 0.4% error in the measured dose. Incorporating the uncertainties in the stopping powers and cross sections used for the Monte-Carlo-derived k_Q factors and expected uncertainties in the chamber measurement and charge correction factors results with a final uncertainty within the absorbed dose of less than 2% at the 95% confidence level.

1.6 CONCLUSION

A substantial amount of work and effort has gone into developing the current absorbed dose-to-water protocol that is based in a rich historical context reaching as far back as the 1960s. What the modern-day medical physicist has inherited is a streamline process for determining a very complex and arbitrary quantity that is a major focus of their practice in medical physics. However, the physicist should be warned that the simplicity and eloquence of the TG-51 formalism can be misleading as so much of the fundamental physics is hidden. A prudent reader will observe that the TG-51 formalism is not exhaustive; it is only well defined for a single set of reference conditions. Any changes to these specified conditions will require us to look critically at how our absorbed dose-to-water estimates could be affected from the assumptions we have made, forcing us to rethink the way we determine dose at the physical level. These influences may manifest from changes from CPE conditions, volume averaging effects, and how our detectors behave from our initial calibration conditions, which will be the focus of our attention later on. The modern role of the practicing medical physicist is far from being considered *ad nauseam*. As the practice of medical physics expands, so must the knowledge and abilities of the physicist to solve tomorrow's problems.

REFERENCES

1. R. Cox, J. Hendry, A. Kellerer, C. Land, C. Muirhead, D. Preston, J. Preston, E. Ron, K. Sanlaranarayanan, R. Shor, R. Ullrich, Biological and Epidemiological Information on Health Risks Attributable to Ionizing Radiation: A Summary of Judgements for the Purposes of Radiological Protection of Humans. International Commission on Radiological Protection, Committee 1 Task Group Report: C1 Foundation Document (2005).

2. H. Staub, H. Bethe, J. Ashkin, N. Ramsey, *Experimental Nuclear Physics* (John Wiley and Sons, Inc., New York, NY, 1953).

3. F. Attix, *Introduction to Radiological Physics and Radiation Dosimetry* (WILEY-VCH Verlag GmbH & Co., KGaA, 2004).

4. E. Podgorsak, *Radiation Physics for Medical Physicists* (Springer International Publishing, New York, NY, 2016).

5. W. Heitler, *The Quantum Theory of Radiation* (DOVER Publications, INC., New York, NY, 1954).

6. O. Klein and Y. Nishina, "On the scattering of radiation by free electrons according to the new relativistic quantum dynamics of Dirac," *Zeitschrift für Physik* **52**, 853–868 (1929).

7. H. Bethe and W. Heitler, "On the stopping of fast particle and on the creation of positive electrons," *Proceedings of the Royal Society of London, A* **146** (1934).

8. L. A. DeWerd and B. R. Smith, "Ionization Chamber Instrumentation," in *Radiation Therapy Dosimetry: A Practical Handbook*, edited by A. Darafsheh (CRC Press, Boca Raton, FL, 2021), Chapter 2, pp. 19–30.

9. R. Schulz, P. Almond, J. Cunningham, J. Holt, R. Loevinger, N. Suntharalingam, K. Wright, R. Nath, G. Lempert, "Task Group 21, Radiation Therapy Committee: A protocol for the determination of absorbed dose from high-energy photon and electron beams," *Medical Physics* **10**, 741–771 (1983).

10. P. Almond, P. Biggs, B. Coursey, W. Hanson, M. Huq, R. Nath, D. Rogers, "AAPM's TG-51 protocol for clinical reference dosimetry of high-energy photon and electron beams," *Medical Physics* **26**, 1847–1870 (1999).

11. L. Gray, "An ionization method for the absolute measurement of γ-ray energy," *Proceedings of the Royal Society of London, A* **156**, 578–596 (1936).

12. L. Spencer and F. Attix, "A theory of cavity ionization," *Radiation Research*, **3**, 239–254 (1955).

13. B. Muir and D. Rogers, "Monte Carlo calculations of k_Q, the beam quality conversion factor," *Medical Physics* **37**(11), 5939–5950 (2010).

14. M. McEwen, "Measurement of ionization chamber absorbed dose k_Q factors in megavoltage photon beams," *Medical Physics* **37**(5), 2179–2193 (2010).

15. B. Muir, M. McEwen, D. Rogers, "Measured and Monte Carlo calculated k_Q factors: Accuracy and comparison," *Medical Physics* **38**(8), 4600–4609 (2011).

16. M. McEwen, L. DeWerd, G. Ibbott, D. Followill, D. Rogers, S. Seltzer, J. Seuntjens, "Addendum to the AAPM's TG-51 protocol for clinical reference dosimetry of high-energy photon beams," *Medical Physics* **41**, 1–20 (2014).

CHAPTER 2

Ionization Chamber Instrumentation

Larry A. DeWerd

University of Wisconsin
Madison, Wisconsin

Blake R. Smith

University of Iowa
Iowa City, Iowa

CONTENTS

2.1 INTRODUCTION TO IONIZATION CHAMBERS

It was recognized shortly after the discovery of x-rays that there was a need for a method to measure the exposure from radiation. Initially, the measurement was done by observing radiation-induced erythema (reddening of the skin) on the individual receiving the exposure. Since there is variation among individuals, this was a subjective measurement technique. The erythema method also involved risk to the patient if there was an excessive exposure. Today, radiation fields are usually quantified by the amount of ionization generated in air or in water under standard conditions. The application of theoretical principles describing the absorption of the radiation with a series of correction factors allows the measured radiation field to be related to the dose in human tissue. This radiation measurement is the scientific field of dosimetry and is the basis for much of medical physicists' contribution to the use of radiation, especially in radiation therapy. The process begins with the measurement by an ionization chamber, which is a precise instrument. An ionization chamber with a known volume may also be accurate enough to be used as a standard in a primary laboratory. Even without a known volume, the ionization chamber is an essential device for the measurement of radiation. The ionization chamber can be calibrated to give an accurate and precise measurement. There are three components involved for this type of measurement device. The ionization chamber together with the electrometer and connecting (triax) cable makes up the dosimetry system. Useful reviews on this topic can be found in works by Attix, Podgorsak, and Andreo [1–3].

2.1.1 Ionization Chambers

Ionization chambers are a generalized class of instruments that also include proportional counters and Geiger–Mueller (GM) counters. These instruments are primarily classified by the voltage applied to the chamber. As shown in Figure 2.1, the response of a generalized chamber is dependent on the voltage applied that generates an electric field to collect the ionization. The ionization chamber region that is the topic of interest for the rest of this chapter is the ion chamber region or the saturation region, or the range of voltages such that the initial charges released in the chamber are collected with near 100% efficiency (the plateau in Figure 2.1). As the voltage increases, the proportional counting region is reached. At voltages in and above this level, specialized gases are generally used. In the proportional counting region, the initial charge released is accelerated enough to create additional ionization, but the total charge collected at a given voltage is still proportional to the incident radiation dose. As the voltage is continually increased, the characteristics change to the point of continuous discharge, or a pulse of current, for each radiation event; this is called the GM region. There are a number of characteristics involved in each particular chamber that need to be considered for a dosimetry system.

2.1.2 Types of Ionization Chambers

There are three main ionization chamber designs for radiation dosimetry, the thimble chamber, parallel-plate

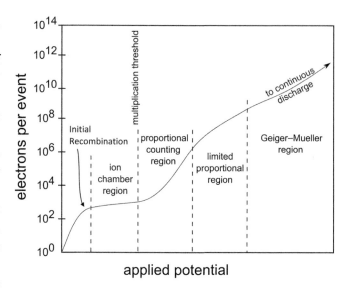

FIGURE 2.1 Voltage curve for a typical ionization chamber.

chamber, and the spherical chamber. These chambers range in size to accommodate various radiation field sizes and dose rates. Each chamber requires calibration to a primary standard for good accuracy. Primary standards are used to calibrate ionization chambers for measurements of x-ray beams as well as megavoltage electron and photon beams. Calorimetric standards are further explained in Chapter 3 of this book. For medium- and low-energy x-ray beams (<300 kVp), the primary standard is a free-air chamber. For megavoltage photon and electron beams, the quantity measured is absorbed dose to water. Within the United States, a ^{60}Co beam, air kerma, and absorbed-dose standards are available. Air kerma is measured with a chamber that has a known volume and wall composition (e.g., graphite). For absorbed dose to water, the standard is a water calorimeter.

The ionization chambers commonly used clinically are cavity ionization chambers, which are well-characterized by Bragg–Gray/Spencer–Attix cavity theory. These chambers consist of a known solid material, called the wall, or shell, containing an air volume. Standard materials used for the wall are C552 (air equivalent material) or graphite. Thus, the effect of the wall necessitates a correction factor known as K_{wall} for in-air measurements or P_{wall} for measurements made in a phantom. Generally, K_{wall} is determined via Monte Carlo computational methods, since experimental techniques have been shown to produce incorrect results [4]. If the chamber is used in air, the walls need to be thick enough to provide full buildup, which is a thickness of material equal to the range of an electron for a given energy.

In most cases, a build-up cap is placed on the chamber, and the cap combined with the chamber wall provides full buildup. The build-up cap has the added advantage of protecting the thimble.

Manufacturer tolerances can affect variations in the volume of the cavity, and thus the response of the chambers is affected. The size of the signal received for a given radiation dose is related to the volume of the cavity. Therefore, a given chamber can be precise in its response, but the signal magnitude can vary among a given model. Thus, each chamber needs calibration related to a primary standard. The correction of the readout, via a calibration coefficient, relates the measurement value to the conventional true value of a primary standard. The calibration coefficient relates the charge reading, in coulombs, to the radiological quantity, air kerma or absorbed dose to water. This calibration coefficient is determined at an American Association of Physicists in Medicine Accredited Dosimetry Calibration Laboratory (AAPM ADCL) by an intercomparison of a reference standard chamber with the chamber being calibrated. In places other than North America, chambers are calibrated by a Secondary Standards Dosimetry Laboratory (SSDL) or a primary laboratory, and Technical Reports Series (TRS) 398 [5] is used. The density of the contained air is always corrected to a standard temperature and pressure.

2.2 BASIC COMPONENTS OF IONIZATION CHAMBERS

The components of an ionization chamber are the shell (also called the wall, thimble, or window), the guard, and the collecting electrode (collector). A schematic of a thimble chamber connected to an electrometer via a triaxial cable (triax) is shown in Figure 2.2. An electric field is established between the ionization chamber components in order to stabilize the ionization collected in the air of the cavity. The electric field is typically generated by a high-voltage power supply or battery within the readout device (electrometer). The schematic of Figure 2.2 is a charge measuring setup; for current measurement, the capacitor is replaced by a resistor. Further details on the electrometer and the triaxial cable are given later. Electrical insulation is provided between each of the ionization chamber components by materials such as polyethylene or PTFE®. The purpose of these insulators is to support the ionization chamber components structurally and to prevent leakage currents. However, even when effective insulators are used, leakage currents may occur when dirt, a hair, or some other type of fiber bridges the insulator. Skin oil is a significant cause of leakage; thus, touching an insulator should be avoided. In general, the signal to noise for a given chamber system always needs to be considered. Ideally, the signal-to-noise ratio should be ≥1000.

Many of the less expensive electrometers, and those no longer meeting original specifications, can have high leakage around 10^{-13} A [6]. Thus, if the signal is from a small volume chamber, then the ideal signal-to-noise ratio of 1000 may not be achievable. The leakage should be subtracted, or the system should be repaired when the noise becomes a significant portion of the measured signal.

The guard of a typical ionization chamber has two main purposes: it electrically shields the collector current through the triaxial cable to the electrometer, and

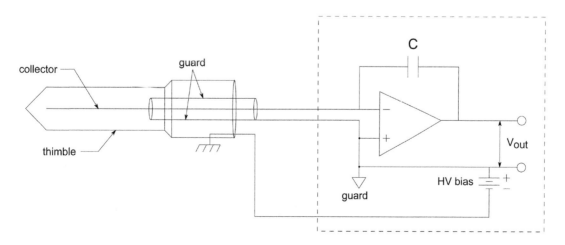

FIGURE 2.2 Schematic of a thimble chamber connected to a charge reading electrometer (a feedback coulombmeter).

it defines the shape of electric field in the ionization chamber collecting volume. In addition, since the guard is at the same potential as the high-impedance collector, it provides a low-impedance path for leakage current, thus reducing the leakage current collected. Improved electric field uniformity in the sensitive volume of the chamber is also provided by the guard. Good field uniformity can minimize fluence perturbations when measurements are performed in electron beams or pulsed beams. The volume is determined by the guard and is more uniformly defined if the chamber is "fully guarded." A chamber is fully guarded when the termination of the guard extends into the air volume of the chamber. A partial or minimal guard has its terminus at the beginning of the air volume or prior to the air cavity. The extent and quality of the guard may cause a difference between the physical and effective volumes of a chamber. Voltage differences between guard and collector [7] or dead spots can occur in the field near the corners where the electric field lines from the collector and guard come together. Dead spots also can occur where the cone meets the cylinder in the classic, Farmer-type chamber. Thus, the design of the fully guarded chamber becomes important for constancy of the volume and determination of the charge collected. A rule of thumb is that the width of the guard should be greater than or equal to three times the width of the gap for a fully guarded parallel-plate chamber.

Although all ionization chambers have the same essential components, chambers are readily defined by their geometric differences. Whether thimble, parallel-plate, or spherical, the size of the chamber signal for a given radiation dose is related to the volume of the cavity. Thus, the volumes of the different geometrical shapes become important since the volume of the collecting region in the chamber generally determines the response of the chamber.

2.2.1 Thimble Chambers

The thimble-type chamber can be used as a model to describe the essential elements of an ionization chamber as well as the characteristics of chambers. Figure 2.2 shows a schematic of a thimble chamber connected to the electrometer through a triaxial cable. The thimble chamber was developed by Farmer [8] (Aird and Farmer, 1972) and is commonly called a Farmer chamber. It has been a standard ionization chamber used by many clinics. The electrodes of the chamber (the collector, the guard, and the outer shell or thimble) are connected through the triaxial cable to an electrometer. The collector of the chamber delivers the current resulting from ionization of the air volume in the chamber to an electrometer device. The electrometer registers the current or the charge collected in a given time period. The electrometer also supplies the high voltage that holds the collector at this potential. When a standard voltage of 300 V is applied between the guard and the thimble, the electric field lines remain uniform across the chamber, defining the collecting volume. The thimble wall contains the air volume. Any ionization occurring in the collecting volume of the chamber results in the positive and negative ions being pulled either to the collector or to the thimble wall. Most often, the collector voltage is positive with respect to the outer shell and results in the collection of negative charge. There are corrections necessary for the operation of ionization chambers. A typical volume used for Farmer chambers is 0.6 cm^3, although there is a range of volumes as small as 0.007 cm^3.

2.2.2 Parallel-Plate Chambers

In a parallel-plate ionization chamber, an electric field is generated between the window and the guard and the window and the collector. The cylindrical separation of the collector and the window defines a collecting volume that is filled with air. Figure 2.3 is a schematic of this type of chamber. The guard electrode, which provides a uniform electric field and a defined volume, is a donut-shaped ring in the plane of the collector with a high voltage applied between the guard and the inner surface of the window. Thus, the collector will collect charge from all the ions that are generated in the gas between the plates. Generally, these chambers have a thin window usually composed of a few microns of conductive

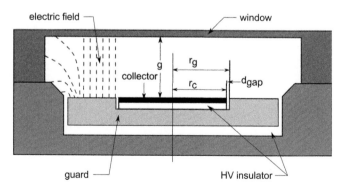

FIGURE 2.3 Schematic of parallel-plate chamber with field lines shown.

Mylar®, Kapton®, or other thin, conductive material. The window would be oriented toward the beam. Most measurement protocols recommend parallel-plate ionization chambers for electron beams with energies below 10 MeV [9]. The plate separation for most of the parallel-plate ionization chambers used for electrons falls between 1 and 2 mm, resulting in a negligible change in beam intensity across the sensitive volume. The small plate separation gives better spatial and depth resolution than cylindrical chambers in beams with a large gradient intensity along the beam axis. Commonly, a parallel-plate chamber is also the type of ionization chamber used for diagnostic exposure measurements. However, the volume of the chamber in this case must be larger since the dose rate is lower. Parallel-plate ionization chambers are calibrated with their plates oriented perpendicular to the beam axis and their window toward the x-ray beam. For diagnostic radiological applications, the electrometer reading is converted from coulombs to an air kerma value measured in units of gray (Gy) [joules per kilogram (J/kg)], roentgens (R), or coulombs per kilogram (C/kg). Again for parallel-plate chambers there is a range of volumes, with 0.62 cm³ being used frequently, and a range of chamber diameters, with 20 mm as an average.

2.2.3 Spherical Chambers

Spherical chambers (such as the one shown in Figure 2.4) with precisely known volumes are used for the primary standard determination of air kerma for ^{60}Co or ^{137}Cs

beams in the United States. (A known volume cylindrical chamber is also used.) Historically, spherical chambers have been used as air-kerma reference chambers in the clinic. Similar characteristics as given for thimble chambers also apply to spherical chambers. A typical volume for spherical chambers is 3.6 cm³. These chambers can become very large, up to 15,700 cm³.

2.2.4 Brachytherapy Well or Re-Entrant Chambers

Brachytherapy is the use of radioactive sources inserted into body cavities or into body tissue. Radiation from brachytherapy sources is measured using a well chamber that has its collector between two ground plates in a cylindrical configuration and a guard at the bottom of the chamber. This type of chamber is elaborated upon in the AAPM 2009 summer school [10]. The radioactive source is inserted into the center of the cylinder, referred to as a well. The strength of a brachytherapy source is specified as the air-kerma rate that the source would produce at one meter in vacuum, called the air-kerma strength, S_K. A complete discussion on the clinical application of these chambers for brachytherapy dosimetry is provided in Chapter 15 of this book [11].

2.3 OTHER COMPONENTS OF THE DOSIMETER SYSTEM

Besides the ionization chamber, the system is composed of the electrometer and the triax cable. Each of these components has its own characteristics that affect the readout.

2.3.1 The Triax Cable

The ionization chamber current is transferred to the electrometer for measurement via a low-noise triaxial cable. A schematic of a triaxial cable is shown in Figure 2.5. A low-noise triaxial cable has insulating qualities that reduce electrical noise from triboelectric and piezoelectric effects in the cable. Triboelectric currents are generated by charges that are created at the interface

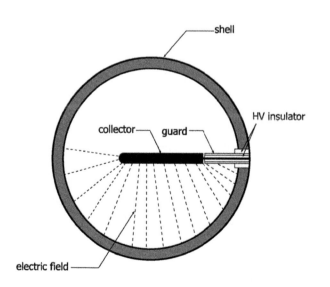

FIGURE 2.4 Schematic of a spherical chamber with field lines shown.

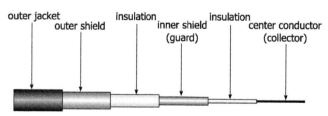

FIGURE 2.5 Schematic of a triaxial cable.

between a conductor and an insulator due to friction when the cable is bent or flexed. Piezoelectric currents are generated when mechanical stress is applied to certain insulating materials. Minimizing the extraneous small signals is of the greatest importance while maintaining applicability. Flexibility of the cable is one of the properties required for convenience. The cable should be positioned in a relaxed state, avoiding twisted coils and sharp bends that induce mechanical stress. Trauma can break the delicate collecting electrode or degrade the insulating qualities. The guard connector provides a contiguous guard throughout the length of the cable. Connections for the cable have to be secure with good insulation between the layers. Also, the cleanliness of the connectors is of extreme importance since one of the major causes of leakage is a dirty connector. Generally, good-quality cables and connectors have very low leakage ($\leq 10^{-15}$ A, when 300 V is applied). A low capacitance per meter is also desirable for a good triaxial cable. Low-capacitance cables allow ionization chambers to have a fast equilibration time following any change in an applied voltage. This fast equilibration time prevents continually increasing or decreasing readings after a change in voltage.

2.3.2 Electrometers

The current produced by ionization chambers for medical physics applications is typically between 0.5 pA and 800 nA. To measure and display such a low-magnitude current, or to integrate the current over time to display a charge collected, requires an instrument with an extremely high input impedance. Starting in the 1970s, electrometers began to be manufactured to meet the requirements of the medical physicist and have been refined to include more capabilities and features. As mentioned, an electrometer is a high impedance measuring device for which low currents or charges are detected such that the device itself does not affect the measurement. There are two basic circuits, one or both of which are found in the preamplifier (front end) of every electrometer. Both circuits are impedance converters in that a high-impedance input is converted to a low-impedance output. Low-level measuring techniques are described in the *Low Level Handbook* [12]. To measure charge, an operational amplifier is configured as an inverting amplifier with a feedback capacitor, as illustrated in Figure 2.2. The amplifier output voltage is a measure of charge transferred from the ionization

process to the feedback capacitor and is related by the capacitance definition shown in Equation 2.1:

$$Q = CV \text{ or } V = \frac{1}{C} \int_0^T I(t)dt = I \times T/C, \text{ if constant current}$$

$$(2.1)$$

where

Q = charge in coulombs
C = capacitance of feedback element in farads
V = voltage across feedback capacitor = V_{OUT}
I = ionization current in amperes
T = integration time in seconds (exposure time)

This coulombmeter will require feedback capacitor values from 0.001 to 1.0 µF depending on the magnitude of charge. Different capacitor values are connected into the circuit by a rotary switch or by an external plug-in module.

For current measurements, an operational amplifier is configured as an inverting amplifier with a feedback resistor. So, replacing the capacitor of Figure 2.2 with a resistor, the amplifier output voltage is a measure of input current related by Ohm's law:

$$V = IR \qquad (2.2)$$

where

V = voltage across feedback resistor = V_{OUT}
I = ionization current in amperes
R = resistance of feedback resistor in ohms

This ammeter will require feedback resistors from 10^7 to 10^{12} ohms, depending on the magnitude of current. Different resistor values are connected into the circuit by a rotary switch, relay, or external plug-in modules. The particular feedback element that is selected in either of the two circuits represents a specific "scale" or range of measurement values. Zero drift current and leakage current values are generally measured only for a charge scale calibration of an analog electrometer (coulombmeter). Most electrometers have the capability of nulling out display offsets caused by the background current. This nulling, or zeroing, process should be performed with the cable and chamber attached, the HV (high-voltage) bias on, and the display stabilized. All electrometers need a "warm-up" time when they are first turned on, so do not zero until after the warm-up has completed. Zero drift current is obtained by first zeroing the charge scale display and then observing the change in reading during a measurement interval. Electrometer leakage current is determined by injecting a half-scale charge value. The source of the charge is then removed, and the decay rate of the reading is determined. A good electrometer should have less than 0.1% change in the reading over the period of one minute. High humidity may increase leakage in some electrometers.

In what can be classified as a "digital" electrometer, the preamplifier voltage is digitally manipulated. For example, a "charge digitizer" electrometer has a coulombmeter preamplifier and digitizes the signal by producing an output pulse for a specific value of input charge (charge bucket). Each time V_{OUT} reaches a certain level, the amplifier is reset, and a pulse is counted. The unit quantifies the charge collected by counting the number of pulses and multiplying by the charge bucket size. The charge measurement resolution is determined by the bucket size. The digital electrometer measures the current by dividing the measured charge by the time interval.

Another type of digital electrometer is a "numerical current integrator" that incorporates the transresistance amplifier configuration. Current is measured by sampling and averaging the preamplifier output. Collected charge is calculated by a microprocessor, which numerically integrates the current. Scale selection determines the feedback resistance and the computing algorithm. Note that each scale of this type of digital electrometer is rate limited. Thus, for a given charge scale, the maximum current, rather than the maximum charge reading, is usually the limiting factor.

The electrometer also applies the high-voltage bias to the ionization chamber through the triaxial cable. As illustrated in Figure 2.2, the high-voltage supply is connected between the outer shield of the triaxial connector and the guard. With the inverting feedback configuration of the preamplifier, the inverting input (collector) is actively forced to maintain equal potential with the noninverting input (guard). Thus, the collector, guard, and all other circuitry "common" to the guard "float" at the high-voltage bias level with respect to the outer shield. If the collecting electrode and the outer shield are shorted together, the input circuitry of the electrometer can be destroyed.

Properly defined voltage requires an indication of the magnitude and the polarity. In the United States,

300 V is the typical high-voltage bias magnitude. In Figure 2.2, the guard (and collector) is +300 V with respect to the outer shell. This perspective is commonly referred to as the "collecting electrode potential." Some electrometer manufacturers reverse the reference point, indicating the outer shell potential with respect to the guard. This perspective is commonly referred to as the "thimble bias." Thus, the thimble bias shown in Figure 2.2 is −300 V. The high-voltage bias configuration in Figure 2.2 will produce a negative charge collection. Unfortunately, the manufacturers do not always label the bias polarity selector switch to indicate whether it is the collecting electrode potential or thimble bias. There may even be inconsistency between models from a particular manufacturer. In addition, some electrometer models display the opposite polarity of the measured charge or current. Generally, the high-voltage bias is measured at the input connector and reported as the guard voltage with respect to the outer shell for each corresponding bias selector setting of the electrometer.

The scales and lowest reading capability of an electrometer are of great importance. Scale selection should ideally provide a reading of four significant digits so that the reading precision is always within 0.1%. Therefore, if the reading is a picoampere, the last digit should be a femtoampere. The lowest end of the measuring range should be at least 200 times larger than the digital resolution, which is the smallest significant increment of the reading. An electrometer may not display the "true" charge or current value. Thus, a "scale-specific" calibration coefficient must be applied to obtain the correct measurement value. A charge scale electrometer calibration coefficient (P_{elec}) has units of coulomb/reading (C/rdg), where "rdg" represents the units of the readout display. A current scale P_{elec} has units of ampere/reading (A/rdg). Although the calibration coefficient is often 1.000 C/rdg or 1.000 A/rdg, it should not be assumed a priori to be unity.

2.4 CHARACTERISTICS OF IONIZATION CHAMBERS

Ionization chambers require some corrections that are related to the basic nature of the chamber. These corrections are made to the measurement value. The ionization chamber should be calibrated by an ADCL to determine a calibration coefficient, $N_{D,w}$. The equation used to determine dose for therapy applications (D_w^Q) is given in Task Group 51 (TG-51) [9, 13]. The TG-51 dose equation is given in Equation 2.3 and M_{raw} is listed in Equation 2.4, with all of the corrections required for the ionization chamber. In addition, the electrometer should also be calibrated and is assigned a calibration coefficient, P_{elec}. This factor, P_{elec}, is usually close to 1.000 C/rdg, but it can vary from 0.995 to 1.005, or more, from unity. In some countries, chamber and electrometers are calibrated as a single system. For diagnostic radiology measurements, the electrometer and chamber are generally calibrated as a system to read out directly in air kerma or exposure. The TG-51 dose-to-water formalism is

$$D_w^Q = M k_Q N_{D,w}^{60Co}, \text{ and} \quad (2.3)$$

$$M = M_{raw} P_{T,P} P_{ion} P_{pol} P_{elec} P_{leak} P_{rP} \quad (2.4)$$

where

k_Q is the conversion from cobalt calibration energy to the energy of the linac

$N_{D,w}^{60Co}$ is the chamber calibration coefficient determined at cobalt energy

M_{raw} is the raw reading of the chamber

P_{TP} is the temperature – pressure or gas density correction factor

P_{ion} is the ion recombination correction factor

P_{pol} is the polarity correction factor

P_{elec} is the electrometer correction factor

P_{leak} is the correction factor from any contribution to the measured reading that is not due to ionization in the chambers collecting volume

P_{rP} is the correction factor accounting for any off-axis variation in the intensity profile of the radiation field over the sensitive volume of the ionization chamber. For further discussions of this correction factor, the reader is referred to the TG-51 addendum [13].

2.4.1 Beam-Quality Dependence (Energy Dependence)

The response of ionization chambers can vary with the energy of the beam, which is accounted for by the beam quality correction, k_Q. Values for the k_Q are given in the addendum [13] or in the TG-51 protocol [9]. This energy dependence is also important for electron beams, orthovoltage and superficial beams (x-ray beams), and for diagnostic x-ray energies. The major requirement for some diagnostic applications is that the air-kerma sensitivity of the chamber varies by less than 3% as a function of beam quality (energy) [14], since a measurement of half value layer can be affected by the variation in air-kerma sensitivity [15].

2.4.2 Pressure/Temperature (Air Density) Effects

The number of ions collected, or the rate of their collection in the gas of ionization chambers, is the signal that is recorded, and this depends on the amount of gas, usually air, in the volume. The calibration coefficient is stated at a standard temperature and pressure (usually 101.3 kPa (760 mm Hg) for pressure and 22°C in North America but 20°C in most of Europe). The density of the air in the chamber must be corrected to these standard conditions during the measurement, since the calibration coefficient assigned to an ionization chamber specifies the charge collected in the chamber per unit quantity of radiation in absence of the chamber. The collecting volume of the chamber is often open to the outside atmosphere. Thus, the mass of air is dependent on conditions of temperature, pressure, and humidity. The temperature and pressure should be measured in the clinic without reliance on outside values. The paper by Rogers and Ross [16] deals with humidity effects and shows that the humidity effect can be considered constant within ±0.15% over a relative humidity range of 30–80%. On the other hand, temperature and pressure effects can be significant since the mass of air in the chamber depends on these atmospheric conditions. The equation (Equation 2.5) expressing this correction, P_{TP}, is derived from the ideal gas law (see [17]). The temperature and pressure must be obtained at the time of measurement. The correction for temperature and pressure is given by:

$$P_{TP} = \left(\frac{T_a + T_0}{T_S + T_0} \right) \left(\frac{P_S}{P_a} \right) \quad (2.5)$$

where

P_a is the ambient pressure in kPa

P_S is the standard pressure (101.3 kPa)

T_a is the ambient temperature in °C

T_0 is the temperature in kelvin at 0°C (273.15 K)

T_S is the standard temperature (22°C in North America)

Higher altitudes have a greater P_{TP} correction due to the lower pressure. It is common to have corrections of more than 10% in mountainous areas. When a calibrated device is used, the physicist must correct the reading using Equation 2.5 to that which would be obtained at standard temperature and pressure. However, for certain chambers with non–air-equivalent components, the standard correction breaks down for low-energy radiation [17–20]. This breakdown occurs when the ranges of the electrons entering the chamber cavity are short with respect to the dimensions of the cavity, and when the photon cross sections differ between the chamber wall and air. La Russa and Rogers [17] demonstrated such an effect for thimble-type chambers in kilovoltage x-rays using Monte Carlo calculations and later confirmed the results with an experimental investigation [20]. The investigators found a smaller deviation than observed for well-type ionization chambers, with a measured 2.5% effect at 0.7 atm for an unfiltered 60 kV x-ray beam with a graphite-walled thimble chamber. An effect of 13% was measured at an air density typical of Mexico City (0.76 atm) for a modified thimble chamber with an aluminum wall, using the same 60 kV x-ray beam. A chamber with an aluminum wall is not common and was used only to demonstrate the effect.

2.4.3 Recombination Effects

Ion recombination effects occur in ionization chambers when ions of opposite charge recombine before they reach the collecting electrode or chamber wall. This leads to an incomplete charge collection, and a correction should be applied for it. Early research in this area was performed by Boag [21, 22], and it has been an area of continued interest for clinical radiation dosimetry (e.g., [7, 23–26]). The recombination effects can affect small-volume chambers in a significant way [7]. Three mechanisms are involved in ion recombination, and all of these have a dependence on the electric field strength in the ionization chamber. The first is initial recombination, in which oppositely charged ions from the same charged particle track meet and recombine. This mechanism is generally only important for high linear energy transfer (LET) radiation such as alpha particles or for gases at high pressure [1, 21]. The second mechanism is ionic diffusion, in which ions diffuse to the collecting electrode or chamber wall against the electric field [27]. The third mechanism is general (or volume) recombination where oppositely charged ions arising from different charged particle tracks meet and recombine. Initial recombination and ionic diffusion have no dependence on the dose

rate, but the occurrence of general recombination will increase with increasing dose rate.

Scott and Greening [28] demonstrated that for a continuous radiation beam (e.g., ^{60}Co or kilovoltage x-rays), a linear relationship exists between the inverse of the collected charge and the inverse of the electric field strength squared when general recombination was the dominant recombination process. Boag and Currant [23] presented a "two-voltage" technique for determining the general recombination for pulsed beams (e.g., linear accelerators), and Weinhous and Meli [29] presented a numerical method to evaluate the general recombination correction for these beams using the two-voltage technique. For measurements in pulsed beams close to saturation, the relationship between the inverse of the collected charge versus the inverse of the electric field strength is close to linear, so the correction from Weinhous and Meli was simplified in the TG-51 report [9] to using a half voltage. The underlying assumption for the two-voltage technique is that the chamber is behaving according to Boag theory [23]. Equation 2.6 shows the relation as recommended in TG-51,

$$P_{ion}(V_H) = \frac{1 - \left(\dfrac{V_H}{V_L}\right)}{\left(\dfrac{M_{raw}^H}{M_{raw}^L}\right) - \left(\dfrac{V_H}{V_L}\right)} \qquad (2.6)$$

where V_H and V_L are the high and low voltages, respectively, and M_{raw}^H and M_{raw}^L are the readings at the high and low voltages, respectively. The TG-51 report states that for a voltage ratio of 2 and a value of P_{ion} less than 1.05, Equation 2.6 should be within 0.2% of the exact equation for pulsed beams and within 0.4% for pulsed swept beams. Since the measured charge is related to the square of the electric field (or voltage) for continuous beams, the general recombination correction for those beams is

$$P_{ion}(V_H) = \frac{1 - \left(\dfrac{V_H}{V_L}\right)^2}{\left(\dfrac{M_{raw}^H}{M_{raw}^L}\right) - \left(\dfrac{V_H}{V_L}\right)^2}. \qquad (2.7)$$

Ionization chambers used for radiation therapy usually use an applied voltage of ±300 V, and the value of P_{ion}

can be up to 5% depending on the type of chamber and dose rate. Larger values of P_{ion} can indicate a malfunctioning ionization chamber and/or electrometer.

Recombination effects can be reduced by increasing the rate of collection of the ionization products produced in the chamber. This collection rate is dependent on the mobility of the ionization products and the electric field strength. A common method to reduce recombination effects is to increase the applied potential so the chamber operates nearer to the saturation region (Figure 2.1). The chamber will asymptotically approach full saturation (complete charge collection) with increasing applied potential, but eventually this can lead to electrical breakdown of the chamber components or gas multiplication in the air. Sometimes, depending on the chamber volume, the voltage should be reduced to be able to operate in the ion chamber region. It is always important to measure at a number of voltages from low to high to be sure the operation of the chamber is in the plateau of the ion chamber region.

2.4.4 Polarity Effects

Ionization chamber polarity effects refer to the difference observed when negative charge versus positive charge is collected. Polarity differences can occur when radiation interactions in the collecting electrode of the ion chamber add or subtract from the measured signal. This effect is known as the Compton current. Current arising from ionizations outside the defined collecting volume, known as extracameral current, can also lead to polarity effects. Kim et al. [30] have associated polarity effects with electric field distortions due to potential differences between the guard electrode and the collecting electrode. In general, this effect is larger for parallel-plate chambers than for thimble chambers and depends on the type of irradiation (i.e., photon beam or electron beam) as well as the depth of measurement [31]. A correction for this effect is determined by performing measurements with positive and negative voltages equal in magnitude. The correction is then determined using the equation from the TG-51 protocol [9]. If sufficient time for the chamber to reach equilibrium after changing voltage is not allowed, the polarity correction would be greater than the chamber actually has, leading to an incorrect measurement. The first polarizing voltage measurement should be repeated to correct for any chamber drift. The polarizing correction is then calculated by Equation 2.8:

$$P_{pol} = \left| \frac{\left(M_{raw}^+ - M_{raw}^- \right)}{2M_{raw}} \right| \tag{2.8}$$

where

M_{raw}^+ is the reading when positive charge is collected

M_{raw}^- is the reading when negative charge is collected

M_{raw} is the reading corresponding to that polarity generally used in the clinic

The type of charge normally collected should be the same as for the chamber calibration. Typical polarity corrections should be between 0.997 and 1.003 (i.e., within ±0.3% of unity). The TG-51 protocol recommends that if users measure a polarity correction outside this range for a photon beam of 6 MV or lower energy, they should determine the polarity correction for the ^{60}Co calibration beam and apply the appropriate correction.

2.4.5 Leakage Currents

The leakage current for the system is measured using the same measurement setup with the accelerator but without the beam on. Although leakage may originate from any component (chamber, electrometer, or cable), leakage generally originates in the cable. If a high leakage is measured, it is suggested to change cables and check it again. The leakage should contribute to less than 0.1% of the signal. At this level or less, leakage is generally ignored and P_{leak} is set to 1.000.

2.5 CONCLUSION

As has been explained in this chapter, a good understanding of ionization chambers and the components involved in the dosimetry system allows the clinical physicist to use them correctly to determine the dose delivered to the patient. Calibration of the ionization chambers and electrometers should be done every 2 years so that they are traceable to a primary laboratory; this could be through an AAPM-accredited ADCL. The clinical physicist should always be aware of the condition of their cable. Special care needs to be taken so that the cable is not kinked, stepped on, or met with other trauma. Such trauma is the major source of leakage current. Various protocols have been developed and should be understood for determinations of dose as well. This chapter has followed TG-51 for the most part, but TG-51 is similar to the IAEA TRS 398. The protocols spell out all the corrections that should be applied for the chamber use.

ACKNOWLEDGMENTS

The authors would like to thank members of the Medical Radiation Research Center at the University of Wisconsin for their support and the customers of the center who provided research support for graduate students. L.A. DeWerd also has a partial interest in Standard Imaging, Inc.

REFERENCES

1. F. H. Attix, *Introduction to Radiological Physics and Radiation Dosimetry* (John Wiley and Sons, Inc., New York, NY, 1986).

2. E. B. Podgorsak (ed), *Radiation Oncology Physics: A Handbook for Teachers and Students* (International Atomic Energy Agency, Vienna, Austria, 2005).

3. P. Andreo, D. T. Burns, A. E. Nahum, J. Seuntjens, F.H. Attix, *Fundamentals of Ionizing Radiation Dosimetry* (Wiley-VCH, Germany, 2017).

4. D. W. O. Rogers and A. F. Bielajew, "Wall attenuation and scatter corrections for ion chambers: Measurements versus calculations," *Physics in Medicine & Biology* **35**, 1065–1078 (1990).

5. IAEA, "*Absorbed Dose Determination in External Beam Radiotherapy,*" Technical Reports Series No. 398 (International Atomic Energy Agency, Vienna, Austria, 2000).

6. L. DeWerd and R. Mackie, "Comment on Comparison of ionization chambers of various volumes for IMRT absolute dose verification [Med Phys 30: 119–123 (2003). Letter to the Editor]," *Medical Physics* **30**, 2264 (2003).

7. J. Miller, R. B. D. Hooten, J. A. Micka, L. A. DeWerd, Polarity effects and apparent ion recombination in microionization chambers, *Medical Physics* **43**, 2141–2152 (2016).

8. E. G. A. Aird and F. T. Farmer, "The design of a thimble chamber for the Farmer dosimeter," *Physics in Medicine & Biology* **17**, 169–174 (1972).

9. P. R. Almond, P. J. Biggs, B. M. Coursey, W. F. Hanson, M. S. Huq, R. Nath, R. D. W. O. Rogers, "AAPM's TG-51 protocol for clinical reference dosimetry of high energy photon and electron beams," *Medical Physics* **26**(9), 1847–1870 (1999). Also available as AAPM Report No. 67.

10. D. W. O. Rogers and J.E. Cygler (ed), Clinical Dosimetry Measurements in Radiotherapy, American Association of Physicists in Medicine, Monograph 34, DeWerd, et al. Chapter 6, pp. 181–204 (2009).

11. C. L. Deufel, W. S. Culberson, M. J. Rivard, F. Mourtada, "Brachytherapy dosimetry," in *Radiation Therapy Dosimetry: A Practical Handbook*, edited by A. Darafsheh (CRC Press, Boca Raton, FL, 2021), Chap. 15, pp. 231–252.

12. Keithley Instruments, Inc., 2016. *Low Level Measurements Handbook: Precision DC Current, Voltage, and Resistance Measurements,* 7th edition (Keithley/Tektronix, Cleveland, OH, 2016).

13. M. McEwen, L. DeWerd, G. Ibbott, D. Followill, D. O. Rogers, S. Seltzer, J. Seuntjens, "Addendum to the AAPM's TG-51 protocol for clinical reference dosimetry of high energy photon beams," *Medical Physics* **41**, 041501-1–041501-20 (2014).

14. L. A. DeWerd and L. K. Wagner, "Characteristics of radiation detectors for diagnostic radiology," *Applied Radiation and Isotopes* **50**, 125–136 (1999).

15. L. A. DeWerd, J. A. Micka, R. W. Laird, D. W. Pearson, M. O'Brien, P. Lamperti. "The effect of spectra on calibration and measurement with mammographic ionization chambers," *Medical Physics* **29**, 2649–2654 (2002).

16. D.W. O. Rogers and C. K. Ross, "The role of humidity and other correction factors in the AAPM TG-21 dosimetry protocol," *Medical Physics* **15**, 40–48 (1988).

17. D. J. La Russa and D. W. O. Rogers. (2006). "An EGSnrc investigation of the PTP correction factor for ion chambers in kilovoltage x rays," *Medical Physics* **33**, 4590–4599.

18. S. L. Griffin, L. A. DeWerd, J. A., Micka, T. D. Bohm, "The effect of ambient pressure on well chamber response: Experimental results with empirical correction factors," *Medical Physics* **32**(3), 700–709 (2005).

19. T. D. Bohm, S. L. Griffin, P. M. DeLuca, Jr., L. A. DeWerd, "The effect of ambient pressure on well chamber response: Monte Carlo calculated results for the HDR 1000 Plus," *Medical Physics* **32**(4), 1103–1114 (2005).

20. D. J. La Russa, M. McEwen, D. W. O. Rogers. (2007). "An experimental and computational investigation of the standard temperature-pressure correction factor for ion chambers and kilovoltage x rays," *Medical Physics* **34**, 4690–4699.

21. J. W. Boag, "Ionization measurements at very high intensities," *The British Journal of Radiology* **23**, 601–611 (1950).

22. J. W. Boag and T. Wilson, "The saturation curve at high ionization intensity," *British Journal of Applied Physics* **3**, 222–229 (1952).

23. J. W. Boag and J. Currant, "Current collection and ionic recombination in small cylindrical ionization chambers exposed to pulsed radiation," *The British Journal of Radiology* **53**, 471–478 (1980).

24. J. W. Boag, E. Hochhäuser, O. A. Balk, "The effect of free-electron collection on the recombination correction to ionization measurements of pulsed radiation," *Physics in Medicine & Biology* **41**, 885–897 (1996).

25. C. Zankowski and E. B. Podgorsak, "Determination of saturation charge and collection efficiency for ionization chambers in continuous beams," *Medical Physics* **25**, 908–915 (1998).

26. F. DeBlois, C. Zankowski, E. Podgorsak, "Saturation current and collection efficiency for ionization chambers in pulsed beams," *Medical Physics* **27**, 1146–1155 (2000).

27. J. Böhm, "Saturation corrections of plane parallel ionization chambers," *Physics in Medicine & Biology* **21**, 754–759 (1976).

28. P. B. Scott and J. R. Greening, "The determination of saturation currents in free-air ionization chambers by extrapolation methods," *Physics in Medicine & Biology* **8**, 51–57 (1963).

29. M. S. Weinhous and J. A. Meli, "Determining P_{ion}, the correction factor for recombination losses in an ionization chamber," *Medical Physics* **11**, 846–849 (1984).

30. Y.-K. Kim, S.-H. Park, H.-S. Kim, S.-M. Kang, J.-H, Ha, C.-E. Chung, S.-Y, Cho, J. K. Kim, "Polarity effect of the thimble-type ionization chamber at a low dose rate," *Physics in Medicine & Biology* **50**, 4995–5003 (2005).

31. B. J. Gerbi and F. M. Khan, "The polarity effect for commercially available plane parallel ionization chambers," *Medical Physics* **14**, 210–215 (1987).

CHAPTER 3

Calorimetry

Larry A. DeWerd

University of Wisconsin
Madison, Wisconsin

Blake R. Smith

University of Iowa
Iowa City, Iowa

CONTENTS

3.1 INTRODUCTION

Calorimetry is used to realize the primary standard for absorbed dose to water. A primary standard is a device that measures a physical quantity directly from the fundamental quantities of substance quantity (mol), length (meter), mass (kilogram), time (second), charge or current (Coulomb or Ampere), temperature (Kelvin), and light intensity (candela). The basis of calorimetry is simple, but its practice is complicated. A good summary is given in Chapter 15 of the AAPM 2009 summer school [1]. Historically, in 1904, Curie and Dewar developed the first cryogenic calorimeter and the history continues to the present day at various temperatures. Calorimetry is the determination of energy absorbed, generally expressed in a temperature rise. For water, the increase in temperature is approximately 2.4×10^{-4} K/Gy or about 0.24 mK/Gy. There are complicating factors involved in water calorimetry that will be addressed later. The quantity of absorbed dose, D_m, for a material, m, is determined from a change in energy, ΔE, that is measured from the material's temperature change

$$\Delta E = \Delta m \, c_p \Delta T \qquad (3.1)$$

where c_p is the specific heat capacity in J kg^{-1} K^{-1} of the material and ΔT is the change in temperature (K) caused by the absorbed radiation. The dose to the material, D_m, is then determined from the known mass, Δm, of the absorbing medium.

$$D_\mathrm{m} = \frac{\Delta E}{\Delta m} = c_p \Delta T \qquad (3.2)$$

The assumption for Equation 3.2 is that all of the energy that is deposited is expressed in terms of heat (temperature rise) with no heat defect (discussed later). If a rate of change in temperature is measured, then the net power (in J s^{-1} or Watt) is expressed by the equation

$$P_{net} = \Delta m \, c_p \frac{dT}{dt} \qquad (3.3)$$

Generally, the temperature is measured with a thermistor, connected via a Wheatstone bridge illustrated in

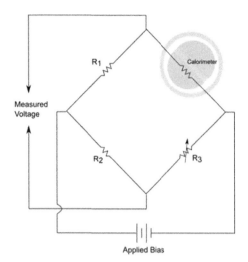

FIGURE 3.1 Diagram of a Wheatstone bridge circuitry commonly used to read out calorimeters. The voltage measured across the bridge provides a measure of the change in resistance from a thermistor placed on the core.

Figure 3.1. A typical bridge output signal is shown in Figure 3.2, with the calorimeter being insulated. The design of all calorimeters is to maximize the ratio of intended signal to the background noise while maintaining simplicity. The thermistor is probably the limiting factor due to self-heating effects. However, control of the environment is also very important; this results in the phantoms and the calorimeter being as isolated as possible, surrounded by well-controlled, insulated phantom materials.

A typical calorimeter consists of three fundamental components: the core, the jacket, and the shield. The core of a calorimeter is the sensitive portion of the detector that the measured change in temperature is referenced. The jacket surrounds and thermally isolates the core from the outside environment. Several thermometers and heaters are placed on the core and jacket to help regulate quasi-adiabatic conditions so that the temperature rise of the core is proportional to the energy absorbed. The shield helps to thermally isolate the jacket and core system. Conventionally, all three components are constructed from the same material and are thermally isolated from one another.

There are three different ways a radiometric calorimeter can operate. The most common method is *heat flow* in which the change in temperature from the absorber to the jacket is monitored. Temperature changes are not expected to be constant and unchanging before or after the irradiation due to the convection of heat from the core to surround material. Specifically, if the system's temperature is linearly changing at some constant rate, then the

change in the system's temperature will result in corresponding changes to their dependent (i.e., temperature) variable intercept on the respective axis. *Adiabatic* techniques fix the absorber and the surrounding material at the same temperature, thereby accounting for any thermal disequilibrium between the core and the surrounding material. The last method, which is arguably the most difficult, is an *isothermal* technique. Isothermal calorimeters electronically cool the core to balance the heat energy imparted. The power used to keep the temperature stable is the measure of the heat energy imparted by the radiation. Graphical illustrations of the calorimeter response are shown in Figure 3.2.

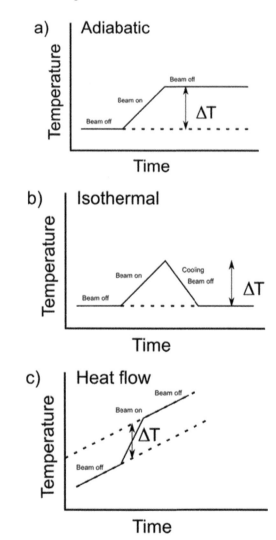

FIGURE 3.2 Schematic of typical bridge output signal for an adiabatic (a), isothermal (b), and heat flow (c) system. The time frame that the beam of radiation is turned on and off, and its impact on the system, is defined into three regions for each case. The measure of temperature change, ΔT, is then used to calculate the dose to the medium.

The two types of calorimeters used in the past by primary laboratories have been a graphite calorimeter or a water calorimeter. Of course, the preferred quantity is absorbed dose to water, but the procedure for water calorimetry is more involved. Water calorimetry has a heat defect that must be taken into account. A heat defect arises when some of the energy deposited in the material is not liberated as heat because of chemical reactions or other effects, exothermic or endothermic [2]. Water can have a significant heat defect, which depends on the impurities and/or dissolved gases. Any perturbations from the surrounding materials must also be taken into account. If graphite is used, there needs to be correction of the absorbed energy from graphite to water. Graphite does not have a significant heat defect [2].

3.2 GRAPHITE CALORIMETERS

The modern era of graphite calorimeters can be traced back to Domen and Lamperti [3]. A number of primary laboratories use this type of calorimeter as their primary standard. This type of calorimeter measures the temperature change of the core and determines the quantity of dose to water through a series of analytical conversions that relate the energy deposition behavior between the core material and the liquid water. This dose can then be used to calibrate ionization chambers. An illustration of a graphite calorimeter initially proposed by Petree and Lamperti [4] is shown in Figure 3.3.

The graphite core is surrounded by a graphite phantom, and this entire arrangement must be thermally isolated to eliminate any extraneous effects. A measure of temperature change in the core is determined relative to the shell that is held at a constant temperature, thus acting in an adiabatic fashion previously discussed. Temperature changes are measured with a set of thermocouples. This is in contrast to water-based calorimeters, which often rely on resistance changes in an electronic system, rather than voltage changes, to detect changes in temperature. To determine dose also necessitates an accurate measurement of the core's mass. Although this primary standard is used in many primary laboratories, the desirable calorimeter is a water calorimeter as no material absorbed dose correction, and thus assumption upon the primary beam of radiation, is required. This type is further explained in the next section.

Graphite calorimeters have also taken on more compact forms and streamline operation. Contemporary graphite and water calorimeters are very fragile, cumbersome, and require specialized equipment and experience to perform a successful measurement. Probe calorimeters have been proposed and shown as a potential integration of absolute graphite dosimetry, which can be performed at a clinic using the resources conventionally available to a medical physicist [5]. The proposed instrument has been operated using both adiabatic and isothermal modes. A Wheatstone bridge is built into the detector electronics, which relates a voltage difference measured using a voltmeter. An illustration of a graphite probe for MV dosimetry is shown in Figure 3.4.

While these compact devices show promise for personalized, absolute dosimetry, they still face several challenges. The sensitive regions (shield, insulation,

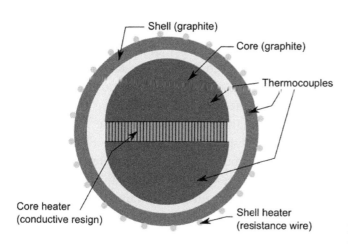

FIGURE 3.3 A cross-sectional diagram of a spherical graphite calorimeter with a solid core. The initial calorimeter design by Petree and Lamperti [4] was 2.05 cm in diameter.

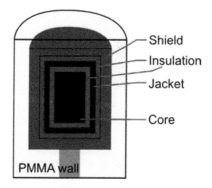

FIGURE 3.4 Diagram schematic of a graphite calorimeter probe proposed by Renaud et al. [5]. The size and sensitive detection volume is close to that of a farmer-type ionization chamber. Several thermistors are placed through the shield, jacket, and core to measure relative temperature changes due to the incident radiation.

and jacket) are layered in insulative materials, epoxy, and thermistors that act as impurities which result in thermal conduction. Accurate measurement of the core mass can also lead to an offset for absolute measurements. Additionally, the expected accuracy of these devices is estimated to be on par with a calibrated ionization chamber rather than their standard calorimeter counterparts due to the limited thermal isolation among components, which can be achieved within these compact devices [5].

3.3 WATER CALORIMETERS

Water calorimeters are the primary standard for Task Group 51 (TG-51) calibrations. There have been a number of varieties built in prior years. At the National Institute of Standards and Technology (NIST), the calorimeter in use is the sealed water calorimeter [6]. This calorimeter uses a sealed glass vessel surrounded by water in the phantom. Sealed water calorimeters are advantageous in that they can reduce impurities and gas exchange with the water, as the medium from which temperature changes are measured is independent of the surrounding calorimeter. As such, a very tight quality control over the water and volume of water within the vessel can be achieved. An example of a glass-vessel water calorimeter is shown in Figure 3.5.

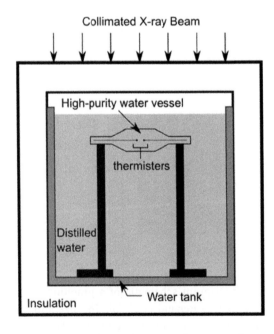

FIGURE 3.5 A sealed-vessel water calorimeter design from Domen [6] that measures the change in temperature from the change in resistance between two thermistors.

Alternatively, the glass vessels are filled with well-known quantities of pure, dissolved gases in addition to high-purity water. In this way, the influences of the impurities within the medium are known and can be accurately corrected. A caveat, though, to sealed vessels is their complex design and introduction of material inhomogeneities within the system.

The temperature of the surrounding phantom is maintained at 4°C to minimize the possibility of convection as liquid water is densest at this temperature. The water phantom dimensions are $30 \times 30 \times 30$ cm^3 containing the vessel at a reference depth. The dose to water is then determined by

$$D_w = \frac{c_w \Delta T}{1 - k_{HD}} k \tag{3.4}$$

And

$$k_{HD} = \frac{E_a - E_h}{E_a} \tag{3.5}$$

where k_{HD} (which can be positive or negative) is the heat defect; $1/(1 - k_{HD})$ is the correction factor. E_a is the energy absorbed by the irradiated water, and E_h is the energy which appears as heat. The other k is the "foreign mass" correction and represents various correction factors such as the effect of the glass in contact with the water [7]. With research about the heat defect, impurities in the water and gasses are found to affect the value of k_{HD} [2]. Water saturated with pure hydrogen or nitrogen gas has a heat defect that is well known and has also been found to reduce the amount of heat defect [2]. This heat defect remains one of the areas for a greater uncertainty in the measurement.

The components composing a water calorimeter are not truly isolated from one another. As such, conduction occurs among the components including the thermistor probe, containment vessel, and the surrounding medium with respect to the relative dose distribution heating the surrounding water at different rates. The thermistor probe is a metallic solid within the calorimeter, which acts as a source of unwanted heat. Likewise, water has a high thermal capacity and will transfer heat energy to any surrounding material with a lower heat capacity. For example, a glass containment vessel has about a 1/5th lower heat capacity than water. The water by itself also experiences some conduction due to the physical dose distribution of the beam. However, the temperature probes for modern calorimeters are able to discriminate dose to a very small region that is nearly negligible relative to the dose gradients resulting from therapeutic x-ray radiation beams but could be larger for nonstandard beams and small fields. Holistically speaking, a calorimeter requires some time to achieve thermal equilibrium due to the heat dissipation, which is dependent upon the location of materials and their abundances relative to one another.

Convection is the tendency for an aqueous material to migrate due to changes in temperature; hot material rises, and cold material sinks with respect to the direction of gravity. The gradients in the dose distribution from the primary beam, both laterally and distally, will cause a gradient in the heat energy imparted to the medium and induce convective currents for fluid media. One of the simplest methods to minimize the effects of convection is to direct the incident radiation from the top of calorimeter's water tank. However, there are instances where this may not be ideal, such as for interferometry-based calorimetry. An alternative, albeit more involved, method is to maintain a water temperature at 4°C. At this temperature, water reaches its maximum density and the influences of convection are negligible.

In addition to thermistor and thermocouple water calorimeters, ultrasound and interferometer-based calorimetry can also be used to measure the heat changes in water from radiation. These alternative calorimeters have an advantage over solid-state calorimeters in that their detector components do not interact within the primary beam. This removes some of the issues regarding component convection previously discussed. An acoustic calorimeter measures the change in the speed of sound wave propagation in water due to the temperature changes in the water. Time-of-flight techniques have been used successfully to measure these heat changes by emitting a tone-burst pulse with a known frequency and comparing the phase change measured with a high-precision ultrasonic phase detector [8]. An illustration of this type of water calorimeter is shown in Figure 3.6.

A tone burst with a well-known phase, φ, is related to the central frequency, f, by

$$\varphi = 2\pi f \tau = 2\pi f \frac{2L}{\bar{v}} \qquad (3.6)$$

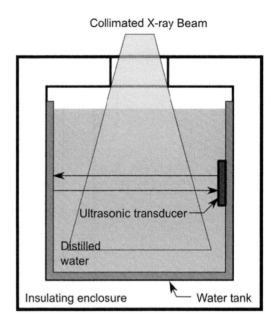

FIGURE 3.6 An ultrasonic calorimeter setup similar to the one published by Malyarenko et al. [8]. A 5 MHz transducer is acoustically coupled to the side of the water tank wall. The transducer tracks changes in the emitted frequency in high-purity water due to temperature changes in the water and wall from the incident radiation.

where the time of flight, τ, is equal to the time it takes the ultrasound pulse to travel an acoustic pathlength, $2L$, at an average velocity of $\bar{v}$ in water. Assuming the water and enclosure are in thermal equilibrium, the increase in temperature will cause an expansion of the walls and increase the acoustic path length by ΔL and also change the acoustic impedance of the water resulting in a change in the speed of the ultrasound pulse by Δv. From Equation 3.6, the change in frequency can be related to the changes in path length and velocity by

$$\frac{\Delta\varphi}{\varphi} = \frac{\Delta L}{L} - \frac{\Delta v}{\bar{v}} \qquad (3.7)$$

The increase in temperature, which caused the changes in path length and velocity, can be approximated by [9],

$$\frac{\Delta L/L}{\Delta v/\bar{v}} \approx 0.036 \frac{\Delta T_{\text{wall}}}{\Delta T_{\text{water}}} \qquad (3.8)$$

Due to the velocity of sound in water, subsequent measurements can be performed on the order of a millisecond, which allows near real-time measurement of dose imparted to water.

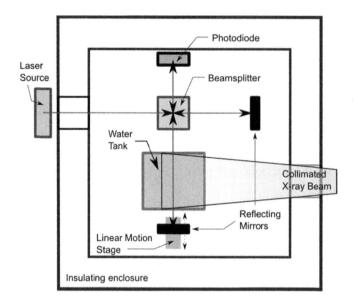

FIGURE 3.7 An interferometer calorimeter design from Flores-Martinez et al. [11]. A He-Ne laser with an optical wavelength of 632.8 nm is split between a stationary reference and measurement arms. The measurement arm was translated using a piezoelectric transducer to create a phase distribution observed using a photodiode.

An alternative, noninvasive calorimetry technique to ultrasonic methods is interferometry. The interferometer was initially proposed by Michelson and Morley [10] in an attempt to detect the luminiferous ether. An interferometer measures the change in phase from an interference pattern observed at the intersection of a single laser split along a fixed and variable-position measurement arm. A diagram for an interferometer-based calorimeter is shown in Figure 3.7.

The refractive index of a material is related to its temperature. An interferometer can therefore be used to measure the dose to a sample of water by measuring the phase shifts, which occur as the water is heated from the incident radiation. There are several empirical and theoretical mathematical models, which exist that predict water's refractive index dependence on temperature which vary in complexity. One such model was proposed by Abbate [12], who measured the change in refractive index and inferred the following relation:

$$\frac{-dn_w}{dT} = B\left[1 - \exp\left(-\frac{T - T_0}{T_k}\right)\right] \quad (3.9)$$

where $B = 26.2 \times 10^{-5}$ K^{-1}, $T_0 = 2.0°C$, and $T_k = 48.5°C$.

While these noninvasive water calorimetry techniques show promise and offer several advantages, they are currently not as accurate as thermistor-based calorimeters and have been the focus of several works to improve their accuracy. For example, these calorimeters are very sensitive to vibrations and thermal instabilities. For interferometer calorimeters in particular, the stability of the laser is a limiting factor in the accuracy of the measurement as it is assumed that the change in intensity measured with the photodiode is due exclusively to the changes in the refractive index of the water. Additionally, these calorimetry techniques do not technically measure dose to a point (or small region) like a thermistor-based calorimeter. Instead, their response is integrated over the entire beam path. As a result, thermal instabilities within the water phantom contribute substantially to the overall uncertainty. Tomographic techniques have been suggested to reconstruct absolute dose distributions but have yet to be demonstrated successfully.

3.4 APPLICATIONS OF CALORIMETERS IN PARTICLE DOSIMETRY

Calorimetry holds a lot of promise as a primary standard for proton and heavy ion beams. Today, clinical ionization chambers used for proton and heavy ion dosimetry are indirectly calibrated in a ^{60}Co beam quality but require several large conversion factors to account for the response of the chamber in a particle beam [13]. These calibrations and conversions are rather arduous and susceptible to large uncertainties from the conversion factors. A direct measure of dose to water from a primary standard in particle beams is challenging. Similar to water calorimetry in photon beams, conduction can induce additional sources of uncertainty, especially for spot-scanning beams [14]. Accounting for heat defect is also difficult to quantify due to an apparent dependence in the ionization density induced along the particle track [14]. This is particularly important to consider for particle beams as the ionization density changes as the beam of radiation traverses through matter. Graphite calorimetry may be a better alternative to water-based calorimetry for particle beams. While graphite calorimetry is not influenced by conduction, there are radiative losses, which must be considered [14]. Finally, the conversion between water and graphite carries its own sources of uncertainty from the stopping powers between water and graphite. As it stands, the $k = 1$ uncertainty for the absorbed dose-to-water measurement with the transferred standard to a clinical ionization chamber is around 2% and 3% for proton and carbon-ion beams, respectively, relative to about 1% for modern medical mega-voltage x-ray beams [13].

3.5 CONCLUSION

A calorimeter thus is the primary standard used to establish absorbed dose to water. Since it is measured in a ^{60}Co beam, there needs to be corrections, k_Q, for chambers when measuring an accelerator beam. The calorimeter could be used in an accelerator beam, but few primary laboratories have a linac. Thus, the standard used is a ^{60}Co source with corrections to the ionization chamber. A water calorimeter is generally the preferred apparatus and is the device used at NIST.

REFERENCES

1. D. W. O. Rogers and J. E. Cygler (ed), *Clinical Dosimetry Measurements in Radiotherapy*, Chapter 15: Primary Standards of Air Kerma for 60Co and X-Rays and Absorbed Dose in Photon and Electron Beams, M. McEwen (Medical Physics Publishing, Madison, WI, 2009).
2. J. Seuntjens and S. Duane, "Photon absorbed dose standards," *Metrologia* **46**(2), S39 (2009).
3. S. R. Domen and P. J. Lamperti, "A heat-loss-compensated calorimeter: Theory, design, and performance," *Journal of Research of the National Bureau of Standards A. Physics and Chemistry* **78A**(5), 595–610 (1974).
4. B. Petree and P. Lamperti, "Comparison of absorbed dose determinations in graphite by cavity ionization measurements and by calorimetry," *Journal of research of the National Bureau of Standards – C. Engineering and Instrumentation* **71C**(1), 19–27 (1966).
5. J. Renaud, A. Sarfehnia, J. Bancheri, J. Seuntjens, "Aerrow: A probe-format graphite calorimeter for absolute dosimetry of high-energy photon beams in the clinical environment," *Medical Physics*, **45**(1), 414–428 (2018).
6. S. R. Domen, "A sealed water calorimeter for measuring absorbed dose," *Journal of Research of the National Institute of Standards and Technology* **99**, 121–141 (1994).
7. C. K. Ross and N. V. Klassen, "Water calorimetry for radiation dosimetry," *Physics in Medicine and Biology* **41**, 1–29 (1996).
8. E. Malyarenko, J. Heyman, H. Chen-Meyer, R. Tosh, "High-resolution ultrasonic thermometer for radiation dosimetry," *Journal of the Acoustical Society of America* **124**(6), 3481–3490 (2008).
9. H. Chandler, R. Bowen, G. Paffenbarger, "Physical properties of a radiopaque denture base material," *Journal of Biomedical Materials Research* **5**, 335–357 (1971).
10. A. Michelson and W. Morley, "On the relative motion of the Earth and the Luminiferous Ether," *American Journal of Science* **34**(203), 333–345 (1887).
11. E. Flores-Martinez, M. Malin, L. DeWerd, "Development and characterization of an interferometer for calorimeter-based absorbed dose to water measurements in a medical linear accelerator," *Review of Scientific Instruments* **87**(11), (2016).
12. G. Abbate, U. Bernini, E. Ragozzino, F. Somma, "The temperature dependence of the refractive index of water," *Journal of Physics D: Applied Physics*, **11**(8), 1167 (1978).
13. IAEA, "Absorbed dose determination in external beam radiotherapy," *An International Practice for Dosimetry on Absorbed Dose to Water*, IAEA Tech. Series No. 398 (2000).
14. L. V. Verhey, A. M. Koehler, J. C. McDonald, M. Goitein, I. Ma, R. J. Schneider, M. Wagner, "The determination of absorbed dose in a proton beam for purposes of charged particle radiation therapy," *Radiation Research* **79**(1), 34–54 (1979).

CHAPTER 4

Semiconductor Dosimeters

Giordano Biasi and Nicholas Hardcastle

University of Wollongong
Wollongong, NSW, Australia
Peter MacCallum Cancer Centre
Melbourne, VIC, Australia

Anatoly B. Rosenfeld

University of Wollongong
Wollongong, NSW, Australia

CONTENTS

4.1 INTRODUCTION

Semiconductor dosimeters are a cornerstone of dosimetry in modern radiation therapy. They can be standalone devices, or fabricated into linear, 2D and quasi-3D arrays to provide high levels of flexibility in measurement geometry. Their radiation-sensitive volume is made of silicon, silicon dioxide, or diamond (Table 4.1); the sensitivity to ionizing radiation is high and the sensitive volume can be small – down to a few micrometers across. Thus, semiconductor dosimeters can measure, with high spatial resolution, in small radiation fields. Further, because they can be read out in real time, they can measure dose distributions that change with time.

At this time, the semiconductor dosimeter family includes silicon diodes, silicon metal-oxide-semiconductor field-effect transistors (MOSFETs), and natural and synthetic diamonds. In brachytherapy and external-beam radiation therapy with electron and photon beams, they are used for integral dosimetry in-phantom or *in vivo*. The dose absorbed in the radiation-sensitive volume over a time span is assessed by measuring radiation-generated charge or current; alternatively, by quantifying the change of the device's electrical characteristics, such

TABLE 4.1   Properties of Silicon, Natural Diamond, and Chemical Vapor Deposition (CVD) Synthetic Diamond

Property at 300 K	Silicon	Diamond	CVD Diamond
Atomic number	14	6	6
Density (g cm^{-3})	2.33	3.51	3.51
Band gap (eV)	1.12	5.47	5.47
Intrinsic resistivity (Ω cm)	2.3×10^5	$>10^{11}$	$\sim10^{13}-10^{14}$
Mean energy to produce e–h pairs (eV)	3.68	13	13
Electron drift mobility (cm^2 V^{-1} s^{-1})	1450	200–2800	~4500
Hole drift mobility (cm^2 V^{-1} s^{-1})	480	~2000	~3800

as its conductivity. In proton therapy and heavy-ion therapy (HIT), semiconductors dosimeters are niche players to verify the energy and range of primary particles, and can also be used to characterize the radiobiological efficiency (RBE) of a therapeutic beam.[1] Further, they have been proposed for dosimetry in experimental radiotherapy techniques such as synchrotron-based micro-beam radiation therapy (MRT).

The reader may complement the present chapter with recent reviews: on the application of semiconductors dosimeters in brachytherapy by Carrara et al. [1] and in particle therapy by Rosenfeld [2]; as well as older ones [3–5].

4.2 DIODES

4.2.1 Construction and Functioning

The life story of each silicon diode starts with a pure silicon crystal – the base, or substrate – in which impurities are introduced. For instance, of phosphorous and aiming for concentrations in the range 10^{14}–10^{16} cm^{-3}, so that valence electrons are donated to the crystal lattice; that results in an n-type base. Alternatively, of boron and aiming for concentrations in the range 10^{15}–10^{17} cm^{-3}, so that holes (i.e., the absence of the covalent bond between atoms of silicon and boron) are created in the lattice; that results in a p-type base. Then, to create a p–n junction (Figure 4.1), one of

the surfaces of the base is heavily doped; here, impurities are of the opposite type and their concentration is of two orders of magnitude, or more, than that used for the base. On the n-side of the junction, electrons are majority and holes are minority charge carriers; on the p-side, the other way around.

As soon as the p–n junction is created, electrons diffuse from the n-side to the p-side; holes diffuse in the opposite direction. Eventually, that diffusion ends up establishing a space-charge region, with a positive charge on n-side and a negative charge on the p-side; across the junction there is an electric field strong enough to prevent further diffusion of majority carriers. The space-charge region has a thickness of about 1 μm and is called the *depleted region*; the electric field has a strength of about 10^3 V/cm and produces the *built-in potential*.

The result is a silicon diode based on a p–n junction with a thin radiation-sensitive volume – the depleted region. The region can be extended by applying a reverse external bias; this extra step, which is necessary to use the diode for spectroscopy of nuclear radiation, is not usually considered for dosimetry.

Take a diode with a p-type base[2] (Figure 4.1). If the diode is exposed to ionizing radiation, ionization generates electron–hole pairs. If the absorbed energy is E, about $N = E/w$ pairs are expected, where w is the mean energy required to produce a single electron–hole pair; for silicon, w is 3.68 eV (Table 4.1). This idea can be expressed by saying that the generation constant of electron–hole pairs in silicon is [6]:

$$g = 4.2 \times 10^{13} \text{ cGy}^{-1} \text{ cm}^{-3} \tag{4.1}$$

The *excess* minority charge carriers generated within one diffusion length from the junction (L_n, for electrons) are swept across the junction by the built-in potential. That results in a radiation-induced current, proportional to the dose rate in silicon; its integral (charge) is

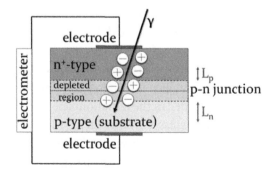

FIGURE 4.1   Cross-sectional view of a silicon diode based on a p–n junction (schematics, not to scale).

proportional to the absorbed dose. Because only charges generated within one diffusion length are collected, the amount of charge Q collected, per cGy to silicon in a p–n junction of 1 cm², is about:

$$Q \propto gL_n \tag{4.2}$$

The diffusion length is a function of the diffusion coefficient D_n and of the mean lifetime of minority carriers τ_n [7]:

$$L_n = \sqrt{D_n \tau_n} \tag{4.3}$$

D_n depends on the resistivity of the silicon substrate – a function of the dopant concentration and of temperature; τ_n is a function of the structure of the generation-recombination (G-R) centers for electron–hole pairs that exists in the crystal lattice. The general expression for τ_n is based on the Read–Shockley recombination theory [8]. Ionizing radiation creates different centers as a function of particle type and energy; each center introduces its own energy level in the forbidden energy-band gap and affects the mean lifetime as a function of injection level[3] and temperature.

Because ionizing radiation produces G-R centers, the concentration of these centers increases with increasing accumulate dose. The result is a degradation of the mean lifetime and, in turn, of the diode sensitivity – the amount of charge collected per unit absorbed dose [9]. That degradation depends on the history of the diode, in terms of previous irradiations, and is non-linear with accumulated dose; also, it plateaus after a certain amount of accumulated dose. For that reason, *priming* has been used to stabilize diode sensitivity; a priming dose can be larger than 10 kGy. An alternative approach to stabilize diode sensitivity introduces suitable G-R centers in the crystal lattice, for example using platinum atoms; their concentration has to be high enough to dominate, over radiation-induced centers, the degradation of the mean lifetime [10]. A more recent tactic to stabilize diode sensitivity requires fixing the dimensions of the radiation-sensitive volume in two directions: laterally, by using guard-rings, and in depth, by growing an epitaxial silicon layer onto a highly conductive substrate. The key is to have a radiation-sensitive volume of dimensions shorter than the expected diffusion length of minority charge carriers, accounting for its degradation with accumulated dose [11].

Ionizing radiation may come in bursts. For example, a medical linear accelerator (linac) delivering radiotherapy with photon beams produces pulses which have a duration in the range 2–6 μs, with a frequency in the range 180–400 Hz. The instantaneous dose rate – the dose per pulse – is in the order of 5.6×10^{-4} Gy/pulse. That number translates to a dose rate of 140 Gy/s, if considering a pulse 4 μs long. At low dose rates, the G-R centers – assuming they have a constant concentration for successive pulses – are mostly empty; a fraction of radiation-generated minority charge carriers is captured and recombines; the remaining charge is collected at the electrodes. As the dose rate in each pulse is increased, more and more centers are filled. The fraction of minority charge carriers that recombine decreases; the fraction of charge collected increases. The dose-rate dependence of a diode's response, first discussed by Rikner and Grusell [12], is thus determined by the G-R centers. It can be controlled by adjusting either the mean lifetime – by priming the diode or by introducing suitable G-R centers – or the resistivity of the silicon substrate [11, 13, 14].

Ionizing radiation may come from different directions. The geometry of a diode is planar and asymmetric (Figure 4.1). If the degree to which a dosimeter perturbs radiation depends on the angle of incidence, a directionally dependent sensitivity is expected. The magnitude of that dependence depends on the design of the diode, compounded by asymmetries in its packaging. It is possible to minimize the directional dependence of a diode – or an array of diodes – by investigating alternative, optimized packaging solutions, e.g., the removal or introduction of high-Z materials [15]; also, geometrical asymmetry can be compensated for by mounting two diodes back-to-back [16]. More recently, an 'edgeless' diode was proposed; the conventional planar design was replaced with a p–n junction that surrounds the base [17, 18]. The directionally independent sensitivity [19] was preserved with a recently developed *drop-in* packaging into a flexible Kapton carrier; lack of high-Z materials prevents perturbations to incident radiation.

But what about perturbations caused by silicon itself? After all, the aim is to use the diode to assess dose-to-water; surely the atomic number of silicon, which is higher than that of water ($Z = 14$ vs. $Z_{eff} \sim 7.42$), requires discussion. In clinical electron beams, the diode is a *sensor* that monitors the electrons; for those electrons, the ratio water-to-silicon of the average mass-collision stopping-powers (S_{el}/ρ) is almost energy-independent. In photon beams, the picture is more nuanced. In megavoltage photon beams, if the diode is small with respect to the range of the electrons (the secondary charged particles), as a first approximation it *senses* the fluence of electrons that would exist in an unperturbed medium[4]; the diode is, again, a sensor that monitors the electrons. In kilovoltage photon beams, however, the electrons originating from the interaction of the photons *inside* the diode contribute a significant portion to the absorbed dose; the diode becomes a sensor that monitors electrons *and* photons. At that point, it is necessary to consider that the ratio water-to-silicon of the average mass energy-absorption coefficients (μ_{el}/ρ) varies with energy; that is explained mainly in terms of a relative increase in cross section for photoelectric effect in the kilovoltage range [20].

In large radiation fields produced by megavoltage photon beams, there is an important component of scattered photons with energies in the kilovoltage range. In that case, it may be desirable to reduce the energy dependence by shielding the diode; a high-Z packaging filters the low-energy component of a photon beam [21]. That shielding is effective was also confirmed with Monte Carlo analysis [22]; a smaller shielding effect is also expected in unshielded diodes, because of the presence of thin layers of materials not intended for shielding. The photon spectrum of megavoltage photon beams gets harder with decreasing field size [23]; in small fields the low-energy component can be neglected. Also, in megavoltage beams, the number of low-energy scattered photons increases with depth; however, a relative over-response moving toward larger depths *may* be offset by an under-response because of dose-rate dependence [24].

The relative temperature instability (RTI) – the dependence on temperature of the diode sensitivity – is a function of the structure of the G-R centers [25]; therefore, it also depends on the history of the diode in terms of previous irradiations. It can be described by [26]:

$$\frac{1}{S}\frac{dS}{dT} = \frac{d\ln S}{dT} = \frac{1}{2}\frac{d\ln(D_n\tau_n)}{dT} \qquad (4.4)$$

where S is the sensitivity, T the temperature, D_n the diffusion coefficient, and τ_n the mean lifetime of

minority carriers. It is important to account for RTI when the diode is used for *in vivo* dosimetry, e.g., placed on the patient's surface for skin dosimetry. It may take up to 5 minutes to reach an equilibrium temperature, with the kinetics of the temperature of the diode depending on its packaging. The internal buildup in commercial diodes typically hinders those measurements anyway.

To conclude, the diode sensitivity depends on accumulated dose (through radiation damage), dose rate, angle of incidence and energy of radiation, and temperature. Yet, at present, diode dosimeters are ubiquitous: they are cost-effective, easy to use and reproducible; they can be operated in passive or active mode (with or without external bias); it is possible to produce free-standing single diode or multidimensional arrays to capture, at once, complex dose distributions; further, they can be read out in real time to measure dose distributions that change with time.

4.2.2 Dosimetry with Diodes

Single diodes were first proposed as dosimeters in photon beams in 1963 [27]; two decades later, Grusell and Rikner pioneered discussions on their advantages and limitations [9, 21, 28, 29]. Diodes have been used in electron beams to measure depth-dose and lateral-profile distributions [30] in place of ionization chambers and don't require depth-dependent corrections [31]. Diodes have also been used for *in vivo* dosimetry in electron and photon beams, typically on the skin [32, 33]. If used for *in vivo* dosimetry, they require the amount of build-up material be minimized, regular calibration against an ionization chamber and the application of correction factors dependent on beam energy, temperature, incident-beam angle, and field size.

At present, single diodes are mainly used for in-phantom integral dosimetry in megavoltage photon beams. Because they can be fabricated of small radiation-sensitive volumes (in the order of 0.1 mm^3), measurements have minimal volume-averaging and, by scanning in water tank, it is possible to achieve sub-millimeter resolutions. Other than lateral profiles, measurements of photon beam output factor, percentage depth-dose and tissue-maximum ratio or tissue-phantom ratio are also possible. Diodes are not used to calibrate the output of a beam. For a bare diode (water is the build-up material) dose-to-water D_{H_2O} can be calculated, using Bragg–Gray cavity theory, from the measured dose-to-silicon D_{Si}; it is necessary to consider the ratio water-to-silicon of the mass-collision stopping powers averaged over the electron-fluence spectrum:

$$D_{H_2O} = D_{Si} \left(\frac{\overline{S_{el}}}{\rho} \right)_{H_2O,Si} P \qquad (4.5)$$

where P accounts for small perturbations, with respect to water, of the fluence of the secondary charged particles.

In small megavoltage photon beams (side or diameter of less than 3 cm), dosimetry gets challenging [34]. Secondary electrons depositing dose in a small silicon volume are, to a large extent, originating from the materials surrounding it [35, 36]. The silicon diode and its packaging (extra-cameral components) perturb, with respect to water, the fluence of electrons; the perturbation is a function of the size of the radiation field. This translates into P in Equation 4.5 being field-size dependent. The formalism required to correct the diode readings with a field-size dependent correction was introduced by Alfonso et al. [37] and later codified in a Code of Practice dedicated to small-field dosimetry [38].

Monte Carlo codes are effective in characterizing diode response in small fields and in calculating field-size dependent corrections; see, for instance [39–42]. Corrections depend on diode design (radiation-sensitive volume and extra-cameral components), linac treatment-head design and measurement conditions. Calculating corrections with Monte Carlo requires knowledge of dosimeter design, which may not be available. Designing a dosimeter maintaining a correction factor close to unity is preferable, and was indeed shown to be possible by introducing appropriate modifications [43, 44].

Shielded diodes are preferable for dosimetry in large fields. They should not be used for small fields; the low-energy photon component is negligible and shielding not only makes photon spectra very different from unperturbed ones, but also perturbs electron spectra [22], which distorts the required corrections. Unshielded diodes, instead, introduce smaller perturbation and have less directional dependence; provided field-size dependent corrections are applied, they measure output factor and high spatial resolution lateral profiles identical to Monte Carlo calculations in water.

One of the most discussed diodes dedicated to small-field dosimetry was the stereotactic-field diode, or SFD diode (IBA Dosimetry, Germany) [45]. Short-term stability and dose-rate dependence were suboptimal,

leading to the development of the Razor diode [46]. The Razor has superior stability, dose linearity and radiation hardness [47]; similarly to its predecessor, it requires field-size dependent corrections [48].

Modern radiotherapy techniques deliver complex intensity-modulated dose distributions. These are created through summation of multiple small fields typically in a dynamic manner, thus are impossible to be measured by scanning a single diode. Fortunately, diodes can be combined into planar arrays. Since 1D arrays of single diodes were first investigated [49], they have gained popularity as stand-alone devices that could be quickly positioned onto the treatment couch, or attached to the linac gantry for measurements in transmission mode. One of the first commercial 1D array was the Profiler (SunNuclear, Melbourne, FL, USA); it had 46 p-type single diodes distributed along 22.5 cm spaced by 5 mm [49]. A further development added a second, orthogonal linear array [50]. The Profiler 2 (SunNuclear) had 83 and 57 single diodes, spaced by 4 mm, in the Y and X directions respectively [51].

In small fields, dosimetry requires a spatial resolution better than about 1 mm, a resolution that is impossible to achieve with arrays of single diodes. Instead, monolithic arrays – diodes are ion-implanted onto a common substrate – can be fabricated in large areas with radiation-sensitive volumes spaced by 1 mm or less. Prototypes have been proposed by several groups [52–54]. For instance, the Dose Magnifying Glass, proposed by the Centre for Medical Radiation Physics (University of Wollongong, Australia), has 128 diodes with a spacing of 0.2 mm [55]. Other than in megavoltage photon beams, the Dose Magnifying Glass was also used for range verification in HIT [56].

While with 1D arrays it is easy to decrease the spacing between diodes, in the case of 2D arrays there is a necessary trade-off between dosimeter area and spatial resolution, to be within the maximum number of readout channels available. The Centre for Medical Radiation Physics has investigated three different compromises (Figure 4.2). The Magic Plate has 512 diodes spaced by 2 mm, uniformly distributed onto a 5×5 cm^2 area [57]; the Duo has the same number of diodes spaced by 0.2 mm, but along two perpendicular arrays [58]; finally, to improve the description of 2D dose distributions, in the Octa diodes are arranged along 4 arrays at 45°, and are spaced by 0.3 mm [59, 60]. The Magic Plate, the Duo and the Octa were used for dosimetry during

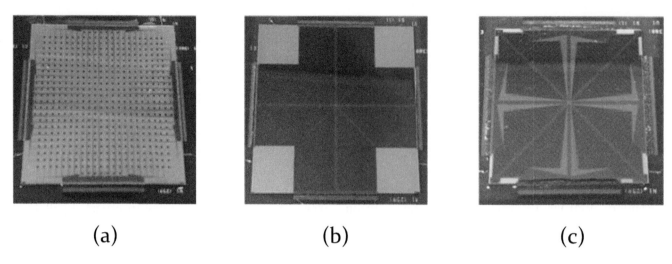

(a)   (b)   (c)

FIGURE 4.2   2D monolithic arrays: (a) the Magic Plate, (b) the Duo, and (c) the Octa [69].

adaptive multi-leaf collimator (MLC)-tracking, including arc deliveries[5] [62]; in transmission mode to verify dose maps at a depth in water [63]; for dosimetry in an environment resembling that of a magnetic resonance (MR)-guided linear accelerator[6] [64]; to verify, in real time, the dosimetry in radiation fields produced with a variable-aperture Iris™ collimator in a CyberKnife® system (Accuray Inc., Sunnyvale, CA, USA) [65]. Also, the design of the Magic Plate, the Duo and the Octa was improved to maintain a correction factor close to unity in small-field dosimetry [66]; it was studied, in terms of uniformity of the diode sensitivity across the large-area array [67], and in terms of dose-rate dependence and directional dependence [68].

When using arrays, there are additional steps to consider to be able to relate readings to dose. For instance, diodes sensitivity may be nonuniform. That may be a result of nonuniformities in the silicon crystal used to fabricate the diodes; also, possible variations involved in the fabrication processes. A nonuniform sensitivity can be compensated for with an equalization procedure – the dosimeter is irradiated with a flat field to calculate equalization factors for each diode.

The introduction of planar arrays was a step forward in dosimetry; however, they have an inherent limitation: 2D information collapses into 1D when radiation comes from a direction parallel with their plane. A workaround is to keep the array always perpendicular to radiation at all times: the dosimeter can be attached to the linac gantry or lodged into a phantom that rotates in synch with the gantry [70, 71]. As an alternative, arrays with quasi-3D geometries have also been also developed. Scaling up

the number of diodes in a 3D arrangement to improve spatial resolution is challenging; clever arrangements were devised. However the more complex the configuration of the radiation-sensitive volumes, the more difficult it becomes to evaluate the results – in particular if the response is directionally dependent.

At present, two quasi-3D diode arrays are commercially available (Figure 4.3): the Delta4 (ScandiDos, Uppsala, Sweden) and the ArcCHECK (SunNuclear). The Delta4 has 1069 cylindrical p-type diodes (diameter 1 mm, thickness 0.05 mm) arranged onto two perpendicular planes; diodes are spaced by 0.5 cm in the 6 × 6 cm^2 central region and 1 cm elsewhere. The ArcCHECK has 1386 SunPoint® n-type diodes (0.8 × 0.8 × 0.03 mm^3) spaced by 1 cm on a HeliGrid™ inserted into a doughnut-shaped phantom. These diode arrays thus allow comparison of calculated doses from treatment planning systems with those measured, typically still on a planar basis. For example the ArcCHECK 'unfolds' the helical array to present the user with a planar dose map. The resultant measurements from these devices can further be processed by interpolating dose between two detectors along a ray line in a given measurement interval, to result in quasi-3D dose distributions.

4.3 MOSFETs

4.3.1 Construction and Functioning

A MOSFET (Figure 4.4) is a device with three terminals.[7] One of the terminals (the gate) controls the conductance of the channel between the other two terminals (the source and the drain).

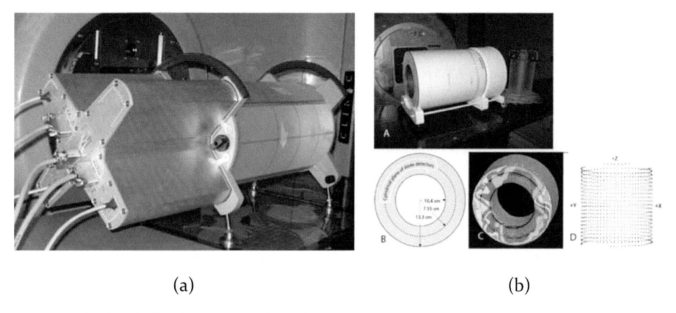

(a)   (b)

FIGURE 4.3   (a) The Delta[4] has diodes on two orthogonal planes. (Image courtesy of Sadagopan et al. [72].) (b) The ArcCHECK has diodes on a HeliGrid™. (Image courtesy of Feygelman et al. [73].)

A MOSFET used for dosimetry is typically a p-MOS-FET (also referred to as p-MOS; p-channel MOSFET; radiation-sensitive FET, RadFET). It is fabricated starting from a pure silicon crystal – the substrate – into which we introduce impurities to create an n-type base (e.g., using phosphorous so that valence electrons are donated to the crystal lattice). Two terminals (the source and the drain) are p+ type; the third (the gate) is placed on top of a silicon-dioxide layer (SiO_2, the radiation-sensitive volume; hereafter, for simplicity: the oxide) in contact with the substrate. The conductive channel is a p-channel formed by the minority charge carriers (holes). When the external bias on the gate (V_g) is zero, relative to the source, the conductance of the channel is very low; we can increase it by increasing the negative bias. Let's say we want a constant current, of our choice, to flow through the channel (I_{ds}, drain–source current);

the threshold voltage (V_{th}) is defined as the bias we must apply on the gate to sustain the flow of that current [7].

Ionizing radiation generates electron–hole pairs in the oxide. The mean energy required to generate a pair is about 18 eV [74]. Electrons either recombine or are swept out of the oxide in a few picoseconds [75]. Holes have lower mobility [76]; if the device is irradiated under positive bias, they disperse (stochastic hopping) toward the interface between oxide and substrate. Their transport is a function of temperature, bias and oxide thickness – and, to a lesser extent, of the processing technique used to grow the oxide and of the concentration of defects [77, 78]. As a first approximation, the fraction of holes that escape recombination (f_E, fractional yield) is a function of the electric field in the oxide and of the initial density of the pairs plasma generated; that density is a function of the type and energy of the incident particle [77, 78].

Holes escaping recombination are trapped in long-term sites – oxide traps and interface-traps – close to the interface between oxide and substrate. They interfere with the conductance of the p-channel, causing an increase in the threshold voltage required to produce any given drain–source current. The change in the threshold voltage (ΔV_{th}, threshold-voltage shift) is proportional to the dose absorbed in the oxide [77, 78]. This idea can be expressed by writing the threshold-voltage shift as [79]:

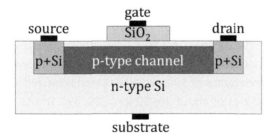

FIGURE 4.4   Cross-sectional view of a p-channel MOSFET (schematics, not to scale).

$$\Delta V_{th} = -\frac{x_{ox} Q_{ox}}{\varepsilon_{ox}} = -\frac{x_{ox} q N_{ox}}{\varepsilon_{ox}} \qquad (4.6)$$

where $q = 1.602 \times 10^{-19}$ C is the electronic charge; x_{ox} is the distance between the gate and the center of the space-charge distribution Q_{ox}, i.e., it is the thickness of the charge-collection region; $\varepsilon_{ox} = 3.365 \times 10^{-11}$ F/m is the electric permittivity of the oxide; N_{ox} is the number of trapped holes per unit area in the oxide. If approximating x_{ox} to the oxide thickness t_{ox}, Equation 4.6 can be rewritten in terms of capacitance per unit area of the oxide ($C_{ox} \overset{\text{def}}{=} \varepsilon_{ox}/t_{ox}$) [80]:

$$\Delta V_{th} \approx -\frac{qN_{ox}}{C_{ox}} \qquad (4.7)$$

Under the same approximation ($x_{ox} \approx t_{ox}$), the number of trapped holes per unit area in the oxide is expressed as [78, 79]:

$$N_{ox} \approx g_0 t_{ox} D A f_E \qquad (4.8)$$

where D is the absorbed dose; $g_0 = 8.1 \times 10^{12}$ cGy^{-1} cm^{-3} is the rate of generation of pairs in the oxide; A is the fraction of holes that escape recombination but end up trapped – it is a function of the electric field in the oxide, of temperature, of the processing technique used to grow the oxide and of the concentration of defects [77, 78]. At this point, it is helpful to define a constant k as:

$$k \overset{\text{def}}{=} \frac{qg_0}{\varepsilon_{ox}} = 39 \; \frac{\text{mV}}{\text{cGy} \; \mu\text{m}^2} \qquad (4.9)$$

and express the MOSFET sensitivity S as:

$$S \overset{\text{def}}{=} \frac{\Delta V_{th}}{D} \approx k t_{ox}^2 f_E \qquad (4.10)$$

It is possible to optimize the MOSFET to operate in a specific dose range; for instance, by maximizing its sensitivity. Equation 4.10 indicates two options. If the requirement is known beforehand, the oxide can be grown of a different thickness; or, the bias can be varied during operation if adapting an existing MOSFET to fit new conditions. There are limitations. Thick oxides (1.0 μm and beyond) have poor reproducibility; they may be exposed to mechanical stress resulting in instability of the threshold voltage. Similarly, a strong electrical field may also inflict mechanical stress.

Those are not the only trade-offs to consider. Any increase in sensitivity comes at the cost of a decrease in the device's lifespan. This is because there is a finite number of available traps in the oxide; eventually, they reach saturation and the threshold voltage plateaus [7]. Also, the threshold voltage shift may increase up to a point where it is too high to be measured with a given reader.[8] There are workarounds; for instance, boron implantation through the gate can be used to reduce the initial threshold voltage and increase the device's lifespan [81]; alternatively, MOSFETs can be annealed [82, 83].

Irradiation in a biased configuration has source, substrate and drain all grounded; a voltage is applied on the gate. The threshold voltage however is measured in a readout configuration: source and substrate are short-circuited, gate and drain short-circuited, and a constant current is forced through the conductive channel. Readout may be affected by *creep-up*, *1/f noise* and *fading*. The *creep-up* effect refers to an instability of the threshold voltage for consecutive queries as a result of charges being injected into the device by the readout electronics; it can be prevented by keeping consecutive queries ~60 s apart. The *1/f noise* introduces instabilities to the threshold voltage [84]; its importance depends on the device's construction, such as processing technique used to grow the oxide, and operating characteristics, such as temperature and applied bias [85]; it also increases with increasing accumulated dose. Finally, *fading* describes a long-term change in the value of the threshold voltage *after* irradiation; the concentration of trapped charges, in the oxide (ΔN_{ot}) and at the interface with the substrate (ΔN_{it}), varies in time depending on the device's construction, and on irradiation and post-irradiation bias [80, 86]. ΔN_{ot} tends to decrease monotonically after irradiation (thus reducing the initial threshold voltage shift), while ΔN_{it} tends to exhibit an initial increase (increasing the initial threshold voltage shift) followed by saturation [85]; in the case of positive bias, the net effect is a negative shift in the threshold voltage. *Fading* is expected to be most pronounced during the first few hours after irradiation, but can be present after days [87].

MOSFETs are fabricated of silicon; the radiation-sensitive volume is of silicon-dioxide. In low-energy photon beams (15–200 keV), MOSFET sensitivity normalized to that in a 6 MV photon beam (effective energy 2 MeV) is larger by a factor of 3–4; sensitivity is maximized at photon energies of about 30–50 keV [88, 89]. In the megavoltage photon range (4–25 MV) and in electron beams in the range 5–21 MeV, MOSFET sensitivity is almost constant [90–92]. The energy-dependence of MOSFET sensitivity depends on the details of the manufacturing process and packaging [89, 93, 94]; because of the relatively small size of the transistor, a relatively high fraction of secondary electrons depositing dose in the oxide is generated in the surrounding packaging (*dose enhancement* effect) [89].

Also, MOSFET sensitivity is influenced by the angle of incidence of radiation. Similarly to diodes, this is explained by, and depends on, asymmetries in the geometry and materials of the transistor and of its packaging. Being device-dependent and measurement-condition dependent, values vary greatly in the literature; see, for instance [90, 95] or, more recently [92, 96].

Finally, because the mobility of the charge carriers in the conductive channel fluctuates with temperature, so does the MOSFET sensitivity. Of course, temperature coefficients are device-dependent; however, they are also negligible because temperature effects are typically within the reproducibility of the MOSFET; they are neglected for clinical measurements [97] expect for *in vivo* skin dosimetry. A typical time required for MOSFETs to reach thermal equilibrium with a patient's skin is 2 minutes [98]. We can minimize temperature dependence by selecting a readout current corresponding to a thermostable point on the I_{sd}–V_g characteristic [99]. Alternatively, we can use an internal p–n junction (drain or source onto substrate) for temperature measurements directly in the channel; or even a dual-MOSFET. In that case, the rationale is that MOSFETs irradiated under different gate bias have different radiation-induced shifts of the threshold voltage, but similar temperature coefficients. The temperature-independent signal proportional to the absorbed dose will result from subtracting the shifts of the threshold voltage of the two MOSFETs [100].

To summarize, MOSFETs dosimeters are robust, light-weight and cost-effective devices. They have advantages over diodes: dose information can be stored permanently and read out, non-destructively, multiple times. A power supply is strictly required only for the readout procedure and for active mode application. Further, the radiation-sensitive volume (the oxide) has dimensions an order of magnitude smaller than that a diode and its sensitivity is adjustable by varying the bias during irradiation. MOSFET sensitivity is also independent of dose rate up to 10^8 Gy/s [101, 102]. However, MOSFET dosimetry has limitations: it is key to account for the energy-dependence, directional-dependence and temperature-dependence of the sensitivity. Also,

creep-up, 1/f *noise* and *fading* may affect the accuracy of the measurements.

4.3.2 Dosimetry with MOSFETs

MOSFETs dosimeters were first proposed in 1970 [103] and later demonstrated to monitor space-radiation on satellites [104, 105]. At present, they have applications in brachytherapy [1]; radiation therapy with kilovoltage [106] and megavoltage [107, 108] photon beams, and with electron beams [109, 110]; also, to monitor radiation in food-irradiation plants [111]. MOSFETs have been proposed for use in intraoperative radiotherapy (IORT) delivered with electron or low-energy photon beams [112, 113]. Despite the small dimensions of the radiation-sensitive volume (the oxide), MOSFETs are not recommended for small-field dosimetry in megavoltage photon beams [38, 114, 115] because of the relatively low measurement precision.

The use of MOSFETs in proton therapy is currently limited by their relatively large dependence on the linear-energy transfer (LET); the height of the Bragg Peak, normalized to the proximal plateau, is almost 40% lower estimated using a MOSFET than with an ionization chamber [116]; however, it can be used to determine the spatial accuracy of the Bragg peak location [117]. MOSFETs can be used for dosimetry in x-ray microbeams, which are relevant to MRT, where beams are separated center-to-center by about 200 μm [118–122].

Most typically, today's MOSFETs are used for *in vivo* dosimetry in external beam radiotherapy. The MobileMOSFET system (Best Medical, Canada) consists of MOSFET detectors, a reader module and Bluetooth wireless transceiver. Multiple detectors can be read out at once, and different detectors are available for various dose ranges. *In vivo* dosimetry, where the detector is placed on the patient surface, is typically used to verify treatment dose when commissioning new techniques, treatment equipment or software, or to verify dosimetry where there may be differences between machine performance and treatment planning calculation. An example of the latter case is verification of junction doses when using matched jaw-defined fields.

MOSFETs have shown particular strengths in skin dosimetry. They are placed on the patient's skin, either at the point of entry (entry-dose measurement) or exit (exit-dose measurement) of incident radiation. Skin dosimetry requires a dosimeter with a reproducible effective-depth close to the 7 mg/cm², i.e., a depth of

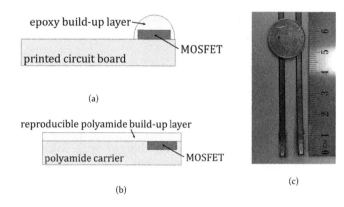

(a)

(b)

(c)

FIGURE 4.5 Packaging of (a) a typical MOSFET dosimeter and (b) a MO*Skin*™. Most manufacturers place the MOSFET on a Kapton film or on a thin printed-circuit-board carrier to which the transistor is wire-bonded; a bubble of epoxy resin is then placed on the top for protection. The MO*Skin*™ (c) replaces that protective bubble with a thin polyamide film.

0.07 mm in tissue,[9] as recommended by the International Commission on Radiological Protection [123].

Initial investigations considered scientific-grade MOSFETs which had their nickel casing removed to leave the bare chip exposed; the transistor was placed in a phantom holder using wax to fill air cavities [124]. Another group [125] adapted for skin dosimetry a commercial MOSFET, which had been previously characterized for *in vivo* dosimetry [95]; the MOSFET was protected by an hemispheric layer of epoxy resin (Figure 4.5) and measured dose at a water-equivalent depth between 0.6 and 1.8 mm, not complying with ICRP recommendations and over-estimating skin dose [125].

The MO*Skin*™ (Figure 4.5) developed at the Centre for Medical Radiation Physics (CMRP, University of Wollongong, Australia) is a MOSFET for real-time *in vivo* dosimetry. The transistor of the MO*Skin*™, which has an oxide of thickness 0.55 μm, is mounted underneath a thin plastic layer with a uniform thickness of 0.07 mm [126]. The use of MO*Skin*™ for skin dosimetry has been reported by various authors; e.g., during breast radiotherapy [126], intensity-modulated radiation therapy (IMRT) for nasopharyngeal carcinoma patients [127] and head & neck cancer patients [128]; for monitoring the eye-lens dose during cerebral angiography [129]. However, the package of the MO*Skin*™ can be adapted to different uses; for instance, to measure dose to the rectal wall in external-beam radiotherapy [130, 131]; by changing the build-up layer and selecting an appropriate water-equivalent depth, to verify dose calculations

at the boundary between low- and high-density media [132]. Brachytherapy applications of the MO*Skin*™ have also been discussed extensively; e.g., see [133, 134].

A further application of MOSFETs to *in vivo* dosimetry uses a MOSFET implanted inside the planning treatment volume (PTV). Commercial implantable wireless MOSFETs (DVS® dose verification system, Sicel Technologies, Morrisville, NC) were investigated by several groups [135–143]. Implanted MOSFETs can be imaged and located with ultrasonography [140, 142], kilovoltage [138–140, 142] or megavoltage [144] computerized tomography (CT) scans. Also, they can be used as fiducial markers in image-guided radiotherapy (IGRT); a feasibility study on their use for extracranial-target treatments delivered with the CyberKnife system was also reported [107].

4.4 DIAMONDS

4.4.1 Construction and Functioning

The crystal of a natural diamond has an energy gap of 5.47 eV and a resistivity in the range 10^{13}–10^{16} Ω cm: it is possible to apply an electric field across the crystal without generating a current.[10] Also, when likened to silicon, the crystal is less prone to radiation damage because the energy required for displacing atoms from their sites in the lattice is larger.

In a crystal of density ρ and volume V, irradiated at a dose rate dD/dt, charge carriers (electron–hole pairs) are generated at a rate (Figure 4.6):

$$\frac{dg}{dt} = \frac{\rho V}{w} \frac{dD}{dt} \qquad (4.11)$$

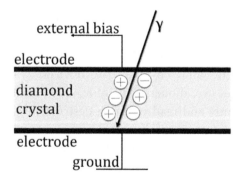

FIGURE 4.6 Ionizing radiation generates electron–hole pairs in the crystal of a diamond; the average energy spent to create a pair is about 13 eV. An external bias inhibits recombination and promotes electrical conductivity; electrons and holes drift toward the electrodes and are collected. Their drift mobility is higher in the crystal of a diamond than in silicon, so that collection time is shorter.

where $w = 13$ eV is the average energy spent to generate a pair; the current I_0 is then:

$$I_0 = e \frac{dg}{dt} \qquad (4.12)$$

where e is the electron charge. An external bias producing an electric field E across the crystal inhibits the recombination of charge carriers and promotes electrical conductivity σ – the crystal is then operated as a resistor in which conductivity is proportional to the dose rate [145]. Charge carriers drift toward the electrodes with a velocity equal to the product of their drift mobility μ and the strength of the electric field:

$$v = \mu E \qquad (4.13)$$

The current I_{out} collected at the electrodes of a crystal of thickness L is:

$$I_{out} = I_0 \frac{\mu \tau}{L} E \qquad (4.14)$$

where τ is the average lifetime of charge carriers. The crystal has a charge-collection efficiency that can be expressed using a gain factor G defined as the ratio of the current I_{out} measured at its electrodes to the current I_o:

$$G \stackrel{def}{=} \frac{I_{out}}{I_0} = \frac{d}{L} \qquad (4.15)$$

where the charge-collection distance d, i.e., the average length the charge carriers drift for, is:

$$d \stackrel{def}{=} \mu \tau E \qquad (4.16)$$

Charge-collection efficiency is maximized with an external bias that saturates the charge-collection distance; however, the bias should not be high enough to cause breakdown[11] or leakage currents.

The crystal of a natural diamond contains impurities such as atoms of nitrogen, boron, and aluminum in various concentrations; those impurities make the crystal n- or p-type, depending on if electrons or holes are donated to the lattice. Impurities with energy levels in the forbidden energy-gap may trap drifting charge carriers. Depending on the time-scale of the trapping process, that may results in a buildup of negative charge close to the positive electrode, and of positive charge close to the negative electrode. The buildups contribute an electric field opposed to that produced by the external bias. For that reason, as accumulated dose increases, at a constant dose rate, the effective field in the crystal decreases. A decreasing effective field also decreases the charge-collection distance in Equation 4.16 and, in turn, the gain in Equation 4.15. This effect is referred to as *polarization* [146]; other than on the effective field, the *polarization* depends on the flux of radiation [147] and the quality of the crystal [148]; also, it becomes more important if the crystal is damaged by radiation [149]. At any rate, eventually, traps are saturated: the *polarization* plateaus and so does the current measured at the electrodes – albeit at a value lower than its initial value at zero accumulated dose.

To minimize the *polarization* in dosimetry, the crystal can be primed, i.e., irradiated before its use. Priming fills deep traps and stabilizes the crystal sensitivity. It must be performed at the operational external bias. Also, the amount of priming required is determined by the quality of the crystal in terms of its impurities; that quality can't be predicted or controlled in the crystal of a natural diamond. A typical priming is in the order of 5–10 Gy.

Too many impurities cause the crystal to be insensitive to radiation, because of a decrease in the recombination time, and to be affected by *polarization*; but impurities are necessary to make the current collected at the electrodes to increase linearly with dose rate. In a parallel-plate biased structure, the current is proportional to the electrical conductivity σ in the crystal. The relationship between that conductivity and the generation rate of pairs is [150]:

$$\sigma \propto \left(\frac{dg}{dt} \right)^{\Delta} \qquad (4.17)$$

where Δ, a dose rate index of linearity, depends on the quality of the crystal. Its value is 0.5 when the recombination of charge-carriers dominates over trapping; for instance, in a pure crystal or when the dose rate is very high. It has values between 0.5 and 1.0 when trapping dominates the lifetime of charge carriers; also, it may be 1.0 or more for particular impurities and cross-sections for capture of charge-carriers [151, 152]. In the literature, Δ has values between 0.49 and 1.16.[12] There is controversy over if Δ varies with the bias, and with the energy and type of radiation [153].

The crystal of a natural diamond is an attractive material for dosimetry because of its quasi-equivalence to human soft-tissue in terms of atomic number ($Z = 6$ vs. $Z_{eff} \sim 7.4$ for human soft-tissue). The ratio diamond-to-water of the mass collision-stopping-powers (S_{el}/ρ) and that of the mass energy-absorption-coefficients (μ_{el}/ρ) are almost energy-independent for a wide range of photon- and electron-energies. That the crystal is energy-independent was confirmed with experiments in megavoltage electron and photon beams [152, 154] and with Monte Carlo calculations [155] – however, the packaging is key because it may include high-Z components. Also, the crystal of a natural diamond has intrinsic symmetry; its sensitivity has a negligible directional dependence; however, the packaging[13] is again key [156, 157].

The crystal of a natural diamond is coveted for its beauty, a property exploited by the jewelry industry; it is also desired, for completely different reasons, in today's dosimetry. Its sensitivity does not depend on angle of incidence and energy of radiation; it can be cut of small size, so that dosimetry with high spatial resolution is possible; it is radiation-hard, i.e., there is negligible radiation damage to the crystal lattice as a function of accumulated dose. True, its sensitivity does depend on accumulated dose and dose rate (through *polarization* and dose-rate dependence), but those effects can be accounted for once characterized. However, the crystal – that selected for dosimetry, at least, if not that for the jewelry industry – is resented because it is *exquisitely* expensive. The quest to overcome that limitation is revealed in the next section.

4.4.2 Dosimetry with Diamonds

Natural diamonds were proposed as dosimeters in 1948 [158]; their commercialization and investigation in clinical photon and electron beams followed shortly after [159]. Although diamonds were discussed as a reliable, small and near-tissue equivalent dosimeters, it was apparent there were challenging in selecting optimal ones from a large batch; those were rare and expensive. Also, they were poorly reproducible, requiring bespoke characterization and priming.

The PTW-60003 (PTW, Freiburg, Germany) is an example of a widely discussed dosimeter based on a natural diamond crystal. Developed in cooperation with the IPTP Institute (Riga, Latvia), it was commercially available in the 1990s. The radius of the crystal varied from 1.0 mm to 2.2 mm and its thickness from 0.2 to

0.4 mm; it required an external bias of 100 V and a priming with up to 5 Gy. Crystals had different and uncontrolled concentrations of impurities and defects; hence, the amount of priming varied between specimens. Also, priming was required after any switching-off of the external bias. Once primed, however, crystals had a stable response, to within 0.1% for short-term stability, over 10 minutes, and to within 1% for long-term stability. Their sensitivity had no temperature dependence; its dose-rate dependence varied between 1.9% and 3.2% for dose rates between 0.9 Gy/min and 4.65 Gy/min [151, 152, 154, 160, 161, 162].

The crystal of a diamond however can also be artificially grown. At present, the chemical vapor deposition (CVD) and the high-pressure high-temperature (HPHT) techniques are widely used to grow synthetic crystals under controlled and reproducible conditions; for instance, groups that have reported on polycrystalline diamonds grown using CVD include [163, 164]; those that have reported on synthetic single-crystal diamonds (SSCDs) grown using CVD or HPHT include [165, 166]. As growth techniques were improved and perfected, synthetic crystals have become more widely available, reproducible, and affordable. Priming, still required [156, 167], and the degree of dose-rate dependence, vary between crystals produced by different groups [165, 166, 168, 169] as a function of the concentration of impurities.[14]

An SSCD in a Schottky-diode configuration (Figure 4.7) was developed at Tor Vergata University (Rome, Italy). An HPHT of area 4×4 mm^2 is covered by a highly conductive film of boron-doped CVD SSCD; a layer 22 μm-thick of nominally intrinsic diamond is grown onto the doped surface; onto that surface, a Schottky junction is then mounted by thermal evaporation of an aluminum contact. The crystal is embedded in a waterproof

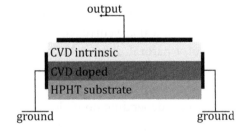

FIGURE 4.7 An SSCD in a Schottky-diode configuration. A film of CVD diamond, p-type doped, is deposited onto a conductive HPHT substrate. A layer of intrinsic diamond is grown onto the doped layer.

cladding made of poly(methyl methacrylate) (PMMA) and epoxy resin. Because of the *built-in* potential at the interface between the metal contact and the layer of diamond, the SSCD works as a Schottky diode and can be operated without an external bias. The radiation-sensitive volume, defined by the depleted region, is about 3.8×10^{-3} mm³, radius 1.1 mm and thickness 1 μm. The sensitivity is stabilized by priming with a few Gy. That SSCD has been made commercially available as the microDiamond PTW-60019 (PTW, Freiburg, Germany); further details can be found in the literature [156, 171–173].

Since its introduction, the microDiamond has been extensively characterized: in photon and electron beams [167, 174–177]; for small-field dosimetry in megavoltage photon beams [176, 178–183]; in x-ray beams produced with a synchrotron [184]; in a magnetic resonance (MR)-guided linear accelerator (linac) [185]; for applicability in a small-animal irradiation unit (SARRP) with x-ray energies up to 220 kV. Those characterizations suggested that the microDiamond, can be scanned in water tank for high spatial resolution dosimetry; if used in the recommended dosimeter-beam orientation, has a sensitivity linear and stable with dose and dose rate, requiring no correction factors; has a directionally dependent sensitivity if the beam is not parallel to the axis of the dosimeter, particularly in the kV energy range; has a low temperature dependence; requires a priming of 10 Gy if used in the kV energy range. The microdiamond is particularly attractive for electron beam scanning dosimetry; since the carbon-water stopping power ratio is nearly constant over typical clinical electron energy ranges, direct dose measurement can be performed [174]. Further, it was proposed that the microDiamond is suitable for *in vivo* dosimetry [156]; for reference dosimetry in radiation fields down to 5 mm in length [186]; for dosimetry in proton and heavy-ion beams – however, its sensitivity has a dependence on the LET in heavy-ion beams, for instance carbon and oxygen [187].

Different groups have proposed alternatives to the microDiamond. For instance, a group has developed an SSCD, referred to as SCDD-Pro (single-crystal diamond dosimeter), for dosimetry in proton beams [188]. The SCDD-Pro is based on a high-quality free-standing SSCD configuration: there is neither substrate nor diamond doping; a bias is required. The sensitivity of the SCDD-Pro exhibits no energy dependence and no dose-rate dependence [188]. Another group investigated a 2D polycrystalline diamond based on CVD [189, 190, 191].

That diamond has dimensions of 2.5×2.5 cm², with a thickness of 300 μm; the front area is divided into 12×12 pixels spaced by 2 mm. Two Schottky barriers are created at the interfaces between the diamond and metallic contacts; it is still necessary to improve the quality of the contacts to ensure a uniform sensitivity [192].

NOTES

1. In that case, devices are referred to as *micro*-dosimeters; we will not cover *micro*-dosimetry applications.
2. For simplicity, the discussion is on p-type diodes; in the base, holes are majority and electrons are minority charge carriers. It is possible to swap around holes and electrons so that the discussion is on n-type diodes.
3. Concentration of excess minority charge carriers with respect to that of equilibrium majority carriers.
4. Unless Bragg–Gray cavity theory breaks down; for instance, in small-field dosimetry.
5. It was necessary to devise a framework for directionally dependent corrections; that methodology relies on real-time information on incident-beam direction [61].
6. The magnetic field affects dose deposition and influences the dosimeter sensitivity.
7. Never mind the fourth terminal (the substrate): it is short-circuited to either source or drain.
8. In other words, readers have a maximum threshold voltage read out.
9. Representing the depth of the radiation-sensitive basal-layer cells.
10. Compare that with the silicon crystal described in Section 3.2. In that case, to apply an electric field across the crystal without generating a current, we need to create a depleted region, formed by a diode based on a p–n junction.
11. In the order of 10^3 V μm⁻¹.
12. For both natural and synthetic crystals.
13. Packaging includes encapsulation, electrodes, cables, etc.
14. The size of the electrodes also impacts the dose-rate dependence [170].

REFERENCES

1. M. Carrara, D. Cutajar, S. Alnaghy, et al., "Semiconductor real-time quality assurance dosimetry in brachytherapy," *Brachytherapy* **17**(1), 133–145 (2018). https://doi.org/10.1016/j.brachy.2017.08.013
2. A. B. Rosenfeld, "Novel detectors for silicon based microdosimetry, their concepts and applications," *Nuclear Instruments and Methods in Physics Research, Section A: Accelerators, Spectrometers, Detectors and Associated Equipment* **809** (February), 156–170 (2016). https://doi.org/10.1016/j.nima.2015.08.059
3. P. D. Bradley, A. B. Rosenfeld, M. Zaider, "Solid state microdosimetry," *Nuclear Instruments and Methods*

in Physics Research, Section B: Beam Interactions with Materials and Atoms **184** (1–2), 135–157 (2001). https://doi.org/10.1016/S0168-583X(01)00715-7

4. A. B. Rosenfeld, "Electronic dosimetry in radiation therapy," *Radiation Measurements* **41** (December), S134–S153 (2007). https://doi.org/10.1016/j.radmeas.2007.01.005

5. A. B. Rosenfeld, "Advanced semiconductor dosimetry in radiation therapy," *AIP Conference Proceedings* **1345**, 48–74 (2011). https://doi.org/10.1063/1.3576159

6. G. C. Messenger and M. S. Ash, *The Effects of Radiation on Electronic Systems* (Van Nostrand Reinhold Company Inc., New York, 1986).

7. S. M. Sze and K. K. Ng, *Physics of Semiconductor Devices*, 3rd ed. (John Wiley & Sons, Inc., Hoboken, NJ, 2006). https://doi.org/10.1002/0470068329

8. W. Shockley and W. T. Read, "Statistics of the recombinations of holes and electrons," *Physical Review* **87**(5), 835–842 (1952). https://doi.org/10.1103/PhysRev.87.835

9. E. Grusell and G. Rikner, "Radiation damage induced dose rate non-linearity in an n-type silicon detector," *Acta Oncologica* **23** (6), 465–469 (1984). https://doi.org/10.3109/02841868409136050

10. B. J. Baliga and E. Sun, "Comparison of gold, platinum, and electron irradiation for controlling lifetime in power rectifiers," *IEEE Transactions on Electron Devices* **24** (6), 685–688 (1977). https://doi.org/10.1109/T-ED.1977.18803

11. M. Bruzzi, "Novel silicon devices for radiation therapy monitoring," *Nuclear Instruments and Methods in Physics Research, Section A: Accelerators, Spectrometers, Detectors and Associated Equipment* **809** (February), 105–112 (2016). https://doi.org/10.1016/j.nima.2015.10.072

12. G. Rikner and E. Grusell, "Effects of radiation damage on the silicon lattice," *Physics in Medicine & Biology* **1261** (28), 1261–1267 (1983).

13. J. Shi, W. E. Simon, T. C. Zhu, "Modeling the instantaneous dose rate dependence of radiation diode detectors," *Medical Physics* **30** (9) (2003). https://doi.org/10.1118/1.1602171

14. P. A. Jursinic, "Dependence of diode sensitivity on the pulse rate of delivered radiation," *Medical Physics* **40** (2) (2013).

15. P. A. Jursinic, "Angular dependence of dose sensitivity of surface diodes," *Medical Physics* **36** (6) (2009).

16. M. Westermark, J. Arndt, B. Nilsson, et al., "Comparative dosimetry in narrow high-energy photon beams," *Physics in Medicine and Biology* **45** (3), 685–702 (2000). https://doi.org/10.1088/0031-9155/45/3/308

17. M. J. Bosma, E. Heijne, J. Kalliopuska, et al., "Edgeless planar semiconductor sensors for a Medipix3-based radiography detector," *Journal of Instrumentation* **6** (11), C11019–C11019 (2011). https://doi.org/10.1088/1748-0221/6/11/C11019

18. X. Wu, J. Kalliopuska, S. Eränen, et al., "Recent advances in processing and characterization of edgeless detectors," *Journal of Instrumentation* **7** (02), C02001–C02001 (2012). https://doi.org/10.1088/1748-0221/7/02/C02001

19. M. Petasecca, S. Alhujaili, A. H. Aldosari, et al., "Angular independent silicon detector for dosimetry in external beam radiotherapy," *Medical Physics* **42** (8), 4708–4718 (2015). https://doi.org/10.1118/1.4926778

20. P. Andreo, D. T. Burns, A. E. Nahum, et al., *Fundamentals of Ionizing Radiation Dosimetry* (John Wiley & Sons, 2017).

21. G. Rikner and E. Grusell, "Selective shielding of a P-Si detector for quality independence," *Acta Oncologica* **24** (1), 65–69 (1985). https://doi.org/10.3109/02841868509134367

22. H. Benmakhlouf and P. Andreo, "Spectral distribution of particle fluence in small field detectors and its implication on small field dosimetry," *Medical Physics* **44** (2) (2017).

23. H. Benmakhlouf, J. Sempau, P. Andreo, "Output correction factors for nine small field detectors in 6 MV radiation therapy photon beams: a PENELOPE Monte Carlo study," *Medical Physics* **41** (4), 041711 (2014). https://doi.org/10.1118/1.4868695

24. M. Heydarian, P. W. Hoban, A. H. Beddoe, "A comparison of dosimetry techniques in stereotactic radiosurgery," *Physics in Medicine and Biology* **41** (1), 93–110 (1996). https://doi.org/10.1088/0031-9155/41/1/008

25. E. Grusell and G. Rikner, "Evaluation of temperature effects in P-type silicon detectors," *Physics in Medicine and Biology* **31** (5), 527–534 (1986). https://doi.org/10.1088/0031-9155/31/5/005

26. A. S. Saini and T. C. Zhu, "Temperature dependence of commercially available diode detectors," *Medical Physics* **29** (4), 622–630 (2002). https://doi.org/10.1118/1.1461842

27. A. R. Jones, "The application of some direct current properties of silicon junction detectors to γ-ray dosimetry," *Physics in Medicine and Biology* **8** (4), 307 (1963). https://doi.org/10.1088/0031-9155/8/4/307

28. G. Rikner and E. Grusell, "General specifications for silicon semiconductors for use in radiation dosimetry," *Physics in Medicine and Biology* **32** (9), 1109–1117 (1987). https://doi.org/10.1088/0031-9155/32/9/004

29. E. Grusell and G. Rikner, "Linearity with dose rate of low resistivity P-type silicon semiconductor detectors," *Physics in Medicine and Biology* **38** (6), 785–792 (1993). https://doi.org/10.1088/0031-9155/38/6/011

30. X. A. Li, C.-M. Ma, D. Salhani, "Measurement of percentage depth dose and lateral beam profile for kilovoltage X-ray therapy beams," *Physics in Medicine and Biology* **42** (12), 2561–2568 (1997). https://doi.org/10.1088/0031-9155/42/12/019

31. R. Tailor, W. Hanson, G. Ibbott, "Response to 'Comment on "Calculated absorbed-dose ratios, TG51/TG21, for most widely used cylindrical and parallel-plate ion chambers over a range of photon and electron energies" '[Med. Phys. 30, 473–477 (2003)]," *Medical Physics* **30** (3), 478–480 (2003). https://doi.org/10.1118/1.1541252

32. D. Marre and G. Marinello, "Comparison of P-type commercial electron diodes for in-vivo dosimetry," *Medical Physics* **31** (1), 50–56 (2004).

33. A. S. Saini and T. C. Zhu, "Energy dependence of commercially available diode detectors for in-vivo dosimetry," *Medical Physics* **34** (5), 1704–1711 (2007). https://doi.org/10.1118/1.2719365

34. P. Andreo, "The physics of small megavoltage photon beam dosimetry," *Radiotherapy and Oncology* **126** (2), 205–213 (2018). https://doi.org/10.1016/j.radonc.2017.11.001

35. H. K. Looe, D. Harder, B. Poppe, "Understanding the lateral dose response functions of high-resolution photon detectors by reverse Monte Carlo and deconvolution analysis," *Physics in Medicine & Biology* **60** (16), 6585–6607 (2015). https://doi.org/10.1088/0031-9155/60/16/6585

36. P. Andreo and H. Benmakhlouf, "Role of the density, density effect and mean excitation energy in solid-state detectors for small photon fields," *Physics in Medicine and Biology* **62** (4), 1518–1532 (2017). https://doi.org/10.1088/1361-6560/aa562e

37. R. Alfonso, P. Andreo, R. Capote, et al., "A new formalism for reference dosimetry of small and nonstandard fields," *Medical Physics* **35** (11), 5179–5186 (2008). https://doi.org/10.1118/1.3005481

38. IAEA, Technical Reports Series No. 483 – Dosimetry of Small Static Fields Used in External Beam Radiotherapy An International Code of Practice for Reference and Relative Dose Determination (2017). http://www-ns.iaea.org/standards/

39. P. Francescon, S. Cora, C. Cavedon, "Total scatter factors of small beams: a multidetector and Monte Carlo study," *Medical Physics* **35** (2), 504–513 (2008). https://doi.org/10.1118/1.2828195

40. G. Cranmer-Sargison, S. Weston, J. A. Evans, et al., "Monte Carlo modelling of diode detectors for small field MV photon dosimetry: detector model simplification and the sensitivity of correction factors to source parameterization," *Physics in Medicine and Biology* **57** (16), 5141–5153 (2012). https://doi.org/10.1088/0031-9155/57/16/5141

41. D. Czarnecki and K. Zink, "Monte Carlo calculated correction factors for diodes and ion chambers in small photon fields," *Physics in Medicine & Biology* **58** (3) (2013). https://doi.org/10.1088/0031-9155/59/3/791

42. C. Moignier, C. Huet, L. Makovicka, "Determination of the correction factors for detectors used with an 800 MU/Min CyberKnife system equipped with fixed collimators and a study of detector response to small photon beams using a Monte Carlo method," *Medical Physics* **41** (7), 071702 (2014). https://doi.org/10.1118/1.4881098

43. P. H. Charles, S. B. Crowe, T. Kairn, et al., "Monte Carlo-based diode design for correction-less small field dosimetry," *Physics in Medicine & Biology* **58** (13), 4501–4512 (2013). https://doi.org/10.1088/0031-9155/58/13/4501

44. T. S. A. Underwood, H. C. Winter, M. A. Hill, et al., "Detector density and small field dosimetry: integral versus point dose measurement schemes," *Medical Physics* **40** (8), 082102 (2013). https://doi.org/10.1118/1.4812687

45. M. Tyler, P. Z.Y. Liu, K. W. Chan, et al., "Characterization of small-field stereotactic radiosurgery beams with modern detectors," *Physics in Medicine and Biology* **58** (21), 7595–7608 (2013). https://doi.org/10.1088/0031-9155/58/21/7595

46. S. Derreumaux, C. Bassinet, C. Huet, et al., "SU-E-T-163: Characterization of the response of active detectors and passive dosemeters used for dose measurement in small photon beams," *Medical Physics* **38** (6 Part 12), 3523–3523 (2011). https://doi.org/10.1118/1.3612113

47. G. Reggiori, P. Mancosu, N. Suchowerska, et al., "Characterization of a new unshielded diode for small field dosimetry under flattening filter free beams," *Physica Medica* **32** (2), 408–413 (2016). https://doi.org/10.1016/j.ejmp.2016.02.004

48. P. Z. Y. Liu, G. Reggiori, F. Lobefalo, et al., "Small field correction factors for the IBA razor," *Physica Medica* **32** (8), 1025–1029 (2016). https://doi.org/10.1016/j.ejmp.2016.07.004

49. T. C. Zhu, L. Ding, C. R. Liu, et al., "Performance evaluation of a diode array for enhanced dynamic wedge dosimetry," *Medical Physics* **24** (7), 1173–1180 (1997). https://doi.org/10.1118/1.598019

50. G. A. Ezzell, J. W. Burmeister, N. Dogan, et al., "IMRT commissioning: multiple institution planning and dosimetry comparisons, a report from AAPM Task Group 119," *Medical Physics* **36** (11), 5359–5373 (2009). https://doi.org/10.1118/1.3238104

51. M. Ahmad, J. Deng, M. W. Lund, et al., "Clinical implementation of enhanced dynamic wedges into the pinnacle treatment planning system: Monte Carlo validation and patient-specific QA," *Physics in Medicine and Biology* **54** (2), 447–465 (2009). https://doi.org/10.1088/0031-9155/54/2/018

52. A. Bocci, M. A. Cortés-Giraldo, M. I. Gallardo, et al., "Silicon strip detector for a novel 2D dosimetric method for radiotherapy treatment verification," *Nuclear Instruments and Methods in Physics Research, Section A: Accelerators, Spectrometers, Detectors and Associated Equipment* **673**, 98–106 (2012). https://doi.org/10.1016/j.nima.2012.01.018

53. E. Pappas, T. G. Maris, F. Zacharopoulou, et al., "Small SRS photon field profile dosimetry performed using a PinPoint air ion chamber, a diamond detector, a novel silicon diode array (DOSI), and polymer gel dosimetry. Analysis and intercomparison," *Medical Physics* **35** (10), 4640–4648 (2008). https://doi.org/10.1118/1.2977829

54. D. Menichelli, M. Bruzzi, M. Bucciolini, et al., "Design and development of a silicon-segmented detector for 2D dose measurements in radiotherapy," *Nuclear Instruments and Methods in Physics Research, Section A: Accelerators, Spectrometers, Detectors and Associated Equipment* **583** (1), 109–113 (2007). https://doi.org/10.1016/j.nima.2007.08.209

55. J. H. D. Wong, M. Carolan, M. L. F. Lerch, et al., "A silicon strip detector dose magnifying glass for IMRT

dosimetry," *Medical Physics* **37** (2), 427–439 (2010). https://doi.org/10.1118/1.3264176

56. E. Debrot, M. Newall, S. Guatelli, et al., "A silicon strip detector array for energy verification and quality assurance in heavy ion therapy," *Medical Physics* **45** (2), 953–962 (2018). https://doi.org/10.1002/mp.12736

57. M. Petasecca, M. K. Newall, J. T. Booth, et al., "MagicPlate-512: A 2D silicon detector array for quality assurance of stereotactic motion adaptive radiotherapy," *Medical Physics* **42** (6), 2992–3004 (2015). https://doi.org/10.1118/1.4921126

58. K. A. Shukaili, M. Petasecca, M. Newall, et al., "A 2D silicon detector array for quality assurance in small field dosimetry: DUO," *Medical Physics* **44** (2), 628–636 (2017). https://doi.org/10.1002/mp.12060

59. G. Biasi, M. Petasecca, S. Guatelli, N. Hardcastle, et al., "A novel high-resolution 2D silicon array detector for small field dosimetry with FFF photon beams," *Physica Medica* **45** (January), 117–126 (2018). https://doi.org/10.1016/j.ejmp.2017.12.010

60. G. Biasi, K. Al Shukaili, M. Petasecca, et al., "Today's monolithic silicon array detector for small field dosimetry: the Octa," *Journal of Physics: Conference Series* **1154** (January), 012002 (2019). https://doi.org/10.1088/1742-6596/1154/1/012002

61. N. Stansook, K. Utitsarn, M. Petasecca, et al., "Technical note: angular dependence of a 2D monolithic silicon diode array for small field dosimetry," *Medical Physics* **44** (8), 4313–4321 (2017). https://doi.org/10.1002/mp.12377

62. M. Duncan, M. K. Newall, V. Caillet, et al., "Real-time high spatial resolution dose verification in stereotactic motion adaptive arc radiotherapy," *Journal of Applied Clinical Medical Physics* **19** (4), 173–184 (2018). https://doi.org/10.1002/acm2.12364

63. K. Utitsarn, G. Biasi, N. Stansook, et al., "Two-dimensional solid-state array detectors: a technique for in vivo dose verification in a variable effective area," *Journal of Applied Clinical Medical Physics* **20** (11), 88–94 (2019). https://doi.org/10.1002/acm2.12744

64. S. J. Alnaghy, T. Causer, M. Gargett, et al., "A feasibility study for high-resolution silicon array detector performance in the magnetic field of a permanent magnet system," *Medical Physics* **46** (9), 4224–4232 (2019). https://doi.org/10.1002/mp.13686

65. G. Biasi, M. Petasecca, S. Guatelli, M. A. Ebert, et al., "CyberKnife® fixed cone and Iris™ defined small radiation fields: assessment with a high-resolution solid-state detector array," *Journal of Applied Clinical Medical Physics* **19** (5), 547–557 (2018). https://doi.org/10.1002/acm2.12414

66. K. Utitsarn, Z. A. Alrowaili, N. Stansook, et al., "The effect of an air gap on a 2D monolithic silicon detector for relative dosimetry," *Journal of Instrumentation* **14** (06), P06018–P06018 (2019). https://doi.org/10.1088/1748-0221/14/06/P06018

67. G. Biasi, J. Davis, M. Petasecca, et al., "On monolithic silicon array detectors for small-field photon beam

dosimetry," *IEEE Transactions on Nuclear Science* **65** (9), 2640–2649 (2018). https://doi.org/10.1109/TNS.2018.2860625

68. G. Biasi, N. Hardcastle, M. Petasecca, et al., "On the instantaneous dose rate and angular dependence of monolithic silicon array detectors," *IEEE Transactions on Nuclear Science* **66** (1), 519–527 (2019). https://doi.org/10.1109/TNS.2018.2885017

69. N. Stansook, G. Biasi, K. Utitsarn, et al., "2D Monolithic silicon-diode array detectors in megavoltage photon beams: does the fabrication technology matter? A medical physicist's perspective," *Australasian Physical & Engineering Sciences in Medicine* **42** (2), 443–451 (2019). https://doi.org/10.1007/s13246-019-00736-7

70. S. Stathakis, P. Myers, C. Esquivel, et al., "Characterization of a novel 2D array dosimeter for patient-specific quality assurance with volumetric arc therapy," *Medical Physics* **40** (7) (2013). https://doi.org/10.1118/1.4812415

71. C. K. McGarry, B. F. O'Connell, M. W. D. Grattan, et al., "Octavius 4D characterization for flattened and flattening filter free rotational deliveries," *Medical Physics* **40** (9) (2013). https://doi.org/10.1118/1.4817482

72. R. Sadagopan, J. A. Bencomo, R. L. Martin, et al., "Characterization and clinical evaluation of a novel IMRT quality assurance system," *Journal of Applied Clinical Medical Physics* **10** (2), 104–119 (2009). https://doi.org/10.1120/jacmp.v10i2.2928

73. V. Feygelman, G. G. Zhang, C. Stevens, et al., "Evaluation of a new VMAT QA device, or the 'X' and 'O' array geometries," *Journal of Applied Clinical Medical Physics* **12** (2), 3346 (2011). https://doi.org/10.1120/jacmp.v12i2.3346

74. G. A. Ausman and F. B. McLean, "Electron–hole pair creation energy in SiO₂," *Applied Physics Letters* **26** (4), 173–175 (1975). https://doi.org/10.1063/1.88104

75. R. C. Hughes, "Charge-carrier transport phenomena in amorphous SiO: direct measurement of the drift mobility and lifetimes," *Physical Review Letters* **30** (26), 1333–1336 (1973). https://doi.org/10.1103/PhysRevLett.30.1333

76. R. C. Hughes, E. P. EerNisse, H. J. Stein, "Hole transport in MOS oxides," *IEEE Transactions on Nuclear Science* **22** (6) (1975).

77. T. R. Oldham and F. B. McLean, "Total ionizing dose effects in MOS oxides and devices," *IEEE Transactions on Nuclear Science* **50** III (3), 483–499 (2003). https://doi.org/10.1109/TNS.2003.812927

78. J. R. Schwank, M. R. Shaneyfelt, D. M. Fleetwood, et al., "Radiation effects in MOS oxides," *IEEE Transactions on Nuclear Science* **55** (4), 1833–1853 (2008). https://doi.org/10.1109/TNS.2008.2001040

79. R. Freeman and A. Holmes-Siedle, "A simple model for predicting radiation effects in MOS devices," *IEEE Transactions on Nuclear Science* **25** (6), 1216–1225 (1978). https://doi.org/10.1109/TNS.1978.4329516

80. G. Ristić, S. Golubović, M. Pejović, "Sensitivity and fading of PMOS dosimeters with thick gate oxide," *Sensors*

and Actuators A: Physical **51** (2–3), 153–158 (1995). https://doi.org/10.1016/0924-4247(95)01211-7

81. A. Jaksic, G. Ristic, M. Pejovic, A. Mohammadzadeh, W. Lane, "Characterisation of radiation response of 400 Nm implanted gate oxide RADFETs," in: *2002 23rd International Conference on Microelectronics. Proceedings (Cat. No.02TH8595)*, vol. **2**, pp. 727–730 (2002). https://doi.org/10.1109/MIEL.2002.1003360

82. S. Alshaikh, M. Carolan, M. Petasecca, et al., "Direct and pulsed current annealing of P-MOSFET based dosimeter: the 'MOSkin.'" *Australasian Physical and Engineering Sciences in Medicine* **37** (2), 311–319 (2014). https://doi.org/10.1007/s13246-014-0261-1

83. G. W. Luo, Z. Y. Qi, X. W. Deng, et al., "Investigation of a pulsed current annealing method in reusing MOSFET dosimeters for in vivo IMRT dosimetry," *Medical Physics* **41** (5), 1–7 (2014). https://doi.org/10.1118/1.4871619

84. D. M. Fleetwood, T. L. Meisenheimer, J. H. Scofield, "1/f noise and radiation effects in MOS devices," *IEEE Transactions on Electron Devices* **41** (11), 1953–1964 (1994). https://doi.org/10.1109/16.333811

85. C. Benson, R. A. Price, J. Silvie, et al., "Radiation-induced statistical uncertainty in the threshold voltage measurement of MOSFET dosimeters," *Physics in Medicine and Biology* **49** (14), 3145–3159 (2004). https://doi.org/10.1088/0031-9155/49/14/009

86. A. Jaksic, G. Ristic, M. Pejovic, A. Mohammadzadeh, C. Sudre, et al., "Gamma-ray irradiation and post-irradiation responses of high dose range RADFETs," *IEEE Transactions on Nuclear Science* **49** (3), 1356–1363 (2002). https://doi.org/10.1109/TNS.2002.1039667

87. M. W. Bower and D. E. Hintenlang, "The characterization of a commercial MOSFET dosimeter system for use in diagnostic X-ray," *Health Physics* **75** (2), 197–204 (1998). https://doi.org/10.1097/00004032-199808000-00013

88. C. R. Edwards, S. Green, J. E. Palethorpe, et al., "The response of a MOSFET, p-type semiconductor and LiF TLD to quasi-monoenergetic x-rays," *Physics in Medicine and Biology* **42** (12), 2383–2391 (1997). https://doi.org/10.1088/0031-9155/42/12/006

89. T. Kron, L. Duggan, T. Smith, et al., "Dose response of various radiation detectors to synchrotron radiation," *Physics in Medicine and Biology* **43** (11), 3235–3259 (1998). https://doi.org/10.1088/0031-9155/43/11/006

90. R. Ramani, S. Russell, P. O'Brien, "Clinical dosimetry using MOSFETS," *International Journal of Radiation Oncology Biology Physics* **37** (4), 959–964 (1997). https://doi.org/10.1016/S0360-3016(96)00600-1

91. R. Ramaseshan, K. S. Kohli, T. J. Zhang, et al., "Performance characteristics of a MicroMOSFET as an in vivo dosimeter in radiation therapy," *Physics in Medicine and Biology* **49** (17), 4031–4048 (2004). https://doi.org/10.1088/0031-9155/49/17/014

92. P. H. Halvorsen, "Dosimetric evaluation of a new design MOSFET in vivo dosimeter," *Medical Physics* **32** (1), 110–117 (2005). https://doi.org/10.1118/1.1827771

93. A. B. Rosenfeld, M. G. Carolan, G. I. Kaplan, et al., "MOSFET dosimeters: the role of encapsulation on dosimetric characteristics in mixed gamma-neutron and megavoltage X-ray fields," *IEEE Transactions on Nuclear Science* **42** (6), 1870–1877 (1995). https://doi.org/10.1109/23.489229

94. B. Wang, C. H. Kim, X. G. Xu, "Monte Carlo modeling of a high-sensitivity MOSFET dosimeter for low- and medium-energy photon sources," *Medical Physics* **31** (5), 1003–1008 (2004). https://doi.org/10.1118/1.1688272

95. P. Scalchi and P. Francescon, "Calibration of a MOSFET detection system for 6-MV in vivo dosimetry," *International Journal of Radiation Oncology Biology Physics* **40** (4), 987–993 (1998). https://doi.org/10.1016/S0360-3016(97)00894-8

96. S. Qin, T. Chen, L. Wang, et al., "Angular dependence of the MOSFET dosimeter and its impact on *in vivo* surface dose measurement in breast cancer treatment," *Technology in Cancer Research & Treatment* **13** (4), 345–352 (2014). https://doi.org/10.7785/tcrt.2012.500382

97. J. Seco, B. Clasie, M. Partridge, "Review on the characteristics of radiation detectors for dosimetry and imaging," *Physics in Medicine & Biology* **59** (20), R303–R347 (2014). https://doi.org/10.1088/0031-9155/59/20/R303

98. T. Cheung, M. J. Butson, P. K. N. Yu, "Effects of temperature variation on MOSFET dosimetry," *Physics in Medicine and Biology* **49** (13), N191–N196 (2004). https://doi.org/10.1088/0031-9155/49/13/N02

99. M. G. Buehler, B. R. Blaes, G. A. Soli, et al., "On-chip p-MOSFET dosimetry (CMOS ICs)," *IEEE Transactions on Nuclear Science* **40** (6), 1442–1449 (1993). https://doi.org/10.1109/23.273520

100. M. Soubra, J. Cygler, G. Mackay, "Evaluation of a dual bias dual metal oxide-silicon semiconductor field effect transistor detector as radiation dosimeter," *Medical Physics* **21** (4), 567–572 (1994). https://doi.org/10.1118/1.597314

101. D. M. Fleetwood, P. S. Winokur, J. R. Schwank, "Using laboratory X-ray and cobalt-60 irradiations to predict CMOS device response in strategic and space environments," *IEEE Transactions on Nuclear Science* **35** (6), 1497–1505 (1988). https://doi.org/10.1109/23.25407

102. J. R. Schwank, S. B. Roeske, D. E. Beutler, et al., "A dose rate independent PMOS dosimeter for space applications," *IEEE Transaction of Nuclear Science* **43** (6), 2671–2678 (1996). https://doi.org/10.1109/23.556852

103. W. Poch and A. G. Holmes-Siedle, "The dosimeter: a new instrument to measure radiation dose," *RCA Engineers* **16** (3), 56–59 (1970).

104. A. Holmes-Siedle, "The space-charge dosimeter," *Nuclear Instruments and Methods* **121** (1), 169–179 (1974). https://doi.org/10.1016/0029-554X(74)90153-0

105. L. Adams and A. Holmes-Siedle, "The development of an MOS dosimetry unit for use in space," *IEEE*

Transactions on Nuclear Science **25** (6), 1607–1612 (1978). https://doi.org/10.1109/TNS.1978.4329580

106. C. Ehringfeld, S. Schmid, K. Poljanc, et al., "Application of commercial MOSFET detectors for in vivo dosimetry in the therapeutic X-ray range from 80 kV to 250 kV," *Physics in Medicine and Biology* **50** (2), 289–303 (2005). https://doi.org/10.1088/0031-9155/50/2/008

107. P. Scalchi, R. Righetto, C. Cavedon, et al., "Direct tumor in vivo dosimetry in highly-conformal radiotherapy: a feasibility study of implantable MOSFETs for hypofractionated extracranial treatments using the Cyberknife system," *Medical Physics* **37** (4), 1413–1423 (2010). https://doi.org/10.1118/1.3315370

108. W. L. Jong, N. M. Ung, A. Vannyat, et al., "'Edge-on' MOSkin detector for stereotactic beam measurement and verification," *Physica Medica* **33**, 127–135 (2017). https://doi.org/10.1016/j.ejmp.2016.12.020

109. E. J. Bloemen-van Gurp, A. W. H Minken, B. J. Mijnheer, et al., "Clinical implementation of MOSFET detectors for dosimetry in electron beams," *Radiotherapy and Oncology* **80** (3), 288–295 (2006). https://doi.org/10.1016/j.radonc.2006.07.002

110. W. L. Jong, N. M. Ung, A. H. L. Tiong, et al., "Characterisation of a MOSFET-based detector for dose measurement under megavoltage electron beam radiotherapy," *Radiation Physics and Chemistry* **144**, 76–84 (2018). https://doi.org/10.1016/j.radphyschem.2017.11.021

111. A. Faigon, J. Lipovetzky, E. Redin, et al., "Extension of the measurement range of MOS dosimeters using radiation induced charge neutralization," *IEEE Transactions on Nuclear Science* **55** (4), 2141–2147 (2008). https://doi.org/10.1109/TNS.2008.2000767

112. R. Consorti, A. Petrucci, F. Fortunato, et al., "In vivo dosimetry with MOSFETs: dosimetric characterization and first clinical results in intraoperative radiotherapy," *International Journal of Radiation Oncology Biology Physics* **63** (3), 952–960 (2005). https://doi.org/10.1016/j.ijrobp.2005.02.049

113. M. Ciocca, V. Piazzi, R. Lazzari, et al., "Real-time in vivo dosimetry using micro-MOSFET detectors during intraoperative electron beam radiation therapy in early-stage breast cancer," *Radiotherapy and Oncology* **78** (2), 213–216 (2006). https://doi.org/10.1016/j.radonc.2005.11.011

114. P. Francescon, S. Cora, C. Cavedon, et al., "Use of a new type of radiochromic film, a new parallel-plate microchamber, MOSFETs, and TLD 800 microcubes in the dosimetry of small beams," *Medical Physics* **25** (4), 503–511 (1998). https://doi.org/10.1118/1.598227

115. O. A. Sauer and J. Wilbert, "Measurement of output factors for small photon beams," *Medical Physics* **34** (6), 1983–1988 (2007). https://doi.org/10.1118/1.2734383

116. R. Kohno, T. Nishio, T. Miyagishi, et al., "Experimental evaluation of a MOSFET dosimeter for proton dose measurements," *Physics in Medicine and Biology* **51** (23), 6077–6086 (2006). https://doi.org/10.1088/0031-9155/51/23/009

117. H.-M. Lu, G. Mann, E. Cascio, "Investigation of an implantable dosimeter for single-point water equivalent path length verification in proton therapy," *Medical Physics* **37** (11), 5858–5866 (2010). https://doi.org/10.1118/1.3504609

118. A. B. Rosenfeld, G. I. Kaplanl, T. Kron, et al., "MOSFET dosimetry of an X-ray microbeam," *IEEE Transactions on Nuclear Science* **46** (6) (1999).

119. G. I. Kaplan, A. B. Rosenfeld, B. J. Allen, et al., "Improved spatial resolution by MOSFET dosimetry of an X-ray microbeam," *Medical Physics* **27** (1), 239–244 (2000). https://doi.org/10.1118/1.598866

120. A. B. Rosenfeld, M. L. F. Lerch, T. Kron, et al., "Feasibility study of online high-spatial-resolution MOSFET dosimetry in static and pulsed X-ray radiation fields," *IEEE Transactions on Nuclear Science* **48** (6), 2061–2068 (2001).

121. A. B. Rosenfeld, E. A. Siegbahn, E. Brauer-Krish, et al., "Edge-on face-to-face MOSFET for synchrotron microbeam dosimetry: MC modeling," *IEEE Transactions on Nuclear Science* **52** (6), 2562–2569 (2005). https://doi.org/10.1109/TNS.2005.860704

122. E. A. Siegbahn, E. Bräuer-Krisch, A. Bravin, et al., "MOSFET dosimetry with high spatial resolution in intense synchrotron-generated x-ray microbeams," *Medical Physics* **36** (4), 1128–1137 (2009). https://doi.org/10.1118/1.3081934

123. ICRP, *The Biological Basis for Dose Limitation in the Skin* (1992).

124. M. J. Butson, A. Rozenfeld, J. N. Mathur, et al., "A new radiotherapy surface dose detector: the MOSFET," *Medical Physics* **23** (5), 655–658 (1996). https://doi.org/10.1118/1.597702

125. P. Scalchi, P. Francescon, P. Rajaguru, "Characterization of a new MOSFET detector configuration for in vivo skin dosimetry," *Medical Physics* **32** (6), 1571–1578 (2005). https://doi.org/10.1118/1.1924328

126. I. S. Kwan, A. B. Rosenfeld, Z. Y. Qi, et al., "Skin dosimetry with new MOSFET detectors," *Radiation Measurements* **43** (2–6), 929–932 (2008). https://doi.org/10.1016/j.radmeas.2007.12.052

127. Z. Y. Qi, X. W. Deng, S. M. Huang, et al., "In vivo verification of superficial dose for head and neck treatments using intensity-modulated techniques," *Medical Physics* **36** (1), 59–70 (2009). https://doi.org/10.1118/1.3030951

128. A. Quinn, L. Holloway, D. Cutajar, et al., "Megavoltage cone beam CT near surface dose measurements: potential implications for breast radiotherapy," *Medical Physics* **38** (11), 6222–6227 (2011). https://doi.org/10.1118/1.3641867

129. M. J. Safari, J. H. D. Wong, K. A. A. Kadir, et al., "Real-time eye lens dose monitoring during cerebral angiography procedures," *European Radiology* **26** (1), 79–86 (2016). https://doi.org/10.1007/s00330-015-3818-9

130. N. Hardcastle, D. L. Cutajar, P. E. Metcalfe, et al., "In vivo real-time rectal wall dosimetry for prostate radiotherapy," *Physics in Medicine and Biology* **55** (13), 3859–3871 (2010). https://doi.org/10.1088/0031-9155/55/13/019

131. S. J. Alnaghy, S. Deshpande, D. L. Cutajar, et al., "In vivo endorectal dosimetry of prostate tomotherapy using dual MOSkin detectors," *Journal of Applied Clinical Medical Physics* 16 (3), 107–117 (2015). https://doi.org/10.1120/jacmp.v16i3.5113

132. E. A. Alhakeem, S. AlShaikh, A. B. Rosenfeld, et al., "Comparative evaluation of modern dosimetry techniques near low- and high-density heterogeneities," *Journal of Applied Clinical Medical Physics* 16 (5), 142–158 (2015). https://doi.org/10.1120/jacmp.v16i5.5589

133. C. Tenconi, M. Carrara, M. Borroni, et al., "TRUS-probe integrated MOSkin detectors for rectal wall in vivo dosimetry in HDR brachytherapy: in phantom feasibility study," *Radiation Measurements* 71, 379–383 (2014). https://doi.org/10.1016/j.radmeas.2014.05.010

134. M. Carrara, C. Tenconi, G. Rossi, et al., "In vivo rectal wall measurements during HDR prostate brachytherapy with MOSkin dosimeters integrated on a trans-rectal US probe: comparison with planned and reconstructed doses," *Radiotherapy and Oncology* 118 (1), 148–153 (2016). https://doi.org/10.1016/j.radonc.2015.12.022

135. C. W. Scarantino, D. M. Ruslander, C. J. Rini, et al., "An implantable radiation dosimeter for use in external beam radiation therapy," *Medical Physics* 31 (9), 2658–2671 (2004). https://doi.org/10.1118/1.1778809

136. R. D. Black, C. W. Scarantino, G. G. Mann, et al., "An analysis of an implantable dosimeter system for external beam therapy," *International Journal of Radiation Oncology Biology Physics* 63 (1), 290–300 (2005). https://doi.org/10.1016/j.ijrobp.2005.05.025

137. T. M. Briere, A. S. Beddar, M. T. Gillin, "Evaluation of pre-calibrated implantable MOSFET radiation dosimeters for megavoltage photon beams," *Medical Physics* 32 (11), 3346–3349 (2005). https://doi.org/10.1118/1.2065447

138. A. S. Beddar, M. Salehpour, T. M. Briere, et al., "Preliminary evaluation of implantable MOSFET radiation dosimeters," *Physics in Medicine and Biology* 50 (1), 141–149 (2005). https://doi.org/10.1088/0031-9155/50/1/011

139. C. W. Scarantino, C. J. Rini, M. Aquino, et al., "Initial clinical results of an in vivo dosimeter during external beam radiation therapy," *International Journal of Radiation Oncology Biology Physics* 62 (2), 606–613 (2005). https://doi.org/10.1016/j.ijrobp.2004.09.041

140. G. P. Beyer, C. W. Scarantino, B. R. Prestidge, et al., "Technical evaluation of radiation dose delivered in prostate cancer patients as measured by an implantable MOSFET dosimeter," *International Journal of Radiation Oncology Biology Physics* 69 (3), 925–935 (2007). https://doi.org/10.1016/j.ijrobp.2007.06.065

141. T. M. Briere, M. T. Gillin, A. S. Beddar, "Implantable MOSFET detectors: evaluation of a new design," *Medical Physics* 34 (12), 4585–4590 (2007). https://doi.org/10.1118/1.2799578

142. G. P. Beyer, G. G. Mann, J. A. Pursley, et al., "An implantable MOSFET dosimeter for the measurement of radiation dose in tissue during cancer therapy," *IEEE Sensors Journal* 8 (1), 38–51 (2008). https://doi.org/10.1109/JSEN.2007.912542

143. C. W. Scarantino, B. R. Prestidge, M. S. Anscher, et al., "The observed variance between predicted and measured radiation dose in breast and prostate patients utilizing an in vivo dosimeter," *International Journal of Radiation Oncology Biology Physics* 72 (2), 597–604 (2008). https://doi.org/10.1016/j.ijrobp.2008.05.058

144. C. Esquivel, P. Rassiah, G. Beyer, et al., "SU-FF-T-203: evaluation and performance characteristics of an implantable dosimeter with the TomoTherapy Hi-Art System," *Medical Physics* 34 (6), 2448 (2007). https://doi.org/10.1118/1.2760864

145. E. A. Burgemeister, "Dosimetry with a diamond operating as a resistor," *Physics in Medicine and Biology* 26 (2), 269–275 (1981). https://doi.org/10.1088/0031-9155/26/2/006

146. S. F. Kozlov, R. Stuck, M. Hage-Ali, et al., "Preparation and characteristics of natural diamond nuclear radiation detectors," *IEEE Transactions on Nuclear Science* 22 (1), 160–170 (1975). https://doi.org/10.1109/TNS.1975.4327634

147. W. Kada, N. Iwamoto, T. Satoh, et al., "Continuous observation of polarization effects in thin SC-CVD diamond detector designed for heavy ion microbeam measurement," *Nuclear Instruments and Methods in Physics Research Section B: Beam Interactions with Materials and Atoms* 331 (July), 113–116 (2014). https://doi.org/10.1016/j.nimb.2013.11.040

148. M. Pomorski, E. Berdermann, W. de Boer, et al., "Charge transport properties of single crystal CVD-diamond particle detectors," *Diamond and Related Materials* 16 (4–7), 1066–1069 (2007). https://doi.org/10.1016/j.diamond.2006.11.016

149. V. Grilj, N. Skukan, M. Jakšić, et al., "Irradiation of thin diamond detectors and radiation hardness tests using MeV protons," *Nuclear Instruments and Methods in Physics Research Section B: Beam Interactions with Materials and Atoms* 306 (July), 191–194 (2013). https://doi.org/10.1016/j.nimb.2012.12.034

150. J. F. Fowler, "Solid state electrical conductivity dosimeters," in: *Radiation Dosimetry: Volume II: Instrumentation*, edited by E. Tochili, F. H. Attix, W. C. Roesch (Academic Press, 1966).

151. P. W. Hoban, M. Heydarian, W. A. Beckham, et al., "Dose rate dependence of a PTW diamond detector in the dosimetry of a 6 MV photon beam," *Physics in Medicine and Biology* 39 (8), 1219–1229 (1994). https://doi.org/10.1088/0031-9155/39/8/003

152. W. U. Laub, T. W. Kaulich, F. Nüsslin, "A diamond detector in the dosimetry of high-energy electron and photon beams," *Physics in Medicine and Biology* 44 (9), 2183–2192 (1999). https://doi.org/10.1088/0031-9155/44/9/306

153. N. Ade and T. L. Nam, "The influence of defect levels on the dose rate dependence of synthetic diamond detectors of various types on exposures to high-energy radiotherapy beams," *Radiation Physics and*

Chemistry **108**, 65–73 (2015). https://doi.org/10.1016/j.radphyschem.2014.11.016

154. C. De Angelis, S. Onori, M. Pacilio, et al., "An investigation of the operating characteristics of two PTW diamond detectors in photon and electron beams," *Medical Physics* **29** (2), 248–254 (2002). https://doi.org/10.1118/1.1446101

155. P. N. Mobit and G. A. Sandison, "An EGS4 Monte Carlo examination of the response of a PTW-diamond radiation detector in megavoltage electron beams," *Medical Physics* **26** (5), 839–844 (1999). https://doi.org/10.1118/1.598593

156. I. Ciancaglioni, M. Marinelli, E. Milani, et al., "Dosimetric characterization of a synthetic single crystal diamond detector in clinical radiation therapy small photon beams," *Medical Physics* **39** (7 Part 1), 4493–4501 (2012). https://doi.org/10.1118/1.4729739

157. S. Kampfer, N. Cho, S. E. Combs, et al., "Dosimetric characterization of a single crystal diamond detector in X-ray beams for preclinical research," *Zeitschrift Für Medizinische Physik* **28** (4), 303–309 (2018). https://doi.org/10.1016/j.zemedi.2018.05.002

158. K. G. Mckay, "Electron bombardment conductivity in diamond," *Physical Review* **74** (11), 1606–1621 (1948). https://doi.org/10.1103/PhysRev.74.1606

159. B. Planskoy, "Evaluation of diamond radiation dosemeters," *Physics in Medicine and Biology* **25** (3), 519–532 (1980). https://doi.org/10.1088/0031-9155/25/3/011

160. M. Heydarian, P. W. Hoban, W. A. Beckham, et al., "Evaluation of a PTW diamond detector for electron beam measurements," *Physics in Medicine and Biology* **38** (8), 1035–1042 (1993). https://doi.org/10.1088/0031-9155/38/8/002

161. P. Bjork, T. Knoos, P. Nilsson, "Comparative dosimetry of diode and diamond detectors in electron beams for intraoperative radiation therapy," *Medical Physics* **27** (11), 2580–2588 (2000). https://doi.org/10.1118/1.1315317

162. A. Fidanzio, L. Azario, R. Miceli, et al., "PTW-diamond detector: dose rate and particle type dependence," *Medical Physics* **27** (11), 2589–2593 (2000). https://doi.org/10.1118/1.1318218

163. A. Fidanzio, L. Azario, R. Kalish, et al., "A preliminary dosimetric characterization of chemical vapor deposition diamond detector prototypes in photon and electron radiotherapy beams," *Medical Physics* **32** (2), 389–395 (2005). https://doi.org/10.1118/1.1851887

164. C. Descamps, D. Tromson, N. Tranchant, et al., "Clinical studies of optimised single crystal and polycrystalline diamonds for radiotherapy dosimetry," *Radiation Measurements* **43** (2–6), 933–938 (2008). https://doi.org/10.1016/j.radmeas.2007.11.080

165. N. Tranchant, D. Tromson, C. Descamps, et al., "High mobility single crystal diamond detectors for dosimetry: application to radiotherapy," *Diamond and Related Materials* **17** (7–10), 1297–1301 (2008). https://doi.org/10.1016/j.diamond.2008.03.025

166. D. Tromson, M. Rebisz-Pomorska, N. Tranchant, et al., "Single crystal CVD diamond detector for high resolution dose measurement for IMRT and novel radiation therapy needs," *Diamond and Related Materials* **19** (7–9), 1012–1016 (2010). https://doi.org/10.1016/j.diamond.2010.03.008

167. W. U. Laub and R. Crilly, "Clinical radiation therapy measurements with a new commercial synthetic single crystal diamond detector," *Journal of Applied Clinical Medical Physics* **15** (6), 92–102 (2014). https://doi.org/10.1120/jacmp.v15i6.4890

168. G. T. Betzel, S. P. Lansley, F. Baluti, et al., "Clinical investigations of a CVD diamond detector for radiotherapy dosimetry," *Physica Medica* **28** (2), 144–152 (2012). https://doi.org/10.1016/j.ejmp.2011.04.003

169. S. Spadaro, G. Conte, M. Pimpinella, et al., "Electrical and dosimetric characterization of a CVD diamond detector with high sensitivity," *Radiation Measurements* **48** (January), 1–6 (2013). https://doi.org/10.1016/j.radmeas.2012.11.017

170. F. Marsolat, D. Tromson, N. Tranchant, et al., "Why diamond dimensions and electrode geometry are crucial for small photon beam dosimetry," *Journal of Applied Physics* **118** (23), 0–14 (2015). https://doi.org/10.1063/1.4937994

171. S. Almaviva, M. Marinelli, E. Milani, et al., "Synthetic single crystal diamond diodes for radiotherapy dosimetry," *Nuclear Instruments and Methods in Physics Research Section A: Accelerators, Spectrometers, Detectors and Associated Equipment* **594** (2), 273–277 (2008). https://doi.org/10.1016/j.nima.2008.06.028

172. S. Almaviva, I. Ciancaglioni, R. Consorti, et al., "Synthetic single crystal diamond dosimeters for intensity modulated radiation therapy applications," *Nuclear Instruments and Methods in Physics Research, Section A: Accelerators, Spectrometers, Detectors and Associated Equipment* **608** (1), 191–194 (2009). https://doi.org/10.1016/j.nima.2009.07.004

173. S. Almaviva, I. Ciancaglioni, R. Consorti, et al., "Synthetic single crystal diamond dosimeters for conformal radiation therapy application," *Diamond and Related Materials* **19** (2–3), 217–220 (2010). https://doi.org/10.1016/j.diamond.2009.10.007

174. C. Di Venanzio, M. Marinelli, E. Milani, et al., "Characterization of a synthetic single crystal diamond Schottky diode for radiotherapy electron beam dosimetry," *Medical Physics* **40** (2) (2013). https://doi.org/10.1118/1.4774360

175. J. E. Morales, S. B. Crowe, R. Hill, et al., "Dosimetry of cone-defined stereotactic radiosurgery fields with a commercial synthetic diamond detector," *Medical Physics* **41** (11), 111702 (2014). https://doi.org/10.1118/1.4895827

176. A. Ralston, M. Tyler, P. Liu, et al., "Over-response of synthetic microdiamond detectors in small radiation fields," *Physics in Medicine and Biology* **59** (19), 5873–5881 (2014). https://doi.org/10.1088/0031-9155/59/19/5873

177. J. M. Lárraga-Gutiérrez, P. Ballesteros-Zebadúa, M. Rodríguez-Ponce, et al., "Properties of a commercial PTW-60019 synthetic diamond detector for the dosimetry of small radiotherapy beams," *Physics in Medicine and Biology* **60** (2), 905–924 (2015). https://doi.org/10.1088/0031-9155/60/2/905

178. G. Azangwe, P. Grochowska, D. Georg, et al., "Detector to detector corrections: a comprehensive experimental study of detector specific correction factors for beam output measurements for small radiotherapy beams," *Medical Physics* **41** (7), 072103 (2014). https://doi.org/10.1118/1.4883795

179. A. Chalkley and G. Heyes, "Evaluation of a synthetic single-crystal diamond detector for relative dosimetry measurements on a CyberKnife™," *The British Journal of Radiology* **87** (1035), 20130768 (2014). https://doi.org/10.1259/bjr.20130768

180. P. Papaconstadopoulos, F. Tessier, J. Seuntjens, "On the correction, perturbation and modification of small field detectors in relative dosimetry," *Physics in Medicine & Biology* **59** (19), 5937–5952 (2014). https://doi.org/10.1088/0031-9155/59/19/5937

181. P. Andreo, H. Palmans, M. Marteinsdóttir, et al., "On the Monte Carlo simulation of small-field micro-diamond detectors for megavoltage photon dosimetry," *Physics in Medicine and Biology* **61** (1), L1–L10 (2015). https://doi.org/10.1088/0031-9155/61/1/L1

182. D. J. O'Brien, L. León-Vintró, B. McClean, "Small field detector correction factors kQclin, Qmsr (fclin, fmsr) for silicon-diode and diamond detectors with circular 6 MV fields derived using both empirical and numerical methods," *Medical Physics* **43** (1), 411 (2016). https://doi.org/10.1118/1.4938584

183. P. Francescon, W. Kilby, J. M. Noll, et al., "Monte Carlo simulated corrections for beam commissioning measurements with circular and MLC shaped fields on the CyberKnife M6 system: a study including diode, microchamber, point scintillator, and synthetic microdiamond detectors," *Physics in Medicine and Biology* **62** (3), 1076–1095 (2017). https://doi.org/10.1088/1361-6560/aa5610

184. J. Livingstone, A. W. Stevenson, D. J. Butler, et al., "Characterization of a synthetic single crystal diamond detector for dosimetry in spatially fractionated synchrotron X-ray fields," *Medical Physics* **43** (7), 4283–4293 (2016). https://doi.org/10.1118/1.4953833

185. D. J. O'Brien, J. Dolan, S. Pencea, et al., "Relative dosimetry with an MR-linac: response of ion chambers, diamond, and diode detectors for off-axis, depth dose, and output factor measurements," *Medical Physics* **45** (2), 884–897 (2018). https://doi.org/10.1002/mp.12699

186. V. De Coste, P. Francescon, M. Marinelli, et al., "Is the PTW 60019 microdiamond a suitable candidate for small field reference dosimetry?" *Physics in Medicine and Biology* **62** (17), 7036–7055 (2017). https://doi.org/10.1088/1361-6560/aa7e59

187. S. Rossomme, M. Marinelli, G. Verona-Rinati, et al., "Response of synthetic diamond detectors in proton, carbon, and oxygen ion beams," *Medical Physics* **44** (10), 5445–5449 (2017). https://doi.org/10.1002/mp.12473

188. C. Moignier, D. Tromson, L. de Marzi, et al., "Development of a synthetic single crystal diamond dosimeter for dose measurement of clinical proton beams," *Physics in Medicine & Biology* **62** (13), 5417–5439 (2017). https://doi.org/10.1088/1361-6560/aa70cf

189. M. Zani, M. Scaringella, C. Talamonti, et al., "Bidimensional polycrystalline CVD diamond detector for intensity modulated radiation therapy pretreatment verifications," *Journal of Instrumentation* **10** (03), C03046–C03046 (2015). https://doi.org/10.1088/1748-0221/10/03/C03046

190. M. Scaringella, M. Zani, A. Baldi, et al., "First dose-map measured with a polycrystalline diamond 2D dosimeter under an intensity modulated radiotherapy beam," *Nuclear Instruments and Methods in Physics Research Section A: Accelerators, Spectrometers, Detectors and Associated Equipment* **796** (8), 89–92 (2015). https://doi.org/10.1016/j.nima.2015.03.023

191. A. Bartoli, I. Cupparo, A. Baldi, et al., "Dosimetric characterization of a 2D polycrystalline CVD diamond detector," *Journal of Instrumentation* **12** (03), C03052–C03052 (2017). https://doi.org/10.1088/1748-0221/12/03/C03052

192. A. Bartoli, M. Scaringella, A. Baldi, et al., "PO-0873: 2D pixelated diamond detector for patient QA in advanced radiotherapy treatments," *Radiotherapy and Oncology* **127** (April), S459–S460 (2018). https://doi.org/10.1016/S0167-8140(18)31183-6

CHAPTER 5

Film Dosimetry

Sina Mossahebi

University of Maryland School of Medicine
Baltimore, Maryland

Nazanin Hoshyar

New York-Presbyterian/Queens Department of Radiation Oncology
Flushing, New York

Rao Khan and Arash Darafsheh

Washington University School of Medicine
St. Louis, Missouri

CONTENTS

5.1 INTRODUCTION

Film dosimeters provide two-dimensional (2D) measurement with high spatial resolution over a large area. There are in general two types of films used as dosimeters: radiographic and radiochromic. Radiographic films have been phased out of most of radiotherapy clinics in the United States and Canada; they have been replaced by 2D flat panel imagers and radiochromic films. Radiochromic films are now widely used in radiation therapy dosimetry and particularly in quality assurance (QA), especially for planar dose distribution comparisons. These films have been very useful for *in vivo* measurements, small field dosimetry as well as dose distribution evaluation of high dose-gradient regions because of their high spatial resolution, water equivalency, and ease of handling [1–6].

A number of clinical applications have benefitted from utilizing radiochromic films. Such applications include brachytherapy, intensity-modulated radiation therapy (IMRT) and volumetric arc therapy (VMAT) commissioning and QA, CyberKnife, linac-based stereotactic radiosurgery (SRS) and stereotactic body radiation therapy (SBRT), and charged particles therapy dosimetry [7–18].

Since the inception of IMRT, the use of small fields for radiotherapy has significantly increased. There has been a steady growth of both techniques and technology such as CyberKnife and linac-based SRS/SBRT where small fields are unavoidable. Small field geometries suffer from the loss of lateral charged particle equilibrium, lower signal and steep dose gradients, which make them difficult to perform accurate dose measurement. Radiochromic films, because of their high spatial resolution and dynamic range, water-equivalent density, signal linearity, dose rate and energy independence in megavoltage (MV) range, are highly desirable for small field dosimetry. The results of film dosimetry are reported to be in close agreement with various detectors for small field sizes of 2×2 cm². Several studies have investigated the small field sizes ($< 2 \times 2$ cm²). Gonzalez-Lopez et al. have shown that for a field size of 1×1 cm² or smaller, due to the volume effect of the chambers, lower agreement (>15% difference) was observed in measuring output factors between EBT2 film and chamber measurement [19]. Morales et al. have observed a <3% difference between relative output factors measured with microdiamond and EBT3 films for cone sizes as small as 4 mm [2].

For radiation dosimetry applications, different models of films have been developed. However, only a few models have survived the evolution; EBT3 and EBT-XD models by GAFchromic™ (Ashland Inc., Bridgewater, NJ) which are currently widely used in external beam therapy applications. EBT-XD film holds the same structure as EBT3 but has a thinner sensitive layer and slightly different chemical composition. In order to use films at higher doses, such as stereotactic radiosurgery applications, EBT-XD films are preferable due to their higher dynamic range (up to 40 Gy).

Historically, MD-V3 film was designed to cover a dose range of 1–100 Gy, and HD-V2 film was later developed to cover higher dose range (10–1000 Gy). These models have smaller thicknesses of sensitive layer compared to the EBT films [16]. In addition to the above models, other models have been developed to measure the low doses of diagnostic radiology examinations at low x-ray energies. One such model is the XR-QA2 film, which has a Bi_2O_3 radiation sensitive layer attached to the reflective polyester surface layer and can measure doses as low as 1 mGy. Several investigators have reported on the use of radiochromic films for dosimetry of low energy x-rays and electronic brachytherapy systems such as percentage depth dose (PDD) measurements of Xoft 50 kVp source. These films require calibration at the appropriate low energies due to variation in response of the film compared to water [20–23].

This chapter provides a summary of film-based dosimetry and its basic principles and applications in radiation dosimetry. Furthermore, the history of film, origin and mechanism of response to radiation, factors influencing the film response, applications, and practical examples are also discussed in more details for radiochromic films.

5.2 RADIOGRAPHIC FILM

In radiation therapy, films (as an inexpensive dosimeter) provide a high-resolution 2D map of doses for photon and electron beams. Radiographic films have been used for over 100 years in radiology, radiation therapy, imaging, and personnel dosimetry since the discovery of radioactivity and x-rays. They have become an integral part of megavoltage x-rays dosimetry for routine QA through verification of beam flatness and symmetry, crossbeam profiles, depth doses in wider areas and also for complex techniques such as dynamic wedges, IMRT, image-guided radiation therapy (IGRT), and small field dosimetry like stereotactic radiosurgery.

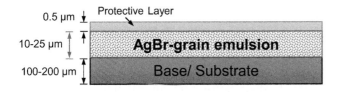

FIGURE 5.1 Films typically use silver bromide contents (30–40% by weight) with 1–2 μm grain diameter at 2 g/cm³ density for x-rays, while higher AgBr contents (70–80% by weight), with grain diameter of about 0.3 μm and 3.3 g/cm³ density are used in nuclear emulsions.

The use of radiographic films for quantitative dosimetry is described in American Association of Physicists in Medicine (AAPM) Task Group (TG) 69 [24]. Figure 5.1 shows a schematic illustration of the radiographic film. The radiographic film is composed of a transparent polyester base, coated on one or both sides with a radiation-sensitive emulsion held in a water-soluble gelatin matrix. The sensitivity of radiographic film to ionizing radiation depends on the composition of the microscopic grains of the silver halide (mostly AgBr), grain size, density, and their microenvironment. Nuclear emulsions, which will not be discussed further here, are used in cosmic-ray research and fast neutrons personnel dosimetry.

Interaction with radiation, ionizes the microscopic grains containing silver bromide; effectively converting ionic silver (Ag⁺) into silver (Ag) atoms. According to Gurney and Mott model of latent image formation [25], this happens in grains containing sensitive centers or 'speck'. Electrons, which produced due to absorption of

radiation in or around grain, make a sensitivity center negatively charged. Mobile silver Ag⁺ rushes to the sensitivity centers to neutralize and form an aggregate of metallic silver and hence the latent image on the film. Only a few metallic silvers primarily along the surface of the grain (containing around ten billion silver bromide) are enough to record the radiation signature.

Following the latent image formation, chemical processing of the films is required to develop the film (as shown in Figure 5.2) while ensuring that the undeveloped silver is eliminated and not exposed due to the processing. The chemical processes convert all of the Ag⁺ in an exposed grain into atoms while all of the bromine is washed away, leaving behind an opaque, microscopic grain of silver.

It is obvious from the Gurney–Mott mechanism, that the formation of opaque metallic silver can be traced back to the impingement of radiation. For dosimetry, the amount and location of elemental silver can be correlated to the absorbed dose by applying an appropriate calibration. The opacity of the film is determined using optical density (OD), which is defined as:

$$OD = -\log_{10} \frac{I}{I_o} \qquad (5.1)$$

where I_o and I, respectively, are the intensities of the incident readout light and the transmitted light through the film. The logarithm of transmitted light can be digitally obtained through the optical sensors. As the image becomes darker, film becomes more opaque. This leads

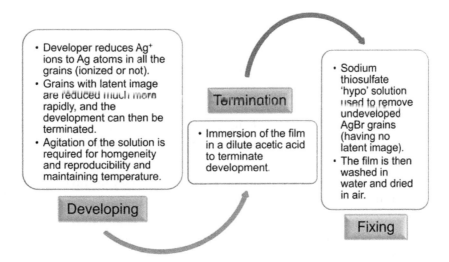

FIGURE 5.2 Chemical processes involved in development of a radiographic film. For dose measurement, the consistency of developing time, temperature, amount of agitation, and developer characteristics are important parameters.

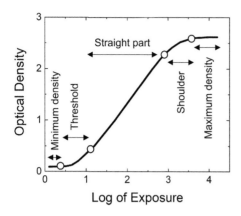

FIGURE 5.3 Plot of optical density of film with exposure is called *characteristics curve, sensitometric curve,* or *H&D curve*. An H&D curve has three main regions; the *toe* or *fog* region defining minimal density threshold, the *gradient* or *latitude* (straight-line portion) which is the exposure range producing acceptable OD, and the *shoulder* which is a region of the curve above the straight-line portion before reaching the *saturation* or maximum density.

to increase of OD indicating higher integrated dose has been delivered to the film. OD is proportional to the electron fluence, and emulsion thickness and hence the absorbed dose. The relationship between absorbed dose and OD is called the *sensitometric curve* or *H&D* curve (named after developers Hurter and Driffield) as shown in Figure 5.3.

Note that the film properties, including fog, contrast, and film speed are functions of developer temperature. For example, the OD of the film increases as the developer temperature is increased.

Energy dependency is one of the major disadvantages of radiographic films. The photoelectric effect in silver bromide grains leads to absorption of x-ray 10–50 times more than that in tissue or air especially for x-ray energies below 100 keV. This energy dependency needs to be accounted for when the ratio of the primary photons to low-energy scattered photons varies, for example, in the penumbra and under the collimator regions when measuring profiles of MV fields. The film over-response can be compensated for by enclosing the film in a high-Z filter to reduce the contribution of low energy x-rays. AgBr films exhibit a significant dose rate effect that is related to the grain composition and processing conditions. Commercial x-ray films are individually packed in a light tight envelope. Sometimes a small amount of air may get trapped inside the packing. The trapped air can create a potentially non-uniform perturbation of

the electron fluence and hence nonuniform dose distribution. Reciprocity law failure, also known as the Schwarzschild effect [26], leads to nonlinear OD especially when exposed to low dose rates.

Despite their non-tissue equivalency ($Z_{eff} = 40$–45), energy dependence, cumbersome chemical processing and development issues, radiographic films played an important role in tumor localization, planning, and dosimetry for both photon and electron beams up until 2010. With the recent introduction of radiochromic films and other advanced dosimeters, the role of radiographic films is diminishing in radiation therapy and radiology.

5.3 RADIOCHROMIC FILM

Although the first reports of radiochromic films were published in the 1960s [27], the predecessors of modern radiochromic film (GAFchromic™) with higher sensitivity were developed in the 1980s. Unlike radiographic films, radiochromic films do not need additional chemicals for wet post-processing. Also, radiochromic films are more water equivalent than radiographic films due to their low effective atomic number of 6.0–6.8 vs. 40–45 [1, 28].

Various film models, such as EBT, EBT2, EBT3, and EBT-XD, have been developed for clinical dosimetry applications over the last two decades. The protective polyester layer differs between EBT2 and EBT3 such that small silicon particles are integrated in the EBT3 polyester layer. The active layer in EBT3 is sandwiched between equal thickness layers of polyester with no adhesive. The structure of both EBT-XD film and EBT3 film comprises of a sensitive layer of lithium pentacosa-10, 12-diynoate (LiPCDA) monomer with 25 and 28 μm thickness, respectively, sandwiched between two symmetric matte-polyester layers each with 125 μm thickness. Different chemical compositions and side-length ratio of the polymer crystal of EBT-3 and EBT-XD creates differences in their active layers. The most recent film model is EBT-XD with smaller sensitive layer thickness which allows for extended dynamic range (up to 40 Gy for EBT-XD vs. 10 Gy for the EBT3). Their higher dynamic range makes them ideal candidate for high dose applications such as SRS and SBRT. The response of both EBT3 and EBT-XD films is reported to be energy-independent for megavoltage photon beams, and electron beams. However, there is still dependency on batches (or batch-to-batch variation) for these models which should be addressed by individual batch calibration [29, 30].

When ionizing radiation interacts with the constituent monomers of radiochromic film, polymers are formed which leads to an observable film coloration, which can be quantified by OD defined by Equation 5.1. A calibration curve allows for establishment of relationship between the film response and the dose received by the film. Flatbed photograph scanners are a common type of readout system used to analyze the films and often used clinically to measure the OD. A broad-band light source and a charge-coupled device (CCD) array of detectors are important components of flatbed scanners which can provide either transmitted or reflected intensity information using red, green, and blue color channels. A mathematical convolution of the densitometer light source emission spectrum, the film absorption spectrum, and the spectral sensitivity of the densitometer's detector will result in the measured OD [31]. An alternate to flatbed scanner is a spectroscopy system which can allow measurement of the absorbance over a broad spectral window or at a single wavelength [32, 33]. The use of radiochromic film for quantitative dosimetry has been described in AAPM's TG 55 [28].

5.3.1 Film Response to Radiation

The polymerization of the constituent monomers of the film occurs upon irradiation. In EBT film series, the irradiated films have a blue appearance. There are two main absorption bands in the net absorbance spectrum of the films, a primary band around 636 nm and a secondary band around 585 nm. Figure 5.4 shows the net absorbance (Net A) spectra of EBT3 and EBT-XD films irradiated with 6 MV photon beams when measured with a fiber coupled optical spectrometer.

As shown in Figure 5.4, the EBT-XD model shows a lower response compared to the EBT3 model due to the fact that the former has been designed to have a wider dynamic range.

In EBT3 film, the primary absorption band reaches to a saturation level earlier than the secondary bands. Therefore, at lower dose level, the primary band (i.e., red channel) can provide higher sensitivity. However, at higher doses from 6 to 10 Gy, the green channel has higher sensitivity compared to the red channel [30].

For absolute dosimetry, a calibration curve is required to convert OD into absorbed dose. The calibration curve is obtained by irradiating several film samples to various dose levels while staying within the dynamic range of the film. The net OD of these samples determined from a film scanner are fit to a polynomial function with the following form [16]:

$$D = a \times \text{Net OD} + b \times \text{Net OD}^n \qquad (5.2)$$

where a, b, and n are the fitting parameters. Figure 5.5 shows a series of EBT3 and EBT-XD films irradiated with 16 MeV electron beam. The corresponding calibration curves are presented in Figure 5.6.

In conventional flatbed scanners, the digital image consists of three-color channels: red, green, and blue. In single channel dosimetry, a calibration curve is established for each color channel. Typically, the red channel is most often used due to its higher sensitivity. However, at doses close to the upper dynamic range of the film, green channel can be used due to its higher sensitivity.

Multichannel film dosimetry as described by Micke et al. [34] varies the dose values until the corresponding scanned OD values for various color channels are matched. The multichannel method balances the color

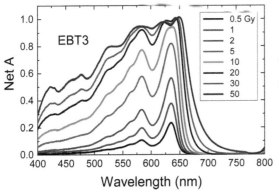
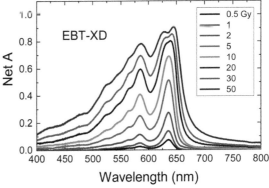

FIGURE 5.4 Net absorbance spectra of EBT3 and EBT-XD films irradiated at different dose levels using a 6 MV beam [29].

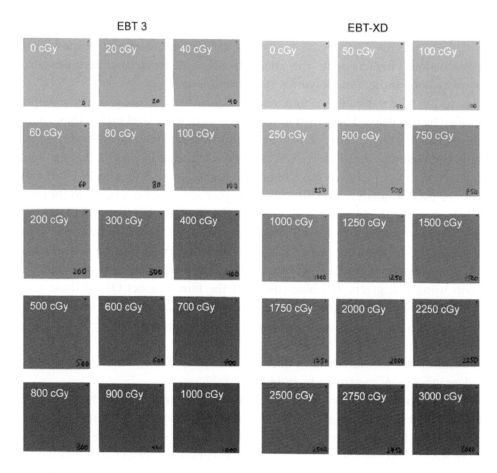

FIGURE 5.5 Series of film samples from EBT3 (left) and EBT-XD (right) irradiated with 16 MeV electron beam to obtain calibration curve for the two EBT film models.

channel with the highest sensitivity and provides the correction of a variety of errors associated with film scan such as film non-uniformities and/or scanner related artifacts as well as covering the entire sensitivity range of the film by the spectral response of the scanner due to different sensitivity of each channel [34]. Multiple studies have shown the benefit of multichannel dosimetry in film-based patient specific QA [17, 18, 35, 36].

It has been shown that the response of EBT3 and EBT-XD films have minimal dose-rate and energy dependence on the megavoltage photon and electron beams, therefore, a calibration obtained at one energy

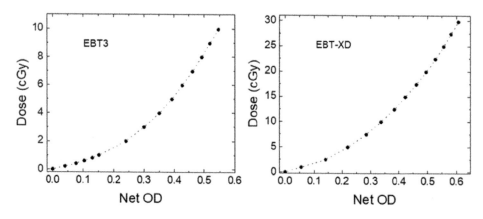

FIGURE 5.6 Calibration curves using the red channel information for EBT3 (left) and EBT-XD (right) films corresponding to images in the previous figure, irradiated by 16 MeV electron beam.

can be used for other energies [29, 30]. However, the film response still remains batch-dependent, which indicates that the calibration must be performed for each new film batch.

5.4 FACTORS AFFECTING RADIOCHROMIC FILM RESPONSE

There are several factors that affect the radiochromic film response to radiation. Type of radiation, energy, dose, and dose rate can influence the film response to varying degrees. Environmental factors such as temperature, humidity, and UV light can also affect the OD of a radiochromic film.

5.4.1 Temperature and Humidity

The effect of temperature on OD of radiochromic film has been thoroughly investigated [37]. When the film temperature increases from 22 to 38°C, a linear change in absorbance causes the fluctuations in hydration of active film components, which affects the location of the absorbance peak and overall sensitivity of the film. It has been reported that the humidity introduces systematic dose errors for radiochromic film dosimetry [38]. A dose error of up to 15% can be introduced when the humidity varies from 80% to 20%. Therefore, it is important to control the storage, humidity, and temperature of the film.

5.4.2 Dose and Dose Rate

Film dosimetry relies on the calibration curve established during the calibration process with a reference beam quality, dose-rate, and under specific conditions. These factors can affect a dosimeter's response.

The dose-rate dependency of radiochromic film has been evaluated in a number of studies across various generations of the film [39]. The EBT films show a small dose-rate dependency (within 1%) when dose rate varies between 16 and 520 cGy/min [39]. In a study of EBT3 film using a Novalis TX linear accelerator, it was found that these films are dose rate independent to within 3% for a wide range of dose rates (50 to 1000 Gy/min) irrespective of total dose delivered [40]. For flattening filter free (FFF) beams, where dose rates of 1200 to 2400 cGy/min are often used, the dose-rate dependency of EBT2 films in FFF beam profiles measurement was found to be negligible considering the uncertainty in scanner reproducibility and film stability [41]. Moreover, multiple studies have shown that the dose-rate dependency of EBT3 and EBT-XD films were insignificant [29, 30].

Additionally, radiochromic films may be useful for dosimetry at ultra-high dose rates (>40 Gy/s), such as FLASH radiation therapy. Overall 5% uncertainty in radiochromic film dosimetry at these dose rates has been reported in the literature for electron beams [42, 43].

5.4.3 Beam Quality and Energy

Several studies evaluated the effect of beam quality on the film; the dependence of calibration curve for megavoltage energies was negligible [17, 18, 29, 30]. However, approximately 5% under-response in kV beams has been reported for EBT3 model compared to megavoltage beam irradiation [20].

5.4.4 Linear Energy Transfer (LET)

In proton therapy beams, and other high LET beams, an under-response of about 10% in the Bragg peak has been reported which makes absolute film dosimetry in proton beams challenging since the film response depends on both dose and LET [44–46]. Figure 5.7 shows PDD

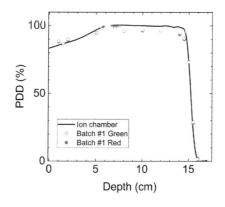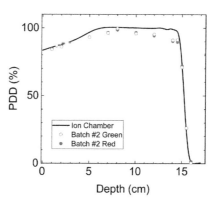

FIGURE 5.7 Percentage depth dose (PDD) measured using two different batches of EBT-XD film samples at discrete depths compared to ionization chamber (solid lines) measurement. Film data were arbitrarily matched at 1.5 cm depth [47].

measurement using EBT-XD films compared with that using an ionization chamber. A matching point near the entrance at an arbitrary depth of 1.5 cm was selected. It can be seen that the film measurements underestimate the dose by ~1–5% and ~7–8% in the spread-out Bragg peak (SOBP) region and at the distal end of the SOBP, respectively. The "quenching" effect is more pronounced in the distal region of the Bragg peak, which has a high LET.

Correcting the quenching is not a trivial task and requires *a priori* knowledge of the beam LET [48, 49]. One approach to correct for quenching is to determine the value of the LET of the beam using Monte Carlo simulations and establish a correction factor as a function of depth. The goal of the correction factor is to correct the dose determination in depth for proton beams. However, this assumes that the Monte Carlo simulation and the determination of the LET are accurate and representative of the beam configuration used.

5.4.5 Batch

Several studies have shown a significant variation of film response between different batches [29, 30] and a minor variation (less than 1%) between the film sheets from the same batch for most types of radiochromic films [1]. The batch dependence can be more than 10% [50], which suggests establishing a separate calibration curve for each batch.

5.4.6 Post Irradiation Time and Temporal Response

Several studies have investigated the temporal response of EBT films and shown a 6–9% increase in post-irradiation OD for EBT films within 12 hours of irradiation in dose range of 1–5 Gy [1]. This increase in OD was higher (13–19%) for the older film models such as MD-55-2, XR type T, and HS radiochromic films.

For EBT3 and EBT-XD films, up to 15% increase in the OD (depending on the dose level and color channel) can occur between the OD measured immediately after the film exposure and that measured at 24 hours post-irradiation. The increase in OD between 12 hours post-irradiation and 24 hours post-irradiation can be as high as 5% [51]. It is recommended to wait at least 24 hours post-irradiation before scanning the films to achieve higher reproducibility in the measurement. If 24-hour post-irradiation wait time is not feasible, a correction function [52, 53] can be used to correct post-irradiation OD. Recent researches have shown that shorter wait time post irradiation may be used [54, 55].

5.4.7 Orientation

Considering the purpose of the measurement, dose level to be determined, film orientation (coronal, sagittal, transversal, etc.) should be carefully selected. After cutting the film strips, it is important to label their orientation clearly to ensure a correct orientation is used during film scanning. Film data are sensitive to film orientation relative to the light polarization direction; therefore, consistency must be followed during calibration, background scanning, and measurement steps [1].

5.4.8 Storage

For optimal accuracy, it is recommended to store radiochromic films in a dark, temperature-controlled environment (ideally in their original box) since they are sensitive to excessive temperatures and UV-light exposures. Pressure and movement during storage should be avoided as it can cause rubbing and scratching of the film pieces. Any visible damage or small scratches can cause significant change in the film response. Moisture, radiation, and magnetic environment should be avoided during storage as it can affect the measurement accuracy and decrease the life of the film. In order to use films properly, they should be kept in the original box in a temperature controlled, dust-, smoke-, and moisture-free area, without any radiation and magnetic field environment [1].

5.4.9 External Magnetic Fields

With the emergence of magnetic resonance image-guided radiation therapy (MRIgRT), that uses MRI for *real time* imaging of the anatomy during the radiation delivery, it is vital to realize to what extent the presence of a magnetic field would impact the response of dosimeters. The presence of an external magnetic field, typically 0.35–1.5 T required in the MRIgRT machines, does not directly affect the primary photon beams, however, their secondary electrons are influenced by the magnetic field through the Lorentz force [56, 57]. The magnetic field will influence the trajectory of electrons and consequently the deposited dose in the medium which can also to some extent impacts the response of routinely used dosimeters.

The existing literature on radiochromic film dosimetry in the context of dosimetry in the presence of an external magnetic field is rather inconsistent (see Table 5.1). However, the common consensus is that below 1.5 T the influence of the magnetic field on the EBT3 and EBT-XD film is within the measurement uncertainty.

TABLE 5.1 Summary of Recent Film Dosimetry Literature in the Context of MRIgRT

Study	Film Model	Clinical MRIgRT	Concurrent B-field and Irradiation	Magnet	B-field (T)	Radiation type (/Reference)	Batch Nos.	Dose Range (Gy)	Result
Reyhan et al. [58]	EBT2	No	No	MRI scanner (GE)	1.5	6 MV	1	0–8	~4% under-measurement of dose.
Reynoso et. al. [59]	EBT2	Yes	Yes	MRIdian (Viewray)	0.35	^{60}Co (/Same system with magnet off)	1	2–18	8.7%, 8%, and 4.3% reduction in the net optical density for red, green, and blue channels, respectively.
Roed et al. [60]	EBT3	No	Yes	Electromagnet	1.5	^{60}Co	1	2–8	<2% under-response, but it was within the measurement uncertainty.
Delfs et al. [61]	EBT3	No	Yes	Electromagnet	0.35–1.42	6 MV	1	1–20	Up to 2.1% under-measurement of dose.
Barten et al. [62]	EBT3	Yes	Yes	MRIdian (Viewray)	0.35	^{60}Co (/Linac for $B = 0$)	1	1–8	Within 1% uncertainty of the measurements. And 1.5% when real-time MRI was performed.
Padilla-Cabal et al. [63]	EBT3	No	Yes	Electromagnet	0.5–1	62.4–252.6 MeV proton	1	0.2–10	No significant change in film response.
Billas et al. [64]	EBT3	No	Yes	Electromagnet	0.5–2	^{60}Co	1	0.75–8	Within measurement uncertainty for $B \leq$ 1.5 T. But, 2.4% over-response at $B = 2$ T.
Volotskova et al. [65]	EBT2, EBT3, EBT-XD	No	No	MRI scanner (Siemens)	1.5–3	6 MV & 6 MV FFF	1	0–20	No significant change in film optical density.
Darafsheh et al. [66]	EBT3, EBT-XD	Yes	Yes	MRIdian (Viewray)	0.35	6 MV FFF (/Linac for $B = 0$)	2	0–20	No significant change in film response.

Source: Adapted from [66].

5.5 OTHER CONSIDERATIONS

5.5.1 Film Handling

Radiochromic films are not overly sensitive to light like their radiographic counterparts, however, they should be kept in a cool, dark, and dry place and never be exposed to UV light, especially sunlight, as it will lead to an ionizing effect on the monomers. The surface of the film should be handled with latex gloves to prevent smudges and fingerprints on the film's surface. A sharp blade or a "guillotine" device should be used to cut the film in order to avoid unnecessary bending or deformation of the film. Since there is a tendency for the separation of the layers of the film, dosimetry should be avoided around the edges of the film where cuts have been made. Thus, care should be taken to include a small buffer area between the regions of interest and the perimeter of the film to prevent unnecessary errors and uncertainties.

Once the film is cut to the desired size, the corner of the film should be marked as the readout of the film is dependent on the polarization of the incident light, and therefore, the orientation of the film. This should be considered in the calibration films, as well for the test films actually used for dosimetry.

5.5.2 Scanners

The scanner should be kept turned on so that the reading electronics is warmed up. This is to stabilize the intensity of the light source. Also, at least some number (typically 10–20) of preview scans should be acquired to adequately warm up the light source. The user can experimentally find the number of sufficient warm up scans.

5.5.3 Calibration Procedures

For each batch of film, a calibration step is necessary to convert the response of the film into a dose distribution, therefore, the model and lot number of the batch for each calibration should be noted. Figure 5.8 demonstrates the steps needed for calibration.

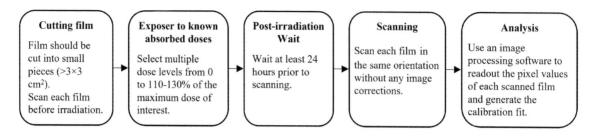

FIGURE 5.8   Radiochromic film calibration process.

The required tools and accessories for film calibration are: the radiochromic films, a readout system which is typically a flatbed scanner, and an analysis software to extract the gray level information of the scanned images. It is recommended to handle the films with gloves and use a frame to place the films at the center of scanner. During the calibration process caution must be exercised to control the ambient temperature during irradiation, storage, and readout in a consistent fashion. Handle the films with care, avoid dust, fingerprints, or over-bending. Store the films in a dry and dark environment; keep them away from light and avoid prolonged exposure to UV light.

About six to eight data points are needed for the calibration curve, covering the range of dose to be determined. Ideally, at least two data points should be higher than the desired maximum dose to be measured. Based on the dynamic range of the film, several dose levels ranging from 0 cGy (the unexposed film) to approximately 110–130% of the maximum dose of interest are selected. Usually an 8″ × 10″ sheet of film is cut into strips of dimensions 1.25″ × 8″. Each strip should be marked to preserve and track its orientation and alignment (portrait or landscape). The orientation directions for both irradiation and readout should be kept the same. Then, each strip is cut into smaller pieces (>3 × 3 cm²). In order to have accurate calibration, each film should be scanned before irradiation to remove the background and establish a baseline level.

Each of the film pieces is exposed to a reference field (e.g., 10 × 10 cm²) of known absorbed dose. The time between irradiation and readout should be consistent, typically waiting for the films to be stabilized (at least 24 hours) prior to scanning them. Shorter waiting times, though, are not recommended but can be used by applying a correction factor or using a separate calibration. Variable waiting-time windows

can lead to larger uncertainties, and may require dose rescaling.

Any flatbed scanner, for example, Epson Expression 10000XL and Epson Perfection V750, can be used to read the films. Films should be positioned in the same orientation (i.e., portrait or landscape) on the scanner close to the center of the scanner using some alignment frame device. The film is scanned in color mode, 48-bit RGB (16 bits per channel), positive film, and at least 72 dpi resolution, without any default image corrections. Finally, an image processing software such as ImageJ (National Institutes of Health, USA) or a commercial product can be used to readout the pixel values of each image and a calibration fit can be generated by using Equation 5.2. The temperature, humidity, post-irradiation time, and film orientation during irradiation and readout are factors that can affect the signal of the radiochromic film and should not change significantly during calibration or test irradiation.

Series of EBT3 and EBT-XD films irradiated with 16 MeV electron beam are shown in Figure 5.5 along with their corresponding calibration curve in Figure 5.6.

5.5.4 Uncertainty Budget

The uncertainty budget of radiochromic films itself depends on several factors such as the film model (EBT, EBT2, EBT3, EBT-XD, etc.), application, dose range, scanning (scanning parameters, orientation, stability), and calibration uncertainties. It should be noted that the total uncertainty in determining the dose depends not only on the dosimetry system (radiochromic film) but also on the known dose measurement uncertainties (i.e., measurement setup, measurement protocol, and output uncertainties). Based on these uncertainties, one can determine error bars or confidence intervals for a particular film measurement protocol.

Several studies have evaluated the uncertainty of radiochromic films used for photon, electron, and

TABLE 5.2 Summary of Selected Literature on Uncertainty Budget in Radiochromic Film Dosimetry

Study	Film Model	Radiation Type	Dose Range (Gy)	Results
van Battum et al. [67]	EBT	6 MV photon beam	0–2.3 Gy	The overall uncertainty in absolute EBT film dose detection 1.8% using a single film and 1.3% for two films. Main factor: Intrinsic film inhomogeneity
Martišíková et al. [68]	EBT	^{60}Co	0.3–1 Gy	Total uncertainty in the net OD determination was estimated to be 1.6% at 0.3 Gy and 0.8% at 1 Gy for ^{60}Co. This leads to a dose uncertainty of 1.6% at 0.3 Gy and 0.9% at 1 Gy.
Aland et al. [69]	EBT2	6 MV photon beam Red and blue channels	2 Gy	Total uncertainty: Worst case (12.6%) and ideal case (2.6%) EBT2 film can be used in routine quality assurance testing for radiotherapy, in situations where a dose uncertainty of up to 2.8% is acceptable.
Palmer et al. [70]	EBT3	Ir-192 and Co-60 HDR brachytherapy Multichannel	0–20 Gy	Film dosimetry fractional uncertainties Calibration dose delivery on a Linac (0.5%), calibration fit (0.5%), film position (0.9%), energy dependence (1.5%), phantom scatter (0.5%), and film scanning (0.2%)
Sorriaux et. al. [71]	EBT3	Photon: 6 MV and 18 MV Electron: 6 MeV Proton: 60 MeV and 100 MeV Red, green, and blue channels	0–10 Gy	The total uncertainties on calibration curve due to film reading and fitting were within 1.5% for photon and proton beams. For electrons, the uncertainty was within 2% for doses larger than 0.8 Gy.
León Marroquin et al. [72]	EBT3	6 MV photon beam Red, green, and blue channels	0–120 Gy	The overall uncertainty in the measured dose for the red, green, and blue channels was 3.2%, 4.9%, and 5.2%, respectively. The relative orientation of the film, the uniformity of response of the scanner, and the response curves and fitting procedure are the main factors in the total uncertainties.
León-Marroquín et al. [29]	EBT-XD	Photon: 6 MV, 6 MV-FFF, 10 MV-FFF, and 15 MV Electron: 6 and 20 MeV Red, green, and blue channels	0.5–50 Gy	The maximum uncertainty in the calculated dose, in ranges 5–40 Gy and 8–50 Gy for red and green channels, respectively, were ~3.1% and ~2.4%. Net OD variations compared to 6 MV: For dose ≤10 Gy: Red (<4%), Green (<5.5%), Blue (10%). For dose >10 Gy: Red (<3%), Green (<3.5%), Blue (5%)

proton beam dosimetry. For megavoltage photon beam using radiochromic film, overall uncertainties from 3% to 5% have been reported. Table 5.2 provides a summary of several studies on the evaluation of the uncertainties in radiochromic film dosimetry systems.

5.6 CONCLUSION

Overall, film dosimeters offer the highest spatial resolution amongst currently available 2D dosimeters and are indispensable to small field dosimetry. A typical protocol for the calibration and the use of radiochromic films can result in overall uncertainties within 2–5%.

REFERENCES

1. I. J. Das, *Radiochromic Film: Role and Applications in Radiation Dosimetry* (CRC Press, Boca Raton, FL, 2017).
2. J. E. Morales, M. Butson, S. B. Crowe, R. Hill, J. V. Trapp, "An experimental extrapolation technique using the Gafchromic EBT3 film for relative output factor measurements in small x-ray fields," *Medical Physics* **43**(8), 4687–4692 (2016).

3. C. Fiandra, M. Fusella, F. R. Giglioli, A. R. Filippi, C. Mantovani, U. Ricardi, R. Ragona, "Comparison of Gafchromic EBT2 and EBT3 for patient-specific quality assurance: Cranial stereotactic radiosurgery using volumetric modulated arc therapy with multiple noncoplanar arcs," *Medical Physics* **40**(8), 082105 (2013).

4. C. Chen, S. Tang, D. Mah, M. Chan, "SU-E-T-286: Dose verification of spot-scanning proton beam using GafChromic EBT3 film," *Medical Physics* **42**(6), 3399–3399 (2015).

5. A. Chu, W. Feng, H. Lincoln, F. Su, R. Nath, "SU-E-T-118: Small dynamic field dosimetry by Gfachromic film (EBT3) and 2D-array diode," *Medical Physics* **40**(6), 230–231 (2013).

6. N. Corradini and S. Presilla, "SU-E-T-58: Submillimeter dose measurement with Gafchromic EBT2 and EBT3 films," *Medical Physics* **40**(6), 216–216 (2013).

7. A. L. Palmer, P. Di Pietro, S. Alobaidli, F. Issa, S. Doran, D. Bradley, A. Nisbet, "Comparison of methods for the measurement of radiation dose distributions in high dose rate (HDR) brachytherapy: Ge-doped optical fiber, EBT3 Gafchromic film, and PRESAGER radiochromic plastic," *Medical Physics* **40**(6), 061707 (2013).

8. D. Cusumano, M. L. Fumagalli, M. Marchetti, L. Fariselli, E. De Martin, "Dosimetric verification of stereotactic radiosurgery/stereotactic radiotherapy dose distributions using Gafchromic EBT3," *Medical Dosimetry* **40**, 226–231 (2015).

9. A. L. Palmer, A. Dimitriadis, A. Nisbet, C. H. Clark, "Evaluation of Gafchromic EBT-XD film, with comparison to EBT3 film, and application in high dose radiotherapy verification," *Physics in Medicine and Biology* **60**(22), 8741–8752 (2015).

10. I. Ali, S. Ahmad, S. Joel, J. F. Williamson, "Optimal densitometry wavelengths that maximize radiochromic film sensitivity while minimizing OD growth and temperature sensitivity artifacts," *Journal of X-Ray Science and Technology* **17**, 61–73 (2009).

11. I. Ali, J. F. Williamson, C. Costescu, J. F. Dempsey, "Dependence of radiochromic film response kinetics on fractionated doses," *Applied Radiation and Isotopes* **62**, 609–617 (2005).

12. Y. Le, I. Ali, J. F. Dempsey, J. F. Williamson, "Prospects for quantitative two-dimensional radiochromic film dosimetry for low dose-rate brachytherapy sources," *Medical Physics* **33**(12), 4622–4634 (2006).

13. J. I. Monroe, J. F. Dempsey, J. A. Dorton, S. Mutic, J. B. Stubbs, J. Markman, J. F. Williamson, "Experimental validation of dose calculation algorithms for the GliaSite™ RTS, a Novel ¹²⁵I liquid-filled balloon brachytherapy applicator," *Medical Physics* **28**(1), 73–85 (2001).

14. Y. Zhu, A. S. Kirov, V. Mishra, A. S. Meigooni, J. F. Williamson, "Quantitative evaluation of radiochromic film response for two-dimensional dosimetry," *Medical Physics* **24**(2), 223–231 (1997).

15. A. Darafsheh, T. Zhao, R. Khan, "Optical spectral analysis of radiochromic films irradiated with

radiation therapy beams," *Proceedings of SPIE* **11231**, 112310I (2020).

16. S. Devic, N. Tomic, D. Lewis, "Reference radiochromic film dosimetry: Review of technical aspects," *Physica Medica* **32**, 541–556 (2016).

17. T. Kairn, N. Hardcastle, J. Kenny, R. Meldrum, W. A. Tomé, T. Aland, "EBT2 radiochromic film for quality assurance of complex IMRT treatments of the prostate: Micro-collimated IMRT, RapidArc, and TomoTherapy," *Australasian College of Physical Scientists in Medicine* **34**(3), 333–343 (2011).

18. L. Marrazzo, M. Zani, S. Pallotta, C. Arilli, M. Casati, A. Compagnucci, C. Talamonti, M. Bucciolini, "GafChromic® EBT3 films for patient specific IMRT QA using a multichannel approach," *Physica Medica* **31**(8), 1035–1042 (2015).

19. A. Gonzalez-Lopez, J. A. Vera-Sanchez, J. D. Lago-Martin, "Small fields measurements with radiochromic films," *Journal of Medical Physics* **40**(2), 61–67 (2015).

20. J. E. Villarreal-Barajas and R. F. H. Khan, "Energy response of EBT3 radiochromic films: Implications for dosimetry in kilovoltage range," *Journal of Applied clinical Medical Physics* **15**(1), 331–338 (2014).

21. S. Devic, L. H. Liang, N. Tomic, H. Bekerat, M. Morcos, M. Popovic, P. Watson, S. Aldelaijan, J. Seuntjens, "Dose measurements nearby low energy electronic brachytherapy sources using radiochromic film," *Physica Medica* **64**, 40–44 (2019).

22. Y. Rong and J. S. Welsh, "Surface applicator calibration and commissioning of an electronic brachytherapy system for nonmelanoma skin cancer treatment," *Medical Physics* **37**(10), 5509–5517 (2010).

23. H. Bekerat, S. Devic, F. DeBlois, K. Singh, A. Sarfehnia, J. Seuntjens, S. Shih, X. Yu, D. Lewis, "Improving the energy response of external beam therapy (EBT) GafChromic™ dosimetry films at low energies (≤ 100 keV)," *Medical Physics* **41**(2), 022101 (2014).

24. S. Pai, I. J. Das, J. F. Dempsey, K. L. Lam, T. J. LoSasso, A. J. Olch, J. R. Palta, L. E. Reinstein, D. Ritt, E. E. Wilcox, "TG-69: Radiographic film for megavoltage beam dosimetry," *Medical Physics* **34**(6), 2228–2258 (2007).

25. R. W. Gurney and N. F. Mott, "The theory of photolysis of silver bromide and the photographic latent image," *Proceedings of the Royal Society of London. Series A, Mathematical and Physical Sciences* **164**(917), 151–167 (1938).

26. A. Djouguela, R. Kollhoff, A. Rubach, D. Harder, B. Poppe, "The Schwarzschild effect of the dosimetry film Kodak EDR 2," *Physics in Medicine and Biology* **50**(21), N317–N321 (2005).

27. W. L. McLaughlin and L. Chalkley, "Measurement of radiation dose distributions with photochromic materials," *Radiology* **84**(1), 124–125 (1965).

28. A. Niroomand-Rad, C. R. Blackwell, B. M. Coursey, K. P. Gall, J. M. Galvin, W. L. McLaughlin, A. S. Meigooni, R. Nath, J. E. Rodgers, C. G. Soares, "Radiochromic

film dosimetry: Recommendations of AAPM Radiation Therapy Committee Task Group 55," *Medical Physics* **25**(11), 2093–2115 (1998).

29. E. Y. León-Marroquín, D. J. Mulrow, A. Darafsheh, R. Khan, "Response characterization of EBT-XD radiochromic films in megavoltage photon and electron beams," *Medical Physics* **46**(9), 4246–4256 (2019).

30. E. Y. León-Marroquín, D. J. Mulrow, R. Khan, A. Darafsheh, "Spectral analysis of the EBT3 radiochromic films for clinical photon and electron beams," *Medical Physics* **46**(2), 973–982 (2019).

31. L. E. Reinstein, G. R. Gluckman, H. I. Amols, "Predicting optical densitometer response as a function of light source characteristics for radiochromic film dosimetry," *Medical Physics* **24**(12), 1935–1942 (1997).

32. J. F. Dempsey, D. A. Low, S. Mutic, J. Markman, A. S. Kirov, G. H. Nussbaum, J. F. Williamson, "Validation of a precision radiochromic film dosimetry system for quantitative two-dimensional imaging of acute exposure dose distributions," *Medical Physics* **27**(10), 2462–2475 (2000).

33. A. Darafsheh, E. Y. León-Marroquín, D. Mulrow, M. Baradaran-Ghahfarokhi, T. Zhao, R. Khan, "On the spectral characterization of radiochromic films irradiated with clinical proton beams," *Physics in Medicine and Biology* **64**(13), 135016 (2019).

34. A. Micke, D. F. Lewis, X. Yu, "Multichannel film dosimetry with nonuniformity correction," *Medical Physics* **38**(5), 2523–2534 (2011).

35. D. Lewis, A. Micke, X. Yu, M. F. Chan, "An efficient protocol for radiochromic film dosimetry combining calibration and measurement in a single scan," *Medical Physics* **39**(10), 6339–6350 (2012).

36. T. Tessonnier, A. Dorenlot, N. Nomikossoff, "Patient quality controls with Gafchromic EBT3 for tomotherapy HI-ART 2: Analysis of the calibration's condition to achieve an absolute dosimetry," *Physica Medica* **29**, e45 (2013).

37. A. Rink, D. F. Lewis, S. Varma, I. A. Vitkin, D. A. Jaffray, "Temperature and hydration effects on absorbance spectra and radiation sensitivity of a radiochromic medium," *Medical Physics* **35**(10), 4545–4555 (2008).

38. F. Girard, H. Bouchard, F. Lacroix, "Reference dosimetry using radiochromic film," *Journal of Applied Clinical Medical Physics* **13**(6), 339–353 (2012).

39. A. Rink, I. A. Vitkin, D. A. Jaffray, "Intra-irradiation changes in the signal of polymer-based dosimeter (GAFCHROMIC EBT) due to dose rate variations," *Physics in Medicine and Biology* **52**(22), N523–N529 (2007).

40. G. Twork and A. Sarfehnia, "SU-E-T-88: Evaluation of the dose-rate dependency of GAFCHROMIC EBT3," *Medical Physics* **40**(6), 223–224 (2013).

41. S. Oyewale, S. Ahmad, I. Ali, "SU-E-T-85: Dose rate and energy dependence of EBT, EBT2, EDR2 films, and Mapcheck2 diode arrays in beam profiles from a Varian TrueBeam system," *Medical Physics* **39**(6), 3722–3722 (2012).

42. M. Jaccard, K. Petersson, T. Buchillier, J. F. Germond, M. T. Duran, M. C. Vozenin, J. Bourhis, F. Bochud, C. Bailat, "High dose-per-pulse electron beam dosimetry: Usability and dose-rate independence of EBT3 Gafchromic films," *Medical Physics* **44**(2), 725–735 (2017).

43. L. Karsch, E. Beyreuther, T. Burris-Mog, S. Kraft, C. Richter, K. Zeil, J. Pawelke, "Dose rate dependence for different dosimeters and detectors: TLD, OSL, EBT films, and diamond detectors," *Medical Physics* **39**(5), 2447–2455 (2012).

44. M. Martišíková and O. Jäkel, "Gafchromic® EBT films for ion dosimetry," *Radiation Measurement* **45**(10), 1268–1270 (2010).

45. M. Martišíková and O. Jäkel, "Dosimetric properties of Gafchromic® EBT films in monoenergetic medical ion beams," *Physics in Medicine and Biology* **55**(13), 3741–3751 (2010).

46. L. Zhao and I. J. Das, "Gafchromic EBT film dosimetry in proton beams," *Physics in Medicine and Biology* **55**(10), N291–N301 (2010).

47. A. Darafsheh, T. Zhao, R. Khan, "Spectroscopic analysis of irradiated radiochromic EBT-XD films in proton and photon beams," *Physics in Medicine and Biology* **65**, 205002 (2020).

48. B. R. Smith, M. Pankuch, C. G. Hammer, L. A. DeWerd, W. S. Culberson, "LET response variability of Gafchromic™ EBT3 film from a ^{60}Co calibration in clinical proton beam qualities," *Medical Physics* **46**(6), 2716–2728 (2019).

49. A. F. Resch, P. D. Heyes, H. Fuchs, N. Bassler, D. Georg, H. Palmans, "Dose- rather than fluence-averaged LET should be used as a single-parameter descriptor of proton beam quality for radiochromic film dosimetry," *Medical Physics* (2020).

50. S. Reinhardt, M. Hillbrand, J. J. Wilkens, W. Assmann, "Comparison of Gafchromic EBT2 and EBT3 films for clinical photon and proton beams," *Medical Physics* **39**(8), 5257–5262 (2012).

51. S. Momin, R. Khan, A. Darafsheh, "Growth kinetics of the EBT3 and EBT-XD films response in radiotherapy beams," *Proceedings of SPIE* **11224**, 112240J (2020).

52. C. Andrés, A. del Castillo, R. Tortosa, D. Alonso, R. Barquero, "A comprehensive study of the Gafchromic EBT2 radiochromic film. A comparison with EBT," *Medical Physics* **37**(12), 6271–6278 (2010).

53. L. Zhao, L. Coutinho, N. Cao, C.-W. Cheng, I. J. Das, "Temporal response of Gafchromic EBT2 radiochromic film in proton beam irradiation," *IFMBE Proceedings* **39**, 1164–1167 (2012).

54. M. F. Chan, *Recent Advancements and Applications in Dosimetry* (Nova Science Publishers, Hauppauge, NY, 2018).

55. M. F. Chan, D. Lewis, X. Yu, "Is it possible to publish a calibration function for radiochromic film?," *International Journal of Medical Physics, Clinical Engineering and Radiation Oncology* **3**(1), 25–30 (2014).

56. B. W. Raaymakers, A. J. Raaijmakers, A. N. Kotte, D. Jette, J. J. Lagendijk, "Integrating a MRI scanner with a 6 MV radiotherapy accelerator: Dose deposition in a transverse magnetic field," *Physics in Medicine and Biology* **49**(17), 4109–4118 (2004).

57. C. Kirkby, T. Stanescu, S. Rathee, M. Carlone, B. Murray, B. G. Fallone, "Patient dosimetry for hybrid MRI-radiotherapy systems," *Medical Physics* **35**(3), 1019–1027 (2008).

58. M. L. Reyhan, T. Chen, M. Zhang, "Characterization of the effect of MRI on Gafchromic film dosimetry," *Journal of Applied Clinical Medical Physics* **16**(6), 325–332 (2015).

59. F. J. Reynoso, A. Curcuru, O. Green, S. Mutic, I. J. Das, L. Santanam, "Technical note: Magnetic field effects on Gafchromic-film response in MR-IGRT," *Medical Physics* **43**(12), 6552–6556 (2016).

60. Y. Roed, H. Lee, L. Pinsky, G. Ibbott, "Characterizing the response of Gafchromic EBT3 film in a 1.5 T magnetic field," *Radiotherapy and Oncology* **123**(Supplement 1), S403 (2017).

61. B. Delfs, A. A. Schoenfeld, D. Poppinga, R.-P. Kapsch, P. Jiang, D. Harder, B. Poppe, H. K. Looe, "Magnetic fields are causing small, but significant changes of the radiochromic EBT3 film response to 6 MV photon," *Physics in Medicine and Biology* **63**, 035028 (2018).

62. D. L. J. Barten, D. Hoffmans, M. A. Palacios, S. Heukelom, L. J. van Battum, "Suitability of EBT3 GafChromic film for quality assurance in MR-guided radiotherapy at 0.35 T with and without real-time MR imaging," *Physics in Medicine and Biology* **63**(16), 165014 (2018).

63. F. Padilla-Cabal, P. Kuess, D. Georg, H. Palmans, L. Fetty, H. Fuchs, "Characterization of EBT3 radiochromic films for dosimetry of proton beams in the presence of magnetic fields," *Medical Physics* **46**(7), 3278–3284 (2019).

64. I. Billas, H. Bouchard, U. Oelfke, S. Duane, "The effect of magnetic field strength on the response of Gafchromic EBT-3 film," *Physics in Medicine and Biology* **64**(6), 06NT03 (2019).

65. O. Volotskova, X. Fang, M. Keidar, H. Chandarana, I. J. Das, "Microstructure changes in radiochromic films due to magnetic field and radiation," *Medical Physics* **46**(1), 293–301 (2019).

66. A. Darafsheh, Y. Hao, B. Maraghechi, J. Cammin, F. Reynoso, R. Khan, "Influence of 0.35 T magnetic field on the spectral response of EBT3 and EBT-XD radiochromic films," *Medical Physics* **47**(9), 4543–4552 (2020).

67. L. J. van Battum, D. Hoffmans, H. Piersma, S. Heukelom, "Accurate dosimetry with GafChromic EBT film of a 6 MV photon beam in water: What level is achievable?," *Medical Physics* **35**(2), 704–716 (2008).

68. M. Martišíková, B. Ackermann, O. Jäkel, "Analysis of uncertainties in Gafchromic® EBT film dosimetry of photon beams," *Physics in Medicine and Biology* **53**(24), 7013–7027 (2008).

69. T. Aland, T. Kairn, J. Kenny, "Evaluation of a Gafchromic EBT2 film dosimetry system for radiotherapy quality assurance," *Australasian Physical & Engineering Sciences in Medicine* **34**, 251–260 (2011).

70. A. L. Palmer, C. Lee, A. J. Ratcliffe, D. Bradley, A. Nisbet, "Design and implementation of a film dosimetry audit tool for comparison of planned and delivered dose distributions in high dose rate (HDR) brachytherapy," *Physics in Medicine and Biology* **58**(19), 6623–6640 (2013).

71. J. Sorriaux, A. Kacperek, S. Rossomme, J. A. Lee, D. Bertrand, S. Vynckier, E. Sterpin, "Evaluation of Gafchromic® EBT3 films characteristics in therapy photon, electron and proton beams," *Physica Medica* **29**(6), 599–606 (2013).

72. E. Y. León Marroquin, J. A. Herrera González, M. A. Camacho López, J. E. Villarreal Barajas, O. A. García-Garduño, "Evaluation of the uncertainty in an EBT3 film dosimetry system utilizing net optical density," *Journal of Applied Clinical Medical Physics* **17**(5), 466–481 (2016).

CHAPTER 6

Thermoluminescence Dosimetry

Tomas Kron and Peta Lonski

Peter MacCallum Cancer Centre
Melbourne, VIC, Australia

CONTENTS

6.1 INTRODUCTION

6.1.1 What is TLD?

Thermoluminescence dosimetry or TLD for short is a dosimetric technique with a lot of history and many applications related to medicine. The name describes the physical principles involved: after irradiation of TLD materials the material is heated which makes it luminesce and the light intensity emitted is proportional to the dose deposited in the material in the first place [1–3]. This is illustrated in Figure 6.1.

This process will be discussed in more detail later and we will explore the variety of materials that can be used for TLD. TLD is a "passive" dosimetric technique: there are no cables or readout mechanism connected to

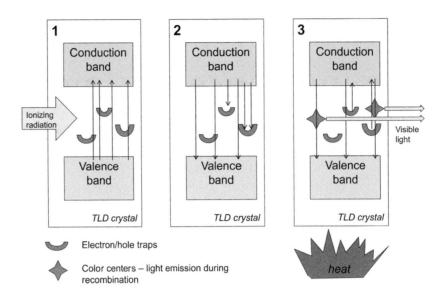

Electron/hole traps

Color centers – light emission during recombination

FIGURE 6.1   Schematic illustration of the thermoluminescence process.

the detectors during irradiation. This allows placement of thermoluminescence (TL) dosimeters in virtually all locations but prevents immediate "real time" readout. The "free standing nature" of TL dosimeters allows taking measurements in many different scenarios of dose levels that can vary by more than 10 orders of magnitude. It is this versatility that makes TLD a useful technique that ensures its continued relevance in radiation medicine.

6.1.2 A Bit of History

TLD has a long history in medicine. Already in 1953 Farrington Daniels, one of the pioneers of TLD, reported the first "medical" application of TLD: "The crystals were swallowed by the patient (who had received an injection containing radioactive isotopes), recovered one or two days later, and the accumulated dosage in roentgens was measured by matching thermoluminescence intensity with that produced in crystals by a known roentgen dosage" [4].

John Cameron, who was the first professor for Medical Physics in Madison, WI, was also a strong promoter of TLD as a tool for medical dosimetry [5]. He established LiF:Mg,Ti (patented by Harshaw Chemicals as TLD-100) as the original "standard" material for TLD. This material and its properties were then characterized extensively by Yigal Horowitz and his school [6–8].

Of particular importance for TLD has been the International Solid State Dosimetry Organisation (ISSDO) holding conferences every 3 years, which bring together many protagonists of TLD. The 20th SSD conference will be held in 2022 in Italy. Since 1980 the conferences publish proceeding in reputable journals [9], which has greatly helped to publicize the science and applications of TLD.

6.1.3 Scope of the Chapter

The present chapter aims to introduce TLD to scientists working broadly in the medical field. This ranges from radiation protection to imaging and radiation therapy. The objective is to link a basic understanding of the underlying science with good practice and useful applications of TLD.

To this end the chapter commences with some theory before covering the instrumentation needed to setup a successful TLD service. As the authors are from a radiotherapy background the focus will be on this; however, the concepts are easily translatable to radiation safety and medical imaging. We will also cover common applications of TLD with particular emphasis on phantom measurements, *in vivo* dosimetry and audits. The chapter will conclude with a summary of limitations of TLD which will need to touch on competing techniques such as optically stimulated luminescence (OSL) dosimetry (Chapter 7) [10] and radiochromic film (Chapter 5) [11]. All this is followed by an attempt to predict the future of a technique, which is still widely used despite having been declared obsolete on many occasions.

6.1.4 Important Literature

This chapter cannot provide a complete summary of TLD in radiotherapy. In addition to the proceedings of the Solid State Dosimetry conferences (e.g., [12–15]), which were mentioned above, the reader is referred to several books that provide more detail on thermoluminescence.

Of particular mention is the three-volume book edited in 1984 by Horowitz [6, 16] and the text books of McKeever et al. [17] and Oberhofer and Scharmann [2]. More clinical applications are covered in review articles by Kron [3, 18], Waligorski [19], and Olko [20] as well as a book chapter by Mobit and Kron [21].

Finally, a recent report of the American Association of Physicists in Medicine (AAPM) task group 191, Clinical Use of Luminescent Dosimeters: TLDs and OSLDs, provides a lot of useful and practical information [22].

6.2 FUNDAMENTAL PRINCIPLES USING LiF:Mg,Ti AS EXAMPLE

6.2.1 Basic Theory

We are attempting a simplified description of the TLD process with the objective to link features of TLD with practical problems and approaches to use it for dosimetry. For a more in depth coverage of TLD physics the reader can be referred to books by Horowitz [6, 23] and McKeever et al. [17].

Thermoluminescence is a physical process occurring in insulating crystals such as salts that feature a large energy gap between valence and conduction band. This is illustrated in Figure 6.1 and large in this context refers to "much larger than kT for room temperature" with k being the Boltzmann constant ($1.38064852 \times 10^{-23}$ m² kg s⁻² K⁻¹) and T the temperature in Kelvin (K). kT for 300 K or 27°C is approximately 0.025 eV and the typical band gap in TLD materials is a few eV. At room temperature all charge carriers such as electrons and electron vacancies (holes) are in the valence band. Radiation can free electrons (for the sake of simplicity "electron" refers to any charge carrier) into the conduction band from which they normally would recombine with their opposite charge to end up again in the valence band.

The trick for TLD materials is that doping them with appropriate impurities creates lattice defects in the crystal where electrons can be trapped. This is typically indicated by naming the crystal and adding the impurities separated by a colon (e.g., LiF:Mg,Ti is a lithium fluoride crystal doped with magnesium and titanium, a common TLD material also referred to as TLD-100).

The energy state of the trap exists between valence and conduction band as illustrated in Figure 6.1. The rate at which electrons are released back into the conduction band is governed by Boltzmann statistics and depends upon the trap depth and temperature of the crystal. Electrons can then recombine with the opposite charge carrier to return to valence band.

Three characteristics of this process are important for dosimetry:

1. The number of trapped electrons is proportional to the radiation dose received.

2. The probability of releasing electrons back into the conduction band depends on temperature. As such increasing the temperature of the TLD material will increase the release of electrons. This can mathematically be described as:

$$-\frac{dn}{dt} = n\, s\, \exp\left(-\frac{E_g}{kT}\right) \tag{6.1}$$

with n the number of electrons in the trap, t the time, s the frequency function associated with a particular trap, E_g the energy gap and kT the Boltzmann constant times the temperature in Kelvin.

3. In appropriately prepared crystals the recombination between electrons in conduction band and missing electrons in valence band can occur with emission of visible light. In this case the intensity of the emitted light is proportional to the number of trapped electrons and as such the initial dose received by the crystal.

6.2.2 Glow Curves

Figure 6.2 shows the emission of light in two TLD materials as a function of the crystal temperature. Several different traps with different abundance and different energy gaps are assumed resulting in several "glow peaks" and the curve shown is often referred to as the glow curve of a TLD material.

In this context some of the complexities of TLD can be demonstrated:

• The number of traps of a certain type depends on the doping and thermal history of the crystal. Doping levels will determine the number of lattice

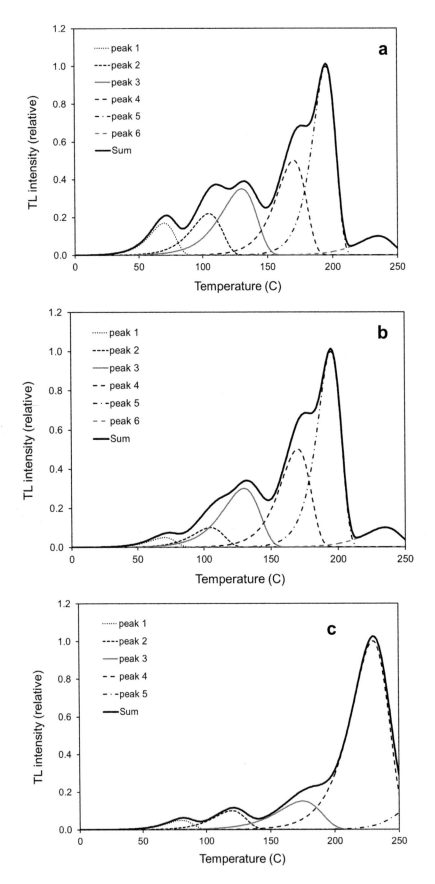

FIGURE 6.2 Schematic glow curves of LiF:Mg,Ti and LiF:Mg,Cu,P. (a) Glow curve of LiF:Mg,Ti; (b) The same glow curve after a pre-read annealing at moderate temperature has been performed; (c) Glow curve for LiF:Mg,Cu,P.

imperfections and as such the number of traps. The thermal history will influence the type of traps. Maintaining a high temperature for a long time will allow the impurities to travel through the crystal where they can start forming dimers, trimers and higher order combinations. In general, the higher the order the deeper the trap. As such the distribution of impurities and therefore energy levels of the traps depend on the thermal history.

• The exact description of the glow curve is not analytical. However a simplified equation [24] can be used to describe the depletion of trapped electrons as the temperature T is linearly increased, as shown in Figure 6.2:

$$I(T) = I_{max} \exp(1 + W(T - T_{max}) - \exp(W(T - T_{max})))$$
(6.2)

with I the light intensity at temperature T in K and $W = E_{gap}/(kT_{max}^2)$ where E_{gap} is the depth of the trap, k the Boltzmann constant and T_{max} the temperature of the glow curve maximum.

• Heating the TL materials after irradiation to a moderate temperature will empty some of the lower electron traps thereby reducing the intensity of the low peaks during readout as illustrated in Figure 6.2b. This "pre-read annealing" is useful in minimizing fading, the reduction of low energy peaks due to storage of the material at room temperature.

• Evaluation of the TLD signal typically includes only light emitted at a temperature that avoids impact of low energy peaks affected by fading. An alternative is glow curve analysis where a fitting routine is used to fit the known glow peaks to the acquired temperature/light intensity curve [25, 26]. This glow curve analysis provides also an excellent quality assurance tool for the dosimetric process and yields typically better accuracy. However, it does require considerably more work to fine tune the process and usually relies on a rather slow readout with slow ramp up of temperature to resolve individual peaks.

6.2.3 Dose Response

One of the main advantages of TLD is the wide dosimetric range which allows dose measurements over many orders of magnitude. This is illustrated in Figure 6.3 which shows the signal of LiF:Mg,Ti TL material as a function of the actual dose received.

As can be seen dose measurements are possible over at least eight orders of magnitude from 0.1 mGy to several 100s of Gy. In some TLD materials such as LiF:Mg,Ti supralinearity is observed as shown in the Figure 6.3.

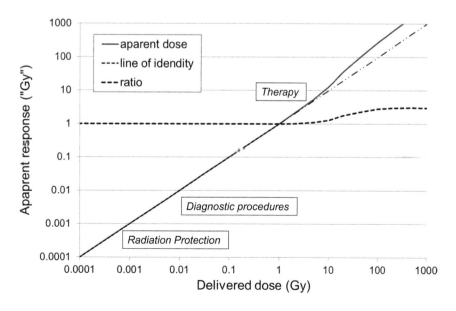

FIGURE 6.3 Dose response curve for LiF:Mg,Ti (solid line). Supralinearity can be seen for doses above approximately 1 Gy. The large range of possible dose measurements from fractions of mGy to 100s of Gy can be appreciated.

At higher doses, there is an overresponse due to the fact that the signal shows a "dual hit" (or quadratic) component due to the microscopic distribution of dose deposition. This is a complex process which depends on the TLD material, the thermal preparation and the radiation quality [27–29].

In practice, either the supralinearity must be characterized during commissioning of the system or standards used for comparison that are exposed to a dose similar to the one in the measurement (compare also Section 6.4).

6.2.4 Other Influences on Dosimetry

Many other factors that affect TLD depend on the material and the readout technique used. As such, some influencing factors such as variation of response with radiation quality will be discussed in Section 6.3. As TLD materials are usually denser than water, additional effects are related to the detector size and shape. In particular at low energy x-rays or other weakly penetrating radiation qualities self-attenuation can have an effect [30, 31].

Size and shape of the dosimeter may also affect angular response. However, as materials are available in different thicknesses and shapes this effect can be generally minimized.

Variation of TLD response with dose rate has not been observed, which makes TLD suitable for some novel very high dose rate (HDR) deliveries such as synchrotron irradiation [32] or FLASH, a novel approach to increase normal tissue sparing in radiotherapy through ultra-high dose rates [33–35].

6.3 TLD MATERIALS

One of the more important features of TLD is the wide variety of materials that exhibit thermoluminescence. Table 6.1 shows a summary of some common TLD materials.

Many more materials exist and the table has been provided more to illustrate certain features than to provide an exhaustive summary of materials. Another important material is Beryllium Oxide (BeO) [41] because of

TABLE 6.1   **TLD Materials Commonly used in Medical Applications and their Characteristics**

TLD Material (Harshaw Designation)	LiF:Mg,Ti (TLD 100)	LiF:Mg,Cu,P (TLD 100H)	$Li_2B_4O_7$:Mn (TLD 800)	Al_2O_3:C (TLD 500)	$CaSO_4$:Dy (TLD 900)
Physical density (g/cm³)	2.64	2.64	2.3	3.9	2.61
Effective atomic number	8.2	8.2	7.4	9.3	15.3
Sensitivity to ^{60}Co radiation (relative to LiF:Mg,Ti)	1	About 30	0.3	40–60	About 15
Energy response 30 keV/1.25 MeV	1.7	1.25	0.9	3.5	12
Temperature of main glow peak (°C)	195	210	200	180	220, 250
Approximate wavelength of emitted light (nm)	400	380	600	420	480, 570
Fading of main glow peak at 20°C	<10% per year	Small	10% per month	5% per year stored in dark	6% in 6 months
Typical annealing procedure	1 h: 400°C 20 h: 80°C	1/4 h: 250°C 2 h: 100°C	1/2 h: 300°C	1 h: 400°C 20 h: 80°C	1/2 h: 400°C
Useful dose range (Gy)	5×10^{-5} to 10^3	$<10^{-6}$ to 100	10^{-4} to 10^4	$<10^{-6}$ to 100	10^{-6} to 10^3
Available physical forms	Powder, rods, crystals, ribbons, Teflon chips	Ribbons, powder	Powder, Teflon-based chips, cards, ribbons	Chips, powder	Ribbons, powder, Teflon-based chips
Toxicity	High if ingested	High if ingested	High if ingested	Low	Low
Remarks	Complex glow curve, most common TLD, available as ^{6}Li and ^{7}Li	Good signal to noise, no supralinearity, low-temperature annealing	Tissue equivalent	Also OSL material; forms also gem stones	Complex glow curve, high sensitivity
Key references	Cameron et al. [5] and Horowitz [23]	Moscovitch and Horowitz, [8], Moscovitch [36], and Duggan et al. [31]	Horowitz [6] and Horowitz et al. [37]	Mehta and Sengupta [38] and Gaza et al. [39]	Horowitz [6] and Souza et al. [40]

Source: Adapted and updated from Kron (1994) [3].

its tissue equivalence and the fact that like Aluminum Oxide it is also useful for OSL dosimetry which extends usability (and scientific interest).

Considering Table 6.1 there are several important considerations that should be taken into account whatever TL material is chosen for use:

- The physical density provided here is the density of a crystal of the material. In practice many TLD crystals are ground to powder and then sintered into various forms such as rods, chips (often called ribbons because the material is pressed into a "ribbon" and then cut in regular intervals) and pellets. This will homogenize the sensitivity of the resulting detectors because it is difficult to grow crystals with consistent doping. The resulting dosimeters often have a different density in particular if binding agents are also used.

- In medical applications dosimetry aims to determine the dose as a quantity related to biological effect, be it detriment or tumor cell kill. This is related to the term "tissue equivalent" which means that a detector has a dose response to different radiation qualities similar to relevant biological tissues. In practice this is often taken as water equivalent. For photon dosimetry the atomic number of the material will significantly affect its response to radiation as different interaction processes (photoelectric effect, Compton interaction and pair production) have vastly different dependencies on radiation quality for different atomic numbers. Figure 6.4a shows the ratio of energy absorption coefficient (broadly related to dose) for different biological tissues as a function of photon energy.

The significant increase in energy absorption at lower energies is due to increasing importance of the photoelectric effect. The same phenomenon is observable for the TL detectors as illustrated in Figure 6.4b. This means that to measure water equivalent doses TLD materials with an effective atomic number around 7.5 would be preferable making LiF and BeO-based materials attractive. On the other hand it is also possible to assess dose to other mediums such as bone using TLD detectors that mimic the variation of response with photon energy of bone. Al_2O_3 is a reasonable choice for this.

Unfortunately, not all of the variation of TL response with energy is determined by atomic number, as can be seen in Figure 6.4b comparing LiF:Mg,Ti and LiF:Mg,Cu,P [31, 42]. The difference in response is in this case due to an anomalous energy response of LiF:Mg,Cu,P described by Olko et al. using microdosimetric analysis [28, 43, 44].

Even more complicated is the response of different detectors to particle irradiation with high linear energy transfer (LET) [27, 29, 37]. Similar to biological cell kill, supralinearity (= a linear quadratic component of dose response) plays a much smaller role than for low LET radiation.

- In some circumstances it is advantageous to combine the TLD powder with a more robust matrix such as Teflon [45, 46]. In this case also, the effective atomic number and physical density of the detector will change and needs to be taken into consideration. A particular interesting application of the concept of mixing TL material with a matrix are "black TLDs." For "black" TLDs, TLD material (with or without Teflon) is mixed with Carbon [47, 48]. Carbon does not affect the dose deposition in a significant way but it will reduce the light that can be emitted from the detector during readout (see next section). By adjusting the carbon loading one can design detectors that report dose up to a depth of approximately 70 μm as is recommended for skin dose measurements. Figure 6.5 shows different TLD materials including "black TLDs."

- An interesting opportunity exists in mixed radiation fields. Given the fact that many TLD materials contain Lithium or Boron, it is possible to modify the response to thermal neutrons by altering the isotope composition. The most common approach is to use pairs of dosimeters based on Li-6 and Li-7. Li-6 has a cross section for thermal neutrons that is several orders of magnitude larger than the one for Li-7. Results obtained using Li-7-based detectors provide information about the photon component of the field, while Li-6-based detectors will exhibit neutron dose signal in addition to this. While this is in principle an attractive way to differentially determine neutron and photon dose it does require the knowledge of the neutron spectrum which is usually not a given [22, 50, 51].

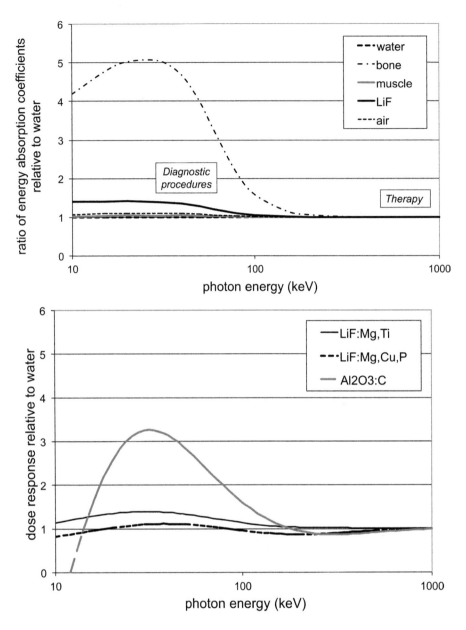

FIGURE 6.4 Variation of energy absorption coefficient with photon energy for different tissue types (a) and different TLD materials (b).

- Table 6.1 also lists a selection of physical forms in which detectors are available. This is another important advantage of TLD where the detector can be designed to suit a particular application such as microcubes and pinworms [52–54] for small field and brachytherapy dosimetry, respectively. In addition to this, the combination of detectors of different physical sizes allows to inter- and extrapolate doses [55, 56] which has been shown to be a useful approach in skin dose measurements [57].

6.4 EQUIPMENT REQUIRED FOR TLD

Equipment and instrumentation used for TLD needs to ensure optimal use in terms of reproducibility, sensitivity, linearity, lack of response variation with radiation quality and incidence direction and of course time and resource requirements. Not surprisingly, accurate temperature control is a key component in every step of TLD. As discussed in Section 6.2, the response of TLD materials may be affected by thermal history, heating and cooling rates, fading, dirt or scratches, and other handling procedures. The success of a TLD system as a

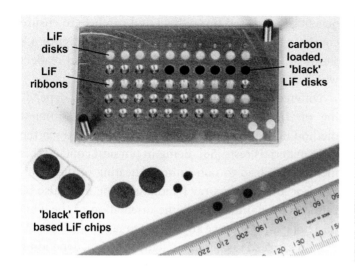

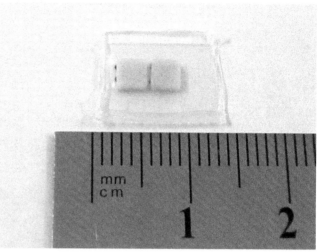

FIGURE 6.5 Several different versions of LiF-based TL detectors (adapted from Kron, 1999 [49]).

FIGURE 6.6 TLD chips pre-packaged in a plastic sleeve using a heat sealer. The use of two detectors improves precision and the plastic sleeve ensures easy handling and labeling (back of the envelope), whilst not significantly attenuating radiation and increasing the size.

whole depends upon careful control of each influencing factor. This section provides an overview of the equipment and processes required to establish a TLD system for radiotherapy.

6.4.1 TLD Handling

Since dirt and scratches can affect the light output from a TLD material, it is imperative that the dosimeter be kept clean and free of damage. Tiny scratches on the surface of a TL dosimeter will not only reduce the amount of physical material available for evaluation, but also reduce light emission for a given radiation exposure. Contaminants such as dirt and oils similarly reduce the light output by blocking light at the surface of the material. Vacuum tweezers can be used for TLD handling, often being inexpensive and an effective means of avoiding damage.

For patient *in vivo* dosimetry, TLDs must be prepared in appropriate packaging suitable for the intended measurement. Often it is desirable to maintain the small physical size of the TLD to ensure patient comfort during measurement and minimize any perturbation of the radiation beam. Packaging procedures must ensure the TLD material does not come into contact with the patient. This can be achieved by encasing TLDs within plastic, which also allows the TLDs to be labeled so they can be correctly identified after treatment. Thin plastic sleeves can be used to secure individual TLDs which can then be labeled. A heat-sealer may be used to secure the edges of the plastic and prevent loss of TLDs, though care should be taken to avoid heat sealing too close to

the TLD. This method provides the flexibility to package multiple TLDs in various configurations and even allows a series of TLDs to be sealed together to form an array. An example of two TLD chips heat-sealed within a plastic sleeve is shown in Figure 6.6. Other TLD encapsulations will be discussed in the next section [51].

Often in medical dosimetry it is necessary to take measurements at multiple locations. It is therefore desirable to group multiple TLDs together in "batches" of 50 or more. TLD disks, rods, and ribbons (chips) are stored in dedicated high-temperature annealing trays which are labeled to allow easy identification of individual TLDs. An example tray with spaces for up to 100 TLD chips is shown in Figure 6.7.

These trays should provide good thermal conductivity throughout the tray, ensuring all TLDs within a batch are exposed to the same temperature/time profile. This is particularly important during high-temperature annealing since any heterogeneity in thermal treatment within a batch can lead to differences in dose response for individual TLDs, which may be unpredictable.

6.4.2 TLD Reader

The role of the TLD reader is to supply thermal energy to the TL material and record subsequent light emissions, usually captured by an inbuilt photomultiplier tube. Commercially available readers vary in complexity, with more advanced models offering automated

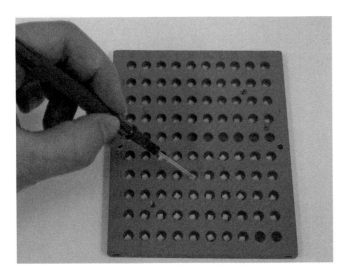

FIGURE 6.7 High-temperature annealing tray with space for up to 100 TLD chips. Vacuum tweezers used for TLD handling are also shown.

readout of multiple chips. Simpler designs require the user to manually read individual TLDs. The benefit of simpler models, besides the reduction in initial cost, is added diversity in the potential types of TLD which can be used. Automated readers are often restricted to a particular form of TLD, requiring a carousel or stacking mechanism which is used to hold many TLDs or TLD cards during readout. Automated readers have the

benefit of increased throughput and can help to ensure consistency in TLD readout. The insert in Figure 6.8 shows an example carousel used in automated readers which can accommodate up to 50 TLD chips.

Different heating methods can be employed depending on the make of the TLD reader. The simplest method of heating utilizes a hot planchet on which the TLD materials rests [58]. Being in physical contact, this method can lead to non-uniform heating of the material. For low energy x-ray irradiation, one must also be cautious to maintain the same orientation of the TLD on a hot planchet, since any self-attenuation during irradiation can lead to changes in light emission, depending on whether the TLD was placed on the planchet in the same orientation as it was during irradiation. An alternative method involves passing a hot inert gas over the TLD as it is held on a vacuum needle. This method helps to ensure uniform heating as well as uniform light emission, and has the advantage that the TLD does not need to sit atop a potentially dirty surface, thereby reducing the risk of contamination.

Automated readers can be controlled by PC-driven software, allowing the user to pre-program customized readout cycles with desired heating rates and readout temperatures. A commercial automated reader and control PC is shown in Figure 6.8. Programmable

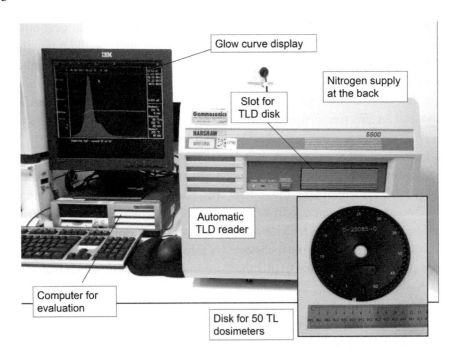

FIGURE 6.8 Automatic programmable TLD reader (Harshaw 5500). The computer screen displays a glow curve with the respective temperature profile indicated by a line. In the insert a carousel which can accommodate up to 50 TLD chips is shown.

readers also allow for material-specific readout procedures which can help to avoid accidental damage from over-heating. The user can define areas of interest in the TL glow curve used for evaluation. Often the area under the curve is used for evaluation since it is usually more robust than the height of a peak. Evaluation of higher energy peaks tends to be more robust than the lower peaks, which can fade quicker after irradiation. In some cases it may be desirable to program a pre-read anneal cycle to eliminate lower, less stable peaks, thereby excluding their contribution to the final glow curve used for analysis.

6.4.3 TLD Evaluation, Quality Assurance, and Analysis

For clinical applications, timely analysis of multiple TLDs including standards is often required. A visual representation of the glow curve should still be available for routine quality assurance and troubleshooting, since the shape of the curve can be useful for interpretation of TLD results. It is desirable to have the raw data exported to a personal computer for further analysis. A simple spreadsheet can be an effective means of analyzing TLD results and templates can be created to aid in maintaining records. For TLD readers which record photomultiplier background signal and the signal from a check light source, regular review of these metrics can be useful as a constancy check for the performance of a TLD reader.

When a system is initially commissioned glow curve analysis can be a useful means of providing insight into the performance of a TLD system and optimize temperature and time settings [24, 26]. In principle, equations such as (6.2) are used to fit the experimentally determined glow curve. Given the finite time required for thermal heating of the crystal to a given temperature, glow curve de-convolution is best done using a slow continuous temperature rise. Glow curve analysis is also useful for defining regions of interest which are to be used for subsequent TLD readings. Commercially available software may be used, or if a glow curve can be exported from the TLD reader, this can be done manually.

6.4.4 Annealing and Reuse

During the readout process most (but not all) of the original signal is lost as trapped electrons are released and interact with recombination centers. Even if the same TLD is read multiple times residual signal will still be present, which can affect subsequent measurements. The remaining signal depends upon the readout temperatures but in practice a high temperature anneal using a programmable oven is required to completely empty remaining traps and ensure the TLD crystal is ready for a new measurement. This is particularly important if a low dose is to be assessed after the same TL chip has received a relatively high dose.

Ensuring reproducible and uniform annealing also helps to control the thermal history of a TLD set, which is a crucial factor to a TLD system as it relies on comparison of TL signal from TLDs irradiated to an unknown dose to the others which have received a known one (typically referred to as standards). An example of a programmable oven used for TLD annealing is shown in Figure 6.9.

TLD readers can also be used to anneal TLDs individually, however this process has some drawbacks. First, only a single TLD can be annealed at a time, leading to a prohibitively long time required to anneal a set. Second, and perhaps most important, the anneal cycle of a TLD reader may not be as reproducible nor can higher temperatures be maintained. TLD readers also cannot achieve longer anneal times which may be required for certain materials (compare Table 6.1).

6.5 APPLICATIONS IN RADIOTHERAPY

Dose measurements are of great importance in radiotherapy as very high and potentially lethal radiation doses are delivered. As a large part of this book deals with a wide range of applications we cover only five areas briefly where TLD has been or continues to be an important dosimetric technique.

Figure 6.10 attempts a summary of dosimetric needs in radiotherapy. They range over at least 8 orders of magnitude in dose from μGy to several tens of Gy and it is beyond the scope of the present chapter to cover all of them in detail. However, as discussed in the previous sections and illustrated in Figure 6.3, TLD would be well suited to cover this range and the related variation in dose rate.

While the accuracy of dose measurements varies between applications it is generally assumed that dosimetry in radiotherapy should yield dose with an uncertainty better than 5% [59]. Quality assurance is an essential part of trying to achieve this and dose measurements and independent audits play an important role.

FIGURE 6.9   High temperature programmable oven used for TLD annealing. A real-time temperature display is shown at the bottom display.

Historically many different systems have been developed to introduce TLD into a radiotherapy clinic. Figure 6.11 shows a general flow of activities which would suit a system using TLD chips in batches as illustrated in the previous section.

6.5.1 Radiation Protection

TLD has been for many years the most important dosimetric technique for personnel dosimetry [60, 61]. This has been now challenged by both OSL (Chapter 7) [10] and radiophotoluminescence (RPL) [62] dosimetric techniques that provide faster and multiple read out. However, for many applications such as finger rings [63] and dosimeters to assess lens dose during diagnostic procedures [64], TLD is still a method of choice particularly where the radiation quality is not well known. A useful reference is the assessment of uncertainty in TL personnel monitoring provided by Veinot [65].

In radiation oncology dose to staff has been decreasing significantly with the replacement of ^{60}Co units with linear accelerators and the introduction of remote afterloading units for brachytherapy. However, for regulatory

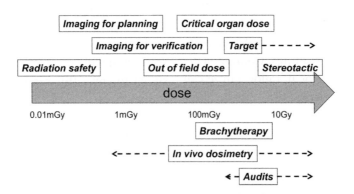

FIGURE 6.10   Application of dosimetry in radiotherapy as a function of dose.

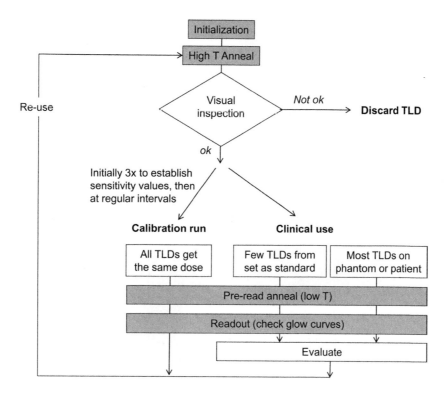

FIGURE 6.11 Typical use cycle for a TLD system based on batches of TLD chips or rods. A batch consists here of some 50 TL detectors which share an identical thermal history. This allows comparing unknown measured dose with known doses from detectors in the same batch that have been exposed to a known dose with the same radiation quality as is measured. The relative response of the different detectors to each other is established using calibration runs in which all detectors in a set are exposed to the same dose.

purposes and best practice in a high-dose environment, personnel dosimetry is usually required.

As integrative dosimeters TLDs are also well suited for environmental dosimetry. In a medical environment, this can be used to verify the shielding of radiotherapy or diagnostic facilities. TL dosimeters can be easily packaged and placed in any location for a suitable period of time. In addition to high sensitivity and small volume they are largely independent of environmental influences. The TLD measurement provides a real-life indication of the dose at a certain location over a given period of time – it verifies the shielding calculation as well as the assumed workload and the actual building and can be performed using a clinical TLD system.

6.5.2 Phantom Measurements

TL dosimeters as passive dosimeters do not require cables or batteries during irradiation. This allows placement in phantoms to mimic the treatment of patients. While this can be done in principle for all types of phantoms, the most important application is the use of TLD in anthropomorphic phantoms representing

the geometry and radiological properties of a human [66–69]. This allows end-to-end (E2E) testing of procedures from imaging to radiotherapy treatment delivery. Figure 6.12 shows an anthropomorphic phantom setup for breast radiotherapy at a linear accelerator.

Anthropomorphic phantoms typically consist of many slices as shown in Figure 6.12. Each slice is a few centimeters thick and includes a grid of pre-drilled holes where detectors can be placed. Holes can be plugged with tissue equivalent material if not in use for dosimeters and it is possible to drill additional holes in areas of anatomical interest as required for particular applications as shown in the insert of Figure 6.12. During the actual dose delivery, TL dosimeters can be placed in selected locations within the phantom and the dose delivery verified.

There are three major applications in radiotherapy for these measurements:

- Commissioning and testing of planning systems

- Development of new treatment techniques

- Quality assurance for individual patients

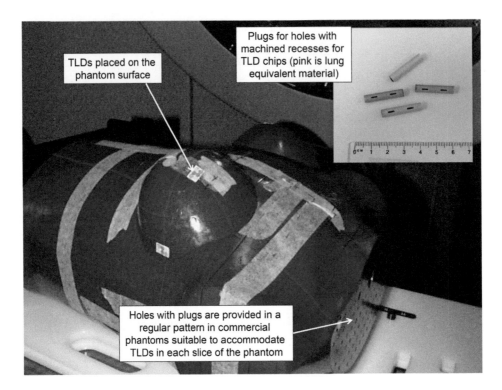

FIGURE 6.12 Anthropomorphic phantom setup for breast radiotherapy. TLDs are used on the surface of the phantom as well as inserted into plugs typically provided in commercial phantoms in a 30 mm or 15 mm pattern. The insert to the figure shows plugs that have been modified to accept TLD chips.

Commissioning of treatment planning systems is one of the most important tasks in setting up a radiotherapy department [70]. It requires beam data acquisition, beam modeling, image manipulation, contouring, as well as geometry and dose calculation assessment. The final important step would be E2E verification where a model representing a patient is imaged, planned and treated in exactly the same way a patient would be. Ideally this would include the creation of relevant structures and calculation of dose volume histograms. Anthropomorphic phantoms are ideal for this purpose and TLDs are well suited for the required measurements as it is not only important to verify the high dose to the target but also the low dose outside of the treated volume which is received by normal tissues. For these out-of-field measurements, TLDs which can be close to tissue equivalence are the ideal dosimeters.

A similar process would be required for the development of new treatment techniques and audits which will be discussed in Section 6.5.5 [71, 72]. New treatment techniques must be verified before put into clinical routine. By their very nature new techniques are unfamiliar to staff and as such an E2E test is useful not only in verifying the radiation dose but also in assessing the workflow and providing training to staff. TLD measurements in anthropomorphic phantoms can provide the dosimetric verification.

Finally, the treatment of individual patients may require verification. While this is rather labor intensive and more commonly done using *in vivo* dosimetry as described in the next section, there is scope to mock up uncommon treatments (e.g., total body irradiations, electron arcs) or unusual patients (e.g., bariatric, pregnant, or pediatric patients) prior to treatment [73].

6.5.3 *In Vivo* Dosimetry

TLD is one of the most commonly used techniques for *in vivo* dosimetry [74]. Its typical applications are the assessment of the dose to critical organs and measurements in difficult geometries. In clinical practice usually LiF-based materials are used. This material has relatively close tissue equivalence ($Z_{eff} = 8.3$), and can be produced in the desired forms such as powder, ribbons or rods with small dimensions of the order of few mm. The typical accuracy reached in clinical practice of about ±3% (2SD) is sufficient for the requirement of *in vivo* dosimetry.

The major advantages of TLD for *in vivo* dosimetry are the small size of the detectors, their stand-alone measurement characteristic and the fact that TL materials typically consist of one material only. Therefore, the reading of the TLD is generally independent of the angular distribution of the radiation. This is an important feature for measurements in complicated geometries where it may be difficult to estimate which direction radiation may come from. TLD detectors are typically prepackaged prior to placing them on the patient's skin in appropriate locations after the setup (compare Figure 6.6). This causes minimal interference with patient setup and treatment.

In radiotherapy, skin dose can mean the dose to the basal cell layer around 0.07 mm deep. While TLDs are typically too thick to give a meaningful result for such a shallow depth, TLD extrapolation or black TLDs can be used [48, 57]. This is relevant for skin toxicity such as desquamation.

On the other hand, the term "skin dose" is often used to describe entrance dose more generally. In megavoltage radiotherapy, this includes the build-up effect and the dosimeters need to be packaged to take this into consideration. Figure 6.13 shows a method to do this using Perspex build-up domes [51]. Similar containers which make handling of the detectors easier and at the same time provide suitable buildup can now also be made using additive manufacturing (3D printing).

For *in vivo* dosimetry, the major disadvantage of TLD is the delayed readout. However, in fractionated radiotherapy it is feasible to obtain the results of measurements after the first fraction of treatment and before the next fraction is delivered to the patient [21].

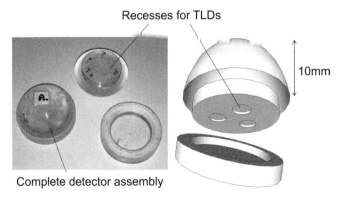

Recesses for TLDs

Complete detector assembly

10mm

FIGURE 6.13 Perspex build-up dome to facilitate entrance dose measurements in megavoltage external beam radiotherapy.

In vivo dosimetry (Chapter 11) [75] can roughly be separated into several different application groups [74]:

1. Routine verification on every patient as is required in some European countries. As these measurements are limited to surface dose assessment they are increasingly difficult to interpret as intensity modulated treatments become more common. Other, real time dosimetric techniques such as diodes, MOSFETs (Chapter 4) [76], or electronic portal dosimetry (Chapter 8) [77] are more suitable for this and as such more commonly employed.

2. Assessment of dose to critical structures, typically outside of the main radiation field. This is usually done for documentation purposes with scrotal and lens dose measurements being common. In the case of scrotal dose measurements these may also be done in the presence of scrotal shielding which make the prediction of the dose using planning systems virtually impossible. In some of these scenarios it is not possible to determine the dose directly. Figure 6.14 is an example for this where the lens dose is estimated from several detectors placed around the lens [21]. This can be done for separate fields if individual fields are used or all together in arc type deliveries.

3. Determination of dose where it is genuinely unknown or very difficult to predict. The most common scenario for this is dose at distance from the treated area. Dose calculation algorithms are limited in their accuracy at distance from the field edge as the dose depends on radiation leakage, internal and external scatter something difficult to model. It also yields a radiation spectrum which is difficult to determine and as such the tissue equivalence and high sensitivity of TL materials such as LiF:Mg,Cu,P provides for highly suitable detectors [51].

6.5.4 Brachytherapy

TLD has some features which make it an ideal dosimeter for brachytherapy, the use of radioactive sources implanted in the patient for treatment of malignancies (Chapter 15) [78]. Brachytherapy implants are characterized by radiation sources of various gamma-ray energies that are very close to tissues resulting in strong dose gradients. The tissue equivalence and good spatial resolution due to small physical size of TL dosimeters make

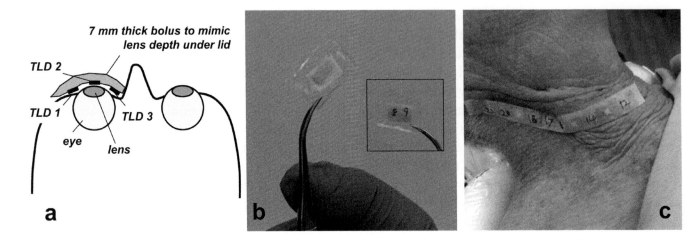

FIGURE 6.14 Suggestion for TLD use for *in vivo* dosimetry. (a) Lens dose measurement using TLDs under buildup that mimics depth of the lens with closed eye lids. Three locations are used that can provide an estimate of lens dose in anterior (TLD2), lateral (TLDs 1 and 3) and multifield/arc deliveries (judicious mix of all three); (b) Label for TLD placed inside a sealed satchel for measurement in the oral cavity of a patient; (c) "Array" of TLDs wrapping around an irregular surface (neck) to investigate early skin reaction.

them a promising tool for dosimetry in these difficult circumstances.

As in external beam therapy, brachytherapy can benefit from *in vivo* dosimetry. For HDR brachytherapy dose rates and fraction sizes are high enough to allow these measurements to be done on skin [79, 80] or in adjacent catheters [81, 82]. Of particular interest are also doses to critical structures such as rectum or bladder which allow for relatively easy insertion of radiation detectors [83, 84]. However, as for other *in vivo* dosimetry the delayed readout of the TL dosimeters is not ideal for brachytherapy where real time dosimetry would be preferential and other active detectors are more commonly used for *in vivo* dosimetry [80, 82, 85].

More recently TLDs have also been used to perform dosimetric intercomparisons and audits for brachytherapy [72, 86].

6.5.5 Audits

TLDs have been used for dosimetric audits for many years by many national and international organizations [87, 88]. The simplest audit is the assessment of dose under reference conditions often referred to as level I audit. Figure 6.15 shows a very simple setup to perform these measurements in water without the need for an expensive scanning water phantom. This measurement can easily be performed in virtually all radiotherapy centers in the world.

Shown in the figure is the dosimetry approach by the WHO/IAEA remote dosimetric audit using TLD capsules filled with TLD powder [88–90].

Audits are often categorized into three groups [91, 92]:

- Level I dosimetric audits verify dose in reference conditions. The setup of Figure 6.15 illustrates such an approach where the dose is verified in a standardized simple scenario.

- Level II dosimetric audits are primarily aimed a treatment planning systems and verify how the planning system handles typical dosimetric scenarios such as off-axis, simple inhomogeneities and different field sizes. Electron output verification is somewhere in between levels I and II for audit purposes as a dose measurement at a point usually also requires an assessment of effective energy to determine dose response [93, 94].

- Level III dosimetric audits finally are an E2E test of a typical (or challenging) treatment scenario, verifying all part of the radiotherapy chain from imaging for planning to dose optimization and calculation to verification and delivery. For this type of dosimetric audit commonly anthropomorphic phantoms as discussed above are used.

TLDs are useful for all levels of audits in particular when the audits are done remotely. In this case the light

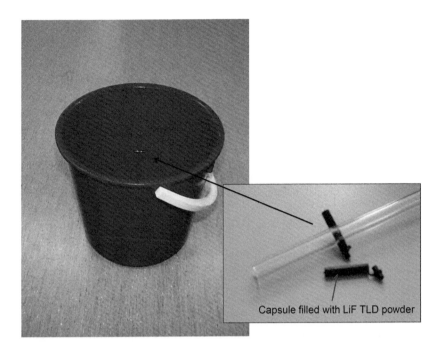

Capsule filled with LiF TLD powder

FIGURE 6.15 Simple setup of capsules filled with LiF powder for remote auditing of radiotherapy beams at 5cm depth in water.

weight, general robustness, and stand-alone nature of TLDs prove to be a significant advantage. As for personnel dosimetry, OSL dosimeters have also become popular for level I dosimetry due to their faster and repeatable readout [93]. For level III applications done remotely TLDs still have the additional advantage of tissue equivalence which is important when dose is assessed also outside of the high-dose region.

In practice, many audits combine measurements of point doses with TLDs with radiochromic film (Chapter 5) [11] which provides for an assessment of not only dose but also dose distribution which has become essential in level III audits of modern radiotherapy techniques.

6.6 LIMITATIONS OF TLD

TL dosimetry has a number of disadvantages that have both stimulated improvements in TL materials and techniques as well as the utilization of other technologies. Some of them are contrasted to the respective advantages in Table 6.2. The most important drawback in a medical environment is probably the fact that TLD is comparatively labor intensive and time consuming. It typically requires several steps (annealing, calibration, packaging, pre-read annealing, readout, signal analysis), many of which need to be tightly controlled to achieve acceptable reproducibility. Automated reader systems can help with this and as "a busy service is a

good service" this in itself can improve the quality of the TLD measurements.

Related to this is the need to meticulous and standardized work, which makes it often necessary to entrust TLD measurements to only few staff members. However, once set up, a TLD service is very versatile and can become useful in areas such as radiation safety, quality assurance and development of new procedures. Being busy also reduces the cost per measurement as the most significant investment is the TLD reader itself [95].

Delayed readout is a limitation of TLD, which is shared amongst all "passive" dosimeters such as radiochromic film and most OSL detectors. This makes TLD not suitable for measurements where real time results are required. In practice the turnaround of TLD measurements is several hours which allows obtaining results from one fraction in radiotherapy for the next.

Information loss after readout is a potential problem for radiation monitoring of staff or the environment. While false-positive results are rare in TLD, a lost reading is not replaceable and an unexpected reading has to be taken at face value and acted upon. Pulsed OSL provides a useful alternative to TL and has replaced TLD in many personal monitoring systems. It is interesting to note that most TLD materials also exhibit OSL properties and the first dual evaluation systems have now become commercially available (https://www.lexsyg.

TABLE 6.2   Summary of Advantages and Disadvantages of TLD for Radiotherapy Applications

Advantages	Disadvantages
Small physical size	Small physical size (difficult handling, fragile)
Tissue equivalence facilitating out-of-field measurements	No absolute dosimetry possible
Integrative, passive dose measurement	Delayed readout
No cables required during measurement	Concerns regarding reliability (black art)
High sensitivity – wide dosimetric range	Concerns regarding accuracy (which can be overcome through meticulous control of all aspects of the dosimeter handling)
Reusable and cheap – multiple measurements can be made at the same time	Expensive readout equipment required
Many forms and materials available (allows optimal choice for particular purpose)	Many forms and materials available (confusing and difficult to find appropriate technique for a particular application)
Mostly independent of environmental conditions such as temperature and pressure	No permanent dose record
Glow curve provides additional information which could be useful to improve accuracy or obtain information on radiation quality	Requires dedicated operator

Source: Adapted from Mobit and Kron (2006) [21].

com/tlosl-reader.html or https://www.nutech.dtu.dk/english/products-and-services/radiation-instruments/tl_osl_reader).

In this context, it also worth mentioning another related dosimetric technique: RPL [62, 96–98]. In RPL ultraviolet light is used for readout of dosimeters. They typically exhibit a small amount of fading and are resistant to most environmental influences. As such they are predominantly used in personnel and accident dosimetry and their applications in radiotherapy are just emerging [97, 99].

6.7 CONCLUSION

A wide range of dosimeters are available to medical physicists to make meaningful assessments of dose in diagnostic and therapeutic procedures. It can be anticipated that TLD will continue to feature as a relevant dose measurement tool in the future as it has some unique characteristics, which will make it difficult to replace it – even if this means accepting the TLD disadvantages of workload and delayed readout.

By comparing advantages and disadvantages listed in Table 6.2, it is possible to identify some areas in medicine where one can expect TLD to maintain a strong presence or find even increasing use in radiotherapy dosimetry:

- Out-of-field dose is difficult to predict from treatment planning but increasingly important as patients live longer and very late effects such as secondary cancers or cardiac toxicity become

more important. Tissue equivalence and high sensitivity are essential for this.

- The same features make TLD an important tool for *in vivo* dosimetry. The stand-alone nature of the measurement as well as the small size help as well.

- Out-of-field and *in vivo* dosimetry often go together. Of increasing relevance is the combined dose from verification imaging and treatment delivery. As radiation quality is even more difficult to predict in this case, TLD with its tissue equivalence has a distinct advantage.

- Remote dosimetric audits can continue to benefit from the relative robustness, long term stability and stand-alone nature of TLDs. It can be expected though that many audits will use TLDs in combination with other dosimeters.

- The same features are relevant for environmental dosimetry. While personnel dosimetry is unlikely to revert back to wide spread use of TLD, assessing new facilities through environmental monitoring will continue to be useful, particular in remote or under-resourced locations.

- Due to reusability TLD is a relatively cheap technique. This helps with environmental dosimetry and audits (if a TLD is lost in action its calibration factor is probably the most costly loss).

The role of TLD to address other emerging problems in radiotherapy such as hadron therapy [100], synchrotron microbeams [101], and FLASH [35] is not clear yet. However, there are many great opportunities to perform meaningful research.

Given this wide range of application for TLD one can anticipate that the technique will continue to be an important dosimetric technique for medical applications. While innovation is typically slow to be widely implemented in clinical practice, the rapid development of new technology and TL materials will offer new opportunities which will sooner or later benefit patients in medicine.

REFERENCES

1. J. Cameron, N. Suntharalingam, G. Kenny, *Thermoluminescence Dosimetry* (University of Wisconsin Press, Madison, 1968).
2. M. Oberhofer and A. Scharmann, *Applied Thermoluminescence Dosimetry* (Adam Hilger, Bristol, 1981).
3. T. Kron, "Thermoluminescence dosimetry and its applications in medicine–Part 1: Physics, materials and equipment," *Australasian Physical and Engineering Sciences in Medicine* 17, 175–199 (1994).
4. F. Daniels, C. Boyd, D. Saunders, "Thermoluminescence as a research tool," *Science* 117, 343–349 (1953).
5. J. Cameron, F. Daniels, N. Johnson, G. Kenny, "Radiation dosimeter utilizing the thermoluminescence of LiF," *Science* 134, 233 (1961).
6. Y. S. Horowitz, "General characteristics of TLD materials," in *Thermoluminescence and Thermoluminescent Dosimetry*, edited by Y. S. Horowitz (CRC Press, Boca Raton, 1984).
7. Y. S. Horowitz, "The annealing characteristics of LiF:Mg,Ti," *Radiation Protection Dosimetry* 20, 219–230 (1990).
8. M. Moscovitch and Y. S. Horowitz, "Thermoluminescent materials for medical applications: LiF: Mg, Ti and LiF: Mg, Cu, P," *Radiation Measurements* 41, S71–S77 (2006).
9. G. Portal, Proceedings of the 6th International Conference on Solid State Dosimetry. *Nuclear Instruments and Methods*, vol. 175 (1980).
10. S. Kry and J. O'Daniel, "Optically stimulated luminescent dosimeters in clinical practice," in *Radiation Therapy Dosimetry: A Practical Handbook*, edited by A. Darafsheh (CRC Press, Boca Raton, 2021), Chapter 7, pp. 97–108.
11. S. Mosahebi, N. Hooshyar, R. Khan, A. Darafsheh, "Film dosimetry," in *Radiation Therapy Dosimetry: A Practical Handbook*, edited by A. Darafsheh (CRC Press, Boca Raton, 2021), Chapter 5, pp. 61–74.
12. Proceedings of the 13th International Conference on Solid State Dosimetry (SSD2001), Athens, Greece, July 9–13, 2001, *Radiation Protection Dosimetry* 100, 1–552 (2002).
13. Solid state dosimetry. Proceedings of the 14th International Conference on Solid State Dosimetry (SSD14), June 27–July 2, 2004, New Haven, Connecticut, USA, *Radiation Protection Dosimetry* 119, 1–517 (2006).
14. Proceedings of the XII International Symposium on Solid State Dosimetry, September 5–9, 2011, Mexico City, Mexico, *Applied Radiation and Isotopes* 71 (Suppl), 1–101 (2012).
15. E. Goldfinch, S. W. S. Mckeever, A. Scharman, *Solid State Dosimetry – Proceedings of the 10th International Conference on Solid State Dosimetry, Held in Washington, DC, July 13–17, 1992* (Nuclear Technology Publishing, Ashford, UK, 1993).
16. Y. S. Horowitz, "TL dose response," in *Thermoluminescence and Thermoluminescent Dosimetry*, edited by Y. S. Horowitz (CRC Press, Boca Raton, 1984).
17. S. W. S. McKeever, M. Moscovitch, P. Townsend, *Thermoluminescence Dosimetry Materials: Properties and Uses* (Nuclear Technology Publishing, Ashford, UK, 1995).
18. T. Kron, "Thermoluminescence dosimetry and its applications in medicine–Part 2: History and applications," *Australasian Physical and Engineering Sciences in Medicine* 18, 1–25 (1995).
19. M. P. R. Waligorski, "What can solid state detectors do for clinical dosimetry in modern radiotherapy?," *Radiation Protection Dosimetry* 85, 361–366 (1999).
20. P. Olko, "Advantages and disadvantages of luminescence dosimetry," *Radiation Measurements* 45, 506–511 (2010).
21. P. Mobit and T. Kron, "Applications of thermoluminescence dosimeters in medicine," in *Microdosimetric Response of Physical and Biological Systems to Low and High LET Radiations*, edited by Y. S. Horowitz (Elsevier, Amsterdam, 2006).
22. S. F. Kry, P. Alvarez, J. E. Cygler, L. A. Dewerd, R. M. Howell, S. Meeks, J. O'Daniel, C. Reft, G. Sawakuchi, E. G. Yukihara, D. Mihailidis, "AAPM TG 191 clinical use of luminescent dosimeters: TLDs and OSLDs," *Medical Physics* 47, e19–e51 (2020).
23. Y. S. Horowitz, "A unified and comprehensive theory of TL dose response of thermoluminescent systems applied to LiF:Mg,Ti," in *Microdosimetric Response of Physical and Biological Systems to Low and High LET Radiations*, edited by Y. S. Horowitz (Elsevier, Amsterdam, 2006).
24. A. Delgado, J. Gomezroz, J. Muniz, J. C. Portillo, "Application of glow curve analysis methods to improve TLD-100 dose reassessment performance," *Health Physics* 62, 228–234 (1992).
25. A. Bos, T. Pithers, J. Gomez Roz, A. Delgado, "An intercomparison of glow curve analysis computer programs: I. Synthetic glow curves," *Radiation Protection Dosimetry* 47, 473–477 (1993).
26. Y. S. Horowitz and D. Yossian, *Computerised Glow Curve Deconvolution: Application to Thermoluminescence Dosimetry* (Nuclear Technology Publishing, Ashford, UK, 1995).

27. P. Olko, P. Bilski, N. A. El-Faramawy, H. Y. Goksu, J. L. Kim, R. Kopec, M. P. Waligorski, "On the relationship between dose-, energy- and LET-response of thermoluminescent detectors," *Radiation Protection Dosimetry* **119**, 15–22 (2006).

28. P. Olko and M. P. Waligorski, "Microdosimetric one hit detector model for calculation of dose and energy response of some solid state detectors," *Radiation Protection Dosimetry* **99**, 381–382 (2002).

29. A. R. Lakshmanan and R. Bhatt, "Supralinearity and LET dependence of thermoluminescence dosimeters," *Physics in Medicine and Biology* **26**, 1157–1163 (1981).

30. N. Nariyama, S. Tanaka, Y. Nakane, Y. Asano, H. Hirayama, S. Ban, H. Nakaskima, Y. Namito, "Responses and glow curves of $Li_2B_4O_7$:Cu, BeO and $CaSO_4$:Tm TLDs to 10–40 keV monoenergetic photons from synchrotron radiation," *Radiation Protection Dosimetry*, **74**, 155–161 (1997).

31. L. Duggan, C. Hood, H. Warren-Forward, M. Haque, T. Kron, "Variations in dose response with x-ray energy of LiF:Mg,Cu,P thermoluminescence dosimeters: Implications for clinical dosimetry," *Physics in Medicine and Biology* **49**, 3831–3845 (2004).

32. G. Grafstrom, B. A. Jonsson, S. E. Strand, P. Spanne, "Dosimetry at 28 keV synchrotron radiation for cell irradiations," *Radiation Protection Dosimetry* **108**, 115–122 (2004).

33. P. G. Jorge, M. Jaccard, K. Petersson, M. Gondre, M. T. Duran, L. Desorgher, J. F. Germond, P. Liger, M. C. Vozenin, J. Bourhis, F. Bochud, R. Moeckli, C. Bailat, "Dosimetric and preparation procedures for irradiating biological models with pulsed electron beam at ultra-high dose-rate," *Radiotherapy and Oncology* **139**, 34–39 (2019).

34. J. Bourhis, W. J. Sozzi, P. G. Jorge, O. Gaide, C. Bailat, F. Duclos, D. Patin, M. Ozsahin, F. Bochud, J. F. Germond, R. Moeckli, M. C. Vozenin, "Treatment of a first patient with FLASH-radiotherapy," *Radiotherapy and Oncology* **139**, 18–22 (2019).

35. M. C. Vozenin, M. Baumann, Coppes, R. P. & J. Bourhis, "FLASH radiotherapy international workshop," *Radiotherapy and Oncology* **139**, 1–3 (2019).

36. M. Moscovitch, New TLD materials and their merits: LiF:Mg,Cu,P and Al_2O_3 (1991).

37. Y. S. Horowitz, J. A. Kalef-Ezra, M. Moskovitch, H. Pinto, "Further studies on the non-universality of the TL-LET response in thermoluminescent LiF and $Li_2B_4O_7$: The effect of high temperature TL," *Nuclear Instruments and Methods* **172**, 479–485 (1980).

38. S. K. Mehta and S. Sengupta, "Al_2O_3 phosphor for thermoluminescence dosimetry," *Health Physics* **31**, 176–177 (1976).

39. R. Gaza, E. G. Yukihara, S. W. McKeever, "The response of thermally and optically stimulated luminescence from Al_2O_3:C to high-energy heavy charged particles," *Radiation Measurements* **38**, 417–420 (2004).

40. J. Souza, L. Darosa, C. Mauricio, "On the thermoluminescence glow curve of $CaSO_4$:Dy," *Radiation Protection Dosimetry* **47**, 103–106 (1993).

41. A. Malthez, M. B. Freitas, E. M. Yoshimura, N. K. Umisedo, V. Button, "OSL and TL techniques combined in a beryllium oxide detector to evaluate simultaneously accumulated and single doses," *Applied Radiation and Isotopes* **110**, 155–159 (2016).

42. B. Obryk, C. Hranitzky, H. Stadtmann, M. Budzanowski, P. Olko, "Energy response of different types of RADOS personal dosemeters with MTS-N (LiF:Mg,Ti) and MCP-N (LiF:Mg,Cu,P) TL detectors," *Radiation Protection Dosimetry* **144**, 211–214 (2011).

43. P. Olko, P. Bilski, E. Ryba, T. Niewiadomski, "Microdosimetric interpretation of the anomalous photon energy response of ultra-sensitive LiF:Mg, Cu, P TL dosemeters," *Radiation Protection Dosimetry* **47**, 31–35 (1993).

44. P. Olko, "Microdosimetric interpretation of photon energy response in TL systems," in *Microdosimetric Response of Physical and Biological Systems to Low and High LET Radiations*, edited by Y. S. Horowitz (Elsevier, Amsterdam, 2006).

45. G. Webb, "Thermal annealing effects in the thermoluminescence of lithium fluoride Teflon dosimeters," *Health Physics* **13**, 814–816 (1967).

46. A. Nash and F. Attix, "Use of LiF Teflon discs in laminated identification cards for personnel accident dosimetry," *Health Physics* **21**, 435–439 (1971).

47. S. Thomas and N. Palmer, "The use of carbon-loaded thermoluminescent dosimeters for the measurement of surface dose in megavoltage x-ray beams," *Medical Physics* **16**, 902–904 (1989).

48. P. Ostwald, T. Kron, C. Hamilton, J. Denham, "Clinical use of carbon-loaded thermoluminescence dosimeters for skin dose determination," *International Journal of Radiation Oncology, Biology, Physics* **33**, 943–950 (1995).

49. T. Kron, "Dose measuring tools," in *Modern Technology of Radiation Oncology*, edited by J. van Dyk (Medical Physics Publishing, Wisconsin, 1999).

50. H. A. Nedaie, H. Darestani, N. Banaee, N. Shagholi, K. Mohammadi, A. Shahvar, E. Bayat, "Neutron dose measurements of Varian and Elekta linacs by TLD600 and TLD700 dosimeters and comparison with MCNP calculations," *Journal of Medical Physics* **39**, 10–17 (2014).

51. P. Lonski, S. Keehan, S. Siva, D. Pham, R. D. Franich, M. L. Taylor, T. Kron, "Out-of-field in vivo dosimetry using TLD in SABR for primary kidney cancer involving mixed photon fields," *Physica Medica* **37**, 9–15 (2017).

52. C. Hood, L. Duggan, S. Bazley, J. Denham, M. Budzanowski, T. Kron, "LiF:Mg,Cu,P 'pin worms': Miniature detectors for brachytherapy dosimetry," *Radiation Protection Dosimetry* **101**, 407–410 (2002).

53. P. Francescon, S. Cora, C. Cavedon, P. Salchi, S. Reccanello, F. Colombo, "Use of a new type of radiochromic film, a new parallel-plate micro-chamber, MOSFETs, and TLD 800 microcubes in the dosimetry of small beams," *Medical Physics* **25**, 503–511 (1998).

54. E. Pantelis, A. Moutsatsos, K. Zourari, W. Kilby, C. Antypas, P. Papagiannis, P. Karaiskos, E. Georgiou, L. Sakelliou, "On the implementation of a recently proposed dosimetric formalism to a robotic radiosurgery system," *Medical Physics* **37**, 2369–2379 (2010).

55. T. Kron, A. Elliott, P. Metcalfe, "The penumbra of a 6MV x-ray beam as measured by thermoluminescence dosimetry and evaluated using an inverse square root function," *Medical Physics* **20**, 1429–1438 (1993).

56. T. Kron, A. Elliott, T. Wong, G. Showell, B. Clubb, P. Metcalfe, "X-ray surface dose measurements using TLD-extrapolation," *Medical Physics* **20**, 703–711 (1993).

57. T. Kron, M. Butson, F. Hunt, J. Denham, "TLD extrapolation for skin dose determination in vivo," *Radiotherapy and Oncology* **41**, 119–123 (1996).

58. T. Kron, M. Butson, T. Wong, P. Metcalfe, "Readout of thermoluminescence dosimetry chips using a contact planchet heater," *Australasian Physical and Engineering Sciences in Medicine* **16**, 137–142 (1993).

59. ICRU, "ICRU report 24: Determination of absorbed dose in a patient irradiated by X or gamma rays in radiotherapy procedures," *ICRU Reports* (International Commission on Radiological Units and Measurements, Bethesda, 1976).

60. P. Olko, L. Currivan, J. W. Van Dijk, M. A. Lopez, C. Wernli, "Thermoluminescent detectors applied in individual monitoring of radiation workers in Europe – A review based on the EURADOS questionnaire," *Radiation Protection Dosimetry* **120**, 298–302 (2006).

61. C. Sneha, S. M. Pradhan, M. M. Adtani, "Study of minimum detection limit of TLD personnel monitoring system in India," *Radiation Protection Dosimetry* **141**, 168–172 (2010).

62. N. Hocine, L. Donadille, C. Huet, C. Itie, I. Clairand, "Personal monitor glass badge: Theoretical dosemeter response calculated with the Monte Carlo transport code MCNPX," *Radiation Protection Dosimetry* **144**, 231–233 (2011).

63. M. Ginjaume, S. Perez, X. Ortega, "Improvements in extremity dose assessment for ionising radiation medical applications," *Radiation Protection Dosimetry* **125**, 28–32 (2007).

64. N. Guberina, M. Forsting, A. Ringelstein, "Efficacy of lens protection systems: Dependency on different cranial CT scans in the acute stroke setting," *Radiation Protection Dosimetry* **175**, 279–283 (2017).

65. K. G. Veinot, "An estimate of the propagated uncertainty for a dosemeter algorithm used for personnel monitoring," *Radiation Protection Dosimetry* **163**, 409–414 (2015).

66. J. A. Bencomo, C. Chu, V. M. Tello, S. H. Cho, G. S. Ibbott, "Anthropomorphic breast phantoms for quality assurance and dose verification," *Journal of Applied Clinical Medical Physics* **5**, 36–49 (2004).

67. C. La Tessa, T. Berger, R. Kaderka, D. Schardt, C. Korner, U. Ramm, J. Licher, N. Matsufuji, C. Vallhagen Dahlgren, T. Lomax, G. Reitz, M. Durante, "Out-of-field dose studies with an anthropomorphic phantom: Comparison of X-rays and particle therapy treatments," *Radiotherapy and Oncology* **105**, 133–138 (2012).

68. A. Molineu, D. S. Followill, P. A. Balter, W. F. Hanson, M. T. Gillin, M. S. Huq, A. Eisbruch, G. S. Ibbott, "Design and implementation of an anthropomorphic quality assurance phantom for intensity-modulated radiation therapy for the Radiation Therapy Oncology Group," *International Journal of Radiation Oncology, Biology, Physics* **63**, 577–583 (2005).

69. M. P. Waligorski, R. Baranczyk, S. Hyodynmaa, J. Eskola, J. Lesiak, B. Rozwadowska-Bogusz, A. Kolodziejczyk, "A TL-based anthropomorphic benchmark for verifying 3-D dose distributions from external electron beams calculated by radiotherapy treatment planning systems," *Radiation Protection Dosimetry* **120**, 74–77 (2006).

70. IAEA, *Commissioning and Quality Assurance of Computerized Planning Systems for Radiation Treatment of Cancer, IAEA Technical report series N430* (International Atomic Energy Agency, Vienna, 2004).

71. E. Adolfsson, P. Wesolowska, J. Izewska, E. Lund, A. C. Tedgren, "End-to-end audit: Comparison of TLD and lithium formate EPR dosimetry," *Radiation Protection Dosimetry* **186**, 119–122 (2019).

72. A. Haworth, L. Wilfert, D. Butler, M. A. Ebert, S. Todd, J. Bucci, G. M. Duchesne, D. Joseph, T. Kron, "Australasian brachytherapy audit: Results of the 'end-to-end' dosimetry pilot study," *Journal of Medical Imaging and Radiation Oncology* **57**, 490–498 (2013).

73. T. Kron, G. Donahoo, P. Lonski, G. Wheeler, "A technique for total skin electron therapy (TSET) of an anesthetized pediatric patient," *Journal of Applied Clinical Medical Physics* **19**, 109–116 (2018).

74. T. Kron, M. Schneider, A. Murray, H. Mameghan, "Clinical thermoluminescence dosimetry: How do expectations and results compare?," *Radiotherapy and Oncology* **26**, 151–161 (1993).

75. D. Bollinger and A. Darafsheh, "Clinical considerations and dosimeters for *in vivo* dosimetry," in *Radiation Therapy Dosimetry: A Practical Handbook*, edited by A. Darafsheh (CRC Press, Boca Raton, 2021), Chapter 11, pp. 151–171.

76. G. Blasi, N. Hardcastle, A. Rosenfeld, "Scintillation fiber optic dosimetry," in *Radiation Therapy Dosimetry: A Practical Handbook*, edited by A. Darafsheh (CRC Press, Boca Raton, 2021), Chapter 4, pp. 39–59.

77. B. Schott, T. Dvergsten, R. Caleb, B. Sun, "EPID-based dosimetry," in *Radiation Therapy Dosimetry: A Practical Handbook*, edited by A. Darafsheh (CRC Press, Boca Raton, 2021), Chapter 8, pp. 109–121.

78. C. L. Deufel, W. S. Culberson, M. J. Rivard, F. Mourtada, "Brachytherapy dosimetry," in *Radiation Therapy Dosimetry: A Practical Handbook*, edited by A. Darafsheh (CRC Press, Boca Raton, 2021), Chapter 15, pp. 231–252.

79. F. Perera, F. Chisela, L. Stitt, J. Engel, V. Venkatesan, "TLD skin dose measurements and acute and late effects after lumpectomy and high-dose-rate brachytherapy only for early breast cancer," *International Journal of Radiation Oncology, Biology, Physics* **62**, 1283–1290 (2005).

80. Z. Jamalludin, W. L. Jong, G. F. Ho, A. B. Rosenfeld, N. M. Ung, "In vivo dosimetry using MOSkin detector during Cobalt-60 high-dose-rate (HDR) brachytherapy of skin cancer," *Australasian Physical and Engineering Sciences in Medicine* **42**, 1099–1107 (2019).

81. D. Adliene, K. Jakstas, B. G. Urbonavicius, "In vivo TLD dose measurements in catheter-based high-dose-rate brachytherapy," *Radiation Protection Dosimetry* **165**, 477–481 (2015).

82. J. Mason, A. Mamo, B. Al-Qaisieh, A. M. Henry, P. Bownes, "Real-time in vivo dosimetry in high dose rate prostate brachytherapy," *Radiotherapy and Oncology* **120**, 333–338 (2016).

83. R. Das, W. Toye, T. Kron, S. Williams, G. Duchesne, "Thermoluminescence dosimetry for in-vivo verification of high dose rate brachytherapy for prostate cancer," *Australasian Physical and Engineering Sciences in Medicine* **30**, 178–184 (2007).

84. E. Jaselske, D. Adliene, V. Rudzianskas, B. G. Urbonavicius, A. Inciura, "In vivo dose verification method in catheter based high dose rate brachytherapy," *Physica Medica* **44**, 1–10 (2017).

85. G. Kertzscher, A. Rosenfeld, S. Beddar, K. Tanderup, J. E. Cygler, "In vivo dosimetry: Trends and prospects for brachytherapy," *The British Journal of Radiology* **87**, 20140206 (2014).

86. A. Roue, J. L. Venselaar, I. H. Ferreira, A. Bridier, J. Van Dam, "Development of a TLD mailed system for remote dosimetry audit for (192)Ir HDR and PDR sources," *Radiotherapy and Oncology* **83**, 86–93 (2007).

87. D. S. Followill, D. R. Evans, C. Cherry, A. Molineu, G. Fisher, W. F. Hanson, G. S. Ibbott, "Design, development, and implementation of the radiological physics center's pelvis and thorax anthropomorphic quality assurance phantoms," *Medical Physics* **34**, 2070–2076 (2007).

88. J. Izewska, P. Bera, S. Vatnitsky, "IAEA/WHO TLD postal dose audit service and high precision measurements for radiotherapy level dosimetry. International Atomic Energy Agency/World Health Organization," *Radiation Protection Dosimetry* **101**, 387–392 (2002).

89. J. Izewska, P. Andreo, S. Vatnitsky, K. R. Shortt, "The IAEA/WHO TLD postal dose quality audits for radiotherapy: A perspective of dosimetry practices at hospitals in developing countries," *Radiotherapy and Oncology* **69**, 91–97 (2003).

90. J. Izewska, D. Georg, P. Bera, D. Thwaites, M. Arib, M. Saravi, K. Sergieva, K. Li, F. G. Yip, A. K. Mahant, W. Bulski, "A methodology for TLD postal dosimetry audit of high-energy radiotherapy photon beams in non-reference conditions," *Radiotherapy and Oncology* **84**, 67–74 (2007).

91. T. Kron, C. Hamilton, M. Roff, J. Denham, "Dosimetric intercomparison for two Australasian clinical trials using an anthropomorphic phantom," *International Journal of Radiation Oncology, Biology, Physics* **52**, 566–579 (2002).

92. T. Kron, A. Haworth, I. Williams, "Dosimetry for audit and clinical trials: Challenges and requirements," *Journal of Physics: Conference Series* **444**, 012014 (2013).

93. P. Alvarez, S. F. Kry, F. Stingo, D. Followill, "TLD and OSLD dosimetry systems for remote audits of radiotherapy external beam calibration," *Radiation Measurements* **106**, 412–415 (2017).

94. A. S. Pradhan, "Influence on heating rate on the TL response of LiF TLD-700, LiF:Mg,Cu,P and Al$_2$O$_3$:C," *Radiation Protection Dosimetry* **58**, 205–209 (1995).

95. K. Kesteloot, A. Dutreix, E. Van Der Schueren, "A model for calculating the costs of in vivo dosimetry and portal imaging in radiotherapy departments," *Radiotherapy and Oncology* **28**, 108–117 (1993).

96. Y. Garcier, G. Cordier, C. Pauron, J. Fazileabasse, "Intercomparison of passive dosimetry technology at EDF facilities in France," *Radiation Protection Dosimetry* **124**, 107–114 (2007).

97. S. Hashimoto, Y. Nakajima, N. Kadoya, K. Abe, K. Karasawa, "Energy dependence of a radiophotoluminescent glass dosimeter for HDR (192) Ir brachytherapy source," *Medical Physics* **46**, 964–972 (2019).

98. H. Vincke, I. Brunner, I., Floret, D. Forkel-Wirth, M. Fuerstner, S. Mayer, C. Theis, "Response of alanine and radio-photo-luminescence dosemeters to mixed high-energy radiation fields," *Radiation Protection Dosimetry* **125**, 340–344 (2007).

99. M. De Saint-Hubert, M. Majer, H. Hrsak, Z. Heinrich, Z. Knezevic, S. Miljanic, P. Porwol, L. Stolarczyk, F. Vanhavere, R. M. Harrison, "Out-of-field doses in children treated for large arteriovenous malformations using hypofractionated gamma knife radiosurgery and intensity-modulated radiation therapy," *Radiation Protection Dosimetry* **181**, 100–110 (2018).

100. C. L. Tessa, T. Berger, R. Kaderka, D. Schardt, S. Burmeister, J. Labrenz, G. Reitz, M. Durante, "Characterization of the secondary neutron field produced during treatment of an anthropomorphic phantom with x-rays, protons and carbon ions," *Physics in Medicine and Biology* **59**, 2111–2125 (2014).

101. R. P. Hugtenburg, A. E. Baker, S. Green, "X-ray synchrotron microdosimetry: Experimental benchmark of a general-purpose Monte Carlo code," *Applied Radiation and Isotopes* **67**, 433–435 (2009).

CHAPTER 7

Optically Stimulated Luminescence Dosimeters in Clinical Practice

Stephen F. Kry

The University of Texas MD Anderson Cancer Center
Houston, Texas

Jennifer O'Daniel

Duke University School of Medicine
Durham, North Carolina

CONTENTS

7.1   DETECTOR AND READER THEORY

Similar to thermoluminescent dosimetry, optically stimulated luminescence dosimetry relies upon a crystalline insulator with added impurities. When these crystals are exposed to ionizing radiation, the resultant ionizations raise a certain number of electrons to a higher energy state (conduction band), leaving an equivalent number of vacancies (holes) in the original energy state (valence band). These electrons and holes will either recombine or become captured in energy traps created by the added crystalline defects. Several types of these traps exist in the crystal that have different depths. Charges in shallow energy traps typically escape spontaneously to recombine within a few minutes. Charges in deeper energy traps can remain caught for years. The number of ionizations and subsequent trapped electrons and/or holes is approximately proportional to the amount of energy incident on the detector. To extract signal from the optically stimulated luminescent dosimeter (OSLD), the electrons and holes in deep energy traps are released by application of light (although they can also be released by heat as with thermoluminescent dosimeters; TLDs). The released electron can combine with a trapped hole, or vice versa, releasing its energy as luminescence. In reality the process is more complex, and competing defects may cause changes in OSLD signals based upon factors such as the amount of radiation dose delivered and upon the detector's exposure history [1, 2].

An excellent overview of the long history of optically stimulated luminescence (OSL) is given by Yukihara and McKeever [3]. There are reports in the literature on OSL phenomena throughout the 19th century. However, the materials studied held the trapped charges at shallow depths, requiring only a small amount of energy to be released. Therefore their signals were unstable at room temperature, and unsuitable for radiation dosimetry. The development of Al_2O_3:C in 1990 provided a sensitive and stable luminescent dosimeter that reacted both to heat and to light in the visible spectrum [4]. As the first OSLD material with sufficient signal stability at room temperature, multiple academic research groups became involved in its development [5, 6]. The advantage of light (as opposed to heat) is that it may be applied in a very precise and therefore reproducible fashion, in terms of wavelength, intensity, and time. By using relatively short bursts of light exposure, only a small portion of the stored energy is released with each reading. Therefore, OSLDs may be read multiple times as compared to the single reading of TLDs via heat exposure. Additionally, using heat requires fairly substantial ancillary equipment (TLD reader, N_2 or Ar gas, annealing equipment) versus a simple desktop photo-multiplier tube setup. At the time of writing this chapter, the Al_2O_3:C form remains the single commercially available OSLD.

In practical OSLD, there are three types of traps: shallow traps that are unstable at room temperature (and make signal unstable for ~10 minutes after irradiation), medium traps that are released when exposed visible light (and are responsible for the dosimetric signal), and deep traps that are practically impossible to empty via light exposure (and remain filled permanently). The number and relative proportion of each kind of trap affects the sensitivity, fading, and reuse of OSLDs. The nature of the simulation light (narrow or broad range of wavelengths, intensity, and duration) also affect how quickly OSLD traps are emptied. A higher OSLD luminescence signal may be obtained by increasing the stimulation intensity or duration, but this will also increase the rate of signal loss. Most commercial readers utilize a 1 second stimulation as a good balance between signal strength and signal loss. Readers may also incorporate different light intensities to stimulate the detector depending on the dose level. For low doses, a higher intensity light is often used so that more signal is generated, but more signal will be lost for subsequent readings.

At the time of writing this chapter and to the best of our knowledge, there is only one commercially available OSLD for clinical use: nanoDot™ by Landauer Inc. (Glenwood, IL). As shown in Figure 7.1, the OSLD powder is shaped into a disk of approximately 4 mm diameter and contained

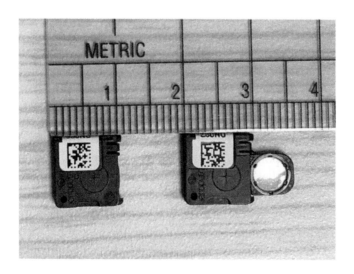

FIGURE 7.1   OSLD both within and without the plastic casing.

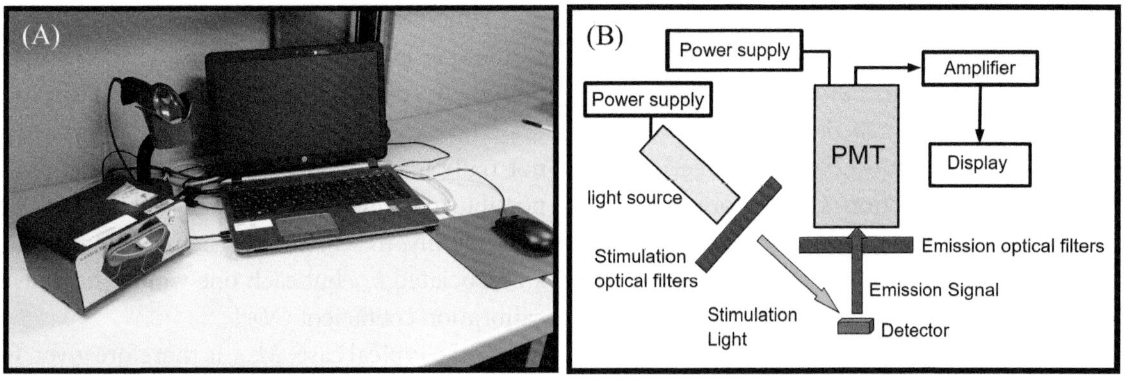

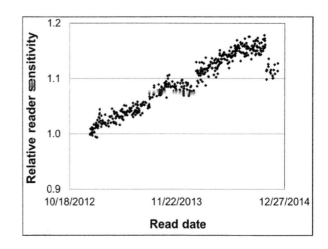

FIGURE 7.2 OSLD reader. (A) Picture of a reader, along with scanner and computer used for processing. (B) Schematic of main OSLD reader components.

within a $1 \times 1 \text{ cm}^2$ light-tight plastic casing. During read-out, the OSLD disk is extruded from the plastic casing.

A current commercial OSLD reader, along with ancillary equipment, is shown in Figure 7.2. The basic components of an OSLD reader include a light source for stimulation, a photomultiplier tube (PMT) for detection, and an amplifier to increase signal strength [3, 7–9]. The light source could be a laser, light-emitting diode (LED), or broadband bulb. Typically optical bandpass filters are used both in front of the light source, to select an appropriate stimulation wavelength as separate from the OSL signal, and in front of the PMT to allow only the OSL signal to pass. For an example, with an $Al_2O_3{:}C$ OSLD the emission light is in the blue spectrum (~420 nm), while the stimulation light is in the green spectrum (~525 nm) [10, 11]. The software associated with the reader will either give the user a simple signal readout in counts accumulated or current integrated per unit interval or will provide a way to calibrate the reader such that the output signal will be converted into a dose reading. The algorithms for such a conversion are discussed later in this chapter. The stability of the reader operation is an essential consideration for the precision of the readings. Routine quality assurance to verify the consistency (or correct for variations) of the stimulation light intensity, duration of stimulation, and light sensitivity of the PMT is essential. This may be done either through direct measurements, using a calibration OSLD to verify reader performance on a per-session basis, or by incorporating a session-specific reader performance correction factor.

OSLD readers are relatively stable over short periods of time. That is, over the course of days or weeks, the stimulation light source produces a consistent light intensity and the PMT converts signal to a consistent number of counts. However, over the period of months this cannot be assumed to be true as readers have been found to not be consistent in terms of how many counts are reported for readout of the same dose. An example of this critical issue is shown in Figure 7.3 for a commercial OSLD reader. Over a period of 2 years, the figure illustrates that the sensitivity of the reader changed by more than 20%. It is therefore important to monitor the performance of the reader on a regular, periodic basis.

7.2 DOSE CALCULATION

Dose (D) is measured with an OSLD based on the number of counts measured by the detector (M) and a calibration coefficient (N) that relates counts to dose:

$$D = M \cdot N \tag{7.1}$$

FIGURE 7.3 Daily assessment of commercial reader sensitivity (counts per dose) over a 2-year period.

Additional corrections are necessary to this simple relationship because the signal is not perfectly proportional to the dose deposited, but is also influenced by factors such as the amount of dose (linearity correction: k_L), time since irradiation (fading correction, k_F), energy (beam quality correction, k_Q), and orientation of the dosimeter to the beam (angular correction, k_θ). The practical equation to determine dose is therefore given by the following:

$$D_w = M_{corr} \cdot N_{D,w} \cdot k_L \cdot k_F \cdot k_Q \cdot k_\theta \qquad (7.2)$$

Each term is this equation is described below.

M_{corr} is the corrected number of counts from the detector. This is the raw number of counts (M_{raw}) corrected for depletion of the signal associated with multiple readouts (k_d) and any background signal (M_{bkg}). Depletion (k_d) is the correction applied to account for the signal lost per reading, and is determined during detector commissioning by reading a dosimeter (or small sample of dosimeters) numerous times and determining the average loss per reading. The signal lost per reading is typically 0.05% for the InLight reader operated in the high-dose mode (i.e., with low stimulation intensity), so k_d is typically 1.0005 [10, 12, 13]. Background counts (M_{bkg}) can be determined by keeping a dosimeter in the same environment as the *experimental* dosimeters, or by reading the *experimentals* right before irradiation. One additional factor is also typically incorporated into M_{corr}: the relative sensitivity of each individual dosimeter. Because different OSLD will have different amounts of actual active material and different trap distributions within the crystal, each OSLD will have a slightly different response. It is therefore common to determine the typical sensitivity of the system, then apply an additional correction factor for each OSLD relative to this central/baseline sensitivity. This is often referred to as using a batch calibration approach, and the relative sensitivity of a given OSLD compared to the baseline sensitivity of the system is given by $k_{s,i}$ (e.g., if the ith detector's sensitivity is lower than the baseline, $k_{s,i}$ corrects its number of counts upwards). How $k_{s,i}$ is actually implemented in practice can vary. Each dosimeter can be characterized and tracked with its own unique $k_{s,i}$ value used during readout; this provides maximum accuracy but requires additional effort (this is the model for a "screened nanoDot™," for example, where the vendor provides detectors with a known sensitivity within a

given uncertainty). Alternately, a group of detectors with similar sensitivities can be considered homogeneous – then the $k_{s,i}$ of each detector need not be tracked, but the uncertainty is larger because the detectors are of course not truly identical. Instead of a batch calibration, it is possible to determine and track each OSLD's sensitivity individually, in which case each OSLD would not have an associated $k_{s,i}$, but each one would then have its own calibration coefficient (N).

For the typical case, M_{corr} is therefore given for J readings of the ith detector as:

$$M_{corr} = k_{s,i} \left(\frac{\sum_j \left(M_{raw,j} \cdot k_d^{j-1} \right)}{J} - M_{bkg} \right) \qquad (7.3)$$

In Equation 7.3, $k_{s,i}$ can easily exceed 5% or more, and is therefore critical to account for. In contrast, depletion (k_d) for a 1 second read is typically a 0.05% correction (i.e., the detector loses 0.05% of the trapped signal during a 1 second read out). While the user should know and understand the depletion of their reader, this term can most often be neglected. Similarly, unless the doses being studied are very small or the background is high, background signal is typically negligible and can also be neglected. Therefore, Equation 7.3 reduces for most applications to:

$$M_{corr} = k_{s,i} \cdot \overline{M_{raw}} \qquad (7.4)$$

$N_{D,w}$ is the calibration coefficient that relates signal to dose. This is determined by irradiating OSLD to a known dose and determining the associated counts. These OSLD, used to determine the system calibration, are referred to as *standards*. The calibration coefficient is determined from the signal readout with the *standards* by simply rearranging Equation 7.1, where the subscript ($_0$) denotes that it is for the reference condition.

$$N_{D,w} = \frac{D_0}{M_{0,corr}} \qquad (7.5)$$

Equation 7.5 defines the calibration coefficient under a specific condition. As indicated by Equation 7.2, $N_{D,w}$ will depend on beam energy, dose level, and the other corrections that affect the proportionality of N and D. The user must therefore remember that $N_{D,w}$ is a function of these parameters and therefore the conditions with which the *standards* have been irradiated. $N_{D,w}$ can only be used if the *experimentals* (i.e., OSLD with unknown dose irradiated for an experiment) were irradiated under identical conditions, or if the differences are accounted for through the application of correction factors (i.e., as shown in Equation 7.2).

While $N_{D,w}$ can be determined for a specific unique condition (i.e., a specific dose, beam quality, etc.), it is common to determine $N_{D,w}$ at several dose levels to create a calibration curve. Such a curve explicitly incorporates dose level in the relationship between dose and counts. In this case, the product ($N_{D,w} \cdot k_L$) is determined by the *standards* instead of simply $N_{D,w}$. An example of three different calibration curves are shown in Figure 7.4, based on the OSL MicroStar reader (Landauer, Inc., Glenwood, IL, USA). The linear calibration curve is appropriate for low doses, <300 cGy. There are actually two linear calibration curves for this system, one for the low dose mode and one for the high dose mode. Higher levels of stimulation light are required for dose levels below 10 cGy ("Linear Low Dose" curve) while lower levels may be used for doses between 10 and 300 cGy ("Linear High Dose" curve). The MicroStar reader automatically switches to the appropriate level based on a short "pretest" exposure reading. For doses above 300 cGy, the relationship between dose and counts becomes nonlinear ("Non-Linear Dose" curve). The calibration curve (based on a second order polynomial) is shown in contrast to a linear curve (fit up to 300 cGy). Particularly for high doses the different in the calibration curves is apparent. Care is required as well

because the user must manually switch to the nonlinear calibration curve when appropriate.

Regardless of how $N_{D,w}$ is determined, it is essential to consider how often it needs to be evaluated. Because the components of the reader are not perfectly stable, if $N_{D,w}$ is determined at a given point in time it will not necessarily be appropriate at a later time. For maximum precision, $N_{D,w}$ can be determined for every reading session by irradiating *standards* for every reading session. Particularly in the case of a calibration curve, it may not be practical to directly establish the system response for every reading session; in such a case the user should at least verify the suitability of the calibration curve. This can be done by maintaining a "constancy dosimeter": a *standard* that is maintained for a long period of time and read out before any session (correcting for fading and depletion). The relative response of the system to the constancy dosimeter indicates the relative performance of the system compared to historical values, and the calibration coefficient or calibration curve can be scaled according to this.

The calibration coefficient is defined with *standards* for a specific dose level (unless a calibration curve is being used), at a specific time since irradiation, for a specific beam quality, and with a specific orientation of the detector to the radiation field. If the *experimentals* are also irradiated under these conditions then $N_{D,w}$ can be directly applied. Otherwise, the different irradiation conditions of the *experimentals* must be incorporated into the determination of dose. This is done through

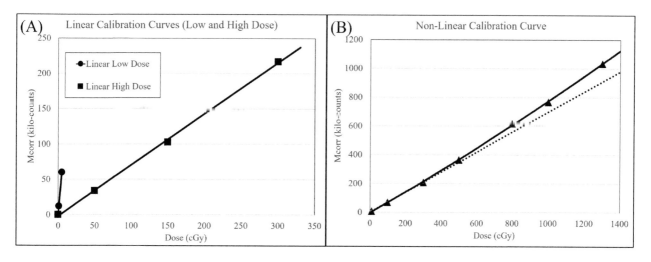

FIGURE 7.4 Calibration curves (a) Linear curves for the low dose (<10 cGy) and high dose (10–300 cGy) scales (b) Non-linear calibration curve for doses above 300 cGy. Dashed line depicts extrapolated calibration curve from the high dose linear region (10–300 cGy), while solid line depicts polynomial calibration curve.

correction factors as described below. These correction factors do not refer to any absolute reference, but simply to the reference defined by the *standards*. For example, application of the fading correction does not account for signal loss since irradiation, but rather accounts for signal loss that is different than that of the *standards*. If both the *standards* and *experimentals* are read out 10 days after irradiation, there is no fading correction necessary because there is no difference in fading between the *experimentals* and the calibration condition defined by the *standards*.

k_L is a correction for non-linearity in the response of the detector to dose, and is necessary because OSLD do not respond uniformly as a function of dose. Instead, as the dose level increases (starting below 1 Gy and increasing with increasing dose), the detector over-responds. By 10 Gy, the detector over-responds by ~15% as compared to its response at low doses [14]. As such, this is an essential correction to be incorporated into the dosimeter response in order to ensure reasonable accuracy. This correction factor can be determined during commissioning by irradiating a group of OSLD to different doses and determining the relationship between signal per dose and dose. Then, for a given *experimental*, the signal can be corrected by the over/under-response at that dose as compared to the dose used to define $N_{D,w}$ by the *standards*.

Alternatively, it is common to establish a calibration curve for OSLD detectors. In such a case, k_L is implicitly incorporated into the calibration coefficient (as a function of dose) of the system and need not be separately determined.

k_F is a correction for loss of detector signal over time (fading) and is necessary because OSLD spontaneously lose signal over time. During the first 10 minutes after irradiation this signal loss is dramatic (more than 40% of signal is lost) as the shallow traps empty spontaneously [10, 15]. It is not possible to achieve reasonable accuracy if reading out during this time period, therefore a minimum of 10 minutes must be waited between irradiation and readout of OSLD. After these shallow traps are emptied, the trapped signal in OSLD is highly stable, and fading is only ~1% per month. If a long-term record is being maintained (or, for example, a constancy dosimeter is being used, see above on $N_{D,w}$) then fading differences between *standards* and *experimentals* may need to be quantified and accounted for. For many applications the magnitude of fading is small and the correction can be neglected with an acceptable increase in the uncertainty budget. Fading can be determined by irradiating a group of dosimeters at different times and reading them out together, or by reading a sample of OSLD repeatedly over an extended period of time (and correcting for depletion).

k_Q is a correction for beam quality and is necessary because OSLDs are not tissue/water equivalent and respond differently to beams with different energies. Al_2O_3:C has different mass attenuation coefficients and stopping powers than tissue or water which affects the response of this dosimeter. Compared to the response at 6 MV, OSLDs respond 2% higher in cobalt, but there is typically less than 1% difference in response across different high-energy photon beams. In electron beams, the response is around 2% lower for all energies as compared to a 6 MV photon beam [9, 10, 15–18]. These values describe the relative response of the detector under reference conditions for different nominal energies. However, beam quality also changes with parameters such as field size, depth, presence of wedges, etc., as happens during measurements in non-reference locations. Within the treatment field, these factors can make up to a 3% difference to the response of the dosimeter. Because the beam quality correction is very difficult to determine for an arbitrary condition, measurements done under non-reference conditions are associated with increased dosimetric uncertainty.

While Al_2O_3:C shows relatively little variation with energy within treatment fields of different energies, as the photon beam energy lowers below the megavoltage range it shows a dramatic over-response. Outside the treatment field from a 6 MV beam, where there is an abundance of scattered radiation and a softer spectrum, OSLD over-respond by up to 30% (compared to an in-field calibration) [19]. In kV applications such as CT dose measurements, OSLD over-respond by as much as a factor of 3.5 compared to the response in an MV beam. This over-response in kV energies is not only sizeable, it also changes dramatically with small changes in beam quality. For applications in CT, for example, kVp, filtration, scan extent, size of the patient, and location of the measurement all impact the spectrum enough to induce measurable effects on the response of the detector [20–23]. This is a major challenge for OSLD application in low energy beams and requires careful attention and calibration.

k_θ is a correction for angular dependence and is necessary because OSLD do not respond the same to en-face irradiation as to edge-on. The disc-shape of the active volume results in the detector under-responding when irradiated edge-on as compared to en-face. A range of values have been reported for the magnitude of this effect in a 6 MV beam, ranging from almost 0% [10] up to 4% [24, 25]; robust and detailed measurements and Monte Carlo simulations found an angular dependence of 2% [26]. Practically, if the irradiations can be controlled such that the *standards* and *experimentals* are irradiated in the same orientation, there is no difference in response and no correction is necessary. In clinical situations where many beams impinge upon the detector it may be impossible to determine the average angle of incidence. In such a case, additional uncertainty must be included in the dose estimate.

7.3 PRECISION

There are three considerations relevant to the precision achievable with OSLD. These same considerations apply to TLD as well, and for the same approach/under the same considerations, very similar uncertainties exist for both OSLD and TLD. These considerations, and detailed uncertainty budgets, are presented in the AAPM TG-191 report [27].

First is the manner in which the OSLDs are calibrated and read. OSLDs can be handled in a range of manners; TG-191 defines a high precision method and a high efficiency method. The high precision approach includes determining $N_{D,w}$ for every reading session, determining $k_{s,i}$ for every detector, and characterizing and applying all correction factors to every reading. It is relatively time intensive (particularly during commissioning of the system when all of the correction factors are determined), but provides the highest precision. In contrast, the high efficiency approach uses a calibration curve to determine $N_{D,w} \cdot k_L$ which is only spot checked and corrected if necessary based on a constancy dosimeter. The relative sensitivity of each detector is provided by the manufacturer within a known window of uncertainty (e.g., "screened" nanoDot™). Fading and beam quality (across MV photons and electrons) are both neglected. This approach is much faster to implement, but has a higher associated uncertainty. At the 1-sigma level, the best reasonable achievable precision with the high precision approach is 1.6%; with the high efficiency approach it is 3.5% (TG-191).

The second consideration relevant to the uncertainty is the control of the irradiation conditions. Under optimal (controlled) conditions, the detectors are irradiated under reference conditions in the radiation field and have a consistent dosimeter orientation relative to the beam. This is the case for beam output audits performed by the Imaging and Radiation Oncology Core (IROC) [28, 29], the International Atomic Energy Agency (IAEA) [30], or the Australian Clinical Dosimetry Services (ACDS) [13]. Under these optimal conditions, the uncertainty may be as low as 1.6% or 3.5% (depending on if a high precision or high efficiency calibration technique has been implemented). Most of the time in clinical practice, measurements are made in "less controlled" conditions. If the dosimeter is used at an arbitrary location in an arbitrary radiation field then the actual beam quality at the measurement location is unknown and will introduce uncertainty. Similarly, the detector may no longer be irradiated under a controlled orientation relative to the beam, introducing angular dependence into the measurement that may be very hard to account for (particularly if multiple beams are incident from a range of different angles). It is also assumed that a less experienced operator is reading the dosimeter. Under such "less controlled" conditions, a high precision approach to OSL dosimetry would be expected to have an uncertainty around 3.9%, while a high efficiency approach to OSL dosimetry would be expected to have an uncertainty around 5.2% (at the 1-sigma level).

The third consideration is the clinical implementation of their use. Riegel et al. (2017) conducted a large clinical study where therapists placed the dosimeter on patients to measure the dose *in vivo* from IMRT treatments [31]. Combining these clinical factors into a high efficiency and less controlled environment led to a sizeable increase in the overall uncertainty. While the dosimeter had approximately a 5.2% uncertainty inherently, the added uncertainty from positioning the dosimeters on patients to generate more than 11,000 measurements resulted in an average (1-sigma) uncertainty of 10.3%.

Achieving good precision with OSLD is feasible and routinely done. However, there is a direct relationship between the effort committed, the control of the irradiation conditions, and the uncertainty in the final dose estimate.

7.4 REUSE

OSLD can be reused by removing trapped electrons and holes from the crystal lattice. This process is called bleaching and is comparable to annealing of TLD except it is based on light. Many different light sources can and have been used to bleach Al_2O_3:C detectors, including fluorescent room lights, LEDs, and tungsten halogen lamps [9, 10, 14, 15]. The time required to bleach the dosimeters depends on the light source, but 12 hours is generally ample time to reduce any residual signal to background levels. For any individual setup, the residual signal after bleaching can be readily checked to ensure it has indeed reached background levels.

As described earlier in this chapter, OSLD have traps of several different depths. Shallow traps spontaneously empty shortly after irradiation, dosimetric traps are highly stable at room temperature but are emptied during the readout process, and deep traps that are so stable that they are not emptied during readout. Similarly, bleaching effectively empties dosimetric traps but does not empty deep traps [32]. Therefore, after bleaching, the OSLD does not return to its original state; rather, the deep traps remain filled. This is important because these deep traps are competitors in the OSL process. The state of occupancy of deep traps affects the sensitivity of the detector by changing the relative trapping and recombination efficiency. In practice, the accumulated dose in OSLDs affects the detector's sensitivity as well as its supralinearity (k_L) [10–12, 33]. As an added complication, the changes in sensitivity are affected by bleaching time, bleaching light spectrum, and accumulated dose [14]. The complex and poorly understood results from the literature regarding changes in sensitivity and supralinearity support using these dosimeters to a maximum accumulated dose of ~10 Gy. Until an accumulated dose of 10 Gy is reached, these dosimeters can be easily bleached and reused without changes to their behavior.

Reusing OSLD has the advantage of repeated use of the detectors and therefore reduced cost. The biggest challenge of reusing OSLD is that, because there is a finite 10 Gy lifespan, the accumulated dose for each dosimeter must be tracked which may be logistically challenging.

7.5 PRACTICAL APPLICATIONS

7.5.1 External Photon Beams

In the setting of external beam radiotherapy, OSLDs may be used to independently verify beam calibration, in anthropomorphic phantoms either on the surface or at depth to verify dose calculations, or on patients to measure *in vivo* doses.

Phantom or *in vivo* measurements are typically performed by placing the OSLD on the surface. The effective point of measurement of a nanoDot™ OSLDs is 0.8 mm below the surface of the plastic case [34]. This places the OSLD in a high dose gradient region where the uncertainty in the dose is high (and if the treatment planning system calculation is being assessed, the accuracy of the treatment planning system is also limited). Therefore, to improve dosimetric accuracy and agreement with the treatment planning system calculation, bolus or buildup caps can be placed over the OSLD to increase its depth of measurement beyond 5 mm. When using bolus, it is important that its lateral extent be at least 5×5 cm^2 to generate lateral charged particle equilibrium [31].

7.5.1.1 Surface Dosimetry

The OSLD may also be used to verify surface/skin dose from photon beams. These measurements are inherently less accurate. The OSLD measurement depth (0.8 mm) will overestimate dose to superficial skin layers (0.07 mm, ICRP 26, 1977 [35]). Users should also expect worse agreement with their treatment planning system calculations, given the difficulty in accurately modeling the high dose gradient region near the surface.

7.5.1.2 Small Fields

The diameter of the nanoDot™ disc is approximately 4 mm, making it suitable for measurements of fields down to 2 cm × 2 cm. For smaller fields, use of nanoDot™ OSLD is not recommended both due to volume-averaging effects due to its size as well as air gaps present within the dosimeter [36]. When performing measurements of small fields, be certain to align the off-centered active area of the nanoDot™ OSLD with the field's central axis rather than centering the dosimeter's casing.

7.5.1.3 Out of Field

Occasionally dose measurements will be needed outside the main field, such as to determine dose to a pacemaker. These measurements are complicated by a few issues. First, nanoDot™ OSLDs that have been calibrated in-field will over-respond by 10–25% due to the lower energy spectrum outside the field [19, 37]. This increases the uncertainty of the dose measurement, although this can be offset by estimating appropriate correction factors from the literature [19]. Second, outside the treatment field the dose is elevated at the surface and decreases down to d_{max}. Unless a surface dose measurement is desired, a bolus of $\sim d_{max}$ thickness should be placed over the dosimeter to avoid measuring this elevated surface dose [38–40]. Third, if the OSLD is out of the primary field but within the imaging field, the OSLD may over-respond to the imaging dose (see Section 7.5.5). This could result in a substantial overestimation of out of field dose. Care should be employed in this scenario, for example by capturing the imaging dose (e.g., through independent measurement or estimation) separately from the therapeutic dose (e.g., by placing another OSLD on the patient after the imaging is performed but before the treatment is delivered).

7.5.2 Electrons

The use of OSLD with electrons is similar to that with photons. The energy dependence varies by less than 2% due to small variations in collisional stopping power from 1 MeV to 20 MeV [10, 15]. In a high efficiency setting, it is acceptable to rely on reader calibrations from a different modality or energy, but using a calibration specific to the measurement's modality and energy improves the measurement accuracy. Note that combining modalities (photon and electron) will increase the measurement uncertainty [41]. Surface dose measurements of electron beams have better agreement with treatment planning dose calculations than in photon beams due to the lower dose gradient. In particular total skin electron therapy has a very small build up region due to additional scatter, and so OSLD measurements without bolus are appropriate. The angular dependence of OSLD with low energy electrons has not been evaluated, which introduces additional uncertainty for oblique measurements.

7.5.3 Protons

Unlike with electrons, the behavior of OSLD changes with protons. First, the greater linear energy transfer (LET) of these particles leads to a higher ionization density along the particle's track. This creates a saturation response in the OSLD resulting in decreased signal per unit dose [31, 42]. This effect is approximately 4% for proton beams relative to photon beams, but is approximately constant for different proton energies. This effect is expected to be greater for carbon beams [15] and may be dependent on the carbon energy and presence of secondary particles. For any charged particle, the distal

fall-off region of the Bragg peak has substantial measurement uncertainty due to the dramatic LET variation, high dose gradient, and the fact that the particle range becomes less than the detector thickness [43]. OSLDs also have less non-linearity with increasing doses when using protons as compared to x-rays, such that the same linearity correction or calibration curve established for photon or electron beams cannot be used. Similar to electrons, surface dose measurements are reasonably accurate due to shallower dose gradients than that of photon beams. Small field measurements have been shown to be accurate down to 5 cm × 5 cm. At the time of writing, only one report for a 2 cm × 2 cm field size was available, which showed the need for a partial irradiation correction of 2% [44]. Users should employ caution when measuring proton fields smaller than 5 cm × 5 cm, and should not use OSLDs below 2 cm × 2 cm field sizes. The angular dependence for protons has been shown to be small ($\leq 0.5\%$) within the spread-out-Bragg-peak [24].

7.5.4 Brachytherapy

Given the small size of OSLDs, they are a good brachytherapy dosimetry option. However, there are several issues to be dealt with in order to optimize accuracy. First, positioning accuracy is crucial, as it is for all dosimeters in a brachytherapy setting, due to the sharp dose gradients (TG-138 [45]). Second, correction for the energy dependence at the lower brachytherapy energies is more complex than with external beam therapy. The calibration could be done either at the brachytherapy source energy, with increased uncertainty in absolute dose, or at an MV energy, with increased uncertainty in k_Q. Both approaches are feasible [46, 47] (Casey et al. 2013). It is important to note that k_Q can vary substantially with source-to-detector distance due to beam hardening. Sharma and Jursinic [46] demonstrated a change of ~10% moving from 2 cm to 10 cm of solid water with ^{192}Ir. The k_Q will also vary based on the source encapsulation. Casey et al. (2013) showed differences of 2.6% between Nucletron and Varian ^{192}Ir sources. Third, the angular dependence has been shown to be ~4% with ^{192}Ir [46]. Finally caution should be used when measuring low energy betas because they may be partially absorbed by the dosimeter (depending on the OSLD thickness and beta energy), resulting in an underestimation of dose. For additional discussion, the reader is referred to the 2009 AAPM summer school [48].

7.5.5 Imaging

In addition to therapeutic applications, OSLDs may also be used to measure dose from imaging procedures. Some recent applications include diagnostic CT [49], cone-beam CT [50, 51], and mammography [52]. With appropriate care taken to characterize the OSLD's response in a kV environment, accurate low dose measurements may be performed. As discussed in the brachytherapy section, the OSLD may be calibrated for low energy kV either via an MV source (with associated large k_Q) or at the low energy itself (with associated uncertainties in absolute dose) [53]. Determining the appropriate energy spectrum for the associated imaging technique is also challenging, and introduces additional sources of uncertainty. Variations in kV from 80 kVp to 140 kVp lead to changes in k_Q of ±10–15% [20]. The presence or absence of a bow-tie filter, amount of material in the beam path, and scan extent also affect k_Q by 4–10% [20, 50, 52, 54]. Due to the large variation in response between kV and MV energies, a single OSLD cannot measure a cumulative imaging dose plus therapeutic dose such as in the setting of image-guided radiation therapy. Separate dosimeters with separate correction factors would be required to appropriately measure each component. In a clinical setting with the high efficiency approach, the relatively small imaging contribution to an in-field therapeutic OSLD measurement may be ignored. If improved accuracy is desired, using two sets of OSLD (one in place only during imaging, the other only during treatment) is necessary.

Angular dependency is also a greater issue with the OSLD under-responding by 30–40% in radiography and 60% in mammography [52] for an edge-on versus en-face irradiation. Interestingly, the angular dependency with the rotating source in a CT scanner is a mere ~5% in comparison [20, 54]. In mammography, self-attenuation of the beam becomes a concern due to the very low energy [55, 56]. Using an en-face position minimizes the attenuation as much as possible. Finally, if a low dose (≤ 5 mGy) is expected, subtracting the background signal will improve accuracy [20].

7.5.6 Quality Assurance

Quality assurance recommendations are discussed in detail in TG-191 [27]. In brief, the physicist needs to verify the reader's functionality (photodetector with dark current and consistent source reading, readout precision, plus an end-to-end test with a range of known standard doses) as well as the dosimeter's functionality (calibration curve or $N_{DW} + k_L$ determination, determine magnitude of k_P, k_d, and k_Q and decide whether/how to correct for them). A simple constancy check based on an OSLD of a known dose should be done for every readout session. This OSLD may be reused for many sessions if the physicist accounts for depletion and fading. Additional QA should be performed on an annual basis and compared to commissioning to determine if there is any drift in the reader functionality.

7.6 SUMMARY

Similar to TLD, OSLD may be used to verify dose both for equipment quality assurance measurements and for *in vivo* dosimetry. Their accuracy and precision is equivalent to that of TLD. OSLD have a clear advantage in terms of having a faster and non-destructive readout process. If the dosimeter will be reused, bleaching of OSLD is also a simpler process than annealing of TLD. While light sensitivity is a conceptual disadvantage of OSLD, the light-tight casing of OSLD ensures this is not a practical problem. Reuse of OSLD is more complicated than for TLD because of the finite lifespan of OSLD (10 Gy) before properties of the detector change. Managing this requires tracking the dose history of each detector.

Ultimately, OSLD dosimeters provide point dose information and are broadly applicable to many clinical situations. The accuracy and precision of OSLD is affected by the calibration and handling practices of these detectors, with a direct trade-off between the accuracy and precision achieved and the effort involved in characterizing the detectors, applying appropriate correction factors, and performing appropriate quality assurance testing. However, it is generally straightforward to achieve good accuracy and precision with these detectors.

REFERENCES

1. R. Chen and S. W. S. McKeever, *Theory of Thermoluminescence and Related Phenomena* (World Scientific, Singapore; River Edge, NJ, 1997).
2. R. Chen and V. Pagonis, *Thermally and Optically Stimulated Luminescence: A Simulation Approach* (Wiley, The Atrium, Southern Gate, Chichester, West Sussex, UK; Hoboken, NJ, 2011).
3. E. G. Yukihara and S. W. S. McKeever, *Optically Stimulated Luminescence: Fundamentals and Applications* (Wiley, Chichester, West Sussex, 2011).

4. M. S. Akselrod and V. S. Kortov, "Thermoluminescent and exoemission properties of new high-sensitivity TLD α-Al$_2$O$_3$:C crystals," *Radiation Protection Dosimetry* **33**, 123–126 (1990).

5. B. G. Markey, L. E. Colyott, S. W. S. Mckeever, "Time-resolved optically stimulated luminescence from α-Al$_2$O$_3$:C," *Radiation Measurements* **24**, 457–463 (1995).

6. S. D. Miller, Composite material dosimeters. U.S. Patent, 5,567,948, issued on 22 October 1996 (1996).

7. L. Bøtter-Jensen, S. W. S. McKeever, et al., *Optically Stimulated Luminescence Dosimetry* (Elsevier, Amsterdam; Boston; London, 2003).

8. S. W. McKeever and M. Moscovitch, "On the advantages and disadvantages of optically stimulated luminescence dosimetry and thermoluminescence dosimetry," *Radiation Protection Dosimetry* **104**(3), 263–270 (2003).

9. E. G. Yukihara, G. Mardirossian, et al., "Evaluation of Al$_2$O$_3$:C optically stimulated luminescence (OSL) dosimeters for passive dosimetry of high-energy photon and electron beams in radiotherapy," *Med Phys* **35**(1), 260–269 (2008).

10. P. A. Jursinic, "Characterization of optically stimulated luminescent dosimeters, OSLDs, for clinical dosimetric measurements," *Medical Physics* **34**(12), 4594–4604 (2007).

11. E. G. Yukihara and S. W. McKeever, "Optically stimulated luminescence (OSL) dosimetry in medicine," *Physics in Medicine and Biology* **53**(20), R351–379 (2008).

12. I. Mrcela, T. Bokulic, J. Izewska, M. Budanec, A. Frobe, Z. Kusic, "Optically stimulated luminescence in vivo dosimetry for radiotherapy: Physical characterization and clinical measurements in (60)Co beams," *Physics in Medicine and Biology* **56**, 6065–6082 (2011).

13. L. Dunn, J. Lye, J. Kenny, J. Lehmann, I. Williams, T. Kron, "Commissioning of optically stimulated luminescence dosimeters for use in radiotherapy," *Radiation Measurements* **51–52**, 31–39 (2013).

14. A. A. Omotayo, J. E. Cygler, G. O. Sawakuchi, "The effect of different bleaching wavelengths on the sensitivity of Al(2)O(3):C optically stimulated luminescence detectors (OSLDs) exposed to 6 MV photon beams," *Medical Physics* **39**, 5457–5468 (2012).

15. C. S. Reft, "The energy dependence and dose response of a commercial optically stimulated luminescent detector for kilovoltage photon, megavoltage photon, and electron, proton, and carbon beams," *Medical Physics* **36**(5), 1690–1699 (2009). [Erratum: *Medical Physics* **39**(9), 5788 (2012)].

16. M. C. Aznar, C. E. Andersen, L. Botter-Jensen, S. A. Back, S. Mattsson, F. Kjaer-Kristoffersen, J. Medin, "Real-time optical-fibre luminescence dosimetry for radiotherapy: Physical characteristics and applications in photon beams," *Physics in Medicine and Biology* **49**, 1655–1669 (2004).

17. V. Schembri and B. J. M. Heijmen, "Optically stimulated luminescence (OSL) of carbon-doped aluminum oxide (Al$_2$O$_3$:C) for film dosimetry in radiotherapy," *Medical Physics* **34**, 2113–2118 (2007).

18. A. Viamonte, L. A. da Rosa, L. A. Buckley, A. Cherpak, J. E. Cygler, "Radiotherapy dosimetry using a commercial OSL system," *Medical Physics* **35**, 1261–1266 (2008).

19. S. B. Scarboro, D. S. Followill, et al., "Energy response of optically stimulated luminescent dosimeters for non-reference measurement locations in a 6 MV photon beam," *Physics in Medicine and Biology* **57**(9), 2505–2515 (2012).

20. S. B. Scarboro, D. Cody, et al., "Characterization of the nanoDot OSLD dosimeter in CT," *Medical Physics* **42**(4), 1797–1807 (2015).

21. P. Mobit, E. Agyingi, G. Sandison, "Comparison of the energy-response factor of LiF and Al$_2$O$_3$ in radiotherapy beams," *Radiation Protection Dosimetry* **119**, 497–499 (2006).

22. C. Ruan, E. G. Yukihara, W. J. Clouse, P. B. Gasparian, S. Ahmad, "Determination of multislice computed tomography dose index (CTDI) using optically stimulated luminescence technology," *Medical Physics* **37**, 3560–3568 (2010).

23. M. C. Aznar, J. Medin, B. Hemdal, A. Thilander Klang, L. Botter-Jensen, S. Mattsson, "A Monte Carlo study of the energy dependence of Al$_2$O$_3$:C crystals for real-time in vivo dosimetry in mammography," *Radiation Protection Dosimetry* **114**, 444–449 (2005).

24. J. R. Kerns, S. F. Kry, et al., "Angular dependence of the nanoDot OSL dosimeter," *Medical Physics* **38**(7), 3955–3962 (2011).

25. D. W. Kim, W. K. Chung, D. O. Shin, M. Yoon, U. J. Hwang, J. E. Rah, H. Jeong, S. Y. Lee, D. Shin, S. B. Lee, S. Y. Park, "Dose response of commercially available optically stimulated luminescent detector, Al$_2$O$_3$:C for megavoltage photons and electrons," *Radiation Protection Dosimetry* **149**, 101–108 (2012).

26. J. Lehmann, L. Dunn, et al., "Angular dependence of the response of the nanoDot OSLD system for measurements at depth in clinical megavoltage beams," *Medical Physics* **41**(6), 061712 (2014).

27. S. F. Kry, P. Alvarez, J. E. Cygler, L. A. DeWerd, R. M. Howell, S. Meeks, J. O'Daniel, C. Reft, G. Sawakuchi, E. G. Yukihara, D. Mihailidis, "AAPM TG 191: Clinical use of luminescent dosimeters: TLDs and OSLDs," *Medical Physics* **47**(2), e19–e51 (2020).

28. T. H. Kirby, W. F. Hanson, D. A. Johnston, "Uncertainty analysis of absorbed dose calculations from thermoluminescence dosimeters," *Medical Physics* **19**, 1427–1433 (1992).

29. P. Alvarez, S. F. Kry, F. Stingo, D. Followill, "TLD and OSLD dosimetry systems for remote audits of radiotherapy external beam calibration," *Radiation Measurements* **106**, 412–415 (2017).

30. J. Izewska, P. Andreo, S. Vatnitsky, K. R. Shortt, "The IAEA/WHO TLD postal dose quality audits for radiotherapy: A perspective of dosimetry practices at hospitals in developing countries," *Radiotherapy & Oncology* **69**, 91–97 (2003).

31. A. C. Riegel, Y. Chen, A. Kapur, L. Apicello, A. Kuruvilla, A. J. Rea, A. Jamshidi, L. Potters, "In vivo dosimetry

with optically stimulated luminescent dosimeters for conformal and intensity-modulated radiation therapy: A 2-year multicenter cohort study," *Practical Radiation Oncology* 7, E135–E144 (2017).

32. E. G. Yukihara, R. Gaza, et al., "Optically stimulated luminescence and thermoluminescence efficiencies for high-energy heavy charged particle irradiation in Al_2O_3:C," *Radiation Measurements* 38(1), 59–70 (2004).

33. P. A. Jursinic, "Changes in optically stimulated luminescent dosimeter (OSLD) dosimetric characteristics with accumulated dose," *Medical Physics* 37, 132–140 (2010)

34. A. H. Zhuang and A. J. Olch, "Validation of OSLD and a treatment planning system for surface dose determination in IMRT treatments," *Medical Physics* 41(8), 081720 (2014).

35. International Commission on Radiological Protection (ICRP). *Recommendations of the International Commission on Radiological Protection. ICRP Publication 26* (Pergamon Press, Oxford, UK, 1977).

36. P. H. Charles, S. B. Crowe, T. Kairn, J. Kenny, J. Lehmann, J. Lye, L. Dunn, B. Hill, R. T. Knifght, C. M. Langton, J. V. Trapp, "The effect of very small air gaps on small field dosimetry," *Physics in Medicine and Biology* 57, 6947–6960 (2012).

37. C. R. Edwards and P. J. Mountford, "Near surface photon energy spectra outside a 6 MV field edge," *Physics in Medicine and Biology* 49(18), N293–N301 (2004).

38. G. Starkschall, F. J. Stgeorge, et al., "Surface dose for megavoltage photon beams outside the treatment field," *Medical Physics* 10(6), 906–910 (1983).

39. S. F. Kry, U. Titt, et al., "A Monte Carlo model for calculating out-of-field dose from a Varian 6 MV beam," *Medical Physics* 33(11), 4405–4413 (2006).

40. S. F. Kry, U. Titt, et al., "A Monte Carlo model for out-of-field dose calculation from high-energy photon therapy." *Medical Physics* 34(9), 3489–3499 (2007).

41. M. J. Lawless, S. Junell, C. Hammer, L. A. DeWerd, "Response of TLD-100 in mixed fields of photons and electrons," *Medical Physics* 40 102103 (2013).

42. G. O. Sawakuchi and E. G. Yukihara, "Analytical modeling of relative luminescence efficiency of Al_2O_3:C optically stimulated luminescence detectors exposed to high-energy heavy charged particles," *Physics in Medicine and Biology* 57(2), 437–454 (2012).

43. J. M. Edmund, C. E. Andersen, et al., "A track structure model of optically stimulated luminescence from Al_2O_3:C irradiated with 10–60 MeV protons," *Nuclear Instruments & Methods in Physics Research Section B-Beam Interactions with Materials and Atoms* 262(2), 261–275 (2007).

44. J. R. Kerns, S. F. Kry, et al., "Characteristics of optically stimulated luminescence dosimeters in the spread-out Bragg peak region of clinical proton beams," *Medical Physics* 39(4), 1854–1863 (2012).

45. L. A. DeWerd, G. S. Ibbott, et al., "A dosimetric uncertainty analysis for photon-emitting brachytherapy

sources: Report of AAPM Task Group No. 138 and GEC-ESTRO," *Medical Physics* 38(2), 782–801 (2011).

46. R. Sharma and P. A. Jursinic, "In vivo measurements for high dose rate brachytherapy with optically stimulated luminescent dosimeters," *Medical Physics* 40(7), 071730 (2013).

47. K. E. Casey, P. Alvarez, S. F. Kry, R. M. Howell, A. Lawyer, D. Followill, "Development and implementation of a remote audit tool for high dose rate (HDR) Ir-192 brachytherapy using optically stimulated luminescence dosimetry," *Medical Physics* 40, 112102 (2013).

48. J. E. Cygler and E. G. Yukihara (eds.), Optically stimulated luminescence (OSL) dosimetry in radiotherapy. *Clinical Dosimetry Measurements in Radiotherapy (AAPM 2009 Summer School)* (Medical Physics Publishing, 2009).

49. C. C. Brunner, S. H. Stern, et al., "CT head-scan dosimetry in an anthropomorphic phantom and associated measurement of ACR accreditation-phantom imaging metrics under clinically representative scan conditions," *Medical Physics* 40(8), 081917 (2013).

50. G. X. Ding and A. W. Malcolm, "An optically stimulated luminescence dosimeter for measuring patient exposure from imaging guidance procedures," *Physics in Medicine and Biology* 58(17), 5885–5897 (2013).

51. T. Giaddui, Y. Cui, et al., "Comparative dose evaluations between XVI and OBI cone beam CT systems using Gafchromic XRQA2 film and nanoDot optical stimulated luminescence dosimeters," *Medical Physics* 40(6), 062102 (2013).

52. R. M. Al-Senan and M. R. Hatab, "Characteristics of an OSLD in the diagnostic energy range," *Medical Physics* 38(7), 4396–4405 (2011).

53. S. Scarboro, D. Cody, F. Stingo, P. Alvarez, D. Followill, L. Court, D. Zhang, M. McNitt-Gray, S. Kry, "Calibration strategies for the use of the nanoDot OSLD in CT applications," *The Journal of Applied Clinical Medical Physics* 20(1), 331–339 (2019).

54. L. Lavoie, M. Ghita, et al., "Characterization of a commercially-available, optically-stimulated luminescent dosimetry system for use in computed tomography," *Health Physics* 101(3), 299–310 (2011).

55. A. Konnai, N. Nariyama, et al., "Energy response of LiF and Mg_2SiO_4 TLDs to 10–150 keV monoenergetic photons," *Radiation Protection Dosimetry* 115(1–4), 334–336 (2005).

56. A. A. Nunn, S. D. Davis, et al., "LiF:Mg,Ti TLD response as a function of photon energy for moderately filtered x-ray spectra in the range of 20–250 kVp relative to ^{60}Co," *Medical Physics* 35(5), 1859–1869 (2008).

57. M. S. Akselrod and S. W. S. McKeever, "A radiation dosimetry method using pulsed optically stimulated luminescence," *Radiation Protection Dosimetry* 81(3), 167–175 (1999).

CHAPTER 8

EPID-Based Dosimetry

Brayden Schott, Thomas Dvergsten, Raman Caleb, and Baozhou Sun

Washington University in St. Louis
St. Louis, Missouri

CONTENTS

8.1 INTRODUCTION

Electronic portal imaging devices (EPIDs) have been developed for verification of patient positioning prior to or during the treatment delivery. Since the introduction of the amorphous silicon (aSi)-based EPIDs, almost every linac is equipped with an EPID panel. Due to its favorable dosimetric characteristics such as high resolution, fast imaging acquisition, and digital format, EPIDs have been used for radiation therapy dose measurements for more than two decades. It has broad dosimetry applications for both patient-specific intensity-modulated radiation therapy/volumetric-modulated arc therapy (IMRT/VMAT) quality assurance (QA) including pre-treatment and *in vivo* patient treatment verification and linac machine QA. It can be used to verify almost every aspect of machine performance. In this chapter we will provide a review of EPIDs dosimetric characteristics and application to both patient-specific and linac QA.

8.2 TECHNOLOGY AND CHARACTERISTICS OF EPID

Beginning in the 1980s, several types of EPID panels emerged as alternatives to radiochromic film for mega-voltage imaging applications. In this chapter we will focus on modern aSi-based EPID technology, which is based on thin-film semiconductor technology. The currently available commercial EPID panel consists of: (1) a thin (typically 1 mm for Varian and Elekta EPID panels) copper plate which acts as a buildup material and absorbs scattered low energy radiation; (2) a scintillating phosphor screen made of Terbium-activated Gadolinium Oxysulfide (Gd_2O_2S:Tb) to convert the

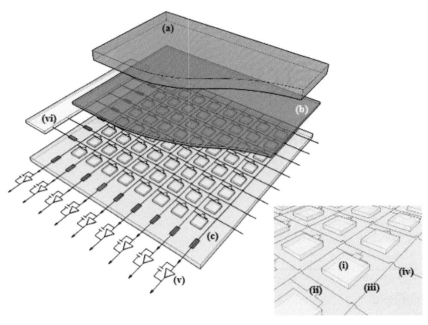

(a) Copper plate
(b) Scintillating phosphor
(c) Active matrix array
    i. a-Si photodiode
    ii. a-Si TFT (switch)
    iii. Data line
    iv. Gate control line
(d) External electronics
    v. Preamplifiers
    vi. Control gate circuitry

FIGURE 8.1 Schematic illustration of a flat panel imaging array. The external electronics controls the conductivity of the pixel FETs and pre-amplify the signal.

incident radiation into optical photons, (3) an array of photodiodes and field-effect transistors (FETs) which detect radiation both directly and indirectly; and (4) read out electronics to translate the charge readings into an image. These parts of the EPID are enclosed in a protective cover. The buildup at the depth of the active imaging layer of the EPID is equivalent to 8 mm of water [1]. Figure 8.1 shows the main components of Varian aS1000 EPID panel.

The image information acquired by the aS1000 is sent as an analog signal to the digitization unit where it is transformed into a digital signal [2]. The newest EPID panel, the aS1200, is an improved version of aS1000 and incorporates a backscatter shield beneath the active area of the EPID [3]. The shielding reduces the amount of backscatter contamination from the support arm and provides a uniform backscatter component to all images. In addition, the readout electronics have been redesigned and integrated to minimize the number of components. The thickness of the aSi-photodiode is reduced by 50%, which reduces image lag giving rise to improved signal linearity.

There are two acquisition modes of interest for dosimetric applications: cine mode and integrated mode. In cine mode, all images are recorded in a fixed time interval and formed in a series of movie images. In contrast, integrated mode creates a sum of all frames from

an acquisition and saves this sum as a single image. The raw EPID image can be processed by using a dark field and a flood field calibration:

$$I = \frac{I_{raw} - I_{dark}}{I_{flood} - I_{dark}} \qquad (8.1)$$

where I_{raw}, I_{dark}, and I_{flood} are the raw, dark field, and flood field images, respectively. Both the dark and flood field are acquired during EPID calibration. The dark field is acquired by imaging without any linac output, giving the electronic noise level of the detector. The flood field is the image of an open field covering as much of the EPID area as possible without irradiating the readout electronics.

Images given from the EPID is stored in DICOM format, with 14, or 16 bit pixel values, where lower values correspond to higher intensities. These images have been corrected with dark field and flood field. To get dosimetric values, all pixel values are subtracted by the maximum pixel value, and then multiplied by a calibration factor and the number of frames in the acquisition, to give so-called calibrated units (CU). The calibration factor is found through the definition of the CU; a 10×10 cm² open field with 100 MU at source-to-detector distance (SDD) of 100 cm should give a response of 1 CU on the central axis (CAX).

A Si-EPID demonstrates excellent dosimetric characteristics including high resolution, fast readout, linearity, reproducibility, high spatial resolution, and no dead time. Varian's aS1000 has the ability to acquire 10 frames/second. The aS1200 can read with the maximum frame rate up to 25 frames per second. The current aSi-EPID panels provide a high spatial resolution compared with most other available dosimeters, with the pixel size smaller than (0.4×0.4 mm^2). The most recent aS1200 utilizes an active matrix of 43×43 cm^2 consisting of 1280×1280 pixels, each one with size of 0.34×0.34 mm^2.

Linearity of EPID response to dose and dose rate have been well studied. EPID demonstrates an excellent linearity of integrated dose and dose rate. Figure 8.2 shows the linearity of EPID with dose rate and integrated dose [4]. The new aS1200 detector showed a significant dosimetric improvement when compared with the previous aS1000 [5]. As shown in Figure 8.3, the aSi-1200 shows an excellent linearity up to 2400 MU/min for 10 MV flattening filter free (FFF) beam on Varian linac [6].

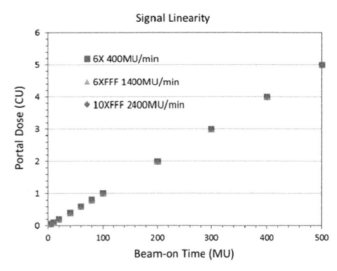

FIGURE 8.3 Central portal doses as a function of dose rate for 100 MU and 20×20 cm^2 field size [6].

The aSi flat-panel EPIDs have been shown to have excellent image stability over time. The long-term reproducibility of the Varian EPID system averaged over all pixels has been shown to be <0.6% over a 3-year period for three EPIDs [7]. Louwe et al. found variations of less than 1% (1 SD) for four aSi EPIDs (Elekta iView-GT) over periods of up to 23 months; this was improved by the application of a dynamic dark field, i.e., a remeasurement of the dark field just prior to image acquisition. With the dynamic dark field correction, the reproducibility was improved to within 0.5% (1 SD) [8].

8.3 USE OF EPID FOR LINAC QA

While the original purpose of EPID use was orientated toward patient QA, the device's utility has recently been expanded to include machine QA. As noted above, an EPID is an integrated imaging system that acquires MV images with high mechanical and dosimetric precision. The seamless integration of these imaging systems not only greatly reduces the setup and processing time that is required for routine film-based or phantom-based QA tests, but also reduces the overall cost of QA practices by minimizing the need for third-party dosimetric instrumentation. Additionally, EPID-based QA grants physicists the freedom to analyze their acquired images by choosing from a variety of well-established commercial analysis software, or by making use of verified in-house software tools. What follows is first a brief overview of common QA

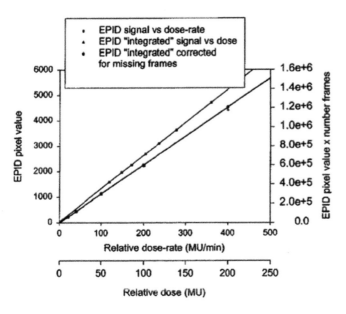

FIGURE 8.2 Linearity of the EPID with dose rate and integrated dose. The dose rate to the EPID was varied by varying the source to detector distance. Ion-chamber readings were recorded at each EPID position in a mini-phantom to determine the relative dose rate. The relative dose rate is expressed in MU/min assuming a nominal output of 400 MU/min at 100 cm from the source. A measurement of integrated dose was obtained by multiplication of the image pixel values by the number of frames averaged, with and without correction for dead time in image acquisition. These were also compared to ion-chamber readings [4].

practices in place to ensure device integrity, followed by a discussion on EPID-based machine QA procedures common in clinical use.

8.3.1 Imager QA

In order for an EPID to be a dependable tool for machine QA, the functional integrity and stability of the imager itself must first be ensured. Due to the numerous integrated components that make up the EPID system (arm, panel, etc.) and the gravitational torque acting on the device, the device and the gantry are susceptible to sag effects, leading to geometric uncertainties. Not only is it critical to minimize sag for the sake of image-guided radiation therapy (IGRT), but it is also necessary for the sake of using the EPID as a machine QA tool. Several AAPM Task Groups (TGs) including AAPM TG-142 [9] and TG-179 [10] provide guidelines for imager QA tests that cover geometric accuracy, image quality, and dosimetric response for MV and kV planar imaging, as well as cone-beam computed tomography (CBCT) imaging. For the remainder of this section, we will primarily focus on MV kV isocenter coincidence.

To perform MV and kV isocenter analysis, physicists usually make use of a phantom with some configuration of BBs embedded within it. A set of 2D kV images at four cardinal gantry angles can be acquired with the EPID and kV imager located at a fixed source-to-imager distance (SID). A set of 2D MV images can be acquired at the same gantry angles as the 2D kV images. The deviation between the appropriate markers and electronic crosshairs (graticule) on each image represents the deviation between the imaging and treatment coordinates.

In addition to simultaneously using MV and kV imagers to determine imager misalignments, some groups have proposed the ability to use a single imager to determine its geometric integrity. For instance, Mao et al. determined it is possible to find the MV isocenter by first irradiating a phantom with embedded BBs at a nonzero gantry angle. One BB is embedded in the center of the phantom while the rest are dispersed strategically along the phantom surface. The projections of the surface BBs are observable in the acquired portal image. The location of ideal BB projections (during full structural integrity) can be calculated. The real location of the BB projections are extracted from the portal image and optimized to determine the functional parameters of the treatment machine, which include the desired MV isocenter result [11].

8.3.2 Machine QA

With the advent of EPID-based machine QA, various AAPM Task Groups have been published to prescribe protocol and tolerance suggestions that encompass both dosimetric and mechanical checks. AAPM TG-142 [9] is the most comprehensive and will be the focus of what follows. In terms of mechanical checks, standard protocols often include multileaf collimator (MLC) positioning, jaw positioning, isocentricity checks, and variable axis angles. In terms of dosimetric checks, output constancy, output factor, profiles, and wedge factor are often taken into account.

8.3.3 MLC Positioning

The leaves of an MLC precisely shape the beam during both static and dynamic fields. Following the introduction of MLC in the mid-90s, several publications offered suggestions to address MLC leaf QA. Most of the suggestions were normalized and summarized in AAPM TG-50 [12], published in 1998. This Task Group, however, did not comprehensively address procedures applicable to static field planning, and were not suitable for dynamic IMRT planning, which was growing in popularity at the time. In response, several publications were released that began to outline dynamic leaf QA tests and procedures, and later, suggestions for dynamic leaf QA were established by AAPM TG-142. The task group suggests "strip test," initially proposed by Chui et al., to be comprehensive in that it is able to acquire leaf positional data as well as transmission information. Now commonly referred to as the picket fence test, Chui et al.'s strip test [13] originally consisted of irradiating 1 mm wide strips on film at known, regular intervals. Initially, the film was visually inspected for positional errors. Visible leaf positional errors may include an abnormally large gap (caused by either the leading leaf stopping too late, the trailing leaf stopping too early, or a combination of the two), a narrow gap (caused by either the leading leaf stopping too early, the trailing leaf stopping too late, or a combination of the two), or a deviated gap where both the leading and trailing leaves either stop too late or too early. Going beyond a simple visual inspection, Sastre-Padro et al. showed it is possible to yield more quantitative results by digitizing the "strip-test" film and relating the relative dose change observed at the leaf junctions to the position and width of each gap [14].

Relying on similar fitting techniques, several publications have highlighted the utility and increased convenience of conducting MLC QA using an EPID [15]. It is possible to obtain sub-pixel accuracy through various fitting techniques. One technique proposed by Mamalui-Hunter et al. made use of a modified Lorentzian function whose parameters described the leaf positions and gap width pattern characteristic to a strip test image [16]. When repeating the acquisition and analysis process five times, the standard deviation of the reproducibility of leaf deviation results was determined to be <0.1 mm [17]. These results give merit to the use of EPID for MLC QA when using the Picket Fence test. It was strongly established that the leaf positional data acquired from an EPID was acceptably comparable to film measurements, and some even suggest the use of an EPID yields more consistent results than using film [17].

8.3.4 Jaw Positioning

In IMRT and SBRT fields, jaws are often used to spare healthy tissue by blocking radiation transmission through MLC leaves, or by defining the extent of a treatment field. Slight misalignments in the jaws positioning may result in significant tissue under or overdosing. Therefore, it is critical to ensure jaw positional accuracy.

Light/Radiation congruence has long been the preferred test for jaw positioning and beam steering. During patient set-up, a therapist often makes use of the light field for alignment. This light field is formed by a light source mounted in the gantry head which shines through the X and Y jaws. The light that passes through the gantry head and onto the patient should be representative of how radiation will travel from the gantry head and onto the patient. Inconsistencies observed between the marked light field and radiation field edge may imply the machine needs to be steered or jaw misalignments are present. As an alternative to using film to conduct the light/radiation congruence test, Prisciandaro et al. suggested an EPID-based method [18]. Instead of film, Prisciandaro et al. utilized a phantom with a diamond cut out in the middle. Each jaw was aligned to a vertex of a diamond using the light field. An open field image was then acquired and the distance between the calculated radiation field edge and a vertex is defined to be the light-radiation field in congruency [18].

An EPID also allows for the quick calculation of open field size, which is often a required parameter for daily linear accelerator (linac) QA. As a part of linac daily QA, Sun et al. showed the feasibility of computing simple field size measurements using an EPID [19]. To acquire these measurements, the open field images were exported from the treatment machine and several signal profiles were extracted from the image in both the inline and crossline directions. The full-width at half-maximum (FWHM) of each profile was converted from pixel values to cm, based upon the EPID's pixel spacing and vertical position. The EPID's sensitivity to jaw positional shifts was tested by recording FWHM measurements when increasing and decreasing the digital jaw positions in 0.1 cm increments. Sun et al. found that the measured field size reflected the digital jaw positional changes [19].

When computing the field size of an FFF beam, one should not simply follow the same method proposed above where the FWHM of each extracted profile is determined. As one may infer from the name, FFF beams are not flat but forward-peaked. In this case, the location of the 50% dose is not representative of the field edges. As a work-around, Fogliata et al. proposed a method that strategically re-normalizes the extracted profiles so that the 50% dose location does align with the field edge [20]. To accomplish this, one must find location in the dose profile that is similar for both flattened beams and FFF beams to normalize to the same dose level. Previously, Pönisch et al. suggested the inflection point of the profile to suffice [21]. Fogliata et al. pointed out, however, that the common step length (1 mm) used to locate the inflection point lacks the precision necessary to accurately locate the point due to the high gradient region of the inflection point [20]. Such imprecision can lead to 10% error in the dose level used for normalization, ultimately potentially leading to a 40% error in central dose level after normalization. As an alternative, Fogliata et al. suggest using what they refer to as the "shoulder point" to renormalize the signal profile. The "shoulder point" is located at the top of the sharp dose gradient defined by the field edge and right before the dose of an FFF beam steadily increases toward the CAX peak [20]. More quantitatively, the "shoulder point" is the maximum of the signal profile's third derivative located inside of the profile's inflection point. This location in the FFF profile is normalized to the same position in the flattened profile, and the relative dose at the FFF CAX is used as the renormalization value. To simplify the procedure, Fogliata et al. developed a function based on field size and EPID measuring depth to compute the renormalization value. Once the renormalization value is determined, it is used to renormalize each FFF profile. The field size of the renormalized FFF profile can then be computed as usual using the FWHM.

Beyond single open field jaw verification, the EPID has shown utility in providing QA for more comprehensive jaw positioning. For instance, Clews et al. developed an EPID-based technique that covers asymmetric jaw QA [22]. When asymmetric jaws are employed, it is often the case that one set of jaws is positioned at the CAX while the other jaw forms the extent of the field. Following, a second field is acquired such that the jaw that was previously positioned at the CAX forms the extent of the field while the opposite jaw is now located at the CAX. If jaw misalignments are present during this configuration, a significantly high dose or low dose line may form at the junction of the abutting CAX lines. Therefore, it is critical to minimize the over/under dose caused by jaw misalignments. In their work, Clews et al. acquired a set of jaw-abutted images and combined them. At the junction point, a dark or bright line may be visible due to underlap or overlap [22]. The ratio between the dose at the CAX and the dose at the field center is computed. Clews et al.'s work showed a linear relationship between the dose at the CAX and the overlap/underlap size (in mm) [22]. Therefore, the ratio of the CAX dose to the field center dose was assumed to be a good indicator of asymmetrical jaw alignment. Moreover, the EPID method proposed yielded comparable results to similar film-based methods, and it was suggested that the EPID-based method added precision while reducing time needed for test completion.

8.3.5 Isocentricity

The motion of a linear accelerator is confined to rotation along three variable axes – gantry rotation, collimator rotation, and couch rotation. It is critical to test the rotation patterns along each variable axes and to quantify how misalignments in each variable axes contribute to the overall machine isocenter size and deviation. A common isocentricity test is known as the Star Shot. Star Shot analysis is used for both 2D and 3D isocenter determination. Two-dimensional analysis is accomplished by placing film perpendicular to a variable axis and irradiating a "strip" of the film at specified angles. The resulting film is made up of multiple "strips," or "rays," that intersect near the center of the film. To define the resulting isocenter size and position, it is common to first define a straight line that represents the center of each ray and to employ a minimization routine to find the optimal intersection point and "wobble" of the defined lines. To find the lines, many groups have first digitized the star shot film image and extracted multiple circular profiles about the estimated star shot center. The resulting pixel profiles yield a series of peaks, which correspond to the irradiated areas. To determine the exact radiation center of each ray, several methods have been proposed including subtracting the mean value of the profile from the profile, finding the zero-crossing points, and finding their respective midpoints [23]. Depuydt et al. suggested simply finding the middle of the FWHM of each ray to define their centers [24]. Once the intersecting lines have been defined, the minimum circle that intersects or is tangent to all the lines is found by some optimization routine that minimizes the distance to each line from some point. The position and size of this circle is defined as the variable axis isocenter size and position [23, 24]. This method can be repeated for each variable axis (gantry, collimator, and couch).

Perhaps a more robust test of isocentricity is the Winston–Lutz test. This test was first introduced in 1988 by Lutz et al. for the purpose of stereotactic radiosurgery QA [25]. The original test consisted of fixing a steel ball at the location of treatment isocenter and attaching a film to the gantry head so that it sits perpendicularly to the beam path. The steel ball is positioned between the gantry head and the film. The film is irradiated at eight distinct gantry and table angle configurations in an open field orientation. The radiation field is larger than the diameter of the ball. If the gantry positioning was solid, the resulting "shadow" of the steel ball embedded in the radiation field (visible on the film) should be located in the exact center of the radiation field. Due to misalignments (wobble), however, deviations occur and are decisive for determining the structural integrity of the linac. The resulting films were visually inspected for isocenter location discrepancies.

Moving beyond a simple visual inspection of the irradiated films, several groups, including Low et al. [26] and Tsai [27], developed minimization algorithms in the mid-1990s to quantitatively describe image offsets that allude to machine or setup misalignments. In more recent years, the effort required for the acquisition and analysis of the Winston–Lutz test has been greatly reduced by making use of an EPID. Following a very similar acquisition plan, Winkler et al. first introduced a procedure of using an EPID for the Winston–Lutz test [28]. Like the original, images were acquired at various gantry, couch, and collimator angles, except instead of irradiating films, the images were acquired digitized by the EPID. The minimization routine employed by Winkler et al. differs from the ones stated above in that it divides up the data into variable axes categories. To quantify table wobble, Winkler et al. asserted that the field center can be taken as a stationary point for each table image and is not subject to change. The coordinates of the tungsten ball center, however, may change due to table wobble. These displacements (between field center and ball center) are plotted about an origin and the minimum inscribed circle about the points is calculated and defined to be the table's wobble radius and position. A similar approach is taken for collimator wobble, except instead of defining the field to be a stationary reference point, the tungsten ball center is now taken as the stationary point. Displacements between the variable field center and ball center are plotted and are also minimized using the minimum inscribed circle technique. For gantry rotation, the tungsten ball center is defined as stationary and the field center is expected to shift with misalignments. The displacement between the ball center and the field center on the image plane is "back-projected," forming a vector normal to the EPID. This is repeated for each image in which gantry positioned differs from the reference position, forming a collection of normal vectors (perpendicular to the x–y plane). If the gantry rotates with no wobble, all lines would intersect at a single point. Since wobble does exist, however, the gantry isocenter is determined by computing the smallest circle that is tangent to or intersects all lines. This is accomplished through a gradient optimization technique which returns the gantry isocenter size and location.

8.3.6 Variable Axis Angles

Apart from using an EPID as a direct replacement for film measurements, the imager also enables physicists to implement entirely new QA methods and procedures. For instance, Wilson et al. proposed a quantitative method to measure true table and collimator rotation angle [29]. Measuring the physical angle of these variable axes is generally completed by using a bubble level and/or the use of physical grid paper. A physicist will align grid paper under the crosshair shadows present from the light field and will rotate the collimator or

table until the crosshair shadow has rotated 90 degrees, as displayed by the grid paper. This value is compared to the digital angle readout to ensure the two are in agreement (within a set tolerance). Film is generally not used for these machine QA measurements.

As one may imagine, using a bubble level and grid paper for variable axis angle measurements can be tedious and is susceptible to user subjectivity. Wilson et al. utilized an EPID and rotation tracking to yield more quantifiable results [29]. To accomplish this, they fabricated a phantom with five embedded BBs. The phantom was made to be fixed to the end of the table and extended out over machine isocenter. The EPID was then positioned directly under the BBs in the phantom. When the table is positioned at 0 degrees, an image was acquired and the BB locations in the images were detected and saved. For each additional angle to measure, the table was rotated to that angle and an image was acquired. Subsequently, the BB locations were detected and stored. The rotational transformation matrix that describes the expected rotation based on three variables – the couch rotation center, the angle of rotation, and an angle scaling factor – was optimized. The resulting angle from the optimization was taken as the physically measured angle. The angle scaling factor and rotation center can be further analyzed for isocenter stability. Wilson et al. proposed the returned information is sufficient for isocenter QA and determined the results are comparable to Star Shot and Winston–Lutz analyses [29]. Similar methods can be followed for collimator rotation. And, with additional care taken to account for EPID tilt and projection geometries, Grelewicz et al. showed a similar method can be followed for measuring gantry angle deviations as well as providing essential isocenter information [30].

8.3.7 Output

Apart from its utility in ensuring the structural integrity of a linac, an EPID has been shown to provide assurance of the dosimetric behavior of a linac. However, the stability and output of an EPID must be verified against ion-chamber readings before it is used for regular dosimetric measurements. Budgell et al. showed there is a linear relationship between measured ion-chamber measurements and EPID measurements normalized to 100 cGy for 100 MU. The MU ranged from 95 MU to 105 MU [31]. Budgell et al. also monitored the output measurements of 100 MU acquired using both an EPID and an ion chamber over the course of a year. From these measurements, it was determined that almost 95% of the EPID measurements were within 1.3% of the associated ion-chamber measurement, which is sufficient agreement for daily output monitoring [31].

Extending beyond simply determining agreement among EPID and ion chamber for CAX output measurements, it is also critical to ensure similar dosimetric properties are detectable for both methods of measurement. One known parameter that affects an ion-chamber reading is field size. To understand how field size affects EPID dosimetric measurements, Sabet et al. acquired relative EPID measurements for field sizes ranging from 5 cm to 19 cm in length [32]. The ion-chamber response was also recorded for each of these field sizes, and the ratio of EPID to ion-chamber measurements was computed. For the two buildup materials (Cu and solid water) and for both energies (6 MV and 18 MV), the difference between relative EPID measurements and ion-chamber response was all within 1.5%. Sabet et al. also evaluated the off-axis response for EPID and ion-chamber measurements by performing a one-dimensional Gamma evaluation on the EPID and ion-chamber measurements [32]. The results showed good agreement for 6 MV and 18 MV, especially when copper buildup was used. Thus, it was determined that an EPID is an acceptable tool for continuous output measurement, especially when comparing output results to a set baseline.

8.3.8 Flatness/Symmetry

As an extension of their investigation on the utility of using an EPID for daily QA, Budgell et al. determined that flatness and symmetry measurements were also possible using an EPID [31]. Both flatness and symmetry are routinely considered as a part of daily machine QA. Flatness is simply the measurement of dose uniformity across an axis (crossline or inline) of an open field image. AAPM TG-45 [33] defines flatness as:

$$Flatness = \frac{M-m}{M+m} \times 100 \qquad (8.2)$$

where M and m are the maximum and minimum signal values within profile width which is bound by the 80% of CAX dose locations. Similarly, symmetry resolves the

symmetry present in an open field beam about the CAX. AAPM TG-45 [33] defines symmetry as:

$$Symmetry = \frac{\max(D_{left} - D_{right})}{D_{center}} \times 100 \qquad (8.3)$$

where D_{left} and D_{right} are dose values at equidistant points to the left and right of the CAX, and D_{center} is the dose at the CAX. Budgell et al. notes that the measurement of flatness and symmetry does not take any additional acquisition if the EPID is used for daily output measurements by means of sampling the CAX dose of an open field image [31]. The same open field images can conveniently be used for flatness and symmetry measurements.

Budgell et al. monitored the flatness and symmetry in terms of deviation from baseline results for 6 MV and 8 MV over the course of a year [31]. During this period, the EPID panel that was used for data acquisition was replaced and the gun and steering currents in the linac were periodically changed. These changes were all reflected in the collected data and were approximately sensitive to changes of half a percent. Such precision is adequate to catch abnormal beam behavior and to track profile trends over time, meaning an EPID is adequate for routine profile monitoring.

8.3.9 Wedge

Physical wedges have long been used in radiation treatment plans to modify the isodose distribution where the isodose lines are oriented at some angle to the line normal to the beam CAX. Using a physical wedge comes with limitations. For instance, physical wedges are manufactured with discrete field size dimensions. Furthermore, physical wedges can be heavy, making it difficult for a therapist to mount to the gantry head. Such weight and positioning also presents a risk of injury to the patient. To obtain the same dosimetric effect, it is now much more common to use a combination of a sweeping jaw motion and a varying dose rate such as is used in Varian's enhanced dynamic wedge (EDW) protocol. Using this method as opposed to a physical wedge allows for the incorporation of a wedge at any angle and any field size and spares any risk of injury to the patient. In order to ensure the accuracy of a wedge profile, it is common to use an ion-chamber array, a diode array, or film to compare the profile to that in the treatment planning system. Perhaps unsurprisingly, this method is slowly being replaced with EPID-based methods, adding convenience and resolution while saving time and cost. Despite it having a linear dose response and a highly reproducible CAX dose, using an EPID for wedge methods presents some otherwise nonexistence problems. For instance, scattering effects differ from an EPID to those of solid water and backscatter from the imager arm is present. Moreover, the phosphor layer of the EPID yields an energy-dependent response. In order to quantify the extent to which these limitations pose an issue to measuring a wedge factor, Greer et al. compared the EDW profiles obtained using an EPID to those obtained using ion-chamber measurements while monitoring the stability of an EPID and developed a correction method to account for using an EPID for EDW measurements [34].

Greer et al. developed a quadratic function to correct for anomalies present in EPID measurements. The function was used for 30-degree and 60-degree wedge measurements for both 6 MV and 18 MV. The wedge factors derived using an EPID were all within 0.6% of the ion-chamber reference factors. The largest deviation in EPID wedge factors over the course of three weeks was 0.9%. Furthermore, Greer monitored the ratio of the EPID dose to the ion-chamber dose over the course of over 200 days and found the standard deviation among 6 MV measurements to be 0.4% and the standard deviation among 18 MV measurements to be 0.3%, which strongly supports the long-term reproducibility of EPID wedge measurements. To correct for the backscatter present due to the imager arm positioning, Greer et al. suggests only taking EDW measurements in the crossline direction. Greer et al. noted that measurement acquired in the inline direction differ with EPID distance from the source due to beam broadening caused by divergence. If an institution uses the Y Jaws for EDW measurements, Greer et al. simply suggest rotating the collimator to 90 degrees in order to minimize the effect of imager arm backscatter [34].

8.4 USE OF EPID FOR PATIENT QA

The advantageous dosimetric properties of EPIDs have inspired exploration into their utilization as patient-specific QA tools, with a variety of clinical implementations reported. In addition to their ease of use and setup reproducibility, gantry-mounted EPIDs offer great flexibility with respect to how such treatment verification is performed. EPID images can be taken before (pretreatment verification) or during (*in vivo* verification) a patient's treatment as well as with (transit dosimetry) or without (non-transit dosimetry) an attenuating medium between the source and EPID. Further, the analysis can be performed either at the level of the EPID or that of the phantom/patient, if applicable. The resulting iterations of these measurement options allow centers to verify the deliverability of treatment plans in a way which suits their workflow while providing the desired treatment information.

EPID-level strategies generally involve evaluating the measured portal image with reference to a predicted image derived from the planned fluence [35, 36]. Either the raw EPID response itself or the dose-to-water at the plane of the EPID may be predicted, with the latter approach requiring conversion of the raw image into a portal dose image. Although portal dose predictions may not represent a direct check of the method used by the treatment planning system to generate the clinical plan's dose distribution, appropriate calibrations do enable the comparison of measured vs. predicted portal dose image to be done on an absolute scale [35, 37]. Further, if the calculation of predicted portal dose is done by the treatment planning system as opposed to an in-house developed software, EPID-level dose prediction can serve as a broader pre-treatment check of the treatment planning system. Measuring and predicting the EPID response with a phantom in the beam path is a similar method of pre-treatment verification which more closely matches a clinical environment [37]. Though both techniques can assess the deliverability of plans, the communication between planning system and treatment machine, and the general integrity of the planning system, neither can identify errors unique to the treatment setting [36]. That shortcoming is addressed when EPID-level transit measurements are performed during an actual treatment, with the EPID image taken after the beam has passed through the patient [36].

Even with the use of daily image guidance from kV sources, it can be difficult to identify soft-tissue anatomy changes and visualize the impact of setup uncertainties or organ motion. With *in vivo* EPID-level dosimetry, these issues lead to an easily identifiable discrepancy between the measured and predicted images which also enables an evaluation of the error's severity [36, 38]. However, the two-dimensional nature of the analysis makes it difficult to precisely identify the source of any such discrepancy (doing so would likely require an additional simulation scan), and this method's focus on the EPID rather than the patient inhibits an understanding of how any error impacts the dose distribution within the patient. Furthermore, performing EPID measurements during treatment requires additional care to avoid collisions between the imager and couch/patient [36]. Still, the method's relative simplicity, the availability of commercial portal dose prediction/analysis tools, and the possibility of combining multiple approaches make EPID-level verification an attractive means of patient-specific QA.

Two significant limitations of EPID-level analyses not resolved with an *in vivo* approach are their inherent lack of a third dimension and inability to directly validate the dose calculation method utilized by the treatment planning system. The three-dimensional and tissue-like construction of many phantoms, meanwhile, allows the use of phantom-level EPID approaches to build upon the benefits of pre-treatment EPID-level verification while addressing the above drawbacks. Generally, phantom-level methods involve utilizing EPID measurements to reconstruct the (two- or three-dimensional) dose within a phantom and comparing the result with a predicted distribution [39–44]. With a tissue-like phantom, the same dose calculation algorithm used to create the treatment plan can be used to generate this prediction, enabling centers to use phantom-level methods to verify the dose calculation model utilized for treatment planning. Further, as long as the reconstruction technique takes advantage of the three-dimensional nature of a phantom, phantom-level approaches can provide better localization of dose discrepancies than two-dimensional techniques. A common phantom-level method utilizes transit EPID images measured with a phantom in the beam's path to back-calculate the dose from the EPID to within the phantom [39, 40]. In this approach, determining the beam transmission through the phantom is required, as it enables an estimation of the EPID dose contribution due to phantom scatter. While this can be achieved with a second non-transit EPID image, modeling the transmission is a reported alternative which obviates the need for another measurement [41]. It is important to note that the phantom may or may not be physically present in the beam's path during the measurement, as the categorization of 'phantom-level' does not require that dose be delivered to an actual phantom, just that the phantom is the site of comparison with the prediction. Indeed, one can also use EPID images from a pre-treatment, non-transit delivery

to derive the fluence of a clinical plan and use that as input to a dose calculation engine [42–44]. The resulting phantom dose distribution, whether two- or three-dimensional, can then be compared to that computed in the treatment planning system.

Although phantom-level approaches expand upon the benefits of pre-treatment EPID-level methods, they still share some of the same disadvantages. For example, even if differences between the measured and predicted three-dimensional dose distributions are observed in a phantom, the effect of these discrepancies on the patient would not be immediately apparent, instead requiring further investigation. This limitation, in addition to the increasing complexity of radiation therapy treatments, has made dose computations at the level of the patient more and more valuable. While such reconstruction has previously been performed using fluence measured with film, EPID dosimetry is an alternative which allows for the possibility of easier, less time-consuming *in vivo* dosimetry as well as automated treatment verification [45, 46]. Consequently, there has been much exploration into dose reconstruction, both two- and three-dimensional, within the patient using EPIDs. One such pre-treatment strategy is the use of EPID-measured fluence as input to a Monte Carlo dose calculation with the patient's planning CT scan. Comparing the resultant distribution with that of the treatment plan facilitates visualization of any differences and how they manifest within the patient's anatomy. Further, as long as the reconstruction algorithm operates independently of the treatment planning system, this method can serve as a verification of the treatment plan's dose computation model. Still, like other pre-treatment approaches, such *in vivo* measurements on their own can only reveal dose discrepancies owing to more systematic errors, like issues with plan deliverability or the treatment planning system. When calculated with treatment-day CBCT scans instead of the planning CT scan, however, this same general strategy can also illuminate dosimetric errors from anatomy changes or setup uncertainties without the added complications, such as scatter corrections and possible collisions, of collecting transit portal images during treatment, thus equipping centers with a unique tool for evaluating the necessity of plan adaptation.

The above method's main limitation, which it shares with other pre-treatment verification procedures, is that it is not based on a patient's actual treatment. Consequently, transit dosimetry, with EPID images taken during a normal treatment session, has also been explored as a means of acquiring an *in vivo* dose distribution [41, 46–50]. Because these transit measurements are performed with the patient in the beam's path, the reconstructed dose necessarily takes into consideration errors unique to the treatment setting, in addition to the more general concerns of plan deliverability, communication between treatment machine and planning system, etc. The reconstruction itself can be performed using a back-projection algorithm, though the information the technique provides depends on whether it makes use of the planning CT scan or a treatment-day CBCT scan. For example, while a change in the patient's anatomy may very well cause a discrepancy between the planned and measured dose distributions when using the planning CT scan for reconstruction, identifying that change as the source of the difference could be difficult without additional imaging or the results of pre-treatment verification. Further, if an anatomy change or setup error was indeed the cause, the reconstructed distribution would not be fully valid. Thus, the use of the planning CT scan in reconstruction can still reveal mistakes from a wide range of sources, but it makes difficult a precise identification of the source and an assessment of the dosimetric impact. Utilizing a treatment-day CBCT scan for the reconstruction, on the other hand, would more clearly show any change from the planning CT scan while providing a fully up-to-date dose distribution so that the effect of any discrepancies with the treatment plan can be readily evaluated [49, 50]. Moreover, while performing patient-level dosimetry during (as opposed to before) treatment does present some complications, such as the need to avoid collisions between the couch/patient and imager, it also has ramifications for patient safety. This technique affords the opportunity to implement consistent online treatment verification which, if a major dose discrepancy is detected, could automatically halt treatments [46]. Still, the lifetime of an imager used in that manner would be another concern, so such a procedure may need to be limited to hypofractionated treatments.

Although EPIDs present many possibilities for patient-specific treatment verification, their utility would be reduced if measurement and analysis techniques were unable to accommodate newer, more complex treatment methods. Instead, techniques such as VMAT and adaptive radiation therapy have been explored with EPID dosimetry, demonstrating its continued relevance. For example, EPID verification of VMAT plans can vary from the confirmation of leaf positions to three-dimensional *in vivo* dosimetry [46, 51]. Furthermore, the ability to use CBCT scans for dose reconstruction makes EPID dosimetry an appealing tool for plan adaptation [52]. Ultimately, EPID dosimetry offers many possibilities for verification and continues to be applicable to the evolving techniques in the field.

REFERENCES

1. H. Gustafsson, P. Vial, Z. Kuncic, C. Baldock, P. B. Greer, "EPID dosimetry: Effect of different layers of materials on absorbed dose response," *Medical Physics* **36**, 5665–5674 (2009).
2. V. M. S. Inc, *PortalVision aS1000: The State of the Art in Electronic Portal Imaging* (2016).
3. B. W. King and P. B. Greer, "A method for removing arm backscatter from EPID images," *Medical Physics* **40**, 071703 (2013).
4. P. B. Greer and C. C. Popescu, "Dosimetric properties of an amorphous silicon electronic portal imaging device for verification of dynamic intensity modulated radiation therapy," *Medical Physics* **30**, 1618–1627 (2003).
5. V. Mhatre, S. Pilakkal, P. Chadha, K. Talpatra, "Dosimetric comparison of a-Si 1200 and a-Si 1000 electronic portal imager for intensity modulated radiation therapy (IMRT)," *Journal of Nuclear Medicine & Radiation Therapy* **9**, 2 (2018).
6. Z. Xu, J. Kim, J. Han, A. T. Hsia, S. Ryu, "Dose rate response of Digital Megavolt Imager detector for flattening filter-free beams," *Journal of Applied Clinical Medical Physics* **19**, 141–147 (2018).
7. B. King, L. Clews, P. Greer, "Long-term two-dimensional pixel stability of EPIDs used for regular linear accelerator quality assurance," *Australasian Physical & Engineering Sciences in Medicine* **34**, 459–466 (2011).
8. R. J. Louwe, L. N. McDermott, J. J. Sonke, R. Tielenburg, M. Wendling, M. van Herk, B. J. Mijnheer, "The long-term stability of amorphous silicon flat panel imaging devices for dosimetry purposes: Stability of EPID response," *Medical Physics* **31**, 2989–2995 (2004).
9. E. E. Klein, J. Hanley, J. Bayouth, F. F. Yin, W. Simon, S. Dresser, C. Serago, F. Aguirre, L. Ma, B. Arjomandy, Task Group 142 report: Quality assurance of medical accelerators," *Medical Physics* **36**, 4197–4212 (2009).
10. J. P. Bissonnette, P. A. Balter, L. Dong, K. M. Langen, D. M. Lovelock, M. Miften, D. J. Moseley, J. Pouliot, J. J. Sonke, S. Yoo, "Quality assurance for image-guided radiation therapy utilizing CT-based technologies: A report of the AAPM TG-179," *Medical Physics* **39**, 1946–1963 (2012).
11. W. Mao, M. Speiser, P. Medin, L. Papiez, T. Solberg, L. Xing, "Initial application of a geometric QA tool for integrated MV and kV imaging systems on three image guided radiotherapy systems," *Medical Physics* **38**, 2335–2341 (2011).

12. A. Boyer, P. Biggs, J. Galvin, E. Klein, T. LoSasso, D. Low, K. Mah, C. Yu, "Basic applications of multileaf collimators. Report of the AAPM Radiation Therapy Committee Task Group No. 50," *Medical Physics* **72**, 16–40 (2001).

13. C. S. Chui, S. Spirou, T. LoSasso, "Testing of dynamic multileaf collimation," *Medical Physics* **23**, 635–641 (1996).

14. M. Sastre-Padro, U. A. van der Heide, H. Welleweerd, "An accurate calibration method of the multileaf collimator valid for conformal and intensity modulated radiation treatments," *Physics in Medicine & Biology* **49**, 2631 (2004).

15. P. Rowshanfarzad, M. Sabet, M. P. Barnes, D. J. O'Connor, P. B. Greer, "EPID-based verification of the MLC performance for dynamic IMRT and VMAT," *Medical Physics* **39**, 6192–6207 (2012).

16. M. Mamalui-Hunter, H. Li, D. A. Low, "MLC quality assurance using EPID: A fitting technique with subpixel precision," *Medical Physics* **35**, 2347–2355 (2008).

17. S. J. Baker, G. J. Budgell, R. I. Mackay, "Use of an amorphous silicon electronic portal imaging device for multileaf collimator quality control and calibration," *Physics in Medicine & Biology* **50**, 1377 (2005).

18. J. Prisciandaro, M. Herman, J. Kruse, "Utilizing an electronic portal imaging device to monitor light and radiation field congruence," *Journal of Applied Clinical Medical Physics* **4**, 315–320 (2003).

19. B. Sun, S. M. Goddu, S. Yaddanapudi, C. Noel, H. Li, B. Cai, J. Kavanaugh, S. Mutic, "Daily QA of linear accelerators using only EPID and OBI," *Medical Physics* **42**, 5584–5594 (2015).

20. A. Fogliata, R. Garcia, T. Knöös, G. Nicolini, A. Clivio, E. Vanetti, C. Khamphan, L. Cozzi, "Definition of parameters for quality assurance of flattening filter free (FFF) photon beams in radiation therapy," *Medical Physics* **39**, 6455–6464 (2012).

21. F. Pönisch, U. Titt, O. N. Vassiliev, S. F. Kry, R. Mohan, "Properties of unflattened photon beams shaped by a multileaf collimator," *Medical Physics* **33**, 1738–1746 (2006).

22. L. Clews and P. B. Greer, "An EPID based method for efficient and precise asymmetric jaw alignment quality assurance," *Medical Physics* **36**, 5488–5496 (2009).

23. A. Gonzalez, I. Castro, J. Martínez, "A procedure to determine the radiation isocenter size in a linear accelerator," *Medical Physics* **31**, 1489–1493 (2004).

24. T. Depuydt, R. Penne, D. Verellen, J. Hrbacek, S. Lang, K. Leysen, I. Vandevondel, K. Poels, T. Reynders, T. Gevaert, "Computer-aided analysis of star shot films for high-accuracy radiation therapy treatment units," *Physics in Medicine & Biology* **57**, 2997 (2012).

25. W. Lutz, K. R. Winston, N. Maleki, "A system for stereotactic radiosurgery with a linear accelerator," *International Journal of Radiation Oncology Biology Physics* **14**, 373–381 (1988).

26. D. A. Low, Z. Li, R. E. Drzymala, "Minimization of target positioning error in accelerator-based radiosurgery," *Medical Physics* **22**, 443–448 (1995).

27. J.-S. Tsai, "Analyses of multi-irradiation film for system alignments in stereotactic radiotherapy (SRT) and radiosurgery (SRS)," *Physics in Medicine & Biology* **41**, 1597 (1996).

28. P. Winkler, H. Bergmann, G. Stuecklschweiger, H. Guss, "Introducing a system for automated control of rotation axes, collimator and laser adjustment for a medical linear accelerator," *Physics in Medicine & Biology* **48**, 1123 (2003).

29. B. Wilson and E. Gete, "Machine-specific quality assurance procedure for stereotactic treatments with dynamic couch rotations," *Medical Physics* **44**, 6529–6537 (2017).

30. Z. Grelewicz, H. Kang, R. D. Wiersma, "An EPID based method for performing high accuracy calibration between an optical external marker tracking device and the LINAC reference frame," *Medical Physics* **39**, 2771–2779 (2012).

31. G. J. Budgell, R. Zhang, R. I. Mackay, "Daily monitoring of linear accelerator beam parameters using an amorphous silicon EPID," *Physics in Medicine & Biology* **52**, 1721 (2007).

32. M. Sabet, P. Rowshanfarzad, F. W. Menk, P. B. Greer, "Transit dosimetry in dynamic IMRT with an a-Si EPID," *Medical & Biological Engineering & Computing* **52**, 579–588 (2014).

33. R. Nath, P. J. Biggs, F. J. Bova, C. C. Ling, J. A. Purdy, J. van de Geijn, M. S. Weinhous, "AAPM code of practice for radiotherapy accelerators: Report of AAPM Radiation Therapy Task Group No. 45," *Medical Physics* **21**, 1093–1121 (1994).

34. P. B. Greer and M. P. Barnes, "Investigation of an amorphous silicon EPID for measurement and quality assurance of enhanced dynamic wedge," *Physics in Medicine & Biology* **52**, 1075 (2007).

35. A. van Esch, T. Depuydt, D. P. Huyskens, "The use of an aSi-based EPID for routine absolute dosimetric pre-treatment verification of dynamic IMRT fields," *Radiotherapy and Oncology* **71**, 223–234 (2004).

36. S. L. Berry, C. Polvorosa, S. Cheng, I. Deutsch, K. C. Chao, C.-S. Wuu, "Initial clinical experience performing patient treatment verification with an electronic portal imaging device transit dosimeter," *International Journal of Radiation Oncology Biology Physics* **88**, 204–209 (2014).

37. S. L. Berry, R. D. Sheu, C. S. Polvorosa, C. S. Wuu, "Implementation of EPID transit dosimetry based on a through-air dosimetry algorithm," *Medical Physics* **39**, 87–98 (2012).

38. M. Kroonwijk, K. L. Pasma, S. Quint, P. C. Koper, A. G. Visser, B. J. Heijmen, "In vivo dosimetry for prostate cancer patients using an electronic portal imaging device (EPID); demonstration of internal organ motion," *Radiotherapy and Oncology* **49**, 125–132 (1998).

39. M. Wendling, R. J. Louwe, L. N. McDermott, J. J. Sonke, M. van Herk, B. J. Mijnheer, "Accurate two-dimensional IMRT verification using a back-projection EPID dosimetry method," *Medical Physics* **33**, 259–273 (2006).

40. L. McDermott, M. Wendling, B. van Asselen, J. Stroom, J. J. Sonke, M. van Herk, B. Mijnheer, "Clinical experience with EPID dosimetry for prostate IMRT pre-treatment dose verification," *Medical Physics* **33**, 3921–3930 (2006).

41. R. Pecharromán-Gallego, A. Mans, J. J. Sonke, J. C. Stroom, Í. Olaciregui-Ruiz, M. van Herk, B. J. Mijnheer, "Simplifying EPID dosimetry for IMRT treatment verification," *Medical Physics* **38**, 983–992 (2011).

42. B. Warkentin, S. Steciw, S. Rathee, B. Fallone, "Dosimetric IMRT verification with a flat-panel EPID," *Medical Physics* **30**, 3143–3155 (2003).

43. W. J. van Elmpt, S. M. Nijsten, R. F. Schiffeleers, A. L. Dekker, B. J. Mijnheer, P. Lambin, A.W. Minken, "A Monte Carlo based three-dimensional dose reconstruction method derived from portal dose images," *Medical Physics* **33**, 2426–2434 (2006).

44. W. J. van Elmpt, S. M. Nijsten, A. L. Dekker, B. J. Mijnheer, P. Lambin, "Treatment verification in the presence of inhomogeneities using EPID-based three-dimensional dose reconstruction," *Medical Physics* **34**, 2816–2826 (2007).

45. W. D. Renner, M. Sarfaraz, M. A. Earl, C. X. Yu, "A dose delivery verification method for conventional and intensity-modulated radiation therapy using measured field fluence distributions," *Medical Physics* **30**, 2996–3005 (2003).

46. B. J. Mijnheer, P. González, I. Olaciregui-Ruiz, R. A. Rozendaal, M. van Herk, A. Mans, "Overview of 3-year experience with large-scale electronic portal imaging device–based 3-dimensional transit dosimetry," *Practical Radiation Oncology* **5**, e679–e687 (2015).

47. P. Francois, P. Boissard, L. Berger, A. Mazal, "In vivo dose verification from back projection of a transit dose measurement on the central axis of photon beams," *Physica Medica* **27**, 1–10 (2011).

48. M. Wendling, L. N. McDermott, A. Mans, J.-J. Sonke, M. van Herk, B. J. Mijnheer, "A simple backprojection algorithm for 3D in vivo EPID dosimetry of IMRT treatments," *Medical Physics* **36**, 3310–3321 (2009).

49. L. N. McDermott, M. Wendling, J. Nijkamp, A. Mans, J.-J. Sonke, B. J. Mijnheer, M. van Herk, "3D in vivo dose verification of entire hypo-fractionated IMRT treatments using an EPID and cone-beam CT," *Radiotherapy and Oncology* **86**, 35–42 (2008).

50. W. van Elmpt, S. Nijsten, S. Petit, B. Mijnheer, P. Lambin, A. Dekker, "3D In Vivo Dosimetry Using Megavoltage Cone-Beam CT and EPID Dosimetry," *International Journal of Radiation Oncology Biology Physics* **73**, 1580–1587 (2009).

51. M. Bakhtiari, L. Kumaraswamy, D. W. Bailey, S. de Boer, H. K. Malhotra, M. B. Podgorsak, "Using an EPID for patient-specific VMAT quality assurance," *Medical Physics* **38**, 1366–1373 (2011).

52. L. C. G. G. Persoon, A. G. T. M. Egelmeer, M. C. Öllers, S. M. J. J. G. Nijsten, E. G. C. Troost, F. Verhaegen, "First clinical results of adaptive radiotherapy based on 3D portal dosimetry for lung cancer patients with atelectasis treated with volumetric-modulated arc therapy (VMAT)," *Acta Oncologica* **52**, 1484–1489 (2013).

CHAPTER 9

Scintillation Fiber Optic Dosimetry

Arash Darafsheh

Washington University School of Medicine
St. Louis, Missouri

CONTENTS

9.1 INTRODUCTION

Using scintillation properties of materials is one of the oldest techniques for ionizing radiation detection [1, 2]. In the context of radiotherapy dosimetry, scintillation fiber optic dosimeters have drawn great attention due to their unique practical advantageous properties including the ability to perform *in vivo*, real-time, and intracavitary measurements with high spatial resolution due to their small physical size and mechanical flexibility. These features make them ideal candidates for many applications in radiotherapy dosimetry, such as in brachytherapy, intensity-modulated radiation therapy, superficial therapy, stereotactic radiosurgery, proton therapy, and small-field dosimetry [3–19].

Working principle of fiber optic dosimeters is based on radioluminescence properties of materials. A visible signal proportional to the absorbed dose is produced as a result of the interaction of the ionizing radiation with the sensitive portion (a scintillator attached to the tip of the fiber) of the fiber. The optical signal is guided by the optical fiber to a detector to measure the dose. However, the main problem with fiber optic dosimetry is that the signal received by the detector through the fiber is "contaminated" with Cherenkov radiation, which may not be directly proportional to the dose. Cherenkov radiation has angular dependency and also depends on the length of the fiber in the radiation field. It is a variable component that cannot be simply subtracted as a calibration factor. Therefore, the total signal must be properly corrected for the contribution of Cherenkov radiation in order to accurately measure the absorbed dose in the scintillator.

A significant issue related to scintillation dosimetry that occurs in proton therapy and other beams with high linear energy transfer (LET) is the non-proportionality between the collected light from the scintillator and the proton dose [20–25]. At low stopping powers the scintillation signal is linear with respect to the energy deposition, however, the scintillation signal "saturates" at high stopping powers. This nonlinearity effect, manifested as under-response of the optical signal to the radiation absorbed dose, is due mainly to the ionization quenching phenomenon resulting from non-radiative de-excitations occurring at high density energy deposition [26–29].

In this chapter a brief overview of scintillation fiber optic dosimetry is given. Working principle of optical fibers, characteristics of scintillators and detectors, and issues related to Cherenkov radiation generated in optical fibers are discussed.

9.2 OPTICAL FIBERS

In fiber optic dosimeter systems, optical fibers are needed to collect and transport the radioluminescent signal from the scintillator to the detector. We briefly describe basic principles and properties of solid-core optical fibers and hollow waveguides (HWGs) in this section.

9.2.1 Solid-Core Fibers

Typical materials of optical fibers are silicate-based glasses and various plastics, such as polymethylmethacrylate (PMMA) and polyethylene (PE) [30]. Conventional optical fibers, fundamentally, consists of two concentric dielectric cylinders: core and cladding. Although the core of the fiber can have other shapes, such as square, most commonly fibers with circular cross section are used. The core region is surrounded by the cladding and has slightly higher refractive index than the cladding ($n_c > n_{cl}$). Usually, there are one or more layers of protective coating outside the cladding, such as a protective plastic buffer layer and a protective jacket. Their main functions are to provide mechanical support, protect the core and cladding against mechanical damage and ambient conditions, such as moisture, and shield against coupling of the ambient light into the fiber. Their effects, however, on the optical properties of the fiber are minimal.

Light is guided in the fiber by total internal reflection at the core-cladding boundary. The acceptance angle of the fiber defines a cone of angles (acceptance cone) within which rays are guided in the fiber. A ray of light incident within the acceptance cone of the fiber on the entrance surface of the core (red ray in Figure 9.1) will impinge on the core-cladding boundary at an angle greater than the critical angle, $\theta_c = \sin^{-1}(n_{cl}/n_c)$, and is "trapped" inside the core, totally internally reflected at

the core-cladding boundary and guided through the core. Rays incident on the entrance face of the fiber with angles greater than the acceptance angle (black ray in Figure 9.1) will impinge on the core-cladding boundary at angles smaller than θ_c, will be only partially reflected at each encounter with the core-cladding boundary and will quickly leak out of the fiber.

The numerical aperture (NA) is an important parameter of optical fibers in quantifying their ability to collect light and radiate outgoing light. NA is related to the critical angle in the fiber that defines a cone of angles within which all rays are guided in the fiber by total internal reflection, as shown in Figure 9.1. Mathematically, $NA = n_o \sin \theta_a = \sqrt{n_c^2 - n_{cl}^2}$, where n_o is the refractive index of the medium in which the fiber's tip is inserted ($n_o = 1$ for air) and θ_a is the half-angle of the fiber's acceptance cone, as illustrated in Figure 9.1. Practically, the NA is defined as the sine of the angle at which the output optical power falls to 5% of the peak value. When two fibers are coupled, it is important to match their NA to optimize the coupling efficiency. Special couplers are designed to optimize the coupling between the fibers by matching their NAs. Coupling loss and reflection losses at both ends of the fiber can be minimized by antireflection coating of the fiber ends.

Depending of the modulation of the refractive index of the core, fibers can have step-index or graded-index profile. In step-index fibers, core has a constant refractive index n_c with radius and total internal reflection occurs at the core-cladding boundary. In graded-index

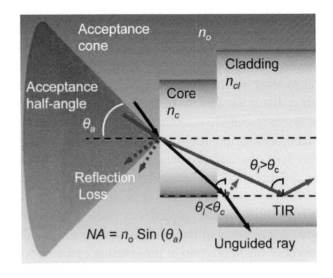

FIGURE 9.1 Schematic of the structure of an optical fiber and its acceptance cone [31].

fibers, the core index $n_c(r)$ varies (gradually decreases) with distance from the central axis of the fiber. In such fibers, rays are gradually bent according to the index profile of the core. Light is guided within all types of fibers in the form of sets of guided electromagnetic waves known as modes of the fiber. Only a certain discrete number of modes can be guided by the fiber. Fibers can be classified as single-mode or multimode. A multimode fiber can sustain many hundreds of modes: $M \approx (4a/\lambda^2)(n_c^2 - n_{cl}^2)$, in which a is the core radius of the fiber and λ is the wavelength of light. For optical power delivery as in fiber optic dosimeters, multimode fibers are used which can be considered as "light pipes" and ray picture of geometrical optics provides a good tool for their analysis. Rigorous analysis of the modal structure of light in the fiber optics is provided elsewhere [32].

The loss in an optical fiber is specified in terms of the absorption coefficient, α, according to the following relation: $\alpha = \frac{10}{L}\log_{10}\frac{1}{T} = \frac{10}{L}\log_{10}\frac{P_{in}}{P_{out}}$, where L is the length of the fiber, T is the transmission, P_{in} and P_{out} are the input and output optical powers, respectively. Transmission loss in optical fibers is extremely small for short-haul applications. For glass fibers the attenuation can be as low as ~0.2 dB/km (>95% transmission after 1 km). For plastic fibers, the attenuation is much higher, typically ~200 dB/km (>95% transmission after 1 m). For practical dosimetry applications, the fiber can be made long enough to accommodate any configuration (including having the readout system in a remote place from the ionizing radiation field or in a separate room), with negligible loss of power in transmission.

9.2.2 Hollow Waveguides

HWGs are generally used for infrared laser power delivery due to their air core. The development of waveguides with hollow cores was a milestone in infrared light transmission, where conventional solid-core fibers dramatically suffer from optical power loss. HWGs structurally are composed of a glass or plastic capillary coated internally with a metal/dielectric layer to enhance the infrared transmission in the waveguide. The structure of a typical HWG is shown in Figure 9.2. A metallic layer of silver (~0.5–2 μm thick) is coated on the inside of the silica tubing and then a dielectric layer of silver iodide is formed over the metal film through an iodization process in which some of the Ag is converted to AgI [33]. The silica tubing has a polymer coating on its outside surface. Since HWGs are typically designed

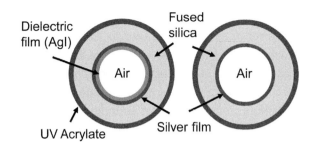

FIGURE 9.2 Schematic of the axial cross section view of a HWG with silver/dielectric coating (left) and with silver-only coating (right).

for infrared light transmission, their transmission in visible range is suboptimal. However, it has been shown that silver-only coated HWGs, i.e., without the dielectric silver iodide layer, have superior transmission in visible range of the spectrum compared with the conventional HWGs with silver/dielectric coating that are optimized for transmission of infrared light. Attenuation in visible spectral region ~0.7 dB/m (~85% transmission after 1 m) has been reported for such HWGs [34]. For visible light transmission, solid-core fibers are preferred over HWGs.

There are two types of sources for loss in HWGs: Propagation loss and bending loss. Propagation loss varies as $1/a^3$, where a is the bore radius. Bending loss varies as $1/R$, where R is the bending radius [33]. HWGs based on plastic and glass tubing have been fabricated with bore diameters ~0.1–2 mm. However, there is a tradeoff between mechanical flexibility, robustness, and throughput. Since the power loss in a HWG is inversely proportional to the cube of the bore radius, increasing the bore diameter reduces the transmission loss. However, there is a practical limit as the waveguide bore size is increased, resulting in a rigid glass rod instead of a relatively flexible HWG. In practice, the largest hollow glass waveguide size that is typically used is ~1 mm inner diameter. HWGs made from plastic tubing are more flexible, however they have higher scattering losses due to their inherently rougher inner surface compared to HWGs made from glass tubing.

9.3 SCINTILLATORS

Radioluminescence or scintillation properties of materials have attracted intense research effort for a myriad of applications in radiation detection and dosimetry, radiotherapy quality assurance, molecular imaging, and so on. Luminescence is defined as spontaneous emission of radiation from an electronically or vibrationally excited species not in thermal equilibrium with its environment [35]. Various types of luminescence, such as photo-, radio-, electro-, thermo-, chemo-, tribo-, etc. are classified according to the mode of excitation [36]. Radioluminescence is an umbrella term for different procedures (fluorescence, phosphorescence, and delayed fluorescence) that lead to light generation as a result of interaction of ionizing radiation with matter. Fluorescence is emission between two states with equal multiplicity (i.e., singlet–singlet or triplet–triplet transitions) whereas phosphorescence results from a transition between two states with different spin multiplicity (e.g., triplet–singlet emission after intersystem crossing). Fluorescence and phosphorescence processes have different time scales due to their allowed and forbidden transitions, typically 0.1–20 ns and 1 ms to 10 s, respectively [32]. Although scintillation and radioluminescence are often used interchangeably, the former can be considered as a subcategory of the latter and usually implies fast emission of light in response to radiation as opposed to phosphorescence.

Scintillators are insulators with wide energy gap between their valence and conduction band. Based on their chemical composition, the scintillators can be divided into two broad categories: organic and inorganic. Main advantage of organic scintillators over inorganic scintillators for radiation therapy dosimetry is that the former has almost water equivalent properties that helps avoid complicated dose calibration process during scintillation dosimetry.

Examples of organic scintillators include conjugated and aromatic compounds, such as aromatic hydrocarbons (naphthalene, anthracene, phenanthrene, pyrene, perylene, porphyrins, phtalocyanins, etc.) and derivatives, dyes (fluorescein, rhodamines, coumarins, and oxazines), polyenes, and diphenylpolyenes. These are employed as solutes in aromatic liquids and polymers to form organic liquid and plastic scintillators, respectively. Examples of inorganic scintillators are lanthanide ions (Eu^{3+}, Tb^{3+}), doped glasses (e.g., with ND, Mn, Ce, Cu), crystals (ZnS, CdS, ZnSe, CdSe, GaS, GaP, ruby: Al_2O_3/Cr^{3+}), semiconductor nanocrystals, metal clusters, carbon nanotubes, and some fullerenes [36]. Inorganic scintillators have higher photon interaction cross section and high light production efficiency. However, due to their high atomic number they require complicated dose calibration process for scintillation dosimetry.

Only a small fraction of the energy absorbed by the scintillator is converted into visible photons and the rest is dissipated non-radiatively by phonons, vibrational modes within large structures. Light yield defined as number of photons per MeV of absorbed radiation is one of the most important parameters of scintillators to quantify their scintillation efficiency. Anthracene has the highest light yield among organic scintillators and often used as a reference for comparison. Its light yield is $\sim 2 \times 10^4$ photons per MeV that corresponds to only ~5% efficiency in the blue region of visible spectrum. The light yield of typical plastic scintillators is ~40–65% of that of anthracene. NaI(Tl) is a typical inorganic scintillator with light yield of ~10%.

Plastic scintillators are composed of a base material, usually polyvinyltoluene (PVT), polystyrene, or acrylic-naphthalene and one or more organic dyes. When fast electrons or other charged particles pass through the scintillator, ~3% of the energy deposited is ultimately converted to scintillation photons in the range of 400–500 nm. A plastic scintillator is relatively transparent to its own light (attenuation length ~4 m), is very fast (<10 ns response time), is highly resistant to radiation damage (~3% loss per 10^4 Gy) and can be drawn into thin fibers or formed into sheets. The scintillation signal is independent of dose rate, total dose, and angle of incidence. Plastic scintillator can be used to form small, rugged, moderately sensitive detectors and with an atomic number that approximately matches that of water.

Emission of observable scintillation is a complex two or three stage process [27]. A simplified description of this process is illustrated in Figure 9.3. The primary process consists of excitation of PVT base molecules (often called solvent molecules), S, by secondary electrons. Only excitation of the weakly bound and delocalized phenyl π-electrons (vs. more tightly bound, inner-shell σ-electrons), can result in fluorescence. Subsequently, an excited base molecule, S*, can dissipate its excitation energy by several mechanisms: self or internal quenching, (S* → S + heat), transfer of its excitation energy to another solvent molecule (S* + S → S + S*), or emission of a fluorescence photon (S* → S + hv). Since the UV photons produced by S* de-excitation in pure plastic base have a low yield and are strongly absorbed by plastic and water-like media, pure plastic scintillator has little practical value. To improve the quantum yield of the process and to shift the scintillation light to longer

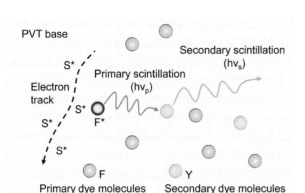

FIGURE 9.3 Simplified description of scintillation mechanism for a ternary plastic scintillator that uses both primary and secondary wave-shifting dyes to produce visible photons in response to ionizing radiation. S* denote excited plastic base molecules, F and F* denote ground state and excited primary dye molecules, respectively, while Y and Y* denote ground and excited states, respectively, of secondary dye molecules.

wavelengths, primary dyes, F, and secondary wave-shifting dyes, Y, are added. The primary dye, F, typically PBD, b-PBD (butyl-PBD), or p-terphenyl in ~1% concentration, absorbs energy from the excited PVT or polystyrene base molecule via dipole–dipole non-radiative or radiative transfer processes and itself enters an excited state: S* + F → S + F*). De-excitation quickly occurs, (F* → F + hv$_p$) accompanied by emission of 320–420 nm "primary photons." This primary process is localized, occurring within 1 mm of the original electron track. The primary photons travel through the plastic with a 0.1–3 mm mean-free path, until absorbed by the secondary dye molecules (typically bisMSB, t-PBD, or d-POPOP in ~0.01% concentration). Energy absorption by the secondary dye molecule (hv$_p$ + Y → Y* → hv$_s$) is followed by reemission of 400–530 nm longer wavelength photons, hv$_s$, which constitute the final and easily observable optical signals.

9.4 DETECTORS

As mentioned in previous section, scintillation is not a very efficient process. However, highly efficient dosimeters can be developed by using photodetectors with high photon collection and efficiency [37]. Commonly used photodetectors in scintillation dosimetry are photomultiplier tubes (PMTs) and photodiodes [38]. When arrays of fibers or a scintillator sheet are used for 2D measurements, multichannel PMTs, photodiode arrays, or charge-coupled device (CCD) and complementary metal-oxide semiconductor (CMOS) cameras can be used as readout systems [39–42].

9.4.1 Photomultiplier Tubes

Conversion of scintillation photons into an electrical signal and amplification of the signal are two functions of a PMT. A PMT consists of an evacuated glass tube with a very thin electrode (called a photocathode) at its entrance, several electrodes (called dynodes) in the interior, and an anode located at the end as the collector of electrons. Scintillation photons eject electrons from the photocathode as a result of photoelectric effect. Approximately one electron is released per five photons incident on the photocathode. The electrons released from the photocathode are accelerated toward the first dynode. Approximately five electrons are ejected as a result of impact of each electron with the dynode. These secondary electrons are then accelerated toward the next dynode (which is held at a more positive potential than the preceding one) which further increases their population upon incident with the dynode. This exponential increase in the population of electrons is repeated several times by letting the electrons pass through a series of dynodes. Typical commercial phototubes may have up to 15 dynodes. The electric field between dynodes is established by applying a successively increasing positive high voltage to each dynode. The voltage difference between two successive dynodes is ~100 V. The end result of this process is an output pulse having amplitude large enough to be easily measured by the associated electronics [37]. The gain of a PMT is strongly dependent on the applied voltage, it can be higher than ~10^6. The electronics should be carefully selected since small changes in the high voltage applied across the cathode and anode can significantly change the output.

PMTs are manufactured for working in ultraviolet, visible, and infrared wavelengths. Important parameters that should be considered in identifying a PMT are their spectral response, quantum efficiency, sensitivity, and dark current. These parameters are determined by the composition of the photocathode. Generally, the long-wavelength cutoff is determined by the photocathode, while the short-wavelength cutoff is determined by the window material. The quantum efficiency of the photon–electron conversion process in the photocathode is low, typically ~20–30% that implies ~80–70% of the scintillation photons are wasted. For optimized results, the spectrum of the scintillator should match the sensitivity of the photocathode. The dark current consists mainly of the thermionic emission of electrons from the photocathode and the first few dynodes, with much smaller contribution from cosmic rays. For example a 5-cm-diameter photocathode may release ~10^5 electrons per second in the dark at room temperature. Cooling of the photocathode greatly reduces the dark current, e.g., by a factor of ~10 for temperature reduction from 20°C to 0°C. However, caution must be exercised to avoid condensation at the PMT window since the moisture will reduce the amount of light incident on the photocathode. In addition, excessive cooling can cause voltage drop across the photocathode. Other ambient conditions influencing the performance of the PMT are changes in the humidity, presence of vibrations and magnetic field.

9.4.2 Photodiodes

A photodiode detector is made of a semiconductor material with appropriate doping optimized to work around the visible spectrum. The scintillation photons produce electron–hole pairs instead of just electrons as in a PMT photocathode. The quantum efficiency of a photodiode can be as high as ~80% which makes it attractive for use in low-level radiation detection. PMTs generally have higher sensitivity than photodiodes. Photodiodes produce more electrical noise than PMTs do but they are smaller and less expensive. Most photodiodes do not amplify the signal. A special kind of photodiodes is avalanche photodiode (APD) where electron–hole pairs produced are multiplied through avalanche process. The process is analogous to the electron multiplication in a PMT except that in this case there are no dynode-like mechanical structures involved. The end result of the electron–hole multiplication is the transformation of a very low-level signal into a pulse having measureable amplitude [37]. The signal amplification in APD is not as much as that in a PMT.

A systematic evaluation of passive (PIN photodiode) and active (APD and PMT) photodetectors performance for plastic scintillation dosimetry based on the expected dose rate (<1 mGy/s, 1–10 mGy/s, >10 mGy/s) and beam energy (kV and MV energy range) is reported in [38]. The main issue associated with the passive photodetectors was their extremely low output signal in ~pA range that required a very sensitive electrometer to obtain reliable measurements. The APD and PMT generated ~2 × 10^3 and ~3 × 10^4 times more current compared to the

PIN system, respectively. Active photodetectors showed higher SNR at low dose rates and low energies. At high dose rates, the SNR of passive photodetectors were on par with active photodetectors. The APD failed to read high dose rates because of the signal saturation. No temperature issues with the PMT were reported; the APD module, however, was affected by the heat. It was recommended to turn on the APD system ~1 hour prior to measurements to avoid background drifting [38].

9.5 CHERENKOV RADIATION IN OPTICAL FIBERS

Cherenkov radiation was discovered by Pavel Čerenkov, for which he shared the 1958 Nobel Prize in Physics with Ilya Frank and Igor Tamm [43–45]. Cherenkov radiation has attracted a considerable amount of recent research interest for its potential applications in life sciences and engineering, such as in molecular imaging, particle detection, ionizing radiation quality assurance, and beam monitoring [46–64]. Cherenkov radiation is a visible light emitted from a dielectric medium when charged particles with velocities greater than the phase velocity of light in that medium, i.e., $v > c/n$, pass through it. The passage of the charged particles induces dipole oscillations through polarization of the medium whose relaxation leads to emission of light when $v > c/n$ due to the constructive interference of the emitted waves.

Cherenkov radiation is a polarized, coherent, and directional emission; its direction is along the surface of a cone that makes the half-angle $\theta = \cos^{-1}(n\beta)^{-1}$ with the particle track, where n is the refractive index of the medium and $\beta = v/c$ is the ratio of the velocity of the particle to that of light. To induce Cherenkov radiation, a charged particle must satisfy the $v > c/n$ condition; the minimum energy required is given as

$$E_{\min} = m_0 c^2 \left(\sqrt{\frac{1}{1-n^{-2}}} - 1 \right) \tag{9.1}$$

where m_0 is the rest mass of the particle.

The threshold electron energies to generate Cherenkov radiation in water ($n = 1.33$), PMMA ($n = 1.5$), and pure silica ($n = 1.55$) are $E_{\min,e} = 264$, 174, and 158 keV, respectively. These energies are far below the energies of megavoltage beams used in modern radiotherapy. The passage of these high energy primary and secondary electrons through the

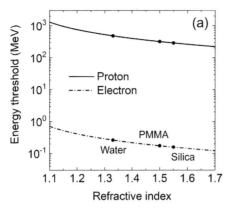
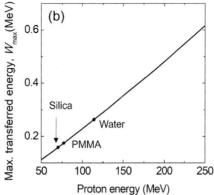

FIGURE 9.4 (a) Minimum energy (E_{min}) of an electron and proton required to generate Cherenkov radiation as a function of the refractive index of the medium. (b) Maximum transferred energy (W_{max}) to an electron in a single collision with a proton as a function of the proton's energy.

fiber optic dosimeters generates Cherenkov light that is the dominant source of the unwanted background signal in the output signal.

The minimum energy of an electron and a proton to generate Cherenkov radiation is plotted in Figure 9.4(a) as a function of the refractive index. It can be seen that a proton requires much higher energy (~1830 times higher) than an electron to generate Cherenkov radiation in a given medium. For example, the threshold energies of a proton to directly generate Cherenkov radiation in water, PMMA, and pure silica are $E_{min,p} = 484$, 320, and 290 MeV, respectively. It should be mentioned that generation of Cherenkov radiation from high energy protons has been observed [65]; however, conventional clinical proton beams have energies below the threshold energy required for directly inducing Cherenkov radiation. Therefore, the incident clinical proton beams cannot directly generate Cherenkov radiation in water. Nevertheless, a proton can transfer enough kinetic energy ($W_{max} > E_{min,e}$) to the electrons in the medium so that the electrons set in motion by the protons can generate Cherenkov radiation. For a particle with rest mass M_0 and kinetic energy E_K, maximum energy transfer (W_{max}) to an electron in a single collision is calculated from Equation 9.2 [66]

$$W_{max} = \frac{2m_e c^2 \beta^2 \gamma^2}{1 + \frac{2\gamma m_e}{M_0} + \left(\frac{m_e}{M_0}\right)^2} \qquad (9.2)$$

where m_e and M_0 are the rest masses of an electron and proton, respectively, and $\gamma = E/(M_0 c^2) = (E_K + M_0 c^2)/(M_0 c^2)$ and $\beta = \frac{v}{c} = \sqrt{1 - \gamma^{-2}}$.

The maximum transferred energy (W_{max}) to an electron in a single collision with a proton as a function of the proton's energy, calculated from Equation 9.2 is plotted in Figure 9.4(b). It can be seen that $W_{max} = E_{min,e}$ in pure silica, PMMA, and water for 70, 77, and 114 MeV proton energy, respectively, indicating that for proton beams with energies greater than the above values Cherenkov radiation can occur from the secondary electrons liberated as a results of columbic interaction of the protons. However, it has been shown that the amount of generated Cherenkov radiation from the secondary electrons liberated as a results of interaction of the proton beam with the medium is not proportional to the absorbed dose in the medium [10, 11].

The number of Cherenkov photons generated by a charged particle with charge ze, where e is the elementary charge, along path length dl in the wavelength region between λ_1 and λ_2 ($\lambda_1 < \lambda_2$) is proportional to λ^{-2} and is given as

$$\frac{dN}{dl} = 2\pi\alpha z^2 \left(\frac{1}{\lambda_1} - \frac{1}{\lambda_2}\right)\left(1 - \frac{1}{n^2\beta^2}\right) \qquad (9.3)$$

where $\alpha = e^2/(4\pi\epsilon_0 \hbar c) \approx 1/137$ is the fine structure constant. The amount of Cherenkov light contamination recorded by the optical fiber dosimeter depends on the angular configuration and spatial position and therefore is not constant, so straightforward subtraction as a calibration constant cannot be done. However, the spectral characteristic of the Cherenkov radiation can be used to decompose the output signal through rigorous spectroscopy. Specifically, Cherenkov radiation has a continuous

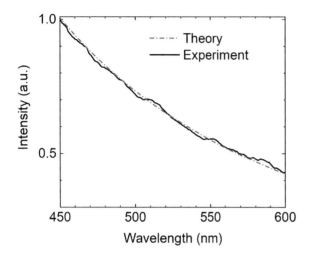

FIGURE 9.5 Theoretical spectrum of Cherenkov radiation over visible spectrum with λ^{-3} intensity dependency on wavelength. A typical spectrum of Cherenkov radiation collected by a silica optical fiber irradiated with a clinical electron beam is plotted for comparison.

spectrum spanning from near-ultraviolet to near-infrared, restricted from both ends of the visible spectrum by the absorption spectrum of the material in which it is generated, with light intensity decreasing proportional to λ^{-3} as the wavelength increases. For comparison, a typical experimental spectrum of Cherenkov radiation observed in an optical fiber irradiated with a clinical electron beam is presented in Figure 9.5, which shows a good agreement between experimental and theoretical spectra.

When optical fibers pass through ionizing radiation fields of high energy, such as in fiber optic dosimetry, Cherenkov radiation generated inside the fibers' core is guided through the fiber if the emitted ray hits the core-cladding boundary with angle greater than the critical angle to satisfy total internal reflection condition required for guided rays. Transmission of Cherenkov radiation is, therefore, dependent on the angle between the particle track and the fiber axis (see Figure 9.6), indicating that the recorded raw signal is not directly related to the absorbed dose by the scintillator.

The intensity of Cherenkov radiation can be ~2 orders of magnitude per mm of transport fiber lower than the intensity of the scintillation signal per mm from the scintillator piece [3]. However, the length of transport fiber in radiation field is much longer than the scintillation part (several cm vs. a few mm), which can lead to a significant Cherenkov contamination in the total signal recorded by the photodetector. In scintillation fiber optic dosimetry, Cherenkov light is an unwanted background signal and separating it from the total optical signal is crucial; using raw signals without removing that leads to a pronounced (up to ~20%) discrepancy. It should be mentioned that in addition to the Cherenkov radiation, the radioluminescence of the bare transport fibers can contribute as an unwanted parasitic signal in photon and electron dosimetry using scintillators [67]. However, their contribution is orders of magnitude lower than the Cherenkov light and can be safely neglected [68].

A comparison study of several transport fiber materials, including polystyrene, PMMA, and silica fibers with low-OH content ("dry" silica) and high-OH content ("wet" silica) irradiated with x-ray beam with <150 kVp energy, showed that bare "dry" silica fibers produce ~4 orders of magnitude stronger radioluminescence than the bare PMMA fibers. The observed signal is the radioluminescence in these fibers because the low energy range selected for irradiation cannot generate Cherenkov radiation in the fibers [67]. The radioluminescence in PMMA fibers reported to be ~1–2 orders

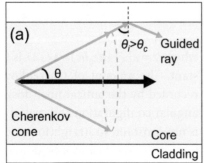
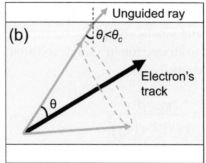

FIGURE 9.6 Schematic illustration of Cherenkov radiation generated in the fiber when it is (a) guided through and (b) leaked from the fiber. θ is the Cherenkov semi-cone angle and θ_i is the incident angle on the core-cladding boundary.

of magnitude weaker than that of polystyrene fibers. PMMA fibers are the preferred choice of transport fiber material in fiber optic scintillation dosimetry due to their mechanical flexibility and lower economic cost (common to plastic fibers).

Several methods have been proposed to minimize the influence of Cherenkov radiation contamination in order to improve the accuracy of fiber optic dosimetry [69–75]. These methods include the (i) Subtraction method based on using a parallel bare fiber identical to the one that is connected to the scintillator piece to produce similar Cherenkov light that can be subtracted from the total signal [76, 77]. However, this technique is not reliable for radiation fields with high-dose gradient. (ii) Optical filtering where a long-wavelength-emitting scintillator in conjunction with a long pass filter is used to selectively measure the signal in longer wavelengths of the spectrum where the intensity of the Cherenkov radiation is weaker due to its λ^{-3} intensity profile [78]. However, this method is not very effective since the filtered signal is still contaminated with Cherenkov radiation due to the fact that the Cherenkov radiation has a continuous spectrum. (iii) Temporal separation that relies on different time scales associated with Cherenkov emission (~ps) and scintillation (~ns or longer) processes [79, 80]. In this methods, gated detection is utilized to collect the signal, since the Cherenkov emission stops immediately (~ps) after the radiation pulse, leaving the radioluminescence of the scintillator to be detected. This method requires fast responding electronics and works only with pulsed radiation fields. (iv) Chromatic removal that requires two different optical filters to measure the signal at two different spectral regions; the dose is then calculated by using coefficients obtained from calibration [81–83]. (v) Rigorous spectral separation based on acquiring the whole spectrum of the transmitted signal and decomposing it into its constituting components using *a priori* knowledge of the spectral shape of the scintillation signal and Cherenkov radiation [73–75]. (vi) Using HWG with air core instead of conventional solid-core fibers would reduce the deteriorating effect of Cherenkov radiation [84] due to the fact that the production of Cherenkov light is minimal in air since its refractive index is very close to 1; such HWGs, however, conventionally have optimal design parameters for transmission of infrared wavelengths [33], whereas the scintillators of interest in fiber optic dosimeters emit primarily visible light. It has been shown that HWGs based on silver-only coating have higher transmission compared to their counterparts with silver/dielectric coating and their performance in scintillation fiber optic dosimetry has been investigated recently [85–87].

9.6 IONIZATION QUENCHING AND LET DEPENDENCIES

As mentioned earlier, a significant issue related to scintillation dosimetry that occurs in proton and other high-LET fields is the non-proportionality between the collected light from the scintillator and the proton dose. The light yield of a scintillating material depends not only on the energy of the ionizing particle, but also on the stopping power of the particle. Birks' semi empirical formula, Equation 9.4, is commonly used to determine the quenching factor of an organic scintillator [26, 27]:

$$\frac{dL}{dr} = \frac{S\frac{dE}{dr}}{1+kB\frac{dE}{dr}} \qquad (9.4)$$

where r is the residual range of the particle, $\frac{dL}{dr}$ is the scintillation yield along the particle's track, S is the absolute scintillation factor, $B\frac{dE}{dr}$ is the density of excitation centers along the track, k is the quenching factor, and $\frac{dE}{dr}$ is the stopping power.

Ionization quenching in plastic scintillators can lead to ~10% under measurement of the dose near the Bragg peak. In principle, quenching correction factors can be empirically determined [21–23]. Such factors, however, depend on the scintillator material and the energy spectrum of the proton beam.

It was suggested that the visible emission from irradiated bare plastic fiber optics, i.e., without the additional scintillator tip, can be used for proton therapy dosimetry [88, 89]. A remarkable distinction of this method compared to conventional fiber scintillators was that the signals from the bare fibers were minimally affected by the radiation quenching effect. The origin of this visible emission was initially attributed to Cherenkov radiation [88, 89]. However, Monte Carlo simulations did not predict direct correlation between the absorbed dose and Cherenkov light production in a typical clinical proton beam [90–92]. The nature of the visible emission responsible for proton dose measurement using PMMA plastic and silica glass bare optical fibers was studied in [10, 11] by performing rigorous luminescent spectroscopic study validated with Monte Carlo simulations.

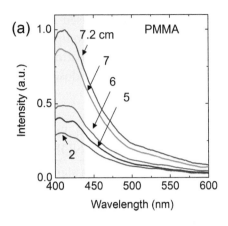 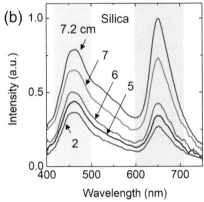

FIGURE 9.7 Radioluminescence spectra of (a) a bare PMMA fiber with peak emission at 410 nm and (b) a silica glass fiber (with high-OH content) with emission peaks at 460 nm and 650 nm, placed at different depths in solid water phantom irradiated with 100 MeV proton beam.

Monte Carlo simulation results, along with the observed experimental spectra, confirmed that Cherenkov radiation cannot be responsible for the proton dose measurement ability of bare optical fibers [93–96].

The luminescent spectroscopy of the visible light generated in proton-irradiated PMMA plastic bare fibers shows a continuous spectrum with a peak at 410 nm with spectral shape different from the spectrum of Cherenkov radiation, see Figure 9.7(a) [10]. Comparing the fiber's spectrum with a typical spectrum of plastic scintillator indicated that the nature of this visible light is the radioluminescence of the fiber material. A good agreement between depth dose measurement using ion chamber array and the fiber was observed. The fact that the acquired signal suffers minimal from quenching effect, which is promising in designing such devices for high-LET beam monitoring, may be qualitatively explained by the lower density of excitation centers in the bare fiber. In the case of the bare fiber, the density of the excitation centers along the track is very low, i.e., $B\frac{dE}{dr} \ll 1$, that leads to $\frac{dL}{dr} \propto S\frac{dE}{dr}$ indicating the minimal effect of ionization quenching effect in the bare fiber.

In contrast to PMMA bare fiber, the emission spectrum of the silica glass fiber irradiated with a proton beam shows two distinct peaks at 460 (blue) and 650 (red) nm, see Figure 9.7(b) [11], whose origin is connected to the silica point defects, namely oxygen-deficiency center (ODC) and non-bridging oxygen hole center (NBOHC) defects, respectively [97–100]. The intensity of the latter correlates with the absorbed dose, whereas that of the former does not directly follow the dose. The fact that the ratio of the signal peak intensities at 460 and 650 nm varies

with the depth in a phantom suggests a dependency on the LET of the beam and can be carefully examined for LET sensing and high-LET beam monitoring. It should be mentioned that in bare PMMA and bare silica fibers, the intensity of the radioluminescent signal is orders of magnitude weaker than the signal from conventional scintillators (e.g., BCF-12).

9.7 CHARACTERISTICS OF A COMMERCIALLY AVAILABLE SYSTEM

The first commercially available fiber optic dosimetry system, Exradin W1 (Standard Imaging, Inc., Middleton, WI), was released in 2013. The fiber scintillator is polystyrene with acrylonitrile butadiene styrene (ABS) plastic enclosure and polyimide stem. The sensitive scintillating piece has density of 1.05 g/cm^3, 1.0 mm diameter, 3.0 mm length (0.0024 cm^3 volume). Housing of the tip has 2.8 mm diameter and 42 mm length. The optical fiber used to transmit the scintillation signal to the photodiode readout system is made of PMMA with polystyrene jacket with 1.0 mm core diameter and 2.2 mm jacket diameter, and 3 m total length. The manufacturer provides a procedure for dose calibration (obtain the "gain" coefficient) and Cherenkov radiation correction. Several groups have studied its characteristics, such as short-term repeatability, dose–response linearity, angular dependency, temperature dependencies, dose-rate dependencies, and long-term stability [101–107].

The Exradin W1 was characterized under standard conditions and short-term reproducibility of 0.3% and dose-rate independence within 0.5% was reported [101].

Consistency of the measure dose per MU was ~0.1% (~2%) when 50–100 MUs (5–100 MUs) were delivered; however it was increased to ~5% for extreme case of 1 MU delivery, indicating that great consideration must be made when using such a system for very low dose measurements. The calibration factor was noted to be beam quality dependent and 1.5% ± 1.4% difference from ion chamber measurement was reported. Variations in gain coefficient ~3% was found when comparing dose-to-water calibration on different days, indicating that the calibration needs to be performed each time that the system is used for absolute dose measurement. Field size dependence was studied by measuring the output factors and 3.3% discrepancy with ion chamber measurements was noted in very large field sizes, indicating that the considered Cherenkov radiation correction method is not very accurate for such large field sizes [101].

In another study [102], maximum angular dependency ~0.34% and decrease in light output with temperature ~0.2–0.3% per °C was reported. The "effective point of measurement" was calculated by comparing the tissue maximum ratio (TMR) curves measured with an ionization chamber and the scintillator system; its location was found to be at 0.6 mm distance upstream with respect to the central axis of the Exradin W1. The loss of the signal per accumulated dose was 0.28% per kGy (for 0–15 kGy) and 0.032% per kGy (for 15–127 kGy) [102–104].

For PDD measurement of photons and electrons, on average 0.5% and 0.6% accuracy is reported comparing to a CC04 (IBA Dosimetry GmbH, Germany) ion chamber, respectively [108]. The PDD measurement in continuous scanning mode using Exradin W1 was not available, and the measurement was done in step-and-shoot method. Average precision ~0.17% and daily reproducibility of ~0.14% was reported for 2 Gy dose measurement.

9.8 SUMMARY

Scintillation fiber optic dosimeters have unique practically advantageous properties including the ability to perform in vivo, real-time, and intracavitary measurements with high spatial resolution due to their small physical size and mechanical flexibility. These features make them ideal candidates for many applications in brachytherapy, intensity-modulated radiation therapy, superficial therapy, stereotactic radiosurgery, proton therapy, and small-field dosimetry. Individual and arrays of fibers can be used for 1D and 2D measurements, respectively. Cherenkov radiation generated in optical fibers is one of the main issues with photon and electron dosimetry and various solutions for dealing with it is available. However the user must have a full understanding of the accuracy of the Cherenkov radiation correction method used in their system, as well as other characteristics of their scintillating fiber optic dosimetry system. For in vivo dosimetry applications, the temperature dependency should be evaluated in advance and properly addressed. Cherenkov radiation generation by proton beams is of less importance, however, ionization quenching effect is the main issue with using scintillators for proton therapy and other high LET beams. There is significant amount of work directed toward making scintillating fiber optic dosimeters for proton therapy, especially those with the ability to simultaneously measure LET and dose.

REFERENCES

1. G. F. Knoll, *Radiation Detection and Measurement* (John Wiley & Sons, Inc., Hoboken, NJ, 2010).
2. N. Tsoulfanidis and S. Landsberger, *Measurement and Detection of Radiation* (CRC Press, Boca Raton, FL, 2015).
3. L. Beaulieu and S. Beddar, "Review of plastic and liquid scintillation dosimetry for photon, electron, and proton therapy," *Physics in Medicine and Biology* **61**(20), R305–R343 (2016).
4. S. Beddar and L. Beaulieu, *Scintillation Dosimetry* (CRC Press, Boca Raton, FL, 2016).
5. J. Lambert, D. R. McKenzie, S. Law, J. Elsey, N. Suchowerska, "A plastic scintillation dosimeter for high dose rate brachytherapy," *Physics in Medicine and Biology* **51**(21), 5505–5516 (2006).
6. B. Mijnheer, S. Beddar, J. Izewska, C. Reft, "*In vivo* dosimetry in external beam radiotherapy," *Medical Physics* **40**(7), 070903 (2013).
7. F. Therriault-Proulx, T. M. Briere, F. Mourtada, S. Aubin, S. Beddar, L. Beaulieu, "A phantom study of an in vivo dosimetry system using plastic scintillation detectors for real-time verification of 192Ir HDR brachytherapy," *Medical Physics* **38**(5), 2542–2551 (2011).
8. L. Wootton, R. Kudchadker, A. Lee, S. Beddar, "Real-time in vivo rectal wall dosimetry using plastic scintillation detectors for patients with prostate cancer," *Physics in Medicine and Biology* **59**(3), 647–660 (2014).
9. I. Veronese, N. Chiodini, S. Cialdi, E. d'Ippolito, M. Fasoli, S. Gallo, S. La Torre, E. Mones, A. Vedda, G. Loi, "Real-time dosimetry with Yb-doped silica optical fibres," *Physics in Medicine and Biology* **62**, 4218–4236 (2017).

10. A. Darafsheh, R. Taleei, A. Kassaee, J. C. Finlay, "The visible signal responsible for proton therapy dosimetry using bare optical fibers is not Čerenkov radiation," *Medical Physics* **43**(11), 5973–5980 (2016).

11. A. Darafsheh, R. Taleei, A. Kassaee, J. C. Finlay, "Proton therapy dosimetry using the scintillation of the silica fibers," *Optics Letters* **42**(4), 847–850 (2017).

12. J. Lambert, Y. Yin, D. R. McKenzie, S. H. Law, A. Ralston, N. Suchowerska, "A prototype scintillation dosimeter customized for small and dynamic megavoltage radiation fields," *Physics in Medicine and Biology* **55**(4), 1115–1126 (2010).

13. D. Létourneau, J. Pouliot, R. Roy, "Miniature scintillating detector for small field radiation therapy," *Medical Physics* **26**(12), 2555–2561 (1999).

14. M. D. Belley, I. N. Stanton, M. Hadsell, R. Ger, B. W. Langloss, J. Lu, O. Zhou, S. X. Chang, M. J. Therien, T. T. Yoshizumi, "Fiber-optic detector for real time dosimetry of a micro-planar x-ray beam," *Medical Physics* **42**(4), 1966–1972 (2015).

15. S. Safai, S. Lin, E. Pedroni, "Development of an inorganic scintillating mixture for proton beam verification dosimetry," *Physics in Medicine and Biology*, **49**(19), 4637–4655 (2004).

16. P. Woulfe, F. J. Sullivan, S. O'Keeffe, "Optical fibre sensors: Their role in in vivo dosimetry for prostate cancer radiotherapy," *Cancer Nanotechnology* **7**(1), 7 (2016).

17. L. M. Moutinho, I. F. C. Castro, L. Peralta, M. C. Abreu, J. F. C. A. Veloso, "Brachytherapy dosimeter with silicon photomultipliers," *Nuclear Instruments and Methods in Physics Research Section A* **787**, 358–360 (2015).

18. M. Eichmann and B. Thomann, "Air core detectors for Cerenkov-free scintillation dosimetry of brachytherapy β-sources," *Medical Physics* **44**(9), 4900–4909 (2017).

19. F. Therriault-Proulx, L. Beaulieu, S. Beddar, "Validation of plastic scintillation detectors for applications in low-dose-rate brachytherapy," *Brachytherapy* **16**(4), 903–909 (2017).

20. A. Hitachi, T. Doke, A. Mozumder, "Luminescence quenching in liquid argon under charged-particle impact: Relative scintillation yield at different linear energy transfers," *Physical Review B* **46**(18), 11463 (1992).

21. L. L. W. Wang, L. A. Perles, L. Archambault, N. Sahoo, D. Mirkovic, S. Beddar, "Determination of the quenching correction factors for plastic scintillation detectors in therapeutic high-energy proton beams," *Physics in Medicine and Biology* **57**(23), 7767–7781 (2012).

22. D. Robertson, D. Mirkovic, N. Sahoo, S. Beddar, "Quenching correction for volumetric scintillation dosimetry of proton beams," *Physics in Medicine and Biology* **58**(2), 261–273 (2013).

23. V. I. Tretyak, "Semi-empirical calculation of quenching factors for ions in scintillators," *Astroparticle Physics* **33**, 40–53 (2010).

24. A. Darafsheh, A. Soldner, H. Liu, A. Kassaee, T. Zhu, J. Finlay, "Phosphor-based fiber optic probes for proton beam characterization," *Medical Physics* **42**(6), 3476 (2015).

25. A. Darafsheh, A. Kassaee, R. Taleei, D. Dolney, J. Finlay, "Fiber optic microprobes with rare-earth-based phosphor tips for proton beam characterization," *Proceedings of SPIE* **9700**, 97000Q (2016).

26. J. B. Birks, "Scintillations from organic crystals: Specific fluorescence and relative response to different radiations," *Proceedings of the Physical Society. Section A* **64**(10), 874–877 (1951).

27. J. B. Birks, *The Theory and Practice of Scintillation Counting* (Pergamon Press, Oxford, 1964).

28. T. J. Gooding and H. G. Pugh, "The response of plastic scintillators to high-energy particles" *Nuclear Instruments and Methods* **7**(2), 189–192 (1960).

29. L. Torrisi, "Plastic scintillator investigations for relative dosimetry in proton-therapy," *Nuclear Instruments and Methods in Physics Research B* **170**, 523–530 (2000).

30. Y. Koike, *Fundamentals of Plastic Optical Fibers* (Wiley-VCH, Weinheim, Germany, 2015).

31. A. Darafsheh, *Optical Super-resolution and Periodical Focusing Effects by Dielectric Microspheres* (Ph.D. dissertation, University of North Carolina at Charlotte, 2013).

32. B. E. A. Saleh and M. C. Teich, *Fundamentals of Photonics* (John Wiley & Sons, Hoboken, NJ, 2007).

33. J. A. Harrington, *Infrared Fibers and Their Applications* (SPIE Press, Bellingham, Washington, 2004).

34. J. E. Melzer and J. A. Harrington, "Investigation of silver-only and silver/TOPAS® coated hollow glass waveguides for visible and NIR laser delivery," *Proceedings of SPIE* **9317**, 93170H (2015).

35. S. Braslavsky, "Glossary of terms used in photochemistry," *Pure and Applied Chemistry* **79**(3), 293–465 (2007).

36. B. Valeur and M. N. Berberan-Santos, *Molecular Fluorescence: Principles and Applications* (Wiley-VCH Verlag & Co. KGaA, Weinheim, Germany, 2012).

37. S. N. Ahmed, *Physics and Engineering of Radiation Detection* (Elsevier, Waltham, MA, 2015).

38. J. Boivin, S. Beddar, M. Guillemette, L. Beaulieu, "Systematic evaluation of photodetector performance for plastic scintillation dosimetry," *Medical Physics* **42**(11), 6211–6220 (2015).

39. M. Guillot, L. Beaulieu, L. Archambault, S. Beddar, L. Gingras, "A new water-equivalent 2D plastic scintillation detectors array for the dosimetry of megavoltage energy photon beams in radiation therapy," *Medical Physics* **38**(12), 6763–6774 (2011).

40. J. Son, S. B. Lee, Y. Lim, S. Y. Park, K. Cho, M. Yoon, D. Shin, "Development of optical fiber based measurement system for the verification of entrance dose map in pencil beam scanning proton beam," *Sensors* **18**(1), 227 (2018).

41. A. M. Frelin, J. M. Fontbonne, G. Ban, J. Colin, M. Labalme, A. Batalla, A. Vela, P. Boher, M. Braud, T. Leroux, "The DosiMap, a new 2D scintillating dosimeter for IMRT quality assurance: Characterization of two Cerenkov discrimination methods," *Medical Physics* **35**(5), 1651 (2008).

42. H. Perera, J. F. Williamson, S. P. Monthofer, W. R. Binns, J. Klarmann, G. L. Fuller, J. W. Wong, "Rapid two-dimensional dose measurement in brachytherapy using plastic scintillator sheet: Linearity, signal-to-noise ratio, and energy response characteristics," *International Journal of Radiation Oncology Biology Physics* **23**(5), 1059–1069 (1992).

43. P. A. Čerenkov, "Visible emission of clean liquids by action of gamma radiation," *Doklady Akademii Nauk SSSR* **2**, 451–454 (1934).

44. I. Frank and I. Tamm, "Coherent visible radiation of fast electrons passing through matter," *Comptes Rendus (Doklady) de l'Académie des Sciences de l'URSS* **14**(3), 109–114 (1937).

45. J. V. Jelly, *Čerenkov Radiation and Its Applications* (Pergamon Press, London, 1958).

46. J. L. Demers, S. C. Davis, R. Zhang, D. J. Gladstone, B. W. Pogue, "Čerenkov excited fluorescence tomography using external beam radiation," *Optics Letters* **38**(8), 1364–1366 (2013).

47. R. Zhang, S. C. Davis, J. L. Demers, A. K. Glaser, D. J. Gladstone, T. V. Esipova, S. A. Vinogradov, B. W. Pogue, "Oxygen tomography by Čerenkov-excited phosphorescence during external beam irradiation," *Journal of Biomedical Optics* **18**(5), 50503 (2013).

48. Y. Helo, I. Rosenberg, D. D'Souza, L. Macdonald, R. Speller, G. Royle, A. Gibson, "Imaging Cerenkov emission as a quality assurance tool in electron radiotherapy," *Physics in Medicine and Biology* **59**(8), 1963–1678 (2014).

49. K. Tanha, A. M. Pashazadeh, B. W. Pogue, "Review of biomedical Čerenkov luminescence imaging applications," *Biomedical Optics Express* **6**(8), 3053–3065 (2015).

50. P. Brůža, H. Lin, S. A. Vinogradov, L. A. Jarvis, D. J. Gladstone, B. W. Pogue, "Light sheet luminescence imaging with Cherenkov excitation in thick scattering media," *Optics Letters* **41**(13), 2986–2989 (2016).

51. R. Robertson, M. S. Germanos, C. Li, G. S. Mitchell, S. R. Cherry, M. D. Silva, "Optical imaging of Cerenkov light generation from positron-emitting radiotracers," *Physics in Medicine and Biology* **54**, N355–N365 (2009).

52. R. S. Dothager, R. J. Goiffon, E. Jackson, S. Harpstrite, D. Piwnica-Worms, "Cerenkov radiation energy transfer (CRET) imaging: A novel method for optical imaging of PET isotopes in biological systems," *PLoS One* **5**(10), e13300 (2010).

53. H. Liu, G. Ren, Z. Miao, X. Zhang, X. Tang, P. Han, S. S. Gambhir, Z. Cheng, "Molecular optical imaging with radioactive probes," *PLoS One* **5**(3), e9470 (2010).

54. A. Ruggiero, J. P. Holland, J. S. Lewis, J. Grimm, "Cerenkov luminescence imaging of medical isotopes," *Journal of Nuclear Medicine* **51**(7), 1123–1130 (2010).

55. A. E. Spinelli, D. D'Ambrosio, L. Calderan, M. Marengo, A. Sbarbati, F. Boschi, "Cerenkov radiation allows in vivo optical imaging of positron emitting radiotracers," *Physics in Medicine and Biology* **55**, 483–495 (2010).

56. J. C. Finlay and A. Darafsheh, "Light sources, drugs, and dosimetry," in *Biomedical Optics in Otorhinolaryngology: Head and Neck Surgery*, edited by B. Wong and J. Ilgner (Springer, New York, NY, 2016), Chap. 19, pp. 311–336.

57. A. Darafsheh, S. Najmr, T. Paik, M. Tenuto, C. Murray, J. C. Finlay, J. S. Friedberg, "Characterization of rare-earth-doped nanophosphors for photodynamic therapy excited by clinical ionizing radiation beams," *Proceedings of SPIE* **9308**, 930812 (2015).

58. G. Lucignani, "Cerenkov radioactive optical imaging: A promising new strategy," *European Journal of Nuclear Medicine and Molecular Imaging* **38**(3), 592–595 (2011).

59. A. Darafsheh, T. Paik, M. Tenuto, S. Najmr, J. S. Friedberg, C. Murray, J. C. Finlay, "Optical characterization of novel terbium-doped nanophosphors excited by clinical electron and photon beams for potential use in molecular imaging or photodynamic therapy," *Medical Physics* **41**(6), 436 (2014).

60. B. Brichard, A. F. Fernandez, H. Ooms, F. Berghmans, "Fibre-optic gamma-flux monitoring in a fission reactor by means of Cerenkov radiation," *Measurement Science and Technology* **18**(10), 3257–3262 (2007).

61. J. S. Cho, R. Taschereau, S. Olma, K. Liu, Y. C. Chen, C. K. Shen, R. M. van Dam, A. F. Chatziioannou, "Cerenkov radiation imaging as a method for quantitative measurements of beta particles in a microfluidic chip," *Physics in Medicine and Biology* **54**(22), 6757–6771 (2009).

62. S. Yamamoto, T. Toshito, S. Okumura, M. Komori, "Luminescence imaging of water during proton-beam irradiation for range estimation," *Medical Physics* **42**(11), 6498–6506 (2015).

63. S. Yamamoto, M. Komori, T. Akagi, T. Yamashita, S. Koyama, Y. Morishita, E. Sekihara, T. Toshito, "Luminescence imaging of water during carbon-ion irradiation for range estimation," *Medical Physics* **43**(5), 2455–2463 (2016).

64. S. Yamamoto, S. Koyama, C. Yamauchi-Kawaura, M. Komori, "Luminescence imaging of biological subjects during X-ray irradiations lower energy than Cerenkov-light threshold," *Optical Review* **24**(3), 428–435 (2017).

65. R. L. Mather, "Čerenkov radiation from protons and the measurement of proton velocity and kinetic energy," *Physical Review* **84**(2), 181–190 (1951).

66. Particle Data Group, "Review of particle physics," *Chinese Physics C* **38**(9), 090001 (2014).

67. R. Nowotny, "Radioluminescence of some optical fibres," *Physics in Medicine and Biology* **52**(4), N67–N73 (2007).

68. F. Therriault-Proulx, L. Beaulieu, L. Archambault, S. Beddar, "On the nature of the light produced within PMMA optical light guides in scintillation fiber-optic dosimetry," *Physics in Medicine and Biology* **58**(7), 2073–2084 (2013).

69. M. A. Clift, R. A. Sutton, D. V. Webb, "Dealing with Cerenkov radiation generated in organic scintillator dosimeters by bremsstrahlung beams," *Physics in Medicine and Biology* **45**(5), 1165–1182 (2000).

70. P. Z. Liu, N. Suchowerska, J. Lambert, P. Abolfathi, D. R. McKenzie, "Plastic scintillation dosimetry: Comparison of three solutions for the Cerenkov challenge," *Physics in Medicine and Biology* **56**(18), 5805–5821 (2011).

71. L. Archambault, L. Beaulieu, S. A. Beddar, "Comment on 'Plastic scintillation dosimetry: Comparison of three solutions for the Cerenkov challenge'," *Physics in Medicine and Biology* **57**(11), 3661–3665 (2012).

72. P. Z. Y. Liu, N. Suchowerska, J. Lambert, P. Abolfathi, D. R. McKenzie, "Reply to the comment on: 'Plastic scintillation dosimetry: comparison of three solutions for the Cerenkov challenge'," *Physics in Medicine and Biology* **57**(11), 3667–3673 (2012).

73. A. Darafsheh, R. Zhang, S. C. Kanick, B. W. Pogue, J. C. Finlay, "Spectroscopic separation of Čerenkov radiation in high-resolution radiation fiber dosimeters," *Journal of Biomedical Optics* **20**(9), 095001 (2015).

74. A. Darafsheh, H. Liu, S. Najmr, M. Tenuto, J. S. Friedberg, C. Murray, T. Zhu, J. C. Finlay, "Phosphor-based fiber optic microprobes for ionizing beam radiation dosimetry," *Proceedings of SPIE* **9317**, 93170R (2015).

75. A. Darafsheh, R. Zhang, S. C. Kanick, B. W. Pogue, J. C. Finlay, "Separation of Čerenkov radiation in irradiated optical fibers by optical spectroscopy," *Proceedings of SPIE* **9315**, 93150Q (2015).

76. S. Beddar, T. R. Mackie, F. H. Attix, "Cerenkov light generated in optical fibres and other light pipes," *Physics in Medicine and Biology* **37**(4), 925–935 (1992).

77. W. J. Yoo, S. H. Shin, D. Jeon, K. T. Han, S. Hong, S. G. Kim, S. Cho, B. Lee, "Development of a fiber-optic dosimeter based on modified direct measurement for real-time dosimetry during radiation diagnosis," *Measurement Science and Technology* **24**(9), 094022 (2013).

78. S. F. de Boer, A. S. Beddar, J. A. Rawlinson, "Optical filtering and spectral measurements of radiation-induced light in plastic scintillation dosimetry," *Physics in Medicine and Biology* **38**, 945–958 (1993).

79. M. A. Clift, P. N. Johnston, D. V. Webb, "A temporal method of avoiding the Cerenkov radiation generated in organic scintillator dosimeters by pulsed mega-voltage electron and photon beams," *Physics in Medicine and Biology* **47**, 1421–1433 (2002).

80. B. L. Justus, P. Falkenstein, A. L. Huston, M. C. Plazas, H. Ning, R. W. Miller, "Gated fiber-optic-coupled detector for *in vivo* real-time radiation dosimetry," *Applied Optics* **43**(8), 1663–1668 (2004).

81. J. M. Fontbonne, G. Iltis, G. Ban, A. Battala, J. C. Vernhes, J. Tillier, N. Bellaize, C. Le Brun, B. Tamain, K. Mercier, J. C. Motin, "Scintillating fiber dosimeter for radiation therapy accelerator," *IEEE Transactions on Nuclear Science* **49**(5), 2223–2227 (2002).

82. A. M. Frelin, J. M. Fontbonne, G. Ban, J. Colin, M. Labalme, A. Batalla, A. Isambert, A. Vela, T. Leroux, "Spectral discrimination of Čerenkov radiation in scintillating dosimeters," *Medical Physics* **32**(9), 3000–3006 (2005).

83. M. Guillot, L. Gingras, L. Archambault, S. Beddar, L. Beaulieu, "Spectral method for the correction of the Cerenkov light effect in plastic scintillation detectors: A comparison study of calibration procedures and validation in Cerenkov light-dominated situations," *Medical Physics* **38**(4), 2140–2150 (2011).

84. J. Lambert, Y. Yin, D. R. McKenzie, S. Law, N. Suchowerska, "Cerenkov-free scintillation dosimetry in external beam radiotherapy with an air core light guide," *Physics in Medicine and Biology* **53**(11), 3071–3080 (2008).

85. A. Darafsheh, J. E. Melzer, J. A. Harrington, A. Kassaee, J. C. Finlay, "Radiotherapy fiber dosimeter probes based on silver-only coated hollow glass waveguides," *Journal of Biomedical Optics* **23**(1), 015006 (2018).

86. A. Darafsheh, H. Liu, J. E. Melzer, R. Taleei, J. A. Harrington, A. Kassaee, T. C. Zhu, J. C. Finlay, "Fiber optic probes based on silver-only coated hollow glass waveguides for ionizing beam radiation dosimetry," *Proceedings of SPIE* **9702**, 970210 (2016).

87. A. Darafsheh, "Fiber optic dosimeters based on silver-only coated hollow waveguides for radiation therapy dosimetry," *Medical Physics* **45**(6), E684–E684 (2018).

88. J. Son, M. Kim, D. Shin, U. Hwang, S. Lee, Y. Lim, J. Park, S. Y. Park, K. Cho, D. Kim, K. W. Jang, M. Yoon, "Development of a novel proton dosimetry system using an array of fiber-optic Cerenkov radiation sensors," *Radiotherapy and Oncology* **117**(3), 501–504 (2015).

89. K. W. Jang, W. J. Yoo, S. H. Shin, D. Shin, B. Lee, "Fiber-optic Cerenkov radiation sensor for proton therapy dosimetry," *Optics Express* **20**(13), 13907 (2012).

90. A. K. Glaser, R. Zhang, D. J. Gladstone, B. W. Pogue, "Optical dosimetry of radiotherapy beams using Cherenkov radiation: The relationship between light emission and dose," *Physics in Medicine and Biology* **59**(14), 3789–3811 (2014).

91. A. Darafsheh, R. Taleei, A. Kassaee, J. C. Finlay, "The connection between Cherenkov light emission and radiation absorbed dose in proton irradiated phantoms," *Medical Physics* **43**(6), 3418 (2016).

92. A. Darafsheh, R. Taleei, A. Kassaee, J. C. Finlay, "On the nature of the background visible light observed in fiber optic dosimetry of proton beams," *Medical Physics* **43**(6), 3500 (2016).

93. A. Darafsheh, R. Taleei, A. Kassaee, J. C. Finlay, "On the origin of the visible light responsible for proton dose measurement using plastic optical fibers," *Proceedings of SPIE* **10056**, 100560V (2017).

94. A. Darafsheh, R. Taleei, A. Kassaee, J. C. Finlay, "Proton therapy dosimetry by using silica glass optical fibers," *Proceedings of SPIE* **10058**, 100580B (2017).

95. A. Darafsheh, R. Taleei, A. Kassaee, J. C. Finlay, "Proton therapy dosimetry by using the scintillation of glass and plastic bare optical fibers," *Medical Physics* **44**(6), 2858 (2017).

96. A. Darafsheh, R. Zhang, A. Kassaee, J. Finlay, "Characterization of the proton irradiation induced luminescence of materials and application in radiation oncology dosimetry," *Proceedings of SPIE* **10478**, 1047815 (2018).

97. H. Zhang, Z. Yuan, J. Zhou, J. Dong, Y. Wei, Q. Lou, "Laser-induced fluorescence of fused silica irradiated by ArF excimer laser," *Journal of Applied Physics* **110**, 013107 (2011).

98. L. Skuja, K. Kajihara, M. Hirano, H. Hosono, "Oxygen-excess-related point defects in glassy/amorphous SiO_2 and related materials," *Nuclear Instruments and Methods in Physics Research Section B* **286**, 159–168 (2012).

99. L. Skuja, "Optically active oxygen-deficiency-related centers in amorphous silicon dioxide," *Journal of Non-Crystalline Solids* **239**(1–3), 16–48 (1998).

100. T. Suzuki, L. Skuja, K. Kajihara, M. Hirano, T. Kamiya, H. Hosono, "Electronic structure of oxygen dangling bond in glassy SiO_2: The role of hyperconjugation," *Physical Review Letters* **90**(18), 186404 (2003).

101. A. R. Beierholm, C. F. Behrens, C. E. Andersen, "Dosimetric characterization of the Exradin W1 plastic scintillator detector through comparison with an in-house developed scintillator system," *Radiation Measurements* **69**, 50–56 (2014).

102. P. Carrasco, N. Jornet, O. Jordi, M. Lizondo, A. Latorre-Musoll, T. Eudaldo, A. Ruiz, M. Ribas, "Characterization of the Exradin W1 scintillator for use in radiotherapy," *Medical Physics* **42**(1), 297–304 (2015).

103. A. R. Beierholm, C. F. Behrens, C. E. Andersen, "Comment on "Characterization of the Exradin W1 scintillator for use in radiotherapy" [Med. Phys. 42, 297–304 (2015)]," *Medical Physics* **42**(7), 4414–4416 (2015).

104. P. Carrasco, N. Jornet, O. Jordi, M. Lizondo, A. Latorre-Musoll, T. Eudaldo, A. Ruiz, M. Ribas, "Response to "Comment on 'Characterization of the Exradin W1 scintillator for use in radiotherapy' [Med. Phys. 42, 297–304 (2015)]," *Medical Physics* **42**(7), 4417–4418 (2015).

105. A. Dimitriadis, I. Silvestre Patallo, I. Billas, S. Duane, A. Nisbet, C. H. Clark, "Characterisation of a plastic scintillation detector to be used in a multicentre stereotactic radiosurgery dosimetry audit," *Radiation Physics and Chemistry* **140**, 373–378 (2017).

106. F. Alsanea, L. Wootton, N. Sahoo, R. Kudchadker, U. Mahmood, S. Beddar, "Exradin W1 plastic scintillation detector for *in vivo* skin dosimetry in passive scattering proton therapy," *Physica Medica* **47**, 58–63 (2018).

107. C. Hoehr, C. Lindsay, J. Beaudry, C. Penner, V. Strgar, R. Lee, C. Duzenli, "Characterization of the exradin W1 plastic scintillation detector for small field applications in proton therapy," *Physics in Medicine and Biology* **63**(9), 095016 (2018).

108. L. Archambault and M. Rilling, "Basic quality assurance: Profiles and depth dose curves," in *Scintillation Dosimetry*, edited by S. Beddar and L. Beaulieu (CRC Press, Boca Raton, FL, 2016), Chap. 6, pp. 87–104.

CHAPTER 10

Cherenkov and Scintillation Imaging Dosimetry

Rachael L. Hachadorian*, Irwin I. Tendler*, and Brian W. Pogue

Dartmouth College,
Hanover, New Hampshire

CONTENTS

10.1 INTRODUCTION

The use of radiotherapy in medicine has been practiced with the concept that the radiation beam and dose delivery could only be visualized by direct use of some measurement probe or device, which samples the field. Imaging via film or external portal imaging devices is most common; however, these inherently sample the field and not the dose delivery to the subject. The use of light emission from radiation can also provide a signal, which can be used in real-time visualization of the beam in 2D and 3D, and when planned carefully can directly report on the deposited dose. With the appropriate time-gating and sensitivity, this can provide the potential for video rate information to the clinical physics or radiotherapy teams. This potential is still developing as optimal camera choices are refined and optical imaging geometries are tested in clinical trials. The choice of the type of radio-luminescence signal to image will alter how the imaging is done, and what the application would be. In this chapter, the use of imaging to capture Cherenkov emission or scintillation emission is discussed from fundamental emission characteristics to how the signal can be used in both quality audit and delivery verification applications.

Perhaps the leading discovery in recent years to make this type of imaging more attractive was that the use of

---

* Equal Contribution Authors

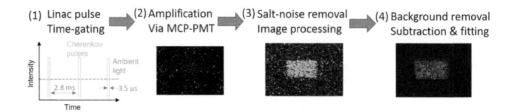

FIGURE 10.1   An illustration of the signal processing steps that go into capturing a cleaned up real-time signal from Cherenkov or scintillation light imaging, starting with (1) time-synchronized detection, (2) followed by amplification in this case by a microchannel plate photomultiplier tube (MCP-PMT), (3) followed by salt noise removal by median filtering in space and time, and finally (4) background subtraction, to suppress any residual ambient light.

camera time-gating synchronized to the linear accelerator (linac) pulses allows for rejection of much of the ambient room lighting and capture of light just during the linac pulse, as illustrated in Figure 10.1. This approach to synchronized imaging, when combined with an amplified system, will allow for real-time video rate capture and could be achieved in ambient room lighting [1]. This discovery has translated into a series of practical tools, which now allow for patient imaging of the radiotherapy beam delivery and accurate dose quantitation [2]. The time synchronization of this and the potential and limitations are outlined here.

Finally, the choice of using Cherenkov or scintillation luminescence to image translates into different applications because the quality of the signal intensity and linearity are different. Cherenkov imaging can provide the signal directly from a patient's tissue, inherently providing a map of the dose directly on the subject, albeit somewhat altered by the tissue optical properties of the patient [3]. Whereas scintillation imaging provides a brighter signal which is known to be highly linear with dose. So, there are strengths and weaknesses associated with both Cherenkov and scintillation imaging, but interestingly, the technology to image these two signals is substantially similar. This is a field that is still emerging, and each tool could turn into either a useful calibration tool or a time-saving clinical verification tool. The range of methods to image and the applications are outlined here.

10.2 CHERENKOV LIGHT

Cherenkov emission is an optical radiation by-product, produced when an electron travels faster than the phase velocity of light in a dielectric medium. In the context of radiation therapy, a linear accelerator delivers the treatment photon beam within the mega-voltage (MV)

range and transfers a substantial fraction of its energy to the secondary electrons within the host medium via Compton scatter interactions. These liberated electrons are scattered into velocities up to $0.75c$, which in a medium with index $n > 1.33$ makes them faster than the speed of light in water. The speed of light in the medium is $c_m = c/n$, where c is the speed of light in a vacuum and n is the index [4].

As high energy electrons travel through a dielectric medium, they polarize and disrupt the local electric fields, causing a torque to the materials polarization, via the valence electrons. The electromagnetic interaction between the relativistic electron velocity and the dipoles of the medium leads to a traveling wave emission, as estimated from semiclassical electromagnetic theory. An optical wavefront is emitted at an angle of approximately 41 degrees from the direction of the traveling electron in water, but this cone of solid angle varies proportionally with the velocity of the particle and the index of the medium [1]. Cherenkov light exists over a wide spectrum of wavelengths from ultraviolet (UV) through to near infrared (NIR), spanning the spectrum over an inverse square relationship ($I = c/\lambda^2$) (Figure 10.2). However self-attenuation by the medium leads to a loss of shorter UV photons, and longer NIR photons, leaving detectable emission largely in the UV-A to visible to short wave NIR wavelengths. By this inverse nature relationship, the shorter wavelengths (blue end) result in the highest intensities. The product is the well-recognized blue glow commonly associated with it and can be seen visually within water in a nuclear reactor pool. It is important to note though that because the light is broadband in nature, the spectrum exiting any medium is distorted by the absorption within that medium and the path of travel to leave it.

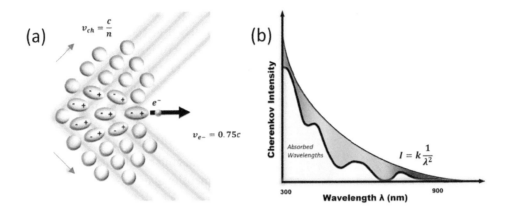

FIGURE 10.2 (a) In radiotherapy, sufficient energy can be transferred to electrons within the tissue to liberate it via Compton scatter, and cause it to travel faster than the velocity of light in water ($v > c_m$). These high-speed electrons polarize the medium transiently, and when the locally disrupted electric field in the medium begins to relax, an optical traveling wave of Cherenkov light is emitted. (b) Cherenkov light is composed of a broad spectrum of wavelengths, from UV through visible to the infrared. This emission spectrum occurs with an inverse square dependence upon wavelength, however within tissue blood absorbs the shorter (blue) wavelengths of light, which substantially shifts the distribution of emitted wavelengths to red and NIR.

Within tissue, the blood in capillaries throughout the volume is a primary absorber of Cherenkov light. Hemoglobin preferentially absorbs out shorter wavelengths of the visible light spectrum, thus serving as a high wavelength passing filter [5]. The smooth, inverse square relationship is radically altered, with the result that the red and NIR wavelengths will make it to the surface of the tissue without being self-attenuated. In some papers, patients reported having observed blue light emitted throughout a stereotactic brain treatment. This was most likely a product of Cherenkov in the vitreous humor of the eye [6–8].

Because the Cherenkov light travels no more than several millimeters *in vivo*, there exists good potential for accurate field verification in clinical imaging, for robust mapping of the exact regions in which the radiation is delivered on the patient's tissue.

10.3 SCINTILLATION LIGHT

Scintillation is a luminescence process – molecular excitation caused by absorption of ionizing radiation results in emission of light during material electronic relaxation. Numerous studies have shown that under appropriate conditions, scintillation signal can be directly proportional to dose [9]. Plastic scintillators used for dosimetry purposes are often composed of a bulk medium and organic fluorophores, where light production is enabled through fluorescence resonance energy transfer between these two components, with

the medium being the conduit for radiation interaction and electron liberation, and the fluorophore being the electron/energy acceptor. A secondary fluorophore (a material that will absorb light at the lower end of the visible light spectrum) is sometimes added to shift the emission spectra to higher wavelengths [10]. Specifically, scintillators emit light when the excited organic fluorophore de-excites via a fluorescence, phosphorescence, or some alternate delayed fluorescence pathway [11].

There exist two main categories of materials used for scintillation dosimetry: organic and inorganic. Popular plastics (organic, usually near water-equivalent) used as base, bulk solvent, materials include polyvinyl toluene (PVT) and polystyrene (PS); these materials are shaped into volumes and fibers. A commercial example of a PVT-based scintillator is Saint Gobain BC-400, while Saint Gobain BCF-12 is PS-based. These scintillators are doped with proprietary organic luminescent agents termed fluors. Some examples of organic fluors used in plastic scintillator dosimeters include anthracene and 3-hydrofluoride. Additionally, certain scintillating materials can be dissolved in solution, thus creating a liquid volume which scintillates. The optical detection of scintillation signal is enabled by the large photon output yield associated with these materials. In the case of liquid scintillators, p-terphenyl and 2,5-diphenyloxazole (PPO) are common organic fluors, which can be incorporated into solution. Other dosimetry systems, such as the Saint Gobain PreLude 420 (cerium-doped lutetium)

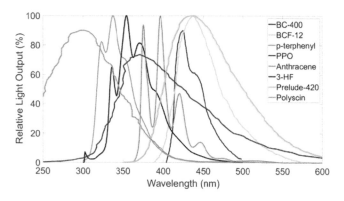

FIGURE 10.3 Emission spectra of various scintillators and organic fluors: plastic (dark and light blue) and crystal (dark and light green) scintillators; solid (gray and black) and liquid (orange and red) organic fluors.

and Polysin NaI(TI) (sodium iodide-doped thallium), utilize inorganic scintillating crystals.

As part of developing an imaging-based scintillator dosimeter, selecting an appropriate scintillating material is highly dependent on the target application. Considering the wavelength of maximum emission is important to ensure high detection efficiency, and here Figure 10.3 shows the range of emission spectra, for example scintillators listed above.

Furthermore, criteria such as scintillator response time (temporal rise and decay of signal) and geometry (dimensions and shape) are other examples of characteristics one would consider when designing a scintillator dosimeter [12]. Examples of scintillator imaging dosimetry systems (1D, 2D, and 3D) are listed below.

10.3.1 1D–Point Measurement

Scintillators have long been used for point dosimetry; many of these scintillators follow a design first suggested by Beddar et al. in 1992: a small ($1 \times 1 \times 1$ mm³)-sized scintillator is coupled to a fiber optic cable. Upon excitation, scintillator light emission is directed into the cable and travels to a receiver, see Figure 10.4A. This signal is converted from analog to digital and undergoes amplification, filtering, and other processes [13, 14]. Modernized versions of this technology (incorporating better Cherenkov elimination techniques and a more robust design compared to earlier versions) have been evaluated for use in proton, photon, and electron dosimetry (both for QA and *in vivo* dosimetry) [15-17]. Optical imaging of scintillator emission for point dose measurements has also recently been utilized within the context

of total skin electron therapy (TSET). Thin scintillator discs attached to the skin surface of patients undergoing TSET; by time-gating an intensified charge-coupled device (ICCD) camera to linac pulses, light output from these plastic discs was captured and related to surface dose remotely and in a rapid manner [18, 19].

10.3.2 2D–Surface and Planar Measurements

Optical imaging of scintillators has been used to conduct two-dimensional dosimetry in a variety of different treatment regimes [20]. For example, it has been shown that 2D visualization of a proton beam is possible using only a commercial digital camera and solid plastic scintillator – this can be useful for computing quenching correction factors and conducting quality assurance beam testing, see Figure 10.4B [21]. Detection of scintillator emission has also been applied to the task of measuring surface dose during EBRT. For example, Jenkins et al. created a flexible, thin, scintillating film using silicone and Gd_2O_2S:Tb (GOS); the sheet was designed to be placed over a patient's skin surface during radiation therapy to track beam movement in real time [22]. Imaging of liquid scintillators has also been used for the purposes of 2D dosimetry. Optical imaging of a light-shielded acrylic tank filled liquid scintillator via electron multiplying charge coupled device has been proven to accurately (compared to treatment plans) measure depth dose lateral profiles and 2D dose distribution for therapeutic photon beams [23].

10.3.3 3D–Volumetric Measurements

Researchers have taken advantage of scintillator volumes for visualizing dose deposition in three dimensions. In the world of proton beam dosimetry, recent innovations have been made to enable to the construction of a real-time 3D system capable of tracking dynamic spot beam delivery and potentially useful for future 3D dose map reconstruction. It should also be noted that this system is capable of high-precision measurements in the low-dose detection regime [24]. To visualize real-time dose deposition in 3D, researchers imaged a volume of scintillator material placed within the beam path. Utilizing a plenoptic camera, linac portal imaging device, and tomographic image reconstruction algorithms, a full, live, 3D reconstruction of dose distribution was achieved [25]. More recently, a plastic scintillator array (with elements extending in 3D) was imaged using a shielded complementary metal-oxide-semiconductor (CMOS) camera,

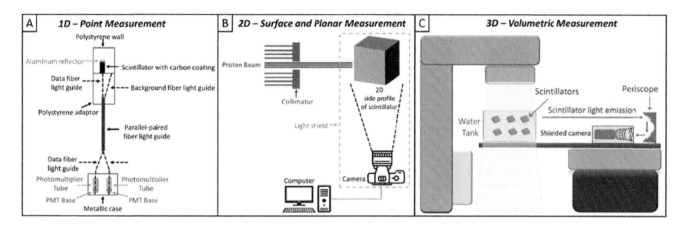

FIGURE 10.4   Examples of 1D [13, 14], 2D [21], and 3D [26] setups used for optical imaging of scintillators. The 1D geometry is for point sampling as used for point dosimetry by Beddar et al. [13, 14]. Planar measurements, B, have been used to profile proton beams for maximum peak depth range analysis by Almurayshid et al. [21]. Finally, 3D imaging can be used with either rotational or angular resolved imaging of the camera, combined with algorithms for tomographic recovery, as shown by a number of groups in very different geometries [24–26].

where light was directed into the camera body using a periscopic mirror and was presented for use in real-time commissioning and verification of dynamic radiotherapy treatments, see Figure 10.4C [26]. Each of these methods is a research study, and it is not obvious, which methodology would gain long range adoption, but each has its strengths and weaknesses depending a lot on the application need.

10.4 LINAC PULSING AND TIME-GATED REAL-TIME IMAGING

While Cherenkov and scintillation imaging has been around for decades, much of the work published has focused around traditional cameras [charge-coupled device (CCD) and CMOS] with frame rates near 1–100 Hz, and long duty cycles [27, 28]. Pulsed or time-gated imaging has allowed Cherenkov and scintillation imaging with an acquisition system (camera) that is synchronized to the linear accelerator pulses. This can be achieved over an optimized time range where the emitted light is emitted from the patient tissue or other source and captured at video frame rates. Considering that a linear accelerator emits ionizing radiation in short 3–5 microsecond pulses at repetition rates driven by the klystron near 100–400 Hz, the camera can be calibrated to acquire images during beam on via a fast triggering mechanism. The fastest image gates tend to be multichannel plate photomultiplier tubes (MCP-PMTs), often called image intensifiers. The image intensifier is turned on and off by electronic gating of

the high voltage, which amplifies the electron signal through the channels. This image intensifier component is critical because it increases the time-resolution capabilities of the optical shutter down to the picosecond scale. Other types of time gates such as mechanical shutters or electronic global shutters operate on the millisecond timescale, and therefore are not sufficient for this fast gating. A unique requirement for these scientific cameras is then to gate during the linac pulses, but then to integrate these pulses on the camera sensor, which images the back end of the MCP-PMT, where a phosphor screen turns the electron image back to an optical image. Depending on the camera/sensor type and software, the user may specify a number of parameters, for example, the number of pulse accumulations to be integrated together, prior to readout of the detector. Triggering of the camera to the linac pulsations is a critical part of the functioning and allows the camera to reject the majority of the ambient light. One Cherenkov or scintillation image therefore can consist of several pulses from the linac, as shown in Figure 10.5, followed by one background exposure. This enables the user to subtract the background light, in real time, thus isolating the Cherenkov light from the image [18]. When attempting to capture Cherenkov or scintillation emission, the goal must be to collect data in a time window directly following a linac pulse, since both are rapid processes. By isolating the same time window over a consistent trigger, the methodology adds reliability and robustness to the clinical acquisition process.

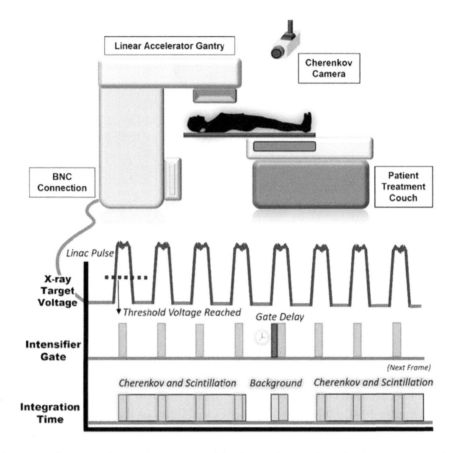

FIGURE 10.5 The linear accelerator is depicted on the top of the figure above, where the gantry rotates about a fixed isocenter. The camera may be mounted to the ceiling for most consistent placement, but may also be mounted to a tripod for temporary use. Below, the camera activity is illustrated in response to the linac pulses. Once a threshold voltage is reached, the intensifier turns on and remains on over the duration of the intensifier gate. The Cherenkov + scintillation image is acquired and both spatial and temporal median-filtered over several frames, followed by one background exposure. Before the background exposure is collected, a short gate delay is employed.

When the camera is triggered using the pulsed radiation delivery of the linac, this is known as an external trigger. The camera requires the signal from the linear accelerator (an external device) in order to initialize acquisition, as previously discussed. However, it is possible for the operator to have control over the time gate over which Cherenkov is acquired. In contrast, this is known as internal triggering, and employs an "always on" or continuous image acquisition mode. Simply put, the camera is turned on by the user; all optical output is detected and recorded, then turned off by the user. This becomes useful for calibrating, focusing, for previewing the field before the Cherenkov image is acquired, as well as for imaging sources of optical output that are not necessarily Cherenkov light.

Many scintillators in the field of imaging dosimetry are selected to have the material property of fast rise and decay times (order of single nanoseconds); this

is advantageous for accurate and precise time-gated imaging. For example, by setting the imaging system to trigger of a rise in klystron voltage, one can capture scintillator emission within nanoseconds after a linac pulse leaving a remaining time gap, which can be used to collect background signal (image collected when there is no active radiation field, e.g., in between linac pulses) [29].

Determining which camera hardware to utilize in conducting imaging-based dosimetry is critical to achieving accurate and precise measurements within a useful timeframe. Examples of camera types available, and commonly used in radiotherapy imaging, are CMOS, CCD, electron multiplying-intensified charge coupled device (EM-ICCD), and ICCD [30]. The cost of these camera systems ranges in magnitude, thus, identifying and selecting the correct imaging criteria is important. For example, one should consider the importance

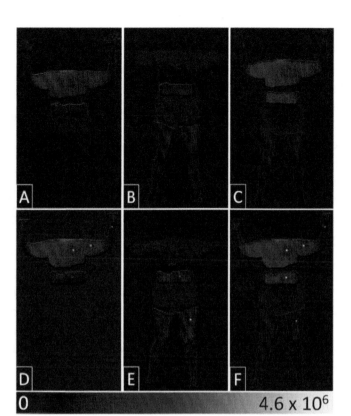

FIGURE 10.6   (A–C) Cherenkov imaging shows the beam delivery in total skin electron therapy to a patient, with the image sequence showing the total dose developed from top and bottom beams and them combined. (D–F) Scintillation imaging of small local plastic scintillating discs show quantitative measures of dose, which can be captured with the same camera set up.

of low light sensitivity, desired frame rate, and background light suppression when choosing equipment for Cherenkov imaging (Figure 10.6) [31].

10.5   STRENGTHS AND WEAKNESSES OF CHERENKOV IMAGING

10.5.1 Strengths

Real-Time Field Verification: The immediate Cherenkov signal from linac pulses allows for image capture in real time. Therefore, the imaging is done while the patient is on the couch, without extra time needed [32]. Conebeam CT or portal imaging provides different diagnostic information during the patient exam, the nature of the information gained with Cherenkov is more related to the deposited dose, instead of position and transmitted beam. Additionally, current software in use allows the team to watch the beam throughout the treatment, and quantitatively assess the treatment immediately postdelivery for patient shifts, day-to-day discrepancies between treatments, and therefore potentially capturing incidents or helping determine when adaptive planning is needed [33, 34].

No Extraneous Dose: Unlike diagnostic beams (such as radiographs and CT scans), which deliver kV-range radiation to large-scale body regions to gain information, Cherenkov emission is a secondary radiation, which essentially yields "free information," and comes with no adverse side effects to the patient, namely dose [35].

Nuclear Medicine Applications: In recent research developments, it has been shown that luminescent compounds can be strategically selected or manufactured such that their excitation wavelengths exist in the Cherenkov spectrum around some of the most abundant Cherenkov emission wavelengths [36]. The phosphorescence can then emitted over a longer time scale, long enough to be imaged much like a radiotracer in Cherenkov-Excited Luminescence Scanned Imaging (CELSI) [37, 38]. The ability to trace a metabolic pathway or any object therein is a common use for nuclear medicine modalities such as PET and SPECT, which require additional modalities. If a patient were concurrently being treated using radiation, further development of CELSI would eliminate the need for the radiation associate with an additional scan [39].

Real-Time in Vivo Dose Verification: A cumulative Cherenkov image is the sum of all the fields of a given delivery. This image is what can be compared to the treatment plan. As previously mentioned, research is being carried out to correct for the discrepancies caused by tissue optical properties, beam entrance and exit differences, and a number of other parameters to establish linearity between dose and Cherenkov light in nonhomogeneous media. Once these parameters have been corrected for, real-time and *in vivo* dose information is feasibly gained. Once finalized, verifying pixel-by-pixel dose immediately after a treatment fraction may become one of the most promising aspects of a cumulative Cherenkov image [40].

10.5.2 Weaknesses

Surface-Limited Dosimetry: The optical wavelengths of Cherenkov emission limit its propagation inside the patient to a few millimeters, by the exponential attenuation with distance. This is its primary weakness because the lower energies have such a high probability of being attenuated out by the time the photon reaches the surface. As a result, the most probable average sampling depth is <10 mm, depending upon the tissue type [41]. As a consequence, real-time field verification (and eventually *in vivo* dose information) can only be used at subcutaneous depths. Therefore, if the tumor or region in question is not at or close to the surface, information may only be gained indirectly, from what is observable at the surface. Thus far, the linearity established between absorbed dose and Cherenkov emission in a homogeneous and transparent medium implies that the imaged surface emission reflects the deposited dose at the surface in homogeneous regions.

Limited to MV Photon and Electron Therapy Beams: While photons and electrons constitute the majority of radiotherapy treatments currently, it neglects proton and heavy ion beams. Although the origin of Cherenkov comes from a relativistic charged particle, the observable effect for proton beams was substantially diminished by comparison to photon and electron beams. Darafsheh et al. demonstrated that the visible signal responsible for proton therapy dosimetry using bare optical fibers is not Cherenkov radiation, yet still the value of beam range finding in this application is compelling [42, 43].

Strengths	Weaknesses
Real-time signal from the actual dose deposition (video rate)	Line-of-sight imaging only
Potential for *in vivo* tissue verification mapping	Surface-limited measurement (<10 mm deep)
Intuitive visualization of procedure for therapy team	Limited to photons and electrons
Molecular sensing potential, excited by Cherenkov *in situ*	Limitations for exceptionally bright illumination rooms

10.6 STRENGTHS AND WEAKNESSES OF SCINTILLATION IMAGING

10.6.1 Strengths

Mature technology of scintillators: The field of scintillator measurement has been advanced now with scintillators that have physical properties that match goal of conducting quantitative optical dosimetry. Due to the fact that large classes of scintillators have been developed as "water equivalent" and can be produced in small in sizes, they can be designed to minimally disturb the applied radiation field. Furthermore, the dose rate independency and linear response to dose for most useful scintillators have been optimized over the years for potential dosimeter use. So, within the scope of imaging-based dosimetry systems, the high photon yield is provided by scintillator emission, and so this matches the concept of remote dosimetry with good signal-to-noise for camera based detection [44].

Stability of signal and linearity with dose: Multiple publications have shown that water-equivalent scintillators are capable of accurately measuring dose in a small-field radiotherapy setting. The ability to prevent field perturbation and avoiding volume averaging effects in converting signal to dose are just a few of the criteria that make optical detection of scintillation effective for small-field dosimetry [45]. These characteristics have also enabled the optical imaging of scintillator emission to be used for dosimetry in brachytherapy treatments [46, 47]. Plastic scintillators have been shown to have strong resistance to radiation damage and, in certain cases, can effectively function independent of temperature (after application of a linear correction factor) temperature independence [15, 48-51].

Insensitivity to optical properties: As compared to Cherenkov imaging, scintillation dosimetry is independent of tissue optical properties; converting light signal to dose is significantly more straightforward when tissue scattering and absorption interactions do not need to be accounted for. As such, direct imaging of scintillators instead of Cherenkov has some significant value [52].

10.6.2 Weaknesses

Complications of mixing Cherenkov & scintillation: Single channel detection for achieving high SNR has been achieved with PIN photodiodes for fast performance at modest cost, while photomultiplier tubes are the optimal choice for high-gradient radiotherapy due to their superior signal sensitivity and larger area [53]. Multiple systems incorporating a small scintillator tip attached to a fiber optic cable for signal transport have been described in the literature (see Chapter 9). The greatest issue these types of dosimetry systems face is generation of Cherenkov within body of the fiber optic cable; in this case, Cherenkov signal is a low-level contamination to the scintillator output. Numerous solutions have been created to combat this issue; these include, but are not limited to: basic optical filtering, two-fiber subtraction, chromatic removal, and multispectral removal [11, 54]. Some groups have attempted to solve the problem of Cherenkov contamination in systems, which use fibers in pulsed radiation environments using removal via temporal filtering; others have focused on removing Cherenkov signal generated in liquid scintillators [55, 56]. While this is a slight weakness, the methods to separate Cherenkov and scintillation have been well developed at this point.

Direct View Limitations: In the case of *in vivo* dosimetry, one of the greatest advantages of imaging scintillator dosimeters lies in the fact that no wires or signal transduction cables are necessary – scintillator output is detected remotely via camera. This fundamental method of signal detection lends well to streamlining and automating dosimetry efforts. However, one of greatest drawbacks is that the dosimeter must be in direct view of the camera. As mentioned previously, many methods avoid this issue by using a fiber optic; however, this eliminates the possibility of wireless signal detection and results in Cherenkov generation within the body of the fiber. Solution employing the use of multiple cameras, mirrors, or unique image reconstruction techniques, and other technical methods has been described in the literature [57, 58]. It should also be noted that potential angular and distance dependencies must be accounted for when dealing with wireless camera-based imaging of scintillator emission.

Strengths	Weaknesses
Linear with dose & dose rate	Line-of-sight imaging
Versatile to different geometries	Cherenkov from fibers in the field can be a confounding signal
Water equivalent scintillators available	Slight angular dependence of signal for remote imaging
Independent of energy for certain scintillators	Distance dependence for remote imaging
Independent of environmental factors such as temperature (following linear correction factor), humidity, and pressure	

10.7 SUMMARY

The use and practical value of scintillation and/or Cherenkov imaging are still being determined as research studies emerge. There is a range of technological realizations of both, which are being evaluated both as devices and as methods for linac QA, pre-treatment QA, and during treatment verification. Much of the device development studies have been around which different camera technologies to use, how to acquire, and to determine if the imaging needs to be done in the setting of normal ambient room lighting. The use of time-gated imaging is quite specialized, and only a few systems allow for this, but it solves the problem of being able to image in the presence of ambient room lighting. The technological systems are still evolving and commercial translation at this point is still embryonic, but this should be changing in the very near future.

Linac or treatment plan QA is the most logical starting point for these technologies, and while the range of devices available for this are many, each has its strengths and weaknesses. The strengths of Cherenkov or scintillation sensing are the real-time nature of the data, and when imaging, the value of remote sensing is important. Human measurement for treatment verification is now widely used for scintillation fiber dosimeters, but beyond that, most efforts are in early-stage clinical trials.

ACKNOWLEDGMENTS

The authors gratefully acknowledge collaboration in all discussions and imaging with other members of the Radiation Oncology and Optics in Medicine research groups at Dartmouth. Additionally, funding from this work comes from NIH grant R01 EB023909 as well as equipment and expertise from DoseOptics LLC.

REFERENCES

1. A. K. Glaser, R. Zhang, S. C. Davis, D. J. Gladstone, B. W. Pogue, "Time-gated Cherenkov emission spectroscopy from linear accelerator irradiation of tissue phantoms," *Optics Letters* **37**, 1193–1195 (2012).

2. C. Snyder, B. W. Pogue, M. Jermyn, I. Tendler, J. M. Andreozzi, "Algorithm development for intrafraction radiotherapy beam edge verification from Cherenkov imaging," *Journal of Medical Imaging* **5**, 015001 (2018).

3. R. Hachadorian, P. Bruza, M. Jermyn, A. Mazhar, D. Cuccia, L. Jarvis, D. Gladstone, B. Pogue, "Correcting Cherenkov light attenuation in tissue using spatial frequency domain imaging for quantitative surface dosimetry during whole breast radiation therapy," *Journal of Biomedical Optics* **24**(7), 071609 (2018).

4. A. K. Glaser, R. Zhang, D. J. Gladstone, B. W. Pogue, "Optical dosimetry of radiotherapy beams using Cherenkov radiation: The relationship between light emission and dose," *Physics in Medicine and Biology* **59**, 3789 (2014).

5. W. G. Zijlstra, A. Buursma, W. P. M. der Roest, "Absorption spectra of human fetal and adult oxyhemoglobin, de-oxyhemoglobin, carboxyhemoglobin, and methemoglobin," *Clinical Chemistry* **37**, 1633–1638 (1991).

6. F. Newman, M. Asadi-Zeydabadis, V. Durairaj, M. Ding, K. Stuhr, B. Kavanagh, "Visual sensations during megavoltage radiotherapy to the orbit attributable to Cherenkov radiation," *Medical Physics* **35**, 77–80 (2008).

7. G. G. Fazio, J. V. Jelley, W. N. Charman, "Generation of Cherenkov light flashes by cosmic radiation within the eyes of the Apollo astronauts," *Nature* **228**, 260–264 (1970).

8. D. T. Blumenthal, B. W. Corn, N. Shtraus, "Flashes of light-radiation therapy to the brain," *Radiotherapy & Oncology* **116**, 331–333 (2015).

9. A. S. Beddar, "Plastic scintillation dosimetry and its application to radiotherapy," *Radiation Measurements* **41**, S124–S133 (2006).

10. S. F. de Boer, A. S. Beddar, J. A. Rawlinson, "Optical filtering and spectral measurements of radiation-induced light in plastic scintillation dosimetry," *Physics in Medicine and Biology* **38**, 945 (1993).

11. L. Beaulieu, S. Beddar, "Review of plastic and liquid scintillation dosimetry for photon, electron, and proton therapy," *Physics in Medicine and Biology* **61**, R305–R343 (2016).

12. I. Tendler, Quantitative Scintillation Imaging for Dose Verification and Quality Assurance Testing in Radiotherapy (Ph.D. dissertation, Dartmouth College, 2020).

13. A. S. Beddar, T. R. Mackie, F. H. Attix, "Water-equivalent plastic scintillation detectors for high-energy beam dosimetry: I. Physical characteristics and theoretical considerations," *Physics in Medicine and Biology* **37**, 1883 (1992).

14. A. S. Beddar, T. R. Mackie, F. H. Attix, "Water-equivalent plastic scintillation detectors for high-energy beam dosimetry: II. Properties and measurements," *Physics in Medicine and Biology* **37**, 1901 (1992).

15. P. Carrasco, N. Jornet, O. Jordi, M. Lizondo, A. Latorre-Musoll, T. Eudaldo, M. Ruiz, M. Ribas, "Characterization of the exradin W1 scintillator for use in radiotherapy: Characterization of exradin W1 for radiotherapy," *Medical Physics* **42**, 297–304 (2016).

16. C. Hoehr, C. Lindsay, J. Beaudry, C. Penner, V. Strgar, R. Lee, C. Duzenli, "Characterization of the exradin W1 plastic scintillation detector for small field applications in proton therapy," *Physics in Medicine and Biology* **63**, 095016 (2018).

17. M. Rahman, P. Brůža, K. M. Langen, D. J. Gladstone, X. Cao, B. W. Pogue, and R. Zhang, "Characterization of a new scintillation imaging system for proton pencil beam dose rate measurements," *Physics in Medicine & Biology* **65**(16), 165014 (2020).

18. P. Bruza, S. Gollub, J. Andreozzi, I. Tendler, B. Williams, L. Jarvis, D. Gladstone, B.W. Pogue, "Time-gated scintillator imaging for real-time optical surface dosimetry in total skin electron therapy," *Physics in Medicine and Biology* **63**, 095009 (2018). https://doi.org/10.1088/1361-6560/aaba19

19. I. Tendler, P. Brůža, J. Andreozzi, M. Jermyn, B. Williams, L. Jarvis, B. Pogue, and D. Gladstone, "Rapid multisite remote surface dosimetry for total skin electron therapy: scintillator target imaging," *International Journal of Radiation Oncology, Biology, Physics* **103**(3), 767–774(2019).

20. D. A. Alexander, R. Zhang, P. Brůža, B. W. Pogue, and D. J. Gladstone, "Scintillation imaging as a high-resolution, remote, versatile 2D detection system for MR-linac quality assurance," *Medical Physics* **47**(9) 3861–3869 (2020).

21. M. Almurayshid, Y. Helo, A. Kacperek, J. Griffiths, J. Hebden, A. Gibson, "Quality assurance in proton beam therapy using a plastic scintillator and a commercially available digital camera," *Journal of Applied Clinical Medical Physics* **18**, 210–219 (2017).

22. C. H. Jenkins, D. J. Naczynski, S.-J. S. Yu, L. Xing, "Monitoring external beam radiotherapy using real-time beam visualization," *Medical Physics* **42**, 5–13 (2015).

23. F. Pönisch, L. Archambault, T. Briere, N. Sahoo, R. Mohan, S. Beddar, M. Gillin, "Liquid scintillator for 2D dosimetry for high-energy photon beams," *Medical Physics* **36**, 1478–1485 (2009).

24. C. D. Darne, F. Alsanea, D. G. Robertson, N. Sahoo, S. Beddar, "Performance characterization of a 3D liquid scintillation detector for discrete spot scanning proton beam systems," *Physics in Medicine and Biology* **62**, 5652–5667 (2017).

25. M. Goulet, M. Rilling, L. Gingras, S. Beddar, L. Beauliue, L. Archambault, "Novel, full 3D scintillation dosimetry using a static plenoptic camera," *Medical Physics* **41**, 082101 (2014).

26. M. W. Jennings, T. P. Rutten, D. J. Ottaway, "Evaluation of the signal quality of an inexpensive CMOS camera towards imaging a high-resolution plastic scintillation detector array," *Radiation Measurements* **104**, 22–31 (2017).

27. B. Robinson, "A new type of Cherenkov imaging detector," *Physica Scripta* **23**, 716 (1981).

28. A. L. Wintenberg, C. G. Moscone, J. P. Jones, G. R. Young, "A CMOS integrating amplifier for the PHENIX ring imaging Cherenkov detector," *IEEE Transactions on Nuclear Science* **45**, 758–763 (1998).

29. M. Ashraf, M. Ramish, P. Bruza, V. Krishnaswamy, D. J. Gladstone, and B. W. Pogue, "Time-gating to medical linear accelerator pulses: Stray radiation detector," *Medical physics* **46**(2), 1044–1048 (2019).

30. J. M. Andreozzi, R. Zhang, A. Glaser, L. Jarvis, B.W. Pogue, D. Gladston, "Camera selection for real-time in vivo radiation treatment verification systems using Cherenkov imaging," *Medical Physics* **42**, 994–1004 (2015).

31. D. A. Alexander, P. Bruza, J. C. M. Farwell, V. Krishnaswamy, R. Zhang, D. J. Gladstone, and B. W. Pogue, "Detective quantum efficiency of intensified CMOS cameras for Cherenkov imaging in radiotherapy," *Physics in Medicine and Biology* **65**(22), 225013(2020).

32. I. Tendler, A. Hartford, M. Jermyn, E. LaRochelle, X. Cao, V. Borza, D. Alexander et al. "Experimentally observed cherenkov light generation in the eye during radiation therapy," *International Journal of Radiation Oncology, Biology, Physics* **106**(2), 422–429 (2020).

33. M. Jermyn, L. Jarvis, S. Gollub, J. Andreozzi, T. Miao, P. Bruza, D. Gladstone, B. Pogue, "Cherenkov video imaging during breast radiation therapy verifies stable beam shapes across treatment days," *International Journal of Radiation Oncology, Biology, Physics* **99**, S230 (2017).

34. L. A. Jarvis, R. A. Hachadorian, M. Jermyn, P. Bruza, D. A. Alexander, I. I. Tendler, B. B. Williams et al. "Initial Clinical Experience of Cherenkov Imaging in EBRT Identifies Opportunities to Improve Treatment Delivery," *International Journal of Radiation Oncology, Biology, Physics* (https://doi.org/10.1016/j.ijrobp.2020.11.013).

35. R. Zhang, D. Gladstone, L. Jarvis, R. Strawbridge, J. Hoopes, O. Friedman, A. Glaser, B.W. Pogue, "Real-time in vivo Cherenkoscopy imaging during external beam radiation therapy," *Journal of Biomedical Optics* **18**, 110504 (2013).

36. X. Cao, S. R. Allu, S. Jiang, J. R. Gunn, C. Yao, J. Xin, P. Bruza et al. "High-Resolution pO2 Imaging Improves Quantification of the Hypoxic Fraction in Tumors During Radiation therapy," *International Journal of Radiation Oncology, Biology, Physics* (https://doi.org/10.1016/j.ijrobp.2020.09.046).

37. R. Zhang, A. D'souza, J. Gunn, T. Esipova, S. Vinogradov, A. Glaser, L. Jarvis, D. Gladstone, B. W. Pogue, "Cherenkov-excited luminescence scanned imaging," *Optics Letters* **40**, 827–830 (2015).

38. X. Cao, S. R. Allu, S. Jiang, M. Jia, J. R. Gunn, C. Yao, E. P. LaRochelle et al. "Tissue pO2 distributions in xenograft tumors dynamically imaged by Cherenkov-excited phosphorescence during fractionated radiation therapy," *Nature Communications* **11**, 573 (2020).

39. P. Brůža, H. Lin, S. Vinogradov, L. Jarvis, D. Gladstone, B. W. Pogue, "Light sheet luminescence imaging with Cherenkov excitation in thick scattering media," *Optics Letters* **41**, 2986–2989 (2016).

40. Hachadorian, R. L., P. Bruza, M. Jermyn, D. J. Gladstone, B. W. Pogue, and L. A. Jarvis, "Imaging radiation dose in breast radiotherapy by x-ray CT calibration of Cherenkov light," *Nature Communications* **11** (2020) 2298.

41. B. W. Pogue, R. Zhang, H. Lin, J. Gunn, A. Glaser, J. Andreozzi, S. Jiang, D. J. Gladstone, L. A. Jarvis, "Molecular Imaging through centimeters of tissue: High resolution imaging with Cerenkov excitation," in *Frontiers in Optics 2015 (2015), paper FW5E.1* (Optical Society of America, 2015). https://doi.org/10.1364/FIO.2015.FW5E.1

42. A. Darafsheh, R. Taleei, A. Kassaee, J. C. Finlay, "The visible signal responsible for proton therapy dosimetry using bare optical fibers is not Čerenkov radiation," *Medical Physics* **43**, 5973 (2016).

43. A. Darafsheh, R. Taleei, A. Kassaee, J. C. Finlay, "Proton therapy dosimetry using the scintillation of the silica fibers," *Optics Letters* **42**(4), 847-850 (2017).

44. L. Beaulieu, M. Goulet, L. Archambault, S. Beddar, "Current status of scintillation dosimetry for megavoltage beams," *Journal of Physics: Conference Series* **444**, 012013 (2013).

45. S. Beddar and L. Beaulieu (eds.), Scintillation Dosimetry (CRC Press, Boca Raton, 2016).

46. G. Kertzscher and S. Beddar, "OC-0172: Development of an inorganic scintillation detector system for in vivo dosimetry for brachytherapy," *Radiotherapy & Oncology* **127**, S89–S90 (2018).

47. F. Therriault-Proulx, L. Beaulieu, S. Beddar, "Validation of plastic scintillation detectors for applications in low-dose-rate brachytherapy," *Brachytherapy* **16**, 903–909 (2017).

48. A. S. Beddar, "Water equivalent plastic scintillation detectors in radiation therapy," *Radiation Protection Dosimetry* **120**, 1–6 (2006).

49. L. Wootton and S. Beddar, "Temperature dependence of BCF plastic scintillation detectors," *Physics in Medicine and Biology* **58**, 2955–2967 (2013).

50. F. Alsanea, L. Wootton, N. Sahoo, R. Kudchadker, U. Mahmood, S. Beddar, "Exradin W1 plastic scintillation detector for in vivo skin dosimetry in passive scattering proton therapy," *Physica Medica, European Journal of Medical Physics* **47**, 58–63 (2018).

51. I. I. Tendler, P. Bruza, M. Jermyn, X. Cao, B. B. Williams, L. A. Jarvis, B. W. Pogue, D. J. Gladstone, "Characterization of a non-contact imaging scintillator-based dosimetry system for total skin electron therapy," *Physics in Medicine and Biology* **64**(12), 125025 (2019).

52. I. I. Tendler, P. Bruza, M. Jermyn, A. Fleury, B. B. Williams, L. A. Jarvis, B. W. Pogue, D. J. Gladstone, "Improvements to an optical scintillator imaging-based tissue dosimetry system," *Journal of Biomedical Optics* **24**(7), 075001 (2019).

53. J. Boivin, S. Beddar, M. Guillemette, L. Beaulieu, "Systematic evaluation of photodetector performance for plastic scintillation dosimetry," *Medical Physics* **42**, 6211–6220 (2015).

54. A. Darafsheh, J. E. Melzer, J. A. Harrington, A. Kassaee, J. C. Finlay, "Radiotherapy fiber dosimeter probes based on silver-only coated hollow glass waveguides," *Journal of Biomedical Optics* **23**(1), 015006 (2018).

55. J. Archer, L. Madden, E. Li, M. Carolan, M. Petasecca, P. Metcalfe, A. Rosenfeld, "Temporally separating Cherenkov radiation in a scintillator probe exposed to a pulsed X-ray beam," *Physica Medica: PM: An International Journal Devoted to the Applications of Physics to Medicine and Biology: Official Journal of the Italian Association of Biomedical Physics (AIFB)* **42**, 185–188 (2017).

56. J. Caravaca, F. B. Descamps, B. J. Land, M. Yeh, G. D. Orebi Gann, "Cherenkov and scintillation light separation in organic liquid scintillators," *European Physical Journal C* **77**, 811 (2017).

57. A. S. Beddar, L. Archambault, D. Robertson, Large-volume scintillator detector for rapid real-time 3-D dose imaging of advanced radiation therapy modalities (2017).

58. J. Dalmasson, G. Gratta, A. Jamil, S. Kravitz, M. Malek, K. Wells, J. Bentley, S. Steven, J. Su, "Distributed imaging for liquid scintillation detectors," *Physical Review D* **97** (2018).

CHAPTER 11

Clinical Considerations and Dosimeters for *In Vivo* Dosimetry

Douglas Bollinger

University of Pennsylvania
Philadelphia, Pennsylvania

Arash Darafsheh

Washington University School of Medicine
St. Louis, Missouri

CONTENTS

11.1 INTRODUCTION

Quality assurance (QA), as defined by American Association of Physicists in Medicine (AAPM)'s Task Group (TG)-100 [1], "confirms the desired level of quality by demonstrating that the quality goals for a task or parameter are met." In practice, this means confirmation that the various aspects of the radiation therapy system are able to perform at the level required to correctly predict and deliver a given dose distribution to a stated target. QA comes in many forms from routine machine maintenance to pre-treatment patient-specific measurements. The form of QA we are concerned with in this chapter is *in vivo* dosimetry. *In vivo* dosimetry is a powerful tool to verify dose delivery accuracy. By placing a radiation detector at a point of interest, we can validate that the expected dose is delivered to the patient (to within some tolerance). However, in order to get an accurate estimate of dose, it is critical to understand the abilities as well as the limitations of different dosimeters. Not all dosimeters are appropriate in all clinical situations and different dosimeters will give the operator different information. For example, the readings from some dosimeters are often represented as point measurements while others lend themselves to two-dimensional (2D) planar data. An alternative method of classifying dosimeters is by the availability of the readings. Some dosimeters provide readings that are instantaneously available (online review) while others produce data that need to be analyzed following treatment in order to extract a dose reading (offline review). This chapter begins by discussing dosimeters that can be used for online review, while the second section is devoted to those designed for offline review. Where possible, a wide variety of clinical examples of the uses of these dosimeters are included.

11.2 ONLINE REVIEW DOSIMETERS

11.2.1 Diodes

Diodes have long been one of the most popular dosimeters for *in vivo* measurements due to their instantaneous read-out ability, small physical size, and high sensitivity. However, their response to radiation suffers from a variety of dependencies, including temperature, angle of incidence of radiation, dose rate, source-to-surface distance (SSD), dose per pulse, integrated dose, radiation type, and energy [2–5]. As such, it is important to evaluate the significance of each of these factors for any application to minimize them or correct for them as needed [6]. In addition to the diode itself, an electrometer is required to read out the diode measurement. Historically, diodes have required a cable connection to an electrometer but some newer models are now able to operate wirelessly. This greatly facilitates their ease of use and opens the way for further applications. In 2005, the AAPM released the TG-62 to provide guidance on the use of diodes for *in vivo* dosimetry. While there have been a number of innovations in diode dosimetry since then, TG-62 remains an excellent resource for understanding the fundamentals of diode dosimetry [7].

There are a large variety of commercially available diodes for *in vivo* dosimetry. Two common classes of diodes are cylindrical and hemispherical; the appropriate geometry will depend on its desired clinical use [8]. Cylindrical diodes have the advantage of not having intrinsic directional dependence perpendicular to the detector axis and so are appropriate when there are likely to be a variety of beam angles impinging on the detector.

If the detector is to be placed perpendicular to the incoming beam(s), however, then a hemispherical diode may be appropriate. Diodes also differ in the material and thickness of the build-up cap. Ideally, there would be a different build-up cap for every clinical energy corresponding to a different depth of d_{max}. This, however, is not practical in many cases and indeed, having a large variety of build-up caps can increase the complexity and potential for error with *in vivo* dosimetry. A common compromise is to have one set of build-up caps for low energy photons (≤12 MV), one for high energy photons (>12 MV), and one for the electron beams.

Diodes must be calibrated before their use for *in vivo* measurement. In addition, as diodes are known to degrade with radiation damage, this calibration should be periodically checked and, if necessary, repeated. Diodes may be used for both photon and electron dosimetry. For electron beams, diodes have the additional advantage of low energy dependence due to the fact that the water to silicon stopping power ratio is relatively independent of electron energy [9].

There are many examples available in the literature for the use of diodes for entrance and exit dose measurements [10–12]. One such example of diodes in clinical practice is provided by Alecu et al., who reported on their use to record the exit and entrance dose for more than 300 pelvic treatments [13]. They calibrated their diodes against ion chamber readings under reference conditions in order to use them for dose reporting to patients. They did not correct for changes in response caused by difference in the SSD and field size between the measurements and the calibration condition. However, they reported that the accuracy of their results could be increased if they apply the appropriate correction factors. Under this system, they found that 95% of measured readings on their diodes agreed to within 5% of the predicted values. Similar results were reported for thoracic measurements. While this study was for 3D plans, diodes may also be used for dose verification of intensity-modulated radiation therapy (IMRT) treatments [14]. In this context, in order to predict the dose to the diode with high accuracy, it is important to use a fine dose grid (e.g., 1 mm) in planning due to the relatively small size (~0.01 mm³) of the detection volume in diodes.

While diodes can be used in a large number of clinical scenarios, extra caution must be exercised for measurements of small fields and at short SSDs. An over-response of 5–10% has been reported in such scenarios in diode measurements [15]. In addition, diodes have a known temperature dependence, which can affect their reading when placed on a patient, the magnitude of which depends on the diode design. One simple solution to circumnavigate this problem is to perform the diode calibration at a higher temperature. Welsh et al. and Saini et al. have published lists of temperature dependencies for a few commercially available diodes (0%/°C to 0.5%/°C) [16, 17]. Accounting for this temperature dependence is important if one wishes to use diodes for skin surface measurement. It is also worth noting that this temperature dependence can vary as a function of beam energy. Another confounding issue for skin measurements is that the effective point of measurement for typical diodes is just under 1 mm, while the International Commission on Radiological Protection (ICRP) definition of skin dose is at a depth of 70 μm [18]. Recently, diodes designed specifically for skin measurements with an effective depth of ~70 μm have been developed [19]; however, their use is not yet widespread.

Due to the photon energy dependence of most diodes, diodes are not commonly used as *in vivo* dosimeters for kilovoltage beams. Work by Saw et al., however, has demonstrated that for some diodes, the energy dependence between 70 kVp and 100 kVp can be negligible as long as the diode is calibrated at a similar low energy [20]. This cannot be assumed for any diode and the user should determine this dependency for themselves if they wish to use a diode in the kilovoltage settings.

Diodes are commonly applied in total body irradiations (TBI). Given the low margin for error, the real-time response of diodes can usefully inform clinical decisions during treatment. There is also no need for large distance (e.g., extended SSD) corrections provided that the diodes are initially calibrated at the corresponding treatment distance. However, if they are calibrated at the isocenter for an extended distance technique, a correction should be applied. In addition, diodes used for TBI dosimetry are typically placed on the patient's skin for extended period of time, which is warmer than the phantoms (solid water, etc.) used for calibration. As such, it is important to factor in the diode's temperature dependence for TBI dosimetry, either by calibrating at a realistic temperature that will be used for treatment or by applying a known temperature correction factor. In addition, due to the large field sizes involved in TBI, a greater length of the diode cable is likely to be present in

the treatment field increasing the impact of leakage on diode readings. This effect is compounded by the longer treatment times common for TBI. This leakage should be evaluated and, if found to be significant, an alternative setup or equipment should be considered. TBI diode dosimetry is made even more complex by the common use of a tray on the gantry. Both this factor and the large field sizes need to be corrected for if high accuracy is desired and the diode is calibrated under standard clinical conditions. As mentioned previously, an alternative approach is simply to calibrate the diode under the condition that mimics the exact TBI treatment geometry.

When placing diodes for TBI *in vivo* dosimetry, unless deliberately accounting for the angular dependence, it is important to orient them as perpendicular to the incoming beam as possible. The exact locations will vary depending on the patient orientation and the anatomical regions we wish to evaluate the dose. If the diodes are used to estimate the dose at the patient's midpoint, they should be left in the same location on the patient for all of the delivered fields. Depending on the size of the build-up cap or bolus thickness, we may need to apply a correction factor to the recorded entrance and/or exit dose before they can be used to calculate the midpoint dose. Popular points of measurement include the head, neck, chest, and abdomen. If a lung block is used, its effect can also be evaluated with diodes measurements.

Like TBI, total skin electron therapy (TSET) treatment typically involves large field sizes at extended SSDs and as such the diode readings must either be corrected to account for that or the calibration must be performed under the TSET conditions. In addition, given the relatively low depth of electron penetration in tissue, unintentionally shielding of a portion of the skin that we wish to treat, with the diode is a real concern. As such, it is important to use minimal buildup on the diodes and correct the readings accordingly. In addition, as TSET is often delivered with patient setup alternating every other fraction, to get a full picture of the delivered dose over the entire course, it is necessary to get readings for at least two subsequent fractions. When doing this, it is critical that the location of the diodes is recorded and reproduced every fraction. As with TBI, the leakage current in TSET can be significant and should be properly evaluated before the diode is used.

As mentioned before, diodes have a known dependence on energy for photon beams. This can make their use to measure out-of-field dose (which has a higher proportion of lower energy scattered photons) challenging. With that said, data presented by Chan et al. show that it is possible to calculate a correction factor as a function of distance from the treatment field [21]. This can be useful for measuring dose to pacemakers or any other areas of concern, such as the genitals.

Volumetric modulated arc therapy (VMAT) can be challenging for diodes due to their directional dependence. Hui et al. conducted a study on the use of diodes in the context of VMAT and while they found a transverse directional dependence for a cylindrical diode can be <1%, they were only able to establish a diode accuracy of 10% for dose to a small target [22]. This must also be considered for units in which a rotation delivery is typical of most or all of its plans, such as TomoTherapy® units and Halcyon™ systems. An additional variable one must consider when performing *in vivo* dosimetry on a TomoTherapy® unit is the pitch of the unit itself. In the aforementioned study, Hui et al. also reported that as pitch increases, diode measurement accuracy decreases.

Recent studies have demonstrated that with the proper calibration, diodes may be used for *in vivo* dosimetry for proton therapy [23, 24]. This introduces a whole host of interesting new applications. One suggested use has been the placement of a diode dosimeter in the esophagus of an anesthetized pediatric craniospinal irradiation (CSI) patient to verify distal range of the proton beam [25]. Another interesting application would be placement of a diode in a rectal balloon as proposed by Gottschalk et al. [26]. One should be careful with the placement of diodes in the context of proton therapy as they will change the range of any proton beam that passes through them, thus impacting the final dosimetry of the plan. In addition, as with photons, the more a diode is used in a proton beam, the more it will degrade. However, this degradation will be more dramatic per unit dose with protons than it is with photons [27].

The use of diodes placed in the patient's rectum for *in vivo* brachytherapy dosimetry was first reported a few decades ago [28]. Rectal dose is of particular interest in both gynecology (GYN) and prostate brachytherapy. In contrast to traditional external beam three-dimensional conformal radiation therapy (3D CRT), a very high level of spatial dose delivery accuracy is critical in brachytherapy. However, compared to conventional external beam radiation, there is a far longer time interval between imaging and treatment. In that time, a patient's anatomy can change dramatically and the applicator

position can shift. As such, an *in vivo* reading by a diode can be a useful QA check. With that said, the positioning uncertainty of the diode within the rectum does lead to a lower level of precision; one study found a range of diode readings for GYN brachytherapy nearly 40% different from those predicted by the treatment planning system (TPS) [29]. The authors were unable to definitively determine whether this discrepancy was due to positional uncertainty surrounding the diode placement or errors in dose delivery; although for an error of this magnitude the latter seems unlikely. Nevertheless, the use of diodes for *in vivo* dosimetry in brachytherapy remains a viable second check provided the user is aware of its potential pitfalls and the accuracy that can be ascribed to it. For example, it could be used to catch gross errors in treatment delivery as demonstrated by Alecu et al., who set a threshold of ±20% for their rectal diode readings [30].

Another confounding factor effecting diodes use in brachytherapy is their demonstrated energy dependence [31]. In the realms of high dose rate (HDR) brachytherapy, Ir-192 has an average energy of 380 keV, significantly below traditional megavoltage energies used in external beam treatments. On the other hand, diodes benefit from having a high sensitivity which is useful in HDR where the dose falls off dramatically as the distance from the source increases. One potential use for diodes in this context is not to measure the dose itself, but instead to assess source position accuracy. For example, one group has looked at Monte Carlo simulations of a diode array embedded in the treatment couch to verify the source position *in vivo* with a reported accuracy of 1 mm [32].

11.2.2 Metal Oxide Semiconductor Field Effect Transistors (MOSFETs)

Basic working principle and design of MOSFET-based dosimeters have been described by multiple authors (e.g., see Chapter 4) [33–36]. Like a diode, a MOSFET-based dosimeter has a relatively small size and can be used to measure dose in real time. Unlike a diode, MOSFETs store dose information, meaning that they can be read multiple times. However, they have a more limited lifetime compared to the diodes; most MOSFET-based dosimeters should not be used once they have recorded more than ~100 Gy (the actual threshold for a given device depends on the manufacturer and model) [37]. Anyone commissioning a MOSFET *in vivo* dosimetry program should be aware of this, lest much of the useful "lifespan" of a MOSFET be used in the commissioning measurements. Even at lower dose levels, as integrated dose to the MOSFET increases, the response of the MOSFET also increases. One study found that for 200 cGy fractions delivered consecutively, the variation in response between one fraction and the next could be higher than 6% [38]. Most MOSFETs come pre-calibrated from the manufacturer. While many clinics will accept the manufacturer calibration during the lifetime of the MOSFET, a higher accuracy can be obtained if the MOSFET is recalibrated prior to every use [39]. Depending upon the measurement conditions, most MOSFETs can be expected to be accurate to within ±5% [40].

MOSFETs can exhibit an angular dependence (though typically less than diodes), which varies in magnitude based on the manufacturer and measurement conditions of the device [41]. Despite this, they have found wide use beyond conventional 3D CRT and IMRT and are frequently utilized in the context of VMAT and TomoTherapy® applications [42]. MOSFETs also have a temperature dependency of the response; however, this is usually minimized through dual biasing (two MOSFETs on a single chip are measured at different levels of bias) [43]. As well as application in conventional external beam radiation therapy (EBRT), MOSFETs have found uses in many of the same applications as diodes ranging from TBI measurements to brachytherapy. Also like diodes, if MOSFETs are intended for entrance or exit dosimetry, a build-up cap should be used. Such a build-up cap can increase the angular dependence of the MOSFET [44] so proper attention must be paid to the obliquity of the incoming beam(s).

It is important that the user be aware that MOSFET dosimeters have a strong energy dependence between the KV and MV energy ranges [45]; they will over-respond to KV photons. If many KV images are to be taken of a patient with the MOSFET in place, then a dose correction may be required. Despite this, MOSFETs have been used in the KV setting, and many groups have reported their potential to measure imaging dose, such as that from a CBCT [46–48]. Such attempts have typically relied on cross-calibrating the MOSFETs with an ion chamber to get suitable dose readings. Reported uncertainties in the referenced publications range from 2.25% to 20% depending on the exact application.

MOSFETs can also be purchased in the form of an array. The RADPOS system (Best Medical Canada) is one such example. What is particularly interesting about

this system is that it has an electromagnetic positioning sensor that can be used to verify the spatial coordinates of the MOSFET as it records dose. This is particularly useful in treatments with high dose gradients, such as in brachytherapy or stereotactic body radiation therapy (SBRT)/stereotactic radiosurgery (SRS). In particular, the CyberKnife® system makes use of circular fields ranging from 5 mm to 60 mm diameter. In such systems, an *in vivo* dosimeter that is both physically small (the dimensions of the sensitive volume of RADPOS are $0.2 \times 0.2 \times 5 \times 10^{-4}$ mm^3) [49] and whose position can be accurately determined is essential. MOSFETs have been evaluated on CyberKnife® for both breast and lung treatments and found to be accurate to within 4% [50]. With that said, CyberKnife® treatments typically involve multiple KV images taken throughout treatment and, as mentioned earlier, this may necessitate a dose correction. In addition, for small field sizes, a correction factor may be necessary for the accurate use of a MOSFET dosimeter.

MOSFETs are also used in the *in vivo* intraoperative radiation therapy (IORT) setting [51]. This often involves physically placing the MOSFETs in the patient's body. In these cases, it is important to consider proper sterilization and it will be necessary to place the MOSFETs in a small sterile container (such as a small plastic bag) before placing them in the patient. The user should be aware that the choice of the container may have an impact on the MOSFET reading, particularly as IORT is often carried out using electrons or low energy photons [52]. Numerous studies have demonstrated the feasibility for this use of MOSFETs [53, 54]. In the IORT setting, it is important to accurately characterize the angular dependence of the MOSFET system beforehand, as it is typically challenging to know the exact angle of the MOSFET in relation to the radiation source.

As mentioned previously, MOSFETs can also be used in the context of TBI. As MOSFETs are typically attached to wires, they can be trickier to handle than other dosimeters in this setting. However, their small sizes combined with their instant readout make them a popular choice for TBI measurements in many clinics. However, as MOSFETs typically come pre-calibrated for more common clinical conditions (lower SSD, smaller field size, etc.), they should be recalibrated if they are intended for TBI dosimetry [55].

Implantable MOSFET devices can be used for *in vivo* dosimetry of patients. These devices usually consist of one or two MOSFETs attached to an antenna, all contained within a small capsule that can be implanted in the patient's breast, prostate, or other sites of clinical interest [56]. When implanted in the prostate, this device not only serves to record the dose but can also be used as a fiducial marker, which helps to minimize the positional uncertainty of the dosimeter placement. Within the context of brachytherapy, MOSFETs may be used to verify skin dose in partial breast treatments [57]. Whenever accurate positioning of the MOSFETs is critical, as is the case in brachytherapy due to steep dose gradients, it is wise to have a method to identify their location on the patient in the TPS beyond rough approximation. A simple solution for patients receiving a CT is to mark the location where the dosimeter will be placed with a BB marker. Then, when the patient is treated, replace the BB with a MOSFET. If a considerable time interval is anticipated before treatment can begin, then one can mark the patient's skin in the location of the BB and use that for MOSFET positioning instead. In addition, as mentioned earlier, MOSFETs have an energy dependence. While the average energy of Ir-192 used in HDR treatments is high enough not to warrant a significant correction for MOSFET response, MOSFETs will over-respond to a number of common LDR sources, such as I-125 or Pd-103. With that being said, their small size makes them attractive for measuring quantities such as urethral dose in a prostate seed implant. When correctly calibrated, such a use has been demonstrated to be feasible by placing the MOSFETs in a urinary catheter [58].

MOSFET dosimeters can also be used in the context of proton therapy [59, 60]. However, they have a strong dependence on the linear energy transfer (LET) of the protons [61]. Available methods to account for this range from simple correction factors to more complex methods relying on Monte Carlo simulations [62]. MOSFETs can be expected to be accurate to within 5% in the context of proton therapy when an appropriate correction method is applied. Without those corrections, errors as great as 26% have been reported [63]. When MOSFETS are used to measure the water equivalent path length, they can be accurate to within 1 mm [64].

11.2.3 Electronic Portal Imaging Devices (EPIDs)

Electronic portal imaging devices (EPIDs) have been in use in radiation therapy departments since the 1980s (see Chapter 8) [65]. The intervening decades have seen dramatic changes and advances in both the technologies in play and their clinical uses, which include, but are not limited to, patient alignment, patient-specific IMRT QA, and, of course, *in vivo* dosimetry [66]. In 2001, AAPM's TG-58 [67] was published, which relates specifically to EPIDs and their use in the clinic. While this is a useful resource for understanding the basic principles upon which EPIDs operate, the modern systems differ significantly from what is described. In addition, many clinics have become significantly more reliant on EPIDs as part of their workflow since the release of TG-58. A perfect example of this is the new Halcyon™ unit (Varian Medical Systems, Palo Alto, CA). This modality has a built-in EPID panel that automatically records the exit fluence from all clinical patient plans, whether or not it is needed. EPIDs can be utilized for *in vivo* dosimetry, in both an online and offline fashion.

Unlike the other dosimeters mentioned thus far, the EPID is primarily concerned with "exit dosimetry." That is to say the dosimetry of particles that have already traveled through the patient. This somewhat limits their application and it is challenging to envision a scenario where they would be useful for brachytherapy or skin dose measurements. Even with these limitations, there are a variety of "pros" to EPID-based dosimetry. These include the convenience of having the device actually attached to the accelerator, lack of angular dependence as the device rotates with the gantry, and a real-time and high-resolution 2D measurement. On the other hand, they have a limited field of view (FOV), although on some modern EPIDs, their detection area can be as large as ~40 × 40 cm². This raises the question of where to place the detector in relation to the patient. If a large FOV is desired, the EPID should be as close to the patient as possible. However, this brings with it a risk of collision, either to the couch or the patient. As such, adequate room should be left to ensure patient safety as well as avoiding mechanical damage.

When using EPIDs for *in vivo* dosimetry, it is important to first decide if they are intended for 2D or 3D dose verification. While 2D dose comparison is inherently and logistically simpler, it is possible to reconstruct a 3D dose based on EPID images. A number of different groups have implemented methods to achieve this [68, 69]. For example, one option is back-projecting the measured fluence and then applying an iterative convolution technique to determine the delivered dose [70]. One of the many advantages of exit dosimetry such as this is that it offers the ability to catch not only errors in delivery but any anatomical changes along the beam path. This, however, can also be a confounding factor. If the user only wishes to evaluate delivery accuracy, independent of setup uncertainty, then positional corrections need to be applied. Alternatively, looser tolerances for *in vivo* measurements can be implemented. To illustrate this, in a study on the use of EPID in *in vivo* breast dosimetry, when setup variations were not accounted for researchers found that only 72% of plans delivered resulted in a match between the predicted and measured dose with an agreement of <5% [71]. When setup variations were incorporated into the results, however, 93% of plans had a predicted dose agreement <5%. Exit dosimetry is also advantageous in that the user does not need to be concerned with perturbation to the dose distribution caused by the placement of the dosimeter in the beam path.

Unlike most of the other dosimeters discussed in this chapter (with the notable exception of film), EPID's typical use in *in vivo* dosimetry results in a full array of measurements to be analyzed instead of more discrete individual points, potentially yielding more information. As such, the EPID is not typically used for individual point dose comparisons in the context of *in vivo* dosimetry. When looking at a 2D image instead of a point, a single dose threshold criterion may no longer be the most appropriate tool for evaluation. Some popular evaluation metrics include looking at the distance to agreement (DTA) or using gamma analysis, although looking at an absolute dose difference map may also be helpful. A full description of gamma analysis is beyond the scope of this chapter but readers who are unfamiliar with the concept are referred to the work of Low et al. [72]. When using gamma analysis, it is important to choose a criterion based on the treatment scheme. For example, SBRT would warrant a stricter tolerance than a more fractionated IMRT regime. In addition, when viewing an array, it is likely that not all measured points will be of equal significance. For example, percent dose deviations in portions of the field where the planned dose is <10% of the max dose are less likely to be clinically significant then percent dose deviations within the target volume. One solution is to implement a dose threshold, below which dose deviations are not evaluated [73].

McDermott et al. have looked into the feasibility of replacing patient-specific IMRT QA with *in vivo* EPID measurements [74]. Using a combination of phantom and *in vivo* patient measurement for 75 prostate plans, they were able to demonstrate that similar results to IMRT QA could be obtained by averaging the *in vivo* dosimetry results of a patient over at least three fractions. They calculated this could reduce the time required for patient specification by a factor of ~3. If EPIDs are to be used in this way, it is important to average the results over more than one fraction. Any single fraction might be subject to random errors and so can give erroneous results when assessing the delivery accuracy of the plan over the entire course of treatment. With that stated, if there is a systematic error in delivery, the more fractions that are averaged to obtain results, the fewer fractions will be left to correct said error.

Many of the early works in *in vivo* EPID dosimetry looked at correlating the EPID signal with some known measurement of dose, such as an ion chamber in a cylindrical phantom, for either a variety of setup conditions that could be interpolated or for patient-specific plans [75]. Obviously, performing a relative EPID calibration for every patient requires significant clinical resources. Recently, there has been an increased effort to extract dose from EPID readings using more advanced techniques. A full mathematical description of these techniques is beyond the scope of this chapter. To give one example, Mans et al. published a case study where through the use of EPID-based *in vivo* dosimetry, their group was able to catch a plan transfer error for a rectum patient [76]. Using the data from their EPID, they predicted an 11.6% under-dose to their PTV unless the problem was rectified. This highlights the benefits of the EPID for evaluating dose to internal target anatomy (using an aforementioned back-projection approach), which is more challenging or often impossible with other online dosimeters.

Recent years have also seen the emergence of commercially available software packages for dose reconstruction, such as EPIgray (DOSIsoft) or iViewDose (Elekta). These software packages not only facilitate the *in vivo* process, but work for a variety of treatment options including 3D CRT, IMRT, and VMAT. When EPID dosimetry is intended for a variety of treatment techniques, it may be a good practice to set different *in vivo* tolerances for different techniques and sites, as average agreement values have been reported to differ between them [77].

As mentioned earlier, EPIDs are not often a practical choice for brachytherapy applications that is not to say that no investigation has been pursued down this avenue. While getting an accurate dose measurement for an *in vivo* brachytherapy source can be challenging, a simpler application of using the EPID to verify source position during treatment has already been implemented. Smith et al. have demonstrated this, using a flat-paneled detector embedded underneath a treatment couch to confirm that their source was traveling to the correct position during multiple HDR prostate treatment deliveries [78]. It is possible to envisage such an application applied to other brachytherapy sites.

11.2.4 Scintillation Fiber Optic Dosimeters

Of all of the dosimeters mentioned in this chapter thus far, scintillators have perhaps experienced the least use in the context of routine *in vivo* dosimetry. Despite this, they possess a number of advantageous traits, not least of which is their water equivalence (see Chapter 9). Also, they have small physical size and mechanical flexibility, do not require a correction for angle, energy, and dose rate, and have the benefit of MR compatibility. With the emergence and growing popularity of real-time MR-guided EBRT this last point could be particularly important [79].

In an *in vivo* study, Wooten et al. measured the rectal wall dose during prostate IMRT by placing fiber optic plastic scintillators on a rectal balloon [80]. Rectal balloons are typically used as immobilization devices during prostate radiation therapy. They also afford a unique opportunity to measure the dose to the rectal wall with the affixation of a dosimeter, such as a scintillator, to the surface of the balloon. The researchers took bi-weekly readings of five patients over the course of radiation therapy resulting in 142 scintillator dose measurements. They found excellent agreement (0.4%) between these measurements and the predicted dose from the TPS. An important takeaway from this study, other than the feasibility of using scintillators in this setting, is that the scintillators themselves were attached to three fiducial markers, thus allowing their localization on CT. As mentioned earlier, one of the advantages of scintillators is their water equivalence; however, due to this very equivalence, they are difficult to detect on CT, which derives its contrast from variations in electron density. As such, when using scintillators for *in vivo* dosimetry, if the placement accuracy is of great importance, the user

should give consideration to using fiducials with known spatial positions in relation to the scintillator. This is particularly the case with SBRT, one of the features of which is typically a very steep dose falloff. If attention is paid to careful positioning of the dosimeter, however, there is no reason that scintillators cannot be used in this context too, as has been demonstrated by the work of Cantley et al. [81] This group also used scintillators on the surface of rectal balloons for *in vivo* measurements, however, within the context of 5 fraction SBRT. Here, they found agreement within 6% between the predicted TPS dose and the measured dose.

The use of scintillators has also been investigated for out-of-field dose measurements [82, 83]. Traditionally, these measurements are a challenge for many dosimeters due to a variety of factors including the relatively low signal recorded (typically <5% of the maximum field dose), the angle of incidence compared to the primary beam, and the difference in energy spectrum from the main beam [84]. For EBRT, this change in energy spectrum not only affects the photon component of the measurement, but there will also be an increase in the relevant proportion of contaminant electrons. Once again, the water equivalence of plastic scintillators helps to overcome some of these challenges. Out-of-field dose is particularly relevant to pregnant patients or patients with an implanted cardiac device, such as a pacemaker. Work has been done using scintillators to estimate the reduction in out-of-field dose to such devices using lead shielding.

Bourgouin et al. [82, 83] looked at using scintillators to measure out-of-field dose for both photon (6 MV and 23 MV) and electron beams (6 MeV and 18 MeV). Placing the detectors 3 cm away from the field edge, they collected percentage depth dose (PDD) curves with both a scintillator and an ion chamber for the purpose of comparison. They also collected a series of readings with both detectors placed at different out-of-field distances. When compared to the TPS, both the readings of the ion chamber and the scintillator showed significant variation (up to 12% for the peripheral dose measurements and up to 80% in the PDD measurements at shallower depths than the d_{max}); however, they agreed more closely with each other. It is important to note that the TPS also struggles with accuracy for out-of-field measurements and so this is not necessarily an indicator of poor dosimeter performance. The ion chamber and the scintillator showed good agreement for all measurements with the exception of the results recorded for the 6 MeV beam. Here, the authors attributed the discrepancy to the relatively large collection diameter of the ion chamber. The other important takeaway from this study is that the scintillator used was able to accurately measure doses as low as ~0.5 cGy for both the photon and electron beam. While further validation is advised, these results do indicate that plastic scintillators may have a use in *in vivo* dosimetry for out-of-field measurements.

Scintillators can also be used for *in vivo* brachytherapy measurements due to their mechanical flexibility and relatively small size. In these cases, it is especially important to consider the potential temperature dependence of the scintillator, which can vary hugely depending on the type of scintillator [85]. The user is advised to consult with the manufacturer, or ideally perform their own measurements, to asses this dependency for their own scintillators. There is considerable debate about the significance of the background signal created by Cherenkov radiation; however, many groups have recommended either for correcting it or avoiding it using specially designed scintillators [86, 87]. If using scintillators for *in vivo* dosimetry in brachytherapy, it is also important to be aware that they can be somewhat mechanically delicate and may degrade quickly if bent too much or placed under too much strain. One example of their use would be to place them in unused catheters for patients undergoing interstitial therapy [88].

Cherenkov radiation generated in optical fibers is one of the main issues with scintillation fiber optic dosimetry of photon and electron fields. Cherenkov radiation generation by proton beams is of less importance; however, ionization quenching effect is the main issue with proton therapy and other high LET beams dosimetry using scintillators. Several methods have been proposed to minimize the influence of Cherenkov radiation contamination in order to improve the accuracy of fiber optic dosimetry [89–91]. However, the user must have a full understanding of the accuracy of the Cherenkov radiation correction method used in their system, as well as other characteristics of their scintillating fiber optic dosimetry system. For *in vivo* dosimetry applications, the temperature dependency should be evaluated in advance and properly addressed as a decrease in light output with temperature ~0.2–0.3%

per °C has been reported [92]. The loss of the signal per accumulated dose in scintillators ~0.28% per kGy (for 0–15 kGy) and 0.032% per kGy (for 15–127 kGy) has also been reported.

11.2.5 Ion Chambers

Ion chambers, while are more typically used for reference dosimetry and machine calibration, have also found a use in *in vivo* programs. Certainly, they are useful for cross-calibration of a number of other dosimeters, such as diodes. With that stated, their requirement for a high voltage bias introduces unique safety concerns that are not present for the other dosimeters discussed here. In addition, this means that the measurement volume must be connected with cables to an electrometer which can make the setup more unwieldy. All of this makes the ion chamber somewhat unusual as a choice for *in vivo* dosimetry. Furthermore, unsealed ion chambers need to be carefully corrected for changes in temperature and pressure. Their ability to measure dose directly, however, means that in certain situations, they may be an appropriate dosimeter.

In using ion chambers for *in vivo* dosimetry, one needs to be aware of the size of the chamber. If being used to measure a region with a high dose gradient, as is often the case in IMRT and VMAT, an ion chamber can suffer from a volume averaging effect if its volume is too large in relation to the dose gradient. Furthermore, the relatively large physical size of ion chambers makes them unsuitable for intracavitary dosimetry applications, for example, internal measurements in brachytherapy, where a dosimeter with a small physical size is needed.

Despite all of these drawbacks, there are a number of published works on *in vivo* dosimetry with ion chambers [93, 94]. For example, Piermattei et al. looked at using a micro ion chamber to directly measure dose to 30 patients treated to the pelvis. The ion chamber was placed behind each patient on the beam's central axis by affixing it to the EPID on the machine. In this way, they were able to relate the recorded signal to the dose predicted by the TPS for simple 3DCRT plans at a variety of beam angles. Their reported accuracy was within 4% for 95% of measured points. It is important to note that this study did not place the ion chamber directly on the patient's surface, thus circumnavigating the potentially hazardous voltage dilemma. If using ion chambers *in vivo*, this is a prudent strategy.

It is also worth mentioning that at the time this study was published, their EPID was not able to accurately measure the dose. In most modern clinical settings, an EPID could easily replace an ion chamber in this context.

Entrance dosimetry is another application of ion chambers that avoids affixing them to or within the patient. Poppe et al. looked at this in the context of IMRT. They used a prototype wire ion chamber positioned in the linac gantry head, below the multi-leaf collimators (MLCs). Transmission-type ion chambers were used to minimize disturbances to the beamline caused by the presence of the ion chamber. This technique of dosimetry cannot account for changes in the skin surface or patient position, however, it does offer an easy way to validate and isolate the accuracy of delivery in terms of radiation leaving the gantry head and MLC movement. Wires were placed along the gantry head such that each wire position corresponded to each MLC pair. During routine patient-specific QA, a baseline reading for each of the wires could be established, and then the measured reading for each wire for each treatment could be compared with this baseline value. Using this technique, the investigators claimed they were able to verify MLC leaf position accuracy during delivery to within 1 mm. If placing an ion chamber in the beam head, it is not only important that it does not disturb the beamline but it also must either be translucent or positioned in such a way as to not block the light field if that is used for patient setup. In addition, it must be robustly attached to prevent it falling during treatment or setup and causing injury to the patient and/or the therapists.

Ion chambers can also find a use in TBI *in vivo* dosimetry. Due to the large field sizes typically employed, it is often possible to place an ion chamber in a solid water box or phantom within the treatment field at the same time as the patient. For example, if the patient is positioned sitting up with their knees bent, the ion chamber within the phantom can be placed under their knees. When used in such a way, if the proper calibrations and corrections (such as temperature and pressure) are applied, the ion chamber can be used to ensure that the expected dose is delivered with each field and MU settings. The downside of this approach is that we are not measuring the dose to the patient themselves and, as such, this technique cannot be used to verify that the patient thickness used for the MU calculations is correct nor that the patient is positioned correctly.

11.3 OFFLINE REVIEW DOSIMETERS

11.3.1 Film Dosimeters

Unlike the other dosimeters discussed up to this point, film dosimeter is used in an offline setting when performing *in vivo* dosimetry. Both radiographic and radiochromic films have been used for *in vivo* dosimetry [95–97]. A full description of their construction as well as the physics behind their interactions is described elsewhere (see Chapter 5) [98, 99]. Interested readers are also referred to AAPM's TG-55 [100] and TG-69 [101] for more information about radiochromic and radiographic films, respectively. Since radiographic films are rarely used nowadays, the focus of this section is on the application of radiochromic films. As far as their practical use is concerned, the user should be aware that the radiographic film requires development in a dedicated film processor unit before it can be read. Radiochromic film is "self-developing" and does not share this requirement. However, the reading on the radiochromic film (e.g., optical density of the film) is known to change based on how much time has elapsed since the radiation exposure. For example, it has been shown that the response of Gafchromic™ EBT-3 and EBT-XD films can change ~8 and ~16% during the first 24 hours post-irradiation [102]. However, a stable response is reached after ~24 hours in both film models. A common recommendation is to wait at least 24–48 hours before reading. What is most critical is that the time between delivery and measurements is the same as that used for the calibration. As such, for *in vivo* dosimetry necessitating immediate and quantitative results, film is not a suitable choice for most scenarios. Radiochromic films have the advantage of being nearly tissue equivalent and so their response is largely independent of energy in the megavoltage range [103, 104]. There are also a number of other benefits to film dosimetry. We will not go into all of them here but chief among them is the high spatial resolution of film (which is most often limited by the choice in scanner settings as opposed to the film itself) and the ability to perform 2D dose measurement using a single film sheet as opposed to a single point dose (1D) measurement offered by most other dosimeters.

In vivo film dosimetry has also found a use in proton therapy. There is a large variation in the reported energy dependence of radiochromic film with some groups reporting variations on the order of 10% [105] while other have claimed it to be minimal [106]. The energy dependence of film also depends on what portion of the proton depth dose curve it is being used to measure and under-doses as high as 20% have been reported in the Bragg peak region [107]. As such, while film may be used for relative dosimetry in a proton beam, care and careful consideration should be applied before using it for any kind of absolute dosimetry [108]. Despite this, the use of film for absolute dose measurements has been investigated with one group claiming dose accuracy to within 5% [109]. If using film for relative measurements, it need not be calibrated in a proton beam and may be calibrated using photons on a linac if that is logistically simpler. Obviously if being used for absolute dosimetry, the film should be calibrated using protons. When scanning the film, LET corrections can also be applied to mitigate the energy dependence [110].

A potential use for film within the context of proton therapy is for range verification. If the film is placed along the proton beam distal to the patient, it can be used to record the distal fall-off of the beam. Using this information, the range of the beam can be calculated and compared to the predicted range from the TPS. Zheng et al. demonstrated the feasibility of this technique for *in vivo* measurements for a variety of proton patient sites, including lung [106]. They reported that the film was accurately able to predict the beam range to within 1 mm. A maximum deviation of 3 mm between the planned and delivered range was reported in their work for the lung treatment cases.

Film dosimetry can be used for electrons, 3DCRT, IMRT, and VMAT plans. It can also be used both externally and internally to the patient; however, due to its larger size, film is not typically applied to as diverse a range of internal applications. Smaller dosimeters like MOSFETs and OSLDs tend to be popular choices for this. When internal use is intended, the film should be wrapped in some type of plastic to protect the film. Since radiographic film typically comes in sealed film sleeves, radiochromic film is the more suitable choice for internal dosimetry as it can be cut into smaller pieces. It is important to exercise caution when cutting radiochromic films as the cutting itself can damage the film and reduce the accuracy of the results around the region of the cut affected by strain artifact. With that stated, Moylan et al. investigated the film size

dependency and concluded that film as small as 5 × 5 mm^2 could be used for *in vivo* dosimetry [111]. When cutting film, a guillotine paper cutter should be used if available.

One example of internal use of film is provided by Chuang et al., who used small squares of radiochromic film placed inside the patients' mouths to verify the delivered dose for a head and neck IMRT plan [112]. Another example is the use of film in the breast IORT setting. Commonly in IORT, a protective disk is placed in the surgical cavity prior to the radiation delivery. This disk serves to protect the healthy tissue beyond the cavity that the operator does not wish to irradiate. Radiochromic film may be affixed to that disk to not only verify the dose delivery accuracy, but also that the disk is functioning as expected. Interested readers are referred to a paper by Severgnini et al. for a more detailed description [113]. IORT is sometimes performed using low energy photons and radiochromic film is also an appropriate dosimeter in this scenario. However, the user should be cautioned that some Gafchromic™ film models (e.g. EBT2, EBT3) can display large energy dependencies in the kilovoltage range. The magnitude of this effect depends on the film itself so the user should be aware of both the manufacturer and model of the film they are using to see if reference to the energy dependence of their film is available in the current literature. Gafchromic™ film has also found a limited use within the confines of brachytherapy. For example, it can be wrapped around the surface of a vaginal cylinder and used to verify dose at the cylinder surface during treatment. Such a technique was utilized by Pai et al., who reported a dose delivery accuracy within 10% [114].

While using films for internal *in vivo* dosimetry, such as the one suggested above, is interesting, in the majority of *in vivo* film dosimetry cases, the films are placed externally to the patients. When used to measure the skin dose, the film is typically placed on the patient's surface with no buildup. However, if the aim of *in vivo* dosimetry is to measure the entrance dose at a certain depth, then appropriate amount of bolus material should be placed over the film. When film is used to measure the skin dose, it is typically placed in the build-up region of the beam. As such, the effective point of measurement may not be at the correct depth to give us the clinically relevant skin dose measurement we want. As such, a correction factor should be applied to get an accurate dose estimate. This correction factor is typically on the order of 15–16% [115]; however, it can vary depending on the manufacturer. It is important as well to avoid stressing the film too much by bending it. Ashburn et al. found a change in film response of >5% when it was significantly bent [116]. A common use of film is verification of junction matching, such as that between the spinal and cranial fields on a CSI case. In such cases, the film is typically placed in a block beneath the patient's head to verify that we are neither under not over-irradiating at the match line.

Film can also be used in the context of TSET *in vivo* dosimetry. In this context, the film is typically cut into small squares, wrapped in plastic or paper sleeves, and placed on the patient at the desired measurement points. Due to the small size (~2 × 2 cm²) of the film pieces needed for this application, a single sheet of Gafchromic™ film can provide enough samples for multiple patients. This raises the possibility of having a sheet-specific (instead of batch-specific) calibration for all films used, if a high level of accuracy is desired. As alluded to previously, TSET is often delivered by placing the patient in six different positions with three positions used per day. As such, the radiation cycles are on a two-day basis and to get an accurate estimate of the dose for the entire course of treatment, film will need to be placed in the same anatomical locations on two consecutive treatment days.

11.3.2 Radiophotoluminescent Dosimeters (RPLDs)

While not as ubiquitous as some of the other dosimeters mentioned in this chapter, RPLDs have been used in radiation therapy for decades, particularly in the context of personnel monitoring [117]. Similar to TLDs and OSLDs, RPLDs have a relatively small physical size (e.g., the GD-301 model is 1.5 × 8.5 mm²). RPLDs are also largely energy-independent in the megavoltage range (provided a suitable energy compensating filter is used) and can be read multiple times [118]. However, they cannot be used in an online setting. They may also be annealed (typically by heating them to a temperature of ~400°C for 1 hour) to allow for repeated use. RPLDs have been investigated for use in the context of IMRT by Hashimoto et al. They found that for small field sizes (<3 × 3 cm²), RPLDs suffer from both a density perturbation and volume averaging effect. As such, a correction factor may be necessary if the user wishes to use RPLDs for *in vivo* dosimetry of small fields [119]. Under IMRT conditions, however, they found that RPLDs were able to accurately measure dose within a PTV region to within 1.2% (compared to an ion chamber reading) and, as such, a correction factor may not be required unless treating a particularly small target, such as with SBRT. Another useful feature regarding RPLDs in the context of *in vivo* measurements is the relative lack of angular dependence for angles <45° on the device, which increases the simplicity of setup on the patient [120].

Use of RPLDs for *in vivo* TBI measurements has been reported by Rah et al. [121] Compared to other popular dosimeters for TBI, such as diodes, RPLDs do not have large angular, SSD, or energy dependence in the MV range, which simplifies the measurement process. Rah et al. used RPLDs along with TLDs and MOSFETs on three TBI patients and estimated an uncertainty in the RPLD readings of <3%. They also found that the readings of the RPLDs agreed with those of the TLDs and MOSFETs to within ~6.5%, which they attributed largely to differences in position of the dosimeters on the patient's surface. When using RPLDs in the context of TBI, it is important to remember to place adequate bolus over the dosimeter to allow for dose buildup and reduce the effect of electron contamination, particularly if the RPLD is placed behind a lung block. Another important takeaway from this study is that Rah et al. found the RPLDs to have a linear response with dose in the range of 0.5 Gy–10 Gy, although other reports go further and claim a linearity all the way up to 500 Gy [122].

As mentioned previously, the response of RPLDs used in the megavoltage range is largely energy-independent. However, when they are used for *in vivo* brachytherapy measurements, a correction is required if the calibration energy is not the same as the source energy [123]. RPLDs are known to over-respond to low energy photons as demonstrated by Hashimoto et al., where they found a 13% increase in dose response at a 7 cm distance from an Ir-192 source compared to a 3 cm distance. They also provide a list of correction factors for detector response and absorbed dose energy dependence as a function of distance for RPLDs calibrated in 4 MV beam to be used in conjunction with Ir-192.

Despite the energy dependence of RPLDs for brachytherapy, they have been used in this application [124]. Takayuki et al. investigated using RPLDs to monitor dose to 61 head and neck intestinal brachytherapy patients. They used a variety of techniques to implant and localize the RPLDs, including sealing with adhesive and suturing to the mouth, sandwiching them between two plastic plates in a template, and placing them within a silicone impression of the mandible. Small radiopaque markers were also attached to the RPLDs to accurately assess their location for dose comparison purposes. Unlike TLDs, RPLD readers are able to distinguish the signal caused by surface contamination from that caused by dose [125]. This is particularly relevant in the context of head and neck treatment verifications due to the presence of saliva. It also reduces (but by no means eliminates) the need for careful handling and the avoidance of fingerprints.

The largest *in vivo* RPLD study carried out at the time of this writing was also performed by Takayuki et al., and consisted of over 1000 measurements taken during pelvis brachytherapy irradiations [126]. This study spanned multiple treatment sites including the prostate as well as gynecological malignancies. What is particularly interesting is that the RPLDs were placed using a variety of different techniques to measure dose in the urethra, rectum, and vagina. For vaginal readings, the RPLDs were placed in a small slot within a vaginal cylinder being used for treatment. For the urethral readings, a "train" of RPLDs was made by placing 10 end-to-end into a single Teflon tube which, in turn, was placed in the patient's urinary catheter. The rectum dose readings were achieved using a similar technique, however, had sutures placed at the ends and middle of the train to hold it in place during treatment. This range of applications in a single study demonstrates the versatility of the RPLD as an *in vivo* dosimeter. It is worth noting that the authors only found an accuracy to within 20% for the measurements of the rectum, which they attributed to organ motion rather than any uncertainties originating in the detector.

11.3.3 OSLDs and TLDs

The physics of OSLDs and TLDs has been described in Chapter 7 and Chapter 6 of this book, respectively. Both OSLDs and TLDs have a long history of successful implementation for *in vivo* dosimetry [127–130]. They have a number of advantages including their small size, ability for reuse, and relatively small energy dependence in the megavoltage range [131]. However, caution must be exercised in their use since most OLSDs and TLDs exhibit supralinearity beyond 10–20 Gy [132]. For some models, the threshold is even lower than 10Gy [133]. In addition, for lower energy applications, such as measuring

the imaging dose, OSLDs, and TLDs are both known to over-respond [134] and significant correction factors must be applied if they are to be used in that setting. Also arising from this effect, if measuring out-of-field photon dose for a megavoltage beam, they will also over-respond to the lower energy scattered photons. As such, careful consideration to appropriate corrections needs to be made if a high accuracy is desired in that scenario. It should be noted that TLDs tend to exhibit less energy dependence than OSLDs as they are typically made of a more tissue-equivalent material (e.g., LiF vs. Al_2O_3:C) [135]. This is also relevant to HDR brachytherapy dosimetry, where the radiation has already lower energy spectrum which would further decrease with depth resulting in the potential need for a depth-dependent correction factor to maintain accuracy [136]. Despite this, the significance of this energy softening has been debated and the argument can be made that for many applications, this correction is negligible compared to far larger factors of experimental uncertainty. When OSLDs and TLDs are used for entrance or exit dose measurement, bolus or a build-up cap should be used to allow for electronic equilibrium. OSLDs may also be used to measure surface dose with no additional buildup [137].

While external applications are the most common, OSLDs and TLDs have also been used internally. For example, one interesting work by Garner et al. looked into using TLDs to measure cardiac and esophageal dose to a canine receiving radiation on a CyberKnife® system [138]. In this case, the intent was assessing the accuracy of the CyberKnife® when delivering dose to targets significantly affected by cardiac motion. Traditional surface dosimeters such as diodes were inappropriate due to the large variety of entrance angles, and their inability to account for cardiac motion. Due to their small size, Garner et al. were able to surgically implant TLD crystals on to the surface of the heart itself where they remained during treatment. The TLDs were visible during simulation of the animals, allowing for an accurate assessment of their spatial position for dose prediction. Despite their small size, the high dose gradients present in CyberKnife® treatment necessitated averaging the expected dose across the entire volume of the TLD. This was achieved by contouring the TLDs on the simulation scan, and recording the average dose in the region of the contour. It is important to note that this technique can be applied to other *in vivo* dosimeters that are scanned with the patient for an accurate estimate of dose. Due to its invasive nature, this particular form of *in vivo* dosimetry is unlikely to be popular with human subjects (the canine was euthanized in order to retrieve the TLDs). However, the use of TLDs in this context is testament to their versatility and highlights a number of considerations that need to be made for many more common clinical scenarios.

Their small size also means that OSLDs and TLDs are popular choice for *in vivo* brachytherapy dosimetry. A number of studies have looked at using them to measure dose within the context of prostate HDR [139–141]. One such example is the work by Anagnostopoulos et al., who used 50 TLDs to record the dose to either the urethra or posterior rectal wall during 18 fractions of HDR prostate brachytherapy. The TLDs were placed in a chain within one of the needles that had been implanted for treatment or, if such needle were not available, within an extra needle placed explicitly for the purposes of *in vivo* dosimetry. There are advantages and disadvantages associated to both techniques. Implanting an extra needle for dosimetry results in increased trauma to the patient. However, using a needle that will also be used for delivery required an interruption in the treatment before the source gets to the needle in question. Using this method, Anagnostopoulos et al. were able to establish a dose delivery accuracy to within a mean of 7% for all treatments. Similar to the work by Garner et al., they also compared their measured dose to the dose predicted to the catheter that contained the TLDs on the simulation CT scan. As the catheter itself was visible, and TLDs ran down its length, no additional markers to indicate the location of the TLDs were necessary. It is of interest to note that in this study, the TLDs were calibrated in a 6 MV photon beam and no correction for the lower energy of Ir-192 was made. Brezovich et al. also used TLDs to measure urethral dose during HDR prostate brachytherapy procedures [142]. However, instead of using needles, they placed the TLD directly in the patient's urethra via a Foley catheter. Looking at three patients, they found the mean urethral dose to be within 4% of what was predicted for each case.

The use of OSLDs within brachytherapy is not limited to the prostate treatment. Another example of their application is in verification of the dose distribution of custom-made applicators [143]. While applicators specializing in skin surface treatments do exist, many clinics will create their own using a variety of techniques and materials, including thermoplastics and wax bolus. The downside of creating such a unique applicator for each patient, other than the time commitment, is the uncertainty in dose distribution. *In vivo* dosimetry with OSLDs in these cases can be a valuable tool to verify the accuracy of the treatment delivery. There are a few ways to go about predicting the dose to the OSLDs in such instances. Perhaps the simplest is to scan the patient with both the custom applicator and OSLDs in place. Alternatively, if the OSLDs cannot be visualized with this technique, the applicator can be scanned again without the patient, but instead with BBs placed at each location where an OSLD will be present during treatment. Assuming the applicator is rigid, this scan can then be registered to the patient scan and a dose estimate to the OSLDs can be calculated. This alternative approach also avoids exposing the OSLDs to imaging dose.

A full history of the clinical implementation of TLDs and OSLDs is beyond the scope of this chapter, and indeed, this book; however, it would be remiss to conclude this brief summary without mentioning the application of OSLDs for both TBI and TSET [144, 145]. Due to their relative ease of use, OSLDs and TLDs are popular dosimeters for verifying dose to TBI and TSET patients. In the case of TBI, they can be placed at various anatomical points of interest during delivery and are often used with bolus to get an estimate of entrance and/or exit dose. Bolus is rarely used in TSET cases due to the sparing effect it would have on the tissue beneath it and instead OSLDs and TLDs (often covered by a plastic wrap) are placed directly on the patient's surface. Most OSLDs and TLDs also have the advantage of low angular dependence which is particularly useful in TBI and TSET where dosimeters are often placed in locations with an oblique angle of incidence to the beam. The user should be aware that the design of the dosimeter and its casing will affect the angular dependence so angular independence should not be freely assumed without evidence for the model in question. If the OSLDs or TLDs used for TBI and TSET measurements are calibrated under standard clinical reference conditions, it may be necessary to apply a correction factor to their readings [146]. This correction factor will depend on the specific dosimeters as well as the clinical setup being used for TBI and TSET.

11.4 SUMMARY

As this chapter shows, there are a wide variety of options available for *in vivo* dosimetry. It is up to the clinical physicist to figure out how to make best use of the resources available to them and, if those resources are insufficient, make recommendations to the hospital administration as to what devices are needed. There is no single dosimeter that will work in all scenarios but similarly, there are few, if any, departments that will need all of the dosimeters discussed here. The needs of a department will also strongly depend on how many specialized procedures they are doing (e.g., TBI, TSET, IORT). Each dosimeter has its advantages and disadvantages, and it is important to be aware of those before beginning a new *in vivo* dosimetry program. For example, if the dosimeter is going to be used to measure treatments in the kV energy range, then its energy dependence needs to be carefully established and/or well documented in the literature. When applied correctly, an *in vivo* dosimetry program can be an incredibly valuable QA tool to help verify the safety and efficacy of dose delivery. However, if such a program is applied with little thought or understanding given to the limitations of the dosimeters used, then any results obtained may be meaningless, or worse, lead to incorrect clinical judgments and treatments. Pros and cons relevant to *in vivo* dosimetry for each of the discussed dosimeters is provided in Table 11.1.

TABLE 11.1 Advantages and Disadvantages of Dosimeters for *In Vivo* Dosimetry Applications

Dosimeter	Pros for *In Vivo* Dosimetry	Cons for *In Vivo* Dosimetry
Diodes	• Small • Highly sensitive • Can be reused • Immediate results • Large body of supporting literature	• Temperature dependence • Angular dependence • Dose rate and Dose per pulse dependence • Energy dependence • Most models require wires and an electrometer • Sensitivity changes with accumulated dose
MOSFET	• Small • Immediate results • Can be reused • Stores dose information permanently • Large body of supporting literature • Minimal temperature dependence with a dual bias	• Angular dependence • Dose per pulse dependence • Energy dependence between KV and MV ranges • Sensitivity changes with accumulated dose
EPIDs	• Immediate results • Can be reused • Doesn't require additional equipment • 2D planar dosimetry • Large body of supporting literature	• Cannot be placed within or on the patient • Not practical for brachytherapy • Not practical for entrance dose measurements • Not practical for TBI or TSET
Scintillators	• Small • Immediate results • Can be reused • Water equivalent • Minimal angular or dose rate dependence • MR compatible	• Less widely used than most other dosimeters • Temperature dependence • Mechanically delicate • Susceptible to "noise" caused by Cherenkov radiation
Ion Chambers	• Immediate results • Can be reused • Can directly measure dose	• Bulky • Requires a high voltage bias • Requires wires and an electrometer • Temperature and pressure dependence • Not practical for brachytherapy
Film Dosimeters	• 2D planar dosimetry • Gafchromic film is tissue equivalent • Large body of supporting literature	• Not appropriate for online review • Reading changes dramatically based on time since exposure • Requires careful handling • Requires use of an appropriately calibrated scanner
RPLDs	• Small • Can be reused • Minimal angular dependence	• Not appropriate for online review • Less widely used than most other dosimeters • Requires a reader
OSLDs and TLDs	• Small • Can be reused • Large body of supporting literature • TLDs are close to tissue equivalent • OSLDs are easy to use	• Not appropriate for online review • TLDs require careful handling • Requires a reader

A physicist should begin the process of dosimeter selection by first asking themselves what is the ultimate goal of the program. Is it to catch gross errors or to observe trends? Is absolute dosimetry needed or is a relative measurement sufficient? Is a single point measurement good enough or is a 2D dose plane needed? Do *in vivo* readings need to be able to identify a problem with delivery immediately or will there be time to process and analyze the results offline? Without having a concrete goal in mind from the outset, it is easy to get confused an overwhelmed by the sheer volume of available choices out there. Hopefully, the review provided by this chapter helps to clarify the roles that different dosimeters play in different treatment scenarios.

REFERENCES

1. M. S. Huq, et al., "The report of Task Group 100 of the AAPM: Application of risk analysis methods to radiation therapy quality management," *Medical Physics* **43**(7), 4209–4262 (2016).
2. R. Alecu, et al., "Guidelines on the implementation of diode in vivo dosimetry programs for photon and electron external beam therapy," *Medical Dosimetry* **24**(1), 5–12 (1999).
3. F. Ghahramani, et al., "Dependency of semiconductor dosimeter responses, used in MDR/LDR brachytherapy, on factors which are important in clinical conditions," *Reports of Practical Oncology & Radiotherapy* **13**(1), 29–33 (2008).
4. P. A. Jursinic, "Angular dependence of dose sensitivity of surface diodes," *Medical Physics* **36**(6 Part 1), 2165–2171 (2009).
5. S. F. Kry, et al., "AAPM TG 158: Measurement and calculation of doses outside the treated volume from external-beam radiation therapy," *Medical Physics* **44**(10), e391–e429 (2017).
6. K. González-Pérez, et al., "Calibration and correction factors of a set of semiconductor diodes for its use in external radiation therapy," *AIP Conference Proceedings* **1747**(1), 040001 (2016).
7. E. Yorke et al., *Diode In Vivo Dosimetry for Patients Receiving External Beam Radiation Therapy*. Report of Task Group 62 of the Radiation Therapy Committee (Medical Physics Publishing, Madison, 2005).
8. X. R. Zhu, "Entrance dose measurements for in-vivo diode dosimetry: Comparison of correction factors for two types of commercial silicon diode detectors," *Journal of Applied Clinical Medical Physics* **1**(3), 100–107 (2000).
9. L. L. W. Wang and D. W. O. Rogers, "Monte Carlo study of Si diode response in electron beams," *Medical Physics* **34**(5), 1734–1742 (2007).
10. T. Loncol, et al., "Entrance and exit dose measurements with semiconductors and thermoluminescent dosimeters: A comparison of methods and in vivo results," *Radiotherapy and Oncology* **41**(2), 179–187 (1996).
11. D. I. Thwaites, et al., "Experience with in vivo diode dosimetry for verifying radiotherapy dose delivery: Practical implementation of cost-effective approaches," *International Atomic Energy Agency (IAEA)* 245–246 (2002).
12. S. Heukelom, et al., "Comparison of entrance and exit dose measurements using ionization chambers and silicon diodes," *Physics in Medicine and Biology* **36**(1), 47–59 (1991).
13. R. Alecu, et al., "A method to improve the effectiveness of diode in vivo dosimetry," *Medical Physics* **25**(5), 746–749 (1998).
14. P. D. Higgins, et al., "In vivo diode dosimetry for routine quality assurance in IMRT," *Medical Physics* **30**(12), 3118–3123 (2003).
15. J. G. Wierzbicki and D. S. Waid, "Large discrepancies between calculated and diode readings for small field sizes and small SSDs of 15 MV photon beams," *Medical Physics* **25**(2), 245–246 (1998).
16. K. T. Welsh and L. E. Reinstein (2001). "The thermal characteristics of different diodes on in vivo patient dosimetry," *Medical Physics* **28**(5), 844–849.
17. A. S. Saini and T. C. Zhu (2002). "Temperature dependence of commercially available diode detectors," *Medical Physics* **29**(4), 622–630.
18. ICRP, "The 2007 Recommendations of the International Commission on Radiological Protection," ICRP Publication 103. Ann. ICRP 37 (2–4) (2007).
19. N. Vicoroski, et al., "Development of a silicon diode detector for skin dosimetry in radiotherapy," *Medical Physics* **44**(10), 5402–5412 (2017).
20. C. B. Saw, et al., "Energy dependence of a new solid state diode for low energy photon beam dosimetry," *Medical Dosimetry* **23**(2), 95–97 (1998).
21. M. Chan, et al., "Estimating dose to ICD outside the treatment field using skin QED diode," *Medical Physics* **33**(6), 2218 (2006).
22. S. Hui, et al., "SU-FF-T-150: In vivo diode dosimetry for helical tomotherapy dose verification," *Medical Physics* **32**(6), 1984–1984 (2005).
23. A. Toltz, et al., "Time-resolved diode dosimetry calibration through Monte Carlo modeling for in vivo passive scattered proton therapy range verification," *Journal of Applied Clinical Medical Physics* **18**(6), 200–205 (2017).
24. U. Sowa, et al., "Dosimetric characteristics of active solid state detectors in a 60 MeV proton radiotherapy beam," *Nukleonika* **57**, 491–495 (2012).
25. A. Toltz, et al., "SU-E-T-524: In-vivo diode dosimetry proton therapy range verification validation study for pediatric CSI," *Medical Physics* **42**(6), 3455–3456 (2015).

26. B. Gottschalk, et al., "Water equivalent path length measurement in proton radiotherapy using time resolved diode dosimetry," *Medical Physics* **38**(4), 2282–2288 (2011).

27. J. R. Srour, et al., "Review of displacement damage effects in silicon devices," *IEEE Transactions on Nuclear Science* **50**(3), 653–670 (2003).

28. K. Tanderup, et al., "In vivo dosimetry in brachytherapy," *Medical Physics* **40**(7), 070902 (2013).

29. M. Allahverdi, et al., "Evaluation of treatment planning system of brachytherapy according to dose to the rectum delivered," *Radiation Protection Dosimetry* **150**(3), 312–315 (2011).

30. R. Alecu and M. Alecu, "In-vivo rectal dose measurements with diodes to avoid misadministrations during intracavitary high dose rate brachytherapy for carcinoma of the cervix," *Medical Physics* **26**(5), 768–770 (1999).

31. A. S. Saini and T. C. Zhu, "Energy dependence of commercially available diode detectors for in-vivo dosimetry," *Medical Physics* **34**(5), 1704–1711 (2007).

32. J. Poder, et al., "HDR brachytherapy in vivo source position verification using a 2D diode array: A Monte Carlo study," *Journal of Applied Clinical Medical Physics* **19**(4), 163–172 (2018).

33. A. S. Kumar, et al., "Characteristics of mobile MOSFET dosimetry system for megavoltage photon beams," *Journal of medical physics* **39**(3), 142–149 (2014).

34. J. H. Koivisto, et al., "Characterization of MOSFET dosimeters for low-dose measurements in maxillofacial anthropomorphic phantoms," *Journal of Applied Clinical Medical Physics* **16**(4), 266–278 (2015).

35. D. J. Gladstone and L. M. Chin, "Automated data collection and analysis system for MOSFET radiation detectors," *Medical Physics* **18**(3), 542–548 (1991).

36. A. Holmes-Siedle, "The space-charge dosimeter: General principles of a new method of radiation detection," *Nuclear Instruments and Methods* **121**(1), 169–179 (1974).

37. D. Shah, *Radiation Effects on MOSFETs* (2016).

38. J. A. Tanyi, et al., "MOSFET sensitivity dependence on integrated dose from high-energy photon beams," *Medical Physics* **35**(1), 39–47 (2008).

39. G. Toncheva, et al., "SU-GG-I-67. Variation of MOSFET calibration factor as a function of dosimeter age," *Medical Physics* **37**(6 Part 3), 3116–3117 (2010).

40. P. H. Halvorsen, "Dosimetric evaluation of a new design MOSFET in vivo dosimeter," *Medical Physics* **32**(1), 110–117 (2005).

41. J. Koivisto, et al., "Characterization of MOSFET dosimeter angular dependence in three rotational axes measured free-in-air and in soft-tissue equivalent material," *Journal of Radiation Research* **54**(5), 943–949 (2013).

42. R. A. Kinhikar, et al., *In vivo dosimetry using MOSFET and TLD for Tomotherapy* (Springer Berlin Heidelberg, Berlin, Heidelberg, 2009).

43. T. Cheung, et al., "Effects of temperature variation on MOSFET dosimetry," *Physics in Medicine and Biology* **49**(13), N191–N196 (2004).

44. S. Qin, et al., "Angular dependence of the MOSFET dosimeter and its impact on in vivo surface dose measurement in breast cancer treatment," *Technology in Cancer Research & Treatment* **13**(4), 345–352 (2014).

45. T. Cheung, et al., "Energy dependence corrections to MOSFET dosimetric sensitivity," *Australasian Physics & Engineering Sciences in Medicine* **32**(1), 16–20 (2009).

46. P. Ravindran, et al., "In vivo dose measurement during IGRT with KV cone beam CT," *Medical Physics* **40**, 188 (2013).

47. M. Amin, et al., "SU-FF-T-387: Performance of a portable MOSFET dosimetry system for in-vivo dose measurements in orthovoltage treatments and kilovoltage cone beam CT," *Medical Physics* **36**(6 Part 15), 2611–2611 (2009).

48. G. X. Ding and C. W. Coffey, "Dosimetric evaluation of the OneDose™ MOSFET for measuring kilovoltage imaging dose from image-guided radiotherapy procedures," *Medical Physics* **37**(9), 4880–4885 (2010).

49. N. Ploquin, et al., "Use of novel fibre-coupled radioluminescence and RADPOS dosimetry systems for total scatter factor measurements in small fields," *Physics in Medicine and Biology* **60**, 1–14 (2014).

50. R. Marants, et al., "Evaluation of the 4D RADPOS dosimetry system for dose and position quality assurance of CyberKnife," *Medical Physics* **45**(9), 4030–4044 (2018).

51. M. Ciocca, et al., "Real-time in vivo dosimetry using micro-MOSFET detectors during intraoperative electron beam radiation therapy in early-stage breast cancer," *Radiotherapy and Oncology* **78**(2), 213–216 (2006).

52. F. W. Hensley, "Present state and issues in IORT physics," *Radiation Oncology* **12**(1), 37 (2017).

53. A. Petoukhova, et al., "In vivo dosimetry with MOSFETs and GAFCHROMIC films during electron IORT for accelerated partial breast irradiation," *Physica Medica* **44**, 26–33 (2017).

54. R. Consorti, et al., "In vivo dosimetry with MOSFETs: Dosimetric characterization and first clinical results in intraoperative radiotherapy," *International Journal of Radiation Oncology, Biology, Physics* **63**(3), 952–960 (2005).

55. T. M. Briere, et al., "Patient dosimetry for total body irradiation using single-use MOSFET detectors," *Journal of Applied Clinical Medical Physics* **9**(4), 200–205 (2008).

56. T. M. Briere, et al., "Implantable MOSFET detectors: Evaluation of a new design," *Medical Physics* **34**(12), 4585–4590 (2007).

57. S. Kim, et al., "In vivo dosimetry of skin dose during HDR breast brachytherapy using balloon catheter," *International Journal of Radiation Oncology, Biology, Physics* **72**(1), S524–S525 (2008).

58. E. J. Bloemen-van Gurp, et al., "In vivo dosimetry using a linear MOSFET-array dosimeter to determine the urethra dose in 125I permanent prostate implants," *International Journal of Radiation Oncology, Biology, Physics* **73**(1), 314–321 (2009).

59. C.-W. Cheng, et al., "Dosimetric characteristics of a single use MOSFET dosimeter for in vivo dosimetry in proton therapy," *Medical Physics* **37**(8), 4266–4273 (2010).

60. G. A. Pablo Cirrone, et al., "Preliminary investigation on the use of the MOSFET dosimeter in proton beams," *Physica Medica* **22**(1), 29–32 (2006).

61. A. B. Rosenfeld, et al., "A new silicon detector for microdosimetry applications in proton therapy," *IEEE Transactions on Nuclear Science* **47**(4), 1386–1394 (2000).

62. R. Kohno, et al., "In vivo proton dosimetry using a MOSFET detector in an anthropomorphic phantom with tissue inhomogeneity," *Journal of Applied Clinical Medical Physics* **13**(2), 3699–3699 (2012).

63. R. Kohno, et al., "Proton dose distribution measurements using a MOSFET detector with a simple dose-weighted correction method for LET effects," *Journal of Applied Clinical Medical Physics* **12**(2), 326–337 (2011).

64. H.-M. Lu, et al., "Investigation of an implantable dosimeter for single-point water equivalent path length verification in proton therapy," *Medical Physics* **37**(11), 5858–5866 (2010).

65. L. E. Antonuk, "Topical review: Electronic portal imaging devices: A review and historical perspective of contemporary technologies and research," *Physics in Medicine and Biology* **47**, R31–R65 (2002).

66. W. van Elmpt, et al., "A literature review of electronic portal imaging for radiotherapy dosimetry," *Radiotherapy and Oncology* **88**(3), 289–309 (2008).

67. M. G. Herman, et al., "Clinical use of electronic portal imaging: Report of AAPM Radiation Therapy Committee Task Group 58," *Medical Physics* **28**(5), 712–737 (2001).

68. M. Wendling, et al., "A simple backprojection algorithm for 3D in vivo EPID dosimetry of IMRT treatments," *Medical Physics* **36**(7), 3310–3321 (2009).

69. B. Mijnheer, et al., "TU-C-BRE-11: 3D EPID-based in vivo dosimetry: A major step forward towards optimal quality and safety in radiation oncology practice," *Medical Physics* **41**(6 Part 26), 457–457 (2014).

70. E. Van Uytven, et al., "Validation of a method for in vivo 3D dose reconstruction for IMRT and VMAT treatments using on-treatment EPID images and a model-based forward-calculation algorithm," *Medical Physics* **42**(12), 6945–6954 (2015).

71. A. Fidanzio, et al., "Breast in vivo dosimetry by EPID," *Journal of Applied Clinical Medical Physics* **11**(4), 249–262 (2010).

72. D. A. Low, et al., "A technique for the quantitative evaluation of dose distributions," *Medical Physics* **25**(5), 656–661 (1998).

73. M. Miften, et al., "Tolerance limits and methodologies for IMRT measurement-based verification QA: Recommendations of AAPM Task Group No. 218," *Medical Physics* **45**(4), e53–e83 (2018).

74. L. N. McDermott, et al., "Replacing pretreatment verification with in vivo EPID dosimetry for prostate IMRT," *International Journal of Radiation Oncology, Biology, Physics* **67**(5), 1568–1577 (2007).

75. K. Slosarek, et al., "EPID in vivo dosimetry in RapidArc technique," *Reports of Practical Oncology and Radiotherapy: Journal of Great Poland Cancer Center in Poznan and Polish Society of Radiation Oncology* **15**(1), 8–14 (2010).

76. A. Mans, et al., "Catching errors with in vivo EPID dosimetry," *Medical Physics* **37**(6 Part 2), 2638–2644 (2010).

77. S. Celi, et al., "EPID based in vivo dosimetry system: Clinical experience and results," *Journal of Applied Clinical Medical Physics* **17**(3), 262–276 (2016).

78. R. L. Smith, et al., "OC-0177: Clinical implementation of in vivo source position verification in high dose rate prostate brachytherapy," *Radiotherapy and Oncology* **115**, S87–S88 (2015).

79. M. van Herk, et al., "Magnetic resonance imaging-guided radiation therapy: A short strengths, weaknesses, opportunities, and threats analysis," *International Journal of Radiation Oncology, Biology, Physics* **101**(5), 1057–1060 (2018).

80. L. Wootton, et al., "Real-time in vivo rectal wall dosimetry using plastic scintillation detectors for patients with prostate cancer," *Physics in Medicine and Biology* **59**(3), 647–660 (2014).

81. J. L. Cantley, et al., "Real-time in vivo dosimetry for SBRT prostate treatment using plastic scintillating dosimetry embedded in a rectal balloon: A case study," *Journal of Applied Clinical Medical Physics* **17**(6), 305–311 (2016).

82. A. Bourgouin, et al., "Estimating and reducing dose received by cardiac devices for patients undergoing radiotherapy," *Journal of Applied Clinical Medical Physics* **16**(6), 411–422 (2015).

83. A. Bourgouin, et al., "Technical Note: Out-of-field dose measurement at near surface with plastic scintillator detector," *Journal of Applied Clinical Medical Physics* **17**(5), 542–547 (2016).

84. A. Skrobala, et al., "Low dose out-of-field radiotherapy, part 2: Calculating the mean photon energy values for the out-of-field photon energy spectrum from scattered radiation using Monte Carlo methods," *Cancer/Radiothérapie* **21**(5), 352–357 (2017).

85. F. Therriault-Proulx, et al., "A method to correct for temperature dependence and measure simultaneously dose and temperature using a plastic scintillation detector," *Physics in Medicine and Biology* **60**(20), 7927–7939 (2015).

86. C. E. Andersen, et al., "Characterization of a fiber-coupled luminescence dosimetry system for online in vivo dose verification during brachytherapy," *Medical Physics* **36**(3), 708–718 (2009).

87. J. Lambert, et al., "A plastic scintillation dosimeter for high dose rate brachytherapy," *Physics in Medicine and Biology* **51**(21), 5505–5516 (2006).

88. C. E. Andersen, et al., "Time-resolved in vivo luminescence dosimetry for online error detection in pulsed dose-rate brachytherapy," *Medical Physics* **36**(11), 5033–5043 (2009).

89. A. Darafsheh, "Scintillation fiber optic dosimetry," in *Radiation Therapy Dosimetry: A Practical Handbook*, edited by A. Darafsheh (CRC Press, Boca Raton, 2021), Chapter 9, pp. 123–137.

90. A. Darafsheh, et al., "Radiotherapy fiber dosimeter probes based on silver-only coated hollow glass waveguides," *Journal of Biomedical Optics* **23**(1), 1–7 (2018).

91. A. Darafsheh, et al., "Spectroscopic separation of Čerenkov radiation in high-resolution radiation fiber dosimeters," *Journal of Biomedical Optics* **20**(9), 1–7 (2015).

92. P. Carrasco, et al., "Characterization of the Exradin W1 scintillator for use in radiotherapy," *Medical Physics* **42**(1), 297–304 (2015).

93. A. Piermattei, et al., "In-vivo portal dosimetry by an ionization chamber," *Physica Medica* **21**(4), 143–152 (2005).

94. B. Poppe, et al., "DAVID – A translucent multi-wire transmission ionization chamber for in vivo verification of IMRT and conformal irradiation techniques," *Physics in Medicine and Biology* **51**, 1237–1248 (2006).

95. G. K. Svensson, "Quality assurance in external beam radiation therapy," *RadioGraphics* **9**(1), 169–182 (1989).

96. H. W. Liu, et al., "Role of in vivo dosimetry with radiochromic films for dose verification during cutaneous radiation therapy," *Radiation Oncology* **10**(1), 12 (2015).

97. M. Williams and P. Metcalfe, "Radiochromic film dosimetry and its applications in radiotherapy," *AIP Conference Proceedings* **1345**(1), 75–99 (2011).

98. S. Mossahebi, N. Hoshyar, R. Khan, and A. Darafsheh, "Film dosimetry," in *Radiation Therapy Dosimetry: A Practical Handbook*, edited by A. Darafsheh (CRC Press, Boca Raton, 2021), Chapter 5, pp. 61–74.

99. S. Devic, "Radiochromic film dosimetry: Past, present, and future," *Physica Medica* **27**(3), 122–134 (2011).

100. A. Niroomand-Rad, et al., "Radiochromic film dosimetry: Recommendations of AAPM Radiation Therapy Committee Task Group 55," *Medical Physics* **25**(11), 2093–2115 (1998).

101. S. Pai, et al., "TG-69: Radiographic film for megavoltage beam dosimetry," *Medical Physics* **34**(6 Part 1), 2228–2258 (2007).

102. A. Darafsheh, T. Zhao, R. Khan, "Spectroscopic analysis of irradiated radiochromic EBT-XD films in proton and photon beams," *Physics in Medicine and Biology* **65**, 205002 (2020).

103. E. Y. León-Marroquín, et al., "Spectral analysis of the EBT3 radiochromic films for clinical photon and electron beams," *Medical Physics* **46**(2), 973–982 (2019).

104. E. Y. León-Marroquín, et al., "Response characterization of EBT-XD radiochromic films in megavoltage photon and electron beams," *Medical Physics* **46**(9), 4246–4256 (2019).

105. L. Zhao and I. J. Das, "Gafchromic EBT film dosimetry in proton beams," *Physics in Medicine and Biology* **55**(10), N291–N301 (2010).

106. Y. Zheng, "SU-E-T-449: In-vivo dosimetry and range verification for proton therapy," *Medical Physics* **40**(6 Part 17), 308–308 (2013).

107. S. Reinhardt, et al., "Comparison of Gafchromic EBT2 and EBT3 films for clinical photon and proton beams," *Medical Physics* **39**(8), 5257–5262 (2012).

108. A. Darafsheh, et al., "On the spectral characterization of radiochromic films irradiated with clinical proton beams," *Physics in Medicine & Biology* **64**(13), 135016 (2019).

109. V. Devaraju and R. Slopsema, "SU-E-T-120: Characterization of EBT2 film for absolute dosimetry in proton beams," *Medical Physics* **38**(6 Part 12), 3513–3513 (2011).

110. D. Kirby, et al., "LET dependence of GafChromic films and an ion chamber in low-energy proton dosimetry," *Physics in Medicine and Biology* **55**(2), 417–433 (2009).

111. R. Moylan, et al., "Dosimetric accuracy of Gafchromic EBT2 and EBT3 film for in vivo dosimetry," *Australasian Physical & Engineering Sciences in Medicine* **36**(3), 331–337 (2013).

112. K. Chuang, et al., "SU-FF-T-237: Gafchromic film in vivo dosimetry for oral radiotherapy," *Medical Physics* **34** (2007).

113. M. Severgnini, et al., "In vivo dosimetry and shielding disk alignment verification by EBT3 GAFCHROMIC film in breast IOERT treatment," *Journal of Applied Clinical Medical Physics* **16**(1), 112–120 (2015).

114. S. Pai, et al., "The use of improved radiochromic film for in vivo quality assurance of high dose rate brachytherapy," *Medical Physics* **25**(7), 1217–1221 (1998).

115. S. Devic, et al., "Accurate skin dose measurements using radiochromic film in clinical applications," *Medical Physics* **33**(4), 1116–1124 (2006).

116. J. Ashburn, "SU-E-T-160: Response of EBT2 Gafchromic film to possible stresses associated with in vivo measures," *Medical Physics* **39**(6 Part 12), 3740–3740 (2012).

117. N. Navab Moghadam, et al., "Response of TLD and RPL personal dosimeters in a national inter-comparison test program," *International Journal of Radiation Research (IJRR)* **14**, 73 (2016).

118. A.-L. Manninen, et al., "The applicability of radiophotoluminescence dosemeter (RPLD) for measuring medical radiation (MR) doses," *Radiation Protection Dosimetry* **151**(1), 1–9 (2012).

119. S. Hashimoto, et al., "Field-size correction factors of a radiophotoluminescent glass dosimeter for small-field and intensity-modulated radiation therapy beams," *Medical Physics* **45**(1), 382–390 (2018).

120. E. H. Silva, et al., "ENERGY and angular dependence of radiophotoluminescent glass dosemeters for eye lens dosimetry," *Radiation Protection Dosimetry* **170**(1-4), 208–212 (2016).

121. J.-E. Rah, et al., "Clinical application of glass dosimeter for in vivo dose measurements of total body irradiation treatment technique," *Radiation Measurements* **46**(1), 40–45 (2011).

122. D. Y. C. Huang and S.-M. Hsu, "Radio-photoluminescence glass dosimeter (RPLGD)," in *Advances in Cancer Therapy* (IntechOpen, 2011).

123. S. Hashimoto, et al., "Energy dependence of a radiophotoluminescent glass dosimeter for HDR 192Ir brachytherapy source," *Medical Physics* **46**(2), 964–972 (2019).

124. T. Nose, et al., "In vivo dosimetry of high-dose-rate brachytherapy: Study on 61 head-and-neck cancer patients using radiophotoluminescence glass dosimeter," *International Journal of Radiation Oncology, Biology, Physics* **61**(3), 945–953 (2005).

125. M. Tsuda, "A few remarks on photoluminescence dosimetry with high energy X-rays," *Igaku Butsuri* **20**(3), 131–139 (2000).

126. T., Nose, et al., "In vivo dosimetry of high-dose-rate interstitial brachytherapy in the pelvic region: Use of a radiophotoluminescence glass dosimeter for measurement of 1004 points in 66 patients with pelvic malignancy," *International Journal of Radiation Oncology, Biology, Physics* **70**(2), 626–633 (2008).

127. J. A. Nieto, "Thermoluminescence dosimetry (TLD) and its application in medical physics," *AIP Conference Proceedings* **724**(1), 20–27 (2004).

128. F. H. Yusof, et al., "On the use of optically stimulated luminescent dosimeter for surface dose measurement during radiotherapy," *PLoS One* **10**(6), e0128544–e0128544 (2015).

129. A. C. Riegel, et al., "In vivo dosimetry with optically stimulated luminescent dosimeters for conformal and intensity modulated radiation therapy: A 2-year multicenter cohort study," *International Journal of Radiation Oncology, Biology, Physics* **96**(2), E605–E606 (2016).

130. N. D. MacDougall, et al., "In vivo dosimetry in UK external beam radiotherapy: Current and future usage," *The British Journal of Radiology* **90**(1072), 20160915–20160915 (2017).

131. R. Ponmalar, et al., "Dosimetric characterization of optically stimulated luminescence dosimeter with therapeutic photon beams for use in clinical radiotherapy measurements," *Journal of Cancer Research and Therapeutics* **13**(2), 304–312 (2017).

132. P. A. Jursinic, "Characterization of optically stimulated luminescent dosimeters, OSLDs, for clinical dosimetric measurements," *Medical Physics* **34**(12), 4594–4604 (2007).

133. Y. R. Ponmalar, et al., "Response of nanodot optically stimulated luminescence dosimeters to therapeutic electron beams," *Journal of Medical Physics* **42**(1), 42–47 (2017).

134. S. B. Scarboro, et al., "Characterization of the nanoDot OSLD dosimeter in CT," *Medical Physics* **42**(4), 1797–1807 (2015).

135. S. B. Scarboro and S. F. Kry, "Characterization of energy response of Al(2)O(3):C optically stimulated luminescent dosimeters (OSLDs) using cavity theory," *Radiation Protection Dosimetry* **153**(1), 23–31 (2013).

136. A. S. Meigooni, et al., "Influence of the variation of energy spectra with depth in the dosimetry of192Ir using LiF TLD," *Physics in Medicine and Biology* **33**(10), 1159–1169 (1988).

137. A. H. Zhuang, and A. J. Olch, "Validation of OSLD and a treatment planning system for surface dose determination in IMRT treatments," *Medical Physics* **41**(8 Part 1), 081720 (2014).

138. E. A. Gardner, et al., "In vivo dose measurement using TLDs and MOSFET dosimeters for cardiac radiosurgery," *Journal of Applied Clinical Medical Physics* **13**(3), 190–203 (2012).

139. G. Anagnostopoulos, et al., "In vivo thermoluminescence dosimetry dose verification of transperineal 192Ir high-dose-rate brachytherapy using CT-based planning for the treatment of prostate cancer," *International Journal of Radiation Oncology, Biology, Physics* **57**(4), 1183–1191 (2003).

140. R. Das, et al., "Thermoluminescence dosimetry for in-vivo verification of high dose rate brachytherapy for prostate cancer," *Australasian Physical and Engineering Sciences in Medicine* **30**(3), 178–184 (2007).

141. W. Toye, et al., "An in vivo investigative protocol for HDR prostate brachytherapy using urethral and rectal thermoluminescence dosimetry," *Radiotherapy and Oncology* **91**(2), 243–248 (2009).

142. I. A. Brezovich, et al., "In vivo urethral dose measurements: A method to verify high dose rate prostate treatments," *Medical Physics* **27**(10), 2297–2301 (2000).

143. E. Ekwelundu, et al., "Custom-designed mouthpiece for HDR brachytherapy of embryonal rhabdomyosarcoma of the soft palate," *Journal of Contemporary brachytherapy* **6**(3), 300–303 (2014).

144. C. Holloway, et al., "Total body irradiation (TBI) optically stimulated luminescence in vivo dosimetry," *Medical Physics* **40**, 223 (2013).

145. T. Kron, et al., "A technique for total skin electron therapy (TSET) of an anesthetized pediatric patient," *Journal of Applied Clinical Medical Physics* **19**(6), 109–116 (2018).

146. P. Alvarez, et al., "TLD and OSLD dosimetry systems for remote audits of radiotherapy external beam calibration," *Radiation Measurements* **106**, 412–415 (2017).

CHAPTER 12

Dosimeters and Devices for IMRT QA

Nesrin Dogan and Matthew T. Studenski
University of Miami
Miami, Florida

Perry B. Johnson
University of Florida Health Proton Therapy Institute
Jacksonville, Florida

CONTENTS

12.1 INTRODUCTION

Intensity-modulated radiation therapy (IMRT) and volumetric modulated arc therapy (VMAT) are sophisticated treatment techniques that are widely available in routine clinical practice [1–5]. In IMRT, treatments are delivered by a series of multiple small beams or beamlets using either static or dynamic multi-leaf collimator (MLC)-based techniques. With VMAT, both treatment planning and delivery are especially complex as patient treatment is delivered while the gantry rotates around the patient, the MLC is dynamically shaping the beam and the dose rate and gantry rotation speed are constantly changing. As compared to conventional 3D conformal radiation therapy (3D CRT), delivery of such complex treatments imposes new demands and requirements for machine and patient-specific quality assurance (QA) and verification methods [6–7].

Currently, there are variety of commercially available dosimeters and measurement devices specifically designed for IMRT QA, including treatment planning as well as machine and patient-specific QA purposes. The selection of a specific dosimeter system depends on the type of measurement and the endpoint of the preferred QA process [8–10].

Many reports and papers on the use of QA devices and techniques have been published since the initial clinical utilization of IMRT. Most of the IMRT techniques published in the early 2000s relied on the use of ionization chambers and radiographic film dosimetry for QA of IMRT fields [11–13]. Use of such traditional methods is time-consuming and limits the analysis of dose to a 2D plane. However, over the last decade, IMRT techniques and tools have rapidly evolved and a variety of new IMRT QA devices such as two-dimensional (2D) array detectors, 3D detectors, electronic portal imaging devices (EPIDs), radiochromic films, and 3D gel dosimetry have become available [14–18]. AAPM Task Group (TG) Report 120 on dosimetry tools and techniques for IMRT [8] provides a comprehensive review of available dosimeters and phantoms for IMRT commissioning and verification of IMRT plans. This report includes how ionization chambers, films, EPIDs, array detectors, and a variety of phantoms (geometric and anthropomorphic) should be used, including limitations and appropriate use of each dosimeter and quantitative comparison techniques for dose distribution evaluation. AAPM TG-142 report [19] indicates the need for dosimetry and measurement tools and provides

updated recommendations for machine-specific QA, which include more stringent tolerances for IMRT. AAPM TG-119 report [12] also provides patient-specific IMRT QA guidelines, including evaluation criteria for QA. The NCS Reports 22 and 23 [20, 21] present a comprehensive review of many QA methods and tests published in the literature and recommended an extensive set of QA techniques and tools (both measurement and non-measurement) for IMRT and VMAT. Recently published AAPM TG-218 presents a comprehensive analysis of performance of tolerance limits and methodologies such as dose difference/distance-to-agreement (DTA) and gamma index metrics and provides specific guidelines on tolerance and action limits for patient-specific IMRT QA [22]. The report suggests a universal tolerance limit of >95% gamma passing rate with 3%/2 mm DTA and 10% dose threshold for evaluation of IMRT QA.

With that knowledge in mind, it is essential to choose suitable dosimetry devices and tools for accurately testing the limitations of different components of an IMRT system such as treatment planning, delivery systems, and accurate comparisons of calculated and measured IMRT dose distributions. Appropriate choice of dosimeters and phantoms is expected to significantly reduce the magnitude and frequency of errors associated with each component in the IMRT QA program. Patient-specific QA is one of the most important aspects of an IMRT QA program since it assures appropriate treatment plans are created, transferred, and delivered correctly for a specific patient. In this chapter, we provide a comprehensive review of the dosimeters and devices needed for proper patient-specific QA, including a description of each device, the acceptance testing and commissioning process, and the performance of periodic QA tests.

12.2 PATIENT-SPECIFIC QA

12.2.1 Rationale

Patient-specific QA, also known as delivery QA, IMRT QA, or measurement-based verification QA, defines a category of tests related to the comparison between the dose planned for a specific patient and the dose measured when that plan is delivered to a QA device. The comparison can be made at a single point, but more often than not is made between spatial dose distributions. Patient-specific QA has been a prevalent aspect of radiation oncology since the introduction of IMRT in the 1990s [23–25]. At the time, there arose the need for the validation of complex dose distributions whose quality could no longer be guaranteed through conventional machine QA and measurement-based, independent dose calculations, that is, hand calculations. In terms of hardware, it also became necessary to assess the deliverability of plans given the mechanical limitations of treatment delivery systems that may influence the physical manifestation of the calculated fluence pattern, for example, leaf speed, leaf positioning, and dose rate. With regard to patient safety, patient-specific QA was also identified as a critical barrier protecting against failures such as plan corruption, missing data, or poor beam modeling [26]. These factors helped lead to the establishment of patient-specific QA as a routine aspect of the pre-treatment review process. This practice was later codified in several society-based recommendations [22, 27–30], and subsequently added by many payers as a requirement for billing codes related to IMRT planning.

In the present, the validation of complex dose distributions has been addressed, to an extent, through the proliferation of software systems capable of providing independent dose calculations for IMRT plans. This has opened a debate as to whether or not measurement-based validation is warranted in lieu of computational alternatives [31, 32]. A key argument in favor of measurement is that with the adoption of advanced modalities such as SBRT, VMAT, and the forthcoming 4π radiotherapy, the physical demands on treatment delivery systems have increased. In order to produce a fluence pattern that matches the one calculated for an advanced modality, modern treatment delivery systems must operate within tighter tolerances while accounting for an increase in the degrees of freedom made available to the treatment planning system. As such, the deliverability of IMRT plans remains a key issue that cannot be addressed by software alone. Another aspect of patient-specific QA, which justifies measurement, is the way in which it validates a variety of factors through a single test, for example, beam model, dose calculation, plan transfer, and deliverability. It could be argued that many of these factors should be addressed through proper IMRT commissioning, yet, a patient-specific measurement goes further by extending these checks to the unique circumstances that exists for a given patient and a given plan. It is the comprehensive nature of the method that helps define its value and maintains its status as one of the most commonly practiced QA measures in the radiation oncology clinic.

12.2.2 Quantitative Evaluation Tools

Dose difference, DTA, and gamma evaluation are the most commonly used tools for the quantitative comparison of dose distributions (e.g., measured versus calculated) for patient-specific QA although there are a number of QA tools for patient-specific IMRT QA reported in the literature [33–35]. The DTA was first used by van Dyk et al. for QA of treatment planning systems [36]. The gamma evaluation, which combines the dose difference and DTA criterion for each point of interest in a single criterion, was first developed by Low et al. for evaluation of 2D dose distributions [34]. The tolerance and acceptance levels of gamma evaluation may be influenced by many factors, including IMRT QA equipment and delivery technique. The DTA is influenced by the spatial resolution of the dose measuring device and the dose calculation grid size. The accurate determination of an appropriate dose tolerance level for gamma evaluation can only be obtained by carrying out an error analysis of the IMRT QA device used for the measurement, treatment planning system, and the choice of 2D versus 3D gamma evaluation. A criterion of at least 3%/3 mm DTA with a 10% dose threshold with minimum pass rate of 90% is generally considered a good agreement between calculated and measured dose using absolute dosimetry [20, 21, 34]. A 3 mm or less dose calculation grid size should be used to achieve 3 mm or better DTA [12]. Recently published AAPM TG-218 report provides a comprehensive review of all IMRT QA tools, including practical considerations using gamma evaluation criteria [22]. The report recommended universal tolerance limits of ≥95% (and 90%) gamma passing rate with 3%/2 mm DTA and 10% dose threshold and local tolerance limits for IMRT QA using gamma evaluation method.

12.2.3 Dosimeters and Devices

12.2.3.1 Point Measurements

Patient-specific QA in its simplest form compares the dose predicted at a single location to the dose delivered to that location when measured using a point detector, that is, a zero-dimensional measurement. The point detector must be capable of providing absolute dose through either a calibration factor traceable to a primary standard or by cross-calibration with a traceable device. The size of the detector must be such that partial volume effects are minimized. This depends on the active volume of the detector, the dose gradient in the area surrounding the measurement location, and the size and shape of the individual IMRT fluence patterns. Other desirable traits include common features such as energy independence, high sensitivity, limited directional dependency, linearity with dose, tissue equivalence, dose rate independence, and low cost.

In the patient-specific QA setting, vented ion chambers meet most of needs described above and have commonly been used for this purpose. As noted, the volume of the ion chamber should be small enough to limit partial volume effects for the given application. An active volume of 0.125 cm³ or less has been previously recommended as a practical cutoff [8, 12]. Consideration must also be given to the size and composition of the material within which the ion chamber is placed, that is, the physical phantom. The application of cavity theory for gas filled detectors is most directly applied when charged particle equilibrium (CPE – or transient CPE) exists. As such, the ideal measurement conditions include a physical phantom with a homogenous elemental composition and density, and a measurement location that avoids regions where CPE may be lost. Examples of the latter include the initial build-up region and any area immediately above or below an imbedded heterogeneity. The physical phantom should closely match the virtual phantom that was used to create the verification plan. Size, shape, and elemental composition should all be considered. Practically speaking, phantoms made from slabs of water-equivalent plastic represent an excellent choice as they meet the conditions described above and are relatively easy to acquire/construct. Liquid water may also be a reasonable option depending upon the delivery technique and whether or not the container holding the liquid water is taken into account. Different geometric and anthropomorphic phantoms have also been offered by several vendors and represent a more realistic interpretation of the clinical setting.

Aside from ion chambers, other point detectors previously used for patient-specific QA include TLDs, OSLDs, and MOSFETs. All three represent small, low cost, integrating dosimeters, which are robust to frequent handling. TLDs and OSLDs have long been used within anthropomorphic phantoms for external dosimetry audits and protocol credentialing [37]. They are reusable, versatile, and capable of 2–3% accuracy when properly calibrated. MOSFETs share many of the same features of TLDs/OSLDs but provide an

immediate readout and offer an improved spatial resolution. MOSFETS, however, show a decreased linearity for low doses (<50 cGy) and can over-respond to phantom scatter. Previous systems utilizing MOSFETs for patient-specific QA implemented patterns of detectors within water equivalent plastic phantoms, both cylindrical and slab-based [38]. In this way, they were able to provide a sample of a 2D or 3D dose distribution. While TLDs, OLSDs, and MOSFETs are commonly used for *in vivo* dosimetry and end-to-end testing, their use as a routine tool for patient-specific QA is limited due to better alternatives as described in the following sections.

12.2.3.2 Measurement Using Film

Historically, the transition from point dose to the measurement of spatial dose distributions came through the combination of ion chambers and film. While a properly calibrated film is capable of providing absolute dose, the accuracy and reproducibility of the measurement are dependent upon the measurement conditions, film processing, and analysis technique [39]. The challenges are such that in the patient-specific QA setting, film is often used to verify the relative dose distribution while a point measurement made with a calibrated ion chamber is used to verify absolute dose. In order to acquire a relative dose distribution, a film must be exposed, processed (depending on film type), scanned, digitized, and calibrated using a sensitometric curve. As opposed to ion chambers, which require extensive corrections in order to accurately measure dose in regions where CPE does not fully exist, film has few limitations in this regard and represents a suitable choice for measuring dose at the surface of a phantom or in the build-up region. For patient-specific QA, the film is most often oriented within a physical phantom at a plane bisecting the high dose region of the dose distribution. If the phantom can accommodate more than one piece of film, concurrent measurements can be made either along different parallel planes and/or in an orthogonal configuration.

Depending upon the sensitivity and corresponding dynamic range of the film, the calculated monitor units may require linear scaling to avoid a saturated response. Early implementation with radiographic film (XV2, Eastman Kodak Co., Rochester, NY) showed a saturation limit near 100 cGy [40]. This was later improved to approximately 600 cGy with the release of extended dose range film (EDR2) by the same manufacturer [41]. The difference between the two products is largely the result of a reduction in the size of AgBr grains, which were decreased by 90% in the EDR2 film. The change in silver content also affected the energy dependence of the film, which had been a previous limitation requiring calibration (optical density to dose) at the same depth as the intended measurement.

To reveal the latent image imparted by a radiation exposure, radiographic film requires chemical processing. Prior to the wide scale adoption of digital imaging, film processors were prevalent in the radiation oncology clinic. Relevant aspects of film processer QA include the periodic assessment of radiographic films exposed to the same dose to assess for constancy and scheduled maintenance/cleaning to avoid streaking and jamming. In the present, digital imaging for patient alignment has removed the primary need for film processors and radiographic film. For QA applications that maintain the utilization of film, the conversion has been made to radiochromic forms [42]. The most common radiochromic film used for patient-specific QA is the commercial product Gafchromic® EBT (International Specialty Products, Wayne, NJ). EBT, like other radiochromic film, is self-developing, largely unaffected by interior room light over short periods of time, relatively energy independent/tissue-equivalent, inherently slow to expose, and provides excellent spatial resolution. Successive versions (EBT, EBT2, and EBT3) have improved uniformity and consistency, and reduced the dependence on scan orientation. The latter is a characteristic of radiochromic film, which is traditionally scanned using a multichannel flatbed scanner. Both a consistent orientation and a centralized position within the flatbed scanner are important when scanning calibration and verification films. It is also important to wait a consistent length of time before scanning, and for the waiting period to be long enough (>24 h) to allow for color formation to reach a plateau. Depending on how the film is made, there will be different absorption patterns for different color spectra. EBT film is typically read out using three color channels, red, green, and blue. The red channel represents the component most sensitive to ionizing radiation and is widely used for single channel dosimetry. Multichannel dosimetry is more complex, as it requires model-based corrections, but affords the opportunity to improve uniformity and/or extend the dynamic range of the film [43].

Once scanned, film requires the use of a sensitometric curve to convert from optical density to dose. A sensitometric a curve is created by exposing blank films to known dose levels. Ideally, calibration films should be exposed/scanned with each QA measurement. To facilitate this process, some institutions have developed calibration plans that simultaneously deliver strip patterns at known dose levels [44]. The sensitometric conversion is the final step in obtaining a relative dose distribution. This distribution can be compared, in relative terms, with a dose plane extracted from the verification plan at the same depth and orientation as the film. As noted previously, an ion chamber is typically utilized to verify absolute dose. The ion chamber is most often located above or below the film near the isocenter. In scenarios where an ion chamber may exhibit partial volume effects, such as the case for stereotactic radiosurgery fields, the calibrated film may be used directly in a comparison between absolute dose distributions. In either case, the measured distribution is aligned with the calculated distribution through registration marks that have been placed on the film. The two distributions can then be compared using various similarity metrics such as the gamma analysis.

12.2.3.3 Detector Arrays

To improve the efficiency of patient-specific QA, several vendors have introduced detector arrays, which allow for easy setup and rapid evaluation of the measured dose. The first iteration of these devices was configured as 2D arrays using either diodes or vented ion chambers (see Table 12.1). The spacing between the detectors is typically 7 mm, though some devices utilize a 10 mm spacing toward the periphery. In either scenario, the spatial resolution is significantly reduced in comparison to film. This may present problems when measuring dose in small fields or fields with very steep dose gradients. One solution is to acquire two measurements with the device shifted during the second acquisition by half the width of the detector spacing. The two measurements can then be combined to double the number of detectors within the field. Recently, detector arrays intended for use in stereotactic applications have been released. The detector spacing for these devices has been reduced to 2.5 mm, which may be suitable for many cases.

Detector arrays are often placed within or beneath slabs of water-equivalent plastic to better simulate patient dosimetry. Several vendors provide device-specific phantoms for this purpose, while others configure their products with inherent buildup. Like film,

TABLE 12.1   Devices and Tools Used for Patient-Specific QA

Manufacturer	Device	Type	Notes
Various	Ion chamber	Point detector	$V < 0.125$ cm^3, location should avoid loss of CPE
LANDAUER	OSLD	Point detector	
BEST medical	MOSFET	Point detector	
Various	TLD	Point detector	
Various	Film	Point detector/spatial distribution	Gafchromic EBT3 is more popular, scanning and calibration required
Sun Nuclear	MapCHECK 3	Detector array (2D planar)	1527 diodes, 26×32 cm^2, 7.07 mm spacing
Sun Nuclear	SRS MapCHECK	Detector array (2D planar)	1013 diodes, 7.7×7.7 cm^2, 2.47 mm spacing
Sun Nuclear	ArcCHECK	Detector array (2D helical)	1386 diodes arranged in a helical pattern
PTW	OCTAVIUS 1500	Detector array (2D planar/rotational)	1405 vented ion chambers, 27×27 cm^2, 7.07 mm spacing, ability to rotate w/gantry
PTW	OCTAVIUS 1000 SRS	Detector array (2D planar)	977 liquid filled ion chambers, 10×10 cm^2, 2.5 mm spacing
ScandiDos	Delta^4 PHANTOM+	Detector array (2D orthogonal)	1069 diodes total (2 planes), 20×20 cm^2, 5 mm central spacing, 10 mm outer spacing
IBA	MatriXX	Detector array (2D planar)	1020 vented ion chambers, 24.4×24.4 cm^2, 7.62 mm spacing
ScandiDos	Delta^4 DISCOVER	Transmission chamber	4040 diodes, 1.5 mm spacing, link with phantom-based IMRT QA
IBA	Dolphin	Transmission chamber	1513 ion chambers, 40×40 cm^2, 5 mm spacing
PTW	DAVID	Transmission chamber	Multiwire ion chamber, real-time, no link to pre-treatment IMRT QA
Varian Medical Systems	Portal dosimetry	EPID based	
Sun Nuclear	PerFRACTION	EPID based/log file analysis	
Mobius Medical Systems	Mobius 3D	Log file analysis	

detector arrays provide a measurement of dose within a phantom analogous to dose within a patient. Another similarity is that calibration is required to convert response to dose, and additionally to ensure a uniform response among the detectors. In contrast to film, 2D arrays (diode versions in particular) have been shown to exhibit nontrivial angular dependence [45, 46]. As a result, the devices are often irradiated using an orientation perpendicular to the beam. This can be achieved by either setting the gantry angles of the verification plan to zero or by rotating the QA device in synchronization with the gantry. The latter technique can further be broken down into methods that affix the 2D array to the gantry or EPID panel, or methods that rotate the array itself using a motor-driven platform (see Octavius 4D).

With the wide scale adoption of arc delivery, new devices have been released to better address this issue while also collecting a broader sample of the 3D dose distribution. The aforementioned Octavius 4D (PTW, Freiburg, Germany) represents one such device, which utilizes a rotating platform to maintain an enface orientation. Then for a given gantry angle, the 3D dose distribution is derived based upon depth-dependent attenuation/scatter factors, which are used to scale the measured dose [47]. Here, the dose is scaled along ray lines, which pass through the 2D measurement plane. The composite dose from all angles provides the full 3D dose distribution. A slightly different approach is utilized by the Delta^4 PHANTOM (ScandiDos, Uppsala, Sweden), which includes two orthogonal detector planes. For this device, measurement-guided correction factors are applied to the dose distribution as calculated by the TPS. The correction factors are based upon the dose measured at each detector and then applied along ray lines, which pass through said detectors [48]. The ArcCHECK (Sun Nuclear, Melbourne FL) represents yet another approach, whereby detectors are arranged in a helical fashion. The helical design maintains a near-perpendicular orientation for arc subfields while also providing a novel method for reconstructing the 3D dose distribution (phantom or patient) based upon entrance and exit dose. Here, the diode measurements are used as corrections factors to scale a pre-calculated, relative dose distribution [49].

12.2.3.4 EPID-Based Analysis

When utilized for patient-specific QA, the EPID available on modern gantry-based delivery systems combines the high spatial resolution of film with the efficiency and convenience of detector arrays [50]. Several versions of EPID-based QA exist including in-house programs, legacy software such as EPIDose™ (Sun Nuclear), vendor-integrated systems such as Portal Dosimetry (Varian Medical Systems, Palo Alto, CA), and new cloud-based platforms such as EPIbeam™ (DOSIsoft, Paris, France) and PerFraction™ Fraction 0 (Sun Nuclear) [51–55]. Like detector arrays, the EPID panel requires calibration in order to ensure a uniform response. Typically, this is achieved via a dark field calibration, which adjusts for background noise and a flood field calibration, which accounts for variation in pixel sensitivity. In contrast to detector arrays, the output of the QA measurement is an indirect indicator of the in-phantom dose distribution. As such, a comparison must be made either with a predicted panel response (Portal Dosimetry), or the panel response must be converted to an in-phantom dose plane (EPIDose™ and EPIbeam™). Both methods require extensive commissioning. A third approach as implemented by PerFraction™ utilizes the detector panel as way to determine the position of individual leaves during a dynamic delivery. This information is then combined with data extracted from machine log files (output, gantry angle, energy, etc.) to reconstruct the incident energy fluence. The fluence can then be forward projected through a CT dataset to calculate a 3D dose distribution.

12.2.3.5 Log File Analysis

For many systems, the log files described above also include time-resolved data related to MLC leaf positioning. This information can be used in a similar fashion to derive the incident energy fluence and subsequent 3D dose distribution. A number of in-house and commercial programs utilize this method (Mobius 3D, Varian Medical Systems, Palo Alto, CA) or provide it as an option (PerFraction™) [54–56]. The use of log files for patient-specific QA has been a topic of much debate for reasons concerning the independent nature of the technique, whether it constitutes a "physical measurement" as required by many payers, and whether or not the log

files capture with high fidelity the actual MLC leaf positioning [32, 57]. The latter point likely depends upon the mechanism being used to report the leaf position, the tolerance within which it operates, and its robustness to failure. Generally speaking, mechanisms that monitor outputs (i.e., optical or magnetic sensors) appear more aptly suited than those that monitor inputs (i.e., stepper motor position). Either way, log file analysis affords many potential benefits including the ability to automate QA, the ability to separate treatment planning errors from treatment delivery errors, and the option to provide continual monitoring over a course of treatment due to the fact that log files are created each time the plan is delivered.

12.2.3.6 Transmission Detectors

Thin transmission detectors are a relatively new class of device designed primarily for treatment monitoring but may also be configured as a tool for patient-specific QA. Currently, there are three commercial systems available. The first system, DAVID (PTW, Freiburg, Germany), utilizes a multiwire ion chamber design where each wire is aligned beneath an individual leaf pair of the MLC [58]. During the initial delivery, the charge accumulated within each wire creates a histogram, which is subsequently used as a baseline for future deliveries. The system, thus, does not perform patient-specific QA, but instead functions as a sophisticated monitoring device to ensure an accurate delivery.

The second system, Delta⁴ DISCOVER (ScandiDos, Uppsala, Sweden), utilizes 4040 p-type diode detectors and operates much like the DAVID system in the way it creates a baseline measurement [59]. The system goes further, however, by providing an option to link the initial baseline to a measurement made using the Delta⁴ PHANTOM. This linking allows the Delta4 DISCOVER to predict phantom measurements for subsequent fractions and analyze the results using the same software tools as used for the initial patient-specific QA.

The third system, Dolphin (IBA, Schwarzenbruck, Germany), incorporates 1513 vented ion chambers with a 5 mm spacing at isocenter [60]. Dolphin serves as both a patient-specific QA device and an on-line monitoring device. In either case, there is no need for a separate QA tool. The Dolphin system is similar to a number of others in the way it performs a forward-based dose calculation using a derived energy fluence. To find said fluence, an ideal energy fluence is modified based upon the measured detector response and the way in which that response differs from the ideal response as predicted by COMPASS, an associated software tool. The derived energy fluence can then be provided as an input, along with the simulation CT, to a dose calculation engine using a collapsed cone convolution algorithm. The calculation results can be compared using a gamma analysis or DVH-based metrics.

12.2.3.7 3D Dosimeters

As noted for many of the devices above, there is an ongoing transition from 2D to 3D dosimetry for the measurement and analysis of patient-specific QA. In each case, a novel approach has been developed to derive the 3D dose distribution based, essentially, on a collection of 2D measurements. These approaches include correction-based interpolation, measurement-guided dose reconstruction, 3D back-projection/calculation, and 3D forward-projected/calculation. In contrast to such methods are 3D dosimeters, which inherently record a high resolution, 3D dose distribution through changes that occur within the composition of a 3D volume. Traditionally known as "gel dosimetry" due to the fact that the volume comprises 80–90% water and 5% gelatin, the type of change which occurs depends upon the remaining material, which may either induce a polymerization process (polymer-based gels) [61] or an oxidation process (Fricke gels) [62]. These changes are actualized through an optical-CT or MRI-based acquisition from which a 3D response map can be derived. The response map can then be converted to dose through a calibration procedure. The advantages of gel dosimetry include the 3D nature of the measurement, energy and dose-rate independence, high sensitivity and linear response, high spatial resolution, a near-tissue equivalence, and the fact that gels can be prepared in anthropomorphic forms with various densities. Limitations include a sensitivity to time, temperature and preparation method, the cost of the materials, the availability of MRI or optical-CT for processing, the need for frequent calibrations, the labor requirements of the process, and the delay between measurement and readout. At this time, the limitations of the method combined with the availability of attractive alternatives prevent 3D dosimeters from achieving widespread use as tool for patient-specific QA. 3D dosimeters do, however, offer unique solutions for the validation of new techniques and may serve a role in the future as a way to validate processes related to the use of adaptive radiotherapy.

12.3 PERIODIC QA

12.3.1 Rationale

In addition to patient-specific QA, periodic tests need to be done to ensure that the linear accelerator components are functioning as they should. While patient-specific QA will catch an error in a particular treatment plan, it can be difficult to determine what aspect of the plan is actually causing the error. For example, is a plan failing due to an output calibration, inconsistent flatness and symmetry or from MLC misalignment? For this reason, the periodic QA on a linac is vital to understanding how your machine is performing over time and for troubleshooting intermittent problems.

For linacs delivering IMRT and VMAT treatment plans, one good resource for periodic QA was published by the NCS in 2015 [21]. This document focuses specifically on different tests for VMAT. In terms of periodic linac QA, AAPM TG-142 is the current standard [19]. As the focus of this chapter is specifically IMRT and VMAT QA, this section will focus on the QA tests and equipment pertinent to IMRT and VMAT QA only.

12.3.2 Acceptance and Commissioning

12.3.2.1 Tests

When a new linear accelerator is installed, it is important to understand what type of procedures will be performed on that machine in order to ensure the proper components are examined. For a linear accelerator that will be used for IMRT (especially VMAT), the number of specific components to check greatly increases from standard 3D conformal therapy due to complexity of the delivery. In addition, this is the time where you can run your initial QA tests to establish baseline values that you can use for your subsequent periodic QA.

The complexity of IMRT delivery requires significant testing before safely treating patients. During VMAT delivery, the gantry is in motion along with the MLC. Therefore, testing these individual components in addition to their interplay with each other is vital. The NCS [21] recommended tests examine the basic mechanical properties of the linac including the accuracy of both the gantry and collimator at all angles. Due to the dynamic delivery in VMAT, they also recommend checking the gantry speed.

Again, due to the dynamic nature of VMAT in both mechanics and output, the linac output must be tested at different gantry angles to ensure proper delivery at all times. First, it is recommended to check that the output is linear and reproducible at the nominal dose rate. The recommended tests for flatness and symmetry are more detailed. Additionally, it is recommended to check at the lowest dose rate at all gantry angles during delivery and at different dose rates to simulate an actual delivery.

The final component recommended to be tested is the MLC. The initial test should be to test the MLC leaf speed. The MLC position during a dynamic delivery should also be tested. Finally, tests that combine the mechanical, output, and MLC should be run. These tests include output checks with varying dose rate while the gantry is rotating and varying gantry speed and dose rate.

12.3.2.2 Equipment

Although the tests for acceptance and commissioning are complex, the equipment needed to run the tests is simple. One essential component is a scanning water tank, usually three-dimensional. The water tank allows for beam characterization and for obtaining point measurements.

For the mechanical acceptance and commissioning tests, a standard level is sufficient to QA the gantry and collimator angles. For the output measurements, an ion chamber and an electrometer are required (discussed in Chapter 2). Depending on the test, a 1D or 3D water tank can be used or solid water is also an option.

The flatness and symmetry tests require some form of a 1D or 2D linear array to acquire profiles. A gantry mount is also helpful in order to acquire the profiles at different gantry angles although the cardinal angles can be acquired by using solid water to support the device on the treatment couch. These linear arrays can be composed of ion chambers or diodes. There are several different commercial options including 2D arrays such as the SunNuclear Profiler (ion chamber or diode), the IBA StarTrack, or the PTW StarCheck. These companies also offer 1D options as well like the SunNuclear WaterProof Profiler, the PTW LA48, and the IBA LDA-99.

The more complex tests require either the use of film or the EPID to acquire an image of the dynamic deliveries. In addition, there are software packages such as Mobius DoseLab, RIT RITG142, or Standard Imaging PiPsPro to process these images or the film or in-house code needs to be written.

12.3.3 Daily QA
12.3.3.1 Tests

The purpose of daily QA is to catch any significant changes in linac functionality that could impact patient treatment. The tests need to be simple and efficient so that treatments can begin on time but they also need to catch the serious errors that could reach the patient. AAPM TG-142 is a good resource for appropriate daily QA tests [19]. The tests are broken down into dosimetry, mechanical, safety, and imaging, and tolerance limits are provided based on the delivery type. The dosimetry test is designed to catch a significant variation in output, meaning a change greater than 3%. The mechanical tests check the laser and ODI accuracy and that the collimator is at the appropriate position. The safety check ensures that the patient can be seen and heard, the radiation monitors are functioning, and that the interlocks are working. The imaging tests are to check that the collision interlocks are functioning and that the imaging and treatment isocenters are coincident to within 2 mm for both MV and kV imaging.

12.3.3.2 Equipment

The daily QA should be quick and efficient. Typically, a detector array containing ion chamber and diodes is used to measure the output, energy, flatness, symmetry, and field size simultaneously. SunNuclear DailyQA3, PTW QuickCheck, and IBA StarTrack are examples of such devices. A separate imaging phantom is used for the imaging tests where an image is taken and compared to a DRR or CT image to assess isocenter coincidence. The QUASAR PentaGuide is an example of such a phantom.

12.3.4 Monthly QA
12.3.4.1 Tests

The purpose of monthly QA is to catch degradations in machine performance over time before they can reach the patient or cause machine malfunctions that interrupt the clinical workflow. Monthly QA tests tend to be more involved and more precise and accurate than daily QA tests. Again, AAPM TG-142 is a good resource along with the NCS report for the appropriate tests to run on a monthly basis [19, 21]. For the dosimetric tests, output constancy should be within 2% of the baseline at all dose rates and the flatness and symmetry should be within 1% of the baseline. The NCS report also recommended to check flatness and symmetry on a monthly basis [21]. The mechanical tests are to ensure that the gantry and collimator rotation is within 1 degree. Specific MLC tests are also recommended for leaf speed and positioning. The leaf speed should be compared to the baseline value acquired during commissioning. A picket fence test can be used for leaf positioning and multiple gantry angles should be tested. The NCS report also recommends checking leaf positioning accuracy on a monthly basis [21]. The imaging tests are the same as the daily QA.

12.3.4.2 Equipment

The equipment needed to run the monthly tests is basically the same that is used for acceptance and commissioning. For the mechanical tests, a level is sufficient to QA the gantry and collimator angles. For the output measurements, an ion chamber, electrometer, and solid water are enough to collect the appropriate information.

The flatness and symmetry tests require some form of a 1D or 2D linear array to acquire profiles. The MLC tests require either the use of film or the EPID to acquire an image of the dynamic deliveries. The leaf speed can be tested by analyzing machine log files or with an ion chamber and a dynamic sweep of the leaves over the chamber.

12.3.5 Annual QA

12.3.5.1 Tests

The purpose of annual QA is to guard against major malfunctions that might occur in machine components not frequently checked before they can reach the patient or cause machine malfunctions that interrupt the clinical workflow. AAPM TG-142 recommends that a full TG-51 protocol is run to bring the absolute output to within 1% [19, 63]. The beam quality should also be within 1% of the baseline value. Output constancy with dose rate and gantry angle and MU linearity should be verified. NCS recommends that linearity beyond 1000 MU be tested. Flatness and symmetry should be within 1% of the baseline and small field output factor should be verified to be within 2% of the commissioned values [21]. For the MLC, the transmission values should be measured to ensure they match the planning system. TG-142 also recommended to do a segmental IMRT

test and a moving window test and four cardinal gantry angles [19].

12.3.5.2 Equipment

The equipment required for annual QA is similar to that for monthly. The only extra piece of equipment would be a 1D or 3D water tank to measure PDDs and to do TG-51 [63].

12.4 SUMMARY

This chapter presents a comprehensive review of the dosimeters and devices needed for proper patient-specific QA, including a description of each device, the acceptance testing and commissioning process, and performing periodic QA tests. Patient-specific QA is a critical component of IMRT and VMAT treatments. Although currently there are many IMRT QA tools, every QA tool has limitations that need to be assessed prior to its use for safe and quality treatments.

REFERENCES

1. A. L. Boyer and C.X. Yu, "Intensity-modulated radiation therapy with dynamic multileaf collimators," *Seminars in Radiation Oncology* **9**(1), 48–59 (1999).
2. D. R. Puri, W. Chou, N. Lee, "Intensity-modulated radiation therapy in head and neck cancers: Dosimetric advantages and update of clinical results," *American Journal of Clinical Oncology* 28(4), 415–423 (2005).
3. K. Otto, "Volumetric modulated arc therapy: IMRT in a single gantry arc," *Medical Physics* **35**(1), 310–317 (2008).
4. J. L. Bedford and A. P. Warrington, "Commissioning of volumetric modulated arc therapy (VMAT)," *Radiotherapy & Oncology* 93(2), 259–265 (2009).
5. D. Rangaraj, S. Oddiraju, B. Sun, L. Santanam, D. Yang, S. Goddu, L. Papiez, "Fundamental properties of the delivery of volumetric modulated arc therapy (VMAT) to static patient anatomy," *Medical Physics* 37(8), 4056–4067 (2010).
6. M. Teoh, C. H. Clark, K. Wood, et al., "Volumetric modulated arc therapy: A review of current literature and clinical use in practice," *British Institute of Radiology* **84**, 967–996 (2011).
7. W. A. Woon, P. B. Ravindran, P. Ekayanake, Y. Y. F. Lim, "Validation of delivery consistency for intensity-modulated radiation therapy and volumetric-modulated arc therapy plans," *Journal of Medical Physics* 43(2), 119–128 (2018).
8. A. D. Low, J. M. Moran, L. Dong, M. Oldham, "Dosimetry tools and techniques for IMRT," *Medical Physics* 38(3), 1313–1338 (2011).

9. L. Masi, F. Casamassima, R. Doro, P. Francescon, "Quality assurance of volumetric modulated arc therapy: Evaluation and comparison of different dosimetric systems," *Medical Physics* **38**, 612–621 (2011).

10. C. De Wagter, "QA-QC of IMRT – European perspective," in *Image-Guided IMRT*, edited by T. Bortfeld, R. Schmidt-Ullrich, W. De Neve, and D. E. Wazer, pp. 117–128 (Springer-Verlag, Berlin, Heidelberg, New York, 2006).

11. R. Kulasekere, J. M. Moran, B. A. Fraass, P. L. Roberson, "Accuracy of rapid radiographic film calibration for intensity-modulated radiation therapy verification," *Journal of Applied Clinical Medical Physics* **7**(2), 86–95 (2006).

12. G. A. Ezzell, et al., "IMRT commissioning: Multiple institution planning and dosimetry comparisons, a report from AAPM Task Group 119," *Medical Physics* **36**(11), 5359–5373 (2009).

13. N. L. Childress, M. Salehpour, L. Dong, C. Bloch, R. A. White, I. I. Rosen, "Dosimetric accuracy of Kodak EDR2 film for IMRT verifications," *Medical Physics* **32**(2), 539–548 (2005).

14. J. Son, T. Baek, B. Lee, D. Shin, S. Y. Park, J. Park, Y. K. Lim, S. B. Lee, J. Kim, M. Yoon, "A comparison of the quality assurance of four dosimetric tools for intensity modulated radiation therapy," *Radiology and Oncology* **49**(3), 307–313 (2015).

15. M. Hussein, E. J. Adams, T. J. Jordan, C. H. Clark, A. Nisbet, "A critical evaluation of the PTW 2D-ARRAY seven29 and OCTAVIUS II phantom for IMRT and VMAT verification," *Journal of Applied Clinical Medical Physics* **14**(6), 274–292 (2013).

16. M. Hussein, P. Rowshanfarzad, M. A. Ebert, A. Nisbet, C. H. Clark, "A comparison of the gamma index analysis in various commercial IMRT/VMAT QA systems," *Radiotherapy & Oncology* **109**(3), 370–376 (2013).

17. J. F. M. Colodro, A. S. Berná, V. P. Puchades, D. R. Amores, M. A. Baños, "Volumetric-modulated arc therapy lung stereotactic body radiation therapy dosimetric quality assurance: A comparison between radiochromic film and chamber array," *Journal of Medical Physics* **42**(3), 133–139 (2017).

18. M. A. Silveira, J. F. Pavoni, O. Baffa, "Three-dimensional quality assurance of IMRT prostate plans using gel dosimetry," *Physica Medica* **34**, 1–6 (2017).

19. E. E. Klein, J. Hanley, J. Bayouth, F. F. Yin, W. Simon, et al., "Task Group 142 report: Quality assurance of medical accelerators," *Medical Physics* **36**(9), 4197–4212 (2009).

20. NCS, "Code of practice for the quality assurance and control for intensity modulated radiotherapy." Netherlands Commission on Radiation Dosimetry Report 22 (http://radiationdosimetry.org) (2013).

21. NCS, "Code of Practice for the Quality Assurance and Control for Volumetric Modulated Arc Therapy." Netherlands Commission on Radiation Dosimetry Report 24 (http://radiationdosimetry.org) (2015).

22. M. Miften, A. Olch, D. Mihailidis, J. Moran, T. Pawlicki, A. Molineu, H. Li, K. Wijesooriya, J. Shi, P. Xia, N. Papanikolaou, D. A. Low, "Tolerance limits and methodologies for IMRT measurement-based verification QA: Recommendations of AAPM Task Group No. 218," *Medical Physics* **45**(4), e53–e83 (2018).

23. X. Wang, S. Spirou, T. LoSasso, et al., "Dosimetric verification of intensity-modulated fields," *Medical Physics* **23**(3), 317–327 (1996).

24. C. S. Chui, S. Spirou, T. LoSasso, "Testing of dynamic multileaf collimation," *Medical Physics* **23**(5), 635–641 (1996).

25. B. Nelms and J. Simon, "A survey on planar IMRT QA analysis," *Journal of Applied Clinical Medical Physics* **8**(3), 76–90 (2007).

26. J. Moran, M. Dempsey, A. Eisbruch, et al., "Safety considerations for IMRT," *Practical Radiation Oncology* **1**(3), 190–195 (2011).

27. A. Hartford, J. Galvin, D. Beyer, et al. "American College of Radiology (ACR) and American Society for Radiation Oncology (ASTRO) practice guideline for intensity-modulated radiation therapy (IMRT)," *American Journal of Clinical Oncology* **35**, 612–617 (2012).

28. ACR practice parameters for intensity modulated radiation therapy (IMRT) (2016).

29. ASTRO model policies, intensity modulated radiation therapy (IMRT) (2015).

30. G. Ezzell, J. Galvin, D. Low, et al., "Guidance document on delivery, treatment planning, and clinical implementation of IMRT: Report of the IMRT subcommittee of the AAPM radiation therapy committee," *Medical Physics* **30**, 2089–2115 (2003).

31. R. Alfredo, C. Siochi, A. Molineu, "Patient-specific QA for IMRT should be performed using software rather than hardware methods," *Medical Physics* **40**(7), 070601 (2013).

32. N. Childress, Q. Chen, Y. Rong, "Parallel/opposed: IMRT QA using treatment log files is superior to conventional measurement-based method," *Journal of Applied Clinical Medical Physics* **16**(1), 4–7 (2015).

33. D. A. Low, W. B. Harms, S. Mutic, J. A. Purdy, "A technique for the quantitative evaluation of dose distributions," *Medical Physics* **25**, 656–661 (1998).

34. D. A. Low, and J. F. Dempsey, "Evaluation of the gamma dose distribution comparison method," *Medical Physics* **30**, 2455–2464 (2003).

35. J. M. Steers, and B. A. Fraass, "IMRT QA: Selecting gamma criteria based on error detection sensitivity," *Medical Physics* **43**(4), 1982–1994 (2016).

36. J. Van Dyk, R. B. Barnett, J. E. Cygler, P. C. Shragge, "Commissioning and quality assurance of treatment planning computers," *International Journal of Radiation Oncology, Biology, Physics* **26**, 261–273 (1993).

37. A. Molineu, N. Hernandez, T. Nguyen, et al., "Credentialing results from IMRT irradiations of an anthropomorphic head and neck phantom," *Medical Physics* **40**(2), 022101 (2013).

38. C. Chuang, L. Verhey, P. Xia, "Investigation of the use of MOSFET for clinical IMRT dosimetric verification," *Medical Physics* **29**(6), 1109–1115 (2002).

39. D. Low, "Quality assurance of intensity-modulated radiotherapy," *Seminars in Radiation Oncology* **12**(3), 219–228 (2002).

40. N. Dogan, L. Leybovich, A. Sethi, "Comparative evaluation of Kodak EDR2 and XV2 films for verification of intensity modulated radiation therapy," *Physics in Medicine and Biology* **47**, 4121–4130 (2002).

41. A. Olch, "Dosimetric performance of an enhanced dose range radiographic film for intensity-modulated radiation therapy quality assurance," *Medical Physics* **29**(90), 2159–2168 (2002).

42. A. Diroomand-Rad, C. Blackwell, B. Coursey, "Radiochromic film dosimetry, recommendations of AAPM radiation therapy committee task group no. 55," *Medical Physics* **25**(11), 2093–2115 (1998).

43. A. Micke, D. Lewis, X. Yu, "Multichannel film dosimetry with nonuniformity correction," *Medical Physics* **38**(5), 2523–2534 (2011).

44. A. Shukla, A. Oinam, S. Kumar, et al., "A calibration method for patient specific IMRT QA using a single therapy verification film," *Reports of Practical Oncology and Radiotherapy* **18**, 235–240 (2013).

45. J. Hosang, V. Keeling, D. Johnson, et al., "Interplay effect of angular dependence and calibration field size of MapCHECK 2 on RapidArc quality assurance," *Journal of Applied Clinical Medical Physics* **15**(3), 80–92 (2014).

46. Y. Shimohigashi, F. Araki, H. Tominaga, et al., "Angular dependence correction of MatriXX and its application to composite dose verification," *Journal of Applied Clinical Medical Physics* **13**(5), 198–214 (2012).

47. B. Allgaier, E. Schule, J. Wurfel, "Dose reconstruction in the OCTAVIUS 4D phantom and in the patient without using dose information from the TPS," in *PTW* (2013).

48. R. Sadagopan, J. Bencomo, R. Martin, et al., "Characterization and clinical evaluation of a novel IMRT quality assurance system," *Journal of Applied Clinical Medical Physics* **10**(2), 104–119 (2009).

49. B. Nelms, D. Opp, J. Robinson, et al., "VMAT QA: Measurement-guided 4D dose reconstruction on a patient," *Medical Physics* **39**(7), 4228–4238 (2012).

50. W. Van Elmpt, L. McDermott, S. Nijsten, et al., "A literature review of electronic portal imaging for radiotherapy dosimetry," *Radiotherapy & Oncology* **88**, 289–309 (2008).

51. A. Esch, T. Depuydt, D. Huyskens, "The use of an aSi-based EPID for routine absolute dosimetric pretreatment verification of dynamic IMRT fields," *Radiotherapy & Oncology* **71**(2), 223–234 (2004).

52. D. Bailey, L. Kumaraswamy, M. Bakhtiari, et al., "EPID dosimetry for pretreatment quality assurance with two commercial systems," *Journal of Applied Clinical Medical Physics* **13**(4), 82–99 (2010).

53. B. Nelms, K. Rasmussen, W. Tome, "Evaluation of a fast method of EPID-based dosimetry for intensity-modulated radiation therapy," *Journal of Applied Clinical Medical Physics* **11**(2), 140–157 (2010).

54. S. Bresciani, M. Poli, A. Miranti, et al., "Comparison of two different EPID-based solutions performing pretreatment quality assurance: 2D portal dosimetry versus 3D forward projection method," *Physica Medica* **52**, 65–71 (2018).

55. E. Hseih, K. Hansen, M. Kent, et al., "Can a commercially available EPID dosimetry system detect small daily patient setup errors for cranial IMRT/SRS?," *Practical Radiation Oncology* **7**, e283–e290 (2017).

56. C. Nelson, B. Mason, K. Robinson, et al., "Commissioning results of an automated treatment planning verification system," *Journal of Applied Clinical Medical Physics* **15**(5), 57–65 (2014).

57. A. Agnew, C. Agnew, M. Grattan, et al., "Monitoring daily MLC positional errors using trajectory log files and EPID measurements for IMRT and VMAT deliveries," *Physics in Medicine and Biology* **59**(9), N49–N63 (2014).

58. G. Karagoz, F. Zorlu, M. Yeginer, et al., "Evaluation of MLC leaf positioning accuracy for static and dynamic IMRT treatments using DAVID in vivo dosimetric system," *Journal of Applied Clinical Medical Physics* **17**(2), 14–23 (2016).

59. T. Li, T. Matzen, F. Yin, et al., "Diode-based transmission detector for IMRT delivery monitoring: A validation study," *Journal of Applied Clinical Medical Physics* **17**(5), 235–244 (2016).

60. J. Thoelking, Y. Sekar, J. Fleckenstein, et al., "Characterization of a new transmission detector for patient individualized online plan verification and its influence on 6 MV X-ray beam characteristics," *Journal of Medical Physics* **26**(3), 200–208 (2016).

61. M. Maryanski, J. Gore, R. Kennan, et al., "NMR relaxation enhancement in gels polymerized and crosslinked by ionizing radiation: A new approach to 3D dosimetry by MRI," *Magnetic Resonance Imaging* **11**, 253–258 (1993).

62. R. Kelly, K. Jordan, J. Battista, "Optical CT reconstruction of 3D dose distributions using the ferrous-benzoic-xylenol (FBX) gel dosimeter," *Medical Physics* **25**, 1741–1750 (1998).

63. P. R. Almond, P. J. Biggs, B. M. Coursey, et al., "AAPM's TG-51 protocol for clinical reference dosimetry of high-energy photon and electron beams," *Medical Physics* **26**(9), 1847–1870 (1999).

CHAPTER 13

Area and Individual Radiation Monitoring

Nisy Elizabeth Ipe

Shielding Design, Dosimetry and Radiation Protection
San Carlos, California

CONTENTS

13.1 INTRODUCTION

Radiation exposure can be external or internal. However, since sealed radioactive sources and particle accelerators are typically used in radiotherapy, all exposures can be considered external. External radiation doses to an individual result from exposure to external sources of ionizing radiation, that is, radiation sources outside the body [1]. Gamma rays, x-rays, high-energy beta particles, protons, and neutrons are considered "penetrating radiation" and present an external exposure hazard. Conversely, low-energy beta particles and alpha particles are relatively "non-penetrating" and present less of an external radiation exposure hazard. Neutrons are produced in particle therapy facilities, therapy linear accelerators (linacs), fast neutron therapy facilities, and boron neutron capture therapy (BNCT) facilities. Particle therapy facilities (heavy ions) produce neutrons with maximum energies in the GeV range [2]. Proton therapy facilities produce neutrons with energies as high as the primary proton beam energy (~250 MeV). Typically, either cyclotrons or synchrotrons are used in these facilities. Fast neutron therapy facilities produce neutrons with energies between 50 MeV and 70 MeV. These neutron therapy beams are produced by reactors, cyclotrons, and linear accelerators. BNCT uses epithermal neutrons for therapy. The neutron source for epithermal radiation is generated from nuclear reactors and accelerator-based neutron sources (ABNS).

For the purposes of this chapter, external radiation monitoring includes the following:

1. Area monitoring – the measurement of radiation levels in adjacent areas outside shielded radiotherapy rooms and accelerators, or outside shielded radioactive sources;

2. Individual monitoring – the measurement of doses received by individuals working outside shielded radiotherapy rooms and accelerators or in the vicinity of the shielded radioactive sources.

This chapter describes the monitoring of external radiation in radiotherapy and does not include radiation monitoring inside therapy treatment rooms during treatment. The radiation monitors used for measuring external radiation levels are typically referred to as area monitors, and include both portable and fixed radiation detectors or survey meters. Fixed radiation detectors are area monitors that are installed at a particular location. The monitors used for recording the dose equivalents received by individuals working with radiation are referred to as personal dosimeters or individual dosimeters. Both area monitors and individual monitors can be either active or passive.

13.2 PROTECTION AND OPERATIONAL DOSE QUANTITIES

External radiation monitoring is performed for the purposes of radiation protection, that is, primarily to demonstrate compliance with the shielding design or

TABLE 13.1   Radiation Weighting Factors Recommended by ICRP Publication 103 [6]

Radiation Type	Energy Range (E)	Radiation Weighting Factor (w_R)
Photons, electrons, and muons	All energies	1
Neutrons	<1 MeV	$2.5 + 18.2 \exp\left[-\dfrac{(\ln(E))^2}{6}\right]$
Neutrons	1 MeV to 50 MeV	$5 + 17 \exp\left[-\dfrac{(\ln(2E))^2}{6}\right]$
Neutrons	>50 MeV	$2.5 + 3.25 \exp\left[-\dfrac{(\ln(0.04E))^2}{6}\right]$
Protons, other than recoil protons	>2 MeV	2
Alpha particles, fission fragments, and heavy nuclei	All energies	20

regulatory dose limits [3]. Thus, the radiation measurements must be expressed in terms of quantities in which the dose limits are defined. The International Commission on Radiological Protection (ICRP) defines the dose quantities. They are expressed in terms of protection quantities measured in the human body. The dose equivalent was specified in ICRP Publication 21 [4]. ICRP Publication 60 [5] introduced the concept of equivalent dose. ICRP Publication 103 [6] modified the weighting factors. The radiation weighting factors, w_R, for the protection quantities as recommended by ICRP Publication 103 [6] are shown in Table 13.1. For neutrons, w_R varies with energy and therefore the computation for the protection quantities is made by integration over the entire energy spectrum.

The **dose equivalent**, H, is the product of D and Q at a point in tissue, where D is the absorbed dose and Q is the quality factor for the specific radiation at that point, thus $H = DQ$.

The unit of the dose equivalent is joule per kilogram ($J\,kg^{-1}$), and its special name is Sievert (Sv).

The **equivalent dose**, H_T, in a tissue or organ is given by $H_t = \sum_R w_R D_{T,R}$, where $D_{T,R}$ is the mean absorbed dose in the tissue or organ, T, due to radiation, R, and w_R is the corresponding radiation weighting factor. The unit of equivalent dose is the Sievert.

ICRP Publications 60 [5] and 103 [6] recommended the use of equivalent dose (H_T) and effective dose (E) as protection quantities. However, these quantities are not directly measurable. For external individual exposure, the accepted convention is the use of the following operational quantities-ambient dose equivalent $H^*(d)$, the directional dose equivalent $H'(d, \Omega)$, and personal dose equivalent $H_p(d)$. The two sets of quantities might

be related to the particle fluence and, in turn, by conversion coefficients to each other. The fluence, Φ, is the number of particles dN (such as photons or neutrons), incident on a sphere divided by the cross-sectional area of the sphere (dA). In other words, it is the total number of particles per unit area with which a material is irradiated. The unit is m^{-2} or cm^{-2}.

Note that the term "dose" might be used in a generic sense throughout this document to refer to the various dose quantities.

The operational quantities - ambient dose equivalent and directional dose equivalent, are defined for area monitoring. The operational quantity - personal dose equivalent, is defined for individual or personal monitoring. The definitions of operational quantities are taken from ICRU Report 51 [7], ICRU Report 57 [8] are as follows:

The **ambient dose equivalent,** $H^*(d)$, at a point in a radiation field, is the dose equivalent that would be produced by the corresponding expanded and aligned field, in the ICRU sphere (diameter = 30 cm, 76.2% O, 10.1% H, 11.1% C, and 2.6% N) at a depth, d, on the radius opposing the direction of the aligned field [7]. The ambient dose equivalent is measured in Sv. In the expanded and aligned field, the fluence and its energy distribution have the same values throughout the volume of interest as in the actual field at the point of reference, but the fluence is unidirectional. The ambient dose equivalent is used for area monitoring.

The **directional dose equivalent,** $H'(d, \Omega)$, at a point in a radiation field, is the dose equivalent that would be produced by the corresponding expanded field in the ICRU sphere at a depth, d, on the radius in a specified direction, Ω [7]. The directional dose equivalent is measured in Sv.

The **personal dose equivalent**, $H_p(d)$, is the dose equivalent in soft tissue, at an appropriate depth, d, below a specified point on the body. The personal dose equivalent is measured in Sv.

For the ambient dose equivalent and personal dose equivalent, a depth of 10 mm is recommended for strongly penetrating radiation. For weakly penetrating radiation (photons below 15 keV and beta particles), a depth of 0.07 mm is recommended for the skin, and a depth of 3 mm is recommended for the lens of the eye.

13.3 ACTIVE AND PASSIVE MONITORING

Radiation monitoring can be active or passive.

13.3.1 Active Monitoring

The purpose of active monitoring is to obtain a current snapshot of the radiation environment [9]. This information may be used to demonstrate compliance with regulatory requirements and verify the shielding integrity of a new facility. It also allows the planning of activities of radiation workers, so that they may be carried out within the regulatory or facility-specified dose limits.

Active monitoring can be used to measure prompt radiation, induced radiation (activation) and radiation from radioactive sources. Active monitors provide a direct display of the radiation levels and are usually equipped with additional functions such as alarm thresholds for dose or dose rate. While most active monitors measure dose rate, some can be used in the "integrate mode" to measure the cumulative dose over a defined period of time. Therefore, such monitors can also be used for passive monitoring.

13.3.2 Passive Monitoring

Passive monitors are used in both area monitoring and individual monitoring. Individual monitors provide estimates of cumulative doses received by the individual over a specified period of time, in terms of the operational quantities used for personnel dosimetry [9]. The purpose of passive monitoring is to determine the cumulative dose over a specified period of time [9].

13.3.2.1 Individual Monitoring

Individual or personal monitoring is the measurement of the radiation doses received by individuals working with radiation [1]. Individual monitoring may also be used to determine accidental exposures.

Typical passive monitors include film dosimeters, thermoluminescent dosimeters (TLDs), or optically stimulated luminescence dosimeters (OSLDs) for photons; and albedo dosimeters, nuclear track etch detectors and bubble detectors for neutrons. Such dosimeters are insensitive to the time structure of the accelerator beam and can be used in pulsed fields. Electronic personal dosimeters (EPDs) may also be used.

The operational quantity for individual monitoring of external exposure is the personal dose equivalent $H_p(d)$, with the recommended depth $d = 10$ mm for strongly penetrating radiation and $d = 0.07$ mm for weakly penetrating radiation (shallow or skin dose). Personal dosimeters are calibrated in these quantities.

13.4 AREA MONITORING

The measurement of radiation levels in adjacent areas, outside shielded radiotherapy rooms or shielded radioactive sources, can be performed with either active or passive radiation monitors. Typical radiation instruments used for active area monitoring are either gas-filled detectors or solid-state detectors such as scintillators and semiconductor detectors [1]. The operational quantities used for area monitoring are ambient dose equivalent and directional dose equivalent.

13.4.1 Calibration of Area Monitors

Area monitors must be calibrated against a reference instrument that is traceable either directly or indirectly to a national standards laboratory. Testing, calibration, and calibration frequency requirements for portable radiation detectors can be found in ANSI Standard N323AB-2013 [10].

Area monitors should be calibrated in terms of the ambient dose equivalent, $H^*(d)$, or the directional dose equivalent, $H'(d, \Omega)$, without the presence of a phantom, that is, "free in air" [11].

The calibration factor, N, is defined as the conventional true value of the quantity the instrument is intended to measure, H, divided by the indication, M (corrected, if necessary) given by the instrument [11].

$$N = H/M \qquad (13.1)$$

The calibration factor is commonly incorporated into an instrument by adjustment of the electronic gain. Several calibration points are evaluated, typically one on each range or decade. An attempt is made to achieve 10%

accuracy over the entire range for photon instruments and 30% for neutron rem meters. The calibration factor is normally only quoted for one reference radiation. If there is no unique factor applicable to the entire measurement range of an instrument, the instrument is said to have a nonlinear response.

The calibration factor N is dimensionless when both the conventional and indicated values have the same units. A calibration factor of 1 implies that the instrument is perfectly accurate.

The reciprocal of the calibration factor is equal to the response under reference conditions. The instrument response is applicable to the prevailing conditions.

The response, R, of the area monitor (with respect to fluence or dose equivalent) typically varies with both the energy and directional distribution of the incident radiation. Therefore, one can consider the response, $R(E,\Omega)$, as a function of the energy E of the incident radiation and of the direction (solid angle), Ω, of the incident monodirectional radiation. $R(E)$ and $R(\Omega)$ describe the "energy dependence" and the "angular dependence," respectively, of the response. The solid angle Ω may also be defined by the angle α between the reference orientation of the instrument and the axis of the incident monodirectional radiation.

13.4.1.1 Photon Calibration

An ionization chamber with a measuring assembly is usually used as a reference instrument for photons [1]. Reference instruments do not directly measure the dose equivalent, H, required for calibration of radiation survey instruments. Instead, basic radiation quantities such as the air kerma (K) in air for photon radiation are measured. Kerma is defined as the quotient of dE_{tr} by dm where dE_{tr} is the sum of the initial kinetic energies of all the charged ionizing particles liberated by uncharged ionizing particles in a volume element of mass dm.

$$K = dE_{tr}/dm \qquad (13.2)$$

$$H = NM_R \qquad (13.3)$$

$$N = hN_R \qquad (13.4)$$

where M_R is the reading of the reference instrument corrected for influence quantities. N_R is the calibration factor (e.g., in terms of air kerma in air or air kerma rate in air) of the reference chamber under reference conditions; h is the appropriate conversion coefficient, that is, dose equivalent per unit kerma. Thus, the dose equivalent H is given by:

$$H = hN_R M_R \qquad (13.5)$$

Reference instruments are calibrated free in air for the range of reference radiation qualities, which are defined by the International Organization for Standardization (ISO)) [12]. These same reference qualities should be used for the calibration of area monitors.

13.4.1.2 Neutron Calibration

The response of an active detector such as a neutron dose-equivalent meter is designed to be proportional to the dose equivalent and is therefore independent of neutron energy. Thus, dose-equivalent meters are useful in radiation fields where the neutron spectrum is not well characterized, or known. The underlying principle of operation for dose-equivalent meters is similar, although they may vary widely in size and geometrical configuration [13].

Neutrons are classified according to their energy as follows:

Thermal: $\bar{E}_n = 0.025$ eV at 20°C. Typically $E_n \leq 0.5$ eV

Intermediate: 0.5 eV $< E_n \leq 10$ keV

Fast: 10 keV $< E_n \leq 20$ MeV

Relativistic: $E_n > 20$ MeV

High-Energy Neutrons: $E_n > 100$ MeV

where E_n is the energy of the neutron and $\bar{E}_n$ is the average energy of the neutron.

A dose-equivalent meter's response is shaped to fit an appropriate fluence-to-dose equivalent conversion coefficient over a particular energy range. The ICRP 21 [4] fluence-to-dose equivalent conversion factors were used in the older designs; and these instruments were referred to as rem-meters. Similar conversion coefficients factors were also recommended by NCRP Report No. 38 [14]. The latter are the basis for the current federal and various state regulations in the United States. The newer designs of dose-equivalent meters comply with the ICRP effective dose recommendations. The operational quantity appropriate for dose meter calibration is the ambient dose equivalent ($H^*(10)$), which is defined for a known neutron spectrum as follows:

$$H^*(10) = \int h_\Phi(E)\, \Phi(E) dE \qquad (13.6)$$

where $h_\Phi(E)$ is the fluence-to-ambient dose equivalent conversion function, and $\Phi(E)$ is the neutron fluence as a function of energy for a given neutron field. The dose-equivalent meter response, R_m, in that field is given by the equation below:

$$R_m = \int Cr_\Phi(E)\, \Phi(E) dE \qquad (13.7)$$

where $r_\Phi(E)$ is the dose-equivalent meter's response function in units of counts per unit fluence, and C is the calibration constant in units of Sievert per count. As long as $r_\Phi(E)$ has a similar energy response to that of $h_\Phi(E)$, the dose-equivalent meter measurement is considered accurate. The traditional energy response of the dose-equivalent meter in terms of counts per unit dose equivalent is given by the ratio $r_\Phi(E)/h_\Phi(E)$. However, $r_\Phi(E)$ does not match $h_\Phi(E)$ over much of the energy range. Therefore, some detectors either over respond or under respond in certain energy regions. Usually, most dose-equivalent meters have a very large over-response in the intermediate energy region. Therefore, they give an adequate measure of dose equivalent between 100 keV and 6 MeV [15]. Thus, it is important to have a rough idea of the spectrum, in order to rely on the instrument readings.

13.4.2 Desirable Properties of Area Monitors or Radiation Survey Meters

13.4.2.1 Sensitivity
The sensitivity S is defined as the inverse of the calibration coefficient N. A high sensitivity is desirable.

13.4.2.2 Energy Dependence
Survey meters are often used in situations in which the radiation field is complex or unknown. Further, they are calibrated at a given energy spectrum or beam quality. Therefore, these survey meters should have a low energy dependence over a wide energy range. In the past, photon survey meters were designed to exhibit a flat energy response that follows exposure or air kerma in air.

A dose-equivalent meter's energy response should vary as the quantity $[H^*(10)/(K_{air})_{air}]$, as shown below:

$$N_{H^*} = [H^*(10)/M] = [H^*(10)/(K_{air})_{air}]/[(K_{air})_{air}/M] \qquad (13.8)$$

13.4.2.3 Directional Dependence
The directional response of the instrument can be studied by rotating the survey meter about its vertical axis. Survey meters should exhibit isotropic response, as required for measuring ambient dose equivalent, within ±60° to ±80° with respect to the reference direction of calibration. They typically have a much better response for higher photon energies, that is >80 keV [1]

13.4.2.4 Measurement Range
The useful dose equivalent range for radiation survey meters is μSv/h to mSv/h. More sensitive instruments are capable of measuring dose equivalents as low as 1 nSv/h.

13.4.2.5 Response Time
The response time of the survey meter is defined as the RC time constant of the measuring circuit. In this case, R is resistance of the decade resistor and C the capacitance of the circuit. High R, hence high RC values are used for low dose equivalent ranges, resulting in slow indicator movements. The meter will stabilize only after about 3–5 time constants.

13.4.2.6 Linearity of Response
The response of survey meters should be linear with the dose rate. Survey meters should read full scale when subjected to dose rates of about ten times the maximum scale range. However, some instruments suffer from saturation as described later.

13.4.2.7 Calibration and Long-Term Stability
Survey meters are required to be calibrated in a standards dosimetry laboratory. In the United States, the regulatory frequency is once a year. They must also be calibrated upon repair or sudden change in response. The long term stability of survey meters should be checked at regular intervals using an appropriate radioactive source.

13.4.2.8 Discrimination in Mixed Fields
The survey meter should be able to distinguish between different types of radiation in mixed radiation fields. For example, end-window Geiger–Mueller (GM) counters

have a buildup cap that can be removed when measuring beta particles; and which can be left in place when measuring gamma rays.

13.4.2.9 Measurement Uncertainties

Measurement uncertainties include uncertainty associated with the calibration factor (Type A); and uncertainties due to energy dependence, angular dependence, and variations from calibration conditions in user fields (Type B). The two types of uncertainties are added in quadrature to obtain the measurement uncertainty.

13.5 PHOTON MONITORING TECHNIQUES

Photon monitoring techniques include active and passive methods.

13.5.1 Active Monitoring

Active methods include the use of gas-filled detectors and solid-state detectors.

13.5.1.1 Gas-Filled Detectors

Gas-filled detectors are typically cylindrical in shape [1]. They are constructed with an outer wall made of plastic and occasionally tissue equivalent composition plastic, and a central electrode. The wall and electrode are insulated from each other. Voltage is applied across the two electrodes. The various regions of operation of a gas-filled detector as a function of the applied voltage are shown in Chapter 2 of this book (see Figure 2.1 in Ref. [16]).

The first region is the recombination region, where the response gradually increases with voltage. Because the electric field in the detector is very weak, the electrons from the cathode are drawn slowly toward the ion chamber. Therefore, the positive ions have an opportunity to recombine with the free electrons before they reach the anode or collecting electrode. In the ion chamber region, the electric field is strong enough to overcome the recombination losses and all the electrons reach the anode. Since all the freed electrons are collected, the resultant current is proportional to the amount of radiation that causes the ionization. In the proportional counting region, the electric field has increased sufficiently so as to accelerate the electrons, which then cause further ionization in the gas. The next region is the limited proportional region. In the GM region, the electric field is so strong that a single ionizing event creates an electron avalanche. The entire gas in the chamber is ionized. The pulse height becomes independent of the primary ionization or the energy of the interacting particles. When the electric field is strong enough to cause a breakdown in the chamber, continuous discharge takes place.

Gas-filled detectors are designed to operate in one of the following regions: The ionization region, proportional region or GM region [1]. Regions of recombination and of limited proportionality are typically not used for survey meters. However, recombination chambers are used to measure the quality factor for radiation fields. Operation in the continuous discharge region will result in damage to detectors.

The design of the survey meter depends upon the type, energy, and nature of the radiation field as well as its specific application [1]. Air ionization survey meters are quite common and are usually vented to the atmosphere. Sealed ionization chambers are typically pressurized and are usually filled with a non-electronegative gas such as argon.

Due to the low mobility of the negative ions formed by electron attachment, these ions would increase the collection time, thus limiting the dose rate that could be monitored. Therefore, noble gases are generally used in these detectors.

Beta-gamma survey meters with thin end windows are used to detect weakly penetrating radiation. The β efficiency is almost 100% for β particles entering the detector; however, the γ efficiency is only a few percent.

The GM counters used to detect gamma rays are much smaller than the ionization chamber detectors, because of their higher sensitivity.

Detectors can operate either in the "current" mode or the "pulse" mode. Proportional and GM counters are normally operated in the "pulse" mode.

13.5.1.1.1 Ionization Chambers The number of primary ions of either sign collected in the ionization region is proportional to the energy deposited by the charged particles in the detector volume [1]. Particle discrimination can be used in this region because of the differences due to the linear energy transfer (LET). Build-up caps are used for high-energy photons (>100 keV) in order to improve the detection efficiency. Some types of ionization chambers have removable caps or covers that enable the measurements of very soft x-rays. The ionization chamber is the most useful photon survey meter because it is almost energy-independent (usually within ± 10% of unity) between 30 keV and a few MeV. However, one cannot measure the dose rates close to the background level, because, the lower detection limit is about 1 μSv/h. Since the ionization chamber survey meter measures a very weak current of the order of femtoamperes (fA) when placed in a field of several μSv/h, it takes several minutes until the detector becomes stable after being switched on [17]. Ionization chamber detectors can operate in the "pulse" mode or in the "current" mode, depending upon the electronics used. Ionization chambers operating in the current mode are more suitable for higher dose rate measurements. They can also be used in pulsed fields such as those encountered with radiotherapy accelerators. However, they may be sensitive to electromagnetic interference caused by stray fields from RF cavities [3].

Ionization chamber survey meters are also the most suitable and reliable detectors for the measurement of ambient dose rate due to residual radioactivity. Detectors are often available with removal ionization chamber windows, to enable the measurement of beta ray dose for the estimation of skin dose [17]. A wide range of dose equivalent rates can be covered with ionization chamber-based survey meters (1 μSv/h–1 Sv/h) through the use of decade resistances, larger detector volumes or detector gases under higher pressures. Figure 13.1 shows the Fluke Biomedical 451P pressurized μR ion chamber survey meter.

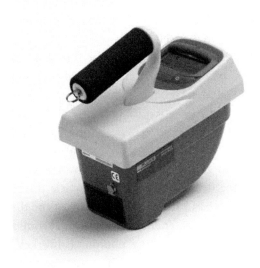

FIGURE 13.1 Fluke Biomedical 451P pressurized μR ion chamber survey meter. (Reproduced with permission from Fluke Biomedical, Everett, Washington.)

The 451P ion chamber is a 230-cubic centimeter volume air ionization chamber pressurized to 6 atmospheres. The plastic chamber wall is 200 mg cm^{-2} thick. The collecting voltage is 90 V DC. It detects beta particles above 1 MeV, and gammas and x-rays above 25 keV. It can measure exposure/dose rate or accumulated exposure/dose using the integrate mode. The operating range for the 451P-RYR (standard chamber) is 0–5 R/h. It is also available as a dose-equivalent chamber (451P-DE-SI-RYR) with the operating range of 0–50 mSv/h.

The energy dependence is within 20% between 50 keV and 1 MeV for 451P-RYR, and within 20% between 30 keV and 1 MeV for 451P-DE-S1-RYR.

Figure 13.2 shows the typical energy dependence for the 451P-RYR.

Figure 13.3 shows the 451B model ion chamber survey mete, which is a 349 cubic centimeter volume air ionization chamber. The phenolic chamber wall is 246 mg cm^{-2} thick. The chamber window is composed of 1.7 mg cm^2 mylar, 0.025 mm thick. The wall is made conductive by the application of graphite. The collection potential is −63 V. The chamber is vented to air through the desiccant. The 451B model has a slide cover made of phenolic 440 mg cm^{-2} thick. The cover is provided as a beta shield and also serves as an equilibrium thickness for photon measurements. It protects the mylar window; and when open allows the detection of alpha particles above 7.5 MeV, beta particles above 100 keV, and gammas above 7 keV. It can measure exposure/dose rate or accumulated exposure/dose using the integrate mode. According to the manufacturer, the operating range for the 451B (standard chamber) is 0–50 R/h, while the operating range for the 451B-DE-SI (dose-equivalent chamber) is 0–500 mSv/h. With the beta shield open, the 451B-DE-SI can measure skin dose at $H^*(0.07)$, and deep dose $H^*(10)$ with the beta shield closed.

Figure 13.4 shows the energy responses of the 451B ion chamber survey meter. The energy dependence is within 20% between 50 keV and 1 MeV for 451B, and within 20% between 30 keV and 1 MeV for 451P-DE-S1-RYR (dose-equivalent ion chamber).

13.5.1.1.2 GM Counters GM survey meters are widely used to measure very low radiation levels because of their very large charge amplification (nine to ten orders of magnitude). They are also suitable for leak testing and detection of radioactive contamination. A GM survey meter with a thin window has almost

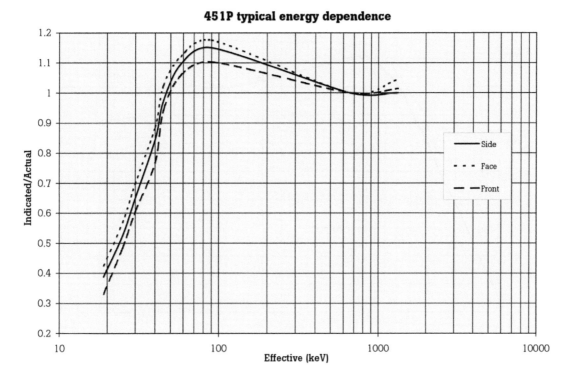

451P typical energy dependence

FIGURE 13.2 Energy response for 451P-RYR. (Reproduced with permission from Fluke Biomedical, Everett, Washington.)

100% sensitivity to the incoming beta rays, and it is very useful in classifying radioactive materials [17]. A survey meter having an extendable rod with a small GM counter installed at its tip can be used for remote high dose rate measurements.

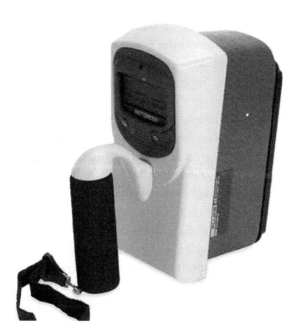

FIGURE 13.3 Fluke 451B ion chamber survey meter with beta slide. (Reproduced with permission from Fluke Biomedical, Everett, Washington.)

GM detectors have a strong energy dependence at low photon energies. They are useful for detecting the presence of radiation, whereas ionization chambers are useful for more precise measurements.

The dead time is defined as the minimum time required between the separations of two events in order that they may be recorded as two pulses [18]. The dead time may be determined by the detector or the electronics. At high counting rates, a true event will be lost because it occurs too quickly following a preceding event. Dead time losses can become severe when high counting rates are encountered. GM detectors have very long dead times, ranging from tens to hundreds of milliseconds. Therefore, GM detectors are not used when accurate measurements are required; and at high count rates (greater than a few hundred counts per second). They cannot be used in pulsed fields unless the count rate of the detector is significantly below the pulse rate; and the dead time is insignificant when compared to the detection rate. Otherwise, they become saturated and count only the repetition rate [19].

13.5.1.1.3 Proportional Counters Proportional counters are always operated in the pulse mode and in the region where, the amplification is about 100 fold [1, 18].

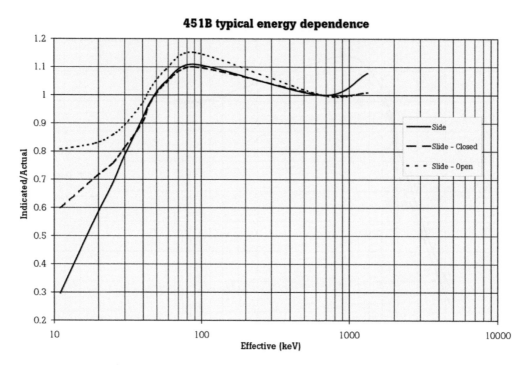

FIGURE 13.4 Energy dependence of 451B ion chamber. (Reproduced with Permission from Fluke Biomedical, Everett, Washington.)

Therefore, proportional counters are more sensitive than ionization chambers and can be used to measure low dose rates. The amount of charge collected from each interaction is proportional to the amount of energy deposited in the gas of the counter by the interaction. Like GM counters, they cannot be used in pulsed fields.

13.5.1.2 Solid-State Detectors

Solid state detectors include scintillators and semiconductors.

13.5.1.2.1 Scintillators Scintillation detectors are based on light emission. Certain phosphors (organic crystals and inorganic crystals doped with activator atoms) emit scintillations upon absorption of radiation [1]. Organic crystals include high-atomic number phosphors, and are primarily used for the measurement of gamma rays; while plastic scintillators are mostly used to measure charged particles, photons, and fast neutrons.

Solid organic materials such as anthracene, stilbene, and plastic scintillators, as well as thallium-activated inorganic phosphors such as sodium iodide (NaI(Tl)) or cesium iodide (CsI(Tl)), are used as scintillators. The light pulse is converted into an electric pulse by a photomultiplier tube (PMT), which is optically coupled to

the scintillator. Some survey meters use photodiodes in place of PMTs. NaI(Tl) scintillation survey meters with correction circuits for energy dependency give accurate results of ambient dose rate, similar to an ionization chamber [17]. The lower detection limits are sufficient for background measurements. They can also be used for the measurement of radioactivity. Sodium iodide and cesium iodide exhibit poor energy response for the measurement of dose equivalent. However, the energy response can be improved through the use of compensation circuits. Scintillation survey meters are usually insensitive to photons with energies below 50 keV and are not appropriate for low-energy x-ray fields.

High-purity germanium (Ge) detectors have an excellent energy resolution and are commonly used for photon spectrometry in research work. Since the Ge detector must be cooled down to liquid-nitrogen temperature, it is not suitable for routine measurements. Handheld scintillation survey meters designed for photon spectral measurements are commercially available, such as InSpector™ 1000[1] [20] and identiFINDER™[2] [21]. Handheld survey meters with cerium-doped lanthanum bromide (LaBr$_3$(Ce)) scintillators are also available. These survey meters have better energy resolution than the conventional thallium-doped sodium iodide (NaI(Tl)) scintillator. Conversion from the light-output distribution to the photon energy spectrum requires an unfolding process.

Scintillation-based systems are more sensitive than GM counters because of higher gamma conversion efficiency and charge amplification (PM tube works by amplifying the electrons generated by a photocathode exposed to a light signal). They are used for surveys at very low dose rates such as contamination monitoring and lost source detection surveys. However, they can also be used at higher dose rates levels because their resolving time is ≤ a few microseconds [1].

Handheld photon spectrometers may be used for nuclide analysis of residual activity [17]. However, complicated spectra cannot be resolved because of their limited energy resolution. High-purity germanium (Ge) detectors can be used to resolve complicated spectra.

Lithium iodide crystals, doped with europium detect thermal neutrons; and therefore, can be used in neutron spectrometers as described later. Plastic scintillation detectors can be used to measure recoil proton spectra in real time, which provide neutron spectral information in the energy range of 2 MeV to 20 MeV [3].

13.5.1.2.2 Semiconductors
A major limitation of scintillators is their relatively poor energy resolution [18]. Semiconductors used as radiation detectors provide a much larger number of carriers for a given incident radiation event, than any other detector. Therefore, they provide the best energy resolution.

Bulk conductivity detectors are formed from intrinsic semiconductors of very high bulk resistivity (e.g., cadmium sulfide (CdS) or cadmium selenide (CdSe)). They act like solid-state ionization chambers on exposure to radiation [1].

Extrinsic (i.e., doped with trace quantities of impurities such as phosphorus or lithium) semiconductors such as silicon or germanium are used to form junction detectors. On application of a reverse bias, these detectors act as a solid-state ionization chamber in detecting radiation.

PIN diodes are diodes with a wide, undoped intrinsic semiconductor region between a p-type semiconductor and an n-type semiconductor region. The p-type and n-type regions are typically heavily doped because they are used for ohmic contacts. In contrast to the ordinary p-n diode, the wide intrinsic region makes a PIN diode a good photo-detector.

The sensitivity of solid-state detectors is about 10^4 times higher than that of gas filled detectors, because of the lower average energy required to produce an ion pair in solid detector materials compared with air (factor of 10); and the higher density of the solid detector materials compared with air (factor of 1000) [1]. The detector dimensions for high-energy electrons or gamma rays can be made much smaller than the equivalent gas-filled detector.

13.5.2 Passive Monitoring
Passive area monitors are similar to those used for individual monitoring. However, they provide an estimate of potential doses received by the individual over a given period of time; and can be used for retrospective dosimetry. The most widely used photon monitors are OSLDs, TLDs, and photographic emulsion film dosimeters. Such dosimeters can be used in pulsed radiation fields because they are insensitive to the time structure of the radiation field.

EPDs are direct-reading dosimeters and show both the instantaneous dose rate and the accumulated dose at any point in time.

13.5.2.1 Individual Monitoring
As previously described, the operational quantity, personal dose equivalent $H_p(d)$ with the recommended depth $d = 10$ mm for strongly penetrating radiation, and $d = 0.07$ mm for weakly penetrating radiation, is used for the individual monitoring of external exposure.

13.5.2.1.1 Calibration   Personal dosimeters are irradiated for calibration while mounted on standard phantoms, in order to account for the backscatter contributions of the human body [1]. The standard phantoms are usually composed of ICRU tissue. In some cases, water phantoms and polymethyl methacrylate (PMAA) phantoms are also used. Different phantom configurations are used for whole body dosimeters (slab), wrist or ankle dosimeters (pillar), and finger dosimeters (rod). The calibration procedure for whole body dosimeters is as follows:

1. Air kerma in air $(K_{air})_{air}$ is measured in a reference field, using a reference ionization chamber that is calibrated by a standards laboratory.

2. The reference instrument reading is converted to $[H_p(d)]_{slab}$ using theoretical values of the ratio, h_{kHp} (available for various beam qualities); where $h_{kHp} = [H_p(d)/(K_{air})_{air}]_{slab}$.

3. The dosimeter to be calibrated is placed at the calibration point on a phantom and its reading M is obtained.

4. The calibration factor in terms of the personal dose equivalent for the dosimeter is given by $N_{Hp} = H_p(d)/M$.

13.5.2.1.2 Desirable Characteristics of Individual Monitors

Personal dosimeters should have high sensitivity and be capable of measuring doses as low as 10–30 μSv [1]. They should not have significant energy dependence. The response should be linear in the dose range of 10 μSv to 10 Sv, thus covering both normal and accidental exposures. They must be iso-directional, that is, the angular response must vary as the directional dose equivalent, $H'(d, \Omega)$. They should also be capable of discriminating against different types of radiation in mixed fields. Unfortunately, no single dosimeter meets all these criteria.

13.5.2.2 Thermoluminescent Dosimeters

Ionizing radiation can produce electron-hole pairs in an inorganic crystal [22]. In TLDs, the crystal material and impurities are such that the electrons and holes remain trapped at the activator sites at room temperature. The number of trapped electrons and holes depends upon the radiation exposure. Upon heating, the trapped electrons and holes migrate and combine emitting photons with energies of a few electron volts. A TLD reader is used to automatically heat the TLD and measure the light output as a function of temperature – referred to as a glow curve. Several peaks appear as traps at different energy levels are emitted. The total light output is given by the area under the glow curve; and can be used to determine the radiation dose. The TLD can be annealed by heating it to high temperatures, so that all the traps are emptied. It can be then reused. Among the many passive dosimeters, TLDs are most commonly used in radiotherapy. The dose response is linear over a wide range of doses used in radiotherapy, but the response increases in the higher dose region, thus exhibiting supralinearity [23]. The most commonly used TLD materials include LiF:Ti,Mg (Lithium fluoride doped with titanium and magnesium), CaSO$_4$:Dy (calcium sulfate doped with dysprosium), and CaF$_2$:Mn (calcium fluoride doped with manganese) [1].

TLDs can be used with filters to discriminate against different types of radiation. Beta, x-ray, and gamma ray doses are determined by measuring the light output under different filters. The results are then compared with results obtained from calibrated dosimeters, which have been exposed to known doses under well-defined conditions.

Dosimeters that have high-atomic number TLDs are not tissue equivalent. Therefore, they require filters to match their energy response to that of tissue. Dosimeters with low-atomic number TLDs do not require such filters. Due to its tissue equivalency, LiF TLD exhibits acceptable energy dependence characteristics. However, CaSO$_4$:Dy shows significant energy dependence and its energy response can be reduced by appropriate dosimeter design.

Due to their small size, TLDs are convenient for use in extremity dosimetry. TLDs exhibit fading, that is, the signal decreases with time after irradiation due to the spontaneous emission of light at room temperature. The dose range covered by TLDs is 100 μSv to 10 Sv [23].

13.5.2.3 Optically Stimulated Luminescence Dosimeters

The principle of optically stimulated luminescence (OSL) is similar to that of thermoluminescence, with the exception that laser light is used instead of heat, to release the trapped energy. Optically stimulated luminescent dosimeters contain a thin layer of aluminum oxide doped with carbon (Al$_2$O$_3$:C). The aluminum oxide is stimulated with selected frequencies of laser light producing luminescence. The light output is proportional to the radiation dose [23].

Figure 13.5 shows the Landauer[3] Luxel+ [24] dosimeter badge which measures radiation exposure due to x-ray, gamma, and beta radiation with OSL technology.

FIGURE 13.5 The Luxel+ dosimeter badge. (Reproduced with permission from Landauer, Inc. Glenwood, IL.)

According to the manufacturer, it is an integrated, self-contained packet. It incorporates a thin strip of Al_2O_3:C (aluminum oxide doped with carbon) sandwiched between a multi-element filter pack that is heat sealed. It is covered by a laminated, light-tight paper wrapper. All of these components are RF-sealed inside a temper-proof plastic blister pack. Luxel+ dosimeter is not affected by heat, moisture, and pressure as long as the clear blister packaging is uncompromised. These dosimeters are very sensitive and can be used to measure photon doses in the range of 10 µSv to 10 mSv over a photon energy range of 5 keV to 40 MeV. They can measure beta ray doses in the range of 100 µSv to 10 mSv over an energy range of 150 keV to 10 MeV. The dosimeters can be reanalyzed several times without losing sensitivity and may be used for up to a period of one year.

The dosimeter can be restimulated with laser numerous times to confirm the accuracy of a radiation dose measurement. A full reanalysis is automatically performed for every measurement yielding a dose in excess of 5 mSv. The badge can also incorporate the neutron detector, Neutrak® 144, described in Section 13.6.2.3.

13.5.2.4 Film Dosimeters

Film badges consist of special emulsion photographic film placed in a light-tight wrapper. The latter is enclosed in a case or holder with windows behind appropriate filters [1]. Film is non-tissue equivalent; therefore, a filter system is used to flatten the energy response. A single filter is adequate for photons of energy above 100 keV, but multiple filters are required for lower energy photons.

The badge holder creates a distinct pattern on the film, which indicates the type and energy of the incident radiation. The film optical density under different filters is read, and the results are compared with calibration films that have been exposed under known conditions (doses and radiation types). The optical density is a measure of the cumulative dose. Film can be used to determine cumulative doses from photons, beta particles, and neutrons (described in Section 13.6.2.1). Film dosimeters can measure doses in the range of 0.1 mSv to 10 Sv.

Films are affected by heat, liquids, and excessive humidity; and the latent image on undeveloped film fades with time. Thus, the usage period is limited to about to three months under ideal conditions.

13.5.2.5 Electronic Personal Dosimeters

EPDs measure both instantaneous dose rates and accumulated doses [1]. Therefore, they are used for supplemental individual monitoring such as:

1. Direct readout of the dose or dose rate at any time.

2. Tracking of daily doses during various activities.

3. Tracking of doses during special situations such as radioactive source loading, surveys, and handling of radiation incidents or emergencies.

EPDs based on miniature GM counters or silicon detectors are available. EPDs with energy-compensated detectors have an energy dependence within ±20% over the photon energy range of 30 keV to 1.3 MeV. They provide an instantaneous display of the cumulative dose at any time. They also have automatic ranging facilities and provide both visual and audio indications, thus monitoring immediate changes in the radiation field.

The RAD-60R[4] EPD uses a 4.5 mm² PIN Si-Diode, energy-compensated with copper filtration. This EPD displays R on the display, but this model actually measures personal dose equivalent, $H_p(10)$ in mrem [25].

The PD-10i[5] EPD utilizes an energy–compensated GM tube. The specified energy response of that unit is tissue-equivalent to within ±25% from 55 keV to 6 MeV. There is a 70% under response at 40 keV. This unit display scales automatically between µR, mR, and R. According to Meier, the EPDs were irradiated with ^{99m}Tc, ^{131}I, and ^{18}F radionuclides with emission energies at 140 keV, 364 keV, and 511 keV, respectively. The authors report that the energy responses of RAD-60R and the PD-10i units significantly diverge from unity at 140 keV, but trend toward unity as the incident photon energy increases to 511 keV. In the energy ranges that were studied, both EPD models were shown to have significant increases in reading in the presence of backscatter from a PMMA phantom.

Thermo Scientific™[6] EPDs measure ionizing radiation in real time [26]. They are equipped with both audible and visual alarms; and can be used to measure gamma radiation, beta radiation, neutron radiation, and x-rays. Various models are available. The EPD-TruDose™ measures $H_p(10)$ within ±5% for ^{137}Cs and $H_p(0.07)$ within ±15% for ^{90}Sr/^{90}Y. It can measure photons in the energy range of 16 keV to

10 MeV. It has a dose rate range of 1 μSv/h to 10 Sv/h and a dose range of 1 μSv–10 Sv.

13.6 NEUTRON MONITORING TECHNIQUES

Neutron monitoring techniques inside treatment rooms have been described in the literature [27, 28]; and are not discussed in this chapter. Only, neutron monitoring outside shielded rooms and accelerators are discussed. Radiotherapy facilities requiring external neutron monitoring include therapy linacs, fast neutron therapy, BNCT and particle therapy facilities.

In neutron monitoring, the quantities of interest are neutron fluence, neutron dose equivalent (usually ambient dose equivalent) or dose equivalent rate, and the neutron spectrum as a function of energy [7].

Neutron monitoring techniques in radiotherapy consist of both active and passive methods. Active methods include the use of dose-equivalent and fluence meters; while passive methods include the use of TLDS, solid-state nuclear track detectors (SSNTDs) and bubble detectors.

13.6.1 Active Neutron Monitoring

Active neutron monitoring is based on slowing down or moderating fast neutrons until they reach thermal energies. The thermal neutrons are then detected by a thermal neutron detector. The geometry and configuration of the instrument is designed to measure either dose equivalent (referred to as dose-equivalent meter) or fluence (fluence meter). Outside the shielded therapy linac rooms, neutron and photon fluences are considerably lower than inside the treatment room [27]. In addition, the neutron pulse is spread over several hundred microseconds due to moderation in the shielding material. Therefore, active monitors may be used as long as photon pulse pile-up and dead times effects on their responses are taken into consideration. Neutron detection for radiotherapy facilities is spread over many decades of energy from thermal energies up to GeV. Unfortunately, no single detector can accurately measure the neutron dose equivalent or fluence over such a wide energy range.

13.6.1.1 Fluence Meters

Neutron fluences can be used to determine dose equivalent; however, this requires the use of fluence-to-dose equivalent conversion coefficients ($h_\Phi(E)$) [8, 29]. Because these coefficients depend strongly on the neutron energy, the neutron spectrum must be known or approximated *a priori* [30]. Typically, most of the neutron detectors are calibrated against standard neutron sources such as Pu-Be (Plutonium–Beryllium) with $\bar{E}n = 4.2$ MeV, Am-Be (Americium–Beryllium) with $\bar{E}_n = 4.5$ MeV or ^{252}Cf with $\bar{E}_n = 2.2$ MeV where $\bar{E}_n$ is the average neutron energy. Therefore, correction factors must be applied to account for the changes in neutron spectra from calibration to actual field conditions.

The most common thermal neutron detectors used in fluence meters are the ^{10}B-enriched BF$_3$ (boron trifluoride) detector, the ^{3}He (helium-3) counter tube, and the ^{6}LiI(Eu) (europium doped lithium-6) scintillator [3]. The reactions that take place in the above-mentioned detectors are as follows:

BF$_3$ proportional counter: ^{10}B(n_{th},α)^{7}Li, $Q = 2.31$ MeV, σ = 3,840 barns,

^{3}He proportional counter: ^{3}He(n_{th},p)^{3}H, $Q = 0.765$ MeV, σ = 5,330 barns,

^{6}Li(Eu) scintillator: ^{6}Li (n_{th},α)^{3}H, $Q = 4.78$ MeV, σ = 940 barns,

where Q is the Q-value (kinetic energy released) and σ is the cross section for the thermal neutron reaction. The cross sections vary roughly as $E_n^{-1/2}$, where E_n is the neutron energy. Therefore, at 1 MeV, the cross section is about four orders of magnitude lower than that at thermal energies. All these bare detectors are sensitive only to thermal neutrons. When used with a moderator, higher energy neutrons can be detected.

The most commonly used neutron fluence meter is a version of the BF$_3$ long counter [27, 31]. The bare BF$_3$ counter detects only thermal neutrons. The sensitivity of the counter is proportional to volume, pressure, and degree of enrichment in ^{10}B. When used in conjunction with a moderator (typically cylindrical), the fast neutrons are thermalized, and detected. By enclosing the moderator in 0.5 mm of cadmium, the thermal neutrons can be completely eliminated. The thickness of hydrogenous moderator is chosen so that a fairly flat response is obtained per unit fluence for neutron energies up to several

MeV. Appropriate fluence-to-dose equivalent factors can be applied to the known fluence, based on an average energy, to obtain dose equivalent [29, 32].

The use of the BF_3 detector requires knowledge of the neutron spectrum. It is particularly suitable for use when neutron dose-equivalent rates are below measurable levels on the dose equivalent meter. Relative variations of the neutron field with time can be monitored with the moderated BF_3 detector. The ratio of the dose-equivalent meter and the moderated BF_3 detector readings provides a rough estimate of the average energy of the neutron spectrum. All the problems associated with dose-equivalent meters (described in the next section) also apply to these detectors.

The ^3He proportional counter is more sensitive and more stable than the BF_3 counter, but it is much more expensive [27]. The $^6Li(Eu)$ scintillator has a very high sensitivity but a very poor photon rejection. Therefore, it is difficult to use in mixed photon–neutron fields such as those encountered at therapy facilities. It is important to note that for all active monitors, manufacturers normally state photon rejection for steady fields and not pulsed fields.

13.6.1.2 Dose-Equivalent Meters

Typically, most commercial dose-equivalent meters consist of a neutron moderator such as polyethylene or some other hydrogenous material, surrounding a thermal neutron detector. Dose-equivalent meters incorporating BF_3 proportional counters are more commonly used outside the shielding in therapy facilities because of their excellent photon rejection and low cost. In the BF_3 detector, the thermal neutrons are captured in the boron via the $^{10}B(n_{th},\alpha)^7Li$ reaction. Both the alpha particle and recoil ^7Li nucleus produce large pulses in the proportional counter. The large pulses being orders of magnitude higher, can be easily discriminated from the smaller pulses produced by photons in mixed fields. The energy response of a moderated detector is determined mainly by its size and geometrical configuration [3].

The commonly used Andersson–Braun (AB) rem-meter has a cylindrical polyethylene moderator surrounding a BF_3 counter tube [33]. Its energy response closely follows the NCRP 38 dose equivalent conversion function (from thermal to about 10 MeV), except at intermediate energies where it over responds.

The counter tube is surrounded by a borated plastic sleeve which minimizes its over-response in the 10 keV to 100 keV range. The response at thermal energies is increased because of holes drilled in the sleeve. However, the directional response of the rem-meter is impacted by its cylindrical moderator geometry. According to Cosack and Leisecki [34], a change in response with instrument orientation of as much as 35% has been observed for neutron energy of 1 MeV. When the rem-meter is exposed with its side oriented to a source of thermal neutrons, a 65% underestimation in the true dose equivalent has been observed. A modified version with a spherical polyethylene moderator and a cadmium layer was designed by Hankins in order to improve the directional dependence [35]. It is currently marketed by Thermo Fisher Scientific as the model NRD. However, the high-energy response of the NRD decreases steadily at energies above 7 MeV. At these energies, the neutron fluence is considerably reduced for therapy linacs. Therefore, the rem-meter can still be used. However, it cannot be used for particle therapy and fast neutron therapy facilities, because of the higher-energy neutrons that are encountered at such facilities.

Figure 13.6 shows a Thermo Eberline ASP/2e NRD Neutron Survey Meter – a portable battery operated instrument. The detector is a 22.9 cm (9 inch) in diameter, cadmium-loaded polyethylene sphere with a BF_3 tube in the center. The instrument is available with display either in rem or Sv. According to the manufacturer, the energy response closely follows the theoretical dose equivalent curve for neutrons over the energy range from 0.025 eV (thermal) to about 10 MeV. It measures dose equivalent rates in the range of 1–100 mSv/h. The BF_3 tube provides gamma rejection up to 500 R/hr depending on the applied voltage (1600–2000 V). The sensitivity is 45 cpm/(mrem/h) or 2700 counts per mrem. The dead time is 10 μs (nominal). It has a directional response within 10%. Its response time is programmable from 0–255 milliseconds. The counting instrument is the Model ASP-2e which has a dual analog/digital display. It has a rate meter that can be used on an integrating or scaler mode. It has a count range of 1–1.3 million counts per minute.

The Fluke Biomedical[7] Neutron Survey Meter Model 190N, is based upon the classical Andersson–Braun rem-meter design. A polyethylene cylinder 24 cm long,

FIGURE 13.6 Thermo Fisher Scientific NRD Asp/2e Neutron Survey Meter.

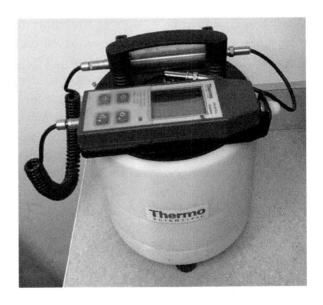

FIGURE 13.7 FHT 762 Wendi-2 Wide Energy Neutron Detector. (Courtesy of Arash Darafsheh.)

and 21.6 cm in diameter surrounds the BF₃ tube. The tube is filled with 96% enriched ^{10}B. The instrument is available with display in either rem or Sv. According to the manufacturer, the instrument has a dose equivalent range from 0 μSv/h to 0.75 Sv/h, and it can integrate doses from 0 μSv to 10 Sv. It has a gamma rejection up to 500 Roentgens/h for ^{137}Cs. The directionality is less than 20% in orthogonal directions.

The maximum neutron energy for a 15 MV linear accelerator is about 10 MeV [27]. Since the response of the neutron dose-equivalent meters (AB, NRD, 190N) falls off rapidly at energies above 6 MeV, these detectors are not necessarily accurate over the energy range of interest in radiotherapy linac facilities, but are adequate for monitoring purposes.

The Andersson–Braun design suffers from angular dependence due to the lack of spherical symmetry [13]. In addition, all instruments using a pure polyethylene moderator suffer from a lack of high-energy response. Because an ordinary rem-meter is practically insensitive to neutrons of energies above 15 MeV, it underestimates the dose equivalent by as much as a factor of 3 when used outside a shielding at particle therapy facilities [17]. Therefore, such rem-meters cannot be used in fast neutron therapy where neutron energies extend to 70 MeV; and particle therapy facilities where maximum neutron energies for particle therapy facilities are in the GeV range. Further, neutron dose

equivalent rates are a strong function of angle at these facilities [36].

Figure 13.7 shows the Wide Energy Neutron Detection Instrument Wendi-2 (also known as WENDI-II) dose-equivalent meter.

The Wendi-2 was originally designed at Los Alamos National Laboratory [13]. A cylindrical polyethylene moderator assembly surrounds a ^{3}He counter. Tungsten powder surrounds the counter at an inner radius of 4 cm, thus, generating neutrons with energies above 8 MeV and absorbing neutrons with energies below several keV. The interaction of high-energy neutrons with the tungsten material causes neutron multiplication and energy degrading reactions such as (n, 2n), thus improving the sensitivity to high-energy neutrons which extends well beyond 20 MeV to 5 GeV. The instrument exhibits increased sensitivity, with a sensitivity to bare ^{252}Cf that is 12 times higher than the NRD sensitivity. Additionally, the energy response for Wendi-2 closely follows the theoretical ambient dose equivalent per unit fluence function [$H^\star(10)/\Phi$] above 0.1 MeV. Its energy response at 500 MeV is approximately 15 times higher than that of the NRD and Andersson–Braun meters.

According to the manufacturer [37], the Wendi-2 has an excellent energy response in the normal energy range up to 15 MeV and closely follows the ambient dose equivalent up to 5 GeV. The linearity of response is within 20%. It can measure neutron ambient dose equivalent in the energy range of thermal to 5 GeV. The dose equivalent rate range is 0.01 μSv/h to 100 mSv/h for ^{252}Cf neutrons. The angular dependence is within 20% in all directions. The neutron sensitivity (^{252}Cf) is 0.84 cps/(μSv/h) and the photon sensitivity is 1 to 5 μSv/h at 100 mSv/h of 662 keV gamma rays from ^{137}Cs. No pulse pile-up needs to be considered for gamma dose rates up to 1 Sv/h.

Figure 13.8 shows the calculated energy response relative to $H^{*}(10)$, normalized for a bare ^{252}Cf calibration, for the following rem meters: WENDI-II (side), Hankins-NRD, and AB (side) [13].

An intercomparison of the performance of various active neutron monitors, including Wendi-2 was carried out in at the CERN-EU reference field (CERF) facility [38]. The CERF mixed radiation field is produced by a positive hadron beam comprised of 2/3 protons and 1/3 positive pions, with a momentum of 120 GeV c^{-1} impinging on a copper target placed inside an irradiation cave [39]. The measurements were made behind an 80-cm thick concrete shield in the CT7 reference exposure location. The neutron spectral fluence at this location is characterized by a low-energy peak with an energy of about 0.4 eV, an intermediate region between the thermal and the evaporation peak located at of about 1 MeV; and a high-energy peak at about 100 MeV.

The results show that the Wendi-2 agrees well within its uncertainties; and within one sigma of the $H^{*}(10)$ value calculated by Monte Carlo calculation using the FLUKA code. The conventional rem counters were in good agreement within their uncertainties and underestimated $H^{*}(10)$ as measured by the extended range instruments and as predicted by FLUKA [40]. The non-neutron part of the stray field accounts for about 30% of the total $H^{*}(10)$.

Measurements performed in quasi-monoenergetic neutron fields indicate that $H^{*}(10)$ measurement results of the Wendi-2 rem-meter can be reproduced satisfactorily by FLUKA for energies of a few hundred MeV [41].

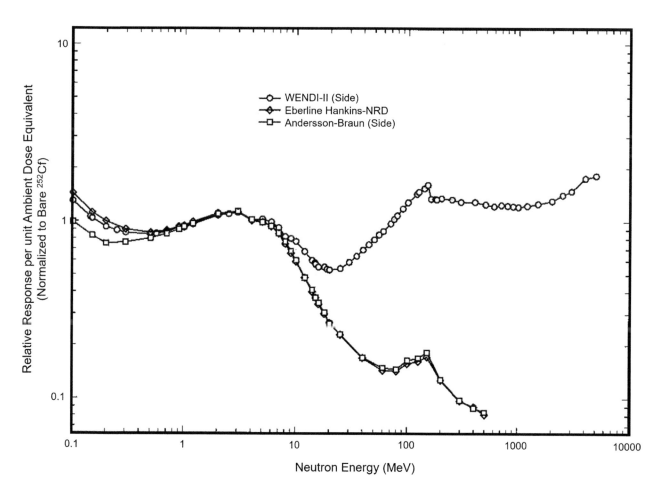

FIGURE 13.8 Energy response relative to $H^{*}(10)$, normalized for a bare ^{252}Cf calibration, for the following rem meters: WENDI-II (side), Hankins-NRD, and AB (side) [13].

The energy response function of the Wendi-2 rem-meter and the ambient dose equivalent per unit fluence as a function of energy showed good agreement. According to the authors, the Wendi-2 can be considered to be an appropriate device to measure $H^*(10)$ neutron dose behind shielding for primary proton beam energies of about 250 MeV and about 400 MeV.

Thus, Wendi-2 can be used for neutron monitoring of fast neutron therapy and particle therapy facilities.

13.6.1.3 Neutron Spectrometers

The neutron spectrum can be determined by using thermal neutron detectors inside a series of hydrogenous spheres of varying diameters [42]. It is possible to calculate the total spectrum by taking all the response functions and folding them into a series of equations, because the amount of moderation varies in each of these spheres. This method is referred to as the Multi-Sphere or Bonner–Sphere method. The process is laborious, as it requires software, calibration, a large number of spheres and long measurements times. Bonner sphere spectrometers suffer from the same energy response limitations that constrain rem-meters due to their pure polyethylene composition.

In addition, successful unfolding of the neutron spectrum requires some knowledge of the actual spectrum since there is no unique solution for a set of integral equations. This approach works pretty well for nuclear power plants but would prove impractical for therapy installations.

The Bubble Technology Industries, Inc., ROSPEC [43] is a rotating neutron spectrometer designed specifically for the spectral measurement of degraded fission neutrons, and detects neutrons in the energy range of thermal to 4.5 MeV. It is capable of generating accurate spectral and dosimetric data simply and routinely in minutes or hours. Data analysis is performed via a notebook computer, with on-line display of individual or all counter pulse height distributions, and an unfolding program for generation of neutron spectra. Fluence, kerma, maximum dose equivalent, ambient dose equivalent $H^*(10)$ and dose rate are calculated, and stored on a hard disk.

Alternatively, a scintillation spectrometer with either a plastic or liquid scintillator can be used [44]. However, the interference from photons in a mixed photon–neutron field can be significant. A considerable amount of time may be required in adjusting the pulse shape discrimination to reject the photon signal. Therefore, the use of a scintillation spectrometer is not practical for routine neutron measurements around radiotherapy facilities. Further, it is also affected by the pulsed nature of the neutron field.

On the other hand, time-of-flight spectroscopy is a laboratory method, not suitable for field applications. The best measurements of neutron spectra are obtained with the time-of-flight spectrometers. When a neutron is created or when it first enters the detector, a signal is produced at the point of creation. The time taken by the neutron to get to a detector at some distance away is measured. The time taken to travel a given distance can then be used to determine the neutron energy. The time-of-flight method may not be applicable for external radiation monitoring at radiotherapy facilities because of the scattering of the majority of neutrons after they are created. It is also not clear that a suitable "start signal" may be obtained in such fields. Apart from being susceptible to pulsed field effects, these are expensive detectors, and require difficult measurements. Therefore, they are not very useful for external neutron monitoring at radiotherapy facilities.

13.6.2 Passive Neutron Monitoring

The various types of passive monitors include film, TLDs, SSNTDs, and bubble detectors. When using passive detectors, it is important to take several measurements with the same detector type at a given location to reduce random errors.

13.6.2.1 Film

Photographic emulsions or film are divided into two categories: radiographic film and nuclear emulsions [18]. In radiographic films, a general darkening of the emulsion occurs due to the cumulative effect of multiple individual interactions. Thus, an image is recorded of the transmitted intensity of a radiation beam. Radiographic film can also be used for thermal neutron monitoring [1]. The cadmium window in the film holder absorbs thermal neutrons. The latent image from capture gamma radiation blackens the film below this window after processing.

Nuclear emulsions are much thicker and differ in composition from radiographic films. They are best suited for recording single particle tracks. Nuclear track emulsions are used for fast neutron monitoring. Recoil protons are produced by neutrons interactions with the hydrogen nuclei in the emulsion and surrounding material. The latent image created by the recoil protons, leads to darkening of the film along their tracks after processing. Nuclear emulsions can be loaded with special target material such as boron or uranium to allow the detection of thermal neutrons.

13.6.2.2 TLDs

Various techniques are available such as using the body as a moderator to thermalize neutrons (similar to albedo dosimeters). The thermal neutrons are then detected using LiF enriched with ^{6}Li. The reaction of thermal neutrons in ^{6}Li provides enhanced sensitivity.

According to Mirion Technologies (GDS), Inc.,[8] the Genesis TLD Dosimeter [45] is a 4-element TLD with three ^{7}LiF:Mg,Cu,P (TLD 700H) and one ^{6}LiF:Mg,Cu,P (TLD 600 H) TLDs. The TLD 700H measures betas, gammas, and x-rays. The TLD 600H responds accurately to betas (0.251 MeV to 5 MeV), gamma rays and x-rays (5 keV to 6 MeV), and neutrons (200 keV to 6 MeV). Neutrons with energies from thermal to 6 MeV are detected by the TLD. The neutron energy can be extended to 20 MeV with the optional use of CR-39 (described in the next section). The response of each element is adjusted by the application of its own unique correction factor. These dosimeters measure deep dose, lens of eye, and shallow doses. Unlike other TLD products, the Genesis Ultra TLD has virtually no fading characteristics. Due to the increase in signal response a minimum reportable dose as low as 0.01 mSv is available for the dosimeter. The useful dose range is 0.01 mSv to 10 Gy.

13.6.2.3 Solid-State Nuclear Track Detectors

SSNTDs such as the polymer CR-39 (poly allyl diglycol carbonate) are used for individual monitoring. Recoil nuclei from neutron interactions with its constituent atoms (hydrogen, carbon, and oxygen) leave submicroscopic damage trails or tracks in the detector. A suitable etching process, such as chemical etching, electrochemical etching, or a combination is used to reveal these tracks. The track density can then be related to the neutron dose equivalent through calibration. The response of CR-39 can be enhanced through the use of a hydrogenous radiator in contact with it. Additional protons are generated within the radiator. CR-39 can detect neutrons over a wide energy range from about 100 keV to 20 MeV when combined with an irradiator. The energy range,

dose equivalent response, and the lower limit of detection depend on the material, the type and thickness of the radiator, and the etching process that is used. These detectors suffer from directional dependence. CR-39 detectors can be used for neutron monitoring inside and outside the therapy linac treatment room but not inside the primary photon beam due to the following photon-induced effects [46, 47].

^{2}D$(\gamma,n)p$, E_{th} = 2.23 MeV, (E_{th} = threshold energy)
(0.02% of hydrogen is deuterium)

^{16}O$(\gamma,\alpha)^{12}$C, E_{th} = 7.2.MeV

^{12}C$(\gamma,\alpha)^8$Be $\rightarrow 2\alpha$, E_{th} = 7.4.MeV

The Landauer[3] Neutrak 144° is one of the commercially available CR-39 detectors suitable for use at radiotherapy facilities as a personnel dosimeter. There are two neutron dosimeters [48]: a fast neutron dosimeter; and a combination fast, intermediate and thermal neutron dosimeter. The fast neutron dosimeter uses a polyethylene radiator with CR-39. Recoil protons resulting from neutron interactions in the dosimeter are recorded. This dosimeter is used for monitoring personnel working with unmoderated or moderately shielded fast neutron sources such as ^{252}Cf and Am-Be. Landauer uses a material that is exclusive to them and is not the typical performance seen by others, which generally has an energy response threshold of 100–150 keV.

The combination neutron dosimeter measures fast, intermediate, and thermal neutrons. The left area of the chip uses a polyethylene radiator for fast neutrons. The right area uses a boron loaded Teflon$^{\circledR}$ radiator for fast, intermediate, and thermal neutrons. The alpha particles resulting from thermal neutron interactions with the boron radiator are recorded in the dosimeter. This dosimeter is used for personnel monitoring outside shielded high-energy accelerators producing neutrons with energy less than or equal to 40 MeV.

The CR-39 is etched in a chemical bath for 15 hours to enlarge the tracks; and the tracks are then counted in an automatic counter. The track density is a measure of the neutron dose equivalent. The fast neutron dose equivalent is measured by counting the tracks generated as a result of the proton recoil with the polyethylene radiator, while the thermal neutron dose equivalent is measured by counting the alpha tracks generated with the boron radiator.

13.6.2.4 Bubble Detectors

Bubble detectors consist of tiny superheated droplets that are dispersed throughout a firm elastic polymer contained in a small sealed tube. The detector can be sensitized by unscrewing the cap. The interaction of neutrons with the droplets results in the production of secondary charged particles. The charged particles deposit energy and cause the droplets to vaporize, producing bubbles which remain fixed in the polymer. The bubbles can be counted either by eye or in an automatic reader, as long as the detector is not saturated with bubbles. A simple method of counting is to project the image of the bubble detector via a video camera on to a TV screen, and then count by eye or score bubbles on a clear transparency sheet placed over the TV screen [46, 49]. The bubble detector should be rotated several times and counted again. The number of bubbles is a measure of the neutron dose equivalent. Another method would be to take photographs of the detector with the bubbles, magnify the photographs, and count the bubbles [27]. Bubble detectors are easy to use, have a range of available sensitivities, are reusable, integrating and allow instant visible detection of neutrons. They also have an isotropic response. Bubble detectors can be used for neutron monitoring inside and outside the therapy linac treatment room but not inside the primary photon beam due to photon-induced effects [46, 47]. The primary disadvantage is that the sensitivities vary within a given batch, and occasionally spurious results are obtained. In addition, there may be a loss of linearity at the higher doses, which according to the manufacturer may be attributed to a change of pressure within the gel when there are a large number of bubbles present. Finally, it is difficult to get good statistics with a single detector. Therefore, it is important to use a minimum of three detectors at each measurement location.

Figure 13.9 shows a photograph of bubble detectors available from Bubble Technology Inc.[9] The bubble detectors are normally calibrated by the manufacturer against an Am-Be source. The BD-PND (Bubble Detector-Personal Neutron Dosimeter) is a temperature-compensated bubble detector and therefore, useful for monitoring, since no temperature corrections are required. According to the manufacturer, the BD-PND [50] has an approximate energy threshold of 100 keV with a reasonably flat dose equivalent response from about 200 keV to greater than 15 MeV. The dose range is from 1 μSv to 5 mSv. There are three sensitive ranges:

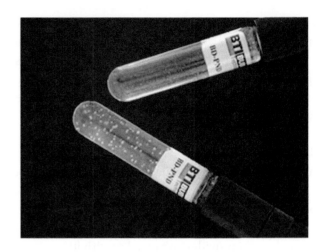

FIGURE 13.9 BD-PND Bubble Detectors. (Courtesy of Bubble Technology Industries Inc., Chalk River, Ontario, Canada.)

Low: 0.093 to 0.019 bubbles/μSv; Medium: 0.94–1.8 bubbles/μSv and High: 1.9–3.7 bubbles/μSv.

The energy response of the BD-PND is shown in Figure 13.10 [51, 52].

By varying the formulation, bubble detectors with different thresholds have been made. The bubble detector spectrometer or the "BDS" can be used for spectral measurements. There are six thresholds – 10 keV, 100 keV, 600 keV, 1 MeV, 2.5 MeV, and 10 MeV. A total of 18 detectors (three of each threshold) are used to make the spectral measurement. The measured spectrum is derived by unfolding the response of the different threshold detectors. A recompression chamber is used to zero the detectors, thus allowing for its reuse.

Figure 13.11 shows the response per unit fluence of the BDS. For the BDS-2500, the response on the graph has been reduced by a factor of 10; hence, the notation 2500/10.

The BDT (bubble detector thermal) uses a ^{6}Li compound dispersed throughout the polymer and a special formulation to detect preferentially α particles from the ^{6}Li$(n,\alpha)^3$H reaction [51]. The thermal to fast neutron ratio is approximately 10:1. The energy range is thermal. The dose range is 1.1–112 μSv. The sensitivity range is 1.8–3.6 bubbles/μSv.

13.7 CONTINUOUS, PULSED, AND MIXED RADIATION FIELDS

In radiotherapy, one encounters both continuous and pulsed radiation fields. Brachytherapy and radioactive sources provide continuous radiation, while therapy

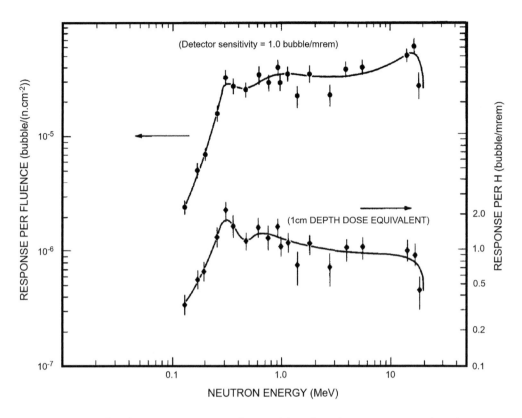

FIGURE 13.10   BD-PND normalized response per unit fluence (closed circles – upper curve) and response per unit dose equivalent (closed diamonds – lower curve). Conversion from fluence-to-dose equivalent based on NCRP Report No. 38 [51].

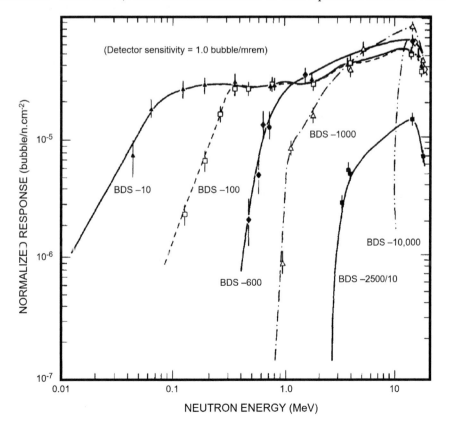

FIGURE 13.11   Normalized response of BDS at 6 different threshold energies. The number 2500/10 means that on the graph, the response of the BDS-2500 has been reduced by a factor of 10 [51].

linear accelerators (linacs) and accelerator-based BNCT, fast neutron therapy, and particle therapy facilities can produce pulsed radiation. In particle therapy facilities, the beams from a synchrotron and a synchrocyclotron are considered pulsed, while the beam from an isochronous cyclotron is considered continuous [53]. However, the extracted beam from the synchrotron is considered continuous, when the particles are extracted slowly [17].

Therapy linacs produce mixed radiation fields comprised of both photons and neutrons [27]. For example, the photon fluence inside the primary beam of a therapy linac is at least 1000 to 4000 times higher than the corresponding neutron fluence. Outside the primary beam, the photon fluence is at least 10 to 100 times higher than the neutron fluence.

Typically, therapy linacs are operated at repetition rates that vary from 100 to 400 Hz with pulse widths of 1 μs to 10 μs [27]. Peak electron currents may vary from 20 to 120 mA per pulse in the photon mode, and 3 to 15 mA per pulse in the electron mode. Higher peak currents may be used in the flattening filter-free mode (FFF).

The fraction of operating time during which the beam is on, is called the duty factor (DF) [19, 54]. The duty factor is the product of the pulse width and the repetition rate or frequency. The peak or instantaneous intensity, I_p, is given by:

$$I_p = \frac{I_{av}}{DF} \qquad (13.9)$$

where I_{av} is the average intensity.

The duty factor for a beam with a pulse width of 1 μs and a repetition rate of 100 Hz is equal to 1×10^{-4}. This very small duty factor imposes a severe limitation on the radiation detector, as in this case, the peak intensity is 10,000 times the average intensity. This intense pulse usually overwhelms any active detector which detects particle events electronically. As a result, the measured readings are only an indication of the repetition rate. A detector that usually responds well to an average dose rate spread out in time may not be able to cope with such high dose rates.

The resolving time or dead time is the time required by the detector to return to its normal state, after registering a pulse. It is the minimum time interval by which two pulses must be separated in order to be detected as separate pulses by the counter and its accessories. Detectors, which have long dead times such as

the Geiger–Muller and proportional counters, tend to become saturated in such fields, and count only the repetition rate. The dead time usually lies between 10^{-8} s and 10^{-4} s.

Ionization chambers are less influenced as long as they are operated with a voltage that is sufficient to overcome recombination losses. Ionization chambers operating in the current mode are more suitable for higher dose rate measurements. Therefore, they are more widely used than GM or proportional counters in pulsed fields.

Scintillation detectors may become non linear at high dose rates in pulsed fields, because photomultipliers cannot handle the high instantaneous dose rates.

For neutron detectors that are moderated, it has been observed that the measured readings are higher than the repetition rate due to scattered radiation within the room; and coupled with the fact that neutron moderation allows an event to be detected even after the pulse has terminated [27]. It may be possible to adjust the accelerator to a low output per pulse, but the photon-to-neutron dose rate may not remain the same at the lower output. Further, in mixed photon–neutron fields, the neutron detectors may have photon-induced reactions [47], which cannot be separated from the neutron interactions. This interference should be considered when measuring neutrons in the primary photon beam of a therapy linac. Thus, only passive neutron detectors should be used in the primary photon beam, provided that they do not experience any photon-induced reactions.

For pulsed radiation, with a pulse repetition rate of P Hz, if the dead time, T_D, is shorter than the pulse width T_P, the corrected counts C_{corr}, is related to the measured counts, C, by the following expression [55]:

$$C_{corr} = \frac{C}{1 - \frac{CT_D}{PT_P}\left(1 - \frac{T_D}{2T_P}\right)} \qquad (13.10)$$

If T_D is >> than T_P

$$C_{corr} = C - P\ln(1 - C/P) \qquad (13.11)$$

For a moderated thermal neutron detector, the effective dead time is determined by the moderation time, and not the pulse width for the neutrons. Therefore, the pulse width is replaced by the moderation time in the above equations. For well-shielded facilities, the neutron dose equivalent rates are fairly low and may be of the order of 1 μSv/h or so. Therefore, integrated measurements of dose equivalent are preferable to instantaneous dose equivalent rate measurements.

13.8 CONCLUSIONS

Photon and neutron area monitoring and individual monitoring are discussed. Examples of some commercial monitors (available at the time of writing this chapter) are provided. However, with time, some of these monitors may become obsolete.

13.9 ACKNOWLEDGMENTS

Sincere gratitude is expressed to Richard H. Olsher and Kamran Vaziri for their review of this chapter and for their suggestions and comments.

NOTES

1 Mirion Technology Inc., 3000 Executive Pkwy #222, San Ramon, CA 94583, USA.

2 ICx Radiation Inc., 100 Midland Road, Oak Ridge, TN 37830, USA.

3 Landauer, Inc., 2 Science Road, Glenwood, IL 60425-1586, USA.

4 RADOS Technology Oy, Turku, Finland.

5 Scientific Applications International Corporation, McLean, VA, USA.

6 Thermo Fisher Scientific, 168 Third Avenue Waltham, MA 02451, USA.

7 Fluke Biomedical, 6045 Cochran Rd., Cleveland, OH 44139, USA.

8 Mirion Technologies (GDS), Inc.2652 McGaw Avenue | Irvine, CA 92614, USA.

9 Bubble Technology Industries Inc., Chalk River, Ontario, Canada.

REFERENCES

1. G. Rajan and J. Izewska, "Radiation monitoring instruments," in *Review of Radiation Oncology Physics. A Handbook for Teachers and Students*, edited by E. Podgorsak (International Atomic Energy Agency, Vienna, 2005), pp. 101–120.

2. N. E. Ipe, "Radiological aspects of particle therapy facilities," in *Shielding Design and Radiation Safety at Particle Therapy Facilities, Particle Therapy Cooperative Group Report 1*, edited by N. E. Ipe (Particle Therapy Cooperative Group, Villigen, 2009), pp. 24–79.

3. NCRP, *Radiation Protection for Particle Accelerators. National Council on Radiation Protection and Measurements Report No. 144* (National Council on Radiation Protection and Measurements, Bethesda, 2003).

4. ICRP, Data for Protection Against Ionizing Radiation from External Sources. Supplement to ICRP Publication 15. *International Commission on Radiation Protection Publication 21*(Pergamon Press, Oxford, 1973).

5. ICRP, *1990 Recommendations of the International Commission on Radiological Protection. ICRP Publication 60. Ann. ICRP 21 (1-3).* (Pergamon Press, UK, 1991).

6. ICRP, *The 2007 Recommendations of the International Commission on Radiological Protection. ICRP Publication 103. Ann. ICRP 37 (2-4)* (Elsevier Science, UK, 2007).

7. ICRU, *Quantities and Units in Radiation Protection Dosimetry. International Commission on Radiation Units and Measurements Report 51* (International Commission on Radiation Units and Measurements, Bethesda, 1993).

8. ICRU, Conversion Coefficients for Use in Radiological Protection against External Radiation. *International Commission on Radiation Units and Measurements Report 57* (International Commission on Radiation Units and Measurements, Bethesda, 1998).

9. L. Moritz, "Radiation measurements and monitoring," in *Topics in Accelerator Health Physics*, edited by J. D. Cossairt, V. Vylet, J. W. Edwards (Medical Physics Publishing, Wisconsin, 2008).

10. ANSI, American National Standard for Radiation Protection Instrumentation Test and Calibration, *Portable Survey Instruments N323AB-2013* (American National Standard Institute, Washington, 2013).

11. IAEA, *Calibration of Radiation Monitoring Instruments. International Atomic Energy Agency Report 16* (International Atomic Energy Agency, Vienna, 2000).

12. ISO, *ISO 4037: Parts 1–3*: Radiological protection - X and gamma reference radiation for calibrating dosemeters and doserate meters and for determining their response as a function of photon energy (International Organization for Standardization, 2019).

13. R. H. Olsher, H. H. Hsu, A. Beverding, et al., "WENDI: An improved neutron rem meter," *Health Physics* **79**, 170–181 (2000).

14. NCRP, *Protection Against Neutron Radiation. National Council on Radiation Protection and Measurements Report No. 38* (National Council on Radiation Protection and Measurements, Washington, 1971).

15. D. Rogers, "Why not to trust a neutron rem meter," *Health Physics* **37**, 735–742 (1979).

16. L. A. DeWerd and B. R. Smith, "Ionization chamber instrumentation," in *Radiation Therapy Dosimetry: A Practical Handbook*, edited by A. Darafsheh (CRC Press, Boca Raton, FL, 2021), Chapter 2, pp. 19–30.

17. Y. Uwamino and G. Fehrenbacher, "Radiation monitoring," in *Shielding Design and Radiation Safety of Charged Particle Therapy Facilities. PTCOG Report 1. Particle Therapy Cooperative Group*, edited by N. Ipe (Particle Therapy Cooperative Group, Villigen, 2010), pp. 176–200.

18. G. Knoll, *Radiation Detection and Measurement. Second Edition* (John Wiley & Sons, New York, 1989).

19. R. C. McCall and N. E. Ipe, "The response of survey meters to pulsed radiation fields," in *Seventh International Congress of the International Radiation Protection Association. Radiation Protection Practice Vol. 1* (International Radiation Protection Association, Sydney, 1988).

20. Mirion Technologies (Canberra) Inc., canberra.com [Online] (2017). Available at: http://www.canberra.com/products/hp_radioprotection/pdf/In1K-SS-C38987.pdf [Accessed 28 October 2018].

21. Thermo Fisher Scientific, *thermo.com/rmp.* [Online] (2007). Available at: http://www.thermo.com.cn/Resources/200802/productPDF_30096.pdf [Accessed 15 March 2020].

22. J. Turner, *Atoms, Radiation, and Radiation Protection* (Pergamon Press, New York, 1986).

23. J. Izewska and G. Rajan, "Radiation dosimeters," in *Review of Radiation Oncology Physics. A Handbook for Teachers and Students*, edited by E. Podgorsak (International Atomic Energy Agency, Vienna, 2005), pp. 71–99.

24. Landauer, Inc., *Landauer.com.* [Online] (2018). Available at: https://www.landauer.com/sites/default/files/product-specification-file/Luxel%2B_1.pdf [Accessed 15 March 2020].

25. J. Meier and S. C. Kappadath, "Characterization of the energy response and backscatter contribution for two electronic personal dosimeter models," *Journal of Applied Clinical Medical Physics* **16**(6), 423–434 (2015).

26. Thermo Fisher Scientific, thermofisher.com. [Online] (2019). Available at: https://www.thermofisher.com/order/catalog/product/EPD161081000 [Accessed 15 March 2020].

27. NCRP, *Structural Shielding Design and Evaluation for Megavoltage X- and Gamma-Ray Radiotherapy Facilities. National Council on Radiation Protection and Measurements Report No. 151* (National Council on Radiation Protection and Measurements, Bethesda, 2005).

28. AAPM, *Neutron Measurements around High Energy Radiotherapy Machines. American Association of Physicists in Medicine Report No. 19* (American Institute of Physics, Inc., New York, 1986).

29. ICRP, *Conversion Coefficients for Radiological Protection Quantities for External Radiation Exposures. International Commission on Radiation Protection Publication 116, Ann. ICRP 40(2–5)* (International Commission on Radiation Protection, Bethesda, 2010).

30. R. Nath, K. W. Price, H. R. Holeman, "Mixed photon-neutron field measurement," in *Proceedings of Conference on Neutrons from Electron Medical Accelerators. April 9–10, 1979. National Bureau of Standards Special Publication 554* (U.S. Government Printing Office, Washington, 1979).

31. A. O. Hanson and J. L. McKibben, "A neutron detector having uniform sensitivity from 10 keV to 3 MeV," *Physics Review* **72**, 673–677 (1972).

32. NCRP, *Neutron Contamination from Medical Electron Accelerators, National Council on Radiation Protection and Measurements Report No. 79* (National Council on Radiation Protection and Measurements, Bethesda, 1984).

33. I. O. Andersson and J. A. Braun, "A neutron rem counter with uniform sensitivity from 0.025 eV to 10 MeV in neutron dosimetry vol. II," in *Proceedings of an International Atomic Agency Symposium. IAEA STI/PUB/69* (International Atomic Energy Agency, Vienna, 1963).

34. M. Cosack and H. Leisecki, "Dose equivalent survey meters," *Radiation Protection Dosimetry* **10**(1–4), 111–119 (1985).

35. D. E. Hankins and J. R. Cortez, *Directional Response and Energy Dependence of Four Neutron Rem meters. Los Alamos Scientific Laboratory Report LA-5528* (Los Alamos Scientific Laboratory, Los Alamos, 1974).

36. N. E. Ipe, "Secondary radiation production and shielding at proton therapy facilities," in *Proton Therapy Physics, 2nd Edition*, edited by H. Paganetti (CRC Press, Boca Raton, 2018), pp. 207–240.

37. Thermo Fisher Scientific, www.thermofisher.com. [Online] (2018). Available at: https://www.thermofisher.com/order/catalog/product/FHT762WENDI2 [Accessed 15 March 2020].

38. M. Carseana, M. Helmecke, J. Kubancak, et al., "Instrument intercomparison in the high-energy mixed field at the CERN-EU reference field (CERF) facility," *Radiation Protection Dosimetry* **161**(1–4), 66–72 (2014).

39. A. Mitaroff and M. Silari, "The CERN-EU high-energy reference field (CERF) facility for dosimetry at commercial flight altitudes and in space," *Radiation Protection Dosimetry* **102**, 7–22 (2002).

40. A. Ferrari, P. Sala, A. Fasso, et al., FLUKA: *A Multi-Particle Transport Code. CERN Yellow Report CERN 2005-10. INFN/TC 05/11. SLAC-R-773*, (CERN, Geneva, 2005).

41. L. Jagerhofer, E. Feldbaum, D. Forkel-Wirth, et al. "Characterization of the WENDI-II REM counter for its application at MedAustron," *Progress in Nuclear Science and Technology* **2**, 258–262 (2011).

42. R. L. Bramblet, R. I. Ewing, T. W. Bonner, "A new type of neutron spectrometer," *Nuclear Instrument Methods* **9**, 1–12 (1960).

43. Bubble Technology Industries, Inc., bubbletech.ca. [Online] (2018). Available at: http://bubbletech.ca/product/rospec/ [Accessed 8 November 2018].

44. R. C. McCall, *Neutron Measurements. Stanford Linear Accelerator Center Publication 2262* (Stanford Linear Accelerator Center, Menlo Park, 1981).

45. Mirion Technology (GDS), Inc., *mirion.com.* [Online] (2014). Available at: https://mirion.app.box.com/s/u55f753y2qv1x940v5vf [Accessed 26 October 2018].

46. N. E. Ipe and D. D. Busick, BD-100: The Chalk River Nuclear Laboratories' Neutron Bubble Detector. *Stanford Linear Accelerator Center PUB 4398* (Stanford Linear Accelerator Center, Menlo Park, 1987).

47. N. E. Ipe, S. Roesler, S. B. Jiang, C. M. Ma, *Neutron Measurements for Intensity Modulated Radiation Therapy.* CD ROM Proc. of 2000 World Congress on Medical Physics and Biomedical Engineering, July 23-28. Also SLAC PUB 8433 (Stanford Linear Accelerator Center, Menlo Park, 2000).

48. Landauer, www.landauer.com. [Online] (2017). Available at: https://www.landauer.com/sites/default/files/product-specification-file/Neutrak_0.pdf [Accessed 15 March 2020].

49. N. E. Ipe, D. D. Busick, R. W. Pollock, "Factors affecting the response of the bubble detector BD-100 and a comparison of its response to CR-39," *Radiation Protection Dosimetry* **23**, 135–138 (1988).

50. Bubble Technology Industries Inc., bubbletech.ca. [Online] (2018). Available at: http://bubbletech.ca/product/bd-pnd/ [Accessed 26 October 2018].

51. H. Ing, R. A. Noulty, T. D. Mclean, "Bubble detectors – A maturing technology," *Radiation Measurements* **27**, 1–11 (1997).

52. B. J. Lewis, M. B. Smith, H. Ing, et al., "Review of bubble detector characteristics and results from space," *Radiation Protection Dosimetry* **150**(1), 1–21 (2012).

53. M. Schippers, "Proton beam production and dose delivery techniques," in *Principles and Practice, American Association of Physicists in Medicine. Medical Physics Monograph No. 37*, edited by I. J. Das and H. Paganetti (Medical Physics Publishing, Madison, 2015), pp. 129–164.

54. N. E. Ipe and R. C. McCall, The Stanford linear accelerator center pulsed X-ray facility. [Also SLAC-PUB-3966 Revised]. *Health Physics* **52**(4), 463–468 (1987).

55. IAEA, Radiological Safety Aspects of the Operation of Electron Accelerators. *International Atomic Energy Agency Technical Report Series No. 188* (International Atomic Agency, Vienna, 1979).

CHAPTER 14

Monte Carlo Techniques in Medical Physics

Ruirui Liu and Tianyu Zhao

Washington University School of Medicine
St. Louis, Missouri

Milad Baradaran-Ghahfarokhi

Washington University School of Medicine
St. Louis, Missouri
Vanderbilt University Medical Center
Nashville, Tennessee

CONTENTS

14.1 INTRODUCTION OF MONTE CARLO SIMULATION

Monte Carlo (MC) simulation, also known as random sampling technique or statistical experimental method, was first introduced in early 17th century when it was realized how to determine the probability of an event based on its frequency. Jacob Bernoulli concluded in his book, *Ars Conjectrandi*, that "this method was not new nor special," and he continued "everyone knows, in order to make prediction of one type of phenomenon, one time or two times observation is not enough, instead, much more times of observation is needed. The more the observation, the less

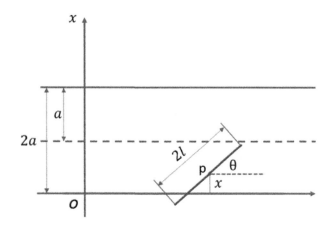

FIGURE 14.1 Position of the needle between the parallel lines.

risk that the goal fails" [1]. The modern form of MC simulation was introduced in the late 1940s by constructing a Markov chain [2] following the development of computers. It was firstly applied in developing of nuclear weapons [3]. In the following paragraph, an example is presented to explain the basic idea of MC simulation.

One well known example for illustrating how MC simulation works is taken from the work by Comte de Buffon (1777) [4], who proposed a MC-like method to determine π by constructing an experiment repeatedly tossing a needle onto a ruled sheet of paper and recording the probability of the needle crossing either of the lines.

Given the coordinate system as shown in Figure 14.1, the center of the needle x and the angle between the needle and the parallel lines θ follow a uniform distribution:

$$f_1(x)=\begin{cases} \dfrac{1}{a}, 0\leq x\leq a \\ 0, \text{ other} \end{cases} \tag{14.1}$$

and

$$f_2(\theta)=\begin{cases} \dfrac{1}{\pi}, 0\leq\theta\leq\pi \\ 0, \text{ other} \end{cases} \tag{14.2}$$

with $x\in[0,a]$ and $\theta\in[0,\pi]$.

The needle crosses the line when $x\leq l\sin\theta$, $0\leq x\leq a$, which can be described by a random variable $g(x,\theta)$,

$$g(x,\theta)=\begin{cases} 1, x\leq l\sin\theta \\ 0, \text{ other} \end{cases} \tag{14.3}$$

The probability of needle crossing the line is given by:

$$\begin{aligned} p &= \iint g(x,\theta)f_1(x)f_2(\theta)dx\,d\theta \\ &= \int_0^\pi \frac{1}{\pi}d\theta \int_0^{l\sin\theta} \frac{1}{a}dx \\ &= \frac{2l}{\pi a} \end{aligned} \tag{14.4}$$

Given the uniform distribution of x and θ, they can be sampled as multiplications of their range and two independent random numbers,

$$\begin{aligned} x &= a\xi_1 \\ \theta &= \pi\xi_2 \end{aligned} \tag{14.5}$$

where ξ_1,ξ_2 are random numbers in $(0,1)$. By the law of large numbers, the integral described in Equation 14.4 can be approximated by taking the arithmetic mean of independent samples of $g(x,\theta)$,

$$p_N = \frac{1}{N}\sum_{i=1}^{N} g(x_i,\theta_i) \tag{14.6}$$

Therefore π can be approximated by

$$\pi \approx \frac{2l}{p_N a}. \tag{14.7}$$

From perspective of probability theory, the idea of MC method is to approximate the true value by running a large number of sampling [5]. Using MC method, the integration could be taken as the mean of the random variable $g(r)$ which follows specific probability density function $f(r)$, which is

$$\langle g\rangle = \int g(r)f(r)dr \tag{14.8}$$

With N samples of parameter r, r_1, r_2, ..., r_N, estimated $\langle g\rangle$ is given by the arithmetic mean of $Z_i = g(r_i)$ as

$$\hat{Z}_N = \frac{1}{N}\sum_{i=1}^{N} Z_i \tag{14.9}$$

By the law of large numbers, $\hat{Z}_N$ converges to $\langle g \rangle$ if $Z_1, Z_2, \ldots, Z_N$ are independent from each other and have finite mean $(E(Z) < \infty)$,

$$\lim_{N \to \infty} P\left(\left| \frac{1}{N} \sum_{i=1}^{N} Z_i - E(Z) \right| < \varepsilon \right) = 1 \quad (14.10)$$

Equation 14.10 implies that, the estimated integration results in converges to a true integration, once the size of sample set is large enough [5].

According to the central limit theorem, if random variables $Z_1, Z_2, \ldots, Z_N$ are independent but follow the same distribution, we have

$$\lim_{N \to \infty} P\left(\frac{\sqrt{N}}{\sigma} \left| \hat{Z}_N - E(Z) \right| < x \right) = \frac{1}{2\pi} \int_{-x}^{x} e^{-\frac{t^2}{2}} dt \quad (14.11)$$

where σ is the standard deviation of variable Z given by

$$\sigma^2 = E(Z - E(Z))^2 = \int (Z(t) - E(Z))^2 f(t) dt \quad (14.12)$$

where $f(t)$ is the distribution of t. It is achievable to reduce the error between the true and estimated value below a desired level with a probability of $1-\alpha$, where α denotes the significance level.

$$P\left(\left| \hat{Z}_N - E(Z) \right| < \frac{x\sigma}{\sqrt{N}} \right) = \frac{1}{2\pi} \int_{-x}^{x} e^{-\frac{t^2}{2}} dt = 1-\alpha \quad (14.13)$$

If $\alpha = 0.05$, then the following inequality holds:

$$\left| \hat{Z}_N - E(Z) \right| < \frac{1.96\sigma}{\sqrt{N}} \quad (14.14)$$

It is worth noting that the error of MC method is a statistical error. Since σ in Equation 14.14 is not known, an estimated value of σ is typically used to quantify the statistical error. The estimated σ can be calculated as

$$\hat{\sigma} = \sqrt{\frac{1}{N} \sum_{i=1}^{N} Z_i^2 - \left(\frac{1}{N} \sum_{i=1}^{N} Z_i \right)^2} \quad (14.15)$$

14.2 APPLICATIONS OF MONTE CARLO SIMULATION IN MEDICAL PHYSICS

14.2.1 Application in Patient Dose Calculation

The MC method has been demonstrated to be a useful tool for dose calculation in radiation therapy. By modeling components of treatment units, radiation sources, and patient structures, MC simulation can accurately estimate the dose distribution in heterogeneous media, including patient geometry [6]. The main practical limitation of using MC simulation for patient dose calculation is the high demand for computational power to achieve acceptable accuracy in the calculated dose. Due to this limitation, MC calculations are mainly used for forward planning (forward dose calculation) [7] as the rapidly increasing computational burden from inverse planning always outpaces the improvement of central processing unit (CPU) performance. Recently, it has been shown that graphic processing units (GPUs) have great potential over conventional CPUs architecture to speed up the dose calculations. Instead of centralized streamline of computation with fewer cores responsible to tasks in a sequential order, GPU, which consists of thousands of small cores performing multiple tasks concurrently, is able to increase the speed of MC simulations dramatically [8].

The modern GPU is a highly data-parallel processor, which is optimized to run excessive floating point arithmetic output for simulations using a single program multiple data (SPMD) model [8]. On a single GPU, the SPMD model operates by launching thousands of threads running the same program (the so-called "the kernel") working on different data. GPU can rapidly switch between high numbers of different threads to ensure the hardware is busy at all the processing times, which can significantly hide memory latency. On the other hand, using CUDA™ (Compute Unified Device Architecture) programs, which are written based on the C programming language, to employ the parallelism of the GPU can provide very fast implementations of standard mathematical functions to calculate patient dose [9].

For photon dose calculations, Badal et al. introduced a GPU-based MC code that simulates photon transport in a voxelized geometry [9]. For their simulations, the accurate physics models were developed from PENELOPE MC package using the CUDA™ programming model (NVIDIA Corporation, Santa Clara, CA). They reported up to 27-fold speedup over a single core CPU for the same calculations.

Fang et al. have presented a parallel MC algorithm accelerated by GPU for modeling time-resolved photon migration in arbitrary 3D turbid media [10]. Similar to Badal et al. the code was implemented in CUDA, and benchmarked under various parameters, such as selection of random number generation (RNG), memory access pattern and thread number. Using a low-cost graphic card in their method, a speed-up ratio over 300 has been reported with 1792 parallel threads over a conventional CPU. In a study on the development of an accurate and efficient dose calculation engine for patient dose calculation designed for online adaptive radiotherapy performed by Gu et al., a finite size pencil beam (FSPB) algorithm was created on GPU with a 3D-density correction method [11]. The developed algorithm was tested for 10 intensity-modulated radiation therapy (IMRT) treatment plans (5 head-and-neck cases and 5 lung cases). Better than 2% of accuracy was reported for the majority of the calculation voxels while the dose calculation was accomplished well within 1 second. Jia et al. have studied their GPU-based MC calculation package for coupled electron–photon transport [12]. Their results demonstrated speedup factors of about 5.0–6.6 times with an NVIDIA Tesla C1060 GPU card over a 2.27 GHz Intel Xeon CPU processor in the range of energies used typically in radiotherapy. They also have discussed about the factors that adversely restrict the efficiency of such simulations, such as the memory access pattern and no cashing of global memory on GPU. Men et al. proposed a GPU-based ultrafast IMRT plan optimization for online adaptive radiotherapy techniques in order to handle the inter-fraction variation of the patient's geometry [13]. Using an NVIDIA Tesla C1060, they achieved speedup factors of 20–40 with accuracy compared to the simulation running on an Intel Xeon 2.27 GHz CPU.

For low-energy brachytherapy, MC simulation has been used widely to determine the dosimetric parameters for various source designs [14]. Compared to the deterministic formalisms which usually take assumptions simplifying patient anatomy, heterogeneity, fiducial seeds orientation and inter-seed attenuation, MC simulation has demonstrated excellent accuracy in the calculated dosimetry [15]. Hissoiny et al. evaluated the capability of GPU-based MC calculations which ultimately resulted in fast and accurate results for routine planning [16]. The study was able to reproduce the dosimetric parameters of the radiation sources to within 1.25% from other literature, depending on the seed type, material, or radionuclide.

GPU-based fast dose calculation in magnetic fields of a MRI–Linac (linear accelerator) has also been studied

by Hissoiny et al. [17]. It was found that influence of Lorentz force from the magnetic field acting on the charged particles, i.e., electrons, should be taken into consideration by the dose calculation engine in the planning stage. According to the Hissoiny et al. GPU-based MC calculations agreed well with measurement on a 2%/2 mm gamma analysis. Moreover, execution time of less than 20 seconds was achieved in a prostate case phantom for a 2% statistical uncertainty, indicating that GPU-based MC calculations is a promising candidate for accurate and fast dose calculations for the hybrid MRI–Linac modality.

On the patient-specific CT/CBCT imaging dose calculation, since patient-specific imaging dose may be proposed for the purpose of dose management, Jia et al. have successfully developed a MC dose calculation code on GPU architecture on the NVIDIA CUDA platform for fast and accurate estimation of the x-ray imaging dose received by a patient during a CT or CBCT scan [18]. They found speedup of about 400 and 76.6 times over EGSnrc MC code when the simulations were performed in a homogeneous water phantom and Zubal phantom, respectively. Despite the inhomogeneities in the Zubal phantom, imaging dose calculation was reported to complete in about 17 s with statistical uncertainty of 0.4%.

For accurate dose calculation in clinical practice, commercial treatment planning systems (TPS) implementing MC dose calculation are available. Cygler et al. implemented clinically a commercial MC-based TPS for electron beams. The software they used was developed by MDS Nucletron based on Kawrakow's VMC++algorithm [19]. The MC module was integrated with Theraplan Plus™ TPS [20]. The adoption of MC for dose calculation has increased over the last decade in some more popular commercial TPSs. Varian's Eclipse electron MC (eMC) deals with electron transport using MC calculations [21]. Elekta's Monaco implements photon and eMC algorithms for patient's dose calculation and Monaco beam models for the Versa HD Linac for routine clinical use [22]. RayStation TPS uses MC simulation for both electron [23], and proton dose calculations [24].

14.2.2 Application in Dosimetry Instruments

MC simulation of dosimetry instruments, i.e., cavity ion chambers, plays a central role in the calculations of ionizing radiation transport in dose determination.

Moreover, beyond different generations of radiation dosimetry protocols, MC techniques have been shown to be the essential part on improving the accuracy of the calculated absorbed dose. Consequently, a large amount of research work has been devoted to these specific applications of MC simulation packages.

In clinical practice, the determination of absorbed dose is performed using a cavity ion chamber placed in water phantom, based on a reference dosimetry setup which is defined by the geometry of interest and radiation source [25, 26]. Over the last decades, efforts have been made to improve the accuracy (0.1% or better) of MC methods to calculate the response of an ion chamber in a radiation beam according to the dosimetry protocols [27–29]. For each dosimetric situation, this level of accuracy can be obtained by assessing different correction factors using MC simulation. Among the dosimetry related factors, beam quality conversion factor has the largest source of uncertainty [26]. Muir et al. performed MC simulations to calculate beam quality conversion factors for large sets of ion chambers used clinically in radiation therapy [30]. They found excellent agreement (up to 0.13% for the entire data set) between measurement and MC calculated factor. This exceptional agreement observed in their study provided great confidence in adopting MC calculated factor for updated dosimetry protocols. Buckley and Rogers focused upon the wall correction factor for thimble ionization chambers [31], which accounts for the dosimetric disturbance from the material difference between the chamber wall and the surrounding phantom, where the Bragg–Gray assumption could break down. MC calculation as an invaluable tool for such investigations showed that, for a cylindrical Farmer type chamber with graphite wall, the factor was significantly sensitive (up to 2.5%) to the depth for a 6 MeV electron beam [31].

Using MC simulations, Bouchard et al. presented a calculation method of dose uncertainties caused by introducing error deliberately, including machine isocenter and detector position in the coordinate system of the dosimetry phantom [32]. This proposed MC method can be beneficial during radiation beam calibration, determination of beam quality correction factors as well as the measurements of output factors.

Some investigators used MC to model response of diode detectors to radiation in dosimetry protocols [33–35]. Compared to the ionization chambers, doped silicon in diodes is much smaller in size with higher sensitivity

to radiation due to its high density, which is approximately 1800 times higher than that of air in the sensitive volume of ion chambers [36]. Using EGSnrc MC code, Wang and Rogers, studied the depth response of diode detector response in both 6 and 18 MeV electron beams [37]. Their results showed that, the water/diode dose ratio does not depend on field size and energy for the broad incident electron beam. Such MC simulations have allowed studies to be performed in dosimetric situations that could not be done by other means, with the same level of accuracy.

Cranmer-Sargison et al. have implemented a MC-based small field dosimetry formalism for a comprehensive set of diode detectors, including IBA (IBA Dosimetry GmbH, Schwarzenbruck, Germany) stereotactic field diode as well as the PTW (PTW The Dosimetry Company, Freiburg, Germany) T60008, T60012, T60016, and T60017 field diodes [36]. Based on their simulations, calculated output factors in water and output ratios at various depths were sensitive to source parameterization in small filed dosimetry.

In another study conducted by Crop et al., the effects of small field sizes, position in penumbra, spot size and measurement depth on perturbation factors such as central electrode, wall, air–wall interface and volume for micro-ionization chambers were investigated [38]. Their results showed that small electron spot size on the target and sharp penumbra led to larger variation for the perturbation factors.

Francescon et al. employed BEAMnrc MC code to calculate total scatter factors in CyberKnife radiosurgery system. In their work, simulations were also performed to calculate correction factors for the microchambers by means of a consistency check between measurement and simulation. They stated that, the MC method represented an "ideal" dosimeter for such studies, since it simulated energy deposition per each radiation particle in a given material and their results could be applied to the Cyberknife radiosurgery systems in clinical practice [39, 40].

14.2.3 Application in Microdosimetry

From the beginning, the aim of microdosimetry has been the study of the stochastic nature in the energy deposition by ionizing radiation and its consequences in the biological response mechanism. The damage induced by radiation in biological structures, such as cells which is in the order of micron scale, depends crucially on the

TABLE 14.1 The Major Characteristics of Different MC Track Structure Codes in Microdosimetry

Code	Interactions	Particle Type	Energy Range
KURBUC	Elastic scattering, excitation, ionization	Proton, alpha	10 eV–10 MeV
PARTRAC	Elastic scattering, excitation, ionization	Electron, photon, and ions	10 eV–10 MeV
NOREC	Elastic scattering, Excitation, ionization	Electron	7.4 eV–1 MeV
Geant4-DNA	Elastic scattering, excitation, ionization	Electron, proton, alpha, and light ions	0 eV–10^6 MeV

spatial distribution of energy deposition events generated by the energetic ions or secondary electrons. Modeling the radiation-induced effects in a microscopic volume, such as cellular structure and DNA molecule, is the domain of microdosimetry or nanodosimetry, where the two terms being often used interchangeably [41]. Microdosimetric quantiles, which describes the energy deposition information in microscopic volumes, are used to explain the model differences of the radiation-induced effects on biological targets at the same average energy deposited per unit mass. Due to the capability of modeling the stochasticity of energy deposition of radiation in cell, MC methods are used to simulate the microdosimetric quantities, such as linear energy and specific energy. Linear energy is the quotient of the energy imparted to the volume, by a single particle or radioactive decay inside it, and the volume mean chord. The specific energy is defined as the energy imparted divided by the mass of the specific region [42]. Both linear energy and specific energy are stochastic variables and commonly are described using probability distribution. MC codes can be coupled to models that describe cell damage and biological effects caused by radiation based on microdosimetric quantities [43]. A MC code addressing this research domain should be capable of simulating the particle track structure over track lengths in the order of nanometers which is compatible with the DNA size. This requirement implies that the cut off energy should go down to 10–100 eV to take soft energy-loss collision into account. It requires detailed knowledge of the electronic properties of the target, to ensure the correct treatment of energy loss occurrences of a magnitude comparable to electronic binding energies [41].

One of the important applications of microdosimetry simulation is quantifying the radiation-induced DNA damage. Here we simply introduce some popular simulation tools for quantifying DNA damages. Event-by-event MC codes such as KURBUC [44], PARTRAC [45], NOREC [46], and Geant4-DNA [47] represent the preferred method of computing distributions of microdosimetric quantities for quantifying radiation-induced

DNA damages. The major characteristics of these codes are tabulated in Table 14.1.

KURBUC was developed for water vapor medium and can handle light ions such as protons (1 keV to 1 MeV) and alpha particles (1 keV to 8 MeV). In addition, the free radical species from water radiolysis in the chemical stage are included in this code. NOREC is a code modified from the original OREC (Oak ridge electron transport code) with updated electron elastic scattering cross sections. The chemical track remains similar as KURBUC but there is no biological phase simulation [48]. PARTRAC was originally developed for track structure of electrons, but the photon and ion can be tracked in the simulation now. The chemical track includes the formation of diverse chemical species similar to KURBUC and NOREC. A DNA double-strand break (DSB) repair module, based on non-homologous end-joining (NHEJ), has been added in the code. Recently, with the new developments of PARTRAC code, heterochromatin and euchromatin structures were modeled to account for variations in DNA damage yields, complexity and repair between these regions. The mitochondrial DNA damage also can be simulated using PARTRAC [49, 50]. Last but not least, Geant4-DNA is an open source simulation toolkit that aims to model the effects of radiation on biological systems at cellular and DNA levels. Geant4-DNA includes physical, chemical and biological models to simulate cellular and subcellular damage induced by ionizing radiation. Geant4-DNA introduces a new set of physical models to increase the definition of physical models at lower particle energies, which is quite beneficial to the DNA damage quantification. Currently, Geant4-DNA handles the transport of electrons, protons, hydrogen nuclei, alpha nuclei (with various charges) and four heavier ions (carbon, nitrogen, oxygen, and iron ions).

Those simulation codes have been adopted as investigation tool in many microdosimetry applications, such as modeling cellular dose, radiation-induced DNA damages of cell, study of radiosensitization of gold nanoparticle aided radiotherapy, etc.

Stewart et al. [51] calculated the single-event distributions, site-hit probabilities and the frequency-mean specific energy per event of a volume in size to a mammalian cell or cell nucleus using PENELOPE. Rollet et al. [52] have calculated frequency distributions of deposited energy in a mini TEPC (with sizes equivalent to 1 and 2 μm) due to photons using the FLUKA MC code. They found a reasonably good agreement between the simulated data with those obtained experimentally. Using Geant4, Cortés-Giraldoand Carabe [53] calculated dose-averaged linear energy transfer (LET) maps in voxelized geometries irradiated with proton beams. They carried out microdosimetry calculations with the aim of deriving reference dose-averaged LET values from microdosimetric quantities. With suitable scoring method corresponding to a step-by-step computation of LET, their simulation gave the best agreement with estimations from microdosimetry calculations on dose-averaged LET.

Due to the fast development of Geant4-DNA project and its open-source oriented characteristics, many works have been done to quantify DNA damages using Geant4-DNA [54–56]. Besides the Geant4-DNA physics, it is worth noting that there are also other physics models for track structure simulation, such as the models discussed in papers [57–60], which also have wide applications in the calculation of DNA damages. In 2011, Francis et al. [61] simulated the DSB using a clustering algorithm DBSCAN (Density-Based Spatial Clustering of Applications with Noise) based on the energy deposition information in cell nucleus obtained by Geant4. Since then, DBSCAN quantification of DNA DSB yield, based on Geant4 radiation transport simulation, has become a popular method in modeling radiation-induced DNA damages [61]. Liu et al. [62] have developed a computational model for quantifying cell dose and DSB number in a multicellular system by simulating the radiation transport in 2D and 3D cell culture. The radiation transport simulation for cells was conducted using Geant4 package with the Geant4-DNA physics to obtain the cellular dose and cellular DSBs yield. Using the method they developed, it is possible to obtain the cellular dose and DNA damage simultaneously. Liu et al. [63] investigated the radiosensitization properties of gold nanoparticles (GNPs) using a simple Geant4 cell model considering a realistic cell geometry and a clustering algorithm to characterize the number of DSBs. In their model, a mixed-physics approach was taken for accurate modeling of low energy photon interactions in different regions of the model employing Geant4-DNA physics within the cell, and Livermore physics within gold, respectively. DBSCAN is used to directly quantify DNA DSBs after irradiation. They obtained the radiosensitization factors of GNP with different sizes and different distances from the cell nucleus. Several carbonatizations of these two conditions were obtained by using their simulation model.

14.2.4 Application in Nuclear Medicine

MC method has been widely used in nuclear medicine [64, 65] with a broad range of applications, such as generating reference S-values in MIRD (Medical Internal Radiation Dose) models, etc. [64]. MC method was initially used for reference dosimetry based on reference models proposed by ICRP or the MIRD committee. The dosimetric parameters, i.e., S-value, were calculated in some dosimetry software like MIRDOSE [66] or OLINDA [67]. In the Oak Ridge National Laboratory, Snyder et al. firstly used MC method to assess the fraction and electron energy emitted from radionuclides in source tissues, deposited in target tissues [68, 69]. According to these simulations, with the anatomic information provided by MRI or CT the three dimensional voxelized patient phantom could be obtained, and with MC simulation the voxel dosimetry introduced in MIRD Pamphlet No.17 could be achieved [68]. Using MC method, the patient dose in radionuclide therapy could be accurately calculated with voxelized phantom or mesh-type computational phantom [70].

14.2.5 Applications for X-ray Imaging

In diagnostic radiology, the use of the MC method to simulate radiation transport has become the most accurate means of predicting the diagnostic x-ray spectra even in complex geometries, owing to incorporation of proper interaction cross section data and more accurate physics modeling. Ay et al. have used MCNP MC code to study x-ray spectra in diagnostic radiology and mammography [71]. They focused on the simulation of various target and tube filter combinations to investigate the effect of tube voltage, target material and filter thickness. They stated that, although using MC to derive x-ray spectra was time consuming, it could provide detailed information about particles' interaction with different parts of a diagnostic x-ray tube, including target and filter. They recommended using MC for x-ray tube design and development of new target/filter combinations to improve image quality in diagnostic radiology and mammography [71]. On the other hand, considering the patient dose in diagnostic radiology, depending on both the attenuation characteristics of the absorbing media and the energy spectrum of the incident x-rays, correction factors are needed to convert dose from model-based dose calculations to absorbed dose-to-medium. Pawlowski et al. have presented a method to accurately account for the medium-dependent effect in model-based dose calculations for diagnostic x-ray [72]. They used MC simulation to derive empirical medium-dependent correction factors as a function of the amount of bone that an x-ray beam must penetrate to reach a voxel. According to their results, their approach resulted in mean dose errors of less than 3% when tested for their patients, providing a good alternative accurate dose calculation method for kV x-rays.

Computed tomography (CT) is a high-dose diagnostic technique that accounts for approximately 3–5% of all diagnostic examinations. However, it has been reported to be responsible for 34–45% of the total collective dose from all diagnostic procedures in western countries [73, 74]. Therefore, calculation of patient CT dose is important, in order to estimate the patient risk for stochastic effects, mainly carcinogenesis, and to optimize the scanning protocols. Avilés Lucas et al., have developed MC computational model of CT to investigate the effect of various physical factors such as bowtie filter as well as the size, shape and position of a patient phantom on the surface air kerma length product, the peak surface air kerma, the air kerma length product within a phantom and the energy imparted [75]. Based on their MC simulations, they highlighted the importance of calculating doses to organs by taking into account their size and position within the CT gantry.

In interventional radiology using fluoroscopically guided procedures, there have been several investigations dealing with estimation of the dose to patient skin and internal organs [76]. Although various methods including direct and indirect measurements have been examined, MC simulations have been shown to provide accurate results for the calculations of the dose to skin and other sensitive organs.

In radiotherapy practice, for complex conformal and IMRT, electronic portal imaging devices (EPIDs) are well proven as important tools for geometric verification [77]. They also provide near-instantaneous important information for 2D mapping of patient position and dosimetric verification of treatments. Chin et al. have developed such a MC-based tool of EPIDs using the BEAMnrc/DOSXYZnrc code [78]. Their model showed good agreement with measured data for an inhomogeneous phantom, where the root-mean-square (RMS) of the difference was under 2% within the field. This MC-based tool was capable of handling arbitrary gantry angles, voxel-by-voxel phantom description and realistic particle transport throughout the geometry. Cone beam CT (CBCT), commonly installed on Linacs for image guided radiation therapy, adds additional imaging dose to the prescribed radiation dose given to patients. Although, the AAPM Task Group 75 report makes recommendations on estimating and reducing imaging dose from CBCT [79], calculation of the imaging dose during treatment planning would allow for a better prediction of the total dose to target and organs at risk. Downes et al. have used MC simulations to calculate patient dose for a kilovoltage CBCT unit [80]. Based on the MC results, they reported that the 3D dose distribution from CBCT in patient was complex. They suggested using full 3D MC calculation, instead of employing numerical factors, to estimate CBCT exposure, especially at regions at the center of the body.

14.3 MONTE CARLO CODES AND SYSTEMS IN THE PUBLIC DOMAIN

Several general-purpose MC codes have been developed for radiation transport calculation, which are commonly used in medical physics, such as, Geant4, MCNP, EGS4/EGSnrc, PENELOPE, and FLUKA. In this section we briefly introduce these MC codes.

3.1 Geant4

Geant4 is a freely available software for performing MC simulations of the interactions of energetic particles in matter [81]. Geant4 is an acronym for "GEometry ANd Tracking." The Geant4 was originally developed by CERN for the purpose of simulating the interaction of high energy particles in large accelerators. The physics process covers a comprehensive range, including electromagnetic, hadronic and optic processes, a large number of particles, materials and elements, over a wide energy range starting from a few eV to TeV [81]. A variety of cross sections are used in Geant4, for instance, EPDL97, EEDL, and EADL for low energy electromagnetic interactions. More detailed information about the physics models used in Geant4 could be referred to their official report [82]. Since its development Geant4 has continued to evolve and expand. The original toolkit implemented methods for handling the fundamental aspects of physical simulation: geometry, materials, particle interactions, particle generation, and particle tracking. Moreover, it is capable of generation and storing event and tracking data, and detector tracking visualization. From these building blocks, the code has expanded widely. Its object-oriented nature allows the end-user to make use of the existing framework by instantiating useful class systems while maintaining a unique and personal application framework. Users can build their own simulation code by using the user's classes provided in Geant4. Since most of the users' classes are designed in base classes, users can directly instantiate the classes for their usage. There are two kinds of user classes including user initialization classes and user action classes. User initialization classes are used during the initialization phase (simulation initialization), while user action classes are used during the simulation (invoked actions for controlling radiation transport). User initialization should call directly G4RunManager class by invoking SetUserInitialization function, while user classes should be defined in G4VUserActionInitialization class. The three necessary initialization classes and five optional user action base classes are listed in Table 14.2 with their description.

Because of its advanced functionality, Geant4 has become a popular simulation toolkit in various domains in the field of medical physics [83]. Geant4 is used in radiotherapy simulations, such as external beam treatment [84–86], brachytherapy [87, 88], and hadron therapy [89–91]. For modeling the patient dose in radiotherapy, Geant4 allows the modeling of complex geometries of treatment equipment and also the patient anatomy by constructing voxelized phantom based on DICOM images [92]. Geant4 also has been used in internal dosimetry simulations [93, 94], and radiodiagnostic applications [95]. In recent years, many simulation applications have been developed for radiation biology studies using Geant4. An international collaborative projects, the so-called Geant4-DNA, has been launched to build an experimentally validated simulation platform for the modeling of DNA damage induced by ionizing radiations, with the help of modern computing tools and techniques [96]. Geant4-DNA, an extension to Geant4 developed under the framework of Geant4-DNA project, which is an open-source simulation toolkit that aims to model the effects of radiation on biological systems at cellular and DNA levels. Geant4-DNA includes physical, chemical and biological models to simulate cellular and subcellular damage induced by ionizing radiation. Geant4-DNA introduces a new set of models to increase the resolution of physical modeling at lower particle energies. Certain processes handled vaguely by standard physics list in Geant4, are treated by a model with increased details. Very low energy electron electromagnetic processes have been introduced, which is quite beneficial to the DNA damage quantification since DNA damages are largely induced by the deposited energies of low energy electrons. More fundamental development and applications of Geant4-DNA were presented in papers [97, 98]. Due to the fast development of Geant4-DNA project and its open-source oriented characteristics, it has been used by many investigators to

TABLE 14.2 Base Classes Used for Building the Geant4 Simulation

Class Type	User Classes	Functionality of Classes
Necessary initialization classes	G4VUserDetectorConstruction	Defining the material and geometrical setup of the detector. Several other properties, such as detector sensitivities and visualization attributes, are also defined in this class.
	G4VUserPhysicsList	Defining all the particles, physics process and cut off parameters.
	G4VUserPrimaryGenerationAction	Generating the primary vertices and particles.
Optional user action base classes	G4UserRunAction	For actions at the beginning and end of every run.
	G4UserEventAction	For actions at the beginning and end of every event.
	G4UserStackingAction	For customizing access to the track stacks.
	G4UserTrackingAction	For actions at the creation and completion of every track.
	G4UserSteppingAction	For customizing behavior at every step.

Source: From [81].

quantify DNA damages [54, 56, 62, 63, 99–101]. Besides the Geant4-DNA physics, it is worth mentioning that there are also other physics models for track structure simulation, such as the models discussed in [57–60], which also have wide applications in the calculation of DNA damages.

There are also some MC simulation packages that are based on Geant4, such as TOPAS (TOlkit for PArticle Simulation) [102] and GATE [95]. The radiation transport simulations in TOPAS and GATE all depend on Geant4, but they have their own methods to implement simulations. TOPAS was originally designed to create a specialized user-friendly tool for proton therapy, and it has been quite widely used in medical physics simulations [103, 104]. TOPAS also has extended its functionality to conduct radiobiology simulation with TOPAS-nBio toolkit [105]. GATE also has wide applications in radiotherapy [86, 106].

14.3.2 MCNP

The MCNP (initially stood for MC Neutron Photon, now stands for Monte Carlo N-Particle transport) code system developed by Group X-6 is the workhorse at the Los Alamos National Laboratory for neutron, photon, and coupled neutron-photon calculations using the MC method [107, 108]. MCNP's history could be dating back to the formulation and first uses of the MC method at Los Alamos during the 1940s by Fermi, Metropolis, Richtmyer, Ulam, and von Neuman [107]. The MCNP is, by far, one of the most widely used MC codes that has been available to the radiation physics community. Most of its users were initially connected to military projects and nuclear power engineering and industry [109]. The cross-section data for photons and neutrons are taken from several sources, some of them dating back to 1970s, expanded with the evaluated data from ENDF, and further completed with the EPDL for high energies. The energy ranges of simulated particles are from 10^{-5} eV to 20 MeV for neutrons with data up to 150 MeV for some nuclides, 1 keV to 1 GeV for electrons, and 1 keV to 100 GeV for photons. In the recent version, MCNP6 includes considerable developments in low-energy electron and photon transport, aiming at simulations down to energies in the few eV range [109]. MCNP has been used successfully to solve many problems in the field of medical physics, and an extensive review is given by Solberg et al. [110]. In radiotherapy, MCNP has been used in many applications, such as in brachytherapy [111, 112],

patient dose calculation [113, 114], internal dosimetry [115, 116], hadron therapy [117, 118]. MCNP is also been used in simulating the microdosimetry of cell, and DNA DSB in order to quantify the relative biological effectiveness (RBE) for photons, neutrons, and light ions [119].

14.3.3 EGS4/EGSnrc

The EGS (Electron Gamma Shower) MC code is a general purpose charged-particle transport simulation system which was originally developed at Stanford Linear Accelerator Center (SLAC) to simulate higher energy electromagnetic cascades, and it was used in large accelerator centers for many years (SLAC, CERN, etc.) [109]. ESG was initially used for high energy electron simulation, later the low-energy electron transport in EGS was improved by inclusion of user-defined restrictions in step sizes. This resulted in a more accurate EGS4 version which added electron transport down to 1 keV and Rayleigh scattering to original photon physics in the code [109]. EGS system evolved into two independent systems, EGSnrc maintained by the National Research Council of Canada (NRCC), and EGS5 maintained by the Higher Energy Accelerator Research Organization (KEK, Japan). EGSnrc is a general purpose MC package for simulation of coupled electron-photon transport in matter, with energy range from 1 keV to 10 GeV [27, 120]. EGS primarily uses cross-section libraries such as EPDL97 [121] and XCOM [122] for photon and electron radiation transport simulation. A very detailed history about EGS system development was given in a report by SLAC [123]. After the release of EGS, it has been used in a wide variety of applications, particularly in medical physics, radiation measurement studies, and industrial development [123]. In medical physics, EGSnrc is one of the most widely used packages in radiation therapy. Particularly, several MC packages were developed based on ESGnrc, such as BEAMnrc for modeling radiotherapy sources [124], DOSXYZnrc for calculating dose distributions in a rectilinear voxel phantom [125]. BEAMnrc can produce a phase-space output of the beam at any specified plain in geometry, and the output data calculated by BEAMnrc can be used as input file for DOSXYZnrc. EGSnrc also has been used in brachytherapy dose calculations for clinical applications [126, 127].

14.3.4 PENELOPE

PENELOPE was originally developed for low energy radiation, and it is maintained at the University of

Barcelona (Spain), and its name is an acronym for "PENetration and Energy Loss of Positrons and Electrons." It simulates electron-photon showers in the energy range from 50 eV to 1 GeV [109]. PENELOPE uses photon cross-section data from the EPDL97 [128]. In PENELOPE, photon transport is simulated by means of the standard, detailed simulation scheme. Electron and positron histories are generated on the basis of a mixed procedure, which combines detailed simulation of hard events with condensed simulation of soft interactions [129]. The main application of PENELOPE includes radiotherapy and nuclear medicine, radiation dosimetry, electron microscopy (SEM, electron-probe microanalysis), detector response, x-ray generators, etc. [130]. In radiotherapy simulations, PENELOPE has been used to conduct Linac simulations [131, 132]. PENELOPE also could be used in simulating the positron emission tomography (PET) imaging [133]. The ability to transport electrons and positions to energies as low as 100 eV makes PENELOPE a suitable MC simulation system for microdosimetry [51].

14.3.5 FLUKA

The FLUKA code is a general purpose MC code for simulation of the interactions and transport of hadrons, heavy ions, and electromagnetic particle from few keV to cosmic ray energies in whichever material [134]. It was developed originally at CERN in the 1960s, its name is an acronym for "FLUktuierende KAskade" [109], and it is used extensively at CERN for all beam-machine interactions, radioprotection calculations and facility design of forthcoming projects [135]. Beside its hadron physics, FLUKA can also handle electromagnetic interaction simulations with a wide energy range: the program can transport photons and electrons from 1 PeV down to 1 keV [134]. FLUKA uses EPDL97 library for photon electron shower [136]. FKUKA is widely used for studies related to both basic research and applications in radiation protection and dosimetry, radiotherapy [137], radiobiology [138], cosmic ray calculations [139].

14.4 LIMITATIONS OF MONTE CARLO SIMULATION

Although MC method can model complex systems by providing flexible and exceptionally powerful techniques, it is important to consider the limitations that constrain its use for simulation of radiation transport. Simulation accuracy of the physics model is one of the most challenging aspects of any MC model. Moreover, having chosen the proper physics model and having finalized the geometry simulation of the problem, appropriate cross-section tables must be selected and assigned to each material in the simulation geometry. Generally, if any of these steps were not performed in detail with reasonable accuracy, the MC method might not be considered as the golden method for dose calculation. Furthermore, in clinical applications, for radiation transport problems with complex geometry and varying material properties, MC-based dose quantification requires long computational times. In other words, detailed simulation of the problem at hand may significantly decrease the efficiency of a calculation. Therefore, there is a compromise between accuracy and efficiency of any MC simulation. For years, MC methods have been considered to be a rather sluggish technique compared with analytical or deterministic methods. Nonetheless, more advanced computing power, aiming at decreasing computational time and simulating realistic geometries, have been recently accessible. Moreover, as stated earlier, developing the general-purpose graphics processing unit (GPU) has been found to be a promising solution by tremendously increasing the speed of MC simulations.

On the other hand, all-purpose MC codes such as MCNP, Geant4, EGS4, and FLUKA usually require either extensive general programming knowledge (for Geant4) or detailed understanding of their specific programming rules (for MCNP, EGS4, etc.), and are not suited well to the inexpert user for medical physics applications. Whereas, MC codes can be much easier to use, when packaged with a user-friendly simulation environment. In this regard, graphical user friendly MC packages such as TOPAS, and GATE can be a valuable tool on making simulation available to the inexperienced programmer, thus facilitating MC use in radiation therapy research.

14.5 SUMMARY

MC simulations are an exceptionally powerful tool not only in external beam radiotherapy but also for modeling brachytherapy and diagnostic imaging systems. In this chapter, we discussed and highlighted the capabilities of Geant4, MCNP, EGS4/EGSnrc, PENELOPE, and FLUKA code for modeling and analyzing radiation transport for clinical applications. Generally, majority of the MC studies in medical physics have been conducted using the MCNP and EGS/EGSnrc-based codes (including the user interface BEAM). However, other codes such as PENELOPE and Geant4 are experiencing an increasing use. Main limitations of MC methods are the lack of efficiency of calculation and storage of large amounts of data. Meanwhile, considering more advanced computing power and also generation of GPU-based desktop computers that are less expensive compared to cluster of CPUs, these limitations are now less of an issue. For implementation of the new technology in radiotherapy and diagnostic imaging, MC-based techniques can play a critical role as speed and accuracy will be of the essence. In the near future, MC methods may replace conventional analytical dose calculation techniques for patient treatment planning and also may be used to face new challenges in patient treatment, such as LET-based dose painting in proton therapy, designing complex beam delivery techniques for modulated electron therapy, and design and dosimetry in MR–Linac systems. These advances in cancer treatment require implementation of commercial MC-based TPS as they provide significant improvement in dose calculation accuracy over the conventional algorithms.

For the studies on detector design and relative dosimeters, for their dose response in advanced complex dose delivery schemes, MC simulations will continue to play an essential role in all these aspects of dosimetry. According to the previous publications, the MC method is the only approach that can accurately model complex dose delivery systems.

REFERENCES

1. L. Pei and X. Zhang, *Monte Carlo Method and its Application in Particle Transport* (Science Press, Beijing, 1980).
2. W. R. Gilks, *Markov Chain Monte Carlo in Practice* (Chapman and Hall/CRC, 1995).
3. A. F. Bielajew, "History of Monte Carlo," in *Monte Carlo Techniques in Radiation Therapy* (CRC/Taylor & Francis, Boca Raton, 2013), pp. 3–15.
4. M. Aigner, G. M. Ziegler, M. Aigner, G. M. Ziegler, "Buffon's needle problem," in *Proofs from THE BOOK* (Springer, Berlin Heidelberg, 2014), pp. 175–178.
5. X. Shuyan, *Application of Monte Carlo Method in Experimental Nuclear Physics* (Atomic. Energy Press, Beijing, 2006).
6. C. M. Ma, et al., "A Monte Carlo dose calculation tool for radiotherapy treatment planning," *Physics in Medicine and Biology* **47**(10), 1671–1689 (2002).
7. E. Alerstam, T. Svensson, S. Andersson-Engels, "Parallel computing with graphics processing units for high-speed Monte Carlo simulation of photon migration," *Journal of Biomedical Optics* **13**(6), 060504 (2008).
8. E. Alerstam, W. C. Yip Lo, T. D. Han, J. Rose, S. Andersson-Engels, L. Lilge, "Next-generation acceleration and code optimization for light transport in turbid media using GPUs," *Biomedical Optics Express* **1**(2), 658 (2010).
9. A. Badaland A. Badano, "Accelerating Monte Carlo simulations of photon transport in a voxelized geometry using a massively parallel graphics processing unit," *Medical Physics* **36**(11), 4878–4880 (2009).
10. Q. Fang and D. A. Boas, "Monte Carlo simulation of photon migration in 3D turbid media accelerated by graphics processing units," *Optics Express* **17**(22), 20178 (2009).
11. X. Gu, U. Jelen, J. Li, X. Jia, S. B. Jiang, "A GPU-based finite-size pencil beam algorithm with 3D-density correction for radiotherapy dose calculation," *Physics in Medicine and Biology* **56**(11), 3337–3350 (2011).
12. X. Jia, X. Gu, J. Sempau, D. Choi, A. Majumdar, S. B. Jiang, "Development of a GPU-based Monte Carlo dose calculation code for coupled electron-photon transport," *Physics in Medicine and Biology* **55**(11), 3077–3086 (2010).
13. C. Men, X. Jia, S. B. Jiang, "GPU-based ultra-fast direct aperture optimization for online adaptive radiation therapy," *Physics in Medicine and Biology* **55**(15), 4309–4319 (2010).
14. R. Nath, L. L. Anderson, G. Luxton, K. A. Weaver, J. F. Williamson, A. S. Meigooni, "Dosimetry of interstitial brachytherapy sources: Recommendations of the AAPM Radiation Therapy Committee Task Group No. 43," *Medical Physics* **22**(2), 209–234 (1995).
15. G. Landry, et al., "Sensitivity of low energy brachytherapy Monte Carlo dose calculations to uncertainties in human tissue composition," *Medical Physics* **37**(10), 5188–5198 (2010).
16. S. Hissoiny, B. Ozell, P. Després, J. F. Carrier, "Validation of GPUMCD for low-energy brachytherapy seed dosimetry," *Medical Physics* **38**(7), 4101–4107 (2011).
17. S. Hissoiny, A. J. E. Raaijmakers, B. Ozell, P. Després, B. W. Raaymakers, "Fast dose calculation in magnetic fields with GPUMCD," *Physics in Medicine and Biology* **56**(16), 5119–5129 (2011).
18. X. Jia, H. Yan, X. Gu, S. B. Jiang, "Fast Monte Carlo simulation for patient-specific CT/CBCT imaging dose calculation," *Physics in Medicine and Biology* **57**(3), 577–590 (2012).
19. I. Kawrakowand M. Fippel, "VMC++, a fast MC algorithm for radiation treatment planning," in *The Use of Computers in Radiation Therapy* (Springer, Berlin Heidelberg, 2000), pp. 126–128.

20. J. E. Cygler, et al., "Clinical use of a commercial Monte Carlo treatment planning system for electron beams," *Physics in Medicine and Biology* **50**(5), 1029–1034 (2005).

21. S. L. Lawrence, N. H. M. van Lieshout, P. M. Charland, "Assessment of Eclipse electron Monte Carlo output prediction for various topologies," *Journal of Applied Clinical Medical Physics* **16**(3), 99–106 (2015).

22. J. E. Snyder, D. E. Hyer, R. T. Flynn, A. Boczkowski, D. Wang, "The commissioning and validation of Monaco treatment planning system on an ElektaVersaHD linear accelerator," *Journal of Applied Clinical Medical Physics* **20**(1), 184–193 (2019).

23. J. Y. Huang, D. Dunkerley, J. B. Smilowitz, "Evaluation of a commercial Monte Carlo dose calculation algorithm for electron treatment planning," *Journal of Applied Clinical Medical Physics* **20**(6), 184–193 (2019).

24. A. N. Schreuder, et al., "Validation of the RayStation Monte Carlo dose calculation algorithm using a realistic lung phantom," *Journal of Applied Clinical Medical Physics* **20**(12), 127–137 (2019).

25. "A protocol for the determination of absorbed dose from high energy photon and electron beams," *Medical Physics* **10**(6), 741–771 (1983).

26. J. E. Rodgers, A. Niroomand-Rad, M. Lundsten, "Comment on clinical implementation of the AAPM's Task Group-51 protocol for calibration of high-energy photon and electron beams," *Medical Physics* **27**(8), 1995–1996 (2000).

27. I. Kawrakow, "Accurate condensed history Monte Carlo simulation of electron transport. II. Application to ion chamber response simulations," *Medical Physics* **27**(3), 499–513 (2000).

28. E. Poon, J. Seuntjens, F. Verhaegen, "Consistency test of the electron transport algorithm in the GEANT4 Monte Carlo code," *Physics in Medicine & Biology* **50**(4), 681 (2005).

29. J. Sempau, E. Acosta, J. Baró, J. M. Fernández-Varea, F. Salvat, "An algorithm for Monte Carlo simulation of coupled electron-photon transport," *Nuclear Instruments and Methods in Physics Research Section B: Beam Interactions with Materials and Atoms* **132**(3), 377–390 (1997).

30. B. R. Muir, M. R. McEwen, D. W. O. Rogers, "Measured and Monte Carlo calculated kQ factors: Accuracy and comparison," *Medical Physics* **38**(8), 4600–1609 (2011).

31. L. A. Buckley and D. W. O. Rogers, "Wall correction factors, Pwall, for thimble ionization chambers," *Medical Physics* **33**(2), 455–464 (2006).

32. H. Bouchard, J. Seuntjens, I. Kawrakow, "A Monte Carlo method to evaluate the impact of positioning errors on detector response and quality correction factors in non-standard beams," *Physics in Medicine and Biology* **56**(8), 2617–2634 (2011).

33. D. T. Burns, "A new approach to the determination of air kerma using primary-standard cavity ionization chambers," *Physics in Medicine and Biology* **51** (4), 929–942 (2006).

34. E. Mainegra-Hing, N. Reynaert, I. Kawrakow, "Novel approach for the Monte Carlo calculation of free-air chamber correction factors," *Medical Physics* **35**(8), 3650–3660 (2008).

35. K. Zink and J. Wulff, "On the wall perturbation correction for a parallel-plate NACP-02 chamber in clinical electron beams," *Medical Physics* **38**(2), 1045–1054 (2011).

36. G. Cranmer-Sargison, S. Weston, J. A. Evans, N. P. Sidhu, D. I. Thwaites, "Implementing a newly proposed Monte Carlo based small field dosimetry formalism for a comprehensive set of diode detectors," *Medical Physics* **38**(12), 6592–6602 (2011).

37. L. L. W. Wang and D. W. O. Rogers, "Monte Carlo study of Si diode response in electron beams," *Medical Physics* **34**(5), 1734–1742 (2007).

38. F. Crop, et al., "The influence of small field sizes, penumbra, spot size and measurement depth on perturbation factors for microionization chambers," *Physics in Medicine and Biology* **54**(9), 2951–2969 (2009).

39. P. Francescon, S. Cora, C. Cavedon, "Total scatter factors of small beams: A multidetector and Monte Carlo study," *Medical Physics* **35**(2), 504–513 (2008).

40. P. Francescon, S. Cora, C. Cavedon, P. Scalchi, "Application of a Monte Carlo-based method for total scatter factors of small beams to new solid state microdetectors," *Journal of Applied Clinical Medical Physics* **10**(1), 147–152 (2009).

41. S. Chauvie, et al., "Geant4 physics processes for microdosimetry simulation: Design foundation and implementation of the first set of models," *IEEE Transactions on Nuclear Science* **54**(6), 2619–2628 (2007).

42. A. Kellerer, "Fundamentals of microdosimetry," in *The Dosimetry of Ionizing Radiation Volume I* (Academic Press, London, 1985), pp. 78–162.

43. T. T. Böhlen, M. Dosanjh, A. Ferrari, I. Gudowska, "Simulations of microdosimetric quantities with the Monte Carlo code FLUKA for carbon ions at therapeutic energies," *International Journal of Radiation Biology* **88**(1–2), 176–182 (2012).

44. T. Liamsuwan, D. Emfietzoglou, S. Uehara, H. Nikjoo, "Microdosimetry of low-energy electrons," *International Journal of Radiation Biology* **88**(12), 899–907 (2012).

45. W. Friedland, M. Dingfelder, P. Kundrát, P. Jacob, "Track structures, DNA targets and radiation effects in the biophysical Monte Carlo simulation code PARTRAC," *Mutation Research – Fundamental and Molecular Mechanisms of Mutagenesis* **711**(1–2), 28–40 (2011).

46. M. Dingfelder, R. H. Ritchie, J. E. Turner, W. Friedland, H. G. Paretzke, R. N. Hamm, "Comparisons of calculations with PARTRAC and NOREC: Transport of electrons in liquid water," *Radiation Research* **169** (5), 584–594 (2008).

47. S. Incerti, et al., "The Geant4-DNA project," *International Journal of Modeling, Simulation, and Scientific Computing* **01**(02), 157–178 (2010).

48. I. El Naqa, P. Pater, J. Seuntjens, "Monte Carlo role in radiobiological modelling of radiotherapy outcomes," *Physics in Medicine and Biology* **57**(11), R75–R97 (2012).

49. W. Friedland, et al., "Comprehensive track-structure based evaluation of DNA damage by light ions from radiotherapy-relevant energies down to stopping," *Scientific Reports* **7**, 45161 (2017).

50. W. Friedland, E. Schmitt, P. Kundrát, G. Baiocco, A. Ottolenghi, "Track-structure simulations of energy deposition patterns to mitochondria and damage to their DNA," *International Journal of Radiation Biology* **95**(1), 3–11 (2018).

51. R. D. Stewart, W. E. Wilson, J. C. McDonald, D. J. Strom, "Microdosimetric properties of ionizing electrons in water: A test of the PENELOPE code system," *Physics in Medicine and Biology* **47**(1), 79–88 (2002).

52. S. Rollet, et al., "Monte Carlo simulation of mini TEPC microdosimetric spectra: Influence of low energy electrons," *Radiation Measurements* **45**(10), 1330–1333 (2010).

53. M. A. Cortés-Giraldoand A. Carabe, "A critical study of different Monte Carlo scoring methods of dose average linear-energy-transfer maps calculated in voxelized geometries irradiated with clinical proton beams," *Physics in Medicine and Biology* **60**(7), 2645–2669 (2015).

54. R. M. Abolfath, D. J. Carlson, Z. J. Chen, R. Nath, "A molecular dynamics simulation of DNA damage induction by ionizing radiation," *Physics in Medicine and Biology* **58**(20), 7143–57 (2013).

55. M. Dos Santos, C. Villagrasa, I. Clairand, S. Incerti, "Influence of the DNA density on the number of clustered damages created by protons of different energies," *Nuclear Instruments and Methods in Physics Research Section B: Beam Interactions with Materials and Atoms* **298**, 47–54 (2013).

56. N. Lampe, et al., "Mechanistic DNA damage simulations in Geant4-DNA part 1: A parameter study in a simplified geometry," *Physica Medica* **48**, 135–145 (2018).

57. W. Friedland, M. Dingfelder, P. Kundrát, P. Jacob, "Track structures, DNA targets and radiation effects in the biophysical Monte Carlo simulation code PARTRAC," *Mutation Research* **711**(1–2), 28–40 (2011).

58. H. Nikjoo, S. Uehara, D. Emfietzoglou, F. A. Cucinotta, "Track-structure codes in radiation research," *Radiation Measurements* **41**(9–10), 1052–1074 (2006).

59. H. Nikjoo, D. Emfietzoglou, T. Liamsuwan, R. Taleei, D. Liljequist, S. Uehara, "Radiation track, DNA damage and response –a review," *Reports on Progress in Physics* **79**(11), 116601 (2016).

60. D. Emfietzoglou, G. Papamichael, H. Nikjoo, "Monte Carlo electron track structure calculations in liquid water using a new model dielectric response function," *Radiation Research* **188**(3), 355–368 (2017).

61. Z. Francis, C. Villagrasa, I. Clairand, "Simulation of DNA damage clustering after proton irradiation using an adapted DBSCAN algorithm," *Computer Methods and Programs in Biomedicine* **101**(3), 265–270 (2011).

62. R. Liu, T. Zhao, M. H. Swat, F. J. Reynoso, K. A. Higley, "Development of computational model for cell dose and DNA damage quantification of multicellular system," *International Journal of Radiation Biology* **95**(11), 1484–1497 (2019).

63. R. Liu, T. Zhao, X. Zhao, F. J. Reynoso, "Modeling gold nanoparticle radiosensitization using a clustering algorithm to quantify DNA double-strand breaks with mixed-physics Monte Carlo simulation," *Medical Physics* **46**(11), 5314–5325 (2019).

64. D. Villoing, S. Marcatili, M. P. Garcia, M. Bardiès, "Internal dosimetry with the Monte Carlo code GATE: Validation using the ICRP/ICRU female reference computational model," *Physics in Medicine and Biology* **62**(5), 1885–1904 (2017).

65. Habib Zaidi, George Sgouros, *Therapeutic Applications of Monte Carlo Calculations in Nuclear Medicine* (CRC/Taylor & Francis, Boca Raton, FL, 2003).

66. M. G. Stabin, "MIRDOSE: Personal computer software for internal dose assessment in nuclear medicine," *Journal of Nuclear Medicine* **37**(3), 538–546 (1996).

67. M. G. Stabin, R. B. Sparks, E. Crowe, "OLINDA/EXM: The second-generation personal computer software for internal dose assessment in nuclear medicine," *Journal of Nuclear Medicine* **46**(6), 1023–1027 (2005).

68. A. Maria, "Monte Carlo simulation in radionuclide therapy dosimetry," *Biomedical Journal of Scientific & Technical Research* **15**(1), 11102–11107 (2019).

69. W. S. Snyder, H. L. Fisher, M. R. Ford, G. G. Warner, "Estimates of absorbed fractions for monoenergetic photon sources uniformly distributed in various organs of a heterogeneous phantom," *Journal of Nuclear Medicine* **3**(Suppl.), 7–52 (1969).

70. L. M. Carter, et al., "PARaDIM – A PHITS-based Monte Carlo tool for internal dosimetry with tetrahedral mesh computational phantoms," *Journal of Nuclear Medicine* **60**(12), 1802–1811 (2019).

71. M. R. Ay, M. Shahriari, S. Sarkar, M. Adib, H. Zaidi, "Monte Carlo simulation of x-ray spectra in diagnostic radiology and mammography using MCNP4C," *Physics in Medicine and Biology* **49**(21), 4897–4917 (2004).

72. J. M. Pawlowski andG. X. Ding, "A new approach to account for the medium-dependent effect in model-based dose calculations for kilovoltage x-rays," *Physics in Medicine and Biology* **56**(13), 3919–3934 (2011).

73. J. T. M. Jansen, P. C. Shrimpton, J. Holroyd, S. Edvean, "Selection of bone dosimetry models for application in Monte Carlo simulations to provide CT scanner-specific organ dose coefficients," *Physics in Medicine and Biology* **63**(12), 125015 (2018).

74. P. C. Shrimpton and S. Edvean, "CT scanner dosimetry," *British Institute of Radiology* **71**(841), 1–3 (1998).

75. P. Avilés Lucas, D. R. Dance, I. A. Castellano, E. Vañó, "Monte Carlo simulations in CT for the study of the surface air kerma and energy imparted to phantoms of varying size and position," *Physics in Medicine and Biology* **49**(8), 1439–1454 (2004).

76. P. Alaei, B. J. Gerbi, R. A. Geise, "Lung dose calculations at kilovoltage x-ray energies using a model-based treatment planning system," *Medical Physics* **28**(2), 194–198 (2001).

77. M. G. Herman, et al., "Clinical use of electronic portal imaging: Report of AAPM Radiation Therapy Committee Task Group 58," *Medical Physics* **28**(5), 712–737 (2001).

78. P. W. Chin, E. Spezi, D. G. Lewis, "Monte Carlo simulation of portal dosimetry on a rectilinear voxel geometry: A variable gantry angle solution," *Physics in Medicine and Biology* **48**(16), (2003).

79. M. J. Murphy, et al., "The management of imaging dose during image-guided radiotherapy: Report of the AAPM Task Group 75," *Medical Physics* **34**(10), 4041–4063 (2007).

80. P. Downes, R. Jarvis, E. Radu, I. Kawrakow, E. Spezi, "Monte Carlo simulation and patient dosimetry for a kilovoltage cone-beam CT unit," *Medical Physics* **36**(9), 4156–4167 (2009).

81. S. Agostinelli, et al., "GEANT4 –a simulation toolkit," *Nuclear Instruments and Methods in Physics Research Section A: Accelerators, Spectrometers, Detectors and Associated Equipment* **506**(3), 250–303 (2003).

82. Geant4Collaboration, *Physics Reference Manual, Release 10.5* (2019). [Online]. Available from: http://geant4-userdoc.web.cern.ch/geant4-userdoc/UsersGuides/PhysicsReferenceManual/fo/PhysicsReferenceManual.pdf (Accessed: 24June2019).

83. L. Aichambault, et al., "Overview of Geant4 applications in medical physics," *IEEE Nuclear Science Symposium Conference Record* **3**, 1743–1745 (2003).

84. B. A. Faddegon, I. Kawrakow, Y. Kubyshin, J. Perl, J. Sempau, L. Urban, "The accuracy of EGSnrc, Geant4 and PENELOPE Monte Carlo systems for the simulation of electron scatter in external beam radiotherapy," *Physics in Medicine and Biology* **54**(20), 6151–6163 (2009).

85. E. Poon and F. Verhaegen, "Accuracy of the photon and electron physics in GEANT4 for radiotherapy applications," *Medical Physics* **32**(6), 1696–1711 (2005).

86. L. Grevillot, T. Frisson, D. Maneval, N. Zahra, J. N. Badel, D. Sarrut, "Simulation of a 6 MV Elekta Precise Linac photon beam using GATE/GEANT4," *Physics in Medicine and Biology* **56**(4), 903–918 (2011).

87. D. Granero, J. Vijande, F. Ballester, M. J. Rivard, "Dosimetry revisited for the HDR 192Ir brachytherapy source model mHDR-v2," *Medical Physics* **38**(1), 487–494 (2011).

88. H. Afsharpour, et al., "ALGEBRA: ALgorithm for the heterogeneous dosimetry based on GEANT4 for BRAchytherapy," *Physics in Medicine and Biology* **57**(11), 3273–3280 (2012).

89. G. A. P. Cirrone, et al., "Hadrontherapy: A Geant4-based tool for proton/ion-therapy studies," *Progress in Nuclear Science and Technology* **2**(0), 207–212 (2011).

90. M. De Napoli, et al., "Carbon fragmentation measurements and validation of the Geant4 nuclear reaction models for hadrontherapy," *Physics in Medicine and Biology* **57**(22), 7651–7671 (2012).

91. G. A. P. Cirrone, et al., "The GEANT4 toolkit capability in the hadron therapy field: Simulation of a transport beam line," *Nuclear Physics B – Proceedings Supplements* **150** (1–3), 54–57 (2006).

92. V. Giacometti, S. Guatelli, M. Bazalova-Carter, A. B. Rosenfeld, R. W. Schulte, "Development of a high resolution voxelised head phantom for medical physics applications," *Physica Medica* **33**, 182–188 (2017).

93. E. Amato, A. Italiano, F. Minutoli, S. Baldari, "Use of the GEANT4 Monte Carlo to determine three-dimensional dose factors for radionuclide dosimetry," *Nuclear Instruments and Methods in Physics Research Section A: Accelerators, Spectrometers, Detectors and Associated Equipment* **708**, 15–18 (2013).

94. S. D. Kost, Y. K. Dewaraja, R. G. Abramson, M. G. Stabin, "VIDA: A voxel-based dosimetry method for targeted radionuclide therapy using Geant 4," *Cancer Biotherapy and Radiopharmaceuticals* **30**(1), 16–26 (2015).

95. S. Jan, et al., "GATE: A simulation toolkit for PET and SPECT," *Physics in Medicine and Biology* **49**(19), 4543–4561 (2004).

96. S. Incerti, et al., "The Geant4-DNA project," *International Journal of Modeling, Simulation, and Scientific Computing* **1**(2), 157–178 (2010).

97. M. A. Bernal, et al., "Track structure modeling in liquid water: A review of the Geant4-DNA very low energy extension of the Geant4 Monte Carlo simulation toolkit," *Physica Medica* **31**(8), 861–874 (2015).

98. S. Incerti, M. Douglass, S. Penfold, S. Guatelli, E. Bezak, "Review of Geant4-DNA applications for micro and nanoscale simulations," *Physica Medica* **32**(10), 1187–1200 (2016).

99. M. Dos Santos, C. Villagrasa, I. Clairand, S. Incerti, "Influence of the DNA density on the number of clustered damages created by protons of different energies,"

Nuclear Instruments and Methods in Physics Research Section B: Beam Interactions with Materials and Atoms **298**, 47–54 (2013).

100. Z. Francis, C. Villagrasa, I. Clairand, "Simulation of DNA damage clustering after proton irradiation using an adapted DBSCAN algorithm," *Computer Methods and Programs in Biomedicine* **101**(3), 265–270 (2011).

101. N. Lampe, et al., "Mechanistic DNA damage simulations in Geant4-DNA Part 2: Electron and proton damage in a bacterial cell," *Physica Medica* **48**, 146–155 (2018).

102. J. Perl, J. Shin, J. Schümann, B. Faddegon, H. Paganetti, "TOPAS: An innovative proton Monte Carlo platform for research and clinical applications," *Medical Physics* **39**(11), 6818–6837 (2012).

103. M. Testa, et al., "Experimental validation of the TOPAS Monte Carlo system for passive scattering proton therapy," *Medical Physics* **40**(12), 1–16 (2013).

104. J. Schümann, H. Paganetti, J. Shin, B. Faddegon, J. Perl, "Efficient voxel navigation for proton therapy dose calculation in TOPAS and Geant4," *Physics in Medicine and Biology* **57**(11), 3281–3293 (2012).

105. A. McNamara, et al., "Validation of the radiobiology toolkit TOPAS-nBio in simple DNA geometries," *Physica Medica* **33**, 207–215 (2017).

106. E. Rault, S. Staelens, R. Van Holen, J. De Beenhouwer, S. Vandenberghe, "Fast simulation of yttrium-90 bremsstrahlung photons with GATE," *Medical Physics* **37**(6), 2943–2950 (2010).

107. R. A. Forster and T. N. K. Godfrey, "MCNP – a general Monte Carlo code for neutron and photon transport," in *Monte-Carlo Methods and Applications in Neutronics, Photonics and Statistical Physics* (Springer-Verlag, 2006), pp. 33–55.

108. J. F. Briesmeister, "MCNPTM–A General Monte Carlo N-Particle Transport Code: Manual," December (2000).

109. P. Andreo, D. Burns, A. Nahum, J. Seuntjens, F. H. Attix, *The Monte Carlo Simulation of the Transport of Radiation Through Matter* (WILEY-VCH, 2017), pp. 349–396.

110. T. D. Solberg, et al., "A review of radiation dosimetry applications using the MCNP Monte Carlo code," *Radiochimica Acta* **89**(4–5), 337–355 (2001).

111. T. D. Bohm, P. M. DeLuca, L. A. DeWerd, "Brachytherapy dosimetry of125I and103Pd sources using an updated cross section library for the MCNP Monte Carlo transport code," *Medical Physics* **30**(4), 701–711 (2003).

112. R. Wang and X. A. Li, "Monte Carlo dose calculations of beta-emitting sources for intravascular brachytherapy: A comparison between EGS4, EGSnrc, and MCNP," *Medical Physics* **28**(2), 134–141 (2001).

113. A. Mostaar, M. Allahverdi, M. Shahriari, "Application of MCNP4C Monte Carlo code in radiation dosimetry in heterogeneous phantom," *Journal of Radiation Research* **1**(3), 143–149 (2003).

114. H. Yoriyaz. et al., "Monte Carlo MCNP-4B – based absorbed dose distribution estimates for patient-specific," *Journal of Nuclear Medicine* **42**(4), 662–669 (2001).

115. A. Bitar, et al., "A voxel-based mouse for internal dose calculations using Monte Carlo simulations (MCNP)," *Physics in Medicine and Biology* **52**(4), 1013–1025 (2007).

116. A. R. Prideaux, et al., "Three-dimensional radiobiologic dosimetry: Application of radiobiologic modeling to patient-specific 3-dimensional imaging-based internal dosimetry," *Journal of Nuclear Medicine* **48**(6), 1008–1016 (2007).

117. S. D. Randeniya, P. J. Taddei, W. D. Newhauser, P. Yepes, "Intercomparision of Monte Carlo radiation transport codes MCNPX, GEANT4, and FLUKA for simulating proton radiotherapy of the eye," *Nuclear Technology* **168**(3), 810–814 (2009).

118. J. Seco and F. Verhaegen, *Monte Carlo Techniques in Radiation Therapy* (CRC/Taylor & Francis, Boca Raton, 2013).

119. R. D. Stewart, et al., "Rapid MCNP simulation of DNA double strand break (DSB) relative biological effectiveness (RBE) for photons, neutrons, and light ions," *Physics in Medicine and Biology* **60**(21), 8249–8274 (2015).

120. I. Kawrakow, "Accurate condensed history Monte Carlo simulation of electron transport. I. EGSnrc, the new EGS4 version," *Medical Physics* **27**(3), 485–498 (2000).

121. D. E. Cullen, J. H. Hubbell, L. Kissel, "EPDL97: The Evaluated Photon Data Library," *UCRL-504006*, 1–35 (1997).

122. M. J. Berger, J. H. Hubbell, S. M. Seltzer, J. Chang, J. S. Coursey, R. Sukumar, D. S. Zucker, K. Olsen, *XCOM: Photon Cross Section Database (NIST Standard Reference Database 8 (XGAM))* (Gaithersburg, MD, 2010).

123. H. Hirayama, et al., *The EGS5 Code System* (2005).

124. D. W. O. Rogers, B. Walters, I. Kawrakow, *BEAMnrc Users Manual* (2009).

125. D. W. O. Rogers, B. Walters, I. Kawrakow, *DOSXYZnrc User Manual* (2005).

126. R. E. P. Taylor, G. Yegin, D. W. O. Rogers, "Benchmarking BrachyDose: Voxel based EGSnrc Monte Carlo calculations of TG-43 dosimetry parameters," *Medical Physics* **34**(2), 445–457 (2007).

127. R. E. P. Taylor and D. W. O. Rogers, "EGSnrc Monte Carlo calculated dosimetry parameters for 192Ir and 169Yb brachytherapy sources," *Medical Physics* **35**(11), 4933–4944 (2008).

128. S. J. Ye, I. A. Brezovich, P. Pareek, S. A. Naqvi, "Benchmark of PENELOPE code for low-energy photon transport: Dose comparison with MCNP4 and EGS4," *Physics in Medicine and Biology* **49**(3), 387–397 (2004).

129. F. Salvat, *PENELOPE2014: A Code System for Monte Carlo Simulation of Electron and Photon Transport* (NEA, Barcelona, Spain, 2015).

130. F. Salvat, J. M. Fernandez-Varea, E. Acosta, J. Sempau, *Penelope - A Code System for Monte Carlo Simulation of Electron and Photon Transport* (Nuclear Energy Agency of the OECD (NEA): Organisation for Economic Co-Operation and Development - Nuclear Energy Agency, Barcelona, Spain, 2001).

131. J. Sempau, A. Badal, L. Brualla, "A PENELOPE-based system for the automated Monte Carlo simulation of clinacs and voxelized geometries-application to far-from-axis fields," *Medical Physics* **38**(11), 5887–5895 (2011).

132. A. Baumgartner, A. Steurer, F. Josef Maringer, "Simulation of photon energy spectra from Varian 2100C and 2300C/D Linacs: Simplified estimates with PENELOPE Monte Carlo models," *Applied Radiation and Isotopes* **67**(11), 2007–2012 (2009).

133. S. Espãa, J. L. Herraiz, E. Vicente, J. J. Vaquero, M. Desco, J. M. Udias, "PeneloPET, a Monte Carlo PET simulation tool based on PENELOPE: Features and validation," *Physics in Medicine and Biology* **54**(6), 1723–1742 (2009).

134. G. Battistoni, et al., "The FLUKA code: Description and benchmarking," *AIP Conference Proceedings* **896** (May), 31–49 (2007).

135. T. T. Böhlen, et al., "The FLUKA Code: Developments and challenges for high energy and medical applications," *Nuclear Data Sheets* **120**, 211–214 (2014).

136. F. Cerutti, A. Ferrari, A. Mairani, P. R. Sala, "New developments in FLUKA," *CERN-Proceedings*, 469–475 (2013).

137. K. Parodi, et al., "The FLUKA code for application of Monte Carlo methods to promote high precision ion beam therapy," *CERN-Proceedings* **2**, 509–516 (2010).

138. A. Mairani, et al., "The FLUKA Monte Carlo code coupled with the local effect model for biological calculations in carbon ion therapy," *Physics in Medicine and Biology* **55**(15), 4273–4289 (2010).

139. Fasso, et al., "The FLUKA code: Present applications and future developments," in *Computing in high Energy and Nuclear Physics Conference Proceedings*, La Jolla, CA, 24-28 March 2003.

II

Brachytherapy

CHAPTER 15

Brachytherapy Dosimetry

Christopher L. Deufel

Mayo Clinic
Rochester, Minnesota

Wesley S. Culberson

University of Wisconsin – Madison
Madison, Wisconsin

Mark J. Rivard

Brown University and Rhode Island Hospital
Providence, Rhode Island

Firas Mourtada

Helen F. Graham Cancer Center and Research Institute
Newark, Delaware

CONTENTS

15.1 INTRODUCTION

Radiotherapy can be delivered using external beam or brachytherapy. The main advantage of brachytherapy is the close proximity of the radiation source to the target (tumor or benign), when this option is viable. Delivery options include interstitial, intracavitary, endoluminal, intraoperative, or surface (i.e., plesiotherapy). Brachytherapy has a long history since the early 1900s and the most recent decade has exploded with new radiation sources, advanced image-guided treatment planning, and improved delivery devices. Cancers of the cervix, prostate, breast, and skin have been most commonly cured using brachytherapy as monotherapy or as a boost to external beam radiotherapy (EBRT) [1–5]. Other disease sites have also benefited from this special radiotherapy modality such as ocular melanoma, endorectal, nasopharynx, lung cancer, and keloids [6–9]. Hence, brachytherapy advancements have enhanced its essential role for the treatment of malignant and nonmalignant disease with radiation. The radiation-emitting sources used for this treatment modality are radionuclide- or electronic-based and may incorporate low (<2 Gy h^{-1}) or high (>1 Gy min^{-1}) dose rates.

There are unique physics challenges that one must realize to ensure accurate and precise dosimetry of such sources. This chapter describes methods to measure and determine the brachytherapy source strength and associated dosimetric parameters while ensuring traceability to the national dosimetry standards. First, we present on the national standards (NIST/PSDL), and the secondary laboratory (ADCL/SSDL/National Metrology Institute) establishments. Then, we describe the required calibration process and requirements at the time of manufacturing by the vendor. Dosimetry tools used for research and development to investigate new sources are presented. End user (clinical physicist) calibrations (assays), measurements, and calculations for photon- and beta- emitting sources are provided, and most current societal recommendations are presented with future directions discussed.

15.2 NIST/PSDL CALIBRATION AND MEASUREMENTS

Source-strength standardization in brachytherapy is crucial for ensuring accurate dose deliveries and clinical trial consistency. The strength of each brachytherapy source used for treatment needs to be characterized by a well-defined and measurable quantity, with high accuracy and precision and a low uncertainty. Since measuring the source strength of a brachytherapy source is not trivial due to generally low photon energies and steep dose gradients, there have been substantial efforts over the past several decades to standardize the quantities, units, and measurement techniques. The organization in the U.S. responsible for maintaining most of the brachytherapy source-strength physical measurement standards is called the National Institute of Standards and Technology (NIST). This is a primary standards dosimetry laboratory (PSDL) and serves as a government-supported laboratory for developing, maintaining, and improving primary radiation calibration standards. Globally, there are many PSDLs in the world and the International Atomic Energy Association (IAEA) and International Bureau of Weights and Measures (BIPM) are responsible for providing support and maintaining their consistency. A PSDL is the top of the calibration chain. One step further down the calibration chain is the secondary standards dosimetry laboratory (SSDL). In the U.S. the SSDLs are not a part of the IAEA network; instead, they are accredited by the American Association of Physicists in Medicine (AAPM) and are called Accredited Dosimetry Calibration Laboratories (ADCLs) instead of SSDLs. In principle, the AAPM ADCLs are similar to SSDLs. The purpose of SSDLs and ADCLs is to disseminate the standards maintained at the PSDLs by way of transferring calibration traceability to end-users.

Traceability is the concept that describes the chain that links a measurement of a quantity in the field to an established standard. This chain of measurements is transferred from the primary standard to a local clinical user by the process of repeated cross calibrations. At the top of the chain are the primary standard measurement devices, maintained by NIST or sometimes by an alternate laboratory as an interim national standard (as is the case for HDR ^{192}Ir) in the U.S. A primary standard is defined as an instrument that does not need a calibration. It directly realizes the quantity of interest. For example, free-air chambers (FACs), calorimeters, and certain well-defined ion chambers may be considered primary standards. In brachytherapy dosimetry, there are several different quantities of interest, determined by a variety of basic measurement devices at NIST. The different source-strength quantities and measurement devices maintained by NIST or interim standards laboratories are reviewed in this section.

Before presenting the different primary brachytherapy standards, it is useful to define the calibration concepts and quantities of interest. Air-kerma strength (S_K) is the quantity used to determine source strength for photon-emitting brachytherapy sources. S_K is the source-strength metric recommended by the AAPM through Task Group (TG) 43 report and its associated updates [10, 11]. S_K is defined by the following equation,

$$S_K = \dot{K}_{air} d^2 \qquad (15.1)$$

where $\dot{K}_{air}$ is the determined air-kerma rate, d is the distance between the source and the "reference distance" of 1 m. The AAPM TG-32 report recommends that the units for S_K are represented by U in μGy m^2 h^{-1} or cGy cm^2 h^{-1} [12]. For beta-emitting sources, the quantity used to determine the source strength is absorbed dose rate to water (ADW). The location of this quantity varies by beta source type. For example, the ADW for planar ophthalmic applicators is on the surface and for beta line sources or line-source arrays it is at a depth of 2-mm in water [13].

15.2.1 Low-Energy LDR Brachytherapy Standards

The primary standard at NIST is a large-volume large-angle free-air ionization chamber (FAC) called the wide-angle free-air chamber (WAFAC) for low-energy low dose rate (LDR) brachytherapy sources such as ^{125}I, [103]Pd, and [131]Cs. The WAFAC is a cylindrically symmetric free-air ionization chamber shown by the basic schematic in Figure 15.1 and detailed in the publication by Seltzer et al. in 2003 [14]. The principal operation of the WAFAC is to determine the air ionization per unit mass, or exposure, which is then converted to air-kerma

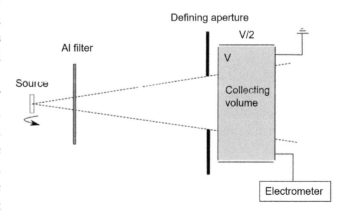

FIGURE 15.1 Basic schematic of the NIST WAFAC showing the source, aluminum filter, defining aperture and collecting volume. V represents the voltage of the polarizing electrode. V/2 represents the voltage of the middle electrode.

rate and then air-kerma strength, S_K, of the source. A beam-defining aperture is placed 30 cm from the source and the center of volume for the ionization chamber is located approximately 50 cm from the source. The measurement is then corrected to the "reference distance" of 1 m. The determination of air-kerma rate is accomplished through a derivative technique where the air volume of the chamber is changed multiple times and the net change in ionization current is measured for each chamber volume change.

It should be noted that determination of the air-kerma strength by means of a large-volume FAC measurement is challenging due to the low signal-to-noise ratio from the weak nature of the LDR sources. Each measurement takes hours to perform, and multiple measurements are acquired and averaged to keep the statistical uncertainties below 1% ($k = 1$). The combined uncertainty estimates for ^{125}I, ^{103}Pd, and ^{131}Cs encapsulated brachytherapy seeds are approximately 1.5%, but depend on each individual measurement and repeatability of the measurements [15].

15.2.2 High-Energy LDR Brachytherapy Standards

For measurement of high-energy LDR brachytherapy sources such as ^{192}Ir and ^{137}Cs with average energies greater than 50 keV, the AAPM and the European Society for Radiotherapy and Oncology (ESTRO) have issued joint recommendations [16]. A procedure was established in 1979 at NIST to utilize known-volume spherical graphite-walled ionization chambers for determining the air-kerma rate at source-to-chamber distances between 50 cm and 100 cm [17]. The wall thickness of each chamber was designed to meet Bragg–Gray cavity theory conditions and the air-kerma rate was determined from cavity theory first principles [18]. The formula to convert the basic quantity of ionization current to air-kerma rate is described in Chapter 16 of the 2009 AAPM Summer School Proceedings by Mitch and Soares [19]. Once the air-kerma rate is determined, S_K is determined similarly as before (Equation 15.1).

Due to the low signal-to-noise ratio in the spherical graphite-walled ionization chambers from the ^{192}Ir and ^{137}Cs sources, sometimes multiple seeds (up to 50) were measured simultaneously [17]. The spherical graphite-walled chamber measurement result is then transferred to a spherical re-entrant ionization chamber. Once the calibration coefficient for this spherical

re-entrant chamber has been established, a ^{226}Ra source (with a half-life of 1,600 years) is used to verify response constancy.

15.2.3 High-Energy HDR Brachytherapy Standards

Presently, there is no NIST standard for the air-kerma strength of high dose rate (HDR) ^{192}Ir sources. For these types of sources, secondary standards are maintained by ADCLs and are referred to as interim standards. There are two basic techniques to determine the air-kerma strength of HDR ^{192}Ir sources as they do not have a suitably shielded facility. The first method is to position a chamber at a fixed distance from the source. Although this sounds straightforward, it is difficult to position the source at a precise location due to its ability to move within the guide tube, resulting in relatively high uncertainties if this method is used. In Europe, this is the preferred method at PSDLs such as Physikalisch-Technische Bundesanstalt (PTB), but in their implementation they use a sophisticated custom source holder without a guide tube. For the ADCLs in the U.S., the preferred method involves measurement of air kerma at multiple distances from the source and solving for unknown quantities of room scatter and positioning offsets by simultaneously solving several equations [20]. The typical number of measurement distances is seven, resulting in the technique being referred to as the seven-distance technique as shown schematically in Figure 15.2. This technique takes advantage of the highly accurate change in distance from the source from a precision lead screw and stepper motor system.

A small (3.6 cm³) spherical ionization chamber, such as an Exradin A3 (Standard Imaging, Middleton, WI)

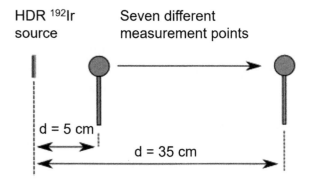

FIGURE 15.2 Schematic of the HDR ^{192}Ir seven-distance measurement technique established at the University of Wisconsin ADCL.

is normally used for HDR ^{192}Ir air-kerma strength measurements [21]. Since a NIST-traceable calibration for ^{192}Ir is not available for these chambers, a linear interpolation method is used to derive the calibration for HDR [192]Ir. The two beam qualities used to interpolate the ^{192}Ir calibration coefficient are ^{137}Cs and M250 x-ray beam, both of which are available at NIST and the ADCLs. This method works well since the average energy of the gamma rays emitted from ^{192}Ir is about midway between the two interpolated energies. The chambers used are selected to have a flat energy response over this range, yielding a variation of only a couple of percent over the range of energies in the interpolation. More details on the specific equations for the determination of air-kerma strength and the equation solver are presented in great detail by Stump et al. in 2002 [20].

15.2.4 Beta Particle Brachytherapy Standards

For many years, the NIST standard for beta particle sources used in medical applications has been an ambient air extrapolation chamber based on the design by Loevinger [22]. Very small gap sizes are utilized, since the range of beta particles is on the order of millimeters in air. Monte Carlo simulations are used to correct each measurement for divergence and chamber materials. Planar beta-particle brachytherapy sources, such as those used for ocular radiotherapy, are measured with a 4-mm diameter electrode. Seed- and line-based sources are measured with a 1-mm diameter electrode and the sources are embedded in water-equivalent plastic phantoms. A larger electrode increases the detector signal, and is therefore used with planar sources whose source strength completely covers the electrode. In-phantom measurement depths for the seed- and line-based sources are usually 2 mm. For all beta-particle measurements at NIST, the source distributions (across the face of the source) are mapped out by stepping the 1-mm electrode around the source to find the region of maximum signal.

15.2.5 Electronic Brachytherapy Standards

Electronic brachytherapy (eBT) involves the treatment of disease with miniature x-ray tubes that mimic the size and geometry of sealed sources. The number of clinical users and manufacturers of eBT devices have increased in the last two decades. Applications of eBT range from surface treatments of skin disease, to internal treatments with applicators for breast, cervical and

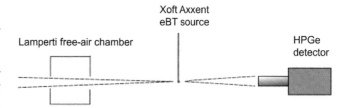

FIGURE 15.3 Schematic of the NIST eBT calibration standard.

vaginal cancers. Although the AAPM has not issued formal recommendations for eBT dosimetry, NIST has developed an air-kerma based eBT standard, which was detailed in a publication by Seltzer et al. in 2014 [23]. A schematic of the basic setup is shown in Figure 15.3. In a specially designed facility at NIST, a FAC and a high-purity germanium detector (HPGe) have been mounted on opposite sides of a source holder that can accommodate a Xoft Axxent (Xoft, a subsidiary of iCAD, Inc, San Jose, CA, USA) eBT source inside its cooling catheter. The spectrum of each source is measured by the HPGe detector and the air-kerma rate is determined from the Lamperti FAC. To average out over source anisotropy, the air-kerma rate is measured at the four cardinal angles around the source (by rotating the source about its long axis) and then averaged to determine a mean value.

15.2.6 Retired Brachytherapy Standards

NIST has retired many of its measurement standards over the years. The reasons for a standard being retired may vary, but are mostly due to sources no longer being manufactured or lack of resources at NIST to maintain the standard. Most notably in recent years was the extrapolation chamber standard for beta-emitting ^{90}Sr/^{90}Y ophthalmic applicators [24, 25]. Consequently, the ADCLs are not able to maintain NIST-traceability and have also discontinued calibrations for ^{90}Sr ophthalmic applicators.

15.3 ADCL AND SSDL CALIBRATIONS AND MEASUREMENTS

NIST maintains the primary standards for brachytherapy sources, however, the demand for source measurements from clinical users far exceed the resources available at NIST or the other PSDLs. It is for this reason that a network of secondary standards laboratories has been established. These labs are called SSDLs or ADCLs in the United States, as described in an earlier section. ADCLs have been an integral part of the calibration chain since their inception in 1971. All three ADCLs currently provide brachytherapy source calibrations to AAPM members. These ADCLs are:

1. University of Wisconsin (Madison, WI)

2. MD Anderson (Houston, TX)

3. K & S Associates (Nashville, TN)

The purpose of the ADCLs is to provide precise calibrations of customer equipment (for the purposes of brachytherapy, these are predominantly well chambers and electrometers), to reduce the time and costs for these calibrations that it would otherwise take if the customer chose to have NIST calibrate their equipment, and to provide technical resources (such as answering questions on equipment selection) for AAPM members. As time has progressed, an AAPM subcommittee, the Calibration Laboratory Accreditation (CLA) Subcommittee, has been developed to oversee the ADCLs. The CLA is comprised of technical experts and representatives from NIST. Members of the CLA serve as auditors for onsite ADCL visits to the clinical users around the Unites States at least once every two years.

The following sections provide an overview of the calibrations provided by the ADCL network. Not every ADCL is accredited to provide every type of calibration, and the accreditation scope for each lab is listed on their website for reference. One unique characteristic of the ADCL network, compared with the IAEA-based SSDLs, is the AAPM requirement for proficiency testing. It is required that all ADCLs periodically inter-compare their calibration results and present their results for CLA review. ADCL source measurement comparisons must agree within 2% for long half-life photon-emitting sources, 3% for short half-life photon-emitting sources, 3% for photon-emitting intravascular brachytherapy (IVBT) sources, and 6% for beta-emitting IVBT sources. The well-chamber calibrations must agree within 3% for photon-emitting calibrations, and 6% for beta-emitting IVBT calibrations. These constancy checks help to ensure standardization within the ADCL network, and distinguishes the ADCLs from SSDLs, which do not have such requirements.

The process of calibrating customer's equipment at the ADCL for brachytherapy sources is similar for all brachytherapy source types. The process involves the determination of a calibration coefficient for the customer's well-type ionization chamber based on a NIST-traceable calibration chain. The specifics of each calibration vary depending on the NIST-traceable quantity and the method of transfer, but ultimately the customer's equipment is returned with a calibration certificate with a unique calibration coefficient for their chamber and source type. Each source model will render a unique calibration coefficient in the customer's well chamber, even if the source radionuclide itself is the same. For example, a model 200 ^{103}Pd seed (Theragenics Corp., Buford, GA) will not yield the same calibration coefficient as the model 2335 ^{103}Pd seed (Best Medical, Springfield, VA). The reason for this is due to the internal characteristics of each seed, yielding differing emission spectra and angular distributions. One should not assume that two different seed models would have the same calibration coefficients simply because they share the same isotope. The next sections provide details on the calibrations of well-type ionization chambers for the different brachytherapy source types. Electrometer calibrations are performed separately from well chamber calibrations and are not described within this chapter.

15.3.1 ADCL Brachytherapy Calibrations

Most recommended brachytherapy source-strength quantities are based on air-kerma calibration standards at NIST. The techniques and measurement standards employed by NIST are extremely complicated and not practical for clinical user measurements. For this reason, well-type ionization chambers (often called well chambers) are used by the clinical user instead. Well chambers are commercially available instruments and are the most common source-strength measurement device used by clinical brachytherapy physicists and calibration laboratories. A well chamber can be calibrated by an ADCL with a high degree of accuracy and precision, and transported easily by mail. Well chambers can be either air-communicating or sealed (pressurized), with the air-communicating type being the more common. A basic schematic of a well chamber is shown in Figure 15.4, where the seed is held securely inside the well chamber with a source holder. The ionization current is read out with an electrometer at the other end of the triaxial cable.

The calibration technique for well chambers is relatively straightforward in an ADCL. The clinical physicist specifies the source model(s) for which they are seeking calibration, and supplies the source holder(s) that they

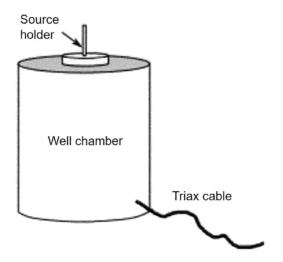

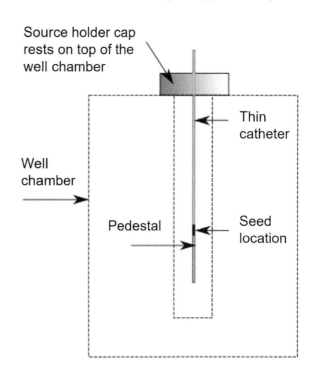

FIGURE 15.4 Basic schematic of a well-type ionization chamber used by a clinical physicist and sent to an ADCL for brachytherapy seed source-strength measurements.

FIGURE 15.5 Schematic of a typical source holder for brachytherapy well chamber calibrations and measurements. The well chamber surrounding the source is indicated by dotted lines.

use. The ADCL calibration should be performed with the end-user's source holder rather than the ADCL's source holder since they can differ structurally and might change the source position, attenuation, or scatter and thereby alter the measured calibration coefficient.

The resulting value from an ADCL well chamber calibration is the air-kerma strength calibration coefficient, N_{SK}, and has units of $\mu Gy\ m^2\ h^{-1}\ A^{-1}$. In North America, N_{SK} is corrected to 22°C and 101.325 kPa. The exception to an air-kerma strength calibration is for eBT source where air-kerma rate at 50 cm in air is the established quantity. The eBT calibration coefficient is $N_{AKR@50cm}$ to reflect the Lamperti FAC-based NIST standard. As stated before, the source holder is an integral part of the reported calibration coefficient. The calibration coefficient assumes that the source is positioned for the maximum response inside the well chamber, also called the *sweet spot*. This position is measured at the time of calibration and is reported on the calibration certificate. The clinical physicist should place the source being measured in the same position as that used for calibration.

A variety of source holders are available for calibration of brachytherapy well chambers, and the correct source holder should be used for each source to ensure reproducible results and safe handling of the sources. The basic design of all holders is similar, with the source being suspended at the cylindrical axis (center) and sweet spot within the well chamber. To accomplish this, a small-diameter plastic catheter (typically Kapton® in order to minimize attenuation) is secured within the

holder, and a standoff spacer is used as a pedestal for the seed to sit upon during the measurement. The seed is dropped into the catheter with a pair of tweezers for LDR sources, while an HDR source is moved into position inside a guide tube. The general design of the source holder is shown in Figure 15.5.

Source holders for eBT calibrations are slightly different since the miniature x-ray tube and its cooling catheter need to be stabilized during the calibration. A specialized source holder with a metal extension tube on top has been developed for eBT air-kerma well chamber measurements. These are commercially available and provided by the manufacturer during eBT acceptance testing and commissioning. As for LDR seeds and HDR sources, these holders should be sent for calibration concurrently with the well chamber.

Uncertainties for each of the ADCL calibration are provided with the calibration certificate and include the NIST-traceable uncertainties as well as any additional uncertainties incurred during the calibration transfer to the end-user. Uncertainties for photon-emitting sources are approximately 2.5% ($k = 1$) and approximately 8% ($k = 1$) for beta-emitting sources (mainly due to uncertainties of the NIST primary measurements) [25, 26].

15.4 SOURCE VENDOR CALIBRATIONS AND MEASUREMENTS

Manufacturers of each brachytherapy source model are responsible for providing end users with a certificate for source strength that is NIST-traceable, accurate, and precise. The manufacturer is essentially at the same level in the traceability chain as the clinical physicist. Manufacturers utilize well-type ionization chambers to determine the source strength and are responsible for calibrating their equipment at least every two years. Each individual source is measured by the manufacturer before sending it to a clinical physicist, and the manufacturer's results are communicated via a source strength certificate. This certificate will inform the end user of the mean source strength or individual source strengths for the many ordered sources, such as would be the case for a permanent seed prostate implant where up to 10^2 seeds might be implanted.

In addition to providing the customer with the mean or individual source strength, the manufacturer is responsible for periodically (approximately once per year) initiating a proficiency test with NIST and the ADCLs to ensure that the manufacturing processes have not changed to a point where the calibration coefficient of a well-type ionization chamber is affected. Proficiency tests first require at least five sources be sent to NIST, and then circulated through the ADCLs, before returning to NIST for a final measurement. Results of each proficiency test are communicated to the ADCL directors as well as the AAPM CLA for review.

15.4.1 Well Chamber Measurement Specifics

For determination of the air-kerma strength of a source by a well chamber measurement, the calibration coefficient for the well chamber is used along with the ionization current and associated corrections according to the following equation:

$$S_K = I \cdot N_e \cdot k_{tp} \cdot N_{SK} \cdot A_{ion} \cdot P_{ion} \qquad (15.2)$$

where I is the measured ionization current, N_e is the electrometer calibration coefficient, k_{tp} is the correction for temperature and pressure, N_{SK} is the air-kerma strength calibration coefficient, A_{ion} is the correction for recombination at the time of calibration, and P_{ion} is the correction for recombination at the time of measurement. Note that this equation involves a current measurement, which is the case for HDR ^{192}Ir measurements where the

ionization currents are high. For well chamber measurements of LDR brachytherapy sources, charge is often measured over a specified time and then divided by a time increment to arrive at the same value (i.e., current).

It is useful for the manufacturer to employ quality assurance (QA) checks of the well chamber to verify its response is stable with time. This is normally achieved by use of another long-lived source such as a ^{137}Cs needle, ^{90}Sr source, or a linear accelerator (irradiation of the entire chamber). In addition to QA checks, leakage of the chamber should be measured to check for instabilities in the electrical signal pathways.

In summary, there is a robust framework of brachytherapy source calibration standards available to end users in North America. The source calibration framework enables traceable calibrations of end-user instrumentation the strength of brachytherapy sources consistently.

15.5 DOSIMETRY INVESTIGATORS MEASUREMENTS AND CALCULATIONS

15.5.1 Rationale

Single-source brachytherapy dose distributions are needed for determining brachytherapy dosimetry parameters as used in clinical brachytherapy treatment planning systems (TPSs). However, evaluation of single-source brachytherapy dose distributions is challenging for medical physicists, and much more difficult than for evaluating EBRT dose distributions. The problem is due to lack of an established dosimetry protocol such as AAPM TG-51 as for EBRT [27], and lack of available resources in a typical radiotherapy clinic to perform such evaluation. Brachytherapy dosimetry is further complicated by its high dose gradients and lower photon energies in comparison to EBRT. Lacking standardized methods and equipment for evaluating brachytherapy dose distributions, physicists must develop their own dosimetry protocol and devise measurement or calculation methods.

Over the years as new brachytherapy sources have entered the marketplace, a cadre of physicists (sometimes termed dosimetry investigators) have embraced this challenge, evaluated single-source brachytherapy dose distributions, and determined the brachytherapy dosimetry parameters as used in TPSs [11]. The field of brachytherapy dosimetry has evolved through scientific advancements learnt from one source that have benefited later investigation of another source, largely from careful documentation of the evaluation techniques in the peer-reviewed literature. Furthermore, the AAPM has provided explicit guidance to dosimetry investigators for performing dose measurements and calculations, see sections V.D. and V.E. of Ref. [11], respectively. Given the necessary initial investment in equipment and knowledge, dosimetry investigators have repeated the process of dose evaluation for new brachytherapy sources as they become available for clinical use. In fact, part of the process in making sources available for general clinical use requires the dosimetry parameters to be determined [28, 29].

From the context of dosimetry, brachytherapy sources may be categorized in a 2-by-2 matrix of low- or high-energy photon emitters or as LDR or HDR sources. As applicable to measuring the dose distribution, sources of lower energy and lower dose rate are harder to evaluate due to increased sensitivity of results because of detector and phantom medium composition and due to lower detector signal-to-noise, respectively. These sources would include LDR seeds of ^{103}Pd, ^{125}I, and ^{131}Cs with photon energies ranging from 0.02 to 0.03 MeV and individual source strengths 10 U or less. Conversely, brachytherapy sources of higher energy and higher dose rate are easier to evaluate due to decreased sensitivity of results because of detector and phantom medium composition and due to higher detector signal-to-noise, respectively. These sources would include HDR sources of ^{192}Ir and ^{60}Co with photon energies ranging from 0.4 to 1 MeV and individual source strengths exceeding 10^4 U.

15.5.2 Measurements

Since dose measurements are sensitive to the aforementioned source differences, dosimetry investigators must customize their techniques and equipment to the type of source being measured [30]. Low-energy sources require special attention since their dose gradients are typically steeper than high-energy sources, due to the fact that they are more easily attenuated, and the high-energy sources are typically used within HDR brachytherapy applicators where patient tissue is positioned farther from the source. Fortunately, low-energy seeds are often used for volume or surface implants where the multi-seed geometry grossly diminishes the dose gradient. For low-energy sources, substantial corrections are needed to adjust the dose measured in a (typically) plastic phantom to the reference material of liquid water. An approach developed over the years is to select a phantom material where the uncertainty in the medium correction is low, even though the magnitude of the medium correction may be greater than a more water-equivalent plastic [31]. For high-energy sources, differences in medium composition make little difference when assessing the dose distribution, and dosimetry investigators will then select the phantom composition based on convenience. However, dosimetry for high-energy sources is more subject to sensitivity of radiation scatter to the experimental setup. Dosimetry investigators need to provide sufficient backscattering material so that full scatter conditions are approximated for dose at a given distance from the source [32]. For LDR sources, seeds having the highest possible source strength are ordered so that the radiation signal is maximized and to provide a long timeframe for which the source can be used since their decay half-lives are 9.7, 17.0, and 59.4 days for ^{103}Pd, ^{125}I, and ^{131}Cs, respectively [11]. For HDR sources, dose at distant locations are sometimes of interest and the dosimetry investigator is challenged trying to characterize its distribution over a large range of distances that may be beyond the dynamic-range capabilities of a single exposure for a radiation detector.

15.5.3 Calculations

The gold standard for calculating single-source brachytherapy dose distributions is through Monte Carlo simulations of radiation transport [33], and is applicable for all brachytherapy source types. Unlike dosimetry measurements, the Monte Carlo technique can simulate things impossible to produce in reality like individual photons from nuclear disintegration (instead of the polyenergetic emissions common to radionuclide decay, for gleaning detector energy-response corrections); exceedingly small detectors (having negligible dose gradients and absolute spatial accuracy); and unusual material dimensions, compositions, and densities (for ascertaining sensitivity of simulation results to assumptions of geometric design).

15.6 SOCIETAL RECOMMENDATIONS
15.6.1 Rationale

The field of brachytherapy dosimetry currently affords global consistency to the brachytherapy community through use of dosimetry parameters in TPSs based on the AAPM TG-43 dose calculation formalism [11]. However, there are subtle issues around which dosimetry parameters to use and how TPSs perform TG-43 dose calculations with specific algorithms [34]. Based on choice of dosimetry parameters or algorithm implementation, dose calculations could vary by more than 50% [35]. Consequently, professional societies like the AAPM, ESTRO, ABS (American Brachytherapy Society), and ABG (Australasian Brachytherapy Group) have issued reports aiming for standardization. These standards include data resolution requirements and data interpolation (and extrapolation) methods.

15.6.2 Consensus Data

Given the dynamic state of the literature on brachytherapy dosimetry data for a given source model, professional societies have developed a process through which publications containing data are reviewed for meeting minimum standards, and then such data are formally disseminated to the community. Currently and largely led by the AAPM, formal committees have examined published dosimetry parameters for new source models. The evaluation methods are described elsewhere in explicit detail [31]. The AAPM works harmoniously with other professional societies so that there are no conflicting consensus data. However, there is not yet an established process to revise the consensus data when new dosimetry parameters are published or improvements are made to the dosimetry techniques.

15.6.3 Data Dissemination

Consensus data are made available online through the posting on the Brachytherapy Source Registry managed jointly by the AAPM and the Imaging and Radiation Oncology Core (IROC) Houston QA Center. The registry is available online and updated regularly given changes to source vendor contact information, inclusion of additional related supporting publications for a specific brachytherapy source model, and addition/subtraction of sources based on compliance with AAPM standards for calibration intercomparison frequency and accuracy [33, 36, 37]. The HDR ^{192}Ir consensus data on the AAPM/IROC Houston registry is actually managed in the form of web links to the ESTRO website, where all data is downloadable and in a standardized format.

15.6.4 Potential Future Directions

There are several new directions that societal recommendations for brachytherapy dosimetry may take in the near future. These could include evaluation of eBT sources and establishing similar dosimetric and calibration standards as for radionuclide-based sources. Such standards would be complicated by the low-energy HDR nature of eBT sources sharing strengths and weaknesses of current standards which are largely low-energy LDR or high-energy HDR. Extension of such standards could also include the applicators associated with the eBT sources. An additional aspect of eBT sources is that they do not explicitly fit the TG-43 dose calculation formulism, which is based on source strength specification. The TPS would require modification or a work-around to fit within the current formulism [38].

Another potential future direction of societal recommendations on brachytherapy dosimetry is adoption of a new source strength metric, dose rate to water at 1 cm on the source transverse plane ($D_{w,1cm}$), to replace the product of air-kerma strength (S_K) and the dose-rate constant (Λ) [39]. This approach would have a smaller uncertainty than the current technique where uncertainties in the $S_K \times \Lambda$ product are determined separately. With $D_{w,1cm}$ being a direct measure of the source output in the medium of interest and for a specific source, not being traced to a measurement (or calculation) performed by a dosimetry investigator at a much earlier date, the uncertainty for LDR seeds and HDR sources could potentially be lower. However, diminishment in the overall uncertainty by this amount is not deemed clinically relevant given it is smaller than the current S_K measurement uncertainty performed by the end-user. The general feeling in the community is that resources should be put toward addressing material heterogeneities for improved brachytherapy dose calculation accuracy [34].

15.7 CLINICAL USER'S CALIBRATIONS (ASSAYS), MEASUREMENTS, AND CALCULATIONS FOR PHOTON EMITTING SOURCES

15.7.1 Clinical User Calibrations

The accurate measurement of LDR and HDR sources is of the utmost importance, and is required or recommended by the US Food and Drug Administration (FDA), AAPM, and ABS. According to AAPM Task Group Reports 40, 56, and 138, the clinical medical physicist has the responsibility to verify the manufacturer-provided source strength measurements using ADCL-traceable instrumentation [13, 40, 41]. Although rare, large deviations in source strength have been reported. In some instances seeds had no radioactivity, and in other instances the source strength was grossly misstated by the vendor calibration certificate. Clerical errors by the manufacturer have also resulted in a clinic receiving the wrong sources or paperwork.

Well chambers are the most common instruments used for air-kerma strength (U) source calibration of LDR and HDR photon sources, as well as for beta-emitting sources. A NIST traceable calibration factor for a well chamber may be obtained by sending the chamber to an ADCL. Alternatively, a clinic may order a source calibrated by an ADCL, insert the source into the chamber, and transfer the ADCL calibration to their well chamber. Nuclear medicine chambers are a less desirable alternative for source calibration, where disadvantages include that they are generally calibrated for source apparent activity (not air-kerma strength), and pre-determined settings are for nuclear medicine radionuclides and not brachytherapy sources. The source geometry may not be adequately affixed and pressurized chambers may experience gas leakage. Furthermore, when a nuclear medicine well chamber is used for both unsealed and sealed radionuclides, there is a risk of contamination or recalibration by the nuclear medicine team.

Commissioning tests for a well chamber include identification of the sweet-spot location and size, measurement of any change in the chamber response as a function of the orientation angle of the seed holder, measurement of the polarity effect, linearity, measurement of recombination effects, and calibration for the sources that will be used clinically. The user should also establish the assay procedure, including performing a constancy check prior to usage with another known source, typically with a source of comparable energy and a long half-life. With a photon energy of approximately 60 keV and half-life of 433 years, ^{241}Am is commonly used for constancy tests. Other long-lived sources such as ^{137}Cs and ^{90}Sr may also be used. Constancy checks are useful for detecting leaks in sealed well chamber systems, as well as problems with the collection region, cables, or electrometer. The well chamber and electrometer system voltage and leakage should also be assessed prior to the measurement.

For LDR sources, the AAPM Report 98 from the Low Energy Brachytherapy Source Calibration Working Group recommended that the end-user calibrate at least 10 sources or 10% of sources, whichever is greater, and preferably all of the sources when small quantities are

to be used (e.g., eye plaques) [37]. When loose seeds are ordered in combination with stranded sources, it is recommended to assay 5% or 5 seeds, whichever is fewer. The clinical user calibration should agree with the manufacturer to within 3%, and an investigation should be performed for greater deviations. It is recommended to contact the vendor if deviations are greater than 5%. Individual seed variations greater than 6% should be discussed with the radiation oncologist. Independent third-party calibration may also be available for all of the sources for a fee. The clinical user should review the third-party results and maintain the documentation for the United States Nuclear Regulatory Commission (NRC) inspection.

For HDR ^{192}Ir sources, the NRC regulation Title 10 CFR requires calibration of the source by the clinical user prior to first use, following repair that requires removal of the source or major repair of the components associated with the source exposure assembly, or at intervals not exceeding 3 months, which is a typical period for source exchanges. The clinical user calibration should agree within 5% with the manufacturer certificate [42]. The typical clinical setup is to place a thin plastic catheter inside of a plastic source holder that fits into the well chamber as shown in Figure 15.6. The source holder helps to center the plastic catheter in the middle of the chamber. The catheter may be secured to the source holder using a small piece of tape. After identifying the "sweet spot" of the chamber, where the chamber response is greatest, through a stepwise measurement (use the afterloader treatment console to step source in ≤5mm steps) of the output as a function of location inside the catheter, the user measures the well chamber current produced by the radioactive source. This current reading is converted to a dose rate using an ADCL calibration coefficient (cGy cm^2 h^{-1} nA^{-1} or U/nA).

The electrometer used in brachytherapy calibrations should have a range that is appropriate to the type of sources being measured. Some low-activity source measurements produce very little ionization current in the well chamber. The electrometer should have a resolution as low as 10 fA for LDR sources or up to 200 nA for HDR sources [43]. The signal resolution should be better than 0.1%. An electrometer with a high and low range is therefore recommended for brachytherapy calibrations, and both settings should be calibrated by the ADCL at a frequency not to exceed two years.

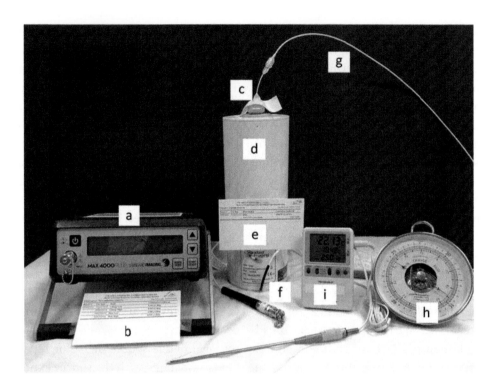

FIGURE 15.6 A typical clinical user calibration setup: (a) Electrometer (model MAX-4000 Plus, Standard Imaging, Inc., Middleton, WI), (b) note card of electrometer ADCL-calibration coefficients, (c) source holder and tape to secure plastic catheter, (d) well chamber (model IVB1000, Standard Imaging, Inc., Middleton, WI), (e) note card of well chamber ADCL-calibration coefficient, (f) triax cable connector for well chamber coupling to electrometer, (g) plastic treatment catheter and transfer tube to HDR remote afterloader unit, (h) aneroid barometer, and (i) electronic thermometer.

15.7.2 Clinical User Measurements

The clinical user may perform dosimetric measurements in order to confirm or elucidate dosimetry, such as to verify the dose calculation parameters or the impact of heterogeneities and treatment geometries, or to perform *in vivo* verification of the treatment dose. In general, there are steep dose gradients in brachytherapy treatments and the clinical user will require high precision phantoms and detectors for these types of measurements. This section highlights examples of the types of experiments that the clinical user might conduct. Additional references are provided to give the reader more comprehensive guidance on measurement techniques and choices of apparatus for using ionization chambers [44–48], radiochromic film [49–54], gel [55, 56], thermoluminescent dosimeters (TLD) [49, 57], optically stimulated luminescence dosimeters (OSLDs) [58], diode [59, 60], metal-oxide-semiconductor field-effect transistors (MOSFET) [61–64], scintillation plastics [57, 65–68], or lithium formate pellets [69, 70].

Verification measurements may be performed to verify dose profiles, depth dose, and the source anisotropy.

For example, the consensus data sets presented in the TG-43 reports were generated using a blend of 2D reference-quality dose distribution data from Monte Carlo and experiments [11]. The experimental verification has an essential role in the consensus process, and provides independent validation prior to clinical use of a new brachytherapy source model. TLD have been most commonly used for such experiments because their diminutive size reduces the uncertainty from steep dose gradients, and well known protocols are available for calibration and energy dependence [71–73]. Although film provides even greater resolution than TLD, care must be taken to properly calibrate the radiochromic film for the resultant photon energies. For low-energy photon-emitting radionuclides, the corrections can exceed 10% [74]. Another example of source parameter verification is the measurement of the profiles and depth dose for a beta-emitting planar source containing ^{32}P. Diodes and radiochromic film have been used to cross-check Monte Carlo results, as well as to evaluate how the dosimetric parameters change when nonstandard geometries are encountered in the clinic

[75, 76]. Radiochromic film is an excellent dosimeter for beta emitters since the energy dependence of the film is smaller than it is for photon emitters [77]. However, the user should take care to properly radiologically scale any phantom materials as well as the materials within the film [75]. Radiological scaling is used to adjust the actual depth in the medium to an equivalent depth in water.

Clinical user measurements may also establish the effect of non-water heterogeneities in the vicinity of the brachytherapy treatment. Heterogeneities may be built into the applicator, present in the patient tissues (or absence of tissues), or even applied external to the treatment site. Experimental measurement offers a method to validate calculations or measure dose under conditions that are difficult to model.

Source applicators are typically constructed from plastics or other materials. The heterogeneity effects are relatively minor when the materials are nearly water-equivalent, but effects can be significant when dense materials such as high-Z metals are involved [78]. For example, a shielded vaginal cylinder that is currently manufactured by both Varian and Nucletron includes tungsten shields ($Z = 74$) for reducing the dose to uninvolved healthy tissues, namely the rectum, bladder, or uninvolved vaginal mucosa. The clinical user may perform measurements in order to experimentally confirm that the shields provide the expected attenuation magnitude and distribution. Radiochromic film has been used to evaluate the dose distribution for the shielded vaginal cylinder with a HDR ^{192}Ir source [79]. A second example of heterogeneous source applicators is an eye plaque as used for temporary LDR treatment of ocular melanoma. COMS eye plaques have non-water equivalent Silastic and Modulay components that have effective atomic numbers greater than water. Eye plaque dose distributions have been measured by several authors [80, 81]. In particular, the work by McCauley et al. demonstrated the use of radiochromic film for commissioning eye plaque therapy and verifying the ~16% attenuation that is due to the Silastic insert that holds ^{125}I seeds within the plaques [81]. McCauley et al. also demonstrated how dosimetric measurements could be used to identify potential deviations in the thickness of the Silastic material that could arise from manufacturing variations.

Patient tissue heterogeneities can also impact delivered dose to the patient, which is more substantial for the lower energy radionuclides and also for tissues with electron densities that differ from water. For example,

an HDR ^{192}Ir planar applicator that is placed upon the skin may have one side facing the tissue and the other side facing air. The lack of tissue causes a reduction in backscatter. Consequently, dose to the treatment site is reduced, with a reduction that varies as a function of depth and applicator geometry. Calcifications, bone, and lung tissues can also produce changes in the dose delivered to the treatment site. Several authors have reported the impact of tissue heterogeneities on the delivered dose [82–89].

Occasionally, external shields are used in brachytherapy to reduce the dose to nearby healthy tissues. For example, lead shields may be used for intraoperative brachytherapy to protect nearby organs that are outside of the treatment field [90]. Similarly, oral shields may be used to protect uninvolved healthy tissue for brachytherapy of the head-and-neck region [91]. Shields may also be used in vaginal cylinders and cervical cancer applicators. Skin applicators may use conical shaped tungsten applicators to focus the radiation and spare uninvolved tissue [92]. The scatter in the applicator and the air cavity between the source and the skin can be difficult to model, and depends on the size and the shape of the conical applicator. The impact of the lead shields is best assessed by experimental measurement using a variety of dosimeters, including diodes, ion chambers, or TLDs [93–96].

In vivo dosimetry is a type of clinical user measurement that provides a record of the dose that is delivered to the patient's tissues. *In vivo* dosimetry is challenging for brachytherapy treatments because of the presence of high dose gradients, and the detector energy sensitivity corrections that may change as a function of distance from the source. The high dose gradients demand precise positioning of the detector active volume with respect to the radiation source(s). If patient dose is to be measured and not simply inferred, the detector must also be inserted into the treatment region. *In vivo* dosimetry may be used in the intraoperative setting to quantifying the dose to the nearby healthy tissues or target [97]. In the intraoperative setting, *in vivo* dosimetry is particularly useful because often a single high-dose treatment fraction is delivered in the OR and it is typically not possible to obtain a CT scan with the applicator in position. Therefore, the physicist is unable to exactly recreate the precise placement and curvature of the planar applicator. *In vivo* dosimetry has also been used to monitor the location and duration of source dwell positions in

HDR brachytherapy in order to prevent gross errors in the execution of the treatment plan, such as an incorrect connection of the treatment unit to the treatment catheters or an incorrect catheter length used in the treatment plan [98, 99]. Another approach is to place an *in vivo* dosimeter directly into the patient tissue, such as the rectum, urethra, or vaginal mucosa to monitor the dose that is delivered near these critical structures during treatments [61, 62, 100–102]. Chapter 11 of this book provides further discussion on the exciting future of *in vivo* dosimetry for brachytherapy.

15.7.3 Dose Calculations

Brachytherapy dose calculations for permanent and temporary implants may be broadly categorized into correction-factor based and model-based approaches. Correction factor methods begin with a calibrated dose under a set of standard conditions, and use correction factors to adjust the dosimetry for the actual conditions encountered in the patient treatment. Model-based calculations are useful when modeling the dosimetric impact of heterogonous materials, and can directly solve the Boltzmann transport equation for photons and electrons using deterministic discrete ordinate methods, collapsed cone convolution, or Monte Carlo radiation transport simulations.

15.7.3.1 Correction-Based Dose Rate Calculations: TG-43

The AAPM TG-43 reports provide the most widely used formalism for brachytherapy dose rate calculations for photons, and can be classified as a correction factor approach. The correction factors are established through a consensus dataset composed of theoretical and experimental results [10, 11]. The reader is encouraged to read the AAPM TG-43 reports for discussion and examples of reference datasets. The dose rate calculation equations for point- and line-source approximation are specified in Equation 15.3 and Equation 15.4, respectively. In Figure 15.7, $\theta_0 = \frac{\pi}{2}$ radians and $r_0 = 1$ cm.

$$\dot{D}(r,\theta) = S_K \Lambda \frac{G_P(r,\theta)}{G_P(r_0,\theta_0)} g_P(r) \phi_{an}(r)$$

(15.3)

(point source approximation)

$$\dot{D}(r,\theta) = S_K \Lambda \frac{G_L(r,\theta)}{G_L(r_0,\theta_0)} g_L(r) F(r,\theta)$$

(15.4)

(line source approximation)

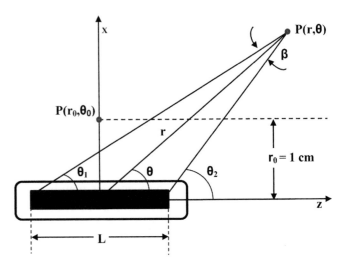

FIGURE 15.7 Coordinate system used for brachytherapy dosimetry calculations. A source of length L is centered at the origin, and the long axis of the source is oriented in the z-direction. A reference point, P, is located at a radial distance, r, and an angle, θ, with respect to the origin. θ_1 and θ_2 represent the angles between the reference point and the ends of the source. The interior angle, β, is also denoted.

The standard conditions in this formalism are represented by the source strength, S_K, as measured by a NIST-traceable calibration, and is intended to reflect the air-kerma rate *in vacuo*, $\dot{K}_\delta(d)$, without radiation scatter and with an energy cutoff that is typically $\delta > 5$ keV for low-energy photon-emitting sources.

$$S_K = \dot{K}_\delta(d)d^2$$

(15.5)

The dose-rate constant, Λ, converts the air-kerma strength into a dose rate to water in water at a distance of 1 cm from the center of the source along the source transverse plane, $\theta = \pi/2$. The dose-rate constant varies with source model, because of differences in the scattering and attenuation in water from various photon energies. The distribution of radioactivity inside the source capsule also influences the value of Λ. Sources that concentrate radioactivity near the capsule center will have higher dose rates at the reference position, and therefore higher Λ values due to the inverse-square law. Also, sources with higher photon energies will generally have higher Λ values.

The geometry function, $G_P(r,\theta)$ or $G_L(r,\theta)$, mitigates the influence of the inverse-square law on dose-rate calculations, and is the dominant factor characterizing the dose-rate distribution. The angles β and θ have units of radians in the calculation.

$$G_P(r,\theta) = \frac{1}{r^2} \qquad (15.6)$$

$$G_L(r,\theta) = \begin{cases} \dfrac{\beta}{Lrsin(\theta)}, & \theta \neq 0 \\[3mm] \dfrac{1}{r^2 - \dfrac{L^2}{4}}, & \theta = 0 \end{cases} \qquad (15.7)$$

While it is convenient in brachytherapy TPSs to use a simple function such as a point- or line-source model for the geometry function, these simplifications do not precisely capture the dose falloff. This is due to simplification of the radioactivity spatial distribution as a point or line segment, as well as how it does not account for radiation attenuation and scatter in water, the reference medium.

The radial dose function is identified as, $g_P(r)$ or $g_L(r)$, depending upon which function is used to model the radioactivity distribution. The radial dose function accounts for scatter and attenuation effects on the source transverse plane. It decreases more rapidly as a function of distance for lower energy sources because they are more easily attenuated in water due to the photoelectric effect.

Brachytherapy source design is meant to be cylindrically symmetric, but asymmetric with respect to polar angle due to the internal components and self-attenuation by the source material itself. Variations in the dose distribution as a function of polar angle around the source are accounted for by the 1D anisotropy function, $\varphi_{an}(r)$, or 2D anisotropy function, $F(r,\theta)$.

TG-43 dose calculations rely on the principle of dose superposition. The dose delivered from a group of sources is simply the sum of the doses delivered by each individual source. This approximation is excellent for a high-energy HDR ^{192}Ir source that is administered with a remote afterloader using a single dwelling source, as well as historically for high-energy LDR sources such as ^{137}Cs tubes and ^{192}Ir ribbons. However, it is important to consider that HDR sources are modeled with a certain amount of source cable, any bending of the cable during treatment delivery can lead to small differences between the ideal dose and the delivered dose at locations proximal to the source. For low-energy brachytherapy where multiple sources are simultaneously placed inside the treatment site, the inter-seed attenuation effects have been reported to have a small effect on the radiation

therapy dose distribution [103–107]. This is due to the properties of a volumetric implant being dominated by dose from the nearest sources, which by their proximity, implies that they are not attenuated by closer sources.

An advantage of the TG-43 formalism is that the correction factors (i.e., dosimetry parameters) may be interpreted to have some physical basis. Additionally, the geometry function dominates for characterizing the dose-rate distribution and is based upon a simple equation that is not subject to interpolation errors. The remaining correction factors exhibit less variation and can be described in small tables with linear interpolation without substantial loss of accuracy. A disadvantage of the TG-43 approach is that it assumes an aqueous medium and does not properly account for material heterogeneities in the brachytherapy applicator or the patient. The assumption of an aqueous medium is good for most LDR and HDR applications; however, the clinical team should carefully evaluate the impact of material heterogeneities when high density (e.g., dense plastics or metals) or low density (e.g., air or lung) media are present near or within the treatment site.

15.7.3.2 Model-Based Dose Calculation Algorithms

The preferred method for calculating dose under realistic conditions that include material heterogeneities is with model-based dose calculation algorithms (MBDCAs). Through varying methods, these approaches solve the linear Boltzmann transport equation to calculate the dose. Monte Carlo methods employ a statistical approach to solve the transport equation for a distribution of initial conditions, and provide a statistical average of the dose based on the sum of microscopic interactions. The quality of Monte Carlo results is statistically dependent on the number of particle histories that are considered. Discrete ordinate approaches are also available, which bin the Boltzmann transport equation into a finite set as a function of predefined spatial grid, angles, and energy histogram and apply radiological equations to the bulk materials. In either case, the quality of the results requires accurate modeling of the source, geometry, and boundary conditions. The accuracy of the discrete ordinates and collapsed cone approaches is dependent on the type and amount of discretization that is used to group the spatial dimensions and energy.

A detailed overview of calculation algorithms is found in a pair of *Medical Physics* Vision 20/20 publications on the use of Model-Based Dose Calculation Algorithms

(MBDCAs) by Rivard et al. [108, 109]. A description of MBDCA methods and recommendations for their clinical implementation is presented in the TG-186 report [110]. Furthermore, a reference set of CT scans and plans is available for the commissioning of a MBDCA [111]. The user may download reference datasets, calculate a dose distribution for an applicator using their TPS, and compare the dose results against the Monte Carlo reference data [112].

15.7.3.3 Dose Calculations for Permanent and Temporary Implants

Once the initial dose rate, $\dot{D}$, to a point in the treatment field is known, the dose to the patient for a permanently implanted source is calculated using the following formula:

$$Dose = \dot{D} \times \tau \quad \text{(Dose from a permanent implant)} \quad (15.8)$$

Where τ represents the average life and is equal to $\frac{t_{1/2}}{\ln(2)}$.

For a temporary implant, the dose calculation is modified by the amount of source strength that remains after completion of the treatment and the sources are removed.

$$Dose = \dot{D} \times \tau \left(1 - e^{-\frac{t}{\tau}}\right) \quad \text{(Dose from a temporary implant)}$$

$$(15.9)$$

15.8 CLINICAL USER'S CALIBRATIONS (ASSAYS), MEASUREMENTS, AND CALCULATIONS FOR BETA EMITTING SOURCES

15.8.1 Clinical User Calibrations

Calibration of beta-emitting sources, such as ^{90}Sr and ^{32}P, is especially challenging due to the steepness of the depth dose curve. The IAEA TECDOC-1274 provides guidance for the calibration of beta emitting sources [43]. The IAEA recommends that beta-emitting planar sources be calibrated at a distance of 1.0 mm measured from the surface of the source. Beta seed and line sources should be calibrated at a distance of 2.0 mm from the center of the source. Beta balloon shells and stents should be calibrated at a distance of 0.5 mm from the surface of the shell or stent. Other detectors may also be used for the calibration of beta-emitting sources. Parallel-plate ionization chambers, diodes, and

radiochromic film may be cross calibrated by a device with a NIST-traceable calibration (e.g., via a linear accelerator with electron energy comparable to the source being assayed) and have been used with success. A depth dose curve should also be obtained for calculation of the dose at depths different than the calibration depth. The IAEA also recommends that the user tests the uniformity of the source in terms of their dose rate at the calibration distance. When the calibration is performed at a depth in phantom and converted to a water-equivalent depth, it is important to properly scale the phantom materials. The IAEA TECDOC-1274 and ICRU Report 56 provide guidance for radiological scaling for beta emitters, and Deufel et al. provides an example for the use of radiological scaling to calibrate a planar ^{32}P source using radiochromic film [43, 75, 113].

15.8.2 Clinical User Measurements

Clinical measurements for beta emitters have a similar role as their photon emitting counterparts. Clinical measurements may be performed in order to verify dosimetry parameters and the impact of nonideal clinical conditions on the delivered dose. Suitable detectors for beta emitters must have high spatial resolution, since depth dose variations of 10% per 0.1 mm change in the depth are possible. Clinical measurements using radiochromic film have been used to assess the dose uniformity, dose profiles, and depth dose as a part of pretreatment QA for ^{32}P planar sources that are used for the treatment of the spinal dura [75]. The impact of source size and material heterogeneities was also evaluated in the same study. Radiochromic film dosimetry has also been applied to the measurement of the dose distribution from a beta source for intravascular brachytherapy [114, 115]. Diodes and microMOSFETs are alternate dosimeters for beta emitters. Diodes have been used to measure the depth dose from planar sources [76]. MicroMOSFETs have been used to evaluate the impact of stents on the dose delivered during intravascular brachytherapy [116, 117].

15.8.3 Dose Calculations

For beta-emitting sources, simple correction-based approaches may be used to calculate the dose rate to a point. Dose profiles and a depth-dose table can be used to estimate the dose at a point from the center of the source. ICRU Report 56 outlines a dose-kernel model that can be used to perform dosimetry for a beta-emitting source in a homogenous medium [113]. The approach assumes a slab geometry and is based on the work by Cross et al. [118]. Janicki et al. extended the work of Cross et al. for a non-homogenous slab medium [119]. The reader is referred to Deufel et al. for a complete description of theoretical calculations and application to ^{32}P planar sources that are used for brachytherapy treatments of the spinal dura [75].

Monte Carlo methods may also be used to calculate the dose for a beta-emitting source. This is essential when the medium cannot be approximated by simple slab geometry or when the source cannot be approximated as planar. Beta-emitting eye plaques, intravascular sources, and planar applicators for spinal dura treatments are good examples of the use of Monte Carlo dosimetry techniques for beta dosimetry [76, 120, 121].

15.9 FUTURE DIRECTIONS

15.9.1 Model-Based vs. TG-43

Each year, computers become faster and less expensive. It is therefore worth considering whether MBDCAs will eventually entirely replace TG-43 as the primary dose calculation algorithm. As noted in Section 15.7.3.2, model-based dose calculations with HDR ^{192}Ir brachytherapy generally result in minor dose differences to soft tissues when compared with the TG-43 standard approach. However larger differences have been noted when there are substantial tissue heterogeneities near the treatment region, for example in an HDR partial breast brachytherapy treatment, the TG-43 will overestimate the dose to skin that is in close proximity to air. The benefits of a more accurate dose distribution will likely mean that MBDCAs will replace TG-43 once high- and low-energy dose distributions can be rapidly calculated and thoroughly validated. Of course, physicians must consider how to modify their prescription doses in circumstances where there are large systematic differences (e.g., eye plaques), akin to the changes in the prescriptions for lung treatments that occurred ~30 years ago when heterogeneity corrections became standard for EBRT. Physicists will also need to unify the method for how the electron density of materials is calculated from CT numbers or assigned for dense materials.

15.9.2 *In Vivo* Quality Control

In vivo measurement of the dose delivered to the patient, which includes real-time quality control (QC) that can detect errors in the treatment delivery, is perhaps the "holy grail" for HDR brachytherapy dosimetry. A *Medical Physics* Vision 20/20 paper by Tanderup et al. highlighted how *in vivo* dosimetry can be used for identifying HDR afterloader malfunction, intra-fraction motion, applicator reconstruction errors, applicator length errors, source step size, and transfer tube connections [122]. *In vivo* dosimetry would also be valuable for the QC of intensity modulated brachytherapy, in which a shielding material is used to asymmetrically block radiation from the source in order to create a dose distribution that is more conformal to the target and thereby better spares healthy tissue. *In vivo* dosimetry could be used to ensure the position and function of any moving parts.

Although researchers have made substantial progress and several prototypes have been used for the types of QC suggested by Tanderup et al., *in vivo* dosimeters have not been incorporated into commercial HDR treatment systems. Flat-panel based *in vivo* dosimetry is a likely candidate for first-to-market for HDR brachytherapy, since it is a straightforward add-on to a brachytherapy system and requires no modifications to the HDR afterloader and placement of additional invasive devices into the patient. The flat panel approach is an indirect measurement of the dose, since it records the treatment dwell locations and time and uses the information to reconstruct the dose inside the patient. Gel, MOSFET, and scintillator detectors have focused on directly measuring the dose in a clinically relevant region, such as the urethra, rectum, or vaginal mucosa. The future of brachytherapy dosimetry will hopefully include *in vivo* tools for evaluating the delivered HDR dose as it is being administered. An immediate challenge that all *in vivo* dosimetry systems face is the interfacing with the treatment delivery system. The *in vivo* system ought to be designed such that it automatically interrupts the treatment when a delivery deviation is detected.

15.9.3 New Dosimeters

New radiation dosimeters are exciting when they provide superior methods of calibration or improve the verification of the dose delivery. Brachytherapy dosimetry is challenging because there are high dose gradients near the treatment region. These high gradients mean that a small uncertainty in the position of the dosimeter can produce a large change in the measured dose. With *in vivo* dosimeters, there is the added challenge that the direct measurement of the dose to a critical structure or target may require a detector that can fit inside of a narrow catheter or similar device.

With regards to calibration dosimetry, a recent publication on graphite calorimetry systems for the dosimetry of therapeutic x-ray beams invites the possibility of a calorimetry-based brachytherapy calibrator [123]. Calorimetry is used by primary standard dosimetry laboratories because it offers a direct measurement of the energy delivered to medium, and therefore relies on fewer correction factors than an ionization chamber, diode, or other dosimeters.

With regards to *in vivo* dosimetry, researchers have been working toward smaller detectors that can be easily placed inside of a catheter. Continued development of MOSFET, diodes, and scintillators will hopefully provide tools for directly measuring the dose in a clinically relevant region, such as the urethra, rectum, or vaginal mucosa. Researchers have also been striving to produce detector systems, internal or external to the patient, that can be used to monitor the location and duration of source dwells. The continued development of new detectors will hopefully make QC provided by *in vivo* dosimetry a routine and straightforward part of the treatment delivery process.

15.9.4 Safety and Learning

At its most fundamental level, the role of the medical physicist is to ensure the safe and accurate delivery of radiotherapy. Recently, incident learning systems (ILS), root cause analyses, and failure modes effect analyses (FMEAs) have been employed to help predict and understand errors in the radiotherapy process.

With regards to brachytherapy dosimetry, ILS is a useful tool for the design of a quality management program. ILS provides a mechanism, for rational, evidence based practice improvement. For example, the ILS can be used to evaluate whether use of the antiquated units of source strength curie (Ci) should be abolished. The IAEA and NRC medical event reports detail how the simultaneous use of Ci and U for clinical applications has been a contributing cause of multiple medical events [43]. Deufel et al. reported on how the comprehensive use of an ILS can be used to reduce risk, improve communication, and improve the quality of written procedures [124]. The ILS described by the authors was used to record each mistake that was made in the radiotherapy treatment process, and the results were used to communicate potential failure modes to team members and to identify ways to improve the process.

Root cause analysis and FMEA are risk-analysis tools for estimating practice risk and identifying weak points in the process. Risk-analysis safety tools can be used to reduce the likelihood or severity of errors in the workflow. The AAPM TG-100 report provides a description of risk-analysis methods that can be used in the design of a quality management program [125]. For brachytherapy dosimetry, this approach might be applied to the design of an LDR source assay procedure, an HDR applicator commissioning procedure, or the design of a planning checklist.

REFERENCES

1. N. Tselis, et al., "High dose rate brachytherapy as monotherapy for localised prostate cancer: Review of the current status," *Clinical oncology (Royal College of Radiologists)* **29**(7), 401–411 (2017).
2. K. K. Lee, et al., "High-dose-rate vs. low-dose-rate intracavitary brachytherapy for carcinoma of the uterine cervix: Systematic review and meta-analysis," *Brachytherapy* **14**(4), 449–457 (2015).
3. K. Ota, et al., "Review: The reemergence of brachytherapy as treatment for non-melanoma skin cancer," *Journal of Dermatological Treatment* **29**(2), 170–175 (2017).
4. J. Skowronek, M. Wawrzyniak-Hojczyk, K. Ambrochowicz, "Brachytherapy in accelerated partial breast irradiation (APBI) – review of treatment methods," *Journal of Contemporary Brachytherapy* **4**(3), 152–164 (2013).
5. L. C. Mendez, et al., "Three-dimensional-guided perineal-based interstitial brachytherapy in cervical cancer: A systematic review of technique, local control and toxicities," *Radiotherapy & Oncology* **123**(2), 312–318 (2017).
6. A. Youroukou, et al., "The potential role of brachytherapy in the irradiation of patients with lung cancer: A systematic review," *Clinical and Translational Oncology* **19**(8), 945–950 (2017).
7. H. Buckley, C. Wilson, T. Ajithkumar, "High-dose-rate brachytherapy in the management of operable rectal cancer: A systematic review," *International Journal of Radiation Oncology, Biology, Physics* **99**(1), 111–127 (2017).
8. J. J. Echegaray, et al., "Iodine-125 brachytherapy for uveal melanoma: A systematic review of radiation dose," *Ocular Oncology and Pathology* **3**(3), 193–198 (2017).
9. I. Goutos and R. Ogawa, "Brachytherapy in the adjuvant management of keloid scars: Literature review," *Scars, Burns & Healing* **3**, 2059513117735483 (2018).
10. R. Nath, et al., "Dosimetry of interstitial brachytherapy sources: Recommendations of the AAPM Radiation Therapy Committee Task Group No. 43. American Association of Physicists in Medicine," *Medical Physics* **22**(2), 209–234 (1995).

11. M. J. Rivard, et al., "Update of AAPM Task Group No. 43 Report: A revised AAPM protocol for brachytherapy dose calculations," *Medical Physics* **31**(3), 633–674 (2004).

12. R. Nath, L. Anderson, D. Jones, C. Ling, R. Loevinger, J. Williamson, W. Hanson, *Specification of brachytherapy source strength: A report by Task Group 32 of the American Association of Physicists in Medicine* (American Institute of Physics, Inc., New York, NY, 1987).

13. R. Nath, et al., "Code of practice for brachytherapy physics: Report of the AAPM Radiation Therapy Committee Task Group No. 56. American Association of Physicists in Medicine," *Medical Physics* **24**(10), 1557–1598 (1997).

14. S. M. Seltzer, et al., "New national air-kerma-strength standards for I-125 and Pd-103 brachytherapy seeds," *Journal of Research of the National Institute of Standards and Technology* **108**(5), 337–358 (2003).

15. W. S. Culberson, et al., "Large-volume ionization chamber with variable apertures for air-kerma measurements of low-energy radiation sources," *Review of Scientific Instruments* **77**(1) (2006).

16. Z. Li, et al., "Dosimetric prerequisites for routine clinical use of photon emitting brachytherapy sources with average energy higher than 50 keV," *Medical Physics* **34**(1), 37–40 (2007).

17. T. P. Loftus, "Standardization of Ir-192 gamma-ray sources in terms of exposure," *Journal of Research of the National Bureau of Standards* **85**(1), 19–25 (1980).

18. F. H. Attix, *Introduction to Radiological Physics and Radiation Dosimetry* (Wiley, New York, NY, 1986), xxi, 607 p.

19. M. G. Mitch and C. G. Soares, "Primary standards for brachytherapy sources," in *Clinical Dosimetry Measurements in Radiotherapy*, edited by D. W. O. Rogers and J. E. Cygler (Medical Physics Publishing, Madison, WI, 2009), Chapter 16, pp. 549–566.

20. K. E. Stump, et al., "Calibration of new high dose rate ^{192}Ir sources," *Medical Physics* **29**(7), 1483–1488 (2002).

21. S. J. Goetsch, et al., "Calibration of Ir-192 high-dose-rate afterloading systems," *Medical Physics* **18**(3), 462–467 (1991).

22. R. Loevinger and N. G. Trott, "Design and operation of an extrapolation chamber with removable electrodes," *International Journal of Applied Radiation and Isotopes* **17**(2), 103–111 (1966).

23. S. M. Seltzer, M. O'Brien, M. G. Mitch, "New national air-kerma standard for low-energy electronic brachytherapy sources," *Journal of Research of the National Institute of Standards and Technology* **119**, 554–574 (2014).

24. J. S. Pruitt, *Calibration of Beta-Particle-Emitting Ophthalmic Applicators* (NBS Measurement Services, Gaithersburg, MD, 1987).

25. J. S. Pruitt, C. G. Soares, and M. Ehrlich, *Calibration of Beta-Particle Radiation Instrumentation and Sources* (NBS Special Publication, Gaithersburg, MD, 1988).

26. S. M. Seltzer, et al., "New national air-kerma-strength standards for I-125 and Pd-103 brachytherapy seeds (vol. 108, p. 337, 2003)," *Journal of Research of the National Institute of Standards and Technology* **109**(2), 301–301 (2004).

27. P. R. Almond, et al., "AAPM's TG-51 protocol for clinical reference dosimetry of high-energy photon and electron beams," *Medical Physics* **26**(9), 1847–1870 (1999).

28. J. Williamson, et al., "Dosimetric prerequisites for routine clinical use of new low energy photon interstitial brachytherapy sources. Recommendations of the American Association of Physicists in Medicine Radiation Therapy Committee. Ad Hoc Subcommittee of the Radiation Therapy Committee," *Medical Physics* **25**(12), 2269–2270 (1998).

29. M. J. Rivard, et al., "Supplement 2 for the 2004 update of the AAPM Task Group No. 43 Report: Joint recommendations by the AAPM and GEC-ESTRO," *Medical Physics* **44**(9), e297–e338 (2017).

30. M. J. Rivard, et al., "Supplement to the 2004 update of the AAPM Task Group No. 43 Report," *Medical Physics* **34**(6), 2187–205 (2007).

31. R. M. Kennedy, et al., "Experimental and Monte Carlo determination of the TG-43 dosimetric parameters for the model 9011 THINSeed brachytherapy source," *Medical Physics* **37**(4), 1681–1688 (2010).

32. J. Perez-Calatayud, et al., "Dose calculation for photon-emitting brachytherapy sources with average energy higher than 50 keV: Report of the AAPM and ESTRO," *Medical Physics* **39**(5), 2904–2929 (2012).

33. L. Beaulieu, et al., "Report of the Task Group 186 on model-based dose calculation methods in brachytherapy beyond the TG-43 formalism: Current status and recommendations for clinical implementation," *Medical Physics* **39**(10), 6208–6236 (2012).

34. M. J. Rivard, J. L. Venselaar, L. Beaulieu, "The evolution of brachytherapy treatment planning," *Medical Physics* **36**(6), 2136–53 (2009).

35. B. R. Thomadsen, et al., "Anniversary paper: Past and current issues, and trends in brachytherapy physics," *Medical Physics* **35**(10), 4708–4723 (2008).

36. L. A. DeWerd, et al., "Procedures for establishing and maintaining consistent air-kerma strength standards for low-energy, photon-emitting brachytherapy sources: Recommendations of the Calibration Laboratory Accreditation Subcommittee of the American Association of Physicists in Medicine," *Medical Physics* **31**(3), 675–681 (2004).

37. W. M. Butler, et al., "Third-party brachytherapy source calibrations and physicist responsibilities: Report of the AAPM Low Energy Brachytherapy Source Calibration Working Group," *Medical Physics* **35**(9), 3860–3865 (2008).

38. L. A. DeWerd, et al., "A modified dose calculation formalism for electronic brachytherapy sources," *Brachytherapy* **14**(3), 405–408 (2015).

39. F. A. Siebert, J. L. Venselaar, M. Paulsen, T. Hellebust, P. Papagiannis, A. Rijnders, M. J. Rivard, "Dose-rate to water calibrations for brachytherapy sources from the end-user perspective," *Metrologia* **49**, S249–S252 (2012).

40. G. J. Kutcher, et al., "Comprehensive QA for radiation oncology: Report of AAPM Radiation Therapy Committee Task Group 40," *Medical Physics* **21**(4), 581–618 (1994).

41. L. A. DeWerd, et al., "A dosimetric uncertainty analysis for photon-emitting brachytherapy sources: Report of AAPM Task Group No. 138 and GEC-ESTRO," *Medical Physics* **38**(2), 782–801 (2011).

42. Commission, U.S.R. NRC 10CFR. [cited 11/15/2020] (2018). Available from: https://www.nrc.gov/reading-rm/doc-collections/cfr/part035/part035-0633.html.

43. International Atomic Energy Agency, IAEA-TECDOC-1274 Calibration of Photon and Beta Ray Sources Used In Brachytherapy (2002).

44. A. Keyvanloo and H. S. Jans, "Technical Note: Minimizing geometrical uncertainties of cylindrical well-type ionization chamber measurements: There is an optimal chamber length," *Medical Physics* **45**(8), 3962–3968 (2018).

45. A. A. Schoenfeld, et al., "Reference conditions for ion-chamber based HDR brachytherapy dosimetry and for the calibration of high-resolution solid detectors," *Zeitschrift für Medizinische Physik* **28**(4), 293–302 (2017). https://pubmed.ncbi.nlm.nih.gov/28969957/

46. I. Iftimia, A. B. McKee, P. H. Halvorsen, "Varian HDR surface applicators – commissioning and clinical implementation," *Journal of Applied Clinical Medical Physics* **17**(2), 231–248 (2016).

47. C. Candela-Juan, et al., "Design and characterization of a new high-dose-rate brachytherapy Valencia applicator for larger skin lesions," *Medical Physics* **43**(4), 1639 (2016).

48. J. Metyko, W. Erwin, S. Landsberger, "Verification of I-125 brachytherapy source strength for use in radioactive seed localization procedures," *Applied Radiation and Isotopes* **112**, 62–68 (2016).

49. M. Aima, et al., "Dosimetric characterization of a new directional low-dose rate brachytherapy source," *Medical Physics* (2018).

50. O. A. Awunor, "Assessment of a source position checking tool for the quality assurance of transfer tubes used in HDR ^{192}Ir brachytherapy treatments," *Brachytherapy* **17**(3), 628–633 (2018).

51. M. Bellezzo, et al., "A novel rectal applicator for contact radiotherapy with HDR ^{192}Ir sources," *Brachytherapy* **17**(6), 1037–1044 (2018).

52. C. L. Deufel, J. P. Mullins, M. J. Zakhary, "A quality assurance device for measuring afterloader performance and transit dose for nasobiliary high-dose-rate brachytherapy," *Brachytherapy* **17**(4), 726–731 (2018).

53. M. Hermida-Lopez and L. Brualla, "Absorbed dose distributions from ophthalmic (106) Ru/(106) Rh plaques measured in water with radiochromic film," *Medical Physics* **45**(4), 1699–1707 (2018).

54. E. Martin, K. Sowards, B. Wang, "Investigation of a source model for a new electronic brachytherapy tandem by film measurement," *Journal of Applied Clinical Medical Physics* **19**(5), 640–650 (2018).

55. E. P. Pappas, et al., "On the use of a novel Ferrous Xylenol-orange gelatin dosimeter for HDR brachytherapy commissioning and quality assurance testing," *Physica Medica* **45**, 162–169 (2018).

56. Y. Watanabe, et al., "Polymer gel dosimetry for the high-dose-rate brachytherapy using Ir-192 source," *Igaku Butsuri* **37**(3), 173–176 (2017).

57. M. D. Belley, et al., "Real-time dose-rate monitoring with gynecologic brachytherapy: Results of an initial clinical trial," *Brachytherapy* **17**(6), 1023–1029 (2018).

58. M. Rejab, et al., "Dosimetric characterisation of the optically-stimulated luminescence dosimeter in cobalt-60 high dose rate brachytherapy system," *Australasian Physical and Engineering Sciences in Medicine* **41**(2), 475–485 (2018).

59. J. Jeong, et al., "Impact of source position on high-dose-rate skin surface applicator dosimetry," *Brachytherapy* **15**(5), 650–660 (2016).

60. N. Chofor, et al., "The mean photon energy EF at the point of measurement determines the detector-specific radiation quality correction factor kQ,M in ^{192}Ir brachytherapy dosimetry," *Zeitschrift für Medizinische Physik* **26**(3), 238–250 (2016).

61. M. Carrara, et al., "In vivo rectal wall measurements during HDR prostate brachytherapy with MOSkin dosimeters integrated on a trans-rectal US probe: Comparison with planned and reconstructed doses," *Radiotherapy & Oncology* **118**(1), 148–153 (2016).

62. J. Mason, et al., "Real-time in vivo dosimetry in high dose rate prostate brachytherapy," *Radiotherapy & Oncology* **120**(2), 333–338 (2016).

63. R. Phurailatpam, et al., "Characterization of commercial MOSFET detectors and their feasibility for in-vivo HDR brachytherapy," *Physica Medica* **32**(1), 208–212 (2016).

64. M. Persson, J. Nilsson, A. Carlsson Tedgren, "Experience of using MOSFET detectors for dose verification measurements in an end-to-end ^{192}Ir brachytherapy quality assurance system," *Brachytherapy* **17**(1), 227–233 (2018).

65. Y. Watanabe, et al., "Automated source tracking with a pinhole imaging system during high-dose-rate brachytherapy treatment," *Physics in Medicine and Biology* **63**(14), 145002 (2018).

66. F. Therriault-Proulx, L. Beaulieu, S. Beddar, "Validation of plastic scintillation detectors for applications in low-dose-rate brachytherapy," *Brachytherapy* **16**(4), 903–909 (2017).

67. G. Kertzscher and S. Beddar, "Inorganic scintillation detectors based on Eu-activated phosphors for ^{192}Ir brachytherapy," *Physics in Medicine and Biology* **62**(12), 5046–5075 (2017).

68. M. Eichmann and B. Thomann, "Air core detectors for Cerenkov-free scintillation dosimetry of brachytherapy beta-sources," *Medical Physics* **44**(9), 4900–4909 (2017).

69. E. Adolfsson, et al., "Measurement of absorbed dose to water around an electronic brachytherapy source. Comparison of two dosimetry systems: Lithium formate

EPR dosimeters and radiochromic EBT2 film," *Physics in Medicine and Biology* **60**(9), 3869–3882 (2015).

70. S. Mishra and T. P. Selvam, "Monte Carlo calculation of beam quality and phantom scatter corrections for lithium formate electron paramagnetic resonance dosimeter for high-energy brachytherapy dosimetry," *Journal of Medical Physics* **42**(2), 72–79 (2017).

71. R. S. Sloboda and G. V. Menon, "Experimental determination of the anisotropy function and anisotropy factor for model 6711 I-125 seeds," *Medical Physics* **27**(8), 1789–1799 (2000).

72. T. D. Solberg, et al., "Dosimetric parameters of three new solid core I-125 brachytherapy sources," *Journal of Applied Clinical Medical Physics* **3**(2), 119–134 (2002).

73. A. S. Meigooni, et al., "Dosimetric characteristics of the bests double-wall ¹⁰³Pd brachytherapy source," *Medical Physics* **28**(12), 2568–2575 (2001).

74. C. G. Hammer, et al., "Experimental investigation of GafChromic((R)) EBT3 intrinsic energy dependence with kilovoltage x rays, (137) Cs, and (60) Co," *Medical Physics* **45**(1), 448–459 (2018).

75. C. L. Deufel, et al., "Experimental and theoretical dosimetry of the RIC-100 phosphorus-32 brachytherapy source for implant geometries encountered in the intraoperative setting," *Brachytherapy* **14**(5), 734–750 (2015).

76. G. N. Cohen, et al., "32P brachytherapy conformal source model RIC-100 for high-dose-rate treatment of superficial disease: Monte Carlo calculations, diode measurements, and clinical implementation," *International Journal of Radiation Oncology, Biology, Physics* **88**(3), 746–752 (2014).

77. J. G. Sutherland and D. W. Rogers, "Monte Carlo calculated absorbed-dose energy dependence of EBT and EBT2 film," *Medical Physics* **37**(3), 1110–1116 (2010).

78. M. J. Price, et al., "Dose perturbation due to the polysulfone cap surrounding a Fletcher-Williamson colpostat," *Journal of Applied Clinical Medical Physics* **11**(1), 68–76 (2010).

79. G. Zwierzchowski, et al., "Film based verification of calculation algorithms used for brachytherapy planning-getting ready for upcoming challenges of MBDCA," *Journal of Contemporary Brachytherapy* **8**(4), 326–335 (2016).

80. P. Saidi, M. Sadeghi, C. Tenreiro, "Experimental measurements and Monte Carlo calculations for (103)Pd dosimetry of the 12 mm COMS eye plaque," *Physica Medica* **29**(3), 286–294 (2013).

81. S. McCauley Cutsinger, C. Deufel, S. Corner, "An experimental method for commissioning and quality assurance of COMS eye plaques," *Brachytherapy* **17**(4, Suppl.), S103 (2018).

82. S. M. Vahabi, M. Bahreinipour, M. Shamsaie Zafarghandi, "Study on the dose modification factor of strut adjusted volume implant (SAVI) with a (169) Yb source using MCNP4C," *Australasian Physical and Engineering Sciences in Medicine* **41**(2), 445–450 (2018).

83. I. Fotina, et al., "A comparative assessment of inhomogeneity and finite patient dimension effects in ⁶⁰Co and ¹⁹²Ir high-dose-rate brachytherapy," *Journal of Contemporary Brachytherapy* **10**(1), 73–84 (2018).

84. D. Jacob, et al., "Clinical transition to model-based dose calculation algorithm: A retrospective analysis of high-dose-rate tandem and ring brachytherapy of the cervix," *Brachytherapy* **16**(3), 624–629 (2017).

85. V. Peppa, et al., "On the impact of improved dosimetric accuracy on head and neck high dose rate brachytherapy," *Radiotherapy & Oncology* **120**(1), 92–97 (2016).

86. C. A. Collins Fekete, et al., "Calcifications in low-dose rate prostate seed brachytherapy treatment: Post-planning dosimetry and predictive factors," *Radiotherapy & Oncology* **114**(3), 339–344 (2015).

87. C. Furstoss, et al., "Monte Carlo study of LDR seed dosimetry with an application in a clinical brachytherapy breast implant," *Medical Physics* **36**(5), 1848–1858 (2009).

88. S. M. Oliveira, et al., "Dosimetric effect of tissue heterogeneity for ¹²⁵I prostate implants," *Reports of Practical Oncology and Radiotherapy* **19**(6), 392–398 (2014).

89. J. K. Mikell, et al., "Impact of heterogeneity-based dose calculation using a deterministic grid-based Boltzmann equation solver for intracavitary brachytherapy," *International Journal of Radiation Oncology Biology Physics* **83**(3), E417–E422 (2012).

90. S. Lloyd, et al., "Intraoperative high-dose-rate brachytherapy: An American Brachytherapy Society consensus report," *Brachytherapy* **16**(3), 446–465 (2017).

91. T. W. Leung, et al., "High dose rate brachytherapy for early stage oral tongue cancer," *Head & Neck.* **24**(3), 274–281 (2002).

92. V. Pastor-Sanchis, et al., "Experimental validation of the Valencia-type applicators developed for the BEBIG HDR afterloader Saginova," *Medical Physics* **44**(6), 3176–3176 (2017).

93. D. Granero, et al., "Radiation leakage study for the Valencia applicators," *Physica Medica* **29**(1), 60–64 (2013).

94. P. T. Finger, et al., "Intraocular radiation blocking," *Investigative Ophthalmology & Visual Science* **31**(9), 1724–1730 (1990).

95. M. Hira, et al., "Measurement of dose perturbation around shielded ovoids in high-dose-rate brachytherapy," *Brachytherapy* **10**(3), 232–241 (2011).

96. A. S. Kirov, et al., "Measurement and calculation of heterogeneity correction factors for an Ir-192 high dose-rate brachytherapy source behind tungsten alloy and steel shields," *Medical Physics* **23**(6), 911–919 (1996).

97. F. W. Hensley, "Present state and issues in IORT Physics," *Radiation Oncology* **12**(1), 37 (2017).

98. A. Espinoza, et al., "The evaluation of a 2D diode array in "magic phantom" for use in high dose rate brachytherapy pretreatment quality assurance," *Medical Physics* **42**(2), 663–673 (2015).

99. J. Poder, et al., "HDR brachytherapy in vivo source position verification using a 2D diode array: A Monte Carlo study," *Journal of Applied Clinical Medical Physics* **19**(4), 163–172 (2018).

100. E. L. Seymour, et al., "In vivo real-time dosimetric verification in high dose rate prostate brachytherapy," *Medical Physics* **38**(8), 4785–4794 (2011).

101. M. Carrara, et al., "Semiconductor real-time quality assurance dosimetry in brachytherapy," *Brachytherapy* **17**(1), 133–145 (2018).

102. D. Wagner, M. Hermann, A. Hille, "In vivo dosimetry with alanine/electron spin resonance dosimetry to evaluate the urethra dose during high-dose-rate brachytherapy," *Brachytherapy* **16**(4), 815–821 (2017).

103. H. Afsharpour, et al., "A Monte Carlo study on the effect of seed design on the interseed attenuation in permanent prostate implants," *Medical Physics* **35**(8), 3671–3681 (2008).

104. J. F. Carrier, et al., "Impact of interseed attenuation and tissue composition for permanent prostate implants," *Medical Physics* **33**(3), 595–604 (2006).

105. J. F. Carrier, et al., "Postimplant dosimetry using a Monte Carlo dose calculation engine: A new clinical standard," *International Journal of Radiation Oncology, Biology, Physics* **68**(4), 1190–1198 (2007).

106. O. Chibani, J. F. Williamson, D. Todor, "Dosimetric effects of seed anisotropy and interseed attenuation for ^{103}Pd and ^{125}I prostate implants," *Medical Physics* **32**(8), 2557–2566 (2005).

107. J. Mason, et al., "Monte Carlo investigation of I-125 interseed attenuation for standard and thinner seeds in prostate brachytherapy with phantom validation using a MOSFET," *Medical Physics* **40**(3), 031717 (2013).

108. M. J. Rivard, L. Beaulieu, F. Mourtada, "Enhancements to commissioning techniques and quality assurance of brachytherapy treatment planning systems that use model-based dose calculation algorithms," *Medical Physics* **37**(6), 2645–2658 (2010).

109. M. J. Rivard, J. L. Venselaar, L. Beaulieu, "The evolution of brachytherapy treatment planning," *Medical Physics* **36**(6), 2136–2153 (2009).

110. L. Beaulieu, et al., "Report of the Task Group 186 on model-based dose calculation methods in brachytherapy beyond the TG-43 formalism: Current status and recommendations for clinical implementation," *Medical Physics* **39**(10), 6208–6236 (2012).

111. Y. Ma, et al., "A generic TG-186 shielded applicator for commissioning model-based dose calculation algorithms for high-dose-rate ^{192}Ir brachytherapy," *Medical Physics* **44**(11), 5961–5976 (2017).

112. Sources, Imaging and Radiation Oncology Core (IROC) Source Registry. Model-Based Dose Calculations. Reference Data Sets (2018) [cited 11/15/2020]; Available from: http://rpc.mdanderson.org/RPC/BrachySeeds/Model_calculations.htm.

113. ICRU, *Dosimetry of External Beta Rays for Radiation Protection* (International Commission on Radiation Units and Measurements, Bethesda, MD, 1997).

114. R. Nath, et al., "Intravascular brachytherapy physics: Report of the AAPM Radiation Therapy Committee Task Group No. 60," *Medical Physics* **26**(2), 119–152 (1999).

115. D. M. Duggan, et al., "Radiochromic film dosimetry of a high dose rate beta source for intravascular brachytherapy," *Medical Physics* **26**(11), 2461–2464 (1999).

116. E. Drud, et al., "Influence of a commonly used stent type on the dose distribution in endovascular brachytherapy," *Radiotherapy and Oncology* **81**, S262–S262 (2006).

117. E. Drud, et al., "Beta dosimetry with microMOSFETs for endovascular brachytherapy," *Physics in Medicine and Biology* **51**(23), 5977–5986 (2006).

118. W. G. Cross, N. O. Freedman, P. Y. Wong, "Beta-ray dose distributions from point sources in an infinite water medium," *Health Physics* **63**(2), 160–171 (1992).

119. C. Janicki, et al., "Dose model for a beta-emitting stent in a realistic artery consisting of soft tissue and plaque," *Medical Physics* **26**(11), 2451–2460 (1999).

120. M. Hermida-Lopez, "Calculation of dose distributions for 12 106Ru/106Rh ophthalmic applicator models with the PENELOPE Monte Carlo code," *Medical Physics* **40**(10), 101705 (2013).

121. J. M. DeCunha and S. A. Enger, "A new delivery system to resolve dosimetric issues in intravascular brachytherapy," *Brachytherapy* **17**(3), 634–643 (2018).

122. K. Tanderup, et al., "In vivo dosimetry in brachytherapy," *Medical Physics* **40**(7), 070902 (2013).

123. I. J. Kim, et al., "Building a graphite calorimetry system for the dosimetry of therapeutic x-ray beams," *Nuclear Engineering and Technology* **49**(4), 810–816 (2017).

124. C. L. Deufel, et al., "Patient safety is improved with an incident learning system—clinical evidence in brachytherapy," *Radiotherapy & Oncology* **125**(1), 94–100 (2017).

125. M. S. Huq, et al., "The report of Task Group 100 of the AAPM: Application of risk analysis methods to radiation therapy quality management," *Medical Physics* **43**(7), 4209 (2016).

III

External Beam Radiation Therapy

CHAPTER 16

Photon Beam Dosimetry of Conventional Medical Linear Accelerators

Francisco J. Reynoso

Washington University in Saint Louis
Saint Louis, Missouri

CONTENTS

16.1 INTRODUCTION

The primary aim of a course of radiation therapy is to cure or control disease while minimizing the damage to normal tissues. The response of tumors and normal tissues to radiation is highly variable and dose response curves can be quite steep. As a result, the optimal window for radiation treatment is quite small and imposes a strong requisite for accurate and consistent radiation dose calculation and delivery. Modern radiotherapy techniques rely on computational algorithms using a treatment planning system (TPS) to calculate dose, and becoming much less frequently dependent on manual dose calculations. It is largely not possible to directly measure doses in patients and physical measurements in phantoms are the foundation of all dose calculation approaches. Phantom-based beam characterization of linear accelerators (linacs) is a staple of Medical Physics practice and forms the basis of dosimetric quantities that characterize how dose changes with depth, along the beam profile, and with the use of beam modifiers.

In the early years of linac manufacturing each machine possessed unique characteristics that made it imperative to carefully characterize each linac separately, with reference data provided as a guide for data comparison. Modern linacs are manufactured to tight specifications with highly reproducible dose characteristics for machines of the same model and energy [1], and manufacturer provided reference data is now routinely used for TPS beam modeling with beam-matched machines sharing the same model [2–5]. The output characteristics of a linac are typically defined relative to an absolute point where the output of the machine is calibrated. The distinction between absolute and relative dosimetry should be made in order to understand the different dosimetric quantities used to define the output of a linac.

Absolute dosimetry protocols like AAPM Task Group 51 (TG-51) [6] and IAEA Technical Report Series TRS-398 [7] carefully define very specific reference conditions to define the absolute dose or dose rate to a point. These reference conditions are important in order to maintain charged particle equilibrium (CPE) and define measurement conditions that are ideal for the accurate measurement of dose at a clinically relevant depth. The dose characteristics throughout the rest of the patient will change significantly as the beam is attenuated and the points get farther away from the source. There are also a wide range of clinical conditions encountered that make it impossible to always define dose or dose rate in absolute terms. Relative external beam dosimetric functions characterize the output in relative terms that scale the calibrated absolute dose under reference conditions to accommodate real clinical conditions. In this chapter, the general dosimetric functions of megavoltage beams are described, general beam parameters are defined, and the details of AAPM's TG-51 are covered.

16.2 PHOTON BEAM DOSIMETRY

Megavoltage beams are characterized for their ability to spare the skin and treat deep lesions by using accelerating potentials that are in the megavoltage range which is now commonly 6–18 MV. This energy requirement is most commonly produced in a linear accelerator that has the general design characteristics similar to those shown in Figure 16.1. A continuous spectrum of megavoltage photons is produced at the target after pulsed electrons are accelerated to the desired energy in the accelerating waveguide. The point-like source of photons from the target produces a diverging field of photons that is collimated within the linac head. This diverging nature

of the field also results in the radial inverse-square law where the photon fluence decreases as $1/r^2$ as the distance from the source increases.

The dose to any point within a phantom or patient contains two main components that determine the total dose: the primary component and the scatter component. The primary photon component originates at the source and is composed of the original photons that were generated at the target. The primary component of dose is mainly dependent on the photon energy and the depth within the phantom, and it is independent of field size. The scatter component is the result of photons scatter within the linac head and the phantom and can be further divided into a collimator scatter component and a phantom scatter component. The collimator scatter originates from photon scatter within the collimator, flattening filter, monitor unit (MU) chamber, and air. The phantom scatter component is the result of scatter within the phantom. The notion of a purely primary dose component is mainly a conceptual one as it is impossible to measure directly as collimator scatter dose is always present. Similarly, phantom scatter only dose is also impossible to measure directly and is a derived quantity from other measurements.

16.2.1 Collimator Scatter Factor

The beam output (dose rate in free space) measured in air depends on the field because as the field size is increased the collimator scatter increases. The scatter component of dose that originates on the linac head and changes mostly with different collimator jaw settings can be quantified with a ***collimator scatter factor*** (S_c) that is measured in air and in the absence of full scatter conditions to avoid the phantom scatter component. The concept of collimator scatter factor is also called an ***in-air output ratio*** and was introduced to characterize how the incident photon fluence per MU changes with jaw settings [8]. Figure 16.1 shows the position of the MU chambers are above both the upper and lower collimator jaws. This results scattered photons from the jaws affecting the charge reading within the MU chambers and nominally changing the amount dose that is deposited per MU. This is the primary effect being characterized by the collimator scatter factor and important parameter that directly changes machine output.

The collimator scatter factor is defined in AAPM TG-74 [8] as the ratio in free-space of primary collision *KERMA* (Kinetic Energy Released in Matter) between a

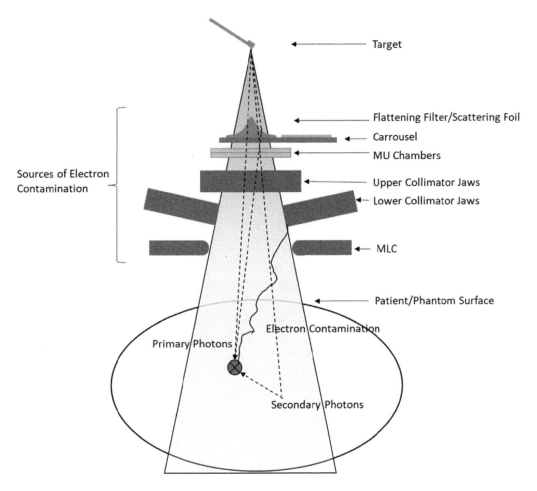

FIGURE 16.1   General schematic of a modern clinical linear accelerator. Dose contributions from different components at any given point within a patient or phantom.

given field size and a reference field size. This is the ratio of the output in air for a given field size and a reference field size typically chosen to be 10 × 10 cm². The ratio can be defined in terms of dose to free space D_{fs} defined as the dose to equilibrium mass of water measured in air under CPE conditions:

$$S_c(FS) = \frac{D_{fs}(r)}{D_{fs}(r_{ref})} \qquad (16.1)$$

where r is the given field size and r_{ref} is the reference field size typically chosen to be 10 × 10 cm². The values are usually measured for a wide range of square fields normalized to 1.0 for the reference field size. The measurement is typically done using an ion chamber with a buildup cap that is large enough to attain CPE (Figure 16.2).

In high-energy applications water equivalent buildup caps can be become prohibitively large for even moderately small field sizes so a high-Z buildup cap with

radiological dimensions that approximate CPE are also acceptable. However, electron contamination and the alteration of the incident spectrum may introduce errors. Therefore, the use of a miniphantom is recommended with dimensions that allow for lateral and longitudinal CPE but also maintain a position for the ion chamber to be deep enough to filter out electron contamination [9]. The values of both D_{fs} and S_c both increase with r because as the jaws open more fluence is seen by the detector and less photons scatter back into the MU chambers. Typical measured values of collimator scatter factors range between 0.90 and 1.05 for most clinical megavoltage beams with values below 1.0 for field sizes below the reference field size (10 × 10 cm²) and above 1.0 for field sizes above. The backscatter from the jaws is dependent on how close each pair of jaws is to the MU chambers so the amount of backscatter differs from lower and upper jaws and results in the so called collimator exchange effect where $S_c(a,b) \neq S_c(b,a)$ for

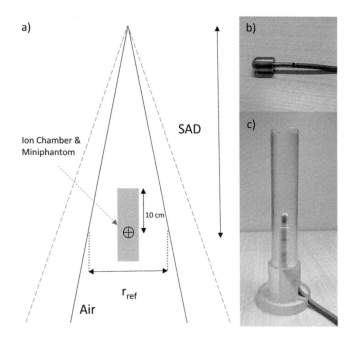

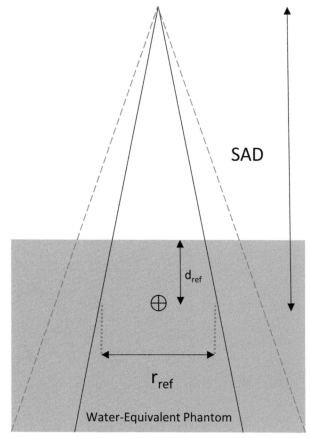

FIGURE 16.2 (a) Collimator scatter factor measurement setup using a miniphantom and ion chamber setup. (b) Brass buildup cap to allow for CPE and the measurement of smaller fields in high-energy applications. (c) Miniphantom and ion chamber setup using a vertical ion chamber alignment.

FIGURE 16.3 Total output factor measurement setup, the depth of measurement is chosen to be deep enough to avoid electron contamination and usually 5 cm for lower energies like 6 MV and 10 MV, and 10 cm for higher energies like 15 MV and 18 MV. The phantom should be large enough to approximate full scatter conditions and capture all relevant phantom scatter.

rectangular fields if the upper and lower jaw settings are inverted. The consequence of this is that the machine output for rectangular fields or blocked fields is not always the same as that of its equivalent square fields and can differ by up to 2%.

16.2.2 Phantom Scatter Factor

The scatter component of dose that originates in the patient itself due to scattered photons can be quantified with a **phantom scatter factor** (S_p), and defined as the ratio of the output in water-equivalent media at a reference depth for a given field size to the same point at the same depth for a reference field size (typically chosen to be 10×10 cm^2) and the same collimator settings such that to capture phantom scatter changes alone. This definition may be conceptually convenient in order to separate collimator and phantom scatter but highly impractical to measure. Under this definition the only way to measure this factor is using a very large collimator setting (e.g., 40×40 cm^2) and changing the size of the phantom to adjust the field size that irradiates the phantom [10] (Figure 16.3).

The most common way to determine phantom scatter factors is indirectly by measuring a **total output factor**

($S_{c,p}$) that contains both the collimator and phantom scatter components, and dividing out the collimator scatter factor (S_c):

$$S_p(FS) = \frac{S_{c,p}(r)}{S_c(r)} \qquad (16.2)$$

where r is the given field size and the measurement of $S_{c,p}(r)$ is taken as the ratio of output for a reference field size taken at the same depth for a reference field size typically chosen to be 10×10 cm^2. The measurement is taken using a source-to-axis (SAD) setup where the measurement is taken at the machine isocenter at a depth that is deep enough to avoid electron contamination. For 6 MV and 10 MV beams the depth is typically 5 cm whereas for higher energy beams like

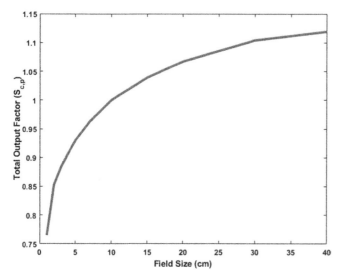

FIGURE 16.4 Total output factor $\left(S_{c,p}\right)$ for a 6 MV beam measured at 5 cm depth and 95 cm SSD. The output factor is normalized for a 10×10 cm^2 and monotonically increases or decreases with a change in field size due to a change in scattering conditions.

15 MV and 18 MV beams a 10 cm depth is more appropriate. Figure 16.4 shows the monotonically increasing function of total output factor for a 6 MV photon beam measured at 5 cm depth at 95 cm source-to-surface distance (SSD). The total output changes rapidly for small fields starting out at around 0.765 for a 1×1 cm^2 field and quickly increasing to its normalized value of 1.00 for a 10×10 cm^2 field. For larger fields, the total output factor changes much more slowly and reaching a maximum of 1.119 for a 40×40 cm^2 field. The measurement of output factors for small fields where there is a lack of lateral CPE ($\sim 3 \times 3$ cm^2) is much more challenging than those for larger fields due not only to the lack of lateral CPE but also to the detector volume averaging, high dose gradients and source occlusions. Large variations in measured output of over 10% have been measured for small field and as such great care must be taken when measuring small field outputs [11,12]. The field size where this becomes an issue depends on the photon beam energy but typically understood to be below $\sim 3 \times 3$ cm^2 or when the detector volume is comparable to the field size. Detailed recommendations and measurement were made by the joint IAEA/AAPM working group on small field dosimetry and published on the IAEA TRS-483 [13]. The formalism is based on the published technique by Alfonso et al. [14].

16.2.3 Percent Depth Dose Curve

The quantities defined so far characterize different components of the dose and how it changes with collimator settings. The way that the dose changes with depth can be addressed with several different functions and the most complete and intuitive to understand is the PDD curve. PDD is defined as the percentage of the maximum dose at a depth d_{max} as:

$$PDD(d,r) = \frac{D(d)}{D(d_{max})} \times 100 \qquad (16.3)$$

Figure 16.5 shows a representative depth dose curve and how dose changes with depth for a 6 MV beam entering a phantom. PDD exhibits a dose buildup region where the dose increases monotonically until it reaches a maximum at d_{max}. As photons enter the patient or phantom, high-speed electrons are ejected from the surface and subsequent layers. These electrons can deposit their energy a significant distance away from their site of origin. This means that the electron fluence, and hence absorbed dose, increase with depth until the rising slope due to buildup of charged particles is balanced by the descending slope due to the attenuation of the photon beam. At greater depths, CPE is attained and the absorbed dose decreases as the photon beam is attenuated. Beyond the depth of maximum dose, PDD values increase with photon energy because higher energy beams have greater penetrating power. The field size dependence of PDD can be attributed to the increase in scattered radiation that contributes to the absorbed dose. Because the scattered dose is greater at larger depths than at d_{max}, the PDD increases with increasing field size. The dependence on SSD of PDD is a direct result of the inverse square dependence of photon fluence emitted by a radiation source. In addition to the inverse square law, PDD variation with depth is also governed by exponential attenuation and scattering. Therefore, the variation of the PDD with depth is given by:

$$PDD(d,r,SSD) \sim \left(\frac{SSD + d_{max}}{SSD + d} \right)^2 e^{-\mu(d-d_{max})} \quad (16.4)$$

where μ is the linear attenuation coefficient and shows the exponential attenuation of the beam beyond d_{max}, and the dependence on SSD [10].

At beam entrance, kinetic energy is released in the medium as secondary electrons are released and as such

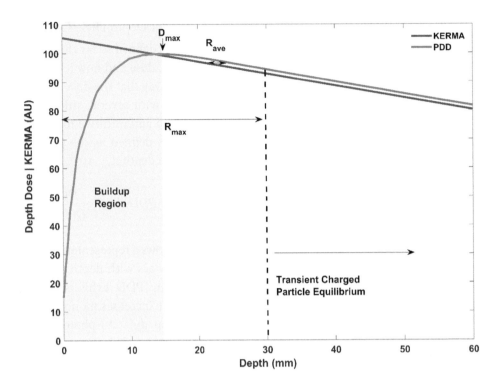

FIGURE 16.5   Percent depth dose curve for a typical 6 MV photon beam in water equivalent media. The shaded area represents the buildup region characteristic of indirectly ionizing radiation where secondary charge particles build up to deposit dose. KERMA decays exponentially as the beam fluence attenuates within the medium and the two decreases by the same rate.

KERMA is at a maximum. The beam fluence attenuates exponentially with depth and KERMA decreases by the same rate. On the other hand, the dose absorbed in the material behaves differently due to the indirectly ionizing nature of photon beams. At the surface, the dose from a megavoltage beam is very low and would be almost zero if it weren't for electron contamination from the head. This low entrance dose gives high-energy megavoltage beams its skin sparing property that is typically below 50% under most conditions. As secondary charged particles are released in the material, the absorbed dose increases and the dose reaches a dose maximum (D_{max}) where the buildup of secondary charged particles is countered by the attenuation of the beam. For the 6 MV beam shown in Figure 16.5, this depth of maximum dose d_{max} occurs at around 1.5 cm. A few centimeters beyond D_{max} the PDD and KERMA curves become parallel and this occurs at a depth equal to the maximum range R_{max} of secondary electrons released in the medium. In the case of 6 MV beam like the one shown in Figure 16.5 this happens at around 3.0 cm for the maximum range of a 6 MeV electron being released through photoelectric interaction from the low yield of 6 MeV photons present in the 6 MV photon spectrum. At these deeper

depths where the PDD and KERMA curves remain mostly parallel the absorbed dose is proportional to KERMA under transient charged particle equilibrium (TCPE). The distance between the PDD and KERMA curves beyond d_{max} represents the average distance that secondary electrons travel through the medium and denoted R_{ave} in Figure 16.5 [15].

16.2.4 Tissue Air Ratio

One of the oldest functions used to characterize how dose changes with depth is the **tissue-air ratio (TAR)**. TAR was first introduced in 1953 to simplify dose calculation for rotational therapy in orthovoltage beams [16]. TAR can be defined as the ratio of dose to some depth d in water-equivalent tissue to the dose in free space D_{fs} for the same field size under SAD conditions.

$$TAR(FS) = \frac{D(d, r_d)}{D_{fs}(r_d)} \tag{16.5}$$

where r_d is the field size at depth. The quantity is convenient because it is measured under SAD conditions and not dependent on the SSD. However, similar to the collimator scatter factor it is difficult to measure D_{fs}

particularly for higher energies because of lack of CPE for small fields and the added uncertainty of using high-Z buildup caps and miniphantoms. Therefore, TARs are typically more useful for low-energy beams with an isocentric treatment geometry. TAR increases with field size because the increased phantom scatter component of the dose measurement at depth increases faster than the increased collimator scatter component of the dose in a free space measurement.

16.2.5 Tissue Phantom Ratio and Tissue Maximum Ratio

Tissue phantom ratios (TPRs) overcome the limitations of the TAR for high-energy photon measurement. TPR is defined in an isocentric geometry as the ratio of dose at a given depth d and field size at depth r, to the dose at a reference depth d_{ref} and reference field size at depth r:

$$TPR(FS) = \frac{D(d,r)}{D(d_{ref},r)} \quad (16.6)$$

The special case where $d_{ref} = d_{max}$ give rise to the tissue maximum ratio (TMR). Isocentric measurements like TAR and TMR are insensitive to changes in SSD because the scatter contribution is not dependent on the divergence on the beam, and only of the actual field size at depth. This is a main difference with PDD measurements that depend on the field size at the surface. This behavior makes TMR a convenient choice for MU calculations because blocked field size don't need to be corrected for divergence at depth. TPR and TMR decrease with depth as the beam is attenuated through tissue, and increases with field size because the increased phantom scatter component of the dose at depth increases (Figure 16.6).

TPR and TMR can be cumbersome to measure directly given the requirement to maintain an isocentric geometric with increasing depth. This can be accomplished in a water tank that can be raised while simultaneously moving the ion chamber deeper into the phantom to maintain an isocentric geometry. Other techniques may include using a solid water phantom and adding different layers of solid water to increase the depth of measurement. Modern scanning tank equipment may include software and specially designed tank platforms that can make TPR and TMR measurements much more streamlined. TPR and TMR data are also commonly derived from PDD measurements correcting for divergence and phantom scatter.

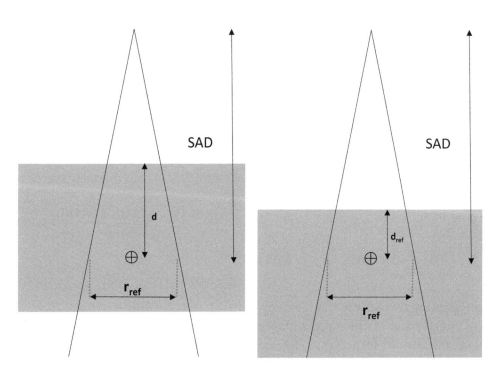

FIGURE 16.6 Isocentric measurement geometry for TPR and TMR measurements in a water equivalent phantom. The reference depth d_{ref} is taken to be d_{max} for TMR measurements.

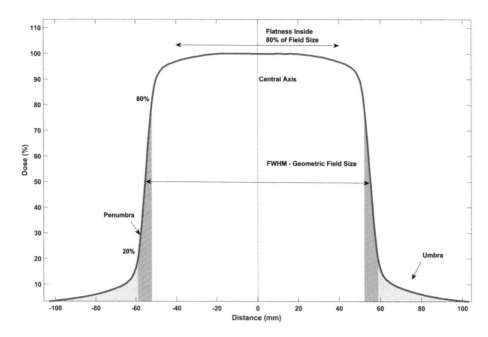

FIGURE 16.7   A 6 MV beam dose profile at 10 cm depth for a 10 × 10 cm² field size at the surface. The beam regions of a megavoltage beam consist of the in-field region, the penumbra and the umbra. The typical width of the penumbral region for a 6 MV field is on the order of 6 mm. It should be noted flattening filters are typically optimized to flatten the field at 10 cm depth like the profile shown.

16.2.6 Beam Dose Profiles and Off-Axis Ratios

All quantities defined up to this point deal with how dose changes with jaw collimator settings or with depth on the central axis (CAX). Dose variation away from the CAX is characterized with dose profile measurements and off-axis output factors ratios (OARs). Dose profiles are an important part of TPS beam model commissioning and validation, and characterize all off-axis aspects of the beam with respect to the CAX values like PDD and output factors. The beam profile consists of three main regions: the in-field region, the penumbra, and the umbra. The central region typically refers to the region of the field that is within 80% of the beam width, i.e., for a 10 × 10 cm² field the central 8 × 8 cm² is considered central region (Figure 16.7).

$$F = \frac{D_{max} - D_{min}}{D_{max} + D_{min}} \times 100 \qquad (16.7)$$

Field flatness and symmetry are two important parameters that are good indicators of machine performance and quality assurance metrics. The flatness (F) is defined as a percentage difference of the maximum (D_{max}) and minimum (D_{min}) values across a profile within the central region of the field, defined as the inner 80% of the

FWHM. Flattening filters are typically designed and optimized to deliver a flat field at 10 cm depth. At shallower depths beam profiles may exhibit *dose horns* where the dose off axis is higher than that at the CAX. This effect is more pronounced for higher energies at d_{max}. Typical flatness values for modern linacs should stay below 3% and typically below 2% for flattened beams. Symmetry is another important beam parameter that reflects the beam symmetry by comparing the area under to profile for each side of the curve. This quantity can typically be directly measured using most modern scanning tank software. Typical symmetry values for most modern linacs are below 2% but routinely stay below 1%.

$$S = \frac{Area_L - Area_R}{Area_L + Area_R} \times 100 \qquad (16.8)$$

16.3 MONITOR UNIT CALCULATIONS

16.3.1 SSD Calculations

Machines are usually calibrated to deliver 1 cGy/MU at d_{max} for a reference field 10 × 10 cm² at 100 SSD. Taking into account all correction factors in an SSD setup, the number of monitor units needed for a specific dose is given by:

$$MU = \frac{Dose}{D_0 \cdot S_c \cdot S_p \cdot Inv^2 \cdot \frac{PDD}{100} \cdot F \cdot TF \cdot WF \cdot OAF} \quad (16.9)$$

where $D_0 = 1$ cGy/MU, Inv^2 is the inverse square factor, F is the Mayneord F factor, TF is the tray factor, WF is the wedge factor and OAF is the of axis factor. Any other correction factors can be included here too. It should be noted that the Mayneord F factor is needed to correct for any changes in SSD. Using Equation 16.4 one can determine how to correct the percent depth does for a change in SSDs. Taking the ratio of PDD at two different SSDs we can obtain the following expression:

$$\frac{PDD(d, SSD_2)}{PDD(d, SSD_1)} = \left(\frac{SSD_2 + d_{max}}{SSD_1 + d_{max}} \right)^2 \left(\frac{SSD_1 + d}{SSD_2 + d} \right)^2 \quad (16.10)$$

The factor on the right-hand side is the Mayneord F factor and corrects the PDD for a change in SSD. It should be noted that the Mayneord F factor works reasonably well for small field sizes since scattering is minimal under such conditions. However, it overestimates the change in PDD for large field sizes and very large SSD. This dependence on SSD makes PDD an inconvenient choice in the clinic because corrections for SSD changes can become quite cumbersome to apply routinely in a clinical setting. The Inv^2 is needed to correct the dose rate, hence the output factor, due to the change in fluence. Therefore, for an SSD setup:

$$Inv^2 = \left(\frac{100 + d_{max}}{SSD + d_{max}} \right)^2 \quad (16.11)$$

$$F = \left(\frac{SSD + d_{max}}{100 + d_{max}} \right)^2 \left(\frac{100 + d}{SSD + d} \right)^2 \quad (16.12)$$

The collimator and phantom scatter factors S_c and S_p take into account different aspects of scatter and are handled differently for SSD MU calculations using PDD. The collimator scatter factor S_c is determined using the jaw collimator setting, while the phantom scatter factor S_p is determined by the portion of the field that contributes scatter dose to the point of interest or the blocked field size at the surface. This can differ from the jaw delimited field based on SSD, machine design, the use of multi-leaf collimators, or other types of photon blocks.

16.3.2 SAD Calculations

In an SAD setup there is no dependence on SSD and hence the Mayneord F factor is not needed. As such the number of monitor units needed for a specific dose is given by:

$$MU = \frac{Dose}{D_0 \cdot S_c \cdot S_p \cdot Inv^2 \cdot TMR \cdot TF \cdot WF \cdot OAF} \quad (16.13)$$

The Inv^2 factor is still needed to correct the output factor, therefore for an SAD setup the Inv^2 is given by:

$$Inv^2 = \left(\frac{100}{SSD + d} \right)^2 \quad (16.14)$$

The collimator and phantom scatter factors S_c and S_p are determined in a similar fashion to MU calculations using PDDs and an SSD geometry. However, the isocentric geometry means that the phantom scatter factor S_p is determined at depth and highlights the convenience of an isocentric geometry that does not dependent to SSD and does not require further correction from the Mayneord F factor.

16.4 ABSOLUTE DOSIMETRY PROTOCOL: AAPM TG-51

AAPM's TG-51 clinical reference dosimetry protocol was prescribed for photon beams with nominal energies between ^{60}Co and 50 MV, and electron beams between 4 and 50 MeV. This is the third generation of such a protocol and was introduced in 1999 replacing TG-21 protocols. TG-21 protocols improved errors in beam calibrations of up to 5% at the expense of much more complex calculations which represent an increased potential for errors in the clinic. A major advantage in TG-51 is that it requires absorbed dose to water calibration factors, making it simpler to implement and understand than earlier protocols. Standards for absorbed dose to water have an uncertainty of less than 1% in ^{60}Co and bremsstrahlung beams up to 25 MV. TG-51 takes advantage of this improved accuracy as it is based on the use of ion chambers calibrated in terms of absorbed dose to water in a ^{60}Co beam. TG-51 protocols must be performed in a water phantom to ensure simplicity and accuracy because the quantity of interest is absorbed dose to water.

The main purpose of the TG-51 protocol is to ensure uniformity of clinical reference dosimetry in external radiation therapy with high-energy photons and electrons. This is accomplished with a common starting point of an ion chamber calibration factor, which is directly traceable to national standards of absorbed dose to water maintained by Primary Standards Laboratories (National Institute of Standards and Technology, NIST). This traceability is maintained via calibration factors obtained from an Accredited Dosimetry Calibration Laboratory (ADCL). Furthermore, all data used in the TG-51 protocols apply only under certain, well-defined reference conditions and are specified individually where it is relevant. These conditions include depth of measurement, field size, SSD, and number of MUs.

The absorbed dose to water under reference conditions is given by:

$$D_w^Q = MN_{D,w}^Q \tag{16.15}$$

where D_w^Q is the absorbed dose to water in J/kg (Gy) at the point of measurement of the ion chamber when it is absent, M is the fully corrected electrometer reading in coulombs (C) and $N_{D,w}^Q$ is the absorbed-dose to water calibration factor in a beam of quality Q.

16.4.1 Absorbed-Dose Calibration Factor $N_{D,w}^Q$

The absorbed-dose calibration factor is obtained for reference conditions in a ^{60}Co beam and denoted $N_{D,w}^Q$. In order to convert the absorbed-dose calibration factor for a ^{60}Co beam into the calibration factor for an arbitrary beam of quality Q, it is multiplied by the quality conversion factor k_Q such that:

$$N_{D,w}^Q = k_Q N_{D,w}^{60Co} (Gy/C) \tag{16.16}$$

This quality conversion factor, k_Q is chamber specific and a function of the photon component of the photon beam percentage-depth dose at 10 cm in a 10×10 cm^2 field on the surface of a water phantom at an SSD of 100 cm or $\%dd(10)_X$. This notation specifies that this is the contributions due to photons only and any electron contamination should be removed when applicable (>10 MV). It has been shown that electron contamination can be reduced to a negligible level by placing a 1 mm thick lead foil below the accelerator head to determine $\%dd(10)_X$ only. The foil should be removed for all other dose measurement steps which can lead to potential confusion and potential operator error. Therefore, the latest recommendations from the addendum to TG-51 [17] are to

forgo the use of a lead foil for flattened photon beams to reduce the potential for error.

For ^{60}Co beams, $\%dd(10)_X$ is not needed as $k_Q = 1.000$ by definition. In order to obtain the appropriate $\%dd(10)_X$, the point of measurement needs to be corrected for gradient effects and effective depth of measurement. The point of measurement for a cylindrical chamber is on the CAX of the chamber and is always placed at the reference depth when measuring dose at an individual point. The effective point of measurement is upstream of the point of measurement as secondary electrons are predominantly forward directed. As a result, depth-dose data with a cylindrical chamber is shifted to shallower depths by a distance proportional to the radius of the cavity r_{cav}. For photon beams and cylindrical or spherical chambers, the shift is taken as $0.6r_{cav}$. Once the appropriate $\%dd(10)_X$ is obtained, k_Q can be extrapolated for range of $\%dd(10)_X$ values from data available from AAPM's TG-51 or equivalent published data.

16.4.2 Charge Measurements Corrections

The electrometer reading has to be fully corrected for ion recombination, polarity, and electrometer calibration effects, and also corrected to standard environmental conditions of temperature and pressure and given by:

$$M = P_{ion}P_{TP}P_{elec}P_{pol}M_{raw} (C) \tag{16.17}$$

where M_{raw} is the raw ion chamber reading in coulombs (C), P_{TP} is the temperature-pressure correction, P_{ion} corrects for incomplete ion collection efficiency, P_{pol} corrects for any polarity effects, and P_{elec} takes into account the electrometer's calibration factor if the electrometer and ion chamber are calibrated separately.

Polarity effects vary with beam quality and a variety of conditions such as cable position. Thus it is necessary to correct for these effects by making measurements each time clinical reference dosimetry is performed. To calculate P_{pol}, an ion chamber's raw reading is taken for both polarities applied and calculated as:

$$P_{pol} = \left| \frac{M_{raw}^+ - M_{raw}^-}{2M_{raw}} \right| \tag{16.18}$$

Typical values for P_{pol} should be below 1.003 for 6 MV and can be a good indicator to verify the charge collection setup is working properly.

The electrometer correction factor is obtained from an ADCL and $P_{elec} = 1.000$ if the electrometer and ion chamber are calibrated as a unit.

Calibration factors are always given in standard environmental condition of temperature at $T_o = 22°C$ and pressure of $P_o = 760$ mmHg thus electrometer readings must be corrected to standard environmental conditions using:

$$P_{TP} = \frac{273.2 + T}{273.2 + 22} \times \frac{760 \text{ mmHg}}{P} \quad (16.19)$$

where T is the ambient temperature in degrees C in the water near the ion chamber and P is the pressure in mmHg. The correction means that for P_{TP} to change by 1%, the temperature must change by 3°C and pressure must change by 7.6 mmHg. Humidity is assumed to be within 20–80% for the temperature and pressure correction to be within 0.15%.

The standard two-voltage technique is used to determine the ion recombination correction P_{ion} and involves measuring the charge produced by the ion chamber when two different bias voltages are applied to the detector. It should be noted that once voltage has been changed, it is necessary to wait for the chamber readings to come to equilibrium. For pulsed or pulsed-swept beams, P_{ion} is calculated as

$$P_{ion}(V_H) = \frac{1 - \dfrac{V_H}{V_L}}{\dfrac{M_{raw}^H}{M_{raw}^L} - \dfrac{V_H}{V_L}} \quad (16.20)$$

It must be emphasized that P_{ion} is a function of dose per pulse in accelerator beams and will change if either the pulse rate for a fixed dose rate or the dose rate is changed. A properly functioning ion chamber and charge collection system should exhibit a correction factor P_{ion} below 1.05.

Taking into account all conversion and correction factors, a raw electrometer reading M yields an absorbed dose to water as

$$M_{raw} \xrightarrow{P_{ion}P_{TP}P_{elec}P_{pol}} M \xrightarrow{k_Q N_{D,w}^{60Co}} D_w^Q \quad (16.21)$$

16.5 SUMMARY

The characterization of photon beam output from high-energy linacs remains an important part of the field of Radiation Therapy Physics despite the stability and consistency of modern linacs, and the availability of reliable reference datasets. The understanding of how dose behaves inside the field and how it changes with different machine parameters is a fundamental skill to the field of Medical Physics. The relationship between dose and different collimator settings as well as the basic irradiation conditions that are required to measure these quantities were covered in this chapter. The basics of hand calculated MU as well as the standard reference dosimetry protocol from AAPM was also covered.

REFERENCES

1. R. J. Watts, "Comparative measurements on a series of accelerators by the same vendor," *Medical Physics* **26**(12), 2581–2585 (1999).
2. D. Sjöström, U. Bjelkengren, W. Ottosson, C. F. Behrens, "A beam-matching concept for medical linear accelerators," *Acta Oncologica* **48**(2), 192–200 (2009).
3. Z. Chang, Q. Wu, J. Adamson, et al., "Commissioning and dosimetric characteristics of TrueBeam system: Composite data of three TrueBeam machines," *Medical Physics* **39**(11), 6981–7018 (2012).
4. C. Glide-Hurst, M. Bellon, R. Foster, et al., "Commissioning of the Varian TrueBeam linear accelerator: A multi-institutional study," *Medical Physics* **40**(3), 031719 (2013).
5. M. C. Glenn, C. B. Peterson, D. S. Followill, R. M. Howell, J. M. Pollard-Larkin, S. F. Kry, "Reference dataset of users' photon beam modeling parameters for the Eclipse, Pinnacle, and RayStation treatment planning systems," *Medical Physics* **47**(1), 282–288 (2020).
6. P. R. Almond, P. J. Biggs, B. M. Coursey, et al., "AAPM's TG-51 protocol for clinical reference dosimetry of high-energy photon and electron beams," *Medical Physics* **26**(9), 1847–1870 (1999).
7. *Absorbed Dose Determination in External Beam Radiotherapy* (International Atomic Energy Agency, Vienna, 2001).
8. T. C. Zhu, A. Ahnesjö, K. L. Lam, et al., "Report of AAPM Therapy Physics Committee Task Group 74: In-air output ratio, for megavoltage photon beams," *Medical Physics* **36**(11), 5261–5291 (2009).
9. J. J. M. van Gasteren, S. Heukelom, H. J. van Kleffens, R. van der Laarse, J. L. M. Venselaar, C. F. Westermann, "The determination of phantom and collimator scatter components of the output of megavoltage photon beams: Measurement of the collimator scatter part with a beam-coaxial narrow cylindrical phantom," *Radiotherapy and Oncology* **20**(4), 250–257 (1991).
10. F. M. Khan and J. P. Gibbons, *Khan's the Physics of Radiation Therapy* (Wolters Kluwer Health, 2014).
11. I. J. Das, G. X. Ding, A. Ahnesjö, "Small fields: Nonequilibrium radiation dosimetry," *Medical Physics* **35**(1), 206–215 (2008).

12. I. J. Das, M. B. Downes, A. Kassaee, Z. Tochner, "Choice of radiation detector in dosimetry of stereotactic radiosurgery-radiotherapy," *Journal of Radiosurgery* **3**(4), 177–186 (2000).

13. H. Palmans, P. Andreo, M. S. Huq, J. Seuntjens, K. E. Christaki, A. Meghzifene, "Dosimetry of small static fields used in external photon beam radiotherapy: Summary of TRS-483, the IAEA–AAPM international Code of Practice for reference and relative dose determination," *Medical Physics* **45**(11), e1123–e1145 (2018).

14. R. Alfonso, P. Andreo, R. Capote, et al., "A new formalism for reference dosimetry of small and nonstandard fields," *Medical Physics* **35**(11), 5179–5186 (2008).

15. F. H. Attix, *Introduction to Radiological Physics and Radiation Dosimetry* (Wiley, 2008).

16. H. E. Johns and J.R. Cunningham, *The Physics of Radiology* (Charles C. Thomas, 1983).

17. M. McEwen, L. DeWerd, G. Ibbott, et al., "Addendum to the AAPM's TG-51 protocol for clinical reference dosimetry of high-energy photon beams," *Medical Physics* **41**(4) (2014).

CHAPTER 17

Dosimetric Considerations with Flattening Filter-Free Beams

Jessica Lye

Australian Radiation Protection and Nuclear Safety Agency
Olivia Newton-John Cancer Wellness & Research Centre, Austin Health
Melbourne, Australia

Stephen F. Kry

Imaging and Radiation Oncology Core
The University of Texas MD Anderson Cancer Center
Houston, Texas

Joerg Lehmann

Calvary Mater Newcastle
University of Newcastle
Newcastle, NSW, Australia
University of Sydney
Sydney, NSW, Australia

CONTENTS

17.1 INTRODUCTION

Flattening filter-free (FFF) beams have been used in clinical radiation therapy in specialized treatment machines such as Tomotherapy and Cyberknife for many years. Now standard linear accelerators (linacs) can include FFF beams as well as flattened beams. The new FFF beams are sufficiently different from conventional flattened beams to require special considerations and occasionally unique corrections [1, 2]. In particular, FFF beams are spatially non-uniform, have an elevated dose rate, and a larger spectral spread with a greater lower energy component than conventional flattened beams. These issues, described below, require attention during dosimetric measurements.

17.2 CHARACTERISTICS OF FFF BEAMS

17.2.1 Beam Shape

As shown in Figure 17.1, FFF beams are dosimetrically peaked toward the central axis. This arises because of the forward-peaked nature of the bremsstrahlung generated distribution and the lack of a flattening filter to force the distribution to be spatially uniform. However, the forward-peaked shape is largely constant with depth because the spectrum is approximately constant across the field [3, 4]. This is in contrast to flattened beams where the flattening filter differentially hardens the beam which causes the shape of a flattened beam profile to change with depth (producing horns at shallow depths and rounded profile shoulders at depth).

The issue of beam shape is relevant to general dosimetry discussions because, in static fields, it means that there is always a degree of dose gradient. The inherent field gradient is small (particularly for the low FFF beam energies most commonly available in clinical practice: 6 and 10 MV), so it is only relevant for relatively large detectors. The gradient is also relatively linear so that for many applications the non-uniformity will simply average-out and provide a reasonable dose estimate at the center of the detector. Nevertheless, for applications such as primary beam calibration, where the detector is large and the gradient is non-linear, the shape of the FFF beam introduces a dosimetric perturbation to the measurement. A non-uniformity correction may be necessary to correct such conditions.

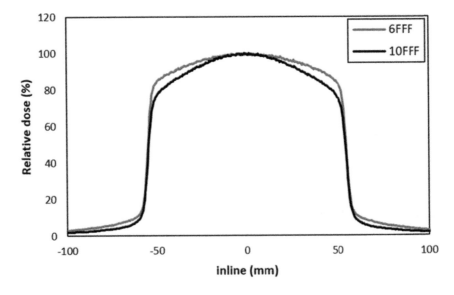

FIGURE 17.1   Typical profile of 10×10 cm^2 FFF beams at 10 cm depth.

17.2.2 Dose Rate

The flattening filter absorbs more than half of the photons generated in the bremsstrahlung target. Removal of the flattening filter therefore dramatically affects the dose rate. It is important to differentiate changes in the instantaneous dose rate (also known as the dose per pulse) from changes in the dose rate that manifest over a longer time period. An example of the latter is changing the pulse repetition rate (rep rate) on the linac; because most linacs change rep rates by pulse dropping (with pulse shaping playing a relatively minor role). Changing the rep rate affects the number of pulses being delivered, not the dose in each pulse and therefore not the instantaneous dose rate. Removal of the flattening filter, on the other hand, increases the instantaneous dose rate as compared to flattened beams. This distinction can be important for dosimetry applications where dose-rate effects manifest. Ion chambers, for example, have a signal collection time on the order of micro-seconds. The dose per pulse therefore affects recombination, but the number/frequency of pulses (e.g., through changing the rep rate) does not.

17.2.3 Beam Quality

Removing the flattening filter changes the energy spectrum of a beam compared to a flattened beam as the filter preferentially removes lower energy photons. For the beams discussed here, the flattening filter has been replaced with a thin sheet of metal in the path of the beam, so some beam hardening occurs [1]. For Varian FFF beams the electron energy is not adjusted with the removal of the flattening filter and the effective energy of the FFF beams are considerably lower than those of the flattened beams with the same nominal energy. For Elekta FFF beams the electron energy is adjusted so the effective energy and beam quality specifier $(TPR_{20,10}/\%dd(10)_x)$ remains similar. This effect, and the magnitude of it, is apparent in Figure 17.2, which shows the $\%dd(10)_x$ specifier for 6 and 10 MV beams of Varian and Elekta accelerators that were equipped with both FFF and flattened beams. This data was self-reported by each institution to IROC in the context of output verification.

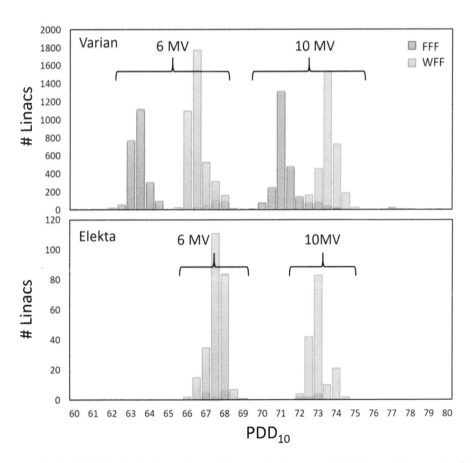

FIGURE 17.2 Beam quality ($\%dd(10)_x$) for different clinical beams for Varian and Elekta accelerators for flattening filter-free (FFF) beams and those with flattening filters (WFF).

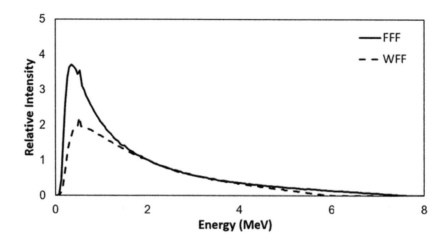

FIGURE 17.3 In-air spectra from modelled Elekta linacs from a beam with the flattening filter (dashed line), and an FFF beam (solid line). All spectra intensities have been normalized to 2 MeV, the approximate average energy.

Even in the case of the Elekta beams where the beam quality specifier is matched between FFF and flattened beams, the spectrum itself is still quite different between the two types of beams as shown in Figure 17.3. The FFF beams have a larger low energy component and the maximum energy is higher in the FFF cases to compensate for this low energy component. The different spectral composition of FFF beams raises questions on the ability of a single beam quality specifier, such as %dd$(10)_x$ to adequately define the energy of the beam and the related beam quality correction factor k_Q in the context of dosimetry.

17.3 REFERENCE DOSIMETRY IN FFF BEAMS
17.3.1 Beam Non-Uniformity

In reference dosimetry with a 0.6 cm³ Farmer type ionization chamber, the pointed shape of the profile requires a geometric correction factor for optimal accuracy. This need arises because the beam profile changes over the length of the collection volume, reducing in all directions from the maximum on the central axis. The non-uniformity correction can be calculated from beam profiles. For a Farmer-type chamber the correction may be approximated by summing the profile over the long axis:

$$k_n = \frac{D_0}{\sum D_i}$$

where D_0 is the central axis dose and D_i is the value at each point i in the scan over the length of the chamber. Depending on the chamber used, a profile with 1 mm resolution will suffice to provide reasonable accuracy on

the correction. Even at that level, the noise will likely require some smoothing before the calculation can be performed. Alternatively a curve can be fitted to the data and used for the calculation.

Correction factors have been reported to range from about 0.4% for 6 MV to about 0.7% for 10 MV FFF beams [5] for Farmer-type ion chambers with a length of ~2.3 cm. This non-uniformity correction is large enough to warrant inclusion in the reference dosimetry, however the non-uniformity correction is insensitive to depth and has no effect on the beam quality measurement [6].

17.3.2 Recombination

A substantial part of the appeal of FFF beams lies in their higher dose rate compared to the same energy flattened beams. As described above, this corresponds to an increase in the dose per pulse, unlike a change in pulse repetition rate where the dose per pulse remains the same.

The increased dose per pulse has potential dosimetric consequences as it results in higher recombination for an ion chamber and therefore necessitates a larger recombination correction factor. This correction is typically around 1% at 10 cm depth [6, 7], but can approach 2% at d_{max} where the dose rate is higher [7].

To date, Farmer chambers evaluated in FFF beams have been found to be "well behaved" in terms of recombination. That is, the recombination can be described accurately following the standard 2-voltage technique which is recommended by calibration protocols such as Task Group 51 (TG-51) [8] and TRS-398 [9]. However, ion chambers do not always behave according to standard recombination theory, and it is possible for the 2-voltage technique to calculate the recombination erroneously. As recommended in the TG-51 addendum [10], for any substantially new beam (e.g., FFF) and a new type of ion chamber, a clinic should (upon commissioning of the system) verify that the ion chamber does in fact follow recombination theory and can have its recombination corrected using the standard 2-voltage technique. This can be verified through the measurement of a Jaffé plot and looking for linearity in the relationship between $1/V$ and $1/Q$ (for a pulsed beam) [11].

17.3.3 Energy Spectrum and Corrections with FFF Beams

Due to the softer beam spectrum, a different k_Q is expected for FFF beams than for flattened beams of the same nominal energy. The interesting question that has arisen is how to calculate k_Q. Initial work into the effect of removing the flattening filter on k_Q was based on Monte Carlo simulations. Xiong and Rogers [12] calculated restricted stopping power ratios where flattening filters were completely removed and found that k_Q factors determined for flattened and FFF beams could lead up to 0.2% errors when following AAPM TG-51 %dd$(10)_x$ or 0.7% when applying TPR$_{20,10}$. This study suggested that TPR$_{20,10}$ was a less accurate beam quality metric than %dd$(10)_x$ for FFF beams. A similar Monte Carlo study reported by Dalaryd et al. [13] was based on more realistic clinical FFF beams where the flattening filter was replaced with a thin metal filter. Their study showed smaller differences for TPR$_{20,10}$ than predicted by Xiong and Rogers and this was supported by Lye et al. [6] who found differences smaller than 0.2% comparing absorbed dose measurements based on TPR$_{20,10}$ (TRS-398) or %dd$(10)_x$ (TG-51) for flattened and FFF beams at 6 MV or 10 MV.

The accuracy of k_Q when following TRS 398 using TPR$_{20,10}$ for commercially available FFF beams was further confirmed by de Prez et al. [14] who performed direct calorimetry measurements and concluded that for farmer type chambers, differences less than 0.23% existed in k_Q factors based on TPR$_{20,10}$ between the corresponding flattened and FFF beams.

In short, k_Q will be similar for an FFF beam and for a flattened beam with the same %dd$(10)_x$. Moreover, it is acceptable to use either %dd$(10)_x$ or TPR$_{20,10}$ to determine k_Q according to the standard methodology used for flattened beams. One consideration for users with

TPR$_{20,10}$ beam quality is there may be a small ~0.3% correction recommended for FFF beams depending on the protocol followed in the user's country [1, 8, 9].

17.4 ROUTINE CLINICAL MEASUREMENTS

Routine clinical measurements, for example, beam scanning, IMRT QA, etc. can be done with a variety of different dosimeters, and measurements are taken at a wide range of locations within the treatment field. These two issues (dosimeter and variable measurement location) are discussed below in the context of the unique characteristics of FFF beams as described in Section 17.2.

17.4.1 Profile Shape

In general, beam non-uniformity corrections are not needed when measurements are performed away from the center of the beam as the approximately linear dose gradient across the detector will cause the measured dose to accurately describe the dose to the center of the detector volume. Even for measurements in the center of the beam, non-uniformity corrections are minimal for typical detector sizes, being less than 0.1% for detectors with the longest dimensions less than 1 cm [9]. Figure 17.4 shows a Farmer 2571 and a CC13 chamber in scale to the profile roll-off. Given that relative dosimetry is typically conducted with such detectors (non-Farmer ion chambers, diodes, thermoluminescent dosimeters (TLD), optically stimulated luminescent dosimeters (OSLD), and planar dosimeters such as film), no correction for profile shape is generally warranted.

17.4.2 Dose Rate and Recombination

The elevated dose rate associated with FFF beams leads to increased recombination in ion chambers. This can reach almost 2% for some chambers [7]. Notably, because the dose rate changes as a function of depth, recombination actually changes with depth as well; for a Farmer-chamber the recombination correction can vary by nearly 0.5% between d_{max} and 10 cm depth [7]. The recombination would similarly vary across the radiation field as the dose rate decreases from a maximum on the central axis.

However, such a large recombination correction, and the corresponding sensitivity in recombination to the measurement location, is largely reserved for just Farmer-type ion chambers. Smaller volume ion chambers, as are more commonly used for relative dose measurements, have less recombination because of their

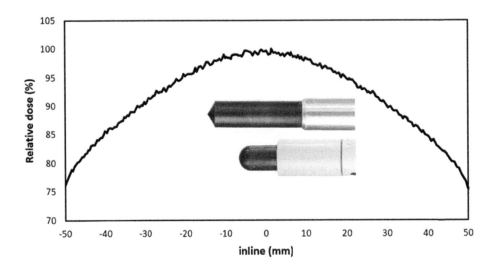

FIGURE 17.4   Comparison of length of a farmer-type 2571 and CC13 thimbles (shown to scale) with respect to the profile roll-off with a 10 FFF beam at 10 cm depth.

smaller volume. Recombination is therefore a smaller concern, and correspondingly, recombination varies less with depth or lateral position.

Other types of detectors typically exhibit smaller recombination effects than ion chambers. TLD and OSLD have negligible dose-rate dependency over a very wide range of dose rates [15]. Similarly, film has no measurable dose-rate dependence [16, 17]. No correction or special consideration is therefore warranted for dose-rate effects for these detectors in FFF beams. On the other hand, diodes may show dose-rate dependence, which is the primary component of the "SSD dependence" of these detectors [18]. Consequently, it is not possible to make universal statements about dose-rate effects. In general, if the dose-rate dependence for given detector is not well known, it is prudent to evaluate this characteristic before using the detector in high dose rate environments including FFF beams. The instantaneous dose rate can be easily reduced by increasing the source-to-detector distance or the depth of the measurement in, for example, a water tank, to reduce the photon flux. The measured dose can also be compared with the measured dose of a dose-rate independent detector. Once the recombination effects are known, a clinical management decision can be made (e.g., effects are corrected for, or the effects are less than the uncertainty in the measurements and are therefore neglected).

17.4.3 Energy Dependence

As discussed previously, FFF beams have a softer spectrum than flattened beams, and are therefore more prone to energy dependence effects. On the favorable side, FFF beams show a uniform spectrum across the treatment field (unlike flattened beams), so there is no variation in response within a given treatment field.

Energy responses of different detectors have been studied at length, including for ion chambers, film, TLD, OSLD, and diodes, although specific FFF versus flattened beam responses have received very little attention so far. Ion chambers show very little energy dependence except for chambers with steel central electrodes (typically microchambers) [19, 20]. Modern radiochromic film shows minimal energy dependence across megavoltage energies, and a flattened calibration would generally suffice for an FFF beam [21–23]. Similarly, LiF-based TLD (e.g., TLD-100) has an effective atomic number very close to water and therefore shows very little energy dependence throughout the MV energy range. Because of this, it is unlikely that TLD would require unique calibration or corrections for measurements in an FFF beam [15]. OSLD (Al_2O_3:C) has a slightly more distinct effective atomic number and may show more variability to FFF beam energies. While a difference in response of up to 1–2% may be expected at different locations in the treatment field [24], both IROC and the Australian Clinical Dosimetry Service (ACDS) use the same beam quality factor for flattened and FFF beams at d_{max} under reference conditions (in their respective Level I remote audits) and achieve high accuracy (within 1%). There are many types of diodes, and these devices often have very pronounced energy effects. Unique calibrations are likely required for these devices when employed in FFF beams [18].

For each of the detectors described above, studying the energy dependence is generally quite challenging. Practically therefore, the most reasonable approach to determine a given detector's response is through a cross calibration in an FFF beam. An ion chamber, having very little energy dependence, is a good choice for performing such a cross calibration. In general, if a detector is being used that has an effective atomic number markedly different from that of water (e.g., diode, MOSFET, etc.), there is a good chance that a specific FFF calibration coefficient (or correction factor) is needed.

17.5 OTHER CLINICAL CONSIDERATIONS

17.5.1 Small Fields

For small field measurements, a similar approach is taken with both FFF beams and conventional flattened beams following the TRS-483 protocol [25]. The small field chamber is cross-calibrated with reference chamber in a reference field with the FFF beam. When measuring the FFF small fields the appropriate field output correction factor $k_{Q_{clin},Q_{msr}}^{f_{clin},f_{msr}}$ is applied determined from the equivalent square field size. This has been specifically evaluated for FFF beams in recent studies of different detectors [26, 27].

17.5.2 Motion and Interplay Effects

For moving targets the high dose rates and associated short treatment times of FFF beams require reconsideration of possible interplay between multileaf collimator (MLC) motion and tumor motion, in particular for hypo-fractionation (SBRT). Interplay effect, which can conceptually lead to over or underdosing if tumor motion is synchronized with MLC motion, was evaluated in the context of the implementation of IMRT and found to be negligible for standard fractionations [28]. Some theoretical and measurement-based studies have been reported on the topic including considerations of the shorter treatment times associated with hypo-fractionated treatments [29–33]. These studies report local dose discrepancies around 10% attributed to interplay. However, such point doses difference need to be put in the context of the other uncertainties of dose delivery to a moving target. The actual impact of interplay effect depends on the MLC pattern, gantry motion, the treatment (beam on) time and the patient motion. It is non-trivial to assess interplay, especially as the actual target and patient motion during the treatment cannot be predicted reliably for a specific patient. Shorter overall treatment (beam on) times (integrated over all fractions) can be associated with increased interplay effect for a given modulation. Single fraction treatments might be of concern depending on the specifics of the plan (length of beam on time) and the patient motion. The current body of published literature is limited and based on this it is impossible to exclude the possibility of the interplay effect at high dose rates (short beam on times) for all possible scenarios.

17.5.3 Electron and Neutron Contamination

Neutron contamination is a concern for patient safety and radiation shielding when beam energies >10 MV are employed. Most commercial implementations of FFF beams are 10 MV or less, so neutron contamination is often not an issue. However, there are some FFF beams with energy >10 MV in clinical operation (e.g., Siemens units [1]). For high energy FFF beams, neutron production has been found to be dramatically reduced as compared to flattened beams [34–36]. This occurs because the photon treatment is delivered more efficiently so the neutron production is reduced (fewer photons are generated for the treatment which create correspondingly fewer neutrons) and also because removing the flattening filter removes a direct source of neutrons. These benefits will be offset if the energy of the FFF beam is raised to restore the PDD as neutrons are produced primarily by the high-energy component of the photon spectrum [37].

In addition to the abundance of low energy x-rays that comprise FFF beams, there is also the potential for increased electron contamination [38]. To minimize the contribution of electrons, filtration of a few mm of metal is placed in the accelerator head (in the same location in the accelerator head as the flattening filter) [1].

17.6 SUMMARY

FFF beams are spatially non-uniform, have an elevated dose rate, and a larger spectral spread with a greater lower energy component than conventional flattened beams. These issues require attention during dosimetric measurements. A geometric correction factor is needed for reference dosimetry with Farmer type chambers. Recombination effects may become an issue in selected ionization chambers. FFF beams have a softer spectrum which is more uniform across the treatment field. The very low energy component as well as the increased electron contamination is generally reduced with a thin metal plate in the beam path. Detectors with energy dependent response should be cross calibrated before use in an FFF beam.

REFERENCES

1. Y. Xiao, S. F. Kry, R. Popple, E. Yorke, N. Papanikolaou, S. Stathakis, P. Xia, S. Huq, J. Bayouth, J. Galvin, et al.,"Flattening filter-free accelerators: A report from the AAPM Therapy Emerging Technology Assessment Work Group," *Journal of Applied Clinical Medical Physics* **16**(3), 5219 (2015).

2. G. Budgell, K. Brown, J. Cashmore, S. Duane, J. Frame, M. Hardy, D. Paynter, R. Thomas, "IPEM topical report 1: Guidance on implementing flattening filter free (FFF) radiotherapy," *Physics in Medicine and Biology* **61**(23), 8360–8394 (2016).

3. O. N. Vassiliev, U. Titt, F. Ponisch, S. F. Kry, R. Mohan, M. T. Gillin, "Dosimetric properties of photon beams from a flattening filter free clinical accelerator," *Physics in Medicine and Biology* **51**(7), 1907–1917 (2006).

4. D. Georg, G. Kragl, S. Wetterstedt, P. McCavana, B. McClean, T. Knoos, "Photon beam quality variations of a flattening filter free linear accelerator," *Medical Physics* **37**(1), 49–53 (2010).

5. A. Sudhyadhom, N. Kirby, B. Faddegon, C. F. Chuang, "Technical Note: Preferred dosimeter size and associated correction factors in commissioning high dose per pulse, flattening filter free x-ray beams," *Medical Physics* **43**(3), 1507–1513 (2016).

6. J. E. Lye, D. J. Butler, C. P. Oliver, A. Alves, J. Lehmann, F. P. Gibbons, I. M. Williams, "Comparison between the TRS-398 code of practice and the TG-51 dosimetry protocol for flattening filter free beams," *Physics in Medicine and Biology* **61**(14), N362– N372 (2016).

7. S. F. Kry, R. Popple, A. Molineu, D. S. Followill, "Ion recombination correction factors (P(ion)) for Varian TrueBeam high-dose-rate therapy beams," *Journal of Applied Clinical Medical Physics* **13**(6), 3803 (2012).

8. P. R. Almond, P. J. Biggs, B. M. Coursey, et al., "AAPM's TG-51 protocol for clinical reference dosimetry of high-energy photon and electron beams," *Medical Physics* **26**(9), 1847–1870 (1999).

9. Technical Report Series No 398, *Absorbed Dose Determination in External Beam Radiotherapy: An International Code of Practice for Dosimetry Based on Standards of Absorbed Dose to Water* (International Atomic Energy Agency, Vienna, 2006).

10. M. McEwen, L. DeWerd, G. Ibbott, D. Followill, D. W. O. Rogers, S. Seltzer, J. Seuntjens, "Addendum to the AAPM's TG-51 protocol for clinical reference dosimetry of high-energy photon beams," *Medical Physics* **41**(4), 041501-1 (2014).

11. M. Boutillon, "Volume recombination parameter in ionization chambers," *Physics in Medicine and Biology* **43**, 2061–2072 (1998).

12. G. Xiong and D. W. Rogers, "Relationship between %dd(10)x and stopping-power ratios for flattening filter free accelerators: A Monte Carlo study," *Medical Physics* **35**(5), 2104–2109 (2008).

13. M. Dalaryd, T. Knoos, C. Ceberg, "Combining tissue-phantom ratios to provide a beam-quality specifier for flattening filter free photon beams," *Medical Physics* **41**(11), 111716-1 (2014).

14. L. de Prez, J. de Pooter, B. Jansen, T. Perik, F. Wittkamper, "Comparison of kQ factors measured with a water calorimeter in flattening filter free (FFF) and conventional flattening filter (cFF) photon beams," *Physics in Medicine and Biology* **63**(4), 045023 (2018).

15. S. F. Kry, P. Alvarez, J. E. Cygler, L. A. DeWerd, R. M. Howell, S. Meeks, J. O'Daniel, C. Reft, G. Sawakuchi, E. G. Yukihara, D. Mihailidis, AAPM TG-191 report: Clinical use of luminescent dosimeters: TLD and OSLD. Report from the American Association of Physicists in Medicine (2018).

16. L. Karsch, E. Beyreuther, T. Burris-Mog, S. Kraft, C. Richter, K. Zeil, J. Pawelke, "Dose rate dependence for different dosimeters and detectors: TLD, OSL, EBT films, and diamond detectors," *Medical Physics* **39**, 2447–2455 (2012).

17. A. Niroomand-Rad, C.R. Blackwell, B.M. Coursey, K.P. Gall, J.M. Galvin, W.L. McLaughlin, A.S. Meigooni, R. Nath, J.E. Rodgers, C.G. Soares, "Radiochromic film dosimetry: Recommendations of AAPM Radiation Therapy Committee Task Group 55," *Medical Physics* **25**, 2093–2115 (1998).

18. AAPM Task Group 62 report: Diode in vivo dosimetry for patients receiving external beam radiation therapy. Medical Physics Publishing/American Association of Physicists in Medicine (2005).

19. B. R. Muir and D. W. O. Rogers, "Monte Carlo calculations of k(Q), the beam quality conversion factor," *Medical Physics* **37**, 5939–5950 (2010).

20. J. R. Snow, J. A. Micka, L. A. DeWerd, "Microionization chamber air-kerma calibration coefficients as a function of photon energy for x-ray spectra in the range of 20–250 kVp relative to Co-60," *Medical Physics* **40**, 041711-1–041711-5 (2013).

21. H. Bekerat, S. Devic, F. DeBlois, K. Singh, A. Sarfehnia, J. Seuntjens, S. Shih, X. Yu, D. Lewis, "Improving the energy response of external beam therapy (EBT) GafChromic™ dosimetry films at low energies (≤100 keV)," *Medical Physics* **41**, 022101 (2014).

22. C. Richter, J. Pawelke, L. Karsch, J. Woithe, "Energy dependence of EBT-1 radiochromic film response for photon (10 kvp–15 MVp) and electron beams (6–18 MeV) readout by a flatbed scanner," *Medical Physics* **36**, 5506–5514 (2009).

23. M. J. Butson, P. K. N. Yu, T. Cheung, H. Alnawaf, "Energy response of the new EBT2 radiochromic film to

x-ray radiation," *Radiation Measurements* **45**, 836–839 (2010).

24. S. B. Scarboro, D. S. Followill, J. R. Kerns, R. A. White, S. F. Kry, "Energy response of optically stimulated luminescent dosimeters for non-reference measurement locations in a 6 MV photon beam," *Physics in Medicine and Biology* **57**, 2505–2515 (2012).

25. Technical Report Series No 483, *Dosimetry of Small Static Fields Used in External Beam Radiotherapy* (International Atomic Energy Agency, Vienna, 2017).

26. W. Lechner, H. Palmans, L. Sölkner, P. Grochowska, D. Georg, "Detector comparison for small field output factor measurements in flattening filter free photon beams," *Radiotherapy & Oncology* **109**(3), 356–360 (2013).

27. T. S. A. Underwood, B. C. Rowland, R. Ferrand, L. Vieillevigne, "Application of the exradin W1 scintillator to determine Ediode 60017 and microdiamond 60019 correction factors for relative dosimetry within small MV and FFF fields," *Physics in Medicine and Biology* **60**, 6669–6683 (2015).

28. T. Bortfeld, S. B. Jiang, E. Rietzel, "Effects of motion on the total dose distribution," *Seminars in Radiation Oncology* **14**(1), 41–51 (2004).

29. R. I. Berbeco, C. J. Pope, S. B. Jiang, "Measurement of the interplay effect in lung IMRT treatment using EDR2 films," *Journal of Applied Clinical Medical Physics* **7**(4), 33–42 (2006).

30. C. L. Ong, M. Dahele, B. J. Slotman, W. F. Verbakel, "Dosimetric impact of the interplay effect during stereotactic lung radiation therapy delivery using flattening filter-free beams and volumetric modulated arc therapy," *International Journal of Radiation Oncology Biology Physics* **86**(4), 743–748 (2013).

31. E. D. Ehler and W. A. Tomé, "Step and shoot IMRT to mobile targets and techniques to mitigate the interplay effect," *Physics in Medicine and Biology* **54**(13), 4311–4324 (2009).

32. M. Rao, J. Wu, D. Cao, T. Wong, V. Mehta, D. Shepard, J. Ye, "Dosimetric impact of breathing motion in lung stereotactic body radiotherapy treatment using intensity modulated radiotherapy and volumetric modulated arc therapy [corrected]," *International Journal of Radiation Oncology Biology Physics* **83**(2), e251–e256 (2012).

33. X. Li, Y. Yang, T. Li, K. Fallon, D. E. Heron, M. S. Huq, "Dosimetric effect of respiratory motion on volumetric-modulated arc therapy-based lung SBRT treatment delivered by TrueBeam machine with flattening filter-free beam," *Journal of Applied Clinical Medical Physics* **14**(6), 4370 (2013).

34. S. F. Kry, R. M. Howell, U. Titt, M. Salehpour, R. Mohan, O. N. Vassiliev, "Energy spectra, sources, and shielding considerations for neutrons generated by a flattening filter-free Clinac," *Medical Physics* **35**(5), 1906–1911 (2008).

35. A. Mesbahi, "A Monte Carlo study on neutron and electron contamination of an unflattened 18-MV photon beam," *Applied Radiation and Isotopes* **67**(1), 55–60 (2009).

36. S. F. Kry, U. Titt, F. Ponisch, et al., "Reduced neutron production through use of a flattening-filter-free accelerator," *International Journal of Radiation Oncology Biology Physics* **68**(4), 1260–1264 (2007).

37. R. M. Howell, S. F. Kry, E. Burgett, N. E. Hertel, D. S. Followill, "Secondary neutron spectra from modern Varian, Siemens, and Elekta linacs with multileaf collimators," *Medical Physics* **36**(9), 4027–4038 (2009).

38. U. Titt, O. N. Vassiliev, F. Ponisch, L. Dong, H. Liu, R. Mohan, "A flattening filter free photon treatment concept evaluation with Monte Carlo," *Medical Physics* **33**(6), 1595–1602 (2006).

CHAPTER 18

Linac-Based SRS/SBRT Dosimetry

Karen Chin Snyder and Ning Wen

Henry Ford Health System
Detroit, Michigan

Manju Liu

William Beaumont Hospital
Royal Oak, Michigan

CONTENTS

18.1　INTRODUCTION TO LINEAR ACCELERATOR-BASED SRS/SBRT

Linear accelerators (linacs) have been used in radiation oncology as medical treatment devices since 1953, when the first patient was treated at the Hammersmith Hospital in London with a stationary linac and a treatment head that could swivel. The original system was simple, allowing for delivery of treatment of open fields, or field blocked to shield normal tissues with devices that were manufactured out of Cerrobend. Linac-based stereotactic surgery occurred in 1980s after Leksell Gamma Knife had been used for several decades. Its implementation was hindered by the mechanical uncertainty and stability of the multiple moving components.

The use of floor frames by several groups mitigated the mechanical uncertainties of the movement of the couch [1, 2]. In 1984, at the joint center for radiation therapy (JCRT) at Massachusetts General Hospital, the Winston–Lutz test was developed that evaluated the accuracy of the mechanical and radiation isocenters [3]. The Winston–Lutz test has still been a common quality assurance test used in linac-based stereotactic radiosurgery (SRS) to determine delivery precision.

18.1.1　Common Treatment Platforms

Several commercial linac platforms exist that are used for radiotherapy treatments as well as some specialized platforms used primarily for radiosurgery. C-arm linac-based SRS and stereotactic body radiotherapy (SBRT) is attractive because of the range and flexibility of treatment options. Whereas Gamma Knife is primarily used to treat single fractionated intracranial lesions, a linac can have a wide variety of field sizes and fractionation schemes used to treat small as well as large lesions. Linacs may also be fitted with stereotactic accessories mounts such as tertiary collimators and cones that can be used to define small fields used in radiosurgery.

18.1.1.1　Novalis Tx

The Novalis Tx (NTx) platform combines a Trilogy machine (Varian Medical Systems, Palo Alto, CA, USA) with an ExacTrac x-ray system (Brainlab AG, Munich, Germany). The system has dual image-guided systems: the Varian on board imaging (OBI) system as well as the Brainlab ExacTrac localization (Figure 18.1(a)). The Varian OBI system can be utilized to acquire cone beam CTs (CBCTs) for volumetric imaging that is useful for imaging 3-dimensional soft tissues. The Brainlab ExacTrac system consists of two kV x-ray generators and two mounted aSi flat panel detectors. They

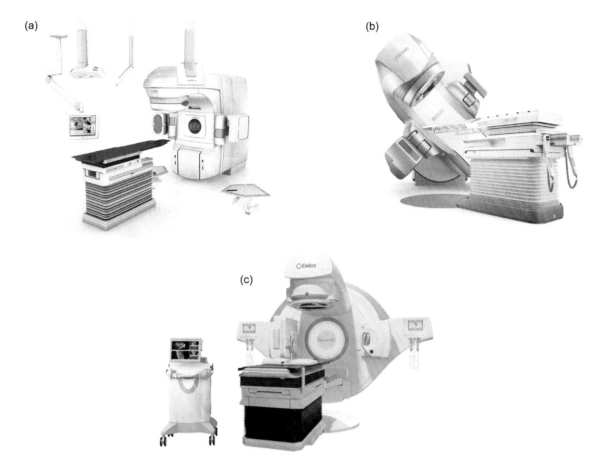

FIGURE 18.1 Common C-arm linear accelerator treatment platforms. (a) Novalis Tx (Image courtesy of Varian Medical Systems.); (b) Varian EDGE (Image courtesy of Varian Medical Systems.); (c) Elekta Versa HD (Image courtesy of Elekta.)

can be utilized for 2D planar x-ray correction as well as patient monitoring during treatment [4].

The Novalis Tx machine can deliver flattened photon beams as well as electron beams. The photon beam used for radiosurgery is a 6X-SRS high dose rate flattened beam. The high dose rate SRS beam has a flattening filter separate from the 6X beam for the high dose rate mode that is limited by field size but is designed for 1000 MU/min dose rate. The Novalis Tx uses either cones or the high definition multi-leaf collimator (MLC) to collimate and shape the beam. The high definition MLC (HD120 MLC) contains 120 MLCs, 60 on each bank, with 32 2.5 mm central leaves and 28 5 mm outer leaves with a maximum field size of 22×40 cm². The cones are manufactured by Brainlab and typically consist of diameters of 4 mm, 6 mm, 7.5 mm, 10 mm, 12.5 mm, and 15 mm.

18.1.1.2 Varian EDGE and Truebeam

The EDGE and Truebeam (Varian Medical Systems, Palo Alto, CA) are streamlined treatment platforms that

integrate a combination of image-guided systems such as kV/MV, CBCT, as well as Calypso and an optical surface monitoring system (OSMS) for image guidance [5]. The CBCT can be used for bony as well as soft tissue localization, and kV planar images can be used to perform 2D/3D matches with 6 degrees of freedom correction.

The EDGE is a dedicated radiosurgery machine based on the Varian Truebeam platform (Figure 18.1(b)). The EDGE has multiple photon energies, 6/10 MV flattened and two high intensity flattening filter-free (FFF) modes, 6X FFF and 10X FFF, as well as a 2.5 MV beam used for imaging. The dose rate for the high dose rate radiosurgery beams are 1400 MU/min for the 6X FFF beam and 2400 MU/min for the 10X FFF beam. The EDGE uses either cones or a high definition MLC to collimate and shape the beam. The high definition MLC (HD120 MLC) contains 120 MLCs, 60 on each bank, with 2.5 mm central leaves and 5 mm outer leaves with a maximum field size of 22×40 cm². The EDGE radiosurgery package includes the Varian Integrated Conical collimator Verification and Interlock (ICVI) system that includes 7 conical collimators of diameters, 4 mm, 5 mm, 7.5 mm, 10 mm, 12.5 mm, 15 mm, and 17.5 mm.

18.1.1.3 Elekta Versa HD

The Elekta Versa HD (Elekta Oncology Systems, Crawly, UK) combines radiosurgery and conventional radiotherapy in a single solution (Figure 18.1(c)). Image guidance on the Versa HD is performed with the Symmetry 4D scan CBCT system which provides 4D image guidance to assist with respiratory tracking and motion management. The Versa HD also utilizes the Clarity Autoscan system which uses ultrasound to monitor target anatomy.

The Elekta Versa can be configured for multiple flattened and FFF photon beams as well as electrons. The high dose rate modes for radiosurgery include a 6X FFF and 10X FFF photon beams with 1400 MU/min and 2400 MU/min, respectively. The Versa HD uses either cones or MLCs to collimate and shape the beam. It is equipped with an Agility MLC high-definition, full field MLC. The MLCs replace the jaws in the orthogonal direction and consist of 160 MLCs, 80 leaves on each bank, with a 5 mm leaf width that spans over the entire 40×40 cm^2 field. Cones of diameters of 5 mm, 7.5 mm, 10 mm, 12.5 mm, and 15 mm are common cone sizes included.

18.1.2 Flattening Filter-Free, High Dose Rate Modes

Linacs have been historically designed with flattening filters for 3D conformal treatments in order to flatten the radiation beam into a useable beam, typically at a reference depth of 10 cm. In radiosurgery, small beams field sizes are often not flat, and peaked in the center, due to the geometric penumbra and source occlusion. Furthermore for intensity-modulated radiation therapy (IMRT) and

volumetric modulated arc therapy (VMAT) planning, the beam does not need to be flat when MLCs modulate the fluence to compensate for the beam profile [6].

The flattening filter is typically designed out of tungsten and copper alloys to filter and attenuate low energy photons. When the flattening filter is removed, the dose rate on the central axis can increase by a factor of 2–5 times the flattened dose rate, given a constant target current [7]. Scatter dose from the head of the linac is also reduced when the flattening filter is removed, which can reduce scatter dose from the head of the machine to the patient [8]. FFF high dose rate beams have become quite attractive in radiosurgery, allowing for delivery of greater amounts of dose in a single or few fractions in a shorter amount of time [9]. High dose rates allow for quicker treatments, thus minimizing the likelihood of patient motion. When the flattening filter is removed, less of the primary beam is attenuated and scattered from the flattening filter, resulting in a lower percentage of low energy photons contributing to out of field dose to the patient, thus reducing the chances of secondary malignancies. However, removal of the flattening filter can lead to challenges in dosimetry due to the change in the beam properties and the response of the dosimeter to the change in energy as well as dose rate across the field. Table 18.1 summarizes characteristic of commercially available C-arm linac FFF beams from the American Association of Physicist in Medicine (AAPM) Therapy Emerging Technology Assessment Work Group for FFF accelerators [6].

18.1.2.1 Ion Recombination Factor

When the flattening filter is removed in the linac, the dose rate is highest on the central axis and decreases

TABLE 18.1 Characteristics of Commercially Available FFF Beams

	Varian		Elekta			Siemens		
Nominal energy (MV)	6 FFF	10 FFF	6 FFF	10 FFF	7 FFF	11 FFF	14 FFF	17 FFF
Bremsstrahlung target material	Tungsten		Tungsten			Tungsten		
Approximate mean electron energy on target (MeV)	6.2	10.5	7	10.5	8.9	14.4	16.4	18.3
Filtration	0.8 mm Brass		2 mm Stainless steel			1.27 mm Al		
D_{max} (cm)	1.5	2.3	1.7	2.4	1.9	2.7	3.0	3.3
Dose at 10 cm depth (%)	64.2	71.7	67.5	73	68.5	74.5	76.5	78
Maximum dose rate on beam axis at D_{max} (cGy/min)	1400	2400	1400	2400	2000	2000	2000	2000
Dose per pulse on beam axis at D_{max} (cGy/pulse)	0.08	0.13	0.06	0.09/0.14[a]	0.13	0.13	0.13	0.13

[a]Feedback/nonfeedback machine.

Note: Dosimetric quantities are given for a 10×10 cm^2 field at 100 cm SSD unless otherwise noted. Dosimetric properties of the beam are specified at depth of maximum dose (D_{max}). Values were provided by the manufacturers and is summarized from "Flattening filter-free accelerators: a report from the AAPM Therapy Emerging Technology Assessment Work Group" [6].

with distance from the central axis. This affects the amount of columnar ion recombination that occurs across the profile of the beam. In the TG-51 absolute dosimetry protocol, ion recombination is corrected using the ion recombination factor, P_{ion}. It is used to correct for incomplete collection of the signal, when ion pairs recombine along an ionization track. The P_{ion} correction factor is clinically measured and corrected using a two-voltage method [10].

The effect of ion recombination in high dose rate flattening filter beams was investigated by Kry et al. using different chambers to verify the validity of the two-voltage method of determining P_{ion} in TG-51 [11]. The ion-recombination correction for FFF beams is larger compared to those of standard flattened beams. The two-voltage technique was within 0.3% of the Jaffe-plot results of determining P_{ion}. The recommended limit for P_{ion}, as suggested by TG-51, is $P_{ion} \leq 1.05$, and still holds for FFF beams [12]. The choice of chamber is important and the P_{ion} for the chamber may exceed the recommended P_{ion} correction. It is recommended for absolute and reference dosimetry, that the detector be evaluated for dose rate dependency.

18.1.2.2 Beam Spectrum

When the flattening filter is removed, the beam has a softer photon spectrum than a beam flattened with a flattening filter. The FFF modes of the Elekta and Siemens machines do not use the same electron beam to create the FFF and corresponding flattened beam. The electron energy at the target is higher for an Elekta 6FFF beam to more closely match the beam spectra at central axis to the Elekta 6X flattened beam. On the Varian linac, the electron energy impinging on the target is the same for the Varian 6FFF as to the Varian 6X flattened beam, and thus the beam spectra of a Varian 6FFF beam is softer relative to the 6X flattened beam.

The beam spectrum is important in characterization of the beam, specifically in its use for absolute dosimetry. AAPM Task Group 51 is a standard calibration protocol used to calibrate the output of a linac [13]. The addendum to AAPM TG-51 gives additional data and guidance for FFF beams, primarily in recommendations in how to characterize the beam spectrum. A softer photon spectrum will result in a different beam quality conversion factor, k_Q. This is due to the difference in water-to-air stopping power ratios as a function of percentage depth dose at 10 cm depth (PDD$_{10}$), or the beam

quality specifier, as well as the differences in the ionization chamber perturbation correction for the specific beam quality. Xiong et al., showed that the relationship between the %dd(10)$_x$ and the stopping power ratios given in TG-51 are acceptable to define beam quality with or without the flattening filter, with a maximum error of 0.4% [14]. Although, it is important to note that this is only valid for chambers with low Z or aluminum electrodes, that chambers with high Z electrodes should not be used for reference dosimetry for FFF beams.

Similarly, for high energy beams greater than 10 MV, TG-51 recommends adding 1 mm of lead, when measuring the beam quality, to reduce electron contamination. However, for FFF beams specified as 10 MV, it is unclear if lead should be used. It is recommended in the TG-51 addendum [12] to measure the %dd(10)$_x$ and %dd(10)$_{Pb}$ for all FFF beams, including those below the 10 MV limit from TG-51.

18.2 PHYSICS OF SMALL FIELD DOSIMETRY

Stereotactic treatments have been found to be effective in treating a large spectrum of intracranial and extracranial lesions. Since lesions treated with SRS/SBRT are typically very small, it brings many challenges to the dosimetric aspects of modeling the beam, calculating and verifying the dose. A small field is typically defined when the field size dimension is smaller than the lateral range of the charged particles, which is also dependent on the density of the medium and energy of the beam in the medium. For example, a field size of less than 3×3 cm^2 in water is considered as small field size for 6 MV beam. Besides lateral charged particle disequilibrium, other challenges in the small field dosimetry include the occlusion of x-ray target due to the collimation and the comparable size of detectors to the beam dimensions. In this section, we will give a brief description of the physics of small field dosimetry and focus on the practical aspects of beam data acquisition of small static fields, the detector consideration, beam modeling and patient specific quality assurance.

18.2.1 Loss of LCPE

Charge particle equilibrium (CPE) condition is defined as the condition when the charged particles entering the cavity or sensitive volume of a detector are equal to the charged particles exiting it. This equilibrium includes the number, energy and angular properties of the charged particles. In small fields, the photon beam loses lateral CPE (LCPE) condition when the beam radius or

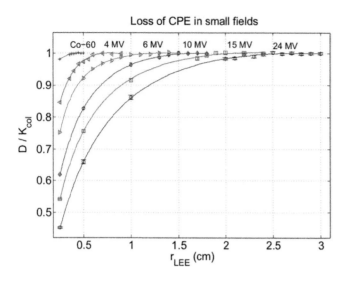

FIGURE 18.2 Ratio of dose to collision KERMA vs. field radius for different energies. This is simulated by Monte Carlo at 5 cm depth. The field radius is defined at 100 cm SSD for 4 MV to 24 MV and 80 cm SSD for ^{60}Co. Loss of LCPE happens at the field sizes that the ratio of dose to collision KERMA is less than 1. (Copied with permission from Dr. Pavlos Papaconstadopoulos [15].)

half beam width is smaller than the maximum lateral range of secondary electrons. r_{LCPE} is defined as the minimum radius of a photon beam field in which collision KERMA equals to absorb dose. r_{LCPE} has been practically used to quantitatively define small fields (Figure 18.2). Papaconstadopoulos performed Monte Carlo simulations to illustrate the relationship between r_{LCPE} and beam quality [15]. The loss of LCPE condition is more pronounced for higher beam energies and in lower density media, in which the charge particle ranges are longer.

18.2.2 Source Size Occlusion

In Monte Carlo beam models, the source is defined as the electron intensity distribution on the target surface, whereas in analytical beam models, the photon fluence distribution below the target, also known as the focal-spot, is defined as the source. The primary photon beam source is on the order of a couple millimeters (less than 5 mm). When the field size is comparable to or smaller than the dimension of the primary photon beam source, the beam output decreases with decreasing field size due to the partial source occlusion effect, where the collimation shields part of the source. The partial source occlusion effect also largely affects the detector response due to the changes of particles spectrum and sharp local absorb dose gradients. The extra-focal source, which

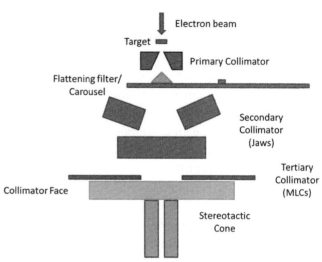

FIGURE 18.3 Schematic demonstrating location of primary collimator relative to the flattening filter, secondary collimators (jaws) and tertiary collimators (MLCs) and stereotactic cones for a Varian system.

models the scatter radiation generated below the target, contributes minimally to beam output in small fields due to the blockage from the collimation.

18.2.3 Field Size Definition for Linac SRS/SBRT

18.2.3.1 Jaws

The radiation field can be defined by multiple collimators on a linac, Figure 18.3. The primary collimator is a conical tungsten block that collimates the beam to the largest circular field size. It is located beneath the target before the flattening filter and monitor chamber. The secondary collimator typically consists of mobile jaws. These are made of thick high-Z material used to collimate the beam into square fields. In Elekta and Siemens machines, the secondary collimator consists of the MLC. In Varian machines, the MLCs are tertiary collimators.

The jaws define the square fields and the corresponding output of the field. The International Electrotechnical Commission (IEC) provides two different definitions of field size: geometric and irradiation field size [16]. The geometrical field size is defined as the geometric projection of the collimator, typically the light field, on a plane perpendicular to the beam axis. The irradiation field size is defined as the dimensions of the full width at half maximum (FWHM) of the irradiated area in a plane perpendicular to the beam axis of the radiation beam.

For large field sizes, where CPE conditions exist, the geometrical field size corresponds to the FWHM of the irradiated field size. The positions of the jaws are usually calibrated using a two point method at larger jaw sizes >5 cm. The position is set at two positions and is linearly interpolated across the entire field. When a linac is used to treat large fields, the field size is historically defined as the full width at half maximum of the radiation field.

At smaller fields the correlation between the geometric field size and irradiation field size begins to deviate, Figure 18.4. When the field size is on the same order of magnitude as the length of the lateral particle track, the machine set jaw size is slightly larger than the machine set field size. This occurs because the penumbra from the field edges overlap with the field [17]. When the machine set jaw size is much smaller, the FWHM of the resulting dose profiles is greatly overestimated. The effect in which the irradiated field size is greater than the machine set field size is called apparent field widening.

The collimator calibration is important in small field, especially for linac-based radiosurgery when the size of the field is used for both large and small field sizes. The accuracy of the jaw position becomes critical at small field sizes since it can drastically change the output and monitor unit settings on the machine. Sharpe et al. showed

a 16% discrepancy in measured output of a 1×1 cm^2 field when the set field size deviated by 2 mm, and decreased to 8% discrepancy when the field size deviation decreased to 1 mm [18].

On C-arm linacs, in which rectangular fields are defined by lower and upper jaws, the measured output factor can change for rectangular fields depending on which jaw delineates the field due to the collimator exchange effect. The collimator exchange effect occurs when photons backscatter off the jaw closest to the monitor chamber, resulting in increased charge collection, thus changing the delivered output. The profiles differ as well since the distance to the source between the X and Y jaws differ and the source has a finite size. Figure 18.5 illustrates the penumbra for an X and Y jaw profile for a square field less than 1 cm.

18.2.3.2 Multi-Leaf Collimators

MLCs are used to shape and modify the radiation treatment beam. The field shape can be defined using static leaf positions, or the MLCs can be modulated in both position and velocity to create the optimal fluence to

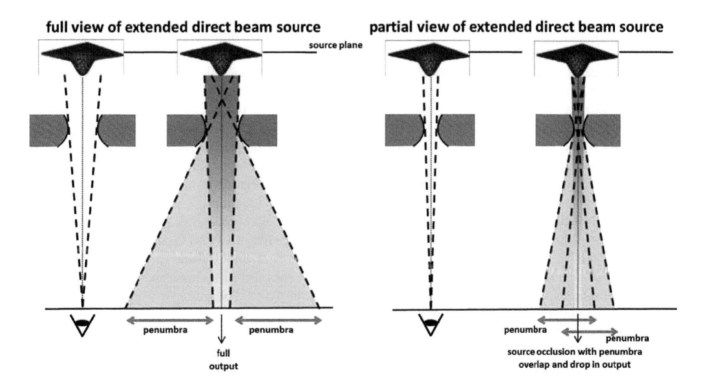

FIGURE 18.4 Demonstration of the change in geometric and irradiation field sizes for large field sizes (right) and for small field sizes when the source is partially blocked and the field size is smaller than lateral charged particle track (left). (Reproduced by kind permission of the Institute of Physics and Engineering in Medicine, York, UK. Small Field MV Photon Dosimetry, IPEM Rep.103, 2010 [19].)

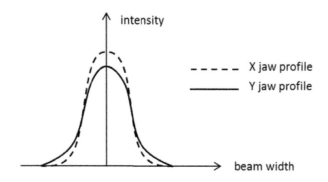

FIGURE 18.5   Illustration of the penumbral effect at very small fields (< 1 cm). Intensity profiles plotted for the same size aperture formed by $X \times Y$ vs. $Y \times X$ jaws. The X (lower) jaws form the smallest aperture [20]. (From Qin et al., Deriving detector-specific correction factors for rectangular small fields using a scintillator detector, JACMP, 17, 379–391, 2016.)

deliver a specified dose distribution [21]. IMRT and VMAT are treatment delivery techniques that are used to create conformal dose distributions. Both IMRT and VMAT use the MLCs to create small segments to vary the intensity across the radiation field. IMRT or VMAT optimization allows the MLCs to conform around irregular shaped lesions, allowing for better sparing of organs at risk. When MLCs are used for IMRT or VMAT, small irregular segments are often created by the MLCs to achieve the desired dose distribution. This can lead to difficulties with measurements and small field measurement techniques are often utilized to verify the delivery of small MLC segments.

MLCs can be secondary collimators or tertiary collimators, succeeding the primary collimator that is used to define the primary radiation field from the target. The MLCs on the radiosurgery platforms discussed in the above section include the Varian High definition MLC (HDMLC), the Elekta Agility MLC, and the Siemens Artiste.

The location of the MLCs can affect the penumbra and transmission values. Example measurement techniques and results of common radiosurgery platforms are given in the following section. When MLCs are the secondary jaws, such as the Elekta and Siemens machine design, the MLCs are located closer to the source and further from the patient surface. Because there is more room in the head of the machine when there is not a tertiary collimation device, the MLCs can be made thicker parallel to the beam, thus reducing the leakage and transmission dose through the MLC. However, the closer the MLC

is to the patient, the sharper the geometric penumbra. Most MLC designs now have curved leaf tips to reduce the penumbra at both large and small field sizes.

Smaller MLCs are preferable for radiosurgery, when small lesions are treated with high doses of radiation. The smaller the leaf, the more conformal the dose distribution can be achieved, especially for small irregular shaped lesions. It has been shown that the conformality of spine and intercranial SRS can be improved using smaller HDMLCs for smaller lesions, relative to the Millennium 120 MLC [22, 23].

18.2.3.3 Stereotactic Cones

Stereotactic cones are the historic method of collimating the linac beam to small, circular field sizes to treat small spherical or ellipsoidal lesions. They are typically made of high density materials such as lead to attenuate the beam and covered with a layer of bronze or aluminum.

Elekta, Brainlab, and Varian have focused cones in which the conical aperture in the center is focused back to the radiation source in order to minimize the penumbra. The diameter of the cone is defined as the projection of 50% of the dose at the isocenter of the machine.

Several studies show that SRS planning with MLCs is advantageous in many situations, by being able to create more conformal distributions and thus have better ability to spare dose to organs at risk. Using MLCs also has the ability to treat multiple lesions simultaneously. In order to treat non spherical lesions with cones, multiple isocenters are used which leads to increased treatment times. However, cones are still advantageous relative to MLCs for shaping radiation fields in some situations.

An advantage of cones over MLCs includes good geometrical stability where MLC positional variability due to calibration or motor issues can result in dosimetric effects. Cones also have a lower transmission value relative to the MLCs. For greater sparing of normal tissue especially when delivering large quantities of dose such as 80–90 Gy in treatment of trigeminal neuralgia, cones are attractive, since there is less transmission through the cone, resulting in less dose to the brainstem. Cones are typically tertiary devices on a linac mounted to the gantry with a collimator mount. Since the cone is mounted on the collimator, the smaller distance from the isocenter to the collimation device results in a sharper geometric penumbra compared to the MLC.

18.3 DETECTORS USED IN CHARACTERIZATION OF LINEAR ACCELERATOR BEAMS FOR STEREOTACTIC RADIOSURGERY

18.3.1 Absolute Dose Calibration

Absolute dose calibration for linacs can be performed following the AAPM TG-51 [13] protocol for clinical reference dosimetry of high-energy photons and electrons or the International Atomic Energy Agency (IAEA) Technical Reports Series (TRS), IAEA TRS-398 [24]. The AAPM TG51 protocol is appropriate for megavoltage photon energies in Co-60 beams to bremsstrahlung beams up to 25 MV, with an uncertainty of less than 1%. The protocol uses ion chambers with absorbed dose-to-water calibration factors that can be traced back to a primary standard, such as NIST (national institute standards) by the ADCL (accredited dosimetry laboratory) if located in the United States.

Absolute dosimetry should be performed in water. In AAPM TG-51 the photon beam quality is defined as %dd(10)$_x$, which is the photon component of the photon beam percentage depth dose (PDD) at 10 cm depth in a 10×10 cm^2 field at a source to surface distance of 100 cm. The beam is calibrated for a 10×10 cm^2 field size, and dose can be calibrated with either a SSD setup or a SAD setup.

Chambers recommended for absolute calibration should be waterproof, have good stability, as well as leakage, polarity and ionization factors within limits. The specification of ionization chambers used for absolute dosimetry in AAPM TG-51 is summarized in Table 18.2.

18.3.1.1 Beam Quality

The beam quality factor, k_Q, published in the original AAPM TG-51 report was based on photon spectrums of flattened beams. The use of FFF beams in SRS may change the k_Q and should be verified.

The low energy photons scattered from the primary beam are reduced for a small field, therefore resulting in a hardening of the energy spectrum. The change of both photon and electron energy spectra of small fields results in a variation of stopping power ratio used for the dose calculation. However, this variation is generally negligible. Ding et al. found that the differences in the mean energy between 4 mm field and 10×10 cm^2 field were over 20% for photons and 5% for electrons [25]. This translated to a variation within 1% in the water-to-air stopping power ratio between a 4 mm cone and 10×10 cm^2 square field.

18.3.1.2 Perturbation and Volume Averaging Effects

Detectors generally contain materials with density property that differ from water, which causes charged particle fluence perturbation. Non-gaseous detectors are usually used for the small field measurement. Such a detector cannot be treated as a Bragg–Gray cavity since the cavity could disturb the charged particle fluence considering the size and material of the detector. For example, high density detectors such as silicon diodes and diamond detectors over respond when measuring the output. Therefore, perturbation correction factors need further quantification in the small field to determine absorbed dose when applying Bragg–Gray cavity theory based on medium-to-detector stopping power ratios.

Monte Carlo simulations have been used to investigate the small field perturbations factors including the electrodes, the walls, and the detector materials [26–29]. The electrode and wall perturbation factors were reported to be close to unity, whereas the volume averaging effect and density differences between the detector and the medium dominate the detector-specific correction factor. Due to the complexity of understanding individual perturbation correction factors, which may be dependent on other, a

TABLE 18.2 Specifications for Ionization Chambers Used for Absolute Photon Dosimetry

	Upper Limit of Specification
Chamber settling	Should be less than a 0.5% change in chamber reading per monitor unit from beam-on, for a warmed up machine, to stabilization of the ionization chamber
P_{leak}	<0.1% of chamber reading ($0.999 < P_{leak} < 1.001$)
P_{pol}	<0.4% of correction ($0.996 < P_{pol} < 1.004$)
P_{ion}	Linear with dose per pulse
Initial recombination	Within 0.3% of unity
P_{ion}, Polarity dependence	<0.1% difference between opposite polarities should be
Chamber stability	Should exhibit less than 0.3% change in calibration coefficient over the typical recalibration period of 2 years

Note: Values given are upper-limit values at the reference depth, not standard uncertainties.
Source: (Summarized from AAPM TG-51.)

correction factor generated from Monte Carlo simulation is favored. However, there is still scientific interest to study the individual components.

In FFF beams, due to the change in the beam profile, a volume averaging correction factor can be used when using a cylindrical farmer type ionization chamber. The non-uniformity of the beam, where it is peaked on the central axis, can lead to an under-response when volume averaged over the length of the ionization chamber. Correction factors can be applied to mitigate the volume averaging effect. Figure 18.6 from IAEA 483 demonstrates the amount of volume averaging over the length of the ionization chamber for common treatment platforms and corresponding FFF beams.

IAEA 483 [30] recommends the following volume factor correction factor $(k_{vol})_Q^{fref}$ which can be calculated empirically from %dd(10,10)$_x$ from Equation 18.1.

$$\left(k_{vol}\right)_Q^{fref} = 1 + \left(5.9 \times 10^{-5} \cdot \%dd\left(10,10\right)_x - 3.38 \times 10^{-3}\right) \cdot$$

$$\left(\frac{100}{SDD}\right)^2 \cdot L^2$$

$$(18.1)$$

where %dd(10,10)$_x$ is the PDD at 10 cm depth in a water phantom for a field size of 10×10 cm^2 at an SSD of 100 cm. L is the length of the ionization chamber (in cm) and SDD is the source-to-detector distance (which equals the source-to-surface distance, SSD, plus the measurement depth), in cm. The assumption is made that the field size is defined at a distance of 100 cm from the photon source.

18.3.1.3 Alfonso Formalism

Alfonso et al. proposed a recommendation formalism to determine absorbed dose using ionization chambers [31]. A new detector-specific correction factor $k_{Q_{clin},Q_{msr}}^{f_{clin},f_{msr}}$ was introduced, which relates to the perturbation and volume averaging effects. It accounts for the difference between the ion chamber response in f_{msr} (machine specific reference field) and f_{clin} (clinical radiation field that needs dose measurement). The detector-specific correction factor is defined in Equation 18.2.

$$k_{Q_{clin},Q_{msr}}^{f_{clin},f_{msr}} = \frac{D_{w,Q_{clin}}^{f_{clin}}/M_{Q_{clin}}^{f_{clin}}}{D_{w,Q_{msr}}^{f_{msr}}/M_{Q_{msr}}^{f_{msr}}} \qquad (18.2)$$

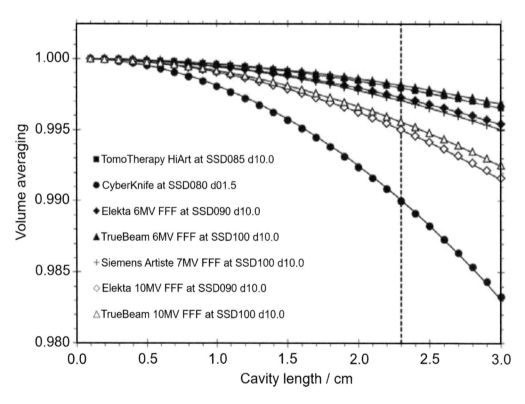

FIGURE 18.6 Volume averaging effect for various FFF beams across the cavity length of the ionization chamber. (Reproduced from IAEA publication: International Atomic Energy Agency, Dosimetry of Small Static Fields Used in External Beam Radiotherapy, Technical Reports Series No. 483, IAEA, Vienna (2017) [30], with permission by the IAEA.)

where $D_{w,Q_{clin}}^{f_{clin}}$ is the absorbed dose to water in the clinical field, f_{clin}, in the clinical beam of quality, Q_{clin}. $M_{Q_{clin}}^{f_{clin}}$ is the measured reading from the detector in the clinical field, f_{clin}, in the clinical beam of quality, Q_{clin}. $D_{w,Q_{msr}}^{f_{msr}}$ is the absorbed dose to water in the machine specific reference field, f_{msr}, in the machine specific beam of quality, Q_{msr}. $M_{Q_{msr}}^{f_{msr}}$ is the measured reading from the detector in the machine specific reference field, f_{msr}, in the machine specific beam of quality, Q_{msr}. The field output factor ($\Omega_{Q_{clin},Q_{msr}}^{f_{clin},f_{msr}}$), Equation 18.3, converts the absorbed dose to water for the machine-specific reference field to absorbed dose to water for the clinical field.

$$\Omega_{Q_{clin},Q_{msr}}^{f_{clin},f_{msr}} = \frac{D_{w,Q_{clin}}^{f_{clin}}}{D_{w,Q_{msr}}^{f_{msr}}} \tag{18.3}$$

The field output factor resembles the definition of traditional output factor but explicitly for dose at a point. The equation for the detector-specific correction factor can be rewritten as in Equation 18.4.

$$k_{Q_{clin},Q_{msr}}^{f_{clin},f_{msr}} = \frac{D_{w,Q_{clin}}^{f_{clin}}/D_{w,Q_{msr}}^{f_{msr}}}{M_{Q_{clin}}^{f_{clin}}/M_{Q_{msr}}^{f_{msr}}} \tag{18.4}$$

As a result, the field output factor can be expressed as a function of the detector-specific correction factor, Equation 18.5.

$$\Omega_{Q_{clin},Q_{msr}}^{f_{clin},f_{msr}} = \frac{M_{Q_{clin}}^{f_{clin}}}{M_{Q_{msr}}^{f_{msr}}} \cdot k_{Q_{clin},Q_{msr}}^{f_{clin},f_{msr}} \tag{18.5}$$

The detector-specific correction factors have been extracted using Monte Carlo for a few linacs and a series of small field detectors [32, 33].

An experimental approach has also been proposed to calculate $k_{Q_{clin},Q_{msr}}^{f_{clin},f_{msr}}$ for different types of detectors. Detectors with minimal corrections can be used as references to help extracting out the detector specific factor for other types of dosimeters. Those approaches were based on the assumption that the detector specific correction for the reference detector is approximately unity, $k_{Q_{clin},Q_{msr}}^{f_{clin},f_{msr}} \approx 1$. The field factor can be calculated by the ratio of reading in the clinical field of beam quality Q_{clin} and machine special field and beam quality. Example reference detectors that have been used in experimental studies include Gafchromic film, thermoluminescence dosimeters, plastic scintillators, and polymer gels [34], all of which are water equivalent and have high resolution.

The advantages of using Monte Carlo are its precision and accuracy when the material composition and geometry are known accurately. Otherwise, errors will be propagated to the calculation of output correction factors. The experimental approach is more practical and includes the detector-related effects even for the ones that are not easily simulated with Monte Carlo. However, the dosimeters that are recommended as reference may not be accessible in the clinic.

18.3.2 Relative Output Factors

Output factors obtained during commissioning of linacs used for SRS include the different output factors for the method in which the field is defined: jaw, MLC, or cones. Typically jaw-based output factors are needed to model the treatment beam.

Detectors may respond and result in different output factors at different field sizes due to either the finite size of the detector or energy response. Relative output factors are measured relative to a reference field size of 10×10 cm². In order to obtain relative output factors, at small field sizes with a relative detector, a daisy chain method is used. The daisy chain method involves measuring the charge detected at an intermediate field size, usually 3×3 or 4×4 cm², and cross calibrated to a measurement at 10×10 cm², Equation 18.6.

$$OF_{detector} = \frac{M_{FS,detector}}{M_{3\times3,detector}} \cdot \frac{M_{3\times3,IC}}{M_{10\times10,IC}} \tag{18.6}$$

It is recommended by the AAPM as well as the Radiosurgery Society (RSS) in the joint Medical Physics Practice Guideline (MPPG) 9.a for SRS-SBRT (AAPM-RSS MPPG 9.a) [35] and several other governing bodies that redundant detectors or multiple detectors be used to measure small field output factors and the output factors compared between the two.

When obtaining relative output factors for small field sizes, it is ideal to scan a cross- and in-line profile at depth to verify that the detector is aligned and positioned at the center of the beam before obtaining output factors.

18.3.2.1 Detectors Used for Relative Dosimetry

Small field dosimetry is difficult due to the challenges in detector choice and setup. The choice of detector is important in accurately characterizing the beam. Many manufacturers have different solutions for small

TABLE 18.3 Different Relative Detectors Characteristics Obtained from the Manufacturers and Published Commissioning Data

Detector	Active Volume	Dimensions		Material/Characteristics		Use
Ion Chambers		Length	Inner Radius	Wall	Electrode	
IBA Dosimetry CC01	0.01 cm³	3.6 mm	1 mm	C552	Steel	Scanning
IBA Dosimetry CC04	0.04 cm³	3.6 mm	2 mm	C552	C552	Scanning, relative output factors
IBA Dosimetry CC13	0.13 cm³	5.8 mm	3 mm	C552	C552	Scanning, ROF, ABS
PTW Pinpoint 31016 3D (new type 31022)	0.016 cm³	2.9 mm	1.45 mm	PMMA/ graphite	Al	Scanning, ROF, ABS
PTW Pinpoint 31015	0.03 cm³	5.0 mm	1.45 mm	PMMA/ graphite	Al	Scanning, ROF
PTW Pinpoint 31014	0.015 cm³	5.0 mm	1 mm	C552	Al	Scanning, ROF
PTW Semiflex 3D 31021	0.07 cm³	4.8 mm	2.4 mm	PMMA/ graphite	Al	Scanning, ROF, ABS
Exradin A16	0.007 cm³	1.65 mm	1.2 mm	C552	NA	ROF
Diodes	**Active Volume**	**Area**	**Thickness**	**n- or p-type**	**Shielding**	**Use**
IBA SFD diode	0.017 mm³	0.6 mm ø	0.06 mm	p-type	No	ROF, Scanning
IBA PFD diode	0.019 mm³	2 mm ø	0.06 mm	p-type	Yes	ROF, Scanning
PTW Diode P 60016	0.03 mm³	1 mm² circular	0.03 mm	p-type	Yes	Scanning, ROF
Sun Nuclear EDGE Detector	0.019 mm³	0.8 mm square	0.03 mm	n-type	Yes	Scanning, ROF
PTW 60012	0.03 mm³	1.13 ø	0.03 mm	p-type	No	Scanning, ROF
PTW 60018 SRS Diode	0.3 mm³	1.13 ø	0.25 mm	p-type	No	ROF
Plastic Scintillator	**Active Volume**	**Diameter ø**	**Thickness**	**Material**		**Use**
Exradin W1 Scintillator	2.4 mm³	1.0 ø	3 mm	Polystyrene		ROF
Diamond	**Active Volume**	**Diameter ø**	**Thickness**	**Material**		**Use**
PTW 60019 microdiamond	0.004 mm³	2.2 ø	0.001 mm	Synthetic single crystal diamond (SCDD)		ROF

Acronyms: ROF = relative output factors, ABS = Absolute dosimetry, PMMA = poly(methyl methacrylate), C552 = C552 Shonka air-equivalent plastic.

stereotactic detectors. Different types of detectors are available commercially. For measuring relative output factors, the volume averaging effect is the major factor for detector selection. The size of a suitable detector is fulfilled when uniform radiation fluence is present across the detector's active volume. The energy dependence of the detector with field size is another factor to consider when measuring output factors. Table 18.3 shows a variety of common detectors, their characteristics, as well as references from institutions that have published their output factor data for small jaw-based linacs. Several studies have done a cross comparison of different detectors [36–39].

18.3.2.1.1 Ion Chambers Ion chambers are tissue equivalent, however due to the air cavity they have less sensitivity than solid state detectors. To compensate for the lack of sensitivity, small field ion chambers may have a high Z central electrode to increase the chamber's sensitivity. However, the high Z electrode results in an over-response at larger field sizes due to the increase in the photoelectric cross section to lower energy photons

from increased phantom and collimator scatter. In small field sizes, the collimator scatter decreases resulting in a harder energy spectrum, thus minimizing the over response found in larger field sizes.

Ion chambers can be mounted for high resolution measurements depending on the size and shape of the ion chamber, the ion chamber can be mounted so that the long axis of the chamber is parallel or perpendicular to the central axis of the beam. Most ion chamber measurements are performed so that the long axis is perpendicular to the axis of the beam. Ion chambers can be used parallel to the central axis of the beam, but the readings may become dependent on polarity effects from irradiation of the stem. Effects such as from stem, cable, polarity and leakage need to be carefully checked and corrected.

18.3.2.1.2 Solid State Detectors Silicon diodes can be advantageous due to their fast response time. They also have higher sensitivity, due to the higher density relative to air, and thus can be made smaller, and have higher spatial resolution than some ion chamber. However some of the disadvantages include dose rate, energy, and depending on the encapsulation angular dependency. Silicon diodes tend to over respond to low-energy scattered radiation due to the larger photo-electric cross section of silicon ($Z = 14$) relative to tissue.

Often times, a metal shield is used in a shielded diode to reduce the over response to low energy photons. However, in FFF beams used in linac-based radiosurgery, the ratio of low energy photons can change with depth. Shielded diodes have shown to over respond to FFF beams due to interaction of the low energy scattered photons; unshielded diodes are recommended to be used in small field dosimetry.

Diode's angular dependency character makes the detector placement critical. It is advised that the detector's axis of symmetry must be aligned parallel to the beam axis. It is also recommended to radiograph the detector at different angles before using in small field measurements. This will help verify the location of sensitive volume against the outside marker provided by the manufacturer for chamber alignment.

Diamond detectors are more tissue equivalent and have a small active volume. However, they are expensive relative to the other commercially available detectors. Natural diamond-based detectors are no longer commercially available although artificial chemical vapor deposition (CVD) diamond-based detectors are commercially available (PTW 60019).

Plastic scintillator detectors (PSD) utilize the light emitted from the scintillating probe during its irradiation. The light is transferred through an optical fiber, and then converted and amplified in the photomultiplier tube (PMT) to signals that are ultimately recorded. The PSDs are nearly water-equivalent with similar density and atomic number to water. They can be manufactured very small while maintaining adequate sensitivity. Their response is linear to the absorbed dose. Owing to the closely matched mass stopping power and mass energy absorption coefficient to water, their response is almost independent of energy, so they can measure in both large and small fields. Several studies have indicated the correction factor for PSDs is close to unity. As discussed in detail in Chapter 9 of this book, the main problem associated with using PSDs is the Cherenkov light generated in the optical fibers, whose effect depends on the length of the irradiated fiber. Various methods have been proposed to solve this problem [40]. One solution is to use a hollow core fiber instead of a plastic transport fiber because the hollow core fiber is not expected to create Cherenkov light [41]. Chromatic removal is another successful method, owing to the fact that the Cherenkov light is mostly concentrated at the lower wavelengths of the visible light spectrum, while the scintillation spectrum peaks at a different wavelength. This method consists of measuring the combined scintillation and Cherenkov light in two different spectral regions; the dose is then calculated by using coefficients obtained from calibration when maximum and minimum portion of the fiber is irradiated [42].

18.3.2.2 Electrometers

Special electrometers may be necessary and the user should verify compatibility when using different detectors. Some small detectors require electrometers with high sensitivity that are suitable for low signal readings. Other detectors, such as the Exradin W1 Scintillator, require an electrometer [42] with two channels in order to subtract the Cherenkov component of the acquired signal.

18.3.2.3 Output Factors from Common Treatment Platforms

Consistency of output factors may not be applicable to very small field sizes. Studies have shown that the size of the source affects dosimetric parameters such as the output factor and beam profiles. This may change on different accelerators and for different energies on the same accelerator model. For the 2×2 cm^2 field size, the standard deviation increased up to 3.5% [43]. The development of a common set of data for small field size output measurements is difficult. Measurements using the correct detector still need to be performed for each individual accelerator; however, published data can be used as a guide to help identify any gross errors in commissioning of small field sizes.

18.3.2.3.1 Jaw and MLC Output Factors   In 2012, the Radiological Physics Center (RPC), now known as the Imaging and Radiation Oncology Core (IROC), published a compilation of small output factors for common treatment platforms [44]. Although it does not include FFF beams, it is a good resource to verify measured output factors during commissioning. The measurements were performed by the RPC on-site as part of on-site audits, and compared their measured values

TABLE 18.4 Summary of Example Small Field RPC-measured and Institution Treatment Planning System-Calculated Small Field for 6 MV Beam from the Varian, Elekta, and Siemens Machine

Field Size (cm × cm)	Varian		Elekta		Siemens	
	RPC	Institution	RPC	Institution	RPC	Institution
10 × 10	1.000	1.000	1.000	1.000	1.000	1.000
6 × 6	0.921	0.929	0.930	0.934	0.914	0.920
	(0.013)	(0.004)	(0.010)	(0.009)	(0.008)	(0.008)
	(0.9%, $n = 64$)		(0.5%, $n = 18$)		(0.7%, $n = 13$)	
4 × 4	0.865	0.874	0.878	0.888	0.855	0.863
	(0.018)	(0.021)	(0.015)	(0.027)	(0.010)	(0.009)
	(1.3%, $n = 64$)		(1.3%, $n = 22$)		(1.1%, $n = 13$)	
3 × 3	0.828	0.841	0.842	0.848	0.820	0.825
	(0.017)	(0.025)	(0.012)	(0.009)	(0.008)	(0.011)
	(1.7%, $n = 62$)		(0.9%, $n = 17$)		(1.3%, $n = 13$)	
2 × 2	0.786	0.796	0.790	0.796	0.764	0.757
	(0.019)	(0.031)	(0.007)	(0.010)	(0.010)	(0.042)
	(2.3%, $n = 55$)		(1.6%, $n = 17$)		(2.8%, $n = 12$)	

Note: The values in square brackets and parentheses are the average absolute percent difference and standard deviation.

Source: Summarized from Followill et al., The Radiological Physics Center's standard dataset for small field size output factors, 13, p. 3962, 2012.

to institution and treatment planning system calculated output factors. Measurements were performed on Varian ($n = 64$), Elekta ($n = 22$), and Siemens ($n = 10$) linacs with varying energies from 6 to 18 MV.

Output factors were acquired for MLC-shaped fields (2×2 cm^2, 3×3 cm^2, 4×4 cm^2, and 6×6 cm^2) with the secondary jaws set at 10×10 cm^2. The measurement setup was 100 cm SSD with the point of measurement at 10 cm depth in water, with relative output factors relative to a 10×10 cm^2 field. The Exradin A16 cylindrical micro ionization chamber (0.007 cm^3 sensitive volume) was used. Across the different institutions, the RPC measured output factors were reproducible from standard deviations ranging from 0.1% to 2.4% (Table 18.4).

18.3.2.3.2 Cone-Based Output Factors Example of measured cone-based output factors is given in Table 18.5.

18.3.2.4 Detector Specific Correction Factors

Extensive research and publications exist for small field measurements for different treatment platforms using a multitude of different detectors. Research has been performed comparing different detectors experimentally, as well comparison of detectors to theoretical methods such as Monte Carlo [48]. Recently the IAEA 483 has compiled a set of small field output correction factors from the literature for selected detectors. In general, solid state detectors tend to overestimate small field output factors, and ionization chambers tend to underestimate small field output factors.

Figure 18.7 shows the detector specific output correction factor $k_{Q_{clin}, Q_{msr}}^{f_{clin}, f_{msr}}$ as a function of square field size for a Sun Nuclear-EDGE shielded diode and a PTW-31014 Micro ionization chamber. Of note is for small field sizes, the square small field correction is less than 1 and approaches 1 around 0.8 cm field size. This shows that

TABLE 18.5 Summary of Example Measured Cone-Based Output Factors for Brainlab, Elekta, and Varian Cones on a Variety of Varian and Elekta Machines for Different Energies, Detectors, and Cone Sizes

Machine	Cone	Energy	Detector	Cone diameter (mm)							
				4	5 (6*)	7.5	10	12.5	15	17.5	20
Varian STx	Brainlab [45]	6X FFF	SNC EDGE diode	0.689	0.790*	0.830	0.871	0.890	0.901	–	–
		10X FFF		0.566	0.699*	0.756	0.826	0.864	0.888	–	–
Novalis Tx	Brainlab [46]	6X-SRS	Scanditronix SFD corrected with MC	–	0.683	0.779	0.836	0.876	0.897	0.911	0.926
Elekta Versa	Elekta [47]	6X FFF	SFD	–	0.564	0.648	0.706	0.740	0.770	–	–
Varian EDGE	Varian [5]	6X FFF	SNC EDGE diode	0.607	0.671	0.755	0.800	0.827	0.848	0.859	–
		10X FFF		0.516	0.589	0.700	0.769	0.815	0.847	0.872	–

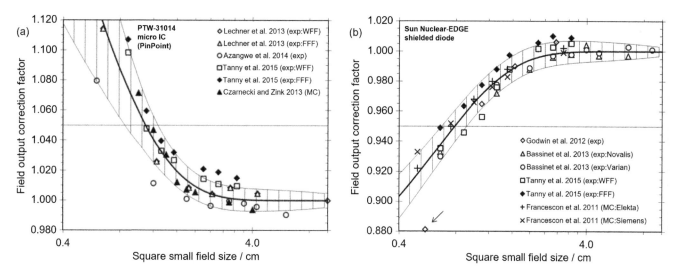

FIGURE 18.7 Plots of detector specific output correction factor as a function of square field size for a (a) PTW-31014 ionization chamber and (b) a Sun Nuclear-EDGE shielded diode. (Reproduced from IAEA publication: International Atomic Energy Agency, Dosimetry of Small Static Fields Used in External Beam Radiotherapy, Technical Reports Series No. 483, IAEA, Vienna (2017), with permission by the IAEA.)

the SunNuclear-EDGE detector over responds at smaller field sizes. The correction factor for the PTW-31014 ionization chamber is greater than 1 and approaches 1 around 0.7 cm. The micro ionization chamber under responds at small field sizes.

The IAEA compiled detector specific output correction factors for fields collimated by an MLC or SRS cone for 6 and 10 MV flattened and FFF beams for commonly used ionization chambers and solid state detectors. Tables 18.6 and 18.7 compile field output correction

TABLE 18.6 Field Output Correction Factors $k_{Q_{clin}, Q_{msr}}^{f_{clin}, f_{msr}}$ for Fields Collimated by an MLC or SRS Cone for 6 MV Flattened and FFF Beams as a Function of Equivalent Square Field Size

Detector	Equivalent Square Field Size (cm)												
Ion Chambers	8.0	6.0	4.0	3.0	2.5	2.0	1.5	1.2	1.0	0.8	0.6	0.5	0.4
IBA Dosimetry CC01	1.002	1.004	1.007	1.008	1.008	1.009	1.011	1.013	1.018	1.027	1.047	–	–
IBA Dosimetry CC04	1.000	1.000	1.000	1.000	1.000	1.002	1.009	1.022	1.041	–	–	–	–
IBA Dosimetry CC13	1.000	1.000	1.000	1.001	1.002	1.009	1.030	–	–	–	–	–	–
PTW Pinpoint 31016 3D	1.000	1.000	1.000	1.001	1.001	1.004	1.013	1.025	1.039	–	–	–	–
PTW Pinpoint 31014	1.000	1.000	1.000	1.002	1.004	1.009	1.023	1.041	–	–	–	–	–
Exradin A16	1.000	1.000	1.000	1.000	1.001	1.003	1.008	1.017	1.027	1.043	–	–	–
Diodes													
IBA SFD diode	1.008	1.017	1.025	1.029	1.031	1.032	1.030	1.025	1.018	1.007	0.990	0.978	0.963
IBA PFD diode	1.000	1.000	0.998	0.995	0.992	0.986	0.976	0.968	0.961	0.952	–	–	–
PTW Diode P 60016	1.000	1.000	0.999	0.995	0.991	0.984	0.970	0.956	–	–	–	–	–
Sun Nuclear EDGE Detector	1.000	1.000	1.000	0.999	0.998	0.994	0.986	0.976	0.966	0.951	–	–	–
PTW 60012	1.005	1.010	1.015	1.017	1.017	1.016	1.010	1.003	0.996	0.985	0.970	0.960	–
PTW 60018 SRS Diode	1.004	1.007	1.010	1.011	1.009	1.006	0.998	0.990	0.983	0.973	0.960	0.952	–
Plastic Scintillator													
Exradin W1 Scintillator	1.000	1.000	1.000	1.000	1.000	1.000	1.000	1.000	1.000	1.000	1.000	1.000	1.000
Diamond													
PTW 60019 microdiamond	1.000	1.000	1.000	1.000	0.999	0.997	0.993	0.989	0.984	0.977	0.968	0.962	0.955

Source: Values summarized from IAEA 483 [30].

TABLE 18.7   Field Output Correction Factors $k_{Q_{clin}, Q_{msr}}^{f_{clin}, f_{msr}}$ for Fields Collimated by an MLC or SRS Cone for 10 MV Flattened and FFF Beams as a Function of Equivalent Square Field Size. Values Summarized from IAEA 483 [30]

Detector	Equivalent squire field size (cm)												
Ion Chambers	8.0	6.0	4.0	3.0	2.5	2.0	1.5	1.2	1.0	0.8	0.6	0.5	0.4
IBA Dosimetry CC01	1.001	1.003	1.004	1.005	1.005	1.006	1.007	1.009	1.014	1.023	1.043	–	–
IBA Dosimetry CC04	1.000	1.000	1.000	1.000	1.000	1.002	1.009	1.022	1.041	–	–	–	–
IBA Dosimetry CC13	1.000	1.000	1.000	1.001	1.002	1.009	1.030	–	–	–	–	–	–
PTW Pinpoint 31016 3D	1.000	1.000	1.000	1.001	1.001	1.004	1.013	1.025	1.039	–	–	–	–
PTW Pinpoint 31014	1.000	1.000	1.000	1.002	1.004	1.009	1.023	1.041	–	–	–	–	–
Exradin A16	1.000	1.000	1.000	1.000	1.001	1.003	1.008	1.017	1.027	1.043	–	–	–
Diodes													
IBA SFD diode	1.005	1.010	1.015	1.018	1.018	1.018	1.015	1.010	1.003	0.992	0.974	0.962	–
IBA PFD diode	1.000	1.000	0.998	0.995	0.992	0.986	0.976	0.968	0.961	0.952	–	–	–
PTW Diode P 60016	1.000	1.000	0.999	0.995	0.991	0.984	0.970	0.956	–	–	–	–	–
Sun Nuclear EDGE Detector	1.000	1.000	1.000	0.999	0.998	0.994	0.986	0.976	0.966	0.951	–	–	–
PTW 60012	1.003	1.006	1.009	1.010	1.010	1.008	1.002	0.994	0.986	0.976	0.960	0.951	–
PTW 60018 SRS Diode	1.002	1.004	1.006	1.006	1.004	1.000	0.992	0.984	0.976	0.966	0.953	–	–
Plastic Scintillator													
Exradin W1 Scintillator	1.000	1.000	1.000	1.000	1.000	1.000	1.000	1.000	1.000	1.000	1.000	1.000	1.000
Diamond													
PTW 60019 microdiamond	1.000	1.000	1.000	1.000	0.999	0.997	0.993	0.989	0.984	0.977	0.968	0.962	0.955

factors for the selected detectors discussed in Section 18.3.2.1 and summarized in Table 18.3.

The relative field output factor can be calculated with the field output correction factor for different detectors utilizing the Alfonso formalism shown in Section 18.3.1.3 in Equation 18.5.

18.3.3 Percentage Depth Dose and Profiles

PDD and profiles (in plane and cross plane) are used to characterize the linac beam. PDD and profiles are obtained using a scanning water phantom. In order to ensure the highest quality results, literature has cited that proper measurement techniques and methods can reduce measurement errors to within 1% [49].

Proper setup and quality assurance (QA) of the scanning water tank include verification of movement of the mechanical arms, including both accuracy and linearity over the entire length of the scanning field. Manufactured scanning systems often offer annual preventative maintenance service which often checks the accuracy and precision of the mechanical movements. Several tests can be performed that are described in AAPM TG-106 to ensure high quality data collection [49]. Scanned profiles for small fields are very sensitive to the setup of the machine and scanning water tank. Khelashvili et al. demonstrated through a Monte Carlo simulation of PDDs of a 5 mm cone that the measured

PDD can change up to 12% when the gantry tilt deviates from the motion of the detector by 0.0 to 1.0 degrees [46]. The deviations in PDD show up in depths greater than 7–8 cm. Since most output factors are typically measured at 5 cm depth, slight misalignment when measuring output factor is less detrimental. However, the alignment can significantly change the magnitude of the PDD for depths greater than 10 cm, Figure 18.8.

The choice of detector is important for small field profile measurements to accurately model the treatment beam. The ideal detector should have high resolution that can accurately determine the penumbra of the beam, be fairly energy independent in response to low energy scattered photons, and have limited dose rate dependency in FFF radiosurgery beams.

Effects from volume averaging can change the shape and size of the penumbra in cross- and in-plane profiles. The use of diodes for scanning can also affect the low dose tail region under the jaw. Kim et al. compared profiles using several detectors, including two diodes (SFD, PFD) and two ion chambers (CC13 and CC01) [4]. For field sizes larger than 6×6 cm^2, the SFD (stereotactic unshielded diode) overestimated the dose by ~2% relative to the CC13. Although the SFD had the highest resolution of the detectors, it tended to over respond to the low energy scattered photons in the tail region of the profiles. The CC13 scanning ion chamber resulted

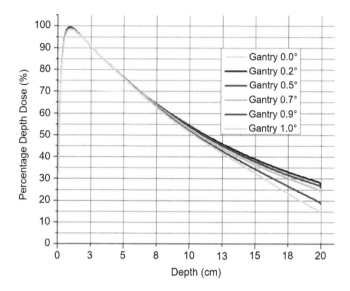

FIGURE 18.8 Dependence of PDD for 5 mm conical cone on alignment of central axis and detector motion. (Figure from Khelashvili et al. [46].)

in a larger blurring of penumbra compared to the other detectors due to the volume averaging effect, Figure 18.9. The smaller scanning ion chamber, the CC01, resulted in sharper penumbra but over responded in the tail regions of the profile, due to the over response of the higher steel electrode to the lower energy scattered photons. When measuring profiles with scanning detectors, the shorter dimension of the active volume is recommended to be placed perpendicular to the central axis to increase the limiting resolution.

Both scanning diodes and ion chambers can be used for measurement of PDDs and profiles, however the type of detector used can impact the measured profiles which in turn can affect the modeling of the beam. Studies of unshielded diodes in FFF beams show over response in the profile of tails due to the relatively high photoelectric cross section of silicon. Thus, shielding is often used to reduce the over response [50].

Diamond detectors [38] have shown significant reduction in response with dose rate, and have been shown to over-respond in the low dose regions in the tail regions of the profile. The Exradin W2 scintillator is an update to the Exradin W1 scintillator which was developed to be used for scanning as well as output factor measurement. The Exradin W1 could not be used for scanning since it was unable to account for instantaneous Cherenkov radiation corrections [51].

A study by Gersh et al. compares the PTW 60012 diode, EDGE diode and IBA CC13 ion chamber for scanning profiles for a 6 MV beam on a Varian Truebeam and the subsequent effect on the anisotropic analytical algorithm (AAA) beam model in the Eclipse treatment planning system [52]. They found that for the different detectors, the size of the penumbra was not a large factor in the beam model. Instead, the dose in the tail region of the affected the beam model created the largest difference in the beam model. When the PTW 60012 diode, an unshielded diode, was used for scanning there was an over response in the tail region and penumbra that became more pronounced for larger field sizes and increased depth where the contribution of low energy scattered photons increased. When the beam profiles of the PTW 60012 were used to create a beam model, the measured over-response resulted in an overestimation of dose to nearby OARs, where scatter dose was the main contributor to the total dose.

18.3.3.1 Reference Detectors

During scanning of PDD and profiles, two detectors are generally used. The primary detector is the scanning detector attached to the mechanical arms of the scanner and measures the profiles. The secondary detector is the reference detector that is used to remove the instantaneous fluctuation in output of the linac.

For small field measurements, the reference detector should be carefully placed to prevent shadowing of the primary scanning detector and perturbation of the primary field. For most fields, the reference detector can be placed on the edge of the primary radiation field above or below the primary detector (see Figure 18.10a). It is also recommended by AAPM TG-106 not to place the reference detector outside the primary beam due to the low signal-to-noise ratio. For very small field measurements, the reference detector can be placed inside the head of the machine, above the secondary and tertiary collimators, but this often requires some disassembling of the head of the machine.

New reference detectors have been designed that minimize the effect of the primary detector and increase efficiency. The Sun Nuclear Out-of-Field Reference detector is mounted on the outside of the linac, on the top of the gantry (see Figure 18.10b). It is a large parallel plate ion chamber, 39 cm³ in volume that measures the head leakage from the linac. The IBA StealthCHAMBER™ is a reference detector that is mounted on the face of the collimator in the primary radiation field. In some linac designs, it attaches to the accessory mount (see Figure 18.10c). It is a transmission

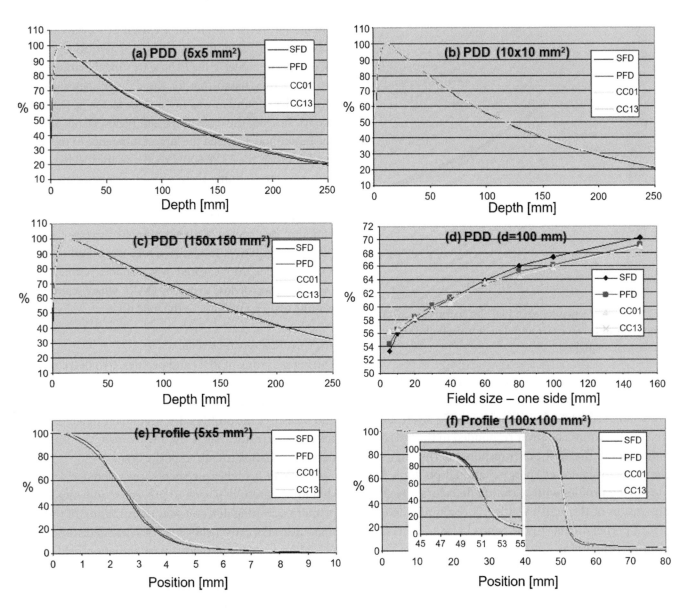

FIGURE 18.9   PDD for (a) 5×5 mm², (b) 10×10 mm², and (c) 150×150 mm² field sizes. (d) PDD at 100 mm depth for field sizes ranging from 5×5 mm² to 150×150 mm². Profiles of (e) 5×5 mm² and (f) 100×100 mm² fields. Measurements were performed with four detectors (SFD, PFD, CC01 and CC13). (Figure from Kim et al. [4].)

detector with an active area 72 mm in diameter. The active area is constructed of carbon fiber and has a total attenuation equivalent to less than 0.5 mm AL.

18.3.3.2 Jaw-Based PDD and Profiles
Commissioning of the linac regardless if it is used for SRS or conventional treatments begin with characterizing the treatment beam. This consists of measuring percent depth dose, cross-plane and in-plane profiles for different field sizes.

Table 18.8 summarizes published data for common SRS/SBRT FFF beams including the Elekta Versa, Varian EDGE, Truebeam and Novalis Tx machines. Machine tolerances from manufacturers are tighter

than they used to be and now machines from the factory are often fairly well matched for a given manufacturer machine type. Although a user's values for basic dosimetric values such as D_{max}, PDD, and penumbras may not match exactly with published data, published data can be used as a baseline to compare measured data for a newly commissioned machine. This can help the physicist verify that the correct detector and measurement techniques are used to help mitigate errors during linac commissioning.

18.3.3.3 Cone-Based Profiles
The relative output factors and thus dosimetric profiles of cones are dependent on the particular configuration

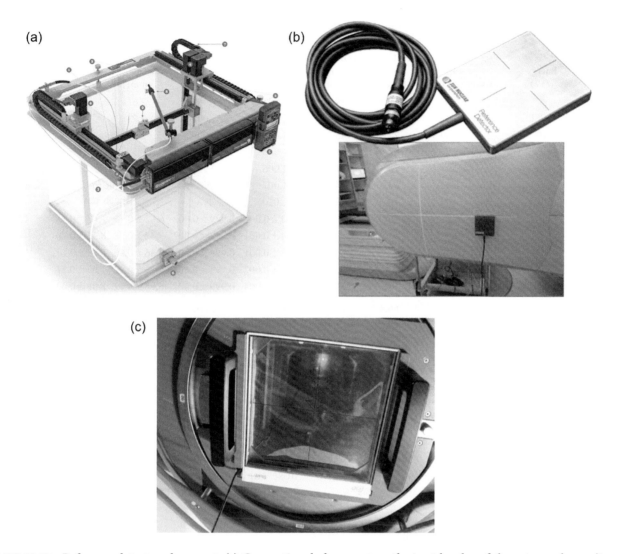

FIGURE 18.10 Reference detector placement. (a) Conventional placement on the in-side edge of the primary beam (Image courtesy of Standard Imaging.) (b). Sun nuclear out-of-field reference detector and mounted location (Image courtesy of Sun Nuclear Corporation.) (c) IBA stealth chamber mounted in accessory mount (Image courtesy of IBA Dosimetry.)

of the linac and cone. Depending on the divergence of the cone, the source occlusion effect on output may become more significant for small field sizes.

Unlike jaw-based fields, PDD and profiles can be measured with unshielded diode. PDD and profiles measured with an unshielded SRS diode (Scanditronix) and were within Monte Carlo simulations of 1% or better [46]. For cones, it is often difficult to avoid volume averaging when using detectors to scan. Film can be used to validate cross- and in-plane profiles.

18.3.4 Tissue Phantom Ratio – TPR

Depending on the treatment planning system, tissue phantom ratio (TPR) may be required in the treatment planning system instead of the PDD. TPRs can be measured directly or can be converted from measured PDDs.

18.3.4.1 TPR Measurement

In a TPR measurement, the detector remains at a fixed distance from the source, while the amount of water between the source and detector is varied. Several commercial add-ons to conventional mechanical scanning water tanks use water-draining systems to directly measure TPR. The system slowly removes water from the tank, and a water sensor monitors the water level tracking the source to surface distance. However, the TPR measurement system is typically an add-on and not always readily available at all clinics.

18.3.4.2 TPR Conversion

When direct TPR measurements are not available, the TPR can be determined from conversion of measured PDD data. Conversion of PDD to tissue maximum ratio

TABLE 18.8   Dosimetric Data for Common SRS/SBRT Platforms

	Energy (MV)	D_{max} (cm)	PDD 10× 10 cm² (5 cm)	PDD 10× 10 cm² (10 cm)	PDD 10× 10 cm² (20 cm)	P_{ion}	P_{pol}	Penumbra (Cross Plane) (mm)	Penumbra (In Plane) (mm)	Detector
Elekta Versa*	6	1.5	87	67.8	39.5	1.003	0.999	7.6	5.5	PTW Semiflex 31010
	6 FFF	1.8	87.4	67.6	39.3	1.005	1.000	7.2	5.1	
	10	2.1	90.8	72.8	45.7	1.003	1.000	8.1	5.6	
	10 FFF	2.4	91.9	72.9	45	1.006	0.998	7.3	5.8	
	18	3.0	95.4	78.3	51.7	1.006	0.999	8.0	6.7	
Varian EDGE†	6 FFF	1.35	84.2	63.0	34.4	1.007	NA	Comparable to Varian Truebeam		Scanditronix CC04
	10 FFF	2.36	90.2	70.6	42.6	1.010	NA			
Varian Truebeam‡	6X	1.5	85.8 ± 0.2	66.2 ± 0.3	38.1 ± 0.3	1.003 ± 0.002	0.999 ± 0.002	6.82 ± 0.13	7.67 ± 0.15	EDGE Detector (UTSW), Scanditronix CC13 (HFHS),
	10X	2.4	91.5 ± 0.1	73.5 ± 0.1	46.4 ± 0.2	1.004 ± 0.001	0.999 ± 0.002	7.42 ± 0.08	8.36 ± 0.20	
	15X	2.9	94.1 ± 0.2	76.6 ± 0.1	49.9 ± 0.2	1.005 ± 0.001	0.998 ± 0.002	7.56 ± 0.06	8.13 ± 0.11	
	6X FFF	1.5	84.3 ± 0.2	63.3 ± 0.1	34.5 ± 0.2	1.006 ± 0.002	0.999 ± 0.001	6.96 ± 0.22	7.65 ± 0.51	
	10 X FFF	2.4	90.4 ± 0.1	71.0 ± 0.1	42.9 ± 0.1	1.010 ± 0.004	0.999 ± 0.001	6.98 ± 0.12	8.08 ± 0.43	
Novalis Tx**	6X-SRS	1.5	86	66.4	37.7	1.000	1.005	4.96	NA	Scanditronix CC13

Note: Including depth of maximum dose (D_{max}), percent depth dose at 5, 10, and 20 cm depth for a 10 × 10 cm² field at 100 SSD, cross-plane and in-plane penumbra widths (mm) from 20% to 80% for 10 × 10 cm² field at 10 cm depth and detector used for scanning.

*   Ganesh et al. University of Texas Health Science Center at San Antonio [53].
†   Wen et al. Henry Ford Health Systems Detroit Michigan [5].
‡   Glide-Hurst et al. Henry Ford Health Systems, University of Colorado, UT Southwestern [54].
** Kim et al. Henry Ford Health Systems Detroit Michigan [4].

(TMR, a special case of TPR) can be calculated for conventional radiotherapy using the formula described in Khan, Equation 18.7.

$$TMR(d,r_d) = P(d,r_s,f)\left(\frac{S_p(r_{t_0})}{S_p(r_d)}\right)\left(\frac{f+d}{f+t_0}\right)^2 \quad (18.7)$$

The method uses the inverse square law and phantom scatter factors to convert PDD to TMR. However, for small fields, the phantom scatter is difficult to determine and can introduce significant errors when converting PDD to TMR. The phantom scatter is determined by calculation from the measured total and collimator scatter factor. However, for small fields the ability to measure small collimator scatter factor becomes problematic.

van Battum et al. proposed a method of converting PDD to TMR using a total scatter factor, in which the total scatter factor is an approximation of the real total scatter factor [55].

TMR can be calculated in the method starting from Equation 18.7, but the phantom scatter S_p is approximated by the PDD and total scatter factor ($S_{c,p}$)

$$TMR(d,r_d) = \left(\frac{D_d}{D_{t_0}}\right) = P(d,r_s,f)\cdot\left(\frac{S_{c,p}(r_s)}{S_{c,p}(r_s^*)}\right)\cdot\left(\frac{f+d}{f+t_0}\right)^2 \quad (18.8)$$

A new term $D_{inf}(d,r_d)$, in Equation 18.9, is the depth dose curve corrected for inverse square law and represents a curve corrected for source-detector distance.

$$D_{inf}(d,r_d) = P(d,r_s,f)\cdot S_{c,p}(r_s)\cdot\left(\frac{f+d}{f+t_0}\right)^2 \quad (18.9)$$

van Battum et al. fitted $D_{inf}(d,r_d)$ for each depth as a function of r, field size, by a double exponential function.

$$D_{inf}(d,r_d) = a_1 + a_2\cdot e^{(-a_3\cdot r)} + a_4\cdot e^{(-a_5\cdot r)} \quad (18.10)$$

And TMR can be calculated combining Equations 18.8–18.10 into Equation 18.11.

$$TMR_{r_d}(d) = \left(\frac{D_{inf}(d,r_d)}{D_{inf}(t_0,r_s^*)}\right) = \left(\frac{D_{inf}(d,r_d)}{S_{c,p}(r_s^*)}\right) \quad (18.11)$$

18.3.5 MLC Characterization

MLCs are characterized dosimetrically in the treatment planning system with several measured parameters. The parameters of the MLC that are input in the treatment planning system to be modeled may include inter- and intra-leaf leakage, transmission, as well as leaf tip transmission and penumbra. Ion chamber and films are often used to characterize the MLC [56]. Due to the small size of MLCs used for radiosurgery, films are useful in accurately characterizing the edge and penumbra regions of the MLC. However, small field detectors as well as multi-detector arrays are often used to verify, validate, and modify different MLC modeling parameters to better match measurements. Different treatment planning systems require different measurements. Depending on the treatment planning system different factors can be tweaked.

18.3.5.1 Varian Eclipse TPS

In the Varian Eclipse treatment planning system, MLCs for the compatible linacs are modeled using two user input parameters, the dosimetric leaf gap (DLG) and leaf transmission. The DLG is the gap between the radiation and light field defined by the MLC. The DLG parameters is used to incorporate the rounded leaf ends of the MLCs, it also incorporates the minimal physical gap between the leaf banks that prevents collisions when the MLCs meet. MLC plans can be obtained from the manufacturer for measurements of these parameters. The Varian provided MLC plans consist of 10 fields and an open field to reference the measurements. MLC transmission is measured with the same field size as the open field with the bank of MLC closed over it, and repeated for the second bank. MLC transmission can be measured using film or ion chamber. For film, a region of interest including transmission through the gap and between the MLC gaps should be averaged. For ion chamber, chamber leakage should be verified, since the typical measured dose in an open field is much higher than the dose through the MLCs. The ion chamber should be oriented parallel to the leaf travel and measured in different positions, between the leaf and through the leaf, and the values averaged.

The DLG is performed after the MLC transmission is measured because the measured MLC transmission is used to correct the DLG measurements for MLC transmission. The DLG measurement uses seven dynamic sliding window fields of varying gaps (2, 4, 6, 10, 14, 16, and 20 mm) [57]. An ion chamber is used for measurement of the DLG. The DLG is defined as the extrapolated gap that would give zero dose for the function $y = ax + b$, where y is the gap and x is the dose. The transmission and DLG is performed for each energy.

18.3.5.2 Monaco TPS

The Monaco treatment planning system uses a vendor-assisted modeling procedure, in which the vendor provides a set of fields to be delivered on the end user's choice of phantom. After the fields are delivered, the information is sent back to the vendor and a vendor-created MLC model is created.

The vendor provides an ExpressQA package for MLC modeling that consists of eight QA fields. Two of the fields simulate head and neck treatment ports, two are simple open fields to check symmetry, and four QA fields are used to obtain specific dosimetric features of the MLC. The four QA fields consist of: 3ABUT – a step-and-shoot plan where three consecutive segments are matched to create a uniform field; DMLC1 – a dynamic sweep field designed to produce the same uniform field as 3ABUT; 7SEGA – similar to 3ABUT but with smaller segments; and FOURL – a four segment step-and-shoot plan that consists of nested L-shape patterns.

Snyder et al. utilized the vendor-provided MLC plans to determine the effect on different parameters on the MLC model [58]. The 3ABUT field was used to investigate changes to the leaf offset, leaf tip leakage, leaf transmission, and corner leakage parameters. The FOURL field was used to investigate changes to the leaf transmission and groove width parameters. And finally, the DMLC1 field (and slight modifications thereof) was used to investigate changes to the interleaf leakage parameter. They found additional adjustments improved the point dose measurements as well as gamma pass rate for TG-119 and IROC phantom irradiation.

18.3.5.3 Pinnacle TPS

MLCs are modeled in the Pinnacle treatment planning system using a Model set table. The parameters are shared by all photon energies. Some of the parameters that can be change in Pinnacle include the MLC offset and rounded leaf tip radius. The MLC offset is the difference in the nominal MLC position measured by the optical light field and the projected physical leaf edge at isocenter. This takes into account the increase in transmission through the rounded leaf ends.

Young et al. utilized film, point dose ion chamber measurements, as well as the SNC ArcCHECK diode array to adjust the rounded leaf end parameters in the Pinnacle TPS to improve the small field dosimetry of VMAT plans with small segments, as well as improve the accuracy of calculated out of field dose [59].

18.4 PATIENT SPECIFIC QA MEASUREMENT DEVICES AND TECHNIQUES

Patient-specific measurements are ideal for SRS and SBRT cases and are recommended by AAPM MPPG 5, since IMRT and VMAT plans can be highly modulated. Measurements can be performed on separate devices to assess the delivered planar or volumetric dose. Several devices are available commercially that are designed to measure planar or 3D dose to verify the delivered dose. Ideal measurement devices for SRS and SBRT should have small detectors and high spatial resolution in order to measure small fields and high dose gradients.

18.4.1 Multi-Detector Arrays

Planar 2D measurement can be evaluated using 2D planar gamma analysis. The gamma analysis is evaluated using a dose difference and distance to agreement criteria that was proposed by Low et al. [60]. Most conventional cases look at 3% dose difference and 3 mm distance criteria. For stereotactic surgery, tighter gamma criteria are usually proposed in order to better evaluate the difference for dose distributions that may be small and have sharp dose gradients. Gamma criteria of 2%/2 mm and 3%/1 mm are commonly used in the literature, to better evaluate the dose difference at sharp dose gradients.

18.4.1.1 Film

Radiochromic film is a common measurement device for SRS and SBRT, due to its high spatial resolution. A common radiochromic film model commercially available is Gafchromic EBT3 film manufactured by Ashland Inc. Gafchromic has been shown in several studies its use for linac-based SRS/SBRT dosimetry. Gafchromic EBT3 film has very good resolution, where the resolution of the film is depending on the scanning resolution, typically 150 dpi. Gafchromic film also is responsive over a large dose range in the clinical range of SRS/SBRT. More recently, a newer Gafchromic EBT-XD model has been introduced by the manufacturer with an extended dynamic range up to 40 Gy.

Gafchromic film is labor intensive and requires more time than other detector arrays for analysis because the readout of the film is not instantaneous. The response of the film changes over the first several hours and plateaus around 10 hours but can be up to 24 hours [61]. Film processing also requires a lot of effort. Studies have shown heterogeneity across the scanner and film, and

corrections should be performed to reduce the uncertainties. Although film dosimetry can be difficult and time consuming, it is a good way to verify the penumbra and profiles for cones, because it has the least amount of volume averaging compared to scanning using a detector or detector array device.

18.4.1.2 Diode Arrays

SRS MapCHECK by Sun Nuclear 2D array for SRS (Figure 18.11(a)). The detector active area is 77×77 mm^2. It has 1013 diodes with a detector spacing of 2.47 mm. Inserted into a stereoPHAN, the 2D array can be used for static, rotational, non-coplanar, FFF, cone and MLC fields. Its predecessor, the MapCheck2, is a 2D array that is often used for conventional radiotherapy treatment that has a larger detector spacing of 7.07 mm and a larger detector array of 32×26 cm^2 and also supports high dose rate FFF modes (Figure 18.11).

ArcCHECK is a 4D array by Sun Nuclear (Figure 18.11(b)). The detectors are located in a concentric circle around the cylindrical phantom, so that the detectors are always perpendicular to the gantry for arc delivery, reducing any angular dependence. Consists of 1386 diodes, 0.019 mm^3 in size, spaced 1 cm apart. The detectors are arranged in a helical pattern so that the detectors are not shadowed in the beams eye view (BEV) and can measure both entrance and exit dose. There are 221 detectors in a 10×10 cm^2 field and 1386 detectors in the full 21×21 cm^2 phantom active area. For small fields for SRS or SBRT, a merge feature can increase the number of detected regions to 442 detectors within a 10×10 cm^2 field. The 4D component of the array is the temporal resolution which allows additional verification of the delivery of dose with respect to time.

A hollow central cavity in the center of the array can be filled with multiple plugs of different density and measurement devices, such as ion chamber and film to measure dose in the lesion. In conjunction with the measurement, 3DVH software can be used to estimate the delivered dose to the patient.

18.4.1.3 Ion Chamber Arrays

The MatriXX FFF array by IBA dosimetry is an air-vented ion chamber array (Figure 18.11(c)). Its predecessor, the MatriXX Evolution, is also an ion chamber array, and although it is suitable, it is not designed to be used in high dose rate flattening filter beams. The MatriXX FFF array consists of 1020 ionization chamber with an active

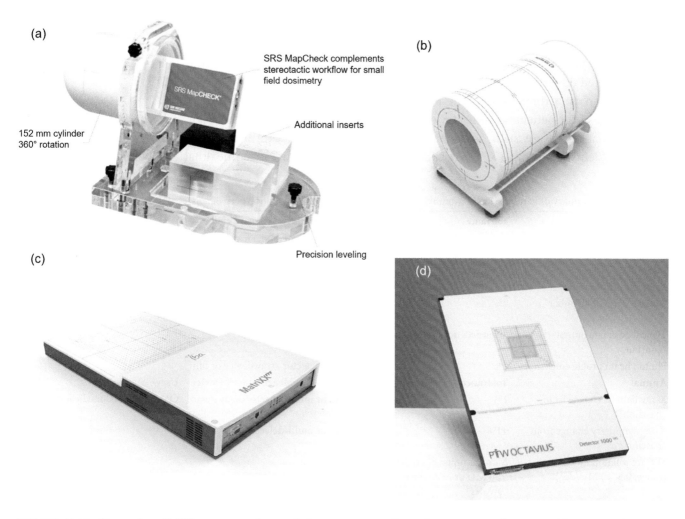

FIGURE 18.11 Examples of different manufactured detector arrays for radiosurgery applications. (a) Sun Nuclear SRS MapCHECK (Image courtesy of Sun Nuclear Corporation.), (b) SunNuclear ArcCHECK (Image courtesy of Sun Nuclear Corporation.), (c) IBA Dosimetry MatrixFFF (Image courtesy of IBA Dosimetry.), and (d) PTW Octavius1000 SRS (Image courtesy of PTW.)

detector area of 24.4 × 24.4 cm². The ion chambers are 0.032 cm³, with a detector spacing of 7.6 mm.

The OCTAVIUS 1000 SRS is a liquid-filled ionization chamber array by PTW (Figure 18.11(d)). The array consists of 977 liquid filled ionization chambers, 0.003 cm³ in size, with a detector spacing of 2.5 mm in the central 5.5 cm region, and 5 mm in the outer area. The total active detector area is 10 × 10 cm². The chambers are filled with liquid ~1000 times denser than air. Thus the liquid ion chambers have very good sensitivity and have good signal relative to an ion chamber of the same volume. The energy response to low scattered photons is much more equivalent and the ionization chambers have been shown to have good stability over time [62]. The arrays have also shown good agreement within 2% of an SRS diode for field sizes down to 1 × 1 cm² [63].

18.4.1.4 Electronic Portal Imaging Device

The electronic portal imaging device (EPID) can also be used to measure the delivered fluence. The EPID is often used to acquire MV images for localization and port verification. Using a back projection algorithm, the measured fluence on the EPID can be converted to dose [64, 65]. High resolution EPID device such as the Varian aS1200 are now available that can be used for measuring high dose rate, FFF beams [66]. The Varian aS1200 is an example of a high resolution EPID device, with an active array of 40 × 40 cm² consisting of 1190 × 1190 pixel array and a pixel pitch of 0.336 mm.

18.4.2 3D Calculation on Patient Geometry

Studies have shown that 2D gamma analysis utilizing a 3%/3 mm passing criteria is too lenient and are unable to catch systematic errors in plan delivery and during commissioning [67, 68]. Other methods to verify and evaluate the deliverability of the plan is to use the measured delivered dose or the machine log files and recalculate the plan using the delivered MLC positions on the CBCT image of the patient from the day of treatment.

Commercial software such as Mobius3D (Mobius Medical Systems, Houston, TX, USA) and SunNuclear Dose check use a separate dose calculation algorithm such as collapsed cone convolution to verify the dose from the primary treatment planning system.

18.4.2.1 Log File Verification

Verification of the plan delivery can also be done through capturing the trajectory log files from the treatment machine that contains the mechanical

TABLE 18.9 Summary of Periodic (Daily, Monthly, Annual) Dosimetric QA for Radiosurgery Beams

Daily	Tolerance	Dosimetric Measurement Device	User
X-ray output constancy (all energies)	±3%	Cross calibrated ion chamber or multi-detector array device	QMP or individual directly supervised by QMP
Monthly	**Tolerance**	**Dosimetric Measurement Device**	**User**
X-ray output constancy	2%	Ion chamber	
Backup monitor chamber constancy	2%	NA	QMP or individual directly supervised by QMP
Typical dose rate output constancy	2% at SRS dose rate, MU	Ion chamber	
Photon beam profile constancy	1%	Multi-detector array device or film	
Annual	**Tolerance**	**Dosimetric Measurement Device**	**User**
X-ray flatness change from baseline	1%	Film or multi-detector array device	QMP (qualified medical physicist)
X-ray symmetry change from baseline	±1%	Film or multi-detector array device	
SRS arc rotation mode (range: 0.5–10 MU/deg)	Monitor units set vs. delivered: 1.0 MU or 2% (whichever is greater)		
X-ray output calibration (TG 51)	±1% (absolute)	NIST traceable ion chamber	
Field size dependent output factors for x-ray (two or more FSs)	±2% from baseline	Detector used for relative dosimetry, small field	
X-ray beam quality (PDD10 or TMR2010)	±1% from baseline	Detector used for relative dosimetry	
X-ray monitor unit linearity (output constancy)	±5% (2–4 MU), ±2% >=5 MU	Ion chamber or detector used for relative dosimetry	
X-ray output constancy vs. dose rate	±2% from baseline	Ion chamber	
x-ray output constancy vs. gantry angle	±1% from baseline	Ion chamber	
X-ray off-axis factor constancy vs. gantry angle	±1% from baseline	Ion chamber or multi-detector array device	
Arc Mode (expected MU/degrees)	±1% from baseline		
MLC transmission (average of leaf and interleaf transmission) all energies	±0.5% from baseline	Ion chamber or film	
Verification of small-field beam data – relative output factors for cones and or/MLC†	±2% from baseline for >1.0 cm apertures, ±5% from baseline for <=1.0 cm apertures	Detector used for relative small field dosimetry	
E2E dosimetric evaluation using SRS frame and or/IGRT system†	±5% measured vs. calculated	Ion chamber	

† From AAPM-RSS MPPG 9.a.

position of the linac during the treatment delivery. Using the trajectory logs, the 3D dose can be calculated to the patient's CT or CBCT, the day of treatment. After calculation of the delivered dose on the CT or CBCT, the dose to the patient such as DVH criteria, 3D gamma analysis between the treatment planned dose vs. the delivered dose, and any delivery errors can be analyzed. Examples of different commercial products that use this technology are MobiusFX and Sun Nuclear PerFraction.

One drawback of this method of using log file verification is that there is no independent MLC or machine measurement since the log files are obtained from the machine's position sensors [69]. Furthermore in according to ASTRO and ACR guidelines a physical measurement is often required for reimbursement of pre-treatment IMRT QA.

18.4.2.2 Transmission Measurement and Back Projection

Other methods for calculation of 3D dose to the patient include using multi-detector arrays such as those described in Section 18.4.1. The detector array is an independent measurement device and does not rely on the machine secondary positional sensors. The dose is measured using the secondary device, and then projected back on the TPCT to verify the dose delivered to the patient.

18.5 RECOMMENDED PERIODICAL QUALITY ASSURANCE

The AAPM TG-142 [70] gives recommendations for periodic QA of medical linacs. It is an update in 2009 to the AAPM TG-40 [71] published in 1994 when the quality, accuracy, and precision of modern linacs have improved greatly allowing for treatment of small lesions with high levels of dose such as in SRS and SBRT.

The goal of periodic QA is to ensure that the machine characteristics, both dosimetric and mechanical, do not deviate significantly from their baseline values obtained during acceptance and commissioning. Table 18.9 summarizes the dosimetric tests that should be performed on a periodic basis and the recommended tolerances for machines used for SRS/SBRT treatments. Although MLCs are a large part of creating conformal dose distributions in SRS, the recommended mechanical tests for the MLCs are not covered in this section; such recommendations were provided elsewhere [56, 72].

In 2017, the AAPM and RSS published the MPPG 9.a for SRS-SBRT [35]. Included in the update are additional annual tests for C-arm linac-based radiosurgery including verification of small field data with different tolerances for fields smaller than 1.0 cm as well as dosimetric tolerances for an end-to-end (E2E) test that should be performed annually.

ACKNOWLEDGMENT

This work was supported by the American Cancer Society under a Research Scholar Grant RSG-15-137-01-CCE.

REFERENCES

1. O.O. Betti, D. Galmarini, V. Derechinsky, "Radiosurgery with a linear accelerator. Methodological aspects," *Stereotactic and Functional Neurosurgery* **57**(1–2), 87–98 (1991).
2. F. Colombo, et al., "Stereotactic radiosurgery utilizing a linear accelerator," *Applied Neurophysiology* **48**(1–6), 133–145 (1985).
3. K. R. Winston and W. Lutz, "Linear accelerator as a neurosurgical tool for stereotactic radiosurgery," *Neurosurgery* **22**(3), 454–464 (1988).
4. J. Kim, et al., "Clinical commissioning and use of the Novalis Tx linear accelerator for SRS and SBRT," *Journal of Applied Clinical Medical Physics* **13**(3), 3729 (2012).
5. N. Wen, et al., "Characteristics of a novel treatment system for linear accelerator-based stereotactic radiosurgery," *Journal of Applied Clinical Medical Physics* **16**(4), 125–148 (2015).
6. Y. Xiao, et al., "Flattening filter-free accelerators: A report from the AAPM Therapy Emerging Technology Assessment Work Group," *Journal of Applied Clinical Medical Physics* **16**(3), 5219 (2015).
7. N. V. Oleg, et al., "Dosimetric properties of photon beams from a flattening filter free clinical accelerator," *Physics in Medicine & Biology* **51**(7), 1907 (2006).
8. J. Cashmore, "The characterization of unflattened photon beams from a 6 MV linear accelerator," *Physics in Medicine & Biology* **53**(7), 1933–1946 (2008).
9. P. F. O'Brien, et al., "Radiosurgery with unflattened 6-MV photon beams," *Medical Physics* **18**(3), 519–521 (1991).
10. M. S. Weinhous and J.A. Meli, "Determining P_{ion}, the correction factor for recombination losses in an ionization chamber," *Medical Physics* **11**(6), 846–849 (1984).
11. S. F. Kry, et al., "Ion recombination correction factors (P(ion)) for Varian TrueBeam high-dose-rate therapy beams," *Journal of Applied Clinical Medical Physics* **13**(6), 3803 (2012).
12. M. McEwen, et al., "Addendum to the AAPM's TG-51 protocol for clinical reference dosimetry of high-energy photon beams," *Medical Physics* **41**(4), 041501 (2014).

13. P. R. Almond, et al., "AAPM's TG-51 protocol for clinical reference dosimetry of high-energy photon and electron beams," *Medical Physics* **26**(9), 1847–1870 (1999).

14. G. Xiong and D. W. Rogers, "Relationship between %dd(10)x and stopping-power ratios for flattening filter free accelerators: A Monte Carlo study," *Medical Physics* **35**(5), 2104–2109 (2008).

15. P. Papaconstadopoulos, *On the Detector Response and the Reconstruction of the Source Intensity Distribution in Small Photon Fields* (McGill theses, McGill University Libraries, 2016).

16. International Electrotechnical Commission, *IEC Technical Report 60788: Medical Electrical Equipment – Glossary of Defined Terms* (2004).

17. I. J. Das, G. X. Ding, and A. Ahnesjo, "Small fields: Nonequilibrium radiation dosimetry," *Medical Physics* **35**(1), 206–215 (2008).

18. M. B. Sharpe, et al., "Monitor unit settings for intensity modulated beams delivered using a step-and-shoot approach," *Medical Physics* **27**(12), 2719–2725 (2000).

19. M. M. Aspradakis, J. P. Byrne, H. Palmans, S. Duane, J. Conway, A. P. Warrington, K. Rosser, *IPEM Report 103: Small Field MV Photon Dosimetry (IAEA-CN-182)* (International Atomic Energy Agency (IAEA), 2010).

20. Y. Qin, et al., "Deriving detector-specific correction factors for rectangular small fields using a scintillator detector," *Journal of Applied Clinical Medical Physics* **17**(6), 379–391 (2016).

21. M. S. Huq, et al., "A dosimetric comparison of various multileaf collimators," *Physics in Medicine & Biology* **47**(12), N159–N170 (2002).

22. Q. J. Wu, et al., "Impact of collimator leaf width and treatment technique on stereotactic radiosurgery and radiotherapy plans for intra- and extracranial lesions," *Radiation Oncology* **4**(1), 3 (2009).

23. J.-Y. Jin, et al., "Dosimetric study using different leaf-width MLCs for treatment planning of dynamic conformal arcs and intensity-modulated radiosurgery," *Medical Physics* **32**(2), 405–411 (2005).

24. International Atomic Energy Agency, *Absorbed Dose Determination in External Beam Radiotherapy*. Technical Reports Series (International Atomic Energy Agency, Vienna, 2000).

25. G. X. Ding and F. Ding, "Beam characteristics and stopping-power ratios of small radiosurgery photon beams," *Physics in Medicine & Biology* **57**(17), 5509–5521 (2012).

26. P. Francescon, et al., "Monte Carlo simulated correction factors for machine specific reference field dose calibration and output factor measurement using fixed and iris collimators on the CyberKnife system," *Physics in Medicine & Biology* **57**(12), 3741–3758 (2012).

27. P. H. Charles, et al., "Monte Carlo-based diode design for correction-less small field dosimetry," *Physics in Medicine & Biology* **58**(13), 4501–4512 (2013).

28. F. Crop, et al., "The influence of small field sizes, penumbra, spot size and measurement depth on perturbation factors for microionization chambers," *Physics in Medicine & Biology* **54**(9), 2951–2969 (2009).

29. A. J. Scott, et al., "Characterizing the influence of detector density on dosimeter response in non-equilibrium small photon fields," *Physics in Medicine & Biology* **57**(14), 4461–4476 (2012).

30. International Atomic Energy Agency, Dosimetry of Small Static Fields Used in External Beam Radiotherapy: An IAEA-AAPM International Code of Practice for Reference and Relative Dose determination/International Atomic Energy Agency. Technical Report Series no. 483 (2017).

31. R. Alfonso, et al., "A new formalism for reference dosimetry of small and nonstandard fields," *Medical Physics* **35**(11), 5179–5186 (2008).

32. G. Cranmer-Sargison, et al., "Monte Carlo modelling of diode detectors for small field MV photon dosimetry: Detector model simplification and the sensitivity of correction factors to source parameterization," *Physics in Medicine & Biology* **57**(16), 5141–5153 (2012).

33. G. Cranmer-Sargison, et al., "A methodological approach to reporting corrected small field relative outputs," *Radiotherapy and Oncology* **109**(3), 350–355 (2013).

34. E. Pantelis, et al., "On the output factor measurements of the CyberKnife iris collimator small fields: Experimental determination of the k(Q(clin),Q(msr)) (f(clin),f(msr)) correction factors for microchamber and diode detectors," *Medical Physics* **39**(8), 4875–4885 (2012).

35. P. H. Halvorsen, et al., "AAPM-RSS Medical Physics Practice Guideline 9.a. for SRS-SBRT," *Journal of Applied Clinical Medical Physics* **18**(5), 10–21 (2017).

36. O. A. Sauer and J. Wilbert, "Measurement of output factors for small photon beams," *Medical Physics* **34**(6), 1983–1988 (2007).

37. W. Lechner, et al., "Detector comparison for small field output factor measurements in flattening filter free photon beams," *Radiotherapy and Oncology* **109**(3), 356–360 (2013).

38. M. Westermark, et al., "Comparative dosimetry in narrow high-energy photon beams," *Physics in Medicine & Biology* **45**(3), 685–702 (2000).

39. S. Dieterich and G. W. Sherouse, "Experimental comparison of seven commercial dosimetry diodes for measurement of stereotactic radiosurgery cone factors," *Medical Physics* **38**(7), 4166–4173 (2011).

40. P. Z. Liu, et al., "Plastic scintillation dosimetry: Comparison of three solutions for the Cerenkov challenge," *Physics in Medicine & Biology* **56**(18), 5805–5821 (2011).

41. A. Darafsheh, et al., "Radiotherapy fiber dosimeter probes based on silver-only coated hollow glass waveguides," *Journal of Biomedical Optics* **23**(1), 015006 (2018).

42. P. Carrasco, et al., "Characterization of the Exradin W1 scintillator for use in radiotherapy," *Medical Physics* **42**(1), 297–304 (2015).

43. D. A. Jaffray, et al., "X-ray sources of medical linear accelerators: Focal and extra-focal radiation," *Medical Physics* **20**(5), 1417–1427 (1993).

44. D. S. Followill, et al., "The Radiological Physics Center's standard dataset for small field size output factors," *Journal of Applied Clinical Medical Physics* **13**(5), 3962 (2012).

45. D. B. Wiant, et al., "Commissioning and validation of BrainLAB cones for 6X FFF and 10X FFF beams on a Varian TrueBeam STx," *Journal of Applied Clinical Medical Physics* **14**(6), 293–306 (2013).

46. G. Khelashvili, et al., "Dosimetric characteristics of the small diameter BrainLab cones used for stereotactic radiosurgery," *Journal of Applied Clinical Medical Physics* **13**(1), 3610 (2012).

47. E. Borzov, et al., "Dosimetric characterization of Elekta stereotactic cones," *Journal of Applied Clinical Medical Physics* **19**(1), 194–203 (2018).

48. G. X. Ding, D. M. Duggan, C. W. Coffey, "Commissioning stereotactic radiosurgery beams using both experimental and theoretical methods," *Physics in Medicine & Biology* **51**(10), 2549–2566 (2006).

49. I. J. Das, et al., "Accelerator beam data commissioning equipment and procedures: Report of the TG-106 of the Therapy Physics Committee of the AAPM," *Medical Physics* **35**(9), 4186–4215 (2008).

50. Z. Yin, R. P. Hugtenburg, A. H. Beddoe, "Response corrections for solid-state detectors in megavoltage photon dosimetry," *Physics in Medicine & Biology* **49**(16), 3691–3702 (2004).

51. P. E. Galavis, et al., "Characterization of the plastic scintillation detector Exradin W2 for small field dosimetry," *Medical Physics* **46**(5), 2468–2476 (2019).

52. J. A. Gersh, R. C. Best, R. J. Watts, "The clinical impact of detector choice for beam scanning," *Journal of Applied Clinical Medical Physics* **15**(4), 4801 (2014).

53. N. Ganesh, et al., "Commissioning an Elekta Versa HD linear accelerator," *Journal of Applied Clinical Medical Physics* **17**(1), 179–191 (2016).

54. C. Glide-Hurst, et al., "Commissioning of the Varian TrueBeam linear accelerator: A multi-institutional study," *Medical Physics* **40**(3), 031719 (2013).

55. L. J. van Battum, M. Essers, P. R. Storchi, "Conversion of measured percentage depth dose to tissue maximum ratio values in stereotactic radiotherapy," *Physics in Medicine & Biology* **47**(18), 3289–3300 (2002).

56. T. LoSasso, C. S. Chui, C. C. Ling, "Physical and dosimetric aspects of a multileaf collimation system used in the dynamic mode for implementing intensity modulated radiotherapy," *Medical Physics* **25**(10), 1919–1927 (1998).

57. W. Yao and J. B. Farr, "Determining the optimal dosimetric leaf gap setting for rounded leaf-end multileaf collimator systems by simple test fields," *Journal of Applied Clinical Medical Physics* **16**(4), 65–77 (2015).

58. M. Snyder, et al., "Modeling the Agility MLC in the Monaco treatment planning system," *Journal of Applied Clinical Medical Physics* **17**(3), 190–202 (2016).

59. L. A. Young, et al., "Rounded leaf end modeling in Pinnacle VMAT treatment planning for fixed jaw linacs," *Journal of Applied Clinical Medical Physics* **17**(6), 149–162 (2016).

60. D. A. Low, et al., "A technique for the quantitative evaluation of dose distributions," *Medical Physics* **25**(5), 656–661 (1998).

61. N. Wen, et al., "Precise film dosimetry for stereotactic radiosurgery and stereotactic body radiotherapy quality assurance using Gafchromic™ EBT3 films," *Radiation Oncology* **11**(1), 132 (2016).

62. B.-G. Jalil, et al., "Long-term stability of liquid ionization chambers with regard to their qualification as local reference dosimeters for low dose-rate absorbed dose measurements in water," *Physics in Medicine & Biology* **46**(3), 729 (2001).

63. M. Markovic, et al., "Characterization of a two-dimensional liquid-filled ion chamber detector array used for verification of the treatments in radiotherapy," *Medical Physics* **41**(5), 051704 (2014).

64. G. Nicolini, et al., "GLAaS: An absolute dose calibration algorithm for an amorphous silicon portal imager. Applications to IMRT verifications," *Medical Physics* **33**(8), 2839–2851 (2006).

65. B. W. King, D. Morf, P. B. Greer, "Development and testing of an improved dosimetry system using a backscatter shielded electronic portal imaging device," *Medical Physics* **39**(5), 2839–2847 (2012).

66. N. Miri, et al., "EPID-based dosimetry to verify IMRT planar dose distribution for the aS1200 EPID and FFF beams," *Journal of Applied Clinical Medical Physics* **17**(6), 292–304 (2016).

67. M. F. Chan, et al., "Using a novel dose QA tool to quantify the impact of systematic errors otherwise undetected by conventional QA methods: Clinical head and neck case studies," *Technology in Cancer Research & Treatment* **13**(1), 57–67 (2014).

68. B. E. Nelms, et al., "Evaluating IMRT and VMAT dose accuracy: Practical examples of failure to detect systematic errors when applying a commonly used metric and action levels," *Medical Physics* **40**(11), 111722 (2013).

69. B. Neal, et al., "A clinically observed discrepancy between image-based and log-based MLC positions," *Medical Physics* **43**(6 Part 1), 2933–2935 (2016).

70. E. E. Klein, et al., "Task Group 142 report: Quality assurance of medical accelerators," *Medical Physics* **36**(9), 4197–4212 (2009).

71. G. J. Kutcher, et al., "Comprehensive QA for radiation oncology: Report of AAPM Radiation Therapy Committee Task Group 40," *Medical Physics* **21**(4), 581–618 (1994).

72. T. Pawlicki, P. Dunscombe, A. J. Mundt, P. Scalliet, "Quality and safety in radiotherapy," in *Imaging in Medical Diagnosis and Therapy*, edited by W.R. Hendee (CRC Press, Boca Raton, FL, 2010), p. 643.

CHAPTER 19

CyberKnife and ZAP-X Dosimetry

Sonja Dieterich

University of California Davis
Sacramento, California

Georg Weidlich

Zap Surgical
San Carlos, California

Christoph Fuerweger

European Cyberknife Center Munich
Munich, Germany

CONTENTS

19.1 INTRODUCTION

The CyberKnife (Accuray Inc., Sunnyvale, CA, USA) and the ZAP-X device (Zap Surgical Systems Inc., San Carlos, CA, USA) are both FDA-approved dedicated radiosurgery devices with linear accelerators (linacs) mounted on robots. While the general small field dosimetry principles discussed in Chapter 18 apply, there are some designs specific to each machine which warrant a separate discussion.

19.2 CYBERKNIFE

The CyberKnife consist of a 6 MV miniature linac mounted on a 6 degree of freedom industrial robot (Kuka, Augsburg, Germany). The beams can be shaped through three collimator systems which can be attached to the linac: fixed cones, an IRIS collimator [1], and the INCISE MLC collimator. Image guidance is provided using orthogonal x-ray cameras mounted at the ceiling with amorphous silicon detectors embedded in the floor. Real-time respiratory motion monitoring is provided using the Synchrony system with surface beacons [2].

The IRIS collimator consists of two leaf banks with six tungsten-alloy segments each, offset at 30 degrees to each other. This dodecagonal treatment field is approximately circular. While the IRIS collimator theoretically can have unlimited numbers of apertures, for treatment planning optimization purposes 12 apertures matching the fixed cones are used.

The INCISE MLC has been released in two versions with different specifications. Version 1 has already been phased out and is no longer in clinical use. Version 2 consists of 26 pairs of flat-sided leaves with a height of 90 mm and is capable of full overtravel and interdigitation. The leaf thickness is 3.85 mm at the reference source-to-axis distance (SAD) of 80 cm, and the maximum field size for clinical use is 11.5×10.1 cm². In the absence of jaws, interleaf leakage is reduced to less than 0.5% by a leaf tilt of 0.5°, which causes asymmetric penumbras in in-plane profiles. Leaf positioning is achieved with a mean accuracy of <0.2 mm and verified by a secondary camera system. As of 2020, the MLC is operated in step-and-shoot mode only.

19.2.1 TG-51 Formalism for CyberKnife

Alfonso et al. [3] developed a formalism for devices which do not have the standard field size required for AAPM Task Group 51 (TG-51) [4] or IAEA TRS-398 [5]. While the newer generation CyberKnife systems offer the option

of an MLC in addition to the cones and IRIS collimator, many existing systems still rely on the Alfonso mechanism of a machine-specific reference (MSR) field. The 60 mm fixed collimator has de facto become the standard for the MSR field.

The first decision in reference dosimetry for CyberKnife is which chamber to choose. As Kawachi et al. discussed, the non-flattened MSR field results in a change of response depending on the length of the chamber [6]. For a traditional Farmer chamber, a correction factor of about 1.7% is required to account for the non-flattened field. An alternative to correcting the chamber reading based on chamber length and field shape is to use a shorter chamber for reference dosimetry. This can either be an ADCL-calibrated chamber, or a chamber cross-calibrated with a Farmer chamber.

The next step is to relate the standard calibration condition of CyberKnife (80 cm SAD using the 60 mm fixed cone as MSR field) to the standard conditions defined in AAPM TG-51 (100 cm SAD, 10×10 cm² field size). In addition, the physicist needs to handle the gradient corrections which are included implicitly in k_Q.

In the first step, the robot is moved up from the water tank to an extended source-to-surface distance (SSD) (for CyberKnife) of 100 cm. In this position, the equivalent square of the 60 mm cone at 100 cm SSD is 6.75×6.75 cm² [7]. The %dd(10, 6.75, 100) is then measured and compared to %dd(10, 6.75, 100) with local reference data from a standard linac or, alternatively, the data published by Jordan in BJR supplement 25 [8]. The corresponding %dd(10,10,100) from the reference dataset will determine k_Q. Based on Figure 2 of Kalach and Rogers' paper [9] comparing flattening-filter free soft beams with flattened beams, k_Q from a flattened beam is valid for a softer unflattened beam as well.

In the second step, the SSD is then changed to the CyberKnife calibration condition (SAD or SSD). $P_{gr}(CK)$ is then determined at the measurement point and the dose is calculated by Equations 19.1–19.3:

$$D_w^Q = \frac{MN_{D,w}^Q}{\%dd \text{ or } TPR},$$ (19.1)

$$M = M_{raw} \prod P,$$ (19.2)

$$N_{D,w}^Q = N_{D,w}^{60Co} k_Q \frac{P_{gr}(CK)}{P_{gr}(10,10,100)}.$$ (19.3)

TABLE 19.1 Sample Set of Output Factors Acquired for a CyberKnife M6 FIM

	Circular		Square/Rectangular	
Field diameter (mm)	Cones	IRIS	Field size (mm²)	MLC
5	0.683	0.507	7.6 × 7.7	0.808
7.5	0.818	0.769	15.4 × 15.4	0.948
10	0.871	0.857	23.0 × 23.1	0.976
12.5	0.908	0.899	30.8 × 30.8	0.988
15	0.934	0.926	38.4 × 38.5	0.994
20	0.960	0.958	46.2 × 46.2	1.000
25	0.973	0.969	53.8 × 53.9	1.005
30	0.979	0.976	69.2 × 69.3	1.013
35	0.984	0.980	84.6 × 84.7	1.020
40	0.987	0.985	100.0 × 100.1	1.025
50	0.993	0.991	115.0 × 100.1	1.026
60	1.000	1.001		

It should be emphasized that a peer-review of the output should be conducted after commissioning and before the first patient is treated, and annually thereafter. External peer review checks are offered by several entities, including the IAEA, IROC Houston, and other dosimetry labs. If this service is not available, an acceptable alternative is to request a peer-review from an outside medical physicist with extensive experience who would be able to bring in their own chambers.

For CyberKnife units equipped with an MLC, the 10×10 cm² field size needed to perform AAPM TG-51 or IAEA TRS-398 reference dosimetry is available. For existing systems that have used the method outlined above for the 60 mm MSR field, the option of retaining that method for consistency and using a separate reference dosimetry method for an MLC which was added later could be discussed. If the reference dosimetry is switched to a traditional 10×10 cm² reference dosimetry, new output factors for both collimators should be compared to the previous output factors, and any changes above 2% be investigated. For new CyberKnife systems who have an MLC available from installation, the reference dosimetry field should be the 10×10 cm² field. Regardless, the absolute dose calibration must be conducted such that the CyberKnife output with the 60 mm fixed cone at 800 mm SAD and 15 mm depth in water is 1 cGy per 1 MU.

19.2.2 Output Factors

The outputs factors (OFs), also called total scatter factors (S_{CP}), are measured relative to the output factor of the 60 mm fixed collimator for three different SSDs (65 cm, 80 cm, and 100 cm) intended to cover the SSD range in clinical use. The output factors are normalized to the 60 mm field size at 80 cm SSD. Accuray recommends detectors optimized for use in small beam dosimetry (e.g., unshielded diodes and synthetic microdiamond detectors). In all cases, Accuray recommends the evaluation and use of published correction factors as appropriate, particularly for output factor determination in the smallest beams. Other detectors such as film, microchambers, liquid filled chambers, and alanine pellets have also been used [10, 11]. Table 19.1 lists a sample set of output factors for all three collimators.

Since the publications of Francescon et al. [12–14], there has been discussion if an energy correction factor should be used to correct the output factors for the smallest collimator. IAEA Technical Report 483 has extensive guidance on implementation of correction factors [15]. For beams and/or detectors which are not covered in IAEA TRS 483, best clinical practice is:

1. Use at least two different suitable detectors, preferably with opposite energy response.

2. Apply correction factors from peer-reviewed literature.

3. Average the results.

4. Compare the result to peer-reviewed literature or use peer review to verify results.

19.2.2.1 Cones

The output factors for the fixed cones are measured with the detector centered on the central axis (CAX) at 1.5 cm depth in water. Output factors are measured at 65 cm, 80 cm, and 100 cm SAD. All output factors are normalized to the 60 mm cone at 80 cm SAD. Measured output factor values have been published for a variety of detectors [16], and energy correction factors are available from Francescon et al. [17, 18].

19.2.2.2 IRIS

Output factors for the IRIS collimator are very similar to the fixed cone sized output factors down to about the 12.5 mm cone. For the three smallest cones (10 mm, 7.5 mm, and 5 mm), differences start to emerge. The reason is the mechanical aperture tolerance of 0.2 mm. To get an average output factor, the collimator opening is changed in between repeat measurements for the three smallest collimators. In clinical practice, the 7.5 mm and 10 mm IRIS collimator size is not heavily used in any treatment plan. Most clinical sites will not use the 5 mm IRIS collimator.

19.2.2.3 MLC

While the smallest clinically used MLC field is 1×1 cm^2, commissioning of the CyberKnife treatment planning system (TPS) includes OF measurement of the 0.76×0.77 cm^2 field. In analogy to the smallest circular fields, diodes can be expected to over-, air-filled microchambers to under-respond for small MLC fields. Recently, Monte-Carlo-based correction factors $k_{\Omega_{clin},\Omega_{msr}}^{f_{clin},f_{msr}}$ specific to the CyberKnife M6 MLC have become available for a range of detectors and should be used for OF determination [19].

Several detectors used to measure output factors, diodes especially, have an energy response as a function of field size. While this response is small over the field size range for fixed and IRIS collimators, it can be considerable over the field sizes achievable with the MLC, depending on the specific detector. In this case, it is recommended to use the same approach as for linacs, i.e., measure the output factors for larger field sizes using an appropriate chamber. Given the requirement to measure output factors for a discrete selection of field sizes for commissioning, the field size of 4.62×4.62 cm^2 is an appropriate choice to daisy-chain chamber-based output factor measurements for larger fields to output factor measurements for smaller fields using a diode or other appropriate detector [16]. Alternatively, the use of MC-based correction factors $k_{\Omega_{clin},\Omega_{msr}}^{f_{clin},f_{msr}}$ can also be considered for the largest fields [19].

For the largest MLC fields, the OF measurement depth of 1.5 cm is well within the build-up region, as the maximum dose point is closer to 1.8 cm. Consequently, exact reproduction of the measurement depth is more critical in comparison to circular or small MLC fields – small errors in measurement depth cause larger differences in detector readings.

19.2.3 Tissue-Phantom Ratios (TPRs)

The CyberKnife TPS currently requires the measurement of tissue-phantom ratios (TPRs) as input. With a regular size water tank, these measurements can be very time consuming, because the robot has to be moved manually for each measurement depth. For earlier CyberKnife versions, this was usually achieved by entering the treatment room. For M6, the Kuka teach pendant can be connected to the robot controller in the equipment room and routed to the control area, which allows for moving the robot from outside the treatment room but requires careful monitoring of the robot using in-room cameras to avoid collision. There is a third, unsupported option applicable to all CyberKnife versions, which uses service tools to move the robot from the delivery console.

The general process can be somewhat abbreviated for the IRIS and the MLC collimator by measuring TPRs at a specific depth for all or at least multiple collimator and field sizes. To make TPR measurements more efficient, smaller water tanks with TMR functionality have been developed [20]. PDD-to-TMR conversions have been attempted but found to fail for the smaller collimators (private communication). Figure 19.1 shows a sample set of TPR data for a range of collimators and field sizes.

19.2.4 Off-Axis Ratios (OARs)

Accurate radiosurgical planning is highly depended on steep dose gradients toward critical structures. Therefore, it is essential that the dose gradients are measured accurately and with the least smearing due to detector size. Figure 19.2 shows a sample set of OAR data for a range of collimators and field sizes.

19.2.4.1 Cone

OARs for the cones are measured using small area detectors such as diodes or diamond detectors. Inline and crossline profiles are measured at several depths as required by the TPS. The data processing is performed similar to taking OARs for external beams.

19.2.4.2 IRIS

Because of the shape of the IRIS collimators, two additional cross profiles must be measured in addition to

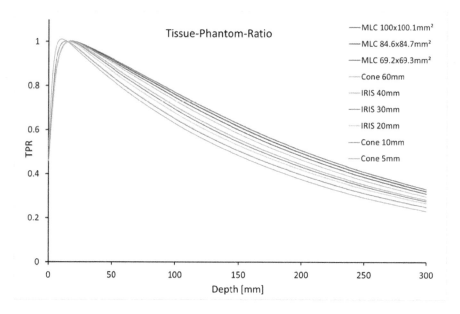

FIGURE 19.1 TPRs acquired for selected field sizes of a CyberKnife M6 FIM.

the inline and crossline profiles. The additional cross-profiles are measured at a 15-degree angle offset to the inline and crossline profiles.

19.2.4.3 MLC

The general procedure of measuring OARs for MLC is similar to fixed cones and IRIS: Beam data is acquired at the same depths, either at a fixed SSD or SAD of 800 mm. However, there are two MLC-specific aspects that require some attention: First, whereas for cones and IRIS, measured OARs are averaged and reduced to a single half beam profile for each field size, full in- and cross-plane profiles of 10 square and one rectangular field are used for MLC commissioning. Consequently, water tank scanning direction and signs matter for MLC only – in- and cross-plane penumbrae are different, and due to the bank tilt, CyberKnife MLC fields will not be perfectly symmetrical even with an optimally tuned beam [21, 22]. Second, correctly aligning the linac to the water phantom is somewhat trickier due to the mechanical degrees

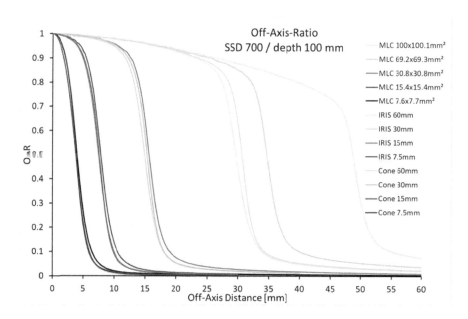

FIGURE 19.2 OARs for selected field sizes of a CyberKnife M6 FIM acquired at SSD 700 mm and depth 100 mm. The graph shows averaged OARs for circular and cross-plane OARs for MLC fields.

of freedom of the robot and the absence of a light field of the MLC, and has to be achieved by a few iterations of measuring off-axis profiles of a narrow field, followed by moving the robot to adjust the linac orientation.

19.3 ZAP SURGICAL UNIT

The ZAP-X is a new, dedicated self-contained and self-shielded radiosurgery system developed and manu-factured by ZAP Surgical Systems, Inc. of San Carlos, California [23–25]. A cross section of the device is shown in Figure 19.3.

This device is intended for stereotactic radiosurgery (SRS) treatment of benign and malignant intracranial and cervical spine lesions. A 3.0 MV S-band linac is the source of therapeutic radiation. Akin to a large gyro-scope, the linac is mounted within a combination of yoked gimbals with attached radiation shielding, each of which accurately rotates around a common isocen-ter. This mechanical construct enables the linac beam crossfire from 2π steradians of solid angle, as is ideally required for cranial SRS.

Accurate therapeutic beam positioning is accom-plished through the two above-mentioned dual-axes, independent rotations of the accelerator, and precise movements of a robotic patient table. Most components needed to produce the beam, such as the radiofrequency power source, waveguiding system, and beam trigger-ing electronics, as well as significant radiation shielding, are mounted on or integrated into the rotating patient

treatment chamber sphere. The patient is supported on a moveable treatment table that extends outside the treat-ment sphere but which itself is also enclosed by addi-tional radiation shielding during radiosurgery. This table shielding consists of a rotary shell and pneumatic door on a steel frame.

The ZAP-X accomplishes three-dimensional (3D) patient registration by means of an integrated planar kilovolt (kV) imaging system that also rotates around the patient's head. Pairs of non-coaxial x-ray images and image-to-image correlation are utilized to determine the location of the patient's anatomy with respect to the machine isocenter, both prior to and during radiosur-gical treatment. The treatment beam is shaped through tungsten collimators.

19.3.1 Radiation Detectors

Radiation detectors appropriate for small field ZAP-X stereotactic radiosurgery fields require small detection volume to prevent averaging of the rapidly changing radiation intensities across the small fields and near the field edges. Acceptable detectors include diodes and ionization chambers. Other detectors with simi-lar detection volumes response are acceptable as well.

19.3.2 Beam Data Acquisition

The ZAP-X TPS utilizes measured and interpolated beam data only by employing the raytracing method with the equivalent path length correction. No modeling

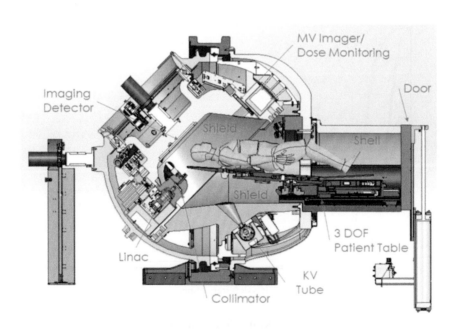

FIGURE 19.3 Cross-sectional view of ZAP-X system. (Figure provided by ZAP Surgical.)

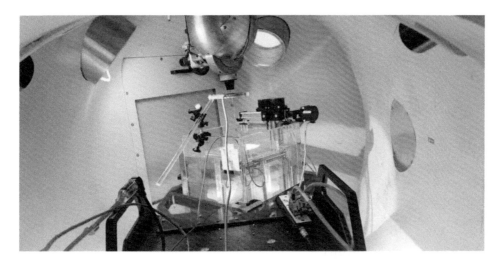

FIGURE 19.4   Setup of 3D PTW water phantom inside the ZAP-X treatment sphere.

or parametrization of the measured beam data is performed. Required ZAP-X beam data include collimator output factors, TPRs, and off-axis ratios (OARs). Percent depth dose (PDD) data shall be acquired and converted to TPR. All beam data should be acquired in a water phantom with a minimum scanning depth of 25 cm and a step size for PDDs and OARs of 0.1–1.0 mm. The SSD should be set to 45 cm for PDD and OAR measurements, while output factors shall be acquired at the calibration point of $d_{\max}$ = 7.0 mm and SSD = 44.3 cm. The unique measurement setup inside the ZAP-X treatment sphere is shown in Figure 19.4.

19.3.3 Reference Dose Calibration of the ZAP-X

The reference dose calibration is performed using AAPM TG-51 [4] protocol with the adaptation for MSR fields as described by Alfonso et al. [3]. Due to the very small field sizes of the ZAP-X, the reference calibration of this system is performed with an ionization chamber mounted with its chamber axis perpendicular to the CAX and a maximum sensitive length of 7 mm. An example of suitable chamber is the PTW 0.125 cm³ semiflex ionization chamber, Model 31010. Accredited Dosimetry Calibration Laboratories can provide calibration factors N_x and $N_{D,W}$ for this type of chamber. The ionization chamber used for reference dose calibration shall be cross calibrated on a conventional linac with a calibration field size of 10×10 cm² and well-established dosimetry. The final calibration of the ZAP-X should be independently verified by TLD or OSLD exposures or independent redundant calibration. The final result of the reference dose calibration will ensure that 1 MU will deliver 1.0 cGy at the calibration conditions of SAD = 45.0 cm, at 7.0 mm depth, for the 25 mm collimator.

The MSR field [3] size is the 25 mm collimator. Because the depth of $d_{\max}$ for the MSR field is 7.0 mm and the ionization chamber has an outer radius of 3.5 mm, the reference calibration is performed at a depth of 5.0 cm. The measured dose is then referenced to calibration conditions (SAD = 45.0 cm, d = 7.0 mm, circular collimator size = 25 mm diameter) using the same methodology as outlined for the CyberKnife in Section 19.2.1. Most suitable for the calculation of dose at the calibration point is the use of TPR ratios. Alternatively, a custom-made acrylic spherical phantom can be attached to the collimator housing via a holder (F-bracket) to provide a known and reproducible measurement geometry and can be used to determine the dose at the calibration point.

19.3.4 Output Factors (OFs)

Output Factors can be measured with small volume detectors such as a diamond detector, pinpoint ionization chamber or SRS diode. Diamond detectors and SRS diodes should be positioned with their stem axis parallel to the central beam axis (CAX) and centered on it. The ionization chamber can be oriented with its stem axis parallel to the central beam axis or perpendicular to it. The output measurements for each collimator are normalized to the 25 mm collimator. Output factors are shown in Figure 19.5.

19.3.5 Percent Depth Dose (PDD)

PDD measurements are best performed with an SRS diode with its stem axis oriented parallel to the ▪ CAX

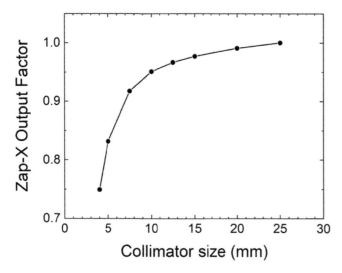

FIGURE 19.5 ZAP-X output factor as a function of field size.

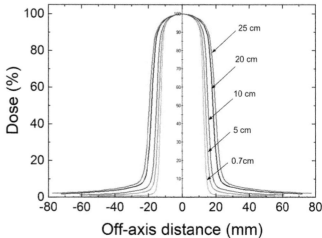

FIGURE 19.7 OAR for the 25 mm collimator at depths 0.7 cm, 5.0, 10.0, 20.0, and 25.0 cm.

and centered on it. Since the collimator sizes are 4, 5, 7.5, 10, 12.5, 15, 20, and 25 mm in diameter at the machine isocenter (45 cm SAD), the use of a small volume reference detector positioned in the primary beam is not practical due to its interference with the radiation measured by the field detector. Instead, it is recommended to use a transmission reference ionization chamber of parallel plate structure with a diameter significantly larger than 25 mm. An example of such a reference chamber is the PTW T-Ref chamber. This chamber should be positioned at least 15 cm above the water surface of the water phantom and should be centered on the CAX. This transmission chamber is considered to be radio-translucent and not to impact the beam intensity in a measurable manner. The surface of the water phantom shall be set at SSD = 45 cm and the PDD is measured from a depth of at least

25 cm to the surface of the water. The PDD graph for the 25 mm collimator is shown in Figure 19.6.

19.3.6 Off-Axis Ratios (OARs)

OAR measurements shall be performed with the SRS diode and the same phantom and reference detector setup as with PDD measurements. OAR profiles shall be measured in the wheel plane (parallel with the collimator wheel) and orthoplane (perpendicular to the wheel plane). Depth of measurement shall be performed at d_{max} = 0.7 cm, 5.0 cm, 10.0 cm, 20.0 cm, and 25.0 cm or maximum achievable depth. OAR data for the 25 mm collimator are shown in Figure 19.7.

19.3.7 Preparation and Processing of Measured Beam Data before Import into the Treatment Planning System

PDD and OAR data will be processed and converted into a standard text format before import into Zap-X TPS. PDD and OAR data are recommended to be smoothed with a low-order polynomial. OAR measurements shall be smoothed, centered, normalized to its CAX value, and a half-beam scan is generated. The processed beam data can now be imported into the TPS.

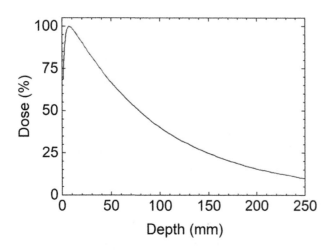

FIGURE 19.6 PDD for the 25 mm collimator size.

REFERENCES

1. G. G. Echner, W. Kilby, M. Lee, E. Earnst, S. Sayeh, A. Schlaefer, B. Rhein, J. R. Dooley, C. Lang, O. Blanck, E. Lessard, C. R. Maurer, W. Schlegel, "The design, physical properties and clinical utility of an IRIS collimator for robotic radiosurgery," *Physics in Medicine and Biology* **54**, 5359–5380 (2009).

2. A. Schweikard, H. Shiomi, J. Adler, "Respiration tracking in radiosurgery," *Medical Physics* **31**, 2738–2741 (2004).

3. R. Alfonso, P. Andreo, R. Capote, M. S. Huq, W. Kilby, P. Kjall, T. R. Mackie, H. Palmans, K. Rosser, J. Seuntjens, W. Ullrich, S. Vatnitsky, "A new formalism for reference dosimetry of small and nonstandard fields," *Medical Physics* **35**, 5179–5186 (2008).

4. P. R. Almond, P. J. Biggs, B. M. Coursey, W. F. Hanson, M. S. Huq, R. Nath, D. W. Rogers, "AAPM's TG-51 protocol for clinical reference dosimetry of high-energy photon and electron beams," *Medical Physics* **26**, 1847–1870 (1999).

5. D. T. B. Pedro Andreo, K. Hohlfeld, M. Saiful Huq, T. Kanai, F. Laitano, V. Smyth, S. Vynckier, *Absorbed Dose Determination in External Beam Radiotherapy: An International Code of Practice for Dosimetry based on Standards of Absorbed Dose to Water* (International Atomic Energy Agency, Report No. TRS-398, Vienna, 2000).

6. T. Kawachi, H. Saitoh, M. Inoue, T. Katayose, A. Myojoyama, K. Hatano, "Reference dosimetry condition and beam quality correction factor for CyberKnife beam," *Medical Physics* **35**, 4591–4598 (2008).

7. M. Day and E. Aird, "Central axis depth dose data for use in radiotherapy: 1996," *British Journal of Radiology. Supplement* **25**, 138–151 (1996).

8. T. Jordan, "Central axis depth dose data for use in radiotherapy," *British Journal of Radiology* **5**, 62–109 (1996).

9. N. Kalach and D. Rogers, "Which accelerator photon beams are "clinic-like" for reference dosimetry purposes?," *Medical Physics* **30**, 1546–1555 (2003).

10. C. Antypas and E. Pantelis, "Performance evaluation of a CyberKnife G4 image-guided robotic stereotactic radiosurgery system," *Physics in Medicine and Biology* **53**, 4697–4718 (2008).

11. P. Francescon, S. Cora, C. Cavedon, P. Scalchi, S. Reccanello, F. Colombo, "Use of a new type of radiochromic film, a new parallel-plate micro-chamber, MOSFETs, and TLD 800 microcubes in the dosimetry of small beams," *Medical Physics* **25**, 503–511 (1998).

12. P. Francescon, S. Cora, C. Cavedon, P. Scalchi, J. Stancanello, "Cyberknife dosimetric beam characteristics: comparison between experimental results & Monte Carlo simulation," in *Robotic Radiosurgery, Vol. 1*, edited by R.F. Mould, pp. 71–80 (CyberKnife Society Press, Sunnyvale, CA, 2005).

13. P. Francescon, W. Kilby, N. Satariano, "Monte Carlo simulated correction factors for output factor measurement with the CyberKnife system—results for new detectors and correction factor dependence on measurement distance and detector orientation," *Physics in Medicine and Biology* **59**, N11 (2014).

14. P. Francescon, W. Kilby, N. Satariano, S. Cora, "Monte Carlo simulated correction factors for machine specific reference field dose calibration and output factor measurement using fixed and IRIS collimators on the CyberKnife system," *Physics in Medicine and Biology* **57**, 3741 (2012).

15. H. Palmans, P. Andreo, K. Christaki, M. Huq, J. Seuntjens, *Dosimetry of Small Static Fields Used in External Beam Radiotherapy: An IAEA-AAPM International Code of Practice for Reference and Relative Dose Determination* (International Atomic Energy Agency, Vienna, 2017).

16. S. Dieterich, G. W. Sherouse, "Experimental comparison of seven commercial dosimetry diodes for measurement of stereotactic radiosurgery cone factors," *Medical Physics* **38**, 4166–4173 (2011).

17. P. Francescon, S. Beddar, N. Satariano, I.J. Das, "Variation of k_{Qclin}, Q_{msr} f_{clin}, f_{msr} for the small-field dosimetric parameters percentage depth dose, tissue-maximum ratio, and off-axis ratio," *Medical Physics* **41**, 101708 (2014).

18. P. Francescon, S. Cora, C. Cavedon, "Total scatter factors of small beams: a multidetector and Monte Carlo study," *Medical Physics* **35**, 504–513 (2008).

19. P. Francescon, W. Kilby, J. Noll, L. Masi, N. Satariano, S. Russo, "Monte Carlo simulated corrections for beam commissioning measurements with circular and MLC shaped fields on the CyberKnife M6 System: a study including diode, microchamber, point scintillator, and synthetic microdiamond detectors," *Physics in Medicine & Biology* **62**, 1076 (2017).

20. N. C. Knutson, M. C. Schmidt, M. D. Belley, N. Nguyen, M. Price, S. Mutic, E. Sajo, H. H. Li, "Equivalency of beam scan data collection using a 1D tank and automated couch movements to traditional 3D tank measurements," *Journal of Applied Clinical Medical Physics* **19**, 60–67 (2018).

21. G. Asmerom, D. Bourne, J. Chappelow, L. Goggin, R. Heitz, P. Jordan, W. Kilby, T. Laing, C. Maurer Jr, J. Noll, "The design and physical characterization of a multileaf collimator for robotic radiosurgery," *Biomedical Physics & Engineering Express* **2**, 017003 (2016).

22. C. Fürweger, P. Prins, H. Coskan, B. J. Heijmen, "Characteristics and performance of the first commercial multileaf collimator for a robotic radiosurgery system," *Medical Physics* **43**, 2063–2071 (2016).

23. G. A. Weidlich, M. B. Schneider, J. R. Adler, "Self-shielding analysis of the Zap-X System," *Cureus* **9**(12), e1917 (2017).

24. G. A. Weidlich, M. B. Schneider, J. R. Adler, "Characterization of a novel revolving radiation collimator," *Cureus* **10**(2), e2146 (2018).

25. G. A. Weidlich, M. Bodduluri, Y. Achkire, C. Lee, J. R. Adler Jr, "Characterization of a novel 3 megavolt linear accelerator for dedicated intracranial stereotactic radiosurgery," *Cureus* **11**(3), e4275 (2019).

CHAPTER 20

Dosimetry in the Presence of Magnetic Fields

Carri Glide-Hurst

University of Wisconsin-Madison
Madison, Wisconsin

Hermann Fuchs and Dietmar Georg

Medical University Vienna
Wien, Austria

Dongsu Du

City of Hope Medical Center
Duarte, California

CONTENTS

20.1 INTRODUCTION

The advent of magnetic resonance-guided radiation therapy (MRgRT) has introduced new considerations for the management of radiation dosimetry in the presence of a magnetic field. This chapter will describe the fundamental physics interactions, quantify the impact on measurement devices, and describe clinical management scenarios to highlight the major considerations.

20.2 PHYSICS INTERACTION CONSIDERATIONS

20.2.1 Physics of Interactions

The external magnetic field surrounding the target area including neighboring organs at risk needs to be considered for MRgRT. Although physics processes, such as particle creation and dose deposition, remain unaffected, the external magnetic field leads to not obvious effects. In general, a moving charged particle in a magnetic field is affected by the Lorentz force. If we disregard electric fields, a particle with charge q moving with speed $\vec{v}$ in a magnetic field $\vec{B}$ experiences the Lorentz force $\vec{F}$

$$\vec{F} = q\vec{v} \times \vec{B} \tag{20.1}$$

The Lorentz force introduces a perpendicular deflection of the particle path, generally resulting in a circular trajectory alongside the particles original track, i.e., a spiralling track. Combining it with the centripetal force and some rearrangement we can introduce the radius of this circular motion, the gyroradius r

$$r = \frac{m\vec{v}_{\neg}}{q\overrightarrow{B_{\neg}}} \tag{20.2}$$

where m is the mass of the particle and $\vec{v}_{\neg}$ and $\overrightarrow{B_{\neg}}$ are the perpendicular components of the respective vectors. The relativistic representation is

$$r = \frac{mv\gamma}{qB} \tag{20.3}$$

where γ is the Lorentz factor, $\gamma = \dfrac{1}{\sqrt{1-\frac{v^2}{c^2}}}$. Rewriting Equation 20.3 leads to

$$r = \frac{mc\sqrt{\gamma^2 - 1}}{qB} \tag{20.3b}$$

Resulting in

$$r = \frac{mc\sqrt{\left(1 + \dfrac{E_{kin}}{mc^2}\right)^2 - 1}}{qB} \tag{20.3c}$$

where c is the speed of light, E_{kin} the kinetic energy.

20.2.2 Considerations for Photons

In MR-guided photon therapy, the primary beam itself is not affected by the magnetic field. However, the secondary electrons which are responsible for dose deposition are affected. This may lead to distorted dose maps compared to conventional photon therapy and may require being accounted for during treatment planning. The most notable effects are observed in proximity to interfaces such as a material with low atomic number Z and the corresponding density gradient that may cause significant changes in the remaining particle range. For example, an electron exiting a dense material into a less dense material such as air, may have sufficient remaining range to turn around and re-enter the previous material. This "electron return effect" [1] (ERE) leads to significantly increased or reduced entry and exit doses especially at tissue surfaces having been reported up to 40% [2–4]. The ERE also impacts air-filled dosimeters [5–9] and selection of measurement phantom geometries [10].

20.2.3 Considerations for Protons

MR-guided proton beam therapy is still in its infancy. In contrary to the effects in therapeutic photon beams described above, the primary proton therapy beam will be affected. The bending of the treatment beam results in a lateral displacement of the Bragg-peak of up to 3 cm for a 250 MeV proton beam in a 1 T magnetic field [11–13]. Furthermore, this bending of the beam path leads to a small reduction in penetration depth. In addition, proton beams experience an effect also encountered in spectrometers, an energy separation occurs and consequently the lateral beam shape gets deformed. This holds true also for initially mono-energetic beams due to the energy/range straggling occurring inside a patient.

In proton therapy, the energy distribution of the generated secondary electrons is generally lower compared to photons, except for low energetic protons close to the Bragg-peak as shown in Figure 20.1. This leads to a reduced range and consequently a reduced ERE. To date, Monte Carlo simulations performed by different groups and limited experimental measurements indicate that electron return specific dose differences at material surfaces are on the order of 2–8% and depend on the magnetic field strength and orientation [4, 11, 14].

20.2.4 Electron Return Effect Dependencies

The magnetic field effect on dose distribution has been primarily studied through Monte-Carlo based dose

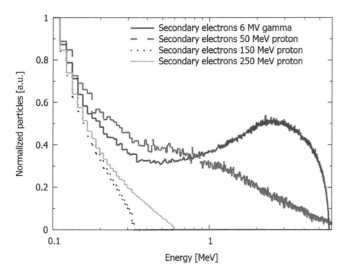

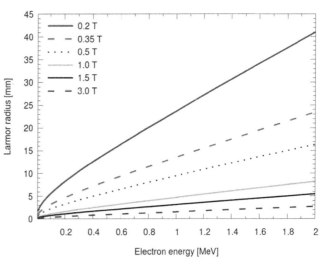

FIGURE 20.1 Electron spectra of a 6 MV photon beam, a 50, 150, and 250 MeV proton beam in water, corresponding to ranges in water of 2.2, 15.8, and 38 cm, respectively. The electron spectra were scored at 2 cm depth.

FIGURE 20.3 Electron return trajectory radius (mm) in vacuum, depends on the magnetic field strength and electron energy. (Based on data from Raaijmakers et al. [13].)

simulation [15–18]. These studies found that magnetic field induced dose effects are dependent on the field orientation, field strengths [1] and beam energy. As described above, the Lorentz force will act on traveling charged particles such as secondary electrons. Figure 20.2 shows the Lorentz force direction when the electron motion is perpendicular to magnetic field.

Generally speaking, a higher energy beam has a larger return electron trajectory's radius and as a result, a lessened magnetic field dose effect. Another important fact

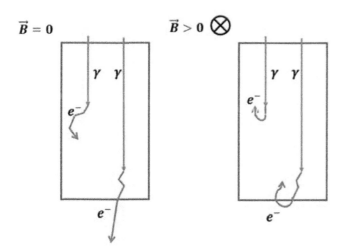

FIGURE 20.2 (Left) Typical interactions of photons and secondary electrons with tissue in the absence of a magnetic field. (Right) Electron return effect due to the presence of the magnetic field perpendicular to the page.

which will affect magnetic field dose is the field strength. Raaijmakers et al. [13] have investigated the ERE dose changes with different field strength in four situations (1) ERE at distal side of the beam, (2) the lateral ERE, (3) ERE in cylindrical air cavities, and (4) ERE in the lungs. Comparing the ERE for four different field strengths ($B = $ 0.2, 0.75, 1.5, and 3 T), the ERE has minimal impact for $B = 0.2$ T while considerable dose increases were observed for all the other fields. Figure 20.3 (using data derived from reference [13]) shows the radius of electron return trajectory depends on the energy of the electron and the magnetic field strength.

Another major factor in the impact of magnetic field on dosimetry is orientation, including that of the magnetic field (inline vs. perpendicular). Here, inline is defined as the MRI scanner and linac axes of symmetry coinciding while perpendicular involves the treatment beam being perpendicular to the main magnetic field [19]. Advantages and disadvantages of each orientation are described by Keall et al. in detail [20]. The overall impact of geometry depends on field strength, radius of gyration, and tissue type. Inline orientations tend to use dose perturbation in terms of electron contamination being focused by the MRI field whereas perpendicular orientation tends to be linked to ERE. Another orientation consideration is the surface angle of entry (both exit and entry sides) [21]. Monte Carlo simulations of a 6 MV beam in a transverse 3 T magnetic field showed that entry surface angles ranging from −30° to −60° yielded comparable entry skin doses to simulations performed with no magnetic field [21].

20.3 DOSIMETRY EQUIPMENT CONSIDERATIONS AND PERFORMANCE

When considering dosimetry equipment in magnetic fields, several key features need to be considered: MR safe or conditional considerations, orientation with respect to the magnetic field, and the overall impact of the magnet on the dosimeter performance. Table 20.1 summarizes these key characteristics for published literature pertaining to dosimeter utility for MRgRT.

Table 20.2 summarizes key detectors that have been tested in MRgRT environments, their chief findings, and related references. Key takeaways from the summary of the literature is that detector performance depends on several factors and orientation as listed, thus end users should be cognizant of the magnetic field dose–response characterization for each individual detector in clinical use as summarized in Table 20.2.

An area of emerging interest is gel dosimetry, where the dosimeter is MR-visible, near-water equivalent, and offers potential for 3D measurement and subsequent analysis. One such example is the radiochromic plastic PRESAGE® dosimeter (Heuris Pharma, LLC, Skillman, NJ) that exhibits optically visible changes occurring with radiation exposure. Another experimental dosimeter developed at MD Anderson Cancer Center is the radiochromic gel FOX, similar to a Fricke dosimeter, that is prepared in-house. Both gel dosimeters can be prepared to include air cavities and demonstrated the ERE that agreed well with Monte Carlo simulations. Commercially available options are currently under development. One such example is a radiochromic gel dosimeter (ClearView, Modus Medical Device Inc.), consisting of gellan gum, propylene glycol, deionized water, and BNC tetrazolium prepared in a transparent vessel (15 cm diameter, 12 cm effective length) as shown in Figure 20.4. Optical readout was conducted using a Vista 16 cone-beam

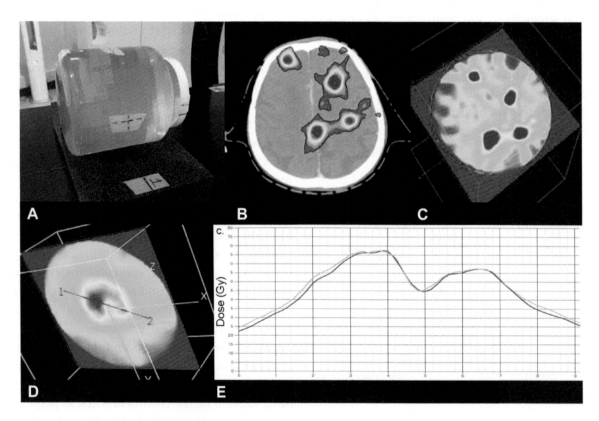

FIGURE 20.4 (A) Radiochromic gel container immobilized and marked in CT using lasers, (B) axial slice of a single isocenter multi-target brain treatment plan generated for a 0.35 T MR-linac, (C) corresponding gel dosimetry plane, (D) measured dose distribution of an intraprostatic dominant lesion prostate plans with the (E) corresponding dose profile through the dominant lesion and urethra sparing for the measured (blue) and calculated (orange) dose distribution of the prostate plan showing good agreement for the complex geometry. Gel experiments performed in conjunction with Modus Medical Devices, Inc., London, ON.

TABLE 20.1 Performance Consideration for Main Dosimeters Evaluated in the Presence of a Magnetic Field

Dosimeter	Conditions Tested	Impact of magnet on performance	Reference
Ion chamber Array, IC PROFILER	Study using 1.5 T MR-Linac with parallel plate chambers within profiler. Measurement results are normalized to the center reference chamber.	The IC PROFILER linearity, reproducibility, pulse rate frequency dependence, are unaffected by the magnetic field. The ionization chamber measurement in magnetic field has some dependence on the ion chamber shape and orientation. Therefore, the setup orientation require attention when normalize the reading to central ion chamber.	Smit et al. [22]
Cylindrical diode array, ArcCHECK-MR	Study using 1.5 T MR-Linac (transverse magnetic field) and a regular linac.	Short-term reproducibility, dose linearity, dose rate dependence, field size dependence, dose per pulse dependence, and inter-diode variation have very little difference with and without magnetic field.	Houweling et al. [23]
Thermoluminescent dosimeters (TLDs)	Inserted in lung and head & neck phantom TLD measurements	– Magnetic field results vs. 0 T for 1.5 T MR-linac and MR-Co^{60} were 0.5% and 0.6%, respectively.	Steinmann et al. [24]
Gafchromic EBT2 (Ashland, Inc., Covington, KY)	Irradiation conducted for a Co-60 MR-IGRT system ($B = 0.35$ T) with/without magnetic field under same conditions.	– Presence of magnetic field may affect crystal orientation and polymerization within active layer of radiochromic film. – NetOD decreased by 8.7%, 8.0%, and 4.3% in red, green, and blue channels, respectively. Underresponse was dose-dependent and differed by up to 15% at 17.6 Gy.	Reynoso et al. [25]
Gafchromic EBT3 (Ashland, Inc., Covington, KY)	*Photons*: Compared 0 and 0.35 T beam profiles generated via Monte Carlo to differences in distance to agreement between 0 and 0.35 T beam profiles measured on EBT3 film.	Negligible absolute differences in DTA were compared at 80%, 50%, and 20% profiles (maximum average difference between 0 and 0.35 T was 0.6 mm for film and 0.4 mm for Monte Carlo), no shift/shape change in beam profiles.	Steinmann et al. [24]
	Protons: Transversal film Irradiation in PMMA with clinical proton beams; $B = 0$–1 T.	Film performance not impacted significantly, dose impact <1% at all positions.	Padilla-Cabal et al. [26]
Electronic Portal Imaging Device (EPID)	aS500 EPID modified with 0.1 cm copper plate to function in 0.5 T parallel magnetic field at 6 MV.	– Comparable flood field, dark field, linearity, uncertainty, radiation profiles, and lag found with and without magnetic field. – Modified EPID responds to 6 MV photons similarly with and without a magnetic field in parallel orientation to the radiation beam.	Rathee et al. [27]

optical computed tomography scanner a 3D image of the dose was reconstructed using an iterative process from 500 projections taken by step-wise rotation around the central axis of the jar a single isocenter multi-target (SIMT) cranial stereotactic radiosurgery plan with 5 metastatic lesions (0.1–0.7 cm³), each receiving 18 Gy.

20.4 TREATMENT PLANNING CONSIDERATIONS

20.4.1 Dose Calculation/Optimization Considerations

For photon-based treatments, the primary beam is not affected by the magnetic field. However, dose is deposited by secondary particles. This leads to asymmetric

TABLE 20.2   Commonly Used Detectors in Magnetic Fields, Testing Conditions, and Correction Factor Considerations Based on Current Literature

Detectors	Conditions Tested	Correction factor	Correction Factor Dependence	Reference
PTW 30013, PTW 31006, PTW 31010, Extradin A1SL	Measurement in different magnetic field strength. (0.75 T, 1.0 T, 1.5 T, 1.75 T transverse).	Only calculated $M_{1.5T}/M_0$ (not correction factor): 1.05–1.09	Depends on the field strength and air gap position around the chamber.	Agnew et al. [28]
PTW 30016 PTW 30015 NE 2571 PTW 30013	Simulation study in both 0.35 T and 1.5 T magnetic field (transverse and parallel).	Correction factor: 0.968–0.977 (0.35 T) 0.954–0.98 (1.5 T)	Correction factor is within 0.99 for parallel magnetic field.	Pojtinger et al. [29]
PTW 30013 and 5 custom build farmer type chamber	Measurement in water tank. Magnetic field ranging from 0 T to 1.1 T. And Monte Carlo simulation study from 0 T to 3 T.	PTW 30013 ~0.97 (0.3 T) ~0.955 (1.5 T) in transverse field.	Depends on chamber radius and magnetic field orientation.	Spindeldreier et al. [30]
NE2571 and PR06C ionization chambers	Monte Carlo simulation dose–response in air and water tank with different magnetic field and detector orientations.	Transverse magnetic field (perpendicular to beam) had maximum of 11% response change near 1.0 T. Longitudinal magnetic field (parallel to the beam) produced slight increase in dose–response (2% at 1.5 T).	Dose–response and correction factor depends on detector type, energy, magnetic field strength and orientation.	Reynolds et al. [31]
PTW 60003 diamond detector, IBA PFD diode detector	Monte Carlo simulation on PTW 60003 diamond detector and IBA PFD diode detector. Simulated the dose–response in air and in water tank with different magnetic field and detector orientations.	Significant dose–response in transverse field geometries (up to 20% at 1.5 T) and dependence on orientation in magnetic field, energy, and detector. Longitudinal magnetic fields show little dose–response, rising slowly with magnetic field, and reaching 0.5–1% at 1.5 T regardless of orientation.	Correction factors higher in transverse magnetic field for both PTW 60003 and IBA PFD diode detector.	Reynolds et al. [32]

dose deposition perpendicular to the magnetic field or for larger field sizes to a small increase in the beam penumbra. In addition, a shift of the depth dose curve occurs. The changes were found to be dependent on magnetic field strengths and orientation, showing notable deviations from symmetric profiles [1, 33–35]. Consequently, employing existing semi-analytic dose calculation algorithms is not possible, and thus MRgRT requires dedicated dose calculation systems as well as beam modeling that takes the presence of the magnetic field into account. The currently available MR-guided photon and Cobalt 60 therapy units employ Monte Carlo based dose calculation engines for treatment planning and optimization [36–38].

In proton beam therapy, treatment planning based on semi-analytical dose calculation algorithms, so-called pencil beam algorithms, remains the clinical standard, although first Monte Carlo based solutions are entering the market. The magnetic field related effects require new dose calculation algorithms. Such semi-analytical dose calculation algorithms taking magnetic field related effects into account have already been developed, showing good agreement with full Monte Carlo simulations [39–41]. In addition to the dose calculation itself, the bent proton beams potentially require new optimization strategies for optimal efficiency. The Bragg-peak displacement is energy dependent, requiring adapted spot positions for each energy layer. Consequently, simple linear ray casting to determine the initial spot positions needs to be adapted. Furthermore, due to the bending of the primary beam, the commonly employed methods to determine beam angles are not well-suited [41, 42].

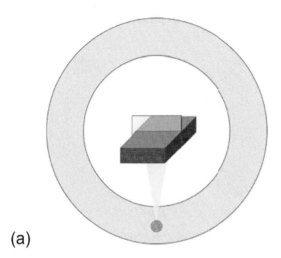

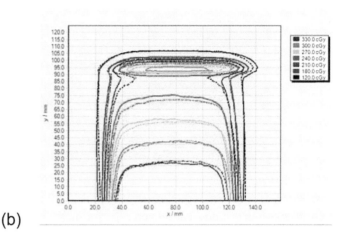

(a)　　　　　　　　　　　　　　　　　(b)

FIGURE 20.5　(a) Film sandwiched in a water slab in an MR-guided Co-60 system and (b) planned and delivered dose distributions. The dashed lines represent the film measurement and the solid line represent dose calculation with magnetic field confirming agreement with the treatment planning system.

20.4.2　Quantifying the Electron Return Effect

An example experimental setup to measure the ERE for a Co-60 MRgRT machine (ViewRay, Oakwood Village, OH) and validate it against the planning system is shown in Figure 20.5a. A 10×10 cm posterior–anterior beam was delivered to a Gafchromic™ film sandwiched by a water slab phantom. The dose was also calculated in the ViewRay Treatment Planning System (TPS) using a Monte Carlo based dose calculation algorithm that models the ERE. The resultant isodose lines (Figure 20.5b) from the calculated dose distribution (solid line) and film measurement (dashed line) are shown. The measured dose distribution agreed well with the calculated dose with a gamma index passing rate of 99.87% using passing criteria of 3%/3 mm, confirming that the ERE at the distal end of the phantom was accurately modeled in ViewRay TPS.

Wooten et al. performed measurements in a Cobalt-60 with a low field 0.35 T magnet in a custom-designed heterogeneous phantom [43]. Open field and IMRT plans were generated on a CT scan of the phantom, plans were delivered, and radiographic film and ionization chamber measurements were obtained. Ionization chamber and film measurements made with the phantom suggested that the dosimetric effect was minimal, and the TPS predicts dose reasonably well for complex heterogeneous scenarios.

20.5　POTENTIAL CLINICAL IMPACT

While few clinics have had the ability to measure the impact of the magnetic field on dose due to the magnet remaining on at all times once the program is clinical, many groups have conducted simulation and validation studies in the presence of a magnetic field. The chief concerns are dose enhancement occurring at transitions from high to low density regions (such as soft tissue to lung) and the reduction in dose in the transition between low to higher density tissue. However, the impact has been shown to vary based on the beam arrangement, number of fields, and orientation between the magnetic field and beam, which will be covered in the following sections.

20.5.1　Beam Orientation and Cavity Size

The MRIdian Linac system (Viewray, Ohio, US) consists of a double-donut superconducting wide (80 cm) split-bore magnet with 0.35 T field strength and a 6 MV flattening filter-free (FFF) Linac. The Linac components are spread among five cylindrical ferromagnetic compartments around a magnetically shielded ring located between the double donuts to avoid magnetic field interference while the shielding also contains carbon fiber to absorb radiofrequency energy. To evaluate the impact of the 0.35 T magnet on treatment planning, a series of anthropomorphic phantom calculations were conducted using a novel pelvic end-to-end phantom (PETE) [44] with rectal air volumes ranging from 30 to 120 cm³. The ViewRay TPS uses a convex, nonlinear programming model for IMRT optimization and fast Monte Carlo dose calculation (source model simulating particle transport in the linear accelerator head (excluding the MLC) and patient model which performs the

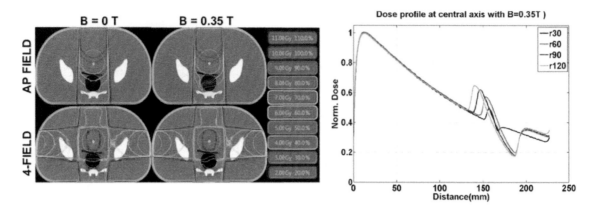

FIGURE 20.6   In this dose calculation experiment, the main 0.35 T superconducting split magnet provides the static B_0 field in the horizontal field direction. (Left) Axial cross section of a pelvis phantom with simulated rectal air volumes for a single anterior–posterior (AP) field (top) and 4-field box (bottom) highlighting the reduction in dose difference with added beams. (Right) Dose profile of a single anterior 6 MV FFF beam traversing simulated rectal gas cavities ranging from 30 to 120 cm^3 (r30–r120). (Figure credit: Laza Rakotondravohitra, Ph.D.)

transport toward the MLC opening and patient geometry) based on VMC++ [45]. The end user can determine whether the 0.35 T magnetic field is implemented in the optimization or dose calculation. For this experiment, it was determined that increasing the rectal air volume yielded a local increase of dose due to the ERE based on cavity size with subsequent decrease after the cavity as shown in Figure 20.6. However, increasing the number of treatment beams decreased the overall impact of the presence of the air and related dose perturbation.

20.5.2 Optimization of Intensity-Modulated Radiation Therapy

Chen et al. performed step-and-shoot intensity-modulated radiation therapy (IMRT) and volumetric-modulated arc therapy (VMAT) plan optimizations in a 1.5 T transverse magnetic field for several disease sites with air/tissue interfaces including pancreas, lung, breast, and head and neck cancer cases [46]. Treatment plans were generated for the following scenarios: (1) no magnetic field, (2) with recalculation (no optimization using the magnetic field), and (3) full reoptimization including the impact of the magnetic field. Standard clinical dosimetric endpoints were considered and all calculations were performed with an Elekta Agility linac head 6 MV flat beam model with fast graphic processing unit (GPU)-based Monte Carlo dose calculation (1% statistical uncertainty/1 mm grid size). Of note is that the Elekta Agility linac implemented in this work is very similar to the 160-leaf multi-leaf collimator used in the Elekta Unity MRI-linac (Elekta AB, Stockholm, Sweden) with

the beam axis perpendicular to the magnet's 1.5 T main field [47]. With the exception of 1 lung case, all planning target volume differences with the transverse magnetic field were <4% different from no magnetic field calculations. Their results also suggested that incorporated many beam angles as used in IMRT, and particularly for VMAT, reduce the impact of the magnetic field on clinical plan quality. Nevertheless, the authors found that by incorporating the transverse magnetic field into the optimization, the maximum doses and dose to 1 cm^3 were reduced, particularly for skin dose.

Overall, the current literature suggests that accounting for the presence of the magnetic field in the optimization and subsequent dose calculation is appropriate for MRgRT. Furthermore, the impact can be lessened by adding beams and counteracting local effects.

20.6 UNMET NEEDS AND NEW DIRECTIONS

While clinical treatments with MRgRT have been underway in the past decade, several unmet needs still exist. MR-guided adaptive radiation therapy, particularly for the online setting where the treatment plan is changed on daily basis using updated anatomy of the day, is an emerging area of clinical trial development. This requires significant needs for benchmarking and workflow development to ensure high confidence in the delivered plan. Dosimetric calibration protocols in the presence of magnetic fields are currently under development at standardized laboratories but have yet to be implemented into standard clinical practice. Recently, real-time MRI guidance in proton or particle therapy

(MRIgPT) has been introduced [48] and a proof of concept system has been used to acquire MR images of a phantom during proton-beam delivery [49], thus this modality appears to be the next generation of MRgRT. Future work will likely encompass dosimetric and planning studies of particle and proton therapy in the presence of magnetic fields.

REFERENCES

1. A. J. Raaijmakers, B. W. Raaymakers, J. J. Lagendijk, "Magnetic-field-induced dose effects in MR-guided radiotherapy systems: Dependence on the magnetic field strength," *Physics in Medicine and Biology* **53**(4), 909–923 (2008).

2. A. E. Rubinstein, Z. Liao, A. D. Melancon, et al., "A Monte Carlo study of magnetic-field-induced radiation dose effects in mice," *Medical Physics* **42**(9), 5510–5516 (2015).

3. B. Oborn, P. E. Metcalfe, M. Butson, A. B. Rosenfeld, "High resolution entry and exit Monte Carlo dose calculations from a linear accelerator 6 MV beam under the influence of transverse magnetic fields," *Medical Physics* **36**(8), 3549–3559 (2009).

4. B. W. Raaymakers, A. J. Raaijmakers, J. J. Lagendijk, "Feasibility of MRI guided proton therapy: Magnetic field dose effects," *Physics in Medicine & Biology* **53**(20), 5615 (2008).

5. D. J. O'Brien, J. Dolan, S. Pencea, N. Schupp, G. O. Sawakuchi, "Relative dosimetry with an MR-linac: Response of ion chambers, diamond, and diode detectors for off-axis, depth dose, and output factor measurements," *Medical Physics* **45**(2), 884–897 (2018).

6. V. N. Malkov and D. Rogers, "Monte Carlo study of ionization chamber magnetic field correction factors as a function of angle and beam quality," *Medical Physics* **45**(2), 908–925 (2018).

7. D. O'Brien, D. Roberts, G. Ibbott, G. Sawakuchi, "Reference dosimetry in magnetic fields: Formalism and ionization chamber correction factors," *Medical Physics* **43**(8 Part 1), 4915–4927 (2016).

8. I. Meijsing, B. W. Raaymakers, A. Raaijmakers, et al., "Dosimetry for the MRI accelerator: The impact of a magnetic field on the response of a Farmer NE2571 ionization chamber," *Physics in Medicine & Biology* **54**(10), 2993 (2009).

9. M. Reynolds, S. Rathee, B. G. Fallone, "Ion chamber angular dependence in a magnetic field," *Medical Physics* **44**(8), 4322–4328 (2017).

10. D. J. O'Brien and G. O. Sawakuchi, "Monte Carlo study of the chamber-phantom air gap effect in a magnetic field," *Medical Physics* **44**(7), 3830–3838 (2017).

11. H. Fuchs, P. Moser, M. Gröschl, D. Georg, "Magnetic field effects on particle beams and their implications for dose calculation in MR-guided particle therapy," *Medical Physics* **44**(3), 1149–1156 (2017).

12. S. M. Schellhammer, S. Gantz, A. Lühr, B. M. Oborn, M. Bussmann, A. L. Hoffmann, "Experimental verification of magnetic field-induced beam deflection and Bragg peak displacement for MR-integrated proton therapy," *Medical Physics* **45**(7), 3429–3434 (2018).

13. J. Hartman, C. Kontaxis, G. Bol, et al., "Dosimetric feasibility of intensity modulated proton therapy in a transverse magnetic field of 1.5 T," *Physics in Medicine & Biology* **60**(15), 5955 (2015).

14. A. Lühr, L. N. Burigo, S. Gantz, S. M. Schellhammer, A. L. Hoffmann, "Proton beam electron return effect: Monte Carlo simulations and experimental verification," *Physics in Medicine and Biology* **64**(3), 035012 (2019).

15. Y. M. Yang, M. Geurts, J. B. Smilowitz, E. Sterpin, B. P. Bednarz, "Monte Carlo simulations of patient dose perturbations in rotational-type radiotherapy due to a transverse magnetic field: A tomotherapy investigation," *Medical Physics* **42**(2), 715–725 (2015).

16. B. M. Oborn, P. E. Metcalfe, M. J. Butson, A. B. Rosenfeld, P. J. Keall, "Electron contamination modeling and skin dose in 6 MV longitudinal field MRIgRT: Impact of the MRI and MRI fringe field," *Medical Physics* **39**(2), 874–890 (2012).

17. B. M. Oborn, Y. Ge, N. Hardcastle, P. E. Metcalfe, P. J. Keall, "Dose enhancement in radiotherapy of small lung tumors using inline magnetic fields: A Monte Carlo based planning study," *Medical Physics* **43**(1), 368 (2016).

18. J. M. Park, S. Y. Park, H. J. Kim, H. G. Wu, J. Carlson, J. I. Kim, "A comparative planning study for lung SABR between tri-Co-60 magnetic resonance image guided radiation therapy system and volumetric modulated arc therapy [published online ahead of print 2016/07/13]," *Radiotherapy & Oncology* **120**(2), 279–285 (2016).

19. D. E. Constantin, R. Fahrig, P. J. Keall, "A study of the effect of in-line and perpendicular magnetic fields on beam characteristics of electron guns in medical linear accelerators," *Medical Physics* **38**(7), 4174–4185 (2011).

20. P. J. Keall, M. Barton, S. Crozier, "The Australian magnetic resonance imaging-linac program [published online ahead of print 2014/06/17]," *Seminars in Radiation Oncology* **24**(3), 203–206 (2014).

21. B. Oborn, P. E. Metcalfe, M. Butson, A. B. Rosenfeld, "Monte Carlo characterization of skin doses in 6 MV transverse field MRI-linac systems: Effect of field size, surface orientation, magnetic field strength, and exit bolus," *Medical Physics* **37**(10), 5208–5217 (2010).

22. K. Smit, J. Kok, J. Lagendijk, B. Raaymakers, "Performance of a multi-axis ionization chamber array in a 1.5 T magnetic field," *Physics in Medicine & Biology* **59**(7), 1845 (2014).

23. A. Houweling, J. De Vries, J. Wolthaus, et al., "Performance of a cylindrical diode array for use in a 1.5 T MR-linac," *Physics in Medicine & Biology* **61**(3), N80 (2016).

24. A. Steinmann, D. O'Brien, R. Stafford, et al., "Investigation of TLD and EBT3 performance under

the presence of 1.5 T, 0.35 T, and 0 T magnetic field strengths in MR/CT visible materials," *Medical Physics* **46**(7), 3217–3226 (2019).

25. F. J. Reynoso, A. Curcuru, O. Green, S. Mutic, I. J. Das, L. Santanam, "Magnetic field effects on Gafchromic-film response in MR-IGRT," *Medical Physics* **43**(12), 6552–6556 (2016).

26. F. Padilla-Cabal, P. Kuess, D. Georg, H. Palmans, L. Fetty, H. Fuchs, "Characterization of EBT3 radiochromic films for dosimetry of proton beams in the presence of magnetic fields," *Medical Physics* **46**(7), 3278–3284 (2019).

27. S. Rathee, B. G. Fallone, S. Steciw, "EPID's response to 6 MV photons in a strong, parallel magnetic field," *Medical Physics* **46**(1), 340–344 (2019).

28. J. Agnew, F. O'Grady, R. Young, S. Duane, G. J. Budgell, "Quantification of static magnetic field effects on radiotherapy ionization chambers," *Physics in Medicine & Biology* **62**(5), 1731 (2017).

29. S. Pojtinger, O. S. Dohm, R.-P. Kapsch, D. Thorwarth, "Ionization chamber correction factors for MR-linacs," *Physics in Medicine & Biology* **63**(11), 11NT03 (2018).

30. C. Spindeldreier, O. Schrenk, A. Bakenecker, et al., "Radiation dosimetry in magnetic fields with Farmer-type ionization chambers: Determination of magnetic field correction factors for different magnetic field strengths and field orientations," *Physics in Medicine & Biology* **62**(16), 6708 (2017).

31. M. Reynolds, B. Fallone, S. Rathee, "Dose response of selected ion chambers in applied homogeneous transverse and longitudinal magnetic fields," *Medical Physics* **40**(4), 042102 (2013).

32. M. Reynolds, B. Fallone, S. Rathee, "Dose response of selected solid state detectors in applied homogeneous transverse and longitudinal magnetic fields," *Medical Physics* **41**(9), 092103 (2014).

33. A. Raaijmakers, B. W. Raaymakers, J. J. Lagendijk, "Experimental verification of magnetic field dose effects for the MRI-accelerator," *Physics in Medicine & Biology* **52**(14), 4283 (2007).

34. H. K. Looe, B. Delfs, D. Poppinga, D. Harder, B. Poppe, "Magnetic field influences on the lateral dose response functions of photon-beam detectors: MC study of wall-less water-filled detectors with various densities," *Physics in Medicine & Biology* **62**(12), 5131 (2017).

35. S. J. Woodings, J. Bluemink, J. de Vries, et al., "Beam characterisation of the 1.5 T MRI-linac," *Physics in Medicine & Biology* **63**(8), 085015 (2018).

36. B. Raaymakers, I. Jürgenliemk-Schulz, G. Bol, et al., "First patients treated with a 1.5 T MRI-Linac: Clinical proof of concept of a high-precision, high-field MRI guided radiotherapy treatment," *Physics in Medicine & Biology* **62**(23), L41 (2017).

37. S. B. Ahmad, A. Sarfehnia, M. R. Paudel, et al., "Evaluation of a commercial MRI Linac based Monte Carlo dose calculation algorithm with geant 4," *Medical Physics* **43**(2), 894–907 (2016).

38. M. R. Paudel, A. Kim, A. Sarfehnia, et al., "Experimental evaluation of a GPU-based Monte Carlo dose calculation algorithm in the Monaco treatment planning system," *Journal of Applied Clinical Medical Physics* **17**(6), 230–241 (2016).

39. R. Wolf and T. Bortfeld, "An analytical solution to proton Bragg peak deflection in a magnetic field," *Physics in Medicine & Biology* **57**(17), N329 (2012).

40. F. Padilla-Cabal, D. Georg, H. Fuchs, "A pencil beam algorithm for magnetic resonance image-guided proton therapy," *Medical Physics* **45**(5), 2195–2204 (2018).

41. S. M. Schellhammer and A. L. Hoffmann, "Prediction and compensation of magnetic beam deflection in MR-integrated proton therapy: A method optimized regarding accuracy, versatility and speed," *Physics in Medicine & Biology* **62**(4), 1548 (2017).

42. O. Jäkel and J. Debus, "Selection of beam angles for radiotherapy of skull base tumours using charged particles," *Physics in Medicine & Biology* **45**(5), 1229 (2000).

43. H. Wooten, O. Green, H. Li, V. Rodriguez, S. Mutic, "Measurements of the electron-return-effect in a commercial magnetic resonance image guided radiation therapy system. WE-G-17A-4. Paper presented at: AAPM Annual Meeting (Austin, TX, 2014).

44. J. M. Cunningham, E. A. Barberi, J. Miller, J. P. Kim, C. K. Glide-Hurst, "Development and evaluation of a novel MR-compatible pelvic end-to-end phantom," *Journal of Applied Clinical Medical Physics* **20**(1), 265–275 (2019).

45. F. Hasenbalg, M. K. Fix, E. J. Born, R. Mini, I. Kawrakow, "VMC++ versus BEAMnrc: A comparison of simulated linear accelerator heads for photon beams," *Medical Physics* **35**(4), 1521–1531 (2008).

46. X. Chen, P. Prior, G. P. Chen, C. J. Schultz, X. A. Li, "Dose effects of 1.5 T transverse magnetic field on tissue interfaces in MRI-guided radiotherapy," *Medical Physics* **43**(8 Part 1), 4797–4802 (2016).

47. M. Glitzner, P. Woodhead, P. Borman, J. Lagendijk, B. Raaymakers, "MLC-tracking performance on the Elekta unity MRI-linac," *Physics in Medicine & Biology* **64**(15), 15NT02 (2019).

48. G. P. Liney, B. Whelan, B. Oborn, M. Barton, P. Keall, "MRI-linear accelerator radiotherapy systems [published online ahead of print 2018/09/10]," *Clinical Oncology (Royal College of Radiologists (Great Britain))* **30**(11), 686–691 (2018).

49. S. Schellhammer, L. Karsch, J. Smeets, et al., "OC-0605: First in-beam MR scanner for image-guided proton therapy: Beam alignment and magnetic field effects," *Radiotherapy and Oncology* **127**, S318–S319 (2018).

CHAPTER 21

Helical Tomotherapy Treatment and Dosimetry

Reza Taleei

Sidney Kimmel Medical College at Thomas Jefferson University
Philadelphia, Pennsylvania

Sarah Boswell

Accuray Incorporated
Madison, Wisconsin

CONTENTS

21.1 INTRODUCTION

The concept of helical tomotherapy was formed in the late 1980s with a series of projects defined by Mackie's research group at the University of Wisconsin [1]. The idea started by investigating the delivery of non-uniform optimized beams introduced by Brahme [1, 2]. The project evolved into designing a linac delivery system on a CT ring gantry with helical delivery [3]. The helical TomoTherapy® System (Accuray Incorporated) is an intensity-modulated radiation therapy (IMRT) delivery system capable of treating targets with thousands of coplanar beamlets as the patient continuously translates through the beam in the longitudinal direction (see Figures 21.1 and 21.2).

The following list is summarized and freely adapted from the *Physics Essentials Guide* with permission by Accuray Incorporated [4]. The helical tomotherapy system features:

- A patient couch that travels at a constant velocity (per treatment plan) in the longitudinal direction. The target, which can be less than 1 cm to more than 1 m in length, is treated from the superior end to the inferior end as the couch translates through the beam.

- A gantry that rotates at a constant velocity (per treatment plan) in the clockwise direction.[1] The combined effect of the couch translation and gantry rotation is a helical delivery pattern on the patient, with 3264 coplanar beamlets available for optimization at each gantry rotation. See Figures 21.3 and 21.4 for sample plan dose distributions.

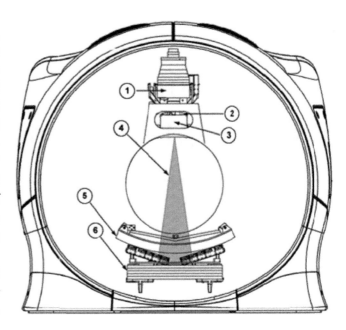

FIGURE 21.2 Major components of the helical tomotherapy system: (1) linac; (2) jaws; (3) MLC; (4) beam; (5) CT detector; (6) beam stop. (Image reprinted from [4]; used by permission of Accuray Incorporated.)

- A standing wave linear accelerator (6 MV), mounted on the rotating gantry with source-to-axis distance 85 cm. See Figure 21.5. The same linear accelerator can be tuned to a lower fluence rate and energy to produce the beam for megavoltage computed tomography (MVCT) imaging.

- Transmission monitor chambers, to provide feedback for achieving a stable dose rate during the delivery. See Figure 21.6.

- Fixed tungsten collimator that produces a fan beam. The beam size is 40 cm in the transverse direction at isocenter.

- A single pair of moveable jaws that collimate the beam to 1 cm, 2.5 cm, or 5 cm in the longitudinal direction at isocenter. To sharpen the superior/inferior dose fall-off, the jaws gradually open as the superior end of the target enters the beam, and gradually close as the inferior end of the target exits the beam. See Figure 21.7.

- A 64-leaf, binary multi-leaf collimator (MLC) for intensity modulation in the transverse direction. See Figure 21.8.

- A single-slice MVCT detector mounted opposite the linac.

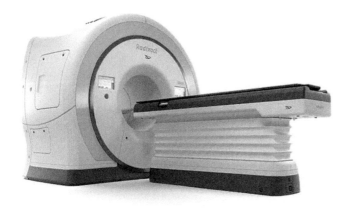

FIGURE 21.1 Helical tomotherapy system. (Image used by permission of Accuray Incorporated.)

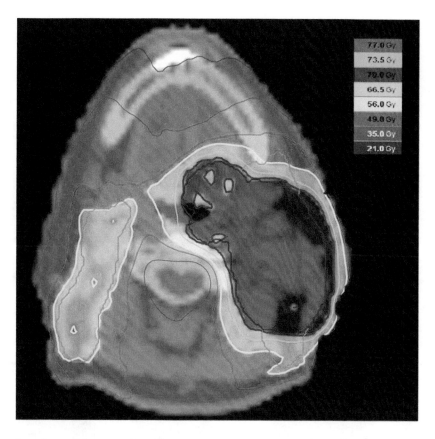

FIGURE 21.3   Head and neck tomotherapy plan. The prescription to PTV1 is 70 Gy in 35 fractions. Fraction delivery time is 285 s.

- Lead beam stop, opposite the linac and beyond the MVCT detector, to attenuate the primary beam to the extent that room shielding calculations are dominated by head leakage and patient scatter.

- The tomotherapy gantry is enclosed by gantry covers. No accessories can be attached to the linear accelerator. There is no light field or front pointer.

In addition, some systems include motion tracking equipment:

- A kV imaging tube and flat panel detector for acquiring radiographs of the patient at intervals during treatment. See Figure 21.9.

- An optical camera for continuously monitoring the respiratory waveform, by tracking light-emitting markers attached to the patient surface. See Figures 21.10.

The helical tomotherapy system is an FFF (Flattening Filter Free) system; see Figure 21.11. All tomotherapy treatments are delivered with moving leaves to achieve a uniform dose distribution. For simple cases that do not require sophisticated IMRT optimization, a 3D Conformal planning option is available. The delivery geometry of a 3D Conformal plan is identical to its IMRT counterpart (helical or fixed angle), but the MLC delivery instructions are determined differently for a 3D Conformal plan, without user-provided optimization inputs [4].

21.2 SYSTEM DETAILS

21.2.1 Beam Production and Monitoring

The high voltage power supply generates direct current to charge a solid-state modulator. The solid-state modulator acts as a switch and provides pulsed high voltage direct current to the magnetron (300 Hz for treatment, 134 Hz for imaging). The magnetron generates the radio-frequency power used by the linac to accelerate electrons. The standing wave linac accelerates electrons from the electron gun to the transmission-style tungsten target, where photons are produced via the bremsstrahlung effect. When operated in treatment mode, the linac produces a 6 MV energy spectrum, as shown in Figure 21.12.

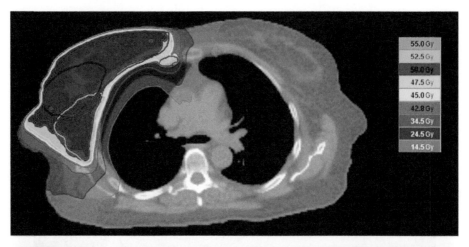

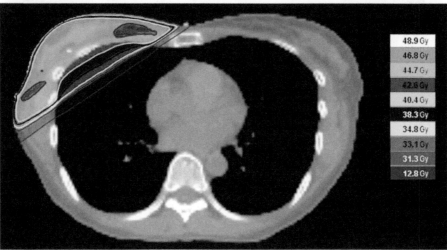

FIGURE 21.4 Above: Helical tomotherapy IMRT breast plan. The prescription is 50 Gy in 25 fractions. Fraction delivery time is 650 s. Below: 3D breast plan with two fixed gantry angles. The prescription is Canadian Fractionation of 42.56 Gy in 16 fractions. Fraction delivery time is 325 s.

Dose monitor chambers are located downstream of the linac and upstream of the jaws and MLC; see Figure 21.6. The monitor chambers are composed of two layers of independently sealed, ceramic, parallel plate chambers. The chambers are transmission chambers: the beam passes through the chambers before entering the collimation system. The monitor chambers provide feedback to the Dose Control System, which is used to adjust the magnitude of the magnetron pulse that establishes the acceleration field of the linac. The Dose Control System stabilizes the dose rate so that it does not fluctuate cyclically with gantry rotation. An additional feedback loop on the gun current helps to stabilize the injector current, which affects the beam energy.

Tomotherapy treatments are planned and delivered based on time, given a known, fixed dose rate. In this aspect, a tomotherapy treatment is like radioactive source treatment, such as high dose rate (HDR) brachytherapy or cobalt external beam therapy. (By contrast, in conventional linac-based systems, the treatment plan specifies a number of monitor units, and the beam turns off when the exact number of monitor units has been delivered). The time-based planning and delivery approach is well-suited to tomotherapy treatments, for which the couch translates at a constant velocity, and targets are treated from the superior end to the inferior end. Each voxel within the target passes through the beam plane during a limited amount of time, and is affected only by the dose delivered while that voxel is in the beam plane (or within scatter's reach of the beam plane).

Although tomotherapy delivery is not controlled by monitor units, monitor units are displayed for informational purposes. The conversion from raw monitor counts to monitor units can be calibrated by the physicist

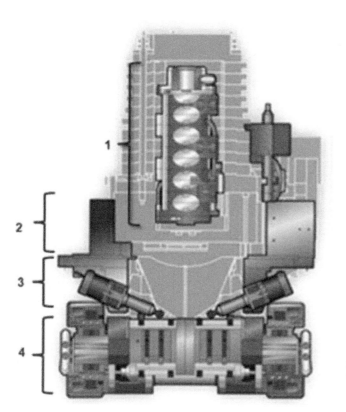

FIGURE 21.5 A side view of the beamline, showing (1) linac; (2) primary collimator and transmission monitor chambers; (3) jaws; (4) MLC. (Image reprinted from [4]; used by permission of Accuray Incorporated.)

so that the display is meaningful, i.e., the display reads 1 cGy/MU under some reference conditions chosen by the user. The monitor unit display calibration does not control the beam delivery.

FIGURE 21.7 Jaws. The jaw setting determines the field size in the longitudinal direction. (Image reprinted from [4]; used by permission of Accuray Incorporated.)

During treatment, the raw counts on the monitor chambers are continually monitored. If the count rate falls outside system tolerances, an interlock will stop the treatment. A make-up procedure can then be delivered.

21.2.2 Beam Collimation

The linac is supported by a tungsten fixture that contains a rectangular aperture to collimate the radiation

FIGURE 21.6 Transmission-style dose monitor chambers. (Image reprinted from [4]; used by permission of Accuray Incorporated.)

FIGURE 21.8 MLC. There are 66 leaves, but the outermost two leaves always remain closed. MLC leaves move along the longitudinal direction. The amount of time that the leaves stay open controls the intensity distribution in the transverse direction. (Image reprinted from [4]; used by permission of Accuray Incorporated.)

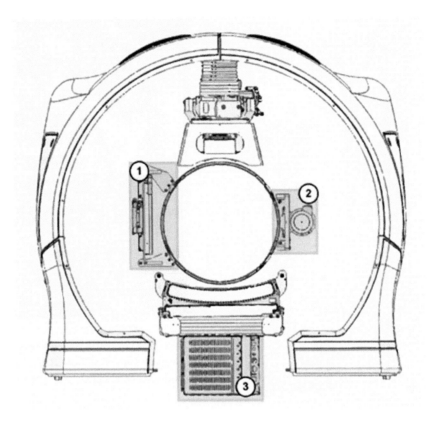

FIGURE 21.9   kV detector (1), tube (2), and generator (3) mounted on gantry for motion tracking. (Image reprinted from [4]; used by permission of Accuray Incorporated.)

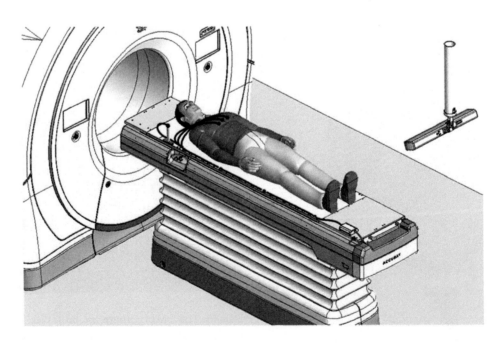

FIGURE 21.10   Motion synchronization. To monitor the patient's breathing, an optical camera mounted to the ceiling tracks the position of light-emitting markers attached to the patient surface. To continuously monitor the target position, the positions of the markers are correlated with intermittent kV radiographs collected during treatment delivery. (Figure by Sam Durst of Accuray Incorporated. Image reprinted from [27]; used by permission of Accuray Incorporated.)

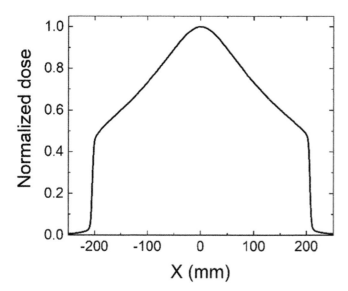

FIGURE 21.11 Sample transverse profile of the tomotherapy beam. The unflattened fluence distribution of photons produced by bremsstrahlung interactions in the target is roughly twice as high in the center as it is on the edges. This profile was measured in a water tank at 85 cm SSD, with the 5 cm jaw size and all leaves open. (Image used by permission of Accuray Incorporated.)

cone beam produced by the linac into a fan beam. This rectangular aperture cannot be adjusted.

The tomotherapy system has only one set of jaws, which are made of a tungsten alloy, and located downstream of the rectangular aperture and upstream of the

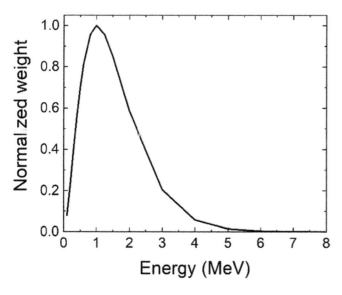

FIGURE 21.12 Example energy spectrum for the tomotherapy treatment beam. (Image reprinted from [4]; used by permission of Accuray Incorporated.)

MLC; see Figure 21.7. The jaws control the width of the radiation beam in the longitudinal direction; there is no collimator rotation. The field width is selected during treatment planning. Available field widths include: 5.0 cm, 2.5 cm, and 1.0 cm; these are nominal full-width at half-maximum (FWHM) values projected to isocenter.

Treatments can be delivered with either fixed or moving (dynamic) jaws. When using fixed jaws, the 5-cm field width has the advantage of the shortest treatment time, but the superior/inferior penumbra is sharpest for the 1-cm beam. Moving the jaws dynamically during treatment allows the planner to achieve treatment times similar to the 5-cm beam, without compromising the superior/inferior penumbra.

Figure 21.13 illustrates a dynamic jaws treatment with the 5-cm field width. The optimizer opens beamlets superior and inferior to the target, where necessary to achieve a uniform target dose. The first beamlets available for optimization occur when the superior end of the target crosses the inferior edge of the beam, and the last beamlets available for optimization occur when the inferior edge of the target crosses the superior edge of the beam. By narrowing the jaw opening at the superior and inferior ends of the target, the penumbra is improved. By using a larger field width internal to the target, the treatment time is improved.

The MLC is located downstream of the jaws and provides intensity modulation along the transverse direction during treatment (see Figure 21.8). There are 64 binary, moveable leaves (tongue and groove design), arranged in banks of 32 leaves each. The two banks of leaves are interlaced and do not meet in the center of the field; when closed, each leaf covers the full longitudinal extent of the beam. The leaves are 10 cm high, and made of tungsten. The leaves are arranged on a curve with a focus near the radiation source. Leaves are equiangular with an average width of 6.25 mm when projected to a line perpendicular to the beam at 85 cm distance. The maximum transverse width of the beam, when all leaves are open, is 6.25 mm per leaf × 64 leaves = 400 mm at isocenter.

Each leaf is controlled independently and pneumatically (unlike other MLC's, the tomotherapy MLC has no motors to replace). The leaves are binary: they are always programmed to be opened or closed. When opened, the leaves are fully removed from the beam, and the fan beam shape is determined by the primary collimator. When closed, each leaf covers the full longitudinal extent of the beam. MLC leaf-open times are stored as

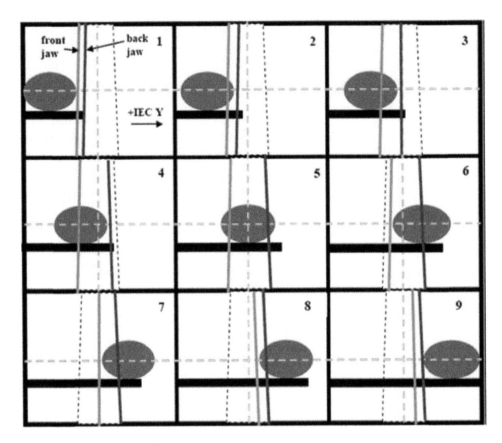

FIGURE 21.13 Stages of a dynamic jaws treatment. The couch travels at a constant velocity in the +IEC Y (longitudinal) direction, into the beam. The intersection of the thicker perpendicular dashed lines marks the radiation isocenter. The thinner dashed line represents the 5-cm field width. The lighter and darker solid lines represent the front and back jaw positions, respectively (the back jaw is always located on the +IEC Y side of the front jaw). As the target enters the beam, the back jaw moves just ahead of the target (beginning with a minimum field width of about 1 cm), continuously expanding until the 5-cm field width is reached. The jaws are fully open to the 5-cm field width while the central portion of the target is being treated, and the leaves modulate during the entire treatment. As the inferior portion of the target begins to move past the beam, the front jaw position narrows to about 1 cm, following the inferior target edge. Discrete jaw positions are illustrated here, but the actual jaw motion is continuous. (Image reprinted from [4]; used by permission of Accuray Incorporated.)

floating-point values, resulting in a very large number of discrete intensity levels (higher precision than deliverable). The number of possible intensity levels for a single beamlet depends on the linac pulse rate and the projection time (see Section 21.4.2 for a discussion of projection time).

To ensure that the leaves open and close at the planned times, each leaf is continuously monitored by two optical position sensors. If the leaves fail to meet their positions, the treatment is interrupted. Newer tomotherapy systems use continuous feedback from the optical sensors to correct the timing of commands to each leaf during treatment. On older systems, the field service engineer uses the CT detector to measure the average response of the MLC to timing commands; this data is used by the planning system to adjust leaf-open time commands to achieve the planned leaf-open times.

MLC leakage can be measured by comparing a "leakage" film exposed with the jaws open and leaves closed, to a "reference" film exposed with the jaws open and leaves open. The maximum rate in the MLC leakage film is less than 2% of the rate in the center of the reference film, and the average rate is less than 0.5% of the average rate in the reference film.

21.2.3 MVCT Imaging

MVCT images are acquired using the treatment linac. The linac parameters are adjusted for imaging to achieve a lower energy (approximately 3 MV spectrum) and lower dose rate than the treatment beam.

The on-board detector is an array of parallel plate ionization chambers (pressurized Xenon-filled cavities separated by stainless steel septa) housed in an aluminum case. The detector consists of about 640 channels, 520 of which are within reach of the imaging beam and used for image reconstruction. The size of the detector channels is approximately 2.5 cm in the longitudinal direction, and approximately 1.24 mm in the transverse direction (since tomotherapy uses a single-slice detector, tomotherapy MVCT imaging is fan beam CT, not cone beam CT).

When a photon interacts in the detector, secondary electrons liberate additional electrons in the gas volume through ionization. Under the influence of an applied electric field (−200 V bias on every other plate), charges flow to the collecting electrode.

Data is acquired helically with a six-second gantry period and 134 Hz pulse rate, providing 804 linac pulses and detector measurements per gantry rotation. The jaw width for imaging projects to approximately 4 mm at isocenter. The couch velocity is determined by the user-selectable acquisition pitch. Couch speeds are 4 mm/rotation (Fine), 8 mm/rotation (Normal), or 12 mm/rotation (Coarse). Thus, a Coarse scan requires one-third of the time and delivers one-third of the patient imaging dose as compared to a Fine scan, but the Fine scan has the best longitudinal resolution. The patient imaging dose for a Fine scan is less than 3 cGy, when measured with an ion chamber near the center of a 30-cm diameter phantom, over a scan distance that spans the entire chamber collection volume.

The imaging system uses two different types of algorithms for image reconstruction: filtered backprojection and iterative reconstruction [5]. While analytical algorithms such as filtered backprojection are based on a single reconstruction, an iterative algorithm uses multiple reconstructions and statistical weighting to reduce the image noise while preserving spatial resolution and image fidelity.

Helical data is interpolated onto discrete slices along the longitudinal axis. Two or four images are reconstructed per gantry rotation, resulting in image reconstruction intervals of 2 mm or 1 mm (Fine), 4 mm or 2 mm (Normal), and 6 mm or 3 mm (Coarse). Reconstruction is performed on a 512 × 512 pixel matrix. The reconstruction field of view is approximately 39 cm.

The spatial resolution in the transverse plane is sufficient to resolve pin holes in a CT resolution plug that are 1.6 mm in diameter, with center-to-center spacing of 3.2 mm. The ability to resolve objects in the longitudinal direction depends on the collimation size, couch motion, and the spacing of the reconstructed slices. Slice sensitivity profiles can be measured using a method described by Polacin et al. [6] and implemented for tomotherapy by Chao et al. [7, 8]: a thin disk is scanned, and images are reconstructed at 0.5 mm intervals (using non-commercial software outside of the treatment system). The average pixel values for the high contrast object are plotted as a function of slice position. The FWHM of the profiles is 3.9 mm ± 1.0 mm for Fine pitch, 5.3 mm ± 1.0 mm for Normal pitch, and 7.2 mm ± 1.0 mm for Coarse pitch.

To help maintain the stability of CT numbers produced by the system, the physicist can run a weekly CT number calibration procedure, which scales the CT numbers by comparing measurements in air and in a phantom against their expected values.

The primary purpose of the MVCT imaging system is to acquire daily pre-treatment images to verify the patient setup. However, MVCT images can also be used for treatment planning, provided the 39-cm field of view is adequate for the patient. A convenient feature of using a megavoltage beam is the lower attenuation through metal, leading to a significant reduction in streak artifacts for patients with high-density features such as prostheses and dental fillings [9, 10] (see Figure 21.14).

In addition, daily MVCT images can be used for adaptive radiation therapy, to monitor changes in the patient anatomy. The daily dose can be calculated on the daily MVCT images, deformed to the planning image and summed across fractions to see the cumulative effect in the patient. This information is helpful to determine if the patient needs to be re-planned due to weight loss, tumor shrinkage, or other anatomical changes.

Finally, MVCT detector data has significant utility as a quality assurance (QA) tool, allowing the physicist to monitor the consistency of the transverse beam profile for various automated machine QA tasks within the TQA® software (Total Quality Assurance, Accuray Incorporated).

21.3 PATIENT POSITIONING AND MOTION SYNCHRONIZATION

Helical tomotherapy calculations on the planning CT image exhibit uniform target coverage and a high degree of critical structure sparing. To successfully deliver the

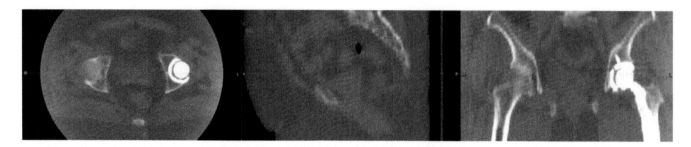

FIGURE 21.14   MVCT images produced on the Radixact® Treatment Delivery System (a form of helical tomotherapy). This patient has a bone prosthesis, but due to the MV energy, the image is not obscured by streak artifacts. The field of view of the MVCT is limited to 39 cm diameter. (Images courtesy of the Radiation Oncology Department at the Froedert and Medical College of Wisconsin.)

plan to the patient, the daily treatment position must be consistent with the planned position.

21.3.1 Patient Positioning

At simulation, the patient receives small tattoo dots or semi-permanent marks.[2] Radiopaque markers are temporarily affixed to the patient surface at the same position of these tattoos so that they will be visible in the CT image. During planning, patient-specific laser positions are assigned so that they correspond to the patient's radiopaque markers on the planning CT.

In preparation for daily treatment, the patient is aligned so that their tattoos correspond to the patient-specific laser positions in the treatment room. Next, a helical MVCT image is used to visualize the patient's internal anatomy in 3D, to ensure that the target is in its planned position. (MVCT images are acquired with all leaves open. Due to MLC leaf modulation, the system cannot reconstruct MVCT images during treatment delivery.)

MVCT images are reconstructed and rigidly registered to the planning kVCT image by automatic or manual methods. Translational adjustments are applied to the couch, prior to delivering the daily treatment. Rotations about the patient's longitudinal axis (roll offsets) are implemented by adjusting the gantry start angle of the daily treatment. The tomotherapy couch does not have rotational degrees of freedom, so if pitch or yaw adjustments are required, the patient is manually adjusted in the treatment room and another MVCT image is acquired to verify the adjustment.

Accurate treatment requires that the target retain its registered position until the daily treatment is complete. Target motion can be minimized through patient education and restraint devices. However, in some treatments, significant target motion is an unavoidable reality, and the physician may choose to expand target volumes to account for motion.

21.3.2 Motion Synchronization

To synchronize the beam delivery with involuntary patient motion such as motion caused by respiration or digestion, some helical tomotherapy systems have a kV imaging source and a flat-panel detector mounted on the rotating gantry, offset 90 degrees from the linac and CT detector; see Schnarr et al. [11]. After initially aligning the patient based on a pre-treatment MVCT image, motion detection is accomplished by collecting 2D kV radiograph images at selected gantry angles while the treatment is in progress (two to six images per gantry rotation). The position of the target or a surgically implanted fiducial set is determined from the radiographs. Motion synchronization is accomplished by continuously updating jaw and leaf positions to address target motion.

Target motion can follow a cyclical pattern (e.g., breathing motion), or the motion can be irregular (e.g., digestive motion). The time between sequential radiographs can range from 1 s to 55 s, depending on the gantry period and the number and placement of imaging angles. The motion tracking system must estimate the target position in between the radiographs:

- To track respiratory motion in between the times of radiograph acquisition, light emitting markers are placed on the patient surface, and their position is monitored by an optical camera system (see Figure 21.10).[3] The system builds a statistical model to correlate the continuously monitored marker positions with the target structure or fiducials in the intermittent radiographs. The jaws and leaves lock onto the target position as determined by this Correlation Model.

- For non-respiratory motion, the target is assumed to maintain the last measured position until the next radiograph is acquired. A statistical Position Model predicts three-dimensional (3D) fiducial positions from sequential, two-dimensional (2D) radiographs. Each time a radiograph is acquired, the collimation system locks onto the updated target position.

A new model is built for each treatment fraction, using only tracking data collected during that session. The beam-on time for treatment with motion tracking is the same as for delivery without motion tracking, since motion correction is accomplished in real-time without gating. Additional time is required for acquiring kV images to build the model prior to treatments with motion tracking, as well as pausing the delivery if adjustments to the motion tracking parameters are needed.

21.4 OPTIMIZATION AND DOSE CALCULATION FOR IMRT PLANS

21.4.1 User-Defined Parameters

Planning CT images are usually acquired on a kV simulator, although an MVCT image from the tomotherapy system may also be used. To account for the attenuation of the treatment couch, the planning CT couch is replaced with an image of the tomotherapy couch.

Mass density is required as an input to the tomotherapy dose calculator. For each CT scanner that supplies images for tomotherapy planning, the customer physicist scans a phantom that contains inserts of various known mass densities, measures the corresponding Hounsfield Unit (HU) values, and creates a table that converts HU values to mass density.

The system calculates dose on all the voxels in the patient CT image (including the patient, restraint devices, couch, air, etc.), so it is important that all the voxels in the CT image accurately represent the density of the patient at the time of treatment. The CT scan is expected to extend a sufficient distance superior and inferior to the target to account for the primary beam as well as scatter. Density override can be used to correct any imaging artifacts, and to remove contrast that will not be present for daily treatments. If the patient extends

laterally beyond the field of view, directional blocking can be used to prevent beam angles from entering through unspecified amounts of tissue.

The planner also specifies the prescription and optimization constraints, and basic information about the plan geometry, including the treatment modality (helical or fixed-angle), jaw size, jaw mode (fixed or dynamic), calculation resolution, pitch, and modulation factor. Pitch and modulation factor are discussed below.

For helical plans, the **pitch** defines the relationship between couch and gantry positions: $pitch = d/W$, where d is the couch travel distance per gantry rotation, and W is the nominal field width projected to isocenter (1.0 cm, 2.5 cm, or 5.0 cm). The maximum allowed pitch is 0.5. The helical pitch may be conceptualized as the "tightness," or amount of overlap, in the helical delivery pattern. A smaller number for the pitch implies more overlap in the rotations. For a smaller pitch number, more rotations are required to cover the same target length.

For a helical plan, as the gantry rotates around an off-axis target, the source-to-target distance varies, impacting the dose rate (inverse square effect), beam size (divergence effect), and fluence rate (transverse profile shape). The overlapping rotations can result in a pattern of ripples in the dose distribution along the longitudinal axis, which are sometimes referred to as the helical thread effect. The ripples are predicted by the dose calculation and are a consequence of the helical delivery pattern. By selecting the pitch value appropriately for the field width and off-axis position of the target, the thread effect can be minimized. See Table II in Mingli Chen et al. [12] for recommended pitch values. In addition to considering the helical thread effect, the appropriate pitch depends on the prescription dose: higher fraction doses require smaller pitch values to provide enough gantry rotations to deliver the dose without exceeding the longest allowed gantry period of 60 s. Conversely, if the selected pitch value is too small, the gantry reaches its shortest allowed gantry period of 11.8 s, and the delivery becomes inefficient. The tomotherapy planning system presents a histogram of leaf open times that can be used to evaluate a plan's selected pitch value for delivery efficiency and reasonable leaf-open times that contribute to better patient QA pass rates [13].

For fixed-angle plans, the pitch is defined as the couch travel in centimeters per sinogram projection. A projection represents an opportunity for each leaf to open and close not more than one time. Thus, the fixed-angle pitch specifies the resolution of the fixed-angle delivery pattern along the longitudinal axis. A smaller number for the pitch implies finer resolution in the delivery pattern. Although the pitch has a different definition in the context of helical or fixed-angle deliveries, similar numerical values are typically selected for planning.

The **modulation factor** is a user-specified parameter that controls the range of leaf-open times in the plan by putting a cap on the ratio between the maximum and average leaf-open times, for the non-zero beamlets. If the modulation factor is exceeded during optimization, the leaf-open times are re-distributed to improve the treatment time. A modulation factor of 1.0 indicates that all leaves, if used, are equally weighted. A modulation factor of 2.0 indicates that the maximum leaf-open time is twice the average leaf-open time; this plan takes twice as long to deliver as a plan with a modulation factor of 1.0.

The physician or planner draws contours on the patient CT to represent targets and avoidance structures. A single plan may include one or more targets, but one target is identified as the prescription target. The entire plan is scaled to meet the prescription, typically specified as a DVH point. Importance values specify the relative priority of achieving the goals for the various contoured structures, and DVH goal points are set with their relative penalty weightings.

21.4.2 Beamlets: The Basic Unit of Optimization

The MLC delivery pattern is the only aspect of the delivery that is optimized by the planning system.

The continuous gantry rotation and couch translation are divided into discrete units called **projections**. For a helical plan, there are 51 projections per 360-degree gantry rotation. Each projection corresponds to a 7.06-degree arc of continuous gantry rotation. For fixed-angle plans, each gantry angle is divided into a number of projections as determined by the user-defined pitch value.

A projection represents an opportunity for each MLC leaf to open and close up to one time. A leaf may be open for the entire projection, closed for the entire projection, or open for some percentage of the projection. If a leaf is open for less than 100% of the projection time, that leaf is opened when the gantry and couch are positioned in the central portion of the projection.

For improved dose calculation accuracy, the plan fluence may be sampled more than once per projection to account for slightly different source positions – a process

called *source super-sampling*. For helical plans, the dose calculation approximates the dose to be delivered from three discrete samples per projection, each corresponding to a 2.35-degree arc. For fixed-angle plans, at least two discrete samples per slice of the planning image are generated. In this context, a "sample" is a distinct source position from which rays are traced, and corresponds to a discrete gantry and couch position.

A **beamlet** represents the leaf-open time or fluence through a given leaf at a given projection position. Beamlets represent the basic units of optimization. For a helical treatment plan, there are $64 \times 51 = 3264$ beamlets per gantry rotation. The number of gantry rotations depends on the target length, field size, and pitch.

A **sinogram** is a two-dimensional data array of MLC delivery instructions, consisting of rows representing projections and 64 columns representing the MLC leaves. See Figure 21.15. Each element in the sinogram represents a beamlet. The gray-level of each beamlet indicates the percentage of fluence or leaf-open time for a particular leaf at a particular range of gantry and couch positions corresponding to that projection.

The vertical axis of the sinogram is a discretized representation of time and couch position. For helical treatments, the vertical axis of the sinogram is also a discretized and cyclical representation of gantry position. Stepping through the rows (projections) of the sinogram in time, the leaves are configured as indicated in each row, while the couch, gantry, and jaws (where applicable) move as specified in the plan. Sinograms get their name from the sinusoidal pattern seen in the beamlet intensity for a helical plan with a target that is offset from the machine isocenter.

21.4.3 Optimization Process

Optimization is the iterative process of adjusting the MLC sinogram to achieve a dose distribution that meets the prescription DVH point and attempts to satisfy other plan objectives indicated by DVH points and maximum/minimum doses.

The optimizer starts by building an initial sinogram that identifies those beamlets that are available for optimization. Beamlets are considered available for optimization if any part of the beamlet passes through a target, and if no part of the beamlet passes through a contour that has been designated by the user as blocked.

Beamlets are projected through the patient to determine if they intersect a target or blocked structure. The

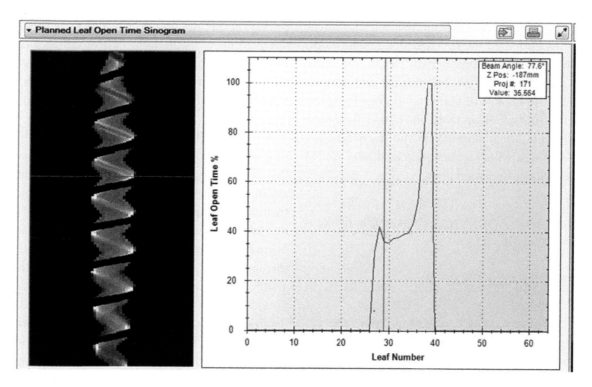

FIGURE 21.15 A sinogram for a prostate case with rectal blocking. The columns represent the 64 MLC leaves and the rows represent the projection angles. White pixels indicate that leaves are open for 100% of the projection, black pixels indicate that leaves are closed, and gray pixels represent open leaves with lower intensity values. The plot on the right illustrates the leaf open times for the projection marked by the vertical line in the sinogram. (Images are from the Delivery Analysis™ software (Accuray Incorporated). Image used by permission of Accuray Incorporated.)

initial sinogram thus determined is an array of zeros and ones. Beamlets with zero intensity are not available for optimization; those beamlets will remain closed. Beamlets that are available for optimization are referred to as "used" beamlets. In the initial sinogram, all the used beamlets are assigned 100% intensity (i.e., leaves open for 100% of their corresponding projections). The intensity weightings for the used beamlets are adjusted during the optimization process.

At each iteration, the dose distribution in the patient is calculated, either by performing a full dose calculation or by using a simple ray-tracing method to predict small changes from the most recent full dose calculation. Then, beamlet weights are adjusted according to plan objectives. Any beamlet weights that exceed the maximum allowed by the modulation factor are truncated. Dose is calculated using the new beamlet weights, the entire plan is scaled to meet the primary prescription point, and the iteration DVH and isodose display is updated.

If the optimization has not been paused by the user and the designated number of iterations has not been exceeded, the beamlet weights are adjusted again. When

optimization is finished, the optimized sinogram is converted into a delivery sinogram to be used by the system. This delivery sinogram discards undeliverable leaf-open times shorter than 18 ms, and generates machine instructions that meet the requirements for the allowed ranges of gantry rotation period, couch speed, jaw speed, etc. A final, full dose calculation is performed on the delivery sinogram and displayed for the planner's evaluation.

21.4.4 Dose Calculation by Convolution/Superposition

The treatment planning system uses a Convolution/Superposition dose calculation algorithm similar to the TomoTherapy one described originally in Ahnesjö [14]. A more recent description of the TomoTherapy algorithm, as implemented on GPU hardware, was published in Chen et al. [15].

Dose calculation is a three-step process:

1. At each projection, the fluence incident on the patient is calculated in a 2D plane.

2. Each 2D fluence plane is projected into the patient volume. Density is interpolated along fluence rays to determine radiological path lengths in water or a bone-like material. Radiological path lengths are used to determine the mass attenuation coefficients from a look-up table. The mass attenuation coefficients are multiplied by the fluence to get the TERMA, which is then projected into the 3D patient dose calculation grid.

3. To determine the dose, TERMA is convolved with a stored scatter kernel, which was originally computed using Monte Carlo.

21.5 QUALITY ASSURANCE

21.5.1 Machine QA

TG-148 [16] is the AAPM Task Group report specifically for quality assurance of helical tomotherapy. This report outlines requirements for daily, monthly, quarterly, annual, and post-service QA. It also provides dose calculation guidance for the non-standard field size (maximum 5×40 cm^2) and 85 cm SAD of the tomotherapy system. Most clinics choose to adopt the TG-148 recommendations for routine QA procedures.

A second task group, TG-306, has been formed to update the TG-148 recommendations, and to account for features that were not available when TG-148 was published in 2010, such as fixed-angle treatments, dynamic jaws, and the dose control system.

As described in Ref. [4], the helical tomotherapy system is provided by the manufacturer with all beam data installed. Thus, the site physicist only needs to check the system routinely to ensure that the machine is performing consistently with the installed beam model. Most system parameters are not user-adjustable, so the site physicist would ask the field service engineer to adjust the machine as needed.

Many system checks can be self-reported, by running the provided procedures and using vendor-provided software to analyze data from the on-board monitor chambers and CT detector. However, since the dose monitor chamber is used to control the beam through the dose control system, it is important to verify the dose with an external ion chamber on a regular basis.

Finally, it is essential to give careful attention to the density models used to convert Hounsfield Units in the planning CT images to mass density. Correct mass density representation is a critical requirement for accurate plan dose calculations. Patient QA measurements, which are performed not in the patient but in a phantom, do not verify that the patient density is properly represented in the treatment plan.

21.5.2 Absolute Dose Measurement and Calibration

In contrast to conventional C-arm gantry-based linacs that deliver output in terms of monitor units, tomotherapy output is delivered based on time. The tomotherapy treatment planning system assumes a constant dose rate which is specified in the beam model as an energy fluence rate. The absolute dose calibration of the tomotherapy system is performed differently from the C-arm gantry-based linacs that follow the TG-51 dosimetry protocol. TG-51 specifies that dose is measured at 10 cm depth for a 10×10 cm^2 static open field in a water tank [17]. However, tomotherapy physical limitations only permit a maximum field size of 5 cm in the longitudinal (y) direction at 85 cm SSD (source to surface distance). The closest field size to the TG-51 protocol is 5×10 cm^2 at 85 cm SSD. Additionally, the lowest available couch position in the tomotherapy bore with 85 cm diameter does not provide sufficient space for measurements at 100 cm SSD and 10 cm depth with appropriate backscatter. In order to overcome the obstacles to implementing the TG-51 protocol for reference dosimetry, a machine-specific reference dosimetry has been suggested for tomotherapy (IAEA-AAPM Joint Committee) [18]. The TG-51 extension for tomotherapy is performed at 85 cm SSD for a field size of 5×10 cm^2. The main component of this protocol is the machine-specific beam quality factor calculated from the beam quality factor for the calibration beam:

$$D_{w,\,Q_{msr}}^{f_{msr}} = M_{Q_{msr}}^{f_{msr}} . N_{D,w,Q_0} . k_{Q_{msr},Q}^{f_{msr}} \tag{21.1}$$

$$k_{Q_{msr},Q}^{f_{msr}} = k_{Q,Q_0} . k_{Q_{msr},Q}^{f_{msr},f_{ref}} \tag{21.2}$$

where, f_{msr} is the machine-specific reference (msr) field size (5×10 cm^2), Q_{msr} is the machine-specific beam quality for 10 cm $\times$ 5 cm and 85 cm SSD, Q is the TG-51 beam quality for a conventional reference 10×10 cm^2 field size and 100 cm SSD, $M_{Q_{msr}}^{f_{msr}}$ is the corrected reading (for temperature, pressure, polarity, ion recombination, and electrometer) for f_{msr} field size, N_{D,w,Q_0} is the absorbed dose to water calibration factor based on TG-51, and $k_{Q_{msr},Q}^{f_{msr}}$ is the beam quality correction factor that takes into account the machine-specific beam quality and field size, k_{Q,Q_0} is the beam quality correction factor for beam quality Q, $k_{Q_{msr},Q}^{f_{msr},f_{ref}}$ corrects the beam quality correction factor for machine-specific beam quality and field size. TG-148 lists the beam quality correction factor for multiple ion chambers including the Exradin A1SL ion chamber that is used by most clinics for tomotherapy beam profile scanning and absolute dosimetry [16, 19, 20].

The machine-specific reference dosimetry is proposed for a fixed beam geometry while tomotherapy treatments are mainly delivered helically. IAEA-AAPM joint committee has proposed a plan-class specific reference (pcsr) dosimetry for rotational output measurements [16, 18]. In this method, a homogenous dose is delivered to an extended and simple target volume. TG-148 suggests a 2 Gy uniform dose to an 8 cm diameter and 10 cm long target in a water-equivalent phantom for pcsr absolute dosimetry. The setup of the water-equivalent phantom is performed with MVCT. The output measurement in the phantom is compared with the treatment planning dose calculation at the position of the ion chamber volume from the CT images. The dose in the ion chamber can be calculated according to the following equation:

$$D_{w,Q_{pcsr}}^{f_{pcsr}} = M_{Q_{pcsr}}^{f_{pcsr}} . N_{D,w,Q_0} . k_{Q_{pcsr},Q}^{f_{pcsr}} \qquad (21.3)$$

$$k_{Q_{pcsr},Q}^{f_{pcsr}} = k_{Q,Q_0} . k_{Q_{msr},Q}^{f_{msr},f_{ref}} . k_{Q_{pcsr},Q_{msr}}^{f_{pcsr},f_{msr}} \qquad (21.4)$$

where, f_{pcsr} is the plan-class specific field size, Q_{pcsr} is the plan-class specific beam quality, $M_{Q_{pcsr}}^{f_{pcsr}}$ is the corrected reading (for temperature, pressure, polarity, ion recombination, and electrometer) for f_{pcsr} field size, and N_{D,w,Q_0} is the absorbed dose to water calibration factor based on TG-51, $k_{Q_{pcsr},Q}^{f_{pcsr}}$ is the beam quality correction factor that takes into account the plan-class specific beam quality and field size. TG-148 lists the beam quality factors for the pcsr dosimetry [16].

21.5.3 Patient Quality Assurance

Patient specific QA is recommended for all tomotherapy treatment plans including 3D plans. Delivery quality assurance (DQA) for tomotherapy, similar to C-arm linacs, is traditionally performed by comparing phantom measurements to dose calculations in a phantom. Vendor-provided software calculates the dose distribution when the patient-specific plan delivery instructions

(gantry, couch, jaws, and leaves) are delivered to a phantom CT image set.

The cylindrical "cheese" phantom has cavities for ion chambers and a place to insert a film. To avoid the inconvenience of film, most users have transitioned to detector array QA devices such as MatriXX [21], ArcCHECK [22], and MapCHECK [23]. In order to set up the phantom for measurement, MVCT of the phantom is normally performed.

The University of Virginia has developed a detector-based tool for patient specific QA [24, 25]. In addition to patient specific QA the toolbox calculates a Monte Carlo (MC) based secondary dose to verify treatment planning dose calculations. The MC calculation uses a tomotherapy simulation model with PENELOPE MC code. The phase-space files from tomotherapy were created from the MC simulation and analytical transfer function for modeling the binary MLC leaves. Patient specific QA is performed through the MLC/log file analysis. The information recorded for the treatment delivery are linac ion chamber measurements (MU1 and MU2), position of gantry and couch, and MVCT exit detector measurements. Tomotherapy log files record the gantry angle and couch position for each projection angle. The MLC positions are derived from the MVCT exit detector measurements. This technique permits 3D dose verification of the treatment plan, and patient specific QA by comparing the dose delivery plan measured with an MVCT exit detector against the treatment plan.

Accuray Incorporated offers a tool called Delivery Analysis™ that analyzes data from the on-board leaf sensors for pre-treatment and in-treatment verification of MLC leaf-open times. Dose is recalculated on the planning CT for comparison with the planned dose. Displayed items include leaf sinograms and histograms, planning CTs with isodose lines, MVCTs, DVHs, and trending data. Linked viewing allows the user to correlate the various data representations for visual analysis. Delivery Analysis verifies the leaf-open times but is not sensitive to other aspects of the delivery; the dose is re-calculated using the beam energy, profiles, and fluence rate from the model data, and the gantry, couch, and jaw positions from the plan.

21.6 SUMMARY

Tomotherapy was invented in the late 1980s to deliver helical IMRT treatments on a CT type rotational platform. Since that time, the system has continued to evolve, most recently to incorporate motion management capabilities. The unique design of tomotherapy enables the delivery of thousands of coplanar beamlets to achieve highly conformal dose distributions, as well as MVCT imaging for patient setup. One of the main advantages of tomotherapy over conventional C-arm treatment linacs is the ability to efficiently treat elongated targets such as total bone marrow irradiation. This chapter reviews the technology of beam delivery, treatment planning, and dosimetry measurements for tomotherapy treatments.

ACKNOWLEDGMENTS

Some text and images were reproduced from the *Physics Essentials Guide* by Accuray Incorporated with their permission. Section 21.2 is summarized and freely adapted from [4] with permission by Accuray Incorporated. Section 21.3.1 is adapted from [26] with permission by Accuray Incorporated. Section 21.3.2 is copied and adapted from [27] with permission by Accuracy Incorporated. Section 21.4 is copied and summarized from [4] with permission by Accuray Incorporated. The authors thank the following people for their input to the *Physics Essentials Guide* and helpful suggestions on this chapter: Eric Schnarr, Andrea Cox, Ed Chao, Dylan Casey, Jacob Shea, Hillary Hausler, Chris Shaw, Michael J. Taylor, and Todd Weston.

CONFLICT OF INTEREST

Sarah Boswell works at Accuray Incorporated.

NOTES

1. For patients that do not require 360 degrees of gantry angles, such as some breast cases for which a pair of tangent beams may be suitable, the tomotherapy system can be used to deliver fixed-angle treatments. For fixed-angle treatments, the gantry is stationary at the user-selected gantry angles, while the couch translates in the longitudinal direction through the beam for each gantry angle [4].

2. Patient external marks are placed in a stable location near the target, but the exact location is arbitrary. The concept of a tumor isocenter is not relevant to tomotherapy treatments, because the couch moves continuously into the bore during treatment.

3. This method of motion synchronization is called Synchrony® Respiratory Tracking™ and is unique to systems produced by Accuray Incorporated: the Radixact® System (a form of helical tomotherapy) and the CyberKnife® System.

REFERENCES

1. T. R. Mackie, "History of tomotherapy," *Physics in Medicine & Biology* **51**(13), R427 (2006).

2. A. Brahme, "Optimization of stationary and moving beam radiation therapy techniques," *Radiotherapy and Oncology* **12**(2), 129–140 (1988).

3. T. R. Mackie, et al., "Tomotherapy: A new concept for the delivery of dynamic conformal radiotherapy," *Medical Physics* **20**(6), 1709–1719 (1993).

4. Accuray Incorporated, *Radixact® Physics Essentials Guide 1062470* (Accuray Incorporated, Madison, 2019).

5. M. Beister, D. Kolditz, W. A. Kalender, "Iterative reconstruction methods in X-ray CT," *Physica Medica* **28**(2), 94–108 (2012).

6. A. Polacin, et al., "Measurement of slice sensitivity profiles in spiral CT," *Medical Physics* **21**(1), 133–140 (1994).

7. E. Chao, et al., "SU-GG-J-120: Longitudinal resolution of the TomoTherapy® MVCT image and potential improvements," *Medical Physics* **35**(6 Part 7), 2706–2706 (2008).

8. E. Chao, et al., "SU-GG-J-32: Evaluation of a thinner-slice MVCT scan mode across several systems," *Medical Physics* **37**(6 Part 9), 3152–3152 (2010).

9. K. Ruchala, et al., "Megavoltage CT on a tomotherapy system," *Physics in Medicine & Biology* **44**(10), 2597 (1999).

10. A. Beavis, "Is tomotherapy the future of IMRT?, " *The British Journal of Radiology* **77**(916), 285–295 (2004).

11. E. Schnarr, et al., "Feasibility of real-time motion management with helical tomotherapy," *Medical Physics* **45**(4), 1329–1337 (2018).

12. M. Chen, et al., "Theoretical analysis of the thread effect in helical TomoTherapy," *Medical Physics* **38**(11), 5945–5960 (2011).

13. D. C. Westerly, et al., "Treatment planning to improve delivery accuracy and patient throughput in helical tomotherapy," *International Journal of Radiation Oncology, Biology, Physics* **74**(4), 1290–1297 (2009).

14. A. Ahnesjö, "Collapsed cone convolution of radiant energy for photon dose calculation in heterogeneous media," *Medical Physics* **16**(4), 577–592 (1989).

15. Q. Chen, M. Chen, W. Lu, "Ultrafast convolution/superposition using tabulated and exponential kernels on GPU," *Medical Physics* **38**(3), 1150–1161 (2011).

16. K. M. Langen, et al., "QA for helical tomotherapy: Report of the AAPM Task Group 148," *Medical Physics* **37**(9), 4817–4853 (2010).

17. P. R. Almond, et al., "AAPM's TG-51 protocol for clinical reference dosimetry of high-energy photon and electron beams," *Medical Physics* **26**(9), 1847–1870 (1999).

18. R. Alfonso, et al., "A new formalism for reference dosimetry of small and nonstandard fields," *Medical Physics* **35**(11), 5179–5186 (2008).

19. S. Thomas, et al., "A Monte Carlo derived TG-51 equivalent calibration for helical tomotherapy," *Medical Physics* **32**(5), 1346–1353 (2005).

20. M. McEwen, "SU-FF-T-218: Evaluation of the Exradin# a19 ion chamber for reference dosimetry in megavoltage photon beams," *Medical Physics* **34**(6 Part 10), 2451–2451 (2007).

21. M. Rao, et al., "Comparison of Elekta VMAT with helical tomotherapy and fixed field IMRT: Plan quality, delivery efficiency and accuracy," *Medical Physics* **37**(3), 1350–1359 (2010).

22. C. Neilson, et al., "Delivery quality assurance with ArcCHECK," *Medical Dosimetry* **38**(1), 77–80 (2013).

23. P. A. Jursinic, R. Sharma, J. Reuter, "MapCHECK used for rotational IMRT measurements: Step-and-shoot, TomoTherapy, RapidArc," *Medical Physics* **37**(6 Part 1), 2837–2846 (2010).

24. L. L. Handsfield, et al., "Phantomless patient-specific TomoTherapy QA via delivery performance monitoring and a secondary Monte Carlo dose calculation," *Medical Physics* **41**(10), 101703 (2014).

25. Q. Chen, et al., "TomoTherapy MLC verification using exit detector data," *Medical Physics* **39**(1), 143–152 (2012).

26. Accuray Incorporated, *1g Patient Positioning Workflows: Simulation, Planning, and Daily Treatment 1059569.B.* In the Radixact® Physics Essentials Course (Accuray Incorporated, Madison, 2017).

27. Accuray Incorporated, *Synchrony® Real-Time Motion Tracking and Correction on the Radixact® Treatment Delivery System 1064193.E* (Accuray Incorporated, Madison, 2019).

CHAPTER 22

Gamma Knife Dosimetry

Nels C. Knutson

Washington University in St. Louis School of Medicine
St Louis, Missouri

CONTENTS

22.1 INTRODUCTION

The following chapter will review important dosimetric measurement techniques used to characterize the Gamma Knife radiosurgery system (Elekta, Stockholm, Sweden). As of writing, the current generation model of the Gamma Knife is the Icon which replaced the Perfexion model which replaced the model 4C. This work will focus on the latest two models as the 4C and previous models are being phased out. This chapter will focus in particular on the dosimetric methods used to characterize the Gamma Knife. This includes absolute dosimetry at the radiation focus and relative dosimetry techniques to further characterize the system. This chapter will focus on techniques used by the author personally as well as some techniques from the literature. Even though attempts to be inclusive will be made, there is always different and new methods being developed, and the reader is encouraged to keep reviewing the literature for new publications. For instance, at the time of writing this chapter, the American Association of Physicists in Medicine (AAPM) task group 178 is preparing for publication of a guidance document on the quality assurance

(QA) of gamma stereotactic units and will hopefully be available to the public soon.

This chapter assumes the reader will have a significant knowledge of general operation and design of the Gamma Knife. We will not include a discussion of previous Gamma Knife models nor their differences compared to the new models. We will only briefly go over the basics of the new design here. The readers are referred to several of the following excellent references for a more in-depth description of the Gamma Knife and differences between models [1, 2]. The Gamma Knife is a specialized radio surgery system that focuses nearly 200 Co-60 sources to a single point. The sources are divided evenly among eight sectors which can be driven independently from a blocked position to any of the three collimator sizes (4 mm, 8 mm, and 16 mm). The location of this focal point is correlated to stereotactic space via either a rigid frame attached to the couch or via a calibrated stereotactic cone beam computed tomography (CBCT) with the Icon model. The patient is immobilized via either a rigid frame or mask system and can be translated to the radiation focus for treatment. The size of a single radiation focus is controlled by the collimation used per sector. Each sector can use the same collimation or consist of different combinations of the collimation used in composite. A single location within the patient of the focal point is often referred to as a shot. Treatment plans can consist of multiple shots with different times to make the desired radiosurgery dose distributions. These distributions are commonly characterized by a high level of conformity and sharp gradient in dose from the prescribed dose level covering the target to the lower dose levels in the surrounding normal tissues.

This chapter will highlight the techniques used during the commissioning of the Gamma Knife. This process though complex and semi-challenging can be done in a relatively short period of time. With proper planning, this commissioning process can typically be completed by a physicist in a single week. Note this chapter focuses on the dosimetry aspects of this process. There are numerous other tests involved in the commissioning of the Gamma Knife that are outside this scope. The readers are referred to other works in the literature for the non-dosimetry commissioning tests used on the Gamma Knife [3–7]. In all sections of this chapter, the author will try to relate their own personal experience to try to help guide the reader. However, the reader should realize the end user is ultimately responsible to investigate all measurement techniques, measurement equipment, and processes themselves, regardless of anyone else's personal experiences. The users are ultimately responsible for the integrity of their own data.

22.2 SAFETY FIRST

As stated, the techniques discussed in this chapter are implemented as part of the commissioning and acceptance process after an installation or source exchange of a Gamma Knife. It is critical that before performing measurements on any machine for the first time, a full radiation safety assessment is completed. This includes a full radiation barrier survey, wipe test, confirmation and accounting of all sources, confirmation of the function of all controls and emergency off switches. Once completed, this will help ensure the machine is safe to operate for further testing.

Though outside of the scope of this chapter, radiation barrier surveys are extremely important for Gamma Knife. Unlike linear accelerators, the Gamma Knife uses cobalt-60 sources that are always emitting gamma rays. They are simply moved from a shielded position to an unshielded position for treatment. The Perfexion and Icon models have 5 potential positions starting from the full retracted position or the home position, then the 8 mm collimator exposed position, then a blocked position, then the 4 mm collimator position, and finally the 16 mm collimator position. The blocked position is used when transitioning the patient between different radiation focus positions. The user should survey the machine in both conditions; with the sources at the home position and fully exposed using the largest collimator size. The user should be sure to conduct surveys using a fast-responding detector such as Geiger counter or scintillator detector to examine shielding barriers for any potential hot spots that could be present due to imperfections in the shielding construction, etc. For the final determination of the dose rate at a survey location a large volume calibrated ionization chamber should be used. For the details on these measurement techniques the user is referred to the following references [8–10].

Also, outside of the scope of this work, before ever using the CBCT on a real patient, it must be commissioned. This includes safety, mechanical, image quality, and positional accuracy tests. This also includes the dosimetric characterization of the CBCT. Even though this is dosimetry, this work will focus on the dosimetry of the treatment beam. The readers are referred to the literature (e.g., Chapter 30) to learn about CBCT dosimetry.

A wipe test is performed to ensure the sealed sources are indeed sealed. In the United States this is done in accordance with title 10 of the code of federal regulations (CFR) part 39.35. Sources with a wipe test greater than 5 nCi must be removed from service. This wipe test can be completed with the sources in the home position and opening the shielding doors in the front of the unit. The service engineer or user can remove the collimator cover with a long suction cup tool. Once removed the user can wipe outside of this cover that was facing the collimator and replace the cover. Instructions for this procedure are provided by the manufacturer and should be followed with care as the cover has pressure sensors that can break if care is not taken in removing or replacing the cover.

Accounting for all sources is important for any Gamma Knife user. A panoramic image using radiochromic film is an easy method for accomplishing this. As described by Cho et al. [11], one can wrap a cylindrical phantom with film and irradiate it to capture an image of the source geometry. This allows for direct accounting for all sources as it may be hard to detect a single missing source out of 192 with other dosimetric methods. This can be done using all three collimator sizes.

Finally, the user should test all controls including all emergency off buttons and door interlock, to ensure they are functional and operate as intended. This ensures in the case of an emergency the user is confident on how to react and the machine will function as intended. This should be tested before each use of the Gamma Knife and is particularly important the very first time the machine is used.

22.3 CONFIRMING THE MEASUREMENT GEOMETRY

For Gamma Knife measurements one cannot visually check that the measurement geometry is sound like one can on a linear accelerator with a field light vs. radiation test. Instead, the user must rely on the correlation of the radiation focus position to the couch positions. With the Gamma Knife Icon one can use the on-board CBCT imaging to confirm this. One can also test this correlation between the couch coordinates and radiation focus position using a film holder with a pin at a fixed position relative to the frame adapter. This is often referred to as the pin prick test. Knowing physical position of the pin relative to the frame, one can irradiate the film and ensure this correlates with center of the radiation focus. This needs to be with 0.3 mm at the center of the frame adapter and within 0.5 mm at the outer positions of the frame adapter. This should be completed in all planes for each collimator. This QA tool can be scanned via CBCT to ensure the CBCT coordinates correlate with the physical positions and can be cross correlated with the radiation focus. This is critical because the detector positioning is critical for the measurement of small radiation fields used in Gamma Knife. If the setup is incorrect this can lead to large differences in the measurement. Anytime a new phantom or measurement technique is being used, the user should confirm the geometry via imaging or physical correlation to the frame/frame adapter via imaging. This can be completed with the frame and fiducial box using CT and for the Icon the phantom can be directly imaged in the measurement position with CBCT. Note that CBCT isocenter location is confirmed on a daily basis via a QA

tool provided by the manufacturer to be within 0.3 mm of the baseline set at the time of installation by the service engineer.

22.4 ABSOLUTE DOSIMETRY

In Gamma Knife, absolute dosimetry is used to determine the dose rate at the radiation focus for the 16 mm collimator at the center of a 16 cm diameter sphere. This calibration is used for all patients, therefore an error in this measurement would propagate across all patients treated. Therefore, it is critical that this measurement be double, and triple checked. Looking in the literature, a round robin study was completed by Drzymala et al. at nine different institutions, using two models of gamma knife, four different phantoms, three dosimetry protocols, and two different ionization chambers [12]. The readers can see how the level of complexity in this type of study can increase rapidly. The reader is referred to this work as it highlights many of the factors for accurate absolute dosimetry in Gamma Knife. This section will look at the individual components of the absolute dosimetry process and will often refer to this round robin study.

There are two different 16 cm diameter spherical phantoms provided by the manufacturer for calibration and will be the focus of the work. The readers are referred to the literature if they are interested in using a third party phantom such as the in air method [13] or water phantom method [14] mentioned in the round robin. Previously the manufacturer provided a gray plastic spherical phantom. Currently this has been replaced with a red colored water equivalent plastic spherical phantom. There are several benefits of the new generation phantom. The new phantom utilizes the frame adapter that is used for the patient treatments. The previous model used a separate adapter to hold the phantom in place which was known to cause just under 1% attenuation of the beam during the measurement [15]. Typically, ionization chambers ranging in volume from 0.007 cm^3 up to 0.125 cm^3 are used for calibration. A list of common ionization chamber models used are shown in Table 22.1, note this is not an all-inclusive list and the reader is encouraged to review other chambers that may be described in the literature. Either phantom has inserts that can be drilled to fit any of these different detectors.

In the United States, the requirements for absolute dosimetry measurement equipment is highlighted in title

TABLE 22.1 Common Ionization Chambers Used for Gamma Knife Absolute Dosimetry

Chamber Model	Manufacturer	Volume (cm³)
Exradin A16	Standard Imaging	0.007
Exradin A14SL	Standard Imaging	0.015
TN31010	PTW	0.125
TN31016	PTW	0.016
PR-05P	Capintec	0.07
PR-05	Capintec	0.14
CC13	IBA	0.13
CC04	IBA	0.04
CC01	IBA	0.01

10 of the code of federal regulations part 35, section 630 (10CFR35.630). Specifically, "(1) The system must have been calibrated using a system or source traceable to the National Institute of Standards and Technology (NIST) and published protocols accepted by nationally recognized bodies; or by a calibration laboratory accredited by the American Association of Physicists in Medicine (AAPM). The calibration must have been performed within the previous 2 years and after any servicing that may have affected system calibration; or (2) The system must have been calibrated within the previous 4 years. Eighteen to thirty months after that calibration, the system must have been intercompared with another dosimetry system that was calibrated within the past 24 months by NIST or by a calibration laboratory accredited by the AAPM. The results of the intercomparison must indicate that the calibration factor of the licensee's system had not changed by more than 2 percent. The licensee may not use the inter comparison result to change the calibration factor. When intercomparing dosimetry systems to be used for calibrating sealed sources for therapeutic units, the licensee shall use a comparable unit with beam attenuators or collimators, as applicable, and sources of the same radionuclide as the source used at the licensee's facility." It is the author's opinion, option 1 is preferred.

Selection of the desired ionization chamber can seem daunting. Further compounding this is the choice of phantom to use in combination with the detector of choice and getting the appropriate plugs drilled for the detector. It is however important to note that the small volume ionization chambers seem attractive due to relatively small fields being measured, however caution should be used as short term stability issues

have been reported for the Exradin A16 chamber in particular [12]. The selection of the dosimetry protocol also impacts the selection of the ionization chamber. Within the limits of what is reasonably achievable to the user, it is advisable for a user to select at least two well-characterized ionization chambers for their calibration. Using a well-established chamber is much easier than trying to fully characterize a new chamber, come up with the appropriate correction factors for the protocol of use, and validate these corrections, which is often beyond the skill set of clinical physicists. At our institution we use the average output as measured by our two relatively large volume chambers (PR-05P and TN31010) as our primary calibration and we have an A16 that can be cross-calibrated for end-to-end to dosimetry testing, but it is not used as a primary calibration due to the stability concerns raised above. Note this is just our experience with one ion chamber and certainly does not represent every A16 chamber, it just highlights the need for using multiple detectors over time to reduce the uncertainties and check for consistency.

This work will discuss two of the primary formalisms used for Gamma Knife absolute dosimetry. The first and longest used is the AAPM TG-21 formalism [16], including revisions made in 1986 [17]. This formalism has the benefit of being very generalized with very few assumptions being made in the measurement geometry. The user must know the material they are measuring in, the construction of the ion chamber, and one can measure dose fairly easily. The first time setting up this calculation requires substantial work as the user will have to look up or calculate approximately fifteen different values found in TG-21 to calculate the output. Once this is completed, the calculation of the output is straightforward. This process must be completed for each phantom and detector combination used. This level of complexity has led many users to seek alternative protocols to determine the output. The formulae below show the different components used in determining the measured dose. Note that user measured inputs are $\bar{M}$ and PDD; values that must be looked up or calculated from TG-21 are A_{wall}, α, $\left(\frac{\bar{L}}{\rho}\right)_{gas}^{wall}$, $\left(\frac{\bar{\mu}_{en}}{\rho}\right)_{wall}^{air}$, $\left(\frac{\bar{L}}{\rho}\right)_{gas}^{cap}$, $\left(\frac{\bar{\mu}_{en}}{\rho}\right)_{cap}^{air}$, $\left(\frac{\bar{L}}{\rho}\right)_{air}^{wall}$, $\left(\frac{\bar{\mu}_{en}}{\rho}\right)_{wall}^{med}$, $\left(\frac{\bar{L}}{\rho}\right)_{air}^{med}$, P_{ion}, P_{repl}, ESC, and $\left(\frac{\bar{\mu}_{en}}{\rho}\right)_{med}^{water}$; values from the ADCL calibration are N_x and A_{ion}; and $k\left(\frac{W}{e}\right)$ is a physical constant.

$$N_{Gas} =$$

$$N_x \times \frac{k\left(\dfrac{W}{e}\right) \times A_{ion} \times A_{wall} \times \beta_{wall}}{\alpha \times \left(\dfrac{\bar{L}}{\rho}\right)^{wall}_{gas} \times \left(\dfrac{\overline{\mu_{en}}}{\rho}\right)^{air}_{wall} + (1-\alpha) \times \left(\dfrac{\bar{L}}{\rho}\right)^{cap}_{gas} \times \left(\dfrac{\overline{\mu_{en}}}{\rho}\right)^{air}_{cap}}$$

$$(22.1)$$

$$P_{wall} = \frac{\left[\alpha \times \left(\dfrac{\bar{L}}{\rho}\right)^{wall}_{air} \times \left(\dfrac{\overline{\mu_{en}}}{\rho}\right)^{med}_{wall} + (1-\alpha) \times \left(\dfrac{\bar{L}}{\rho}\right)^{med}_{air}\right]}{\left(\dfrac{\bar{L}}{\rho}\right)^{med}_{air}} \quad (22.2)$$

$$D_{med} = \bar{M} \times N_{gas} \times \left(\frac{\bar{L}}{\rho}\right)^{med}_{air} \times P_{wall} \times P_{ion} \times P_{repl} \quad (22.3)$$

$$D_{water}\left(at\ d_{max}\right) = \frac{D_{med} \times ESC \times \left(\dfrac{\overline{\mu_{en}}}{\rho}\right)^{water}_{med}}{\dfrac{PDD}{100}} \quad (22.4)$$

In general, the simpler TG-51 [18] formalism has all but replaced TG-21 for typical linear accelerators. This simplicity comes at a cost of less flexibility in the TG-51 formalism. Machines such as Gamma Knife do not meet the criteria of beam quality specification or the reference measurement to use this formalism. The corrections used in the protocol can lead to errors in the calibration if applied blindly. To address these issues Alfonso et al. have proposed a new formalism to extend the TG-51 style easier formalism to non-standard beam geometries such as Gamma Knife [19]. This is the formalism used in IAEA Technical Report Series (TRS) 483 [20]. In this report, the output is tied directly to a corrected reading, the calibration factor from a standards lab, and a conversion from the reference geometry in water, to the measurement geometry in the user's measurement for a chamber. This reduces the number of values the user has to look up to one. However, these values are tabulated for very specific detectors, phantoms, geometries, and machines. Therefore, new machines, phantoms, and chambers still pose an issue to future users as these values will yet to be determined. Again, simplicity comes at a tradeoff for the loss of flexibility. This general equation is shown in Equation 22.5, in which $M^{f_{msr}}_{Q_{msr}}$ is the user measured input, $N^{f_{ref}}_{D,W,Q_o}$ is obtained from the ADCL,

and $k^{f_{msr} \cdot f_{ref}}_{Q_{msr} \cdot Q_o}$ is a single looked up value from Table 14 of TRS 483. Note f indicates field size, Q indicates beam quality, M is charge, k is the correction, N is the calibration factor from the ADCL, Q_o indicates the calibration beam quality, ref indicates the reference field for the calibration, msr indicates a machine specific reference.

$$D^{f_{msr}}_{water,Q_{msr}} = M^{f_{msr}}_{Q_{msr}} \times N^{f_{ref}}_{D,W,Q_o} \times k^{f_{msr} \cdot f_{ref}}_{Q_{msr} \cdot Q_o} \quad (22.5)$$

At our institution during the commissioning of our first Icon unit which was a simple upgrade and reload of an existing Perfexion. We elected to do both formalisms in the new and old phantoms with two chambers. Therefore, 8 measurements of the output were completed and averaged. The maximum difference between two methods was 0.6% and the maximum difference from the average was 0.3%, giving us confidence in our absolute calibration. However, no matter how confident the user is, it is highly recommended the user do an independent audit of the output. This can be completed via the Imaging and Radiation Oncology Core (IROC) at MD Anderson in Houston Texas. Independent audit stated our output was 1.01 × our institution stated output. Post the initial install at our institution the output was measured using the TRS 483 formalism with two chambers in the solid water phantom, in addition to the third party audit on a yearly basis at our institution and compared to the TPS output corrected for decay entered at the time of commissioning.

22.5 RELATIVE DOSIMETRY: OUTPUT FACTORS

After determining the output of the 16 mm collimator, the output of the 4 mm and 8 mm collimators can be determined relative the 16 mm reference. The Gamma Knife treatment planning systems comes with the default values of 1.000, 0.9005, and 0.8140 for the 16 mm, 8 mm, and 4mm collimators, respectively. These values are provided by the manufacturer and are from Monte Carlo simulation. It is the author's opinion that it is best to use measurements to confirm these values instead of using the measurements directly. These measurements are completed in the same reference conditions to the output measurement. This can be accomplished in either of the calibration phantoms discussed previously. Typically, this is done using film. Film has excellent spatial resolution and can easily be placed in either of the manufacturer provided phantoms. Radiochromic film has been widely used in Gamma Knife and reported on in the literature. The convenience of not needing a developer, being nearly tissue equivalent, being largely energy independent, and having a large dosimetric range, as compared to radiographic film has led to radiochromic film to become very popular for Gamma Knife users. As with any detector, radiochromic film is not perfect and has its own drawbacks. There is a time dependency for readout, the user must have a high-quality scanner, there have been reported nonuniformities both per batch and in an individual sheet of film. There are dependencies on the color channel being used. With care and skill however, film can be used to measure Gamma Knife dose distributions with excellent spatial resolution and precise measurements of dose. Consistency is also crucial for film dosimetry so one should take care to be consistent in the time of readout, orientation, scanning protocol, and handle and store all film with care to ensure quality measurement results. These challenges and techniques to help minimize their impacts have been described in the literature [21, 22].

In order to use radiochromic film for relative output factors, the user must irradiate films with different collimator settings for fixed time. There are several schools of thought for this measurement. One school of thought is to convert film optical density to dose via a calibration curve of known doses from no dose to beyond the maximum dose delivered to the output factor films and then directly take ratio of maximum doses for each

collimator for the fixed times to calculate the output factors. This is generally done by using a small region of interest on the scanned film. This technique requires the user to construct a calibration curve for the film first. This is time consuming and any errors (random or systematic) in this process will propagate through to the output factor measurement. To reduce the random error, one can take several measurements for each point on the calibration curve as well as multiple output factor measurements and take the average for each. To reduce the systematic errors, one should follow the best practices for film dosimetry and confirm the output of the Gamma Knife as described in the previous section. An example calibration curve is shown in Figure 22.1.

Another school of thought is to take the ratio of net optical densities (i.e., correcting for the inherent optical density of unirradiated film) for films irradiated with a fixed time for all collimators. This assumes that dose is proportional to net optical density and therefore a simple ratio will cancel this out. This is true of course for small ranges of dose as a user can see from the above-mentioned calibration curve technique (Figure 22.1). It is recommended the user does not use the ratio of net optical densities unless they have well characterized the film they are using and know with a high certainty the dose response is linear in the dose range they are using for the irradiation. The net OD can be read out with a scanner or with a densitometer. Note that Ma et al. proposed a similar method to eliminate the use of a calibration

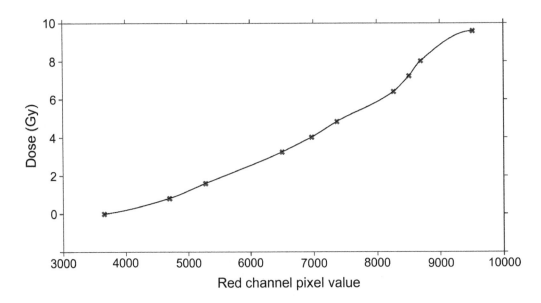

FIGURE 22.1 Example calibration curve obtained with the multiple dose level irradiations in the Gamma Knife for EBT3 Gafchromic film. Note the non-linearity in the curve fit near each end of the calibration curve using a piecewise cubic Hermite interpolating polynomial. This curve would be more useful for measuring profiles with a max dose of 6 Gy.

curve by irradiating a single film with the so-called double shot method [23]. They concluded the uncertainty associated with this assumption to be approximately 3% assuming a 1–2% random uncertainty in the film. They feel this uncertainty is an acceptable trade-off for eliminating the sheet-to-sheet variation and eliminating the need for a calibration curve. The author would point out traditional output factor measurements can typically be done on a single sheet of film anyway, but the double shot technique could be of interest to the reader.

Outside of using film, active detectors have also been proposed. These are typically small diodes, diamond or micro ionization chambers that will give the user charge (M). TRS 483 defines a measured output factor according to Equation 22.6. In which $M_{Q_{clin}}^{f_{clin}}$ and $M_{Q_{msr}}^{f_{msr}}$ are the user measured inputs and $k_{Q_{clin} \cdot Q_{msr}}^{f_{clin} \cdot f_{msr}}$ is obtained from TRS 483. Note f indicates field size, Q indicates beam quality, M is charge, k is the correction, $clin$ indicated the clinical field being measured, msr indicates a machine specific reference:

$$OF_{Q_{clin} \cdot Q_{msr}}^{f_{clin} \cdot f_{msr}} = \frac{M_{Q_{clin}}^{f_{clin}}}{M_{Q_{msr}}^{f_{msr}}} \times k_{Q_{clin} \cdot Q_{msr}}^{f_{clin} \cdot f_{msr}} \qquad (22.6)$$

Detector selection is critical for this measurement due to the small field sizes used in Gamma Knife. It becomes immediately clear that detectors that are used for the 16 mm collimator are not appropriate for use with the 4 mm and 8 mm collimators due to their relatively large size compared to the radiation field size. This is highlighted in Table 25 of TRS 483 where a list of detector correction factors $\left(k_{Q_{clin} \cdot Q_{msr}}^{f_{clin} \cdot f_{msr}} \right)$ for the different clinical field sizes are given based off of the work by Benmakhlouf et al. [24]. Note in TRS 483 there are only five detectors listed as appropriate for measuring all three collimator sizes used in the Gamma Knife. One should also note the correction factors listed become large, on the order of 5%, for the shielded diode. Corrections larger than this should indicate to the user that detector is most likely not suitable for this measurement and should not be used as the sole measurement device. Commonly, the A16 is proposed for use of measurement of the relative output factors. However, due to its relatively large size (1.7 mm length and 2.4 mm diameter) it will have corrections for not only the chamber components (wall, electrode, etc.) but also partial volume averaging. These potential corrections are not given for Gamma Knife in the report, however, for a CyberKnife 6 mm cone the corrections are on the order of 5%.

Just as volume averaging effects are critical in this measurement, geometric accuracy is as well. If the detector holder being used is off even by 1–2 mm, an error in the 4 mm output factor measurement could be well over 10%. Therefore it is highly recommended to confirm this measurement geometry via imaging. This can be done via CT with the fiducial markers or CBCT on the machine. This was described previously but the author would like to re-emphasize this point due to its importance. At our institution we used radiochromic film and a synthetic diamond detector to confirm the output factors at time of installation and during the annual QA. Example CBCT images of measurement geometry is shown in Figure 22.2.

22.6 RELATIVE DOSIMETRY: OFF-AXIS DISTRIBUTIONS

Along with characterizing the output of the Gamma Knife and the relative outputs of the different collimator sizes, the characterization of the radiation profiles is critical. This can and should be done in all planes of the Gamma Knife (Axial, Coronal, Sagittal). The profile data in the TPS come from Monte Carlo simulations and can easily be extracted for comparison to measurements. Looking at the profiles for different collimator settings one will note the very fast fall off. Typically, Gamma Knife plans are prescribed to the 50% isodose to take advantage of this fast fall off. This is of course highly desirable but demands precision to avoid underdosing the target or potentially over dosing OARs.

Much attention has been given to the output and the relative output factors in this chapter, however in radiosurgery the minimal peripheral dose to the target is often just as or more important than the maximum dose delivered to the tumor. This is directly governed by the off-axis distributions of the shots used. Imagine, the positioning errors are zero, however the profile width as defined by the full-width at half-maximum (FWHM) is incorrect by 2 mm. This could mean your prescription dose size is incorrect by 2 mm. In the radiosurgery this difference could lead to underdosing the target if the FHWM is smaller than modeled. Conversely if the FWHM is larger than modeled, one could overdose an OAR. Therefore, it is critical to confirm this FHWM is within 1 mm of the TPS value. The FWHM is not configurable in the Gamma Knife TPS given the fixed geometry of the Gamma Knife. If the user's measurement (assuming there was no measurement error) doesn't match the

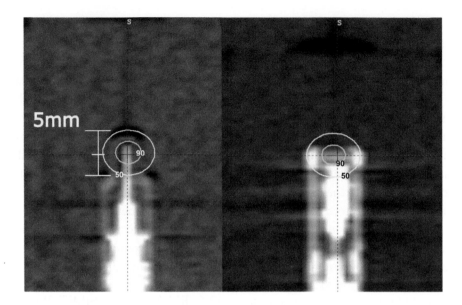

FIGURE 22.2   Examples of confirming the output factor measurement utilizing CBCT. Note position of the shot was found via CBCT. Notice the relative size of the microchamber (left) and synthetic diamond (right) compared to the 4 mm shot. The microchamber is very large compared to the 4 mm shot and the synthetic diamond detector is more appropriate as the amount of volume averaging is reduced due to its size.

expected values this would indicate a manufacturing or installation error which would need to be corrected. Since the Gamma Knife prescription isodose dose does not have to be 50%, and it can be varied significantly, the user should also confirm the 80–20% penumbra of the profiles. In fact the best practice is just to confirm the entire off axis distribution via either dose difference and distance-to-agreement metrics, or using a 1D Gamma test. The user should keep in mind the appropriate metrics for these tests. The user's measurement should be within 3% dose difference or 1 mm distance to agreement. Note that the 1D Gamma test is slightly less strict as it allows for searching the minimum combination of dose difference and distance to agreement for all points on the curve. This does not mean it shouldn't be used, this is simply a word of caution to the reader.

In order to measure the off-axis distribution, the user is mostly limited to using film. The use of 3D gel dosimetry systems have been described in the literature, however, this has yet to see wide spread adoption clinically. Currently, there are no other ways to measure off axis profiles in Gamma Knife. Since film dosimetry is by far the most common method used at the time of writing this chapter, it will be the main focus here. Measuring off axis distributions in Gamma Knife with film is done by constructing a calibration curve of net optical density and dose, as described in the relative output factors section of this chapter and applying it to films irradiated with

the desired collimation and orientations. Again, it can be tempting to just take a profile of optical density from the film and used this as your off-axis ratio. However, caution should be used as this assumes the net optical density is directly proportional to dose. This can be true over a range of doses for radiochromic film; however, it is certainly not true over all ranges. Therefore, one should be sure of the characterization of the film they are using in the dose ranges they are trying to measure.

Film measurements can be noisy. It is tempting to smooth these images using various smoothing kernels. The user should remember that this will inherently limit the spatial resolution of the film, which is needed to characterize the small radiosurgery fields used in Gamma Knife. Therefore, to improve and reduce this random noise in the measurement, the author recommends limiting smoothing techniques and instead using repeated measurements to average out this statistical noise. Example profile measurements compared to the TPS profiles are shown in Figure 22.3.

22.7   PUTTING IT ALL TOGETHER: END-TO-END TESTING

The final portion of this chapter will discuss putting all these principles together in an end-to-end test. This involves taking a phantom and putting it through the entire Gamma Knife treatment process from simulation through to treatment. There are several phantoms that are

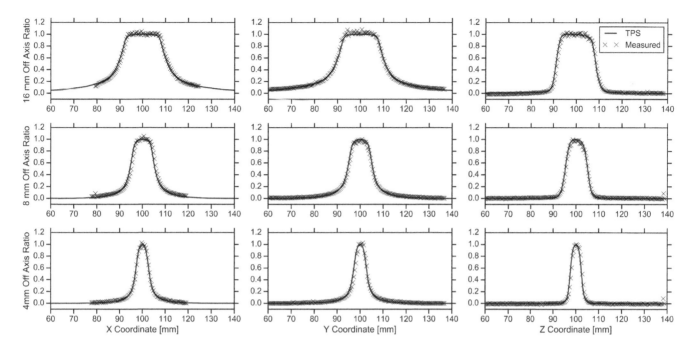

FIGURE 22.3 X, Y, and Z profiles taken with film for 4 mm, 8 mm, and 16 mm collimator compared to the TPS. Note every 4th film data point is plotted for clarity. Note the statistical noise seen in the film measurements.

commercially available that allow the user to end-to-end test their process. This allows the user to simulate, plan, treat and measure the dose delivered via film, gel, or with an active detector. The phantom should be treated just as a patient and go through the entire clinical workflow. The resultant measurements should be completed to confirm both the absolute and relative dosimetry delivered. This should be done over a range of case types that are seen in the clinic, ranging from very simple single shot dose distributions to complex plans consisting of multiple composite shots in an anthropomorphic phantom. Note that these phantoms can make excellent training tools for all parties involved in the Gamma Knife process. This allows the users to work through the processes before trying to treat a patient. An example phantom with different inserts are shown in Figure 22.4.

At the author's institution an end-to-end test is done on an annual basis using both an active detector (using an A16 micro chamber or synthetic diamond detector) and radiochromic film for both a simple spherical target and a complex organic shaped target. For this test we use a 5% dose difference 1 mm distance-to-agreement criteria. Typically, the simpler plans can be measured within 3% or 1 mm distance-to-agreement. Note that during complex target delivery the user should be aware of potential leakage currents in the detector. This can be particularly troublesome if the plan contains numerous couch motions in

the delivery of a complex treatment plan. At our institution, it was not immediately clear why our end-to-end test ion chamber measurements were systemically lower than expected for the complex plans and not for a simple single shot plan. After close inspection it was clear that between shots during couch motions the total charge reading was decreasing due to leakage causing a systematic error in

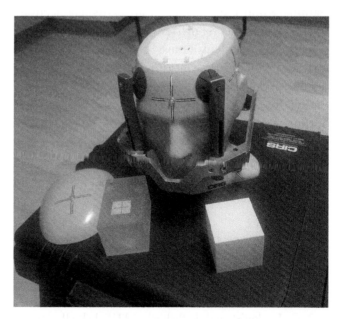

FIGURE 22.4 An example end-to-end phantom in a frame used with multiple detector and target inserts.

the measurement. This was confirmed with the measurement using a second chamber and the original chamber was sent out for repair.

Note this same process can be repeated using an independent third-party phantom (IROC in the United States) that is planned and irradiated by the user but the dosimetry is read out by the third party. This service is an excellent independent test for the user and is highly recommended. This is also required to enroll patients on to many clinical trials. Their phantom uses a combination of film and thermoluminescent dosimeters (TLDs) to report the measured dose in comparison to the planned dose to the user. They use a 1D Gamma comparison on the film profiles and a simple dose difference for the TLD measurements. The criteria to pass this test is currently set to five percent for the point measurements and greater than or equal to 85% of points having a gamma value less than one with a five percent dose difference and three millimeter distance to agreement criteria. In the author's opinion, this is a relatively lax criteria and represents a very minimal requirement a user should be able to meet as Molineu et al. reported only a single failure to meet this criterion in a relatively small subset of 23 irradiations using Gamma Knife [25]. Interestingly the pass rates across all modalities was lower at 81%. At the authors institution this was completed for both frame and frameless SRS modalities.

22.8 CONCLUSIONS

Gamma Knife radiosurgery is an extremely accurate treatment modality for delivering very focused radiation fields. Like any system, it must be validated prior to ever treating a patient. These measurements can be challenging due to the small radiation fields used, but with care the system can be validated in relatively short amount of time to a high degree of accuracy. If the reader is to take anything away from this chapter, please remember, proper planning prevents poor performance. Never rely on a single measurement and always get an independent confirmation of your work!

REFERENCES

1. J. P. Bhatnagar, J. Novotny, M. Saiful Huq, "Dosimetric characteristics and quality control tests for the collimator sectors of the Leksell Gamma Knife® Perfexion™," *Medical Physics* **39**(1), 231–236 (2012).
2. C. Lindquist and I. Paddick, "THE Leksell Gamma Knife Perfexion and comparisons with it's predecessors," *Neurosurgery* **61**(September), 130–141 (2007).
3. N. C. Knutson, et al., "Characterization and validation of an intra-fraction motion management system for masked-based radiosurgery," *Journal of Applied Clinical Medical Physics* **20**(5), 21–26 (2019).
4. I. Aldahlawi, D. Prasad, M. B. Podgorsak, "Evaluation of stability of stereotactic space defined by cone-beam CT for the Leksell Gamma Knife Icon," *Journal of Applied Clinical Medical Physics* **18**(3), 67–72 (2017).
5. W. Li, et al., "Preliminary evaluation of a novel thermoplastic mask system with intra-fraction motion monitoring for future use with image-guided Gamma Knife," *Curēus* **8**(3), e531 (2016).
6. A. Sarfehnia, et al., "Performance characterization of an integrated cone-beam CT system for dedicated gamma radiosurgery," *Medical Physics* **45**(9), 4179–4190 (2018).
7. M. Zeverino, et al., "Commissioning of the Leksell Gamma Knife® Icon™," *Medical Physics* **44**(2), 355–363 (2017).
8. National Council on Radiation Protection and Measurement, *NCRP Report no. 151 Structural Shielding Design and Evaluation for Megavoltage X- and Gamma-Ray Radiotherapy Facilities* (2005).
9. P. H. McGinley, *Shielding Techniques for Radiation Oncology Facilities* (Medical Physics, Madison, WI, 2002).
10. J. K. Shultis, R. E. Faw, K. R. Kase, *Radiation Shielding* (Prentice Hall PTR, Upper Saddle River, NJ, 1996).
11. Y. Bin Cho, M. Van Prooijen, D. A. Jaffray, M. K. Islam, "Verification of source and collimator configuration for Gamma Knife® Perfexion™ using panoramic imaging," *Medical Physics* **37**(3), 1325–1331 (2010).
12. R. E. Drzymala, et al., "A round-robin gamma stereotactic radiosurgery dosimetry interinstitution comparison of calibration protocols," *Medical Physics* **42**(11), 6745–6756 (2015).
13. S. G. Meltsner and L. A. DeWerd, "Air kerma based dosimetry calibration for the Leksell Gamma Knife," *Medical Physics* **36**(2), 339–350 (2009).
14. R. E. Drzymala, R. C. Wood, J. Levy, "Calibration of the Gamma Knife using a new phantom following the AAPM TG51 and TG21 protocols," *Medical Physics* **35**(2), 514–521 (2008).
15. J. P. Bhatnagar, J. Novotny, M. A. Quader, G. Bednarz, M. S. Huq, "Unintended attenuation in the Leksell Gamma Knife Perfexion calibration-phantom adaptor and its effect on dose calibration," *Medical Physics* **36**(4), 1208–1211 (2009).
16. Task Group 21, Radiation Therapy Committee, AAPM, "A protocol for the determination of absorbed dose from high-energy photon and electron beams," *Medical Physics* **10**(6), 741–771 (1983).
17. R. J. Schulz, et al., "Clarification of the AAPM Task Group 21 protocol," *Medical Physics* **13**(5), 755–759 (1986).
18. P. R. Almond, et al., "AAPM's TG-51 protocol for clinical reference dosimetry of high-energy photon and electron beams," *Medical Physics* **26**(9), 1847–1870 (1999).
19. R. Alfonso, et al., "A new formalism for reference dosimetry of small and nonstandard fields," *Medical Physics* **35**(11), 5179–5186 (2008).

20. IAEA, *Dosimetry of Small Static Fields Used in External Beam Radiotherapy* (2017).

21. S. Pai, et al., "TG-69: Radiographic film for megavoltage beam dosimetry," *Medical Physics* **34**(6), 2228–2258 (2007).

22. B. M. Coursey, et al., *Radiochromic Film Dosimetry Radiochromic film dosimetry: Recommendations of AAPM Radiation Therapy Committee Task Group 55* (1998).

23. L. Ma, P. Kjäll, J. Novotny, H. Nordström, J. Johansson, L. Verhey, "A simple and effective method for validation and measurement of collimator output factors for Leksell Gamma Knife Perfexion™," *Physics in Medicine and Biology* **54**(12), 3897–3907 (2009).

24. H. Benmakhlouf, J. Johansson, I. Paddick, P. Andreo, "Monte Carlo calculated and experimentally determined output correction factors for small field detectors in Leksell Gamma Knife Perfexion beams," *Physics in Medicine and Biology* **60**(10), 3959–3973 (2015).

25. A. Molineu, S. Kry, P. Alvarez, N. Hernandez, T. Nguyen, D. Followill, "SU-G-TeP2-12: IROCHouston and MDAPL SRS anthropomorphic phantom results," *Medical Physics* **43**(6 Part 27), 3665 (2016).

CHAPTER 23

Kilovoltage X-Ray Beam Dosimetry

C.-M. Charlie Ma

Fox Chase Cancer Center
Philadelphia, Pennsylvania

CONTENTS

23.1 REVIEW OF KILOVOLTAGE X-RAY DOSIMETRY

Kilovoltage x-ray radiation has been used for medical applications including imaging and cancer therapy ever since Roentgen discovered x-rays in 1895 [1, 2]. X-rays were first used by Emil Herman Grubbe to treat a patient with breast cancer, only one year after the discovery of x-rays [3]. Although the use of kilovoltage x-ray beams for external beam radiation therapy decreased drastically with the development of ^{60}Co teletherapy systems and medical linear accelerators that generate megavoltage electron and photon beams, kilovoltage x-ray therapy is still a viable option for treating superficial targets such as skin cancers and for intraoperative radiation therapy [4–6]. Kilovoltage x-ray imaging has received widespread applications both for diagnostic purposes

and for patient setup and target localization in radiotherapy treatment [7, 8]. On the other hand, kilovoltage x-rays are widely used in radiobiology studies for *in vitro* cell irradiation and *in vivo* animal experiment [9–11].

The dosimetric and biological effects of kilovoltage x-ray beams were investigated in many ways from the early simple visual observations of skin reddening, to photographic or fluorescence methods to measure x-ray intensities, and to primitive chemical and thermal measurements of x-ray absorption. The introduction of the roentgen in 1928 at Stockholm Congress of Radiology marked the beginning of precise physical measurement of the radiation exposure, which was later re-defined at the Chicago Congress of radiology in 1937 for both x- and γ-rays. The device used to measure exposure in accordance with the definition of roentgen was a "free-air" chamber, which has been used in almost all the standards laboratories to establish exposure standards. A free-air chamber must be large enough so that the secondary electrons that produce the ionization can originate and complete their tracks in the air of the chamber. Initial measurements between the standards chambers of several national laboratories showed good agreement (±1%) in the energy range of 100–180 kV as reported by Taylor [12] but were later demonstrated to be untrue; Errors of the order of 1.5% and 2.5% were reported 20 years later [13]. American physicists especially those at the NIST made major contributions to the standardization of x-ray measurements using free-air chambers [14].

Dosimetry protocols play an essential role in the accurate delivery of radiation therapy dose clinically. A survey [15] carried out by the AAPM Task Group 61 (TG61) showed that in North America a variety of dosimetry procedures were used in practice with a combination of conversion and correction factors measured and/or taken from different protocols. The National Council on Radiation Protection and Measurements (NCRP) Report No. 69 [16] provided a formula to calculate the dose to a phantom material at a point in air (with a proper buildup) for tube potential 10 kV through the medium-energy range. Based on NCRP 69 a backscatter factor will be needed to calculate the dose on the phantom surface for clinical dose determination but NCRP 69 did not provide the values of this factor. Therefore, some hospitals adopted the ICRU Report No. 23 [17], which recommended a backscatter method for the low-energy (40–150 kV) range with the backscatter factors taken from BJR suppl. 11 [18], and an in-phantom method for medium-energy (150–300 kV) x-rays. The in-phantom method was for the first time introduced to dose measurement for medium-energy x-rays, which at that time was still widely used for treating deep-seated tumors (therefore the name orthovoltage radiotherapy). The IAEA published a technical report series No. 277 [19] "Absorbed Dose Determination in Photon and Electron Beams; An International Code of Practice," which also recommended the backscatter method for the low-energy range and the in-phantom method for the medium-energy range although the energy ranges were defined slightly different (low-energy: 10–100 kV; medium-energy: 100–300 kV). The backscatter factors for the low-energy formalism were calculated using the Monte Carlo method while the introduction of a chamber related perturbation correction factor for the medium-energy formalism, which was up to 10% at the low-energy end, created some controversies and was later revised. This triggered many studies in the late 1980s and early 1990s on the various aspects of kilovoltage x-ray dosimetry (c.f., Ma and Seuntjens [5]) and several new dosimetry protocols were published [20–22].

The AAPM TG61 was formed in 1994 to evaluate the situation and to produce a dosimetry guide for the users in North America. The TG61 report "AAPM protocol for 40-300 kV x-ray beam dosimetry in radiation therapy and radiobiology" is the only AAPM protocol on kilovoltage x-ray dosimetry [23]. It provided detailed recommendations for reference as well as relative dosimetry. It provided more complete date for dose conversion coefficients, backscatter factors and chamber correction factors, and it also provided guidelines for clinical dosimetry and quality assurance (QA) measurements. Different from previous dosimetry protocols, TG61 recommended the backscatter method for the low-energy (up to 150 kV) range and both the backscatter method and the in-phantom method for the medium-energy (100–300 kV) range. Investigators have compared the dosimetry consistency between the two dosimetry methods for the overlapping x-ray energies and the results showed good agreement for both methods using the TG61 formalism and dosimetric data [24]. No major errors have been reported since the publication of the TG61 report in 2001 [8].

New dose determination methods have been investigated since then to determine the dose to water in the NIST-matched x-ray beams at the University of Wisconsin Accredited Dosimetry Calibration Laboratory (ADCL), e.g., the Monte Carlo-based method and the ^{60}Co-based method [25]. The ^{60}Co method used the absorbed dose-to-water calibration of an ionization chamber acquired from an ADCL and accounted for differences in beam quality with a beam quality correction factor (k_Q), which is analogous to the procedure that is outlined in AAPM TG51 [26]. Both methods showed excellent agreement with the TG61 method although their estimated uncertainty has been much reduced compared to TG61. For example, the Monte Carlo method and the TG61 in-phantom method agreed within 1.8% for all energies, and it also agreed with the ^{60}Co method within 1.8% for all of the beams. The TG61 in-phantom method and the ^{60}Co method showed a maximum discrepancy of 1.0%.

Their results also compare favorably with calorimetry-based standards established at other institutions [27]. Based on a thorough review of the dosimetry protocols currently available, the Kilovoltage Dosimetry Working Group of the Australasian College of Physical Scientists and Engineers in Medicine (ACPSEM) Radiation Oncology Specialty Group (ROSG) recommended the use of the AAPM TG61 as the reference dosimetry code of practice [28].

This chapter suggests the use of AAPM TG 61 for kilovoltage x-ray reference dosimetry for now because it is based on air-kerma standards, which are readily available in most countries and regions, and because it currently contains the widest range of tabulated data with respect to beam characteristics and measurement detectors. The in-air and in-phantom methods, as recommended by TG61, are easy to use and produce accurate results compared with other recent methods that are only available to standards laboratories and a few advanced centers [25, 27, 28]. The TG61 protocol also has a comprehensive discussion on issues related to relative dosimetry, which are useful to the users in the clinical implementation of an x-ray unit. Older reference dosimetry protocols or codes of practice that are based on exposure measurements should not be used because they may contain erroneous data [8]. The measurement methods and dosimetric procedures described in this chapter are primarily based on the recommendations of the TG61 report [23].

23.2 KILOVOLTAGE X-RAY DOSIMETRY PROTOCOL

Kilovoltage x-rays have some unique properties that require special attention in their applications in radiation therapy and radiobiological applications. For example, clinically available 50–300 kV x-rays have a very small electron range (<0.5 mm of water). Their dose distributions include a significant (up to 30%) scatter component, which is energy and field size dependent. Because of the negligible radiative energy loss for this energy range the collision kerma is generally considered to be the same as kerma and the absorbed dose. Commonly used ionization chambers have been generally calibrated as "exposure meters" and used as "photon detectors" since the well-known Bragg–Gray cavity theory no longer applies to this energy range [29]. The following sections describe the AAPM TG61 dosimetry

protocol and the terminology used here is consistent with that defined in the TG61 protocol [23].

23.2.1 Dosimeter Calibration

The TG61 dosimetry formalism is based on the air kerma concept [23]. Air kerma, K_{air}, is defined as the ratio of the kinetic energy, ΔE, transferred from x-rays to charged particles to a given air mass, Δm (i.e., $K_{air} = \Delta E/\Delta m$). If K_{air} is the air kerma at the reference point in air for a given beam quality and M the reading (corrected for temperature, pressure, recombination and polarity effect) of an ionization chamber to be calibrated with its center at the same point, the air kerma calibration factor, N_K, for this chamber at the specified beam quality is defined as:

$$N_K = \frac{K_{air}}{M} \qquad (23.1)$$

Many earlier dosimetry protocols for kilovoltage x-rays were based on the exposure, X, which is defined as the total electric charge, ΔQ, of all ions of one sign (e.g., electrons), produced in air by x-rays in a unit mass of air, Δm (i.e., $X = \Delta Q/\Delta m$). A similar equation can be used to derive the exposure calibration factor, N_X:

$$N_X = \frac{X}{M} \qquad (23.2)$$

We can derive the air kerma calibration factor from the exposure calibration factor from

$$N_K = N_X \left(\frac{W}{e}\right)_{air} (1-g)^{-1} \qquad (23.3)$$

where $\left(\frac{W}{e}\right)_{air}$ has the value 33.97 J/C (or 0.876 cGy per roentgen) for dry air, $(1 - g)$ corrects for the effect of radiative losses by charged particles to bremsstrahlung photons (g is the fractional energy lost to bremsstrahlung photons in air, which is practically zero for x-ray beams below 300 kV).

Both N_K and N_X factors can be obtained from standards laboratories for a number of available x-ray beam qualities to match the actual x-ray beams for an x-ray unit. In the calculation of dose to a patient for a kilovoltage x-ray beam, the accuracy of the chamber calibration determines the final accuracy (see uncertainty analysis in the next section). Each chamber involved in reference dosimetry of the clinical kilovoltage setup must be calibrated, since there can be up to 8% differences in the calibration factor for different chambers of the same type. The chamber must be calibrated for all the beam qualities available for the x-ray unit to minimize the effect of the chamber energy dependence. Well-designed chambers usually exhibit less than 6% energy dependence between 40 kV and 300 kV.

23.2.2 Formalisms for Kilovoltage X-ray Dosimetry

Two major methods have been investigated extensively for kilovoltage x-ray dosimetry determination, i.e., the in-air method and the in-phantom method. The AAPM TG61 recommends the use of the in-air method for low-energy (50–100 kV) x-ray dosimetry for superficial radiotherapy applications. For medium-energy (100–300 kV) x-ray dosimetry determination the TG61 recommends to use either the in-air or the in-phantom methods depending on the application [23]. This recommendation was made based on the fact that the determination of the surface dose can be more accurate using the backscatter method if the point of interest is the skin surface. On the other hand, if the point of interest is at a deeper depth it is more accurate to use the in-phantom method to determine the dose. The uncertainty would be much greater if one derived the dose at a depth based on a surface dose calibration (i.e., using the in-air method) because of the large uncertainties in the depth dose values near the phantom surface, and vice versa.

The In-Air Method: This method can be used for both superficial and orthovoltage x-ray dose determination. In this method the ion chamber is calibrated free in air in a standards laboratory, and used to measure air kerma free in air in a user's beam without the presence of a water phantom. A backscatter factor is needed to account for the effect of photon backscatter from the phantom (Figure 23.1a). The dose to water at the phantom surface can be calculated by

$$D_w = M N_K \left(\frac{\bar{\mu}_{en}}{\rho} \right)_{air}^{w} B_w P_{stem,air} \qquad (23.4)$$

where M is the corrected ionization chamber reading free in air, N_K the chamber air kerma calibration factor at the user's beam quality, $\left(\frac{\bar{\mu}_{en}}{\rho} \right)_{air}^{w}$ the ratio of mass energy absorption coefficients for water to air, evaluated over the x-ray energy spectrum free-in-air, in the absence

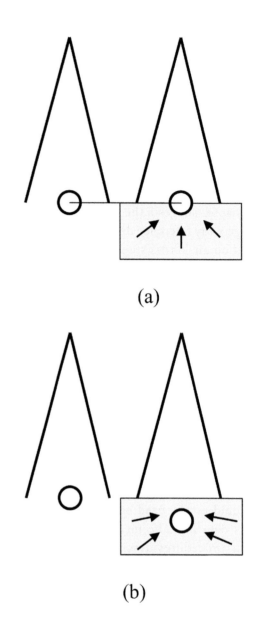

FIGURE 23.1 Schematic diagrams to illustrate (a) the in-air method and (b) the in-phantom method.

of a phantom, B_w the backscatter factor that accounts for the effect of the phantom (water) scatter, and $P_{stem,air}$ the chamber stem correction factor that accounts for the change in photon scatter from the chamber stem between the calibration and measurement due to the difference in field size. Here $\left(\frac{\bar{\mu}_{en}}{\rho} \right)_{air}^{w}$ is independent of the field size used since it is evaluated over the primary beam only [30]. The TG61 recommended values of $\left(\frac{\bar{\mu}_{en}}{\rho} \right)_{air}^{w}$ and B_w for some typical beam qualities are given in Tables 23.1 and 23.2 [23]. $P_{stem,air}$ is taken as unity for a calibrated chamber if, for a given beam quality, the same field size is used in the calibration and the measurement. Otherwise, the $P_{stem,air}$ factor can be measured

TABLE 23.1 Ratios of Average Mass Energy Absorption Coefficients for Water to Air to Convert Air Kerma to Water Kerma, Free-In-Air for Both Superficial and Orthovoltage X-Rays

HVL		$\left(\frac{\bar{\mu}_{en}}{\rho}\right)^{w}_{air}$ (free-in-air)
(mm Al)	(mm Cu)	
0.03		1.047
0.05		1.046
0.08		1.044
0.10		1.044
0.3		1.035
0.5		1.028
0.8		1.022
1.0		1.020
3.0		1.021
5.0		1.029
8.0		1.045
	0.1	1.020
	0.3	1.035
	0.5	1.050
	0.8	1.068
	1.0	1.076
	3.0	1.100
	5.0	1.109

by intercomparing the chamber with unknown $P_{stem,air}$ with a reference chamber for which $P_{stem,air}$ is known. A Farmer type cylindrical chamber with flat response can be used as a reference chamber for the measurement of $P_{stem,air}$ of another chamber since its stem effect varies little with field size (<1%).

The In-Phantom Method: For orthovoltage x-rays, TG61 recommends the in-phantom method if the point of interest is at a depth in the patient [23]. In this method, the chamber is calibrated free in air in a standards laboratory and used to measure the air kerma at the reference depth in water (Figure 23.1b). Assuming the reference depth is 2 cm in water the dose to water for orthovoltage x-rays can be calculated using

$$D_w = MN_K P_{Q,cham} P_{sheath} \left(\frac{\bar{\mu}_{en}}{\rho}\right)^{w}_{air} \quad (23.5)$$

where M is the corrected ionization chamber reading in water, N_K the chamber air kerma calibration factor at the user's beam quality, and $\left(\frac{\bar{\mu}_{en}}{\rho}\right)^{w}_{air}$ the ratio of mass

TABLE 23.2A Water Kerma-Based Backscatter Factors, B_w, for Different Field Diameters (d) and SSD for Superficial X-Rays

SSD (cm)	d (cm)	HVL (mm Al)										
		0.03	0.05	0.08	0.1	0.3	0.5	0.8	1.0	3.0	5.0	8.0
20	1	1.005	1.007	1.011	1.014	1.028	1.036	1.043	1.046	1.061	1.058	1.053
	3	1.005	1.008	1.014	1.019	1.049	1.069	1.092	1.105	1.158	1.165	1.158
	5	1.005	1.008	1.014	1.019	1.054	1.080	1.112	1.131	1.215	1.234	1.236
	10	1.005	1.008	1.014	1.019	1.057	1.088	1.129	1.155	1.291	1.334	1.354
	15	1.006	1.008	1.014	1.019	1.058	1.090	1.133	1.162	1.321	1.380	1.414
	20	1.006	1.008	1.014	1.019	1.058	1.091	1.136	1.165	1.334	1.402	1.444
30	1	1.005	1.007	1.011	1.015	1.027	1.035	1.043	1.047	1.063	1.059	1.053
	3	1.005	1.008	1.014	1.019	1.048	1.069	1.093	1.107	1.164	1.168	1.158
	5	1.005	1.008	1.014	1.019	1.053	1.079	1.111	1.130	1.221	1.242	1.237
	10	1.006	1.008	1.014	1.019	1.057	1.088	1.130	1.157	1.298	1.350	1.367
	15	1.006	1.008	1.014	1.019	1.058	1.091	1.136	1.165	1.332	1.403	1.434
	20	1.006	1.008	1.014	1.019	1.058	1.091	1.138	1.169	1.350	1.428	1.472
50	1	1.005	1.007	1.011	1.014	1.027	1.035	1.042	1.045	1.065	1.059	1.052
	3	1.005	1.007	1.013	1.018	1.049	1.070	1.093	1.106	1.163	1.169	1.160
	5	1.005	1.007	1.013	1.018	1.054	1.081	1.113	1.132	1.226	1.241	1.242
	10	1.006	1.007	1.013	1.018	1.057	1.091	1.134	1.159	1.309	1.352	1.375
	15	1.006	1.007	1.013	1.018	1.058	1.093	1.140	1.169	1.346	1.411	1.448
	20	1.006	1.007	1.013	1.018	1.058	1.094	1.142	1.173	1.363	1.443	1.493
100	1	1.005	1.007	1.011	1.014	1.028	1.036	1.042	1.044	1.062	1.059	1.053
	3	1.005	1.008	1.014	1.019	1.050	1.070	1.092	1.104	1.163	1.169	1.162
	5	1.006	1.008	1.014	1.019	1.055	1.082	1.113	1.131	1.225	1.240	1.243
	10	1.006	1.008	1.014	1.019	1.058	1.091	1.134	1.158	1.311	1.351	1.381
	15	1.006	1.008	1.014	1.019	1.059	1.094	1.140	1.169	1.354	1.417	1.460
	20	1.006	1.008	1.014	1.019	1.059	1.095	1.143	1.172	1.375	1.451	1.508

TABLE 23.2B (continued). Water Kerma-Based Backscatter Factors, B_w, for Different Field Diameters (d) and SSD for Orthovoltage X-Rays

SSD	d	HVL (mm Cu)						
(cm)	(cm)	0.1	0.3	0.5	0.8	1.0	3.0	5.0
20	1	1.061	1.055	1.053	1.048	1.045	1.024	1.018
	3	1.158	1.168	1.155	1.147	1.140	1.082	1.057
	5	1.214	1.242	1.233	1.219	1.209	1.127	1.088
	10	1.290	1.352	1.353	1.339	1.326	1.204	1.141
	15	1.320	1.407	1.415	1.403	1.389	1.251	1.174
	20	1.333	1.434	1.447	1.436	1.421	1.278	1.194
30	1	1.063	1.056	1.052	1.047	1.044	1.024	1.018
	3	1.164	1.169	1.155	1.146	1.139	1.084	1.055
	5	1.220	1.242	1.235	1.221	1.211	1.130	1.087
	10	1.297	1.363	1.367	1.347	1.332	1.214	1.147
	15	1.330	1.417	1.438	1.422	1.405	1.270	1.189
	20	1.348	1.446	1.478	1.464	1.446	1.302	1.213
50	1	1.065	1.054	1.052	1.047	1.045	1.025	1.018
	3	1.163	1.170	1.157	1.148	1.140	1.084	1.057
	5	1.225	1.247	1.240	1.226	1.214	1.131	1.089
	10	1.308	1.367	1.376	1.360	1.344	1.222	1.152
	15	1.345	1.433	1.452	1.446	1.428	1.285	1.195
	20	1.361	1.471	1.499	1.495	1.478	1.325	1.226
100	1	1.062	1.055	1.052	1.047	1.045	1.025	1.018
	3	1.163	1.170	1.160	1.150	1.142	1.085	1.057
	5	1.224	1.245	1.241	1.227	1.217	1.132	1.090
	10	1.310	1.370	1.383	1.369	1.353	1.226	1.155
	15	1.353	1.447	1.463	1.458	1.441	1.291	1.204
	20	1.373	1.490	1.513	1.516	1.499	1.334	1.237

energy absorption coefficients for water to air, evaluated over the x-ray energy fluence at 2 cm depth in water, in the absence of the chamber. The overall chamber correction factor $P_{Q,cham}$ accounts for the effect of the change in chamber response due to photon energy and angular variation between chamber calibration in air and measurement in water, the effect of chamber stem between calibration in air and measurement in water and the effect of displacement of water by the chamber in the measurement in water. The sheath correction factor P_{sheath} is needed when a waterproofing sheath is used, which is not directly related to the individual chamber type. The values of $\left(\frac{\bar{\mu}_{en}}{\rho}\right)_{air}^{w}$, $P_{Q,cham}$, and P_{sheath} for some typical beam qualities are given in Tables 23.3–23.5 [23].

23.3 DOSIMETRIC CONSIDERATIONS FOR KILOVOLTAGE X-RAYS

23.3.1 Detectors and Equipment

Dosimeters: TG61 recommends the use of air-filled ionization chambers for kilovoltage x-ray beam reference dosimetry [23]. Measurements are performed either free in air (in cases where the surface dose is the primary concern) or at a 2 cm depth in water (dose at greater depths is the primary concern). Cylindrical chambers that have a calibration factor varying with the beam quality by less than 3% in this energy range are recommended for the beam output measurement. For superficial x-rays the in-air method is recommended. Cylindrical chambers that have a calibration factor

TABLE 23.3 Ratio of Average Mass Energy Absorption Coefficients for Water to Air at 2 cm Depth in Water for a 100 cm² Field Defined at 50 cm SSD for Orthovoltage X-Rays

HVL		$\left(\frac{\bar{\mu}_{en}}{\rho}\right)_{air}^{w}$
(mm Cu)	(mm Al)	
0.1	2.9	1.026
0.3	6.3	1.037
0.5	8.5	1.046
0.8	10.8	1.055
1.0	12.0	1.060
2.0	15.8	1.081
3.0	17.9	1.094
4.0	19.3	1.101
5.0	20.3	1.105

TABLE 23.4 Overall Chamber Correction Factors $P_{Q,cham}$ for Commonly Used Cylindrical Chambers in Orthovoltage X-Ray Beams

Chamber Type	NE2571	Capintec PR06C	PTW N30001	NE2611 or NE2561
HVL (mm Cu)				
0.10	1.008	0.992	1.004	0.995
0.30	1.023	1.008	1.021	1.017
0.50	1.025	1.010	1.023	1.019
0.80	1.024	1.010	1.022	1.018
1.0	1.023	1.010	1.021	1.017
2.0	1.016	1.007	1.015	1.011
3.0	1.009	1.005	1.010	1.006
4.0	1.004	1.003	1.006	1.003
5.0	1.002	1.001	1.002	1.001

The data applies to the in-phantom method for a chamber at 2 cm depth in water and a 100 cm² field.

varying with the beam quality by less than 5% between 50 and 150 kV are recommended. Extensive studies have been carried out on the correction factors for the commonly used Farmer-type chambers for the in-phantom measurement [5, 31, 32–35]. Other cylindrical chambers may also be used. However, correction factors must then be determined by comparing these chambers with a chamber with known correction factors. If measurements are performed in water, a thin waterproofing sleeve should be used and appropriate correction factors should be applied depending on the material and thickness used [47]. If measurements are performed in air, the thimbles of cylindrical chambers are thick enough, so no build-up cap is required. For x-ray energies below 70 kV, calibrated parallel-plate chambers with a thin entrance window are recommended. Thin plastic build-up foils should be added to the entrance window, if necessary, to provide full electron buildup and to eliminate electron contamination (see Table 23.6). All chamber readings should be corrected for temperature, pressure,

ion recombination and polarity effect before applying Equations 23.1–23.5.

Electrometers: The TG61 protocol [23] recommends that electrometers for kilovoltage x-ray therapy measurements should be capable of reading currents on the order of 0.1 nA, with an accumulated charge of 50–100 nC. They are calibrated by a standards laboratory with proper correction factors applied in the measurement. These correction factors are generally close to 1.000; but can occasionally be 5% different from 1.000. Electrometers and ionization chambers can be calibrated either together or sometimes separately and if so their calibration factors must be combined.

Phantoms: The TG61 protocol recommends the use of a water phantom when the in-phantom method is used in orthovoltage x-ray reference dosimetry [23]. Conversion factors are used to convert from the air kerma in the sensitive chamber cavity to the dose to water. Plastic phantoms are not recommended for in-phantom reference dosimetry for orthovoltage x-rays as

TABLE 23.5 Correction Factors for PMMA Sheaths when Using Cylindrical Chambers for In-water Measurements in Orthovoltage X-Ray Beams

HVL		PMMA (Lucite)			
(mm Cu)	(mm Al)	$t = 0.5$ mm	$t = 1$ mm	$t = 2$ mm	$t = 3$ mm
0.1	3.0	0.998	0.995	0.991	0.986
0.3	6.1	0.998	0.997	0.994	0.991
0.5	8.5	0.999	0.998	0.996	0.994
0.8	11.0	0.999	0.998	0.997	0.996
1.0	12.1	1.000	0.999	0.998	0.997
2.0	15.2	1.000	1.000	1.000	0.999
3.0	17.6	1.000	1.000	1.000	1.000
4.0	19.4	1.000	1.000	1.000	1.000
5.0	20.9	1.000	1.000	1.000	1.000

The data applies to 2 cm depth in water and a 100 cm² field.

TABLE 23.6 Thickness of Build-Up Material for Thin-Window Parallel-Plate Chambers Used for In-Air Calibrations with X-Ray Energy below 100 kV

Generating Potential	Thickness of Foil
(kV)	(mg cm^{-2})
50	1.5
60	3.0
70	4.7
80	6.6
90	8.7
100	10.9

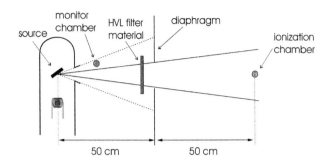

FIGURE 23.2 Recommended experimental setup for HVL measurement including the source (target), the HVL filter (absorber), the diaphragm and the measurement ion chamber.

the chamber correction factors and conversion factors to derive dose at a depth in water for these phantoms are poorly known. Some "water equivalent" plastics are commercially available for routine consistency measurements. Their properties must be investigated before they can be used as water substitutes for reference dosimetry.

23.3.2 Kilovoltage X-Ray Beam Specification

X-ray beams with different energy spectra exhibit penetrating powers in biological media. The quality of an x-ray beam refers to its penetrating power, with high beam quality indicating deeper penetration in a medium or less attenuation by the medium. The exact knowledge of beam quality plays an important role in clinical radiotherapy to ensure proper treatment target coverage and adequate skin sparing. TG61 recommends the use of more than one beam quality parameter to specify an x-ray beam [23]. The usual quantities used are the half-value layer (HVL) and the accelerating potential. A literature review shows that most dosimetric data for kilovoltage x-rays have been given using one beam quality parameter, usually HVL.

The AAPM TG61 defines the HVL as the thickness of an absorber that reduces the air kerma rate of a narrow unidirectional x-ray beam at a point distant from the absorber to 50% of that of the non-attenuated beam [23]. For kilovoltage x-rays, aluminum (Al) and copper (Cu) have been used as the absorbing material for the determination of HVL. The beam quality is expressed in mm Al or mm Cu. For superficial and contact x-ray beams, only mm Al is used to express HVL. For orthovoltage x-rays, both Al and Cu can be used to specify the beam quality although Cu is more frequently used in practice.

The AAPM TG61 protocol recommends the use of an ionization chamber to measure the HVL [23]. A narrow beam setup should be used, in which a collimating aperture is placed half way between the x-ray target and the measurement chamber. The field size is reduced to just enough to encompass the chamber. The absorbing materials (Al or Cu) are placed at least 50 cm from the measurement chamber and close to the collimating aperture. The absorber should be made of high-purity (99.9%) material and the thickness of the absorber should be measured accurately. The thickness that reduces the air kerma value to one half is obtained by interpolation. A monitor detector is used to correct fluctuations of air kerma rate (Figure 23.2). The measurement chamber must have limited beam quality dependence (within 5% between 40 and 300 kV) for accurate HVL measurements. Thin-walled chambers are used for lightly filtered beams to measure the low-energy x-ray components accurately.

It is well known that the peak voltage across an x-ray tube determines the maximum photon energy in an x-ray spectrum. Traditionally, the accelerating potential is expressed in kV for a kilovoltage x-ray beam to indicate that the x-ray beam actually has an energy spectrum with its maximum energy up to its accelerating potential. It should be noted that most often the accelerating potential of a kilovoltage x-ray beam is nominal rather than actual. Higher accelerating potentials produce more penetrating x-rays and, therefore have a direct effect on the beam quality. A review of x-ray equipment for low and medium energies can be found in Ma [6].

23.3.3 Relative Dosimetry Measurement

Output Factors: For kilovoltage x-ray dosimetry, the output factor is defined as the surface dose value for a given source-to-surface distance (SSD) and field size relative to that under the reference conditions. Clinically, output factors are required for all combinations of SSD and field size used for kilovoltage x-ray treatment. Since the scatter contribution from the inside of a cone applicator may be significant, it is not sufficiently accurate to estimate output factors for different applicators using the ratio of the backscatter factors corresponding to the respective field sizes. The output factor for each individual applicator must be measured at each beam quality. If the in-phantom calibration method has been used for orthovoltage x-rays it is necessary to obtain the depth dose values in order to determine the dose at the surface (see Section 23.3.4). This may result in large uncertainties in the output factors since the uncertainties are generally high in the depth dose values near the surface.

Percentage Depth Dose and Lateral Dose Profiles: Percentage depth dose (PDD) and lateral dose profiles are important clinical quantities for treatment planning. The measurement uncertainties are high for kilovoltage x-rays because of the significant variation of x-ray energy (especially for lightly filtered beams) and angular distribution with depth and field size. Care must be exercised in the measurement of PDD and dose profiles for kilovoltage x-ray beams because of their significant depth and field size dependence.

The suitability of x-ray detectors for relative dosimetry measurement has been evaluated extensively [5, 24, 36]. Detectors for the PDD and dose profile measurement should have high spatial resolution and constant energy and angular response. As a general requirement for the evaluation of a specific detector, the relative in-air chamber response and the relative in-phantom response should be compared with a well-behaved cylindrical chamber at depths where reasonable measurements with the cylindrical chamber can be performed. Diamond detectors and the NACP parallel plate chamber (type) have been found to require relatively small depth-dependent correction factors for orthovoltage x-ray beams [5]. Other solid detectors, attractive because of the small size of their sensitive volume (diode detector, TLD, film), usually show significant beam quality dependence or large experimental uncertainties. Well-designed cylindrical chambers have nearly constant energy response in this beam quality range and are suitable for in-phantom measurements. However, the measurement depth is limited to no less than the outer radius of the chamber. Parallel-plate chambers have been used for measurements at smaller depths. Those parallel-plate chambers designed for electron beams usually have a calibration factor varying with beam quality by

20–40% for kilovoltage x-rays. Significant corrections with depth may be required for the PDD measurement with these chambers. Specifically designed parallel-plate chambers for low-energy x-rays usually have a flat energy response in air but not at a depth in a phantom. For instance, more than 10% variations in chamber response have been observed for the Capintec PS-033 chambers. Thus, a depth-related correction factor may be required for these chambers to be used in accurate PDD measurements.

A water tank with a small-volume scanning ionization chamber is recommended for the PDD and profile measurement. A monitor chamber should be used to measure any erratic fluctuations in dose rate. If a thin window parallel plate chamber is used, the chamber must have sufficient buildup material placed over its entrance window (see Table 23.6). Because of the finite size of the ionization chamber in the beam direction, it is necessary to extrapolate the dose to the surface. Since the dose distribution may be nonlinear near the phantom surface, caution should be exercised in extrapolating over the last few millimeters [24, 36].

The PDD can also be measured by placing thin sheets of water-equivalent material over the chamber in a phantom and moving the chamber back by the same amount to maintain a constant SSD. The water-equivalence of the material in the energy range of interest must be verified. PMMA is not suitable for this purpose. Strictly speaking, an ion chamber measures the "depth-ionization" distribution rather than the depth-dose distribution. However, the difference between them is small [5] except for the surface dose measurement. If a suitable detector for relative dosimetry cannot be identified in the clinic it is recommended to use the data from the British Journal of Radiology Supplement 25 [37] or published data that match the kV and HVL values of the user's beams (e.g., Podgorsak et al. [38], Gerig et al. [39], Butson et al. [40], Aukett et al. [41], Kurup and Glasgow [42]).

23.3.4 Dose to Medium Conversion

Dose on the Surface: The AAPM TG61 protocol provides detailed guidelines to determine dose to other biological tissues on the surface or at a depth of a human body for clinical radiotherapy and radiobiology [23].

The surface dose for other media (med) can be calculated from:

$$D_{\mathrm{med},z=0} = C_{\mathrm{w}}^{\mathrm{med}} D_{\mathrm{w},z=0} \qquad (23.6)$$

TABLE 23.7 Free-in-Air Ratios of Mass Energy Absorption Coefficients of Biological Tissue to Water for Application in Conjunction with the In-air Method (The Data Are for SSD = 50 cm)

HVL		"Free-in-Air" Mass Energy Absorption Coefficient Ratio of the Specified Tissues to Water				
(mm Al)	(mm Cu)	Tissue	Muscle	Lung	Skin	Bone
0.3		0.917	1.016	1.031	0.890	4.200
0.4		0.918	1.020	1.035	0.893	4.289
0.5		0.919	1.022	1.037	0.895	4.335
0.6		0.920	1.024	1.039	0.897	4.382
0.8		0.921	1.028	1.043	0.901	4.475
1.0		0.923	1.031	1.046	0.904	4.494
1.2		0.925	1.031	1.046	0.907	4.469
1.5		0.927	1.032	1.047	0.910	4.427
2.0		0.930	1.032	1.047	0.915	4.350
3.0		0.934	1.032	1.045	0.922	4.179
4.0		0.939	1.030	1.042	0.929	3.975
5.0		0.943	1.028	1.039	0.935	3.769
6.0		0.947	1.026	1.036	0.940	3.557
8.0		0.955	1.021	1.030	0.950	3.133
	0.1	0.934	1.032	1.045	0.921	4.209
	0.2	0.942	1.029	1.040	0.934	3.808
	0.3	0.947	1.026	1.036	0.940	3.561
	0.4	0.952	1.023	1.032	0.946	3.314
	0.5	0.956	1.020	1.029	0.952	3.068
	0.6	0.960	1.018	1.026	0.957	2.859
	0.8	0.964	1.015	1.022	0.961	2.657
	1.0	0.967	1.012	1.018	0.965	2.456
	1.5	0.975	1.006	1.009	0.975	1.952
	2.0	0.981	1.001	1.003	0.980	1.637
	3.0	0.986	0.996	0.997	0.985	1.280
	4.0	0.988	0.994	0.994	0.987	1.128
	5.0	0.990	0.992	0.992	0.989	1.026

with the conversion factor from dose-to-water to dose-to-medium given by:

$$C_w^{med} = \frac{B_{med}}{B_w}\left[\left(\frac{\bar{\mu}_{en}}{\rho}\right)_w^{med}\right]_{air} \quad (23.7)$$

where $\left[\left(\frac{\bar{\mu}_{en}}{\rho}\right)_w^{med}\right]_{air}$ represents the ratio of mass energy absorption coefficients medium to water averaged over the primary photon spectrum free-in-air, and $\frac{B_{med}}{B_w}$ the ratio of kerma based backscatter factors medium to water at the surface of the medium versus air.

The AAPM TG61 report gives the ratios of mass energy absorption coefficients averaged over the photon fluence spectrum free-in-air of media of clinical interest relative to water (Table 23.7). The backscatter factor ratios relative to water for these biological tissues, except bone is not differing from unity by more than 1.2% for a large field size and can therefore be ignored. For bone,

Table 23.8 shows the ratios of the backscatter factors, bone to water, for photon beams 50–300 kV (0.875–20.8 mm Al) with different field sizes at various SSD.

Dose at a Depth: The absorbed dose at a depth in a non-water phantom can be calculated by

$$D_{med} = MN_K P_{Q,cham}\left(\frac{\bar{\mu}_{en}}{\rho}\right)_{air}^w C_w^{med} \quad (23.8)$$

where

$$C_w^{med} = \frac{P_{Q,cham}^{med}}{P_{Q,cham}^w}\cdot\left(\frac{\bar{\mu}_{en}}{\rho}\right)_w^{med} \quad (23.9)$$

is the conversion factor. The chamber correction factor for non-water phantoms is not well known but it is expected to be similar to that for water for low-Z materials if their density does not differ significantly from

TABLE 23.8   Ratios of the Kerma-Based Backscatter Factors, Bone to Water, for Photon Beams 50–300 kV (0.05–5 mm Cu) with Different Field Sizes and SSD

SSD	HVL		$\dfrac{B_{bone}}{B_w}$				
(cm)	(mm Cu)	(mm Al)	$1 \times 1 cm^2$	$2 \times 2 cm^2$	$4 \times 4 cm^2$	$10 \times 10 cm^2$	$20 \times 20 cm^2$
10	0.05	1.6	0.958	0.929	0.897	0.861	0.854
	0.1	2.9	0.976	0.945	0.905	0.853	0.838
	0.5	8.5	1.019	1.011	0.974	0.910	0.875
	1	12.0	1.031	1.041	1.026	0.974	0.943
	2	15.8	1.038	1.065	1.077	1.047	1.023
	3	17.9	1.037	1.066	1.092	1.086	1.070
	4	19.3	1.028	1.053	1.082	1.087	1.075
	5	20.3	1.022	1.043	1.074	1.087	1.078
30	0.05	1.6	0.958	0.926	0.889	0.850	0.833
	0.1	2.9	0.976	0.940	0.894	0.837	0.809
	0.5	8.5	1.019	1.011	0.981	0.887	0.833
	1	12.0	1.031	1.042	1.033	0.959	0.902
	2	15.8	1.038	1.067	1.083	1.043	0.989
	3	17.9	1.037	1.067	1.101	1.090	1.047
	4	19.3	1.029	1.055	1.088	1.091	1.065
	5	20.3	1.023	1.045	1.077	1.091	1.079
50	0.05	1.6	0.958	0.927	0.891	0.847	0.827
	0.1	2.9	0.975	0.942	0.897	0.832	0.800
	0.5	8.5	1.018	1.009	0.977	0.881	0.825
	1	12.0	1.031	1.040	1.032	0.958	0.894
	2	15.8	1.038	1.066	1.085	1.047	0.983
	3	17.9	1.036	1.069	1.100	1.095	1.048
	4	19.3	1.028	1.057	1.084	1.094	1.066
	5	20.3	1.022	1.047	1.072	1.094	1.080

unity. Thus, the conversion factor can be given as the ratio of the mass energy-absorption coefficients for the non-water phantom to water averaged over the photon spectrum at the depth z, $\left(\frac{\bar{\mu}_{en}}{\rho}\right)_w^{med}$. The waterproofing sleeve correction factor is not needed for solid phantoms. In general, $\left(\frac{\bar{\mu}_{en}}{\mu}\right)_w^{med}$ is depth, SSD and field-size dependent.

The AAPM TG61 provides the ratios of mass-energy absorption coefficient ratios of tissue to water at depth in water (Table 23.9). It was found that for all the tissues listed in Table 23.9, except for bone, depth dependence of the conversion factors is at most 0.9%, field-size dependence at most 0.8%, and SSD dependence at most 0.1% [48]. Therefore, this table can be, except for bone, applied to any depth, field size and SSD. For bone, the differences were found to be up to 16% (depth dependence), 27% (field size dependence), and 1.5% (SSD dependence), as shown in Figures 23.3, if the field size, depth and SSD deviate significantly from 100 cm², 2 cm and 50 cm, respectively.

23.3.5 Dosimetry Uncertainty

The International Commission on Radiation Units and Measurements (ICRU) report 24 recommended that the uncertainty on the absorbed dose that can be delivered to the treatment target in a clinical situation be preferably better than ±5% [43]. The overall uncertainty on the dose delivered to the target consists of several components, the first part of which occur as a result of uncertainties in the calibration chain linking the calibration of the clinical beam to the standards laboratory (N_K-factor, uncertainties on conversion and correction factors, beam quality specification uncertainties, etc.) and a second part associated with clinical errors in patient setup, immobilization, target localization and dose delivery. Table 23.10 lists several components contributing to the uncertainty (including type A and type B

TABLE 23.9   Ratios of the Mass Energy Absorption Coefficients of Tissue to Water for Converting Dose to Water to Dose to Tissue at an Equivalent Depth

HVL		"In-Phantom" Ratio of Mass Energy Absorption Coefficients of the Specified Tissue to Water			
(mm Al)	(mm Cu)	Tissue	Muscle	Lung	Bone
0.3		0.921	1.030	1.046	4.540
0.4		0.922	1.033	1.049	4.611
0.5		0.922	1.033	1.049	4.615
0.6		0.923	1.034	1.049	4.619
0.8		0.924	1.035	1.050	4.627
1.0		0.925	1.035	1.050	4.585
1.2		0.927	1.035	1.049	4.514
1.5		0.930	1.035	1.049	4.433
2.0		0.932	1.034	1.048	4.345
3.0		0.936	1.033	1.046	4.177
4.0		0.939	1.031	1.044	4.025
5.0		0.942	1.030	1.041	3.873
6.0		0.945	1.028	1.039	3.719
8.0		0.951	1.024	1.034	3.410
	0.1	0.936	1.033	1.047	4.200
	0.2	0.941	1.030	1.042	3.901
	0.3	0.945	1.028	1.039	3.722
	0.4	0.948	1.026	1.036	3.542
	0.5	0.951	1.024	1.033	3.363
	0.6	0.954	1.022	1.031	3.205
	0.8	0.958	1.019	1.027	3.014
	1.0	0.961	1.017	1.024	2.823
	1.5	0.969	1.011	1.016	2.346
	2.0	0.974	1.006	1.011	2.015
	3.0	0.981	1.000	1.002	1.565
	4.0	0.985	0.997	0.997	1.330
	5.0	0.988	0.994	0.994	1.136

Note: The data are for SSD = 50 cm, 100 cm² field size and 2 cm depth.

uncertainties) in the dose to medium, as described in the TG61 report [23] without considering the clinical errors. These uncertainties are fairly large due to both the limited amount of experimental data and the limited computational resources available when it was published. For example, the uncertainties on the dose conversion and detector correction factors can be much reduced [25] and the depth dose measurement can be performed more accurately with well-designed chambers. The estimated uncertainty for kilovoltage x-ray dosimetry following TG61 is likely to be better than 2% at reference points, i.e., at the phantom surface for the in-air method, and at 2 cm depth for the in-phantom method. For dose determination at depths other than the reference points and for medium other than water, the estimated uncertainty is likely to be about 3% following the

TG61 recommendations. This was confirmed by recent studies using different methods with high-precision calculations and measurements at standards laboratories [25, 27].

23.4 CLINICAL IMPLEMENTATION OF KILOVOLTAGE X-RAYS

The clinical implementation of a therapeutic x-ray unit requires acceptance testing, clinical commissioning and a routine QA program to ensure its safe operation. Acceptance testing is required to ensure that the x-ray unit meets the technical specifications set by the manufacturer as well as the local department. Clinical commissioning involves the collection of dosimetry data for beam characterization, treatment planning and QA testing for the routine QA program.

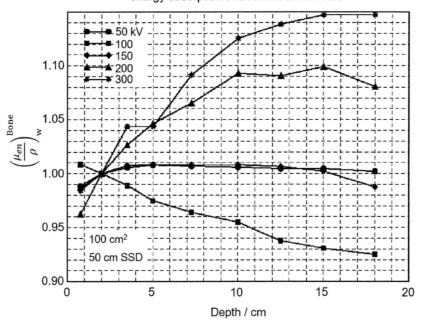

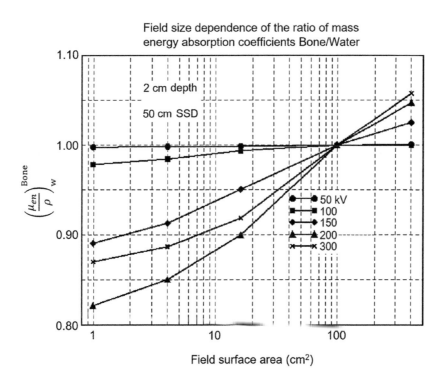

FIGURE 23.3 Depth and field size dependence of the ratio of mass energy absorption coefficients, bone to water. Half value layers for the indicated radiation qualities are: 0.88 mm Al (50 kV), 2.65 mm Al (100 kV), 0.57 mm Al (150 kV), 1.6 mm Cu (200 kV), and 4.3 mm Cu (300 kV).

23.4.1 Acceptance Testing

It is recommended that the user of a kilovoltage x-ray unit should work with the vendor to establish an acceptance testing protocol. The acceptance tests that are required for typical clinical applications include mechanical checks of the x-ray unit, radiation tests, radiation surveys and checking the general operation and functionality of the unit as well as the treatment room.

The following tests are examples for the acceptance testing of a kilovoltage treatment unit as recommended

TABLE 23.10 Components and Typical Values of the 1σ Uncertainty on Determining the Dose to Medium in the Users Beam in the Clinic

Type of Quantity or Procedure	Uncertainty (%)
In-air method (low and medium energies)	
N_K from calibration lab.	0.7
Effect of beam quality difference between calibration and measurement	2.0
Backscatter factor B	1.5
$P_{stem,air}$	1.0
$\left(\frac{\bar{\mu}_{en}}{\rho}\right)_{air}^{water}$	1.5
Free air measurement in clinic	1.5
Overall uncertainty for $D_{w,z=0}$	**3.5**
Transfer to other points in water	3.0
Conversion to dose to tissue	1.0
Overall uncertainty for $D_{med,z}$	**4.7**
In-phantom method (medium energies)	
N_K from calibration lab.	0.7
Effect of beam quality difference between calibration and measurement	2.0
Chamber correction factors	1.5
P_{sheath}	0.5
$\left(\frac{\bar{\mu}_{en}}{\rho}\right)_{air}^{water}$	1.5
In-phantom measurement in clinic	2.0
Overall for $D_{w,z=2}$	**3.6**
Transfer to other points in water	3.0
Conversion to dose to tissue	1.0
Overall uncertainty for $D_{med,z}$	**4.8**

by the ACPSEM kilovoltage dosimetry working group [28]. These tests can be performed along with any tests that are specified by the manufacturer.

Safety tests

- The safety switches and emergency buttons function properly.

- The radiation signs at the door and at the console activate when the beam is on.

- The interlock for the room entrance or door works correctly and disables the beam from being able to be switched unless the door is closed or interlock is armed.

- The patient remote monitoring cameras are functioning properly.

Mechanical tests

- Verification of the alignment of the mounting device that holds the beam applicator to ensure that it is centered about the x-ray source.

- Reproducibility of positioning of the treatment applicators to ensure correct position and treatment distances in the clinical setting.

- Validation of any light field with the mechanical readout for a range of field sizes at an appropriate SSD.

- Confirmation of travel range of the x-ray unit in any direction available for the treatment should be consistent with any indicators display on the stand or on the x-ray tube.

- If the x-ray tube has the ability to be angled in different directions, testing of the accuracy and mechanical rigidity of the system holding the treatment applicator in place is required. In addition, a comparison of the radiation field defined by a standard treatment applicator in a non-rotated position to that in an extremely rotated position should be performed. Ensure the locking mechanism on the unit is functional.

- Testing filter interlocks to verify that the selected filter and kV combination can produce the correct x-ray beam and that if an inappropriate filter is inserted, x-ray will not be produced.

- Check the integrity of each treatment applicator for cracks or dents. Measure lengths and diameter of each treatment applicator to ensure that it meets specifications.

Radiation tests

- Light field versus x-ray field: the congruence between light field and x-ray field should be checked for a number of field sizes at the standard SSD. The comparison should be performed at the FWHM of the x-ray field and the edge of the light field. The two edges should agree to within ±3 mm for field sizes greater than 10 cm and should agree to better than ±2 mm for smaller field sizes. The accuracy of the light field for small field sizes should be carefully evaluated before clinical use.

- X-ray leakage: all sources of leakage should be assessed (head or tube leakage, leakage transmitted through the applicator and the applicator supporting device) and should meet both regulations and specifications. The leakage should be determined using both film and ionization chamber measurements and should be conducted with the highest beam kV and mA available.

X-ray beam performance checks

- Field flatness should be evaluated by measurement of beam profiles in water for the largest treatment field available of each beam to be used clinically. Profiles of the beam should be measured in a scanning water tank both perpendicular to and in the cathode–anode axis. If it is assumed that the beam energy does not change across the field, then radiochromic film in a solid phantom may be used for flatness measurements. For kilovoltage x-ray beams, the uniformity of the field is usually ±5% over the field area.

- Field symmetry may be evaluated using the same beam profile measurement as for flatness. However, differences may be seen in the cathode–anode

beam profile, particularly for the largest fields, giving rise to the "heel effect" produced by the electron interaction onto the angled target.

- Beam quality is usually determined by HVL measurements using a setup consistent with the recommendations in the AAPM TG61 code of practice (see Section 23.3).

Radiation dosimetry Tests

- Focal spot size and position should be determined to ensure that the treatment distances indicated are correct and can be assessed using a pinhole technique to ensure that the treatment applicators are well aligned with the focal spot of the x-ray tube.

- Timer error should be investigated if the x-ray unit does not have a monitor chamber. The accuracy and linearity of the timer should be evaluated. The system should also have a backup timer, in case the primary timer fails. Calculation of the timer error is important due to existence of the "ramp up region" in kilovoltage x-ray beams machines. During the ramp-up, the dose rises from zero to the steady state dose rate over a period of a few seconds. If the steady state dose rate is used for the calculation of dose, it can lead to a dose deficit.

- Dose output linearity as a function of monitor units (MUs) or with treatment time (if using timer mode) should be determined. The stability of the kV and the mA over operating time can be checked with a series of short and long radiation exposure times.

- Dose output reproducibility should be checked by performing a number of dose measurements with the same monitor units (or time).

- Rotation of the x-ray tube should not produce any variation in the exposure rate, beam output, field flatness or symmetry. An ionization chamber may be attached to the applicator, parallel to the anode–cathode axis, and measurements then be performed for a variety of tube angles. Alternatively, film can be attached to the applicator for monitoring beam uniformity at different tube angles.

- Short/long term reproducibility of the dose rate for the clinical treatment techniques should be monitored.

23.4.2 Clinical Commissioning

Clinical commissioning involves the collection of dosimetry data that is used for beam characterization. This is required for both treatment planning and reference dosimetry. Relative dosimetry measurements are also performed for beam characterization and use in treatment planning data. Reference dosimetry is performed for each of the x-ray beams that will be made available on the treatment unit. Finally the QA baselines are established and benchmarks are set which will be used to monitor the performance of the x-ray unit by routine QA tests.

It is recommended that the user of a kilovoltage x-ray unit work with the vendor to establish a comprehensive commissioning protocol based on the specifications of the x-ray unit and its intended clinical applications. The following commissioning tests are examples that were recommended by the ACPSEM kilovoltage dosimetry working group [28], which can be included in the commissioning protocol together with other commissioning tests recommended by the vendor.

HVL measurements

- Measurement of the first and second HVL for each of the radiation beams using narrow beam geometry

Relative dosimetry for each x-ray beam and treatment applicator combination

- Relative output factor and applicator factor
- Percentage depth dose curve
- Lateral profiles in both axes perpendicular to beam direction for at least the largest applicator
- Isodose measurements as required for treatment planning

Validity of inverse square law (ISL) with increasing SSD

- Assess the validity of the ISL correction for increasing distances of stand-off from the end of each applicator; this should be performed for every x-ray beam energy

Planning data

- Relative output factors for standard lead cutouts
- Transmission measurements for lead thickness

Backscatter factor verification

- Check the Back Scatter Factors (BSFs) via direct measurement or comparison with measured output factors

Reference dosimetry

- Perform reference dosimetry measurements for each of the clinical x-ray beams

Independent audit

- Independent audit of the reference dosimetry of each clinical x-ray beam as well as a review of the relative dosimetry data

23.4.3 Routine Quality Assurance

A QA program is required for the safe and accurate delivery of radiation therapy treatments using kilovoltage x-rays. The safe delivery of radiation therapy is ensured by machine interlocks and strategically placed emergency-off buttons. Accurate delivery of radiation treatment is ensured by maintaining the mechanical accuracy within the specification of treatment unit, maintaining the beam quality in its original condition, and maintaining the accuracy of dosimetry calculation and measurements for treatment planning and plan verification. All dosimetry data and mechanical limits are established during the initial machine commissioning period before its clinical use and thereafter annually, or after any change which may significantly alter the dosimetry data and the mechanical limits. Once the baseline is established, maintaining the baseline data becomes the mission of the dosimetry QA program.

The test frequency for a kilovoltage x ray therapy system is determined by the significance of the tests that indicate beam dosimetry changes and mechanical tolerance changes within the limited amount of time. The number of patients that are treated using kilovoltage x-ray beams are much less than for megavoltage radiotherapy. Therefore, the check of the beam dosimetry should depend on the frequency of the beam usage.

Documentation of weekly and monthly spot check measurements is maintained for 2 years. Documentation for each full yearly calibration is maintained for 5 years after the completion of calibration. The TG61 dosimetry QA items and their frequencies [23] are summarized below:

Daily Checks:

- Beam output constancy for energy and filter combinations in use
- Functionality of the audio-visual monitor
- Door and energy interlock circuits and emergency stops

Monthly Checks:

- Items included in the daily checks, and
- Beam flatness and symmetry
- Timer operation
- Light-radiation congruence

Annual Checks:

- Items included in the monthly checks, and
- Dose rate for all energy and filter combinations
- Output factors for each of the applicator (cone)
- Timer accuracy (verification of the timer error)
- Accuracy of the light localizer system
- Accuracy of distance measuring devices
- Beam quality
- Accuracy of depth dose data and isodose charts
- Accuracy of field size dependence data
- Agreement of dose rate with distance from target
- Attenuation in lead for patient block thickness

23.5 SUMMARY

This chapter provides a brief historical review of the kilovoltage x-ray dosimetry and describes the AAPM TG61 code of practice for low- and medium-energy x-rays (40 kV ≤ kVp ≤ 300 kV). TG61 still represents the best kilovoltage x-ray dosimetry protocol for clinical imaging, radiation therapy and radiobiology applications at time of writing of this chapter.

The TG61 report recommends two measurement methods based on ionization chambers calibrated free-in-air in terms of air-kerma. If the point of interest is clinically at or close to the surface, the code recommends one unified approach over the entire energy range to determine absorbed dose to water at the surface of a water phantom based on in-air measurements (the "in-air" method). If the point of interest is at depth, it is recommended to perform an in-phantom ion chamber measurement at 2 cm for kVp values ≥100 kV (the "in-phantom" method) whereas the in-air method is recommended for kVp ≤ 100 kV. The TG61 report provides a comprehensive data set and produces consistent dose values regardless of the method used and presents estimates of uncertainties on the final dose values. Whereas the code provides guidelines for dose measurements using water, it is supplemented with data to calculate dose to other materials of clinical interest.

This chapter does not deal with very low energy x-rays (e.g., 50 kV or lower) such as those used in surface or contact therapy or interstitial radiosurgery [6, 8, 44–46]. For x-ray beams with accelerating potentials less than 40 kV, the AAPM TG61 protocol may be used provided the required data values are present in the associated tables. For values of beam potential down to 10 kV, the in-air method may be used provided the medical physicist can locate appropriate correction factors and associated uncertainties in the literature. It should be noted that for x-ray beams with energies less than 40 kV, the attenuation of the x-rays in air becomes significant. Further dosimetric studies on detector response and dose conversion factors are needed to establish reliable and practical reference and relative dosimetry methods and procedures for the widespread clinical application of these techniques at very low x-ray energies.

REFERENCES

1. O. Glasser (ed.), *The Science of Radiology* (Bailliere, London, 1933).
2. R. Paterson, *The Treatment of Malignant Disease by Radium and X-Rays* (Williams and Wilkins, Baltimore, 1949).
3. E. H. Grubbe, "Priority in the therapeutic use of X-rays," *Radiology* **21**, 156–162 (1933). https://doi.org/10.1148/21.2.156.
4. P. Biggs, C.-M. Ma, K. Doppke, A. Niroomand-Rad, J. Beatty, "Chapter 9: Kilovoltage x-rays," in *The Modern*

Technology of Radiation Therapy, edited by J. Van Dyk (Medical Physics Publishing, Madison, WI, 1999).

5. C.-M. Ma and J. P. Seuntjens (eds.), *Kilovoltage X-ray Beam Dosimetry for Radiotherapy* (Medical Physics Publishing, Madison, WI, 1999).

6. C.-M. Ma, "X-ray therapy equipment, low and medium energy," in *Encyclopedia of Medical Devices and Instrumentation* (John Wiley & Sons, Inc., New York, NY, 2006).

7. C.-M. Ma and K. Paskalev, "In-room CT techniques for image-guided radiation therapy," *Medical Dosimetry* **31**, 30–39 (2006).

8. R. Hill, B. Healy, L. Holloway, Z. Kuncic, D. Thwaites, C. Baldock, "Advances in kilovoltage X-ray beam dosimetry," *Physics in Medicine and Biology* **59**(6), R183 (2014).

9. C. McKerracher and D. I. Thwaites, "Calibration of an x-ray cabinet unit for radiobiology use," *Physics in Medicine and Biology* **51**, 3315 (2006).

10. F. Verhaegen, P. Granton, E. Tryggestad, "Small animal radiotherapy research platforms," *Physics in Medicine and Biology* **56**, R55–R83 (2011).

11. R. Pidikiti, S. Stojadinovic, M. Speiser, K. H. Song, F. Hager, D. Saha, T. D. Solberg, "Dosimetric characterization of an image-guided stereotactic small animal irradiator," *Physics in Medicine and Biology* **56**, 2585–2599 (2011).

12. L. S. Taylor, *Radiology* **18**(99), 627 (1932).

13. H. O. Wyckoff, G. H. Aston, E. E. Smith, "A comparison of x-ray standards," *British Institute of Radiology* **27**, 325 (1954).

14. G. Ibbott, C.-M. Ma, D. W. Rogers, S. M. Seltzer, J. F. Williamson, "Anniversary paper: Fifty years of AAPM involvement in radiation dosimetry," *Medical Physics* **35**, 1418–1427 (2008).

15. C.-M. Ma, C. W. Coffey, L. A. DeWerd, C. Liu, R. Nath, S. M. Seltzer, J. P. Seuntjens, "Status of kilovoltage x-ray beam dosimetry in radiotherapy," in *Kilovoltage X-ray Beam Dosimetry for Radiotherapy and Radiobiology*, edited by C.-M. Ma and J. P. Seuntjens (Medical Physics Publishing, Madison, WI, 1999).

16. NCRP, *NCRP Report 69, Dosimetry of X-ray and gamma ray beams for Radiation Therapy in the Energy Range 10 keV to 50 MeV* (NCRP Publications, Washington, DC, 1981).

17. International Commission on Radiation Units and Measurements (ICRU) Report 23, *Radiation Dosimetry: Measurement of Absorbed Dose in a Phantom Irradiated by a Single Beam of X- or Gamma Rays* (International Commission on Radiation Units and Measurements, Bethesda, MD, 1973).

18. M. Cohen, D. E. A. Jones, D. Greene (Eds.), "Central axis depth dose data for use in radiotherapy," *British Journal of Radiology Supplement 11* (1972).

19. International Atomic Energy Agency (IAEA), *Absorbed Dose Determination in Photon and Electron Beams; An International Code of Practice*, Technical Report Series (IAEA, Vienna, Austria, 1987), vol. 277.

20. S. C. Klevenhagen, R. J. Aukett, R. M. Harrison, C. Moretti, A. E. Nahum, K. E. Rosser, "The IPEMB code of practice for the determination of absorbed dose for x-rays below 300 kV generating potential (0.035 mm Al-4mm Cu HVL; 10–300 kV generating potential)," *Physics in Medicine and Biology* **41**, 2605–2625 (1996).

21. Netherlands Commission on Radiation Dosimetry (NCS), *Dosimetry of Low and Medium Energy X-rays, a Code of Practice for Use in Radiotherapy and Radiobiology*, NCS Report 10 (NCS, Delft, The Netherlands, 1997).

22. International Atomic Energy Agency (IAEA), *Absorbed Dose Determination in External Beam Radiotherapy: An International Code of Practice for Dosimetry Based on Standards of Absorbed Dose to Water*, ISSN 1011-4289 (IAEA, Vienna, 2000).

23. C.-M. Ma, C. Liu, C. Coffey, R. Nath, S. Seltzer, J. Seuntjens, "AAPM protocol for 40–300 kV x-ray beam dosimetry for radiotherapy and radiobiology," *Medical Physics* **28**, 868–893 (2001).

24. C.-M. Ma, X. A. Li, J. Seuntjens, "Consistency study on kilovoltage x-ray beam dosimetry for radiotherapy," *Medical Physics* **25**, 2376–2384 (1998).

25. L. Dimaso, B. Palmer, J. Micka, et al., "Monte Carlo and ^{60}Co-based kilovoltage x-ray dosimetry methods," *Medical Physics* **45**, 5564–5576 (2018).

26. P. Almond, P. J. Biggs, B. M. Coursey, et al., "AAPM's TG-51 protocol for clinical reference dosimetry of high-energy photon and electron beams," *Medical Physics* **26**, 1847–1870 (1999).

27. L. Buermann, A. S. Guerra, M. Pimpinella, et al., "First international comparison of primary absorbed dose to water standards in the medium energy X-ray range," *Metrologia* **53**, 06007 (2016).

28. R. Hill, Healy, D. Butler, D. Odgers, S. Gill, J. Lye, T. Gorjiara, D. Pope, B. Hill, "Australasian recommendations for quality assurance in kilovoltage radiation therapy from the Kilovoltage Dosimetry Working Group of the Australasian College of Physical Scientists and Engineers in Medicine," *Australasian Physical & Engineering Sciences in Medicine* **41**, 781–808 (2018).

29. C.-M. Ma and A. E. Nahum, "Bragg–Gray theory and ion chamber dosimetry in photon beams," *Physics in Medicine and Biology* **36**, 413–428 (1991).

30. A. E. Nahum and R. T. Knight, "Consistent formalism for kV x-ray dosimetry," in *Proceedings of the IAEA International Symposium on Measurement Assurance in Dosimetry* (IAEA, Vienna, Austria, 1994), pp. 451–459.

31. C.-M. Ma and A. E. Nahum, "Monte Carlo calculated stem effect corrections for NE2561 and NE2571 chambers in medium-energy x-ray beams," *Physics in Medicine and Biology* **40**, 63–72 (1995).

32. C.-M. Ma and A. E. Nahum, "Calculations of ion chamber displacement effect corrections for medium-energy x-ray dosimetry," *Physics in Medicine and Biology* **40**, 45–62 (1995).

33. J. P. Seuntjens and F. Verhaegen, "Dependence of overall correction factor of a cylindrical ionization chamber on

field size and depth in medium energy X-ray beams," *Medical Physics* **23**, 1789–1796 (1996).

34. J. P. Seuntjens, H. Thierens, A. Van der Plaetsen, O. Segaert, "Determination of absorbed dose to water with ionisation chambers calibrated in free air for medium energy X-rays," *Physics in Medicine and Biology* **33**, 1171–1185 (1988).

35. J. P. Seuntjens, H. Thierens, U. Schneider, "Correction factors for a cylindrical chamber used in medium energy x-ray beams," *Physics in Medicine and Biology* **38**, 805–832 (1993).

36. X. L. Li, C.-M. Ma, D. Salhani, "Measurement of percentage depth dose and lateral beam profile for kilovoltage x-ray therapy beams." *Physics in Medicine and Biology* **42**, 2561–2568 (1997).

37. British Institute of Radiology (BIR), "Central axis depth dose data for use in radiotherapy," *British Journal of Radiology Supplement 25* (1996).

38. E. B. Podgorsak, M. Gosselin, M. D. C. Evans, "Superficial and orthovotage x-ray beam dosimetry," *Medical Physics* **25**, 1206–1211 (1998).

39. L. Gerig, M. Soubra, D. Salhani, "Beam characteristics of the Therapax DXT-300 orthovoltage therapy unit," *Physics in Medicine and Biology* **39**, 1377–1392 (1994).

40. M. J. Butson, J. Mathur, P. Metcalfe, "Dose characteristics of a new 300 kVp orthovoltage machine," *Australasian Physical & Engineering Sciences in Medicine* **18**, 133–138 (1995).

41. R. J. Aukett, D. W. Thomas, A. W. Seaby, J. T. Gittins, "Performance characteristics of the Pantak DXT-300 kilovoltage X-ray treatment machine," *British Journal of Radiology* **69**, 726–734 (1996).

42. R. G. Kurup and G. P. Glasgow, "Dosimetry of a kilovoltage radiotherapy x-ray machine," *Medical Dosimetry* **18**, 179–86 (1993).

43. International Commission on Radiation Units and Measurements (ICRU) Report 24, *Determination of Absorbed Dose in a Patient Irradiated by Beams of X or Gamma Rays in Radiotherapy Procedures* (International Commission on Radiation Units and Measurements, Bethesda, MD, 1976).

44. M. Dinsmore, K. J. Harte, A. P. Sliski, D. O. Smith, P. M. Nomikos, M. J. Dalterio, A. J. Boom, W. F. Leonard, P. E. Oettinger, J. C. Yanch, "A new miniature x-ray source for interstitial radiosurgery: Device description," *Medical Physics* **23**, 45–52 (1996).

45. Y. Rong and J. S. Welsh, "Surface applicator calibration and commissioning of an electronic brachytherapy system for nonmelanoma skin cancer treatment," *Medical Physics* **37**, 5509–5517 (2010).

46. J. P. Gerard, A. S. Myint, O. Croce, J. Lindegaard, A. Jensen, R. Myerson, J. M. Hannoun-Levi, S. Marcie, "Renaissance of contact x-ray therapy for treating rectal cancer," *Expert Review of Medical Devices* **8**, 483–92 (2011)

47. C.-M. Ma and J. P. Seuntjens, "Correction factors for water-proofing sleeves in kilovoltage x-ray beams," *Medical Physics* **24**, 1507–1513 (1997).

48. C.-M. Ma and J. P. Seuntjens, "Mass energy-absorption coefficient and backscatter factor ratios for kilovoltage x-ray beams," *Physics in Medicine and Biology* **44**, 131–143 (1999).

CHAPTER 24

Electron Dosimetry

John A. Antolak

Mayo Clinic
Rochester, Minnesota

CONTENTS

24.1 INTRODUCTION

The dose deposited by an x-ray beam is primarily due to secondary charged particles (mostly electrons), which implies that any detector that is suitable for measuring dose in an x-ray beam is generally suitable for measuring dose in an electron beam. The goals of this chapter are to provide the reader with (1) a basic understanding of the physics of electron beams that can affect electron dosimetry, (2) recommendations for water-phantom dosimetry of electron beams, (3) guidance for solid phantom measurements using electron beams, and (4) an overview of the use of *in vivo* dosimeters for electron beam radiotherapy.

There are many very good references that are broadly available, and nothing in this chapter is meant to replace the information in those references. The author assumes that the reader has access to the relevant reference materials, avoiding the possibility that transcription errors are introduced that could lead the reader to make mistakes having downstream clinical effects.

24.2 PROPERTIES OF ELECTRON BEAMS

Electrons are fundamental particles with a small mass, approximately 2000 times lighter than a nucleon (proton or neutron). Because they are also charged particles, they readily interact with all of the charged particles that make up normal matter. If the electron comes close enough to the nucleus of an atom, the electric field of the nucleus can deflect the trajectory of the electron, with the amount of deflection depending on the strength of the interaction or how close the electron gets to the nucleus. This interaction is called Coulomb scattering. The number of scattering interactions is approximately 750 per mm of water for >5-MeV electrons [1] and it is not practical to follow each interaction on a microscopic scale. On a macroscopic scale, the deflection of the electron will be the end result of those many interactions and this is called multiple Coulomb scattering. The scattering power of a material is a measure of how much a given material changes the angular distribution of the electrons passing through the material. In some cases, the electron comes close enough to the nucleus that it interacts more strongly with the nucleus and in addition to changing direction, a bremsstrahlung x-ray is emitted as a result of the interaction. This is the basic mechanism for producing x-rays in the anode of an x-ray tube or the target of a linear accelerator. These bremsstrahlung events remove energy from the electron beam, but this energy is not deposited locally and hence contributes very little to the deposited dose.

The most common interaction is with orbital electrons. Even if the incident electron doesn't come very close to the orbital electron, the interaction can lead to a small amount of energy loss and scattering because the masses of the incident and orbital electrons are the same. This is also where most of the energy loss occurs, leading to heating of the matter and ionization of the atoms, the latter of which can eventually lead to biological effects. The mass collisional stopping power is a measure of the energy lost to collisional interactions per unit length divided by the density of the material and the deposited dose is the product of the electron fluence (number of electrons per unit area) and the mass collisional stopping power. The mass collisional stopping power is relatively constant for biological materials. Therefore, the energy deposition is determined primarily by the total electron track length in a given volume, which is equivalent to the electron fluence. The electrons in the beam also redistribute themselves depending on the scattering power of the materials that the beam passes through and calculating the dose deposition due to electron beams is primarily

a problem of calculating the fluence distribution of the electrons in the material.

This is a very simplified description of the electron beam transport, but it is good enough to be able to qualitatively understand some of the basic properties of electron beams used in radiation therapy and how dosimetry of the electron beams might be influenced.

24.3 WATER-PHANTOM DOSIMETRY

The primary purpose of this section is to discuss considerations for measurements commonly carried out during commissioning or annual quality assurance (QA), primarily using water phantoms with automated scanning systems. The American Association of Physicists in Medicine (AAPM) Report 106 (TG-106 [2]) discusses beam commissioning in detail for both photons and electrons, including the use of water phantom systems and AAPM Report 99 (TG-70 [3, 4]) reiterates those recommendation in the context of performing the measurements for electron beams. Most of the recommendations that follow are from those reports, but also includes some preferences of the author based on prior experience.

24.3.1 Reference Dosimetry

Electron beams must be calibrated before they are used clinically. Calibration is the process by which you establish the relationship between the response of the accelerator's monitor chamber and the delivered dose under a specified calibration condition. The most common current calibration protocols are TG-51 (AAPM Report 67 [5, 6]) in the United States and Canada and TRS398 elsewhere [7]. Tailor and Hanson [8] showed that the newer calibration protocols result in 1–2% higher dose in electron beams, depending on the beam quality and ionization chamber used for calibration. The fundamental changes in the latest protocols were that the ion chamber calibration factor is dose to water for a specified beam quality (^{60}Co) and all calibrations must be done in a water phantom. Acceptable dosimeters for these protocols include cylindrical ionization chambers and well-guarded plane-parallel ionization chambers. The dosimeter and electrometer should be calibrated by an Accredited Dosimetry Calibration Laboratory (ADCL) in the United States, or equivalent in other countries. The calibration of the ionization chamber should generally be done every two years or more often if required by regulations. While calibration laboratories will provide a calibration factor for plane-parallel ionization chambers, a process for cross-calibrating such ionization chambers is also described in the reports [5, 7]. The basic process is to perform a calibration measurement in a high-energy electron beam using both ionization chambers and solve for the unknown calibration factor by assuming that the dose at the reference depth is the same for both measurements.

The following general description of the calibration process is based on TG-51, but also generally applies to TRS398 since the protocols are very similar and based on the same underlying research. With respect to electron-beam calibration, the primary difference between the protocols is that while TRS398 requires the use of plane-parallel ionization chambers below 10 MeV ($R_{50} \leq 4.3$ cm), TG-51 only recommends their use for those energies and requires plane-parallel ionization chambers for energies less than 6 MeV ($R_{50} \leq 2.6$ cm). The reader should refer to the original protocol documents to ensure their process is compliant with the protocol and an independent check of the calibration, such as having another person do an independent measurement with different instrumentation, is good practice. In addition, it would be useful to review the article describing common issues found by the Radiological Physics Center (now IROC) as they were helping many centers with their conversion to the TG-51 protocol [9].

The TG-51 protocol for electron beam calibration specifies the reference depth, d_{ref}, for an electron beam as

$$d_{ref} = 0.6\,R_{50} - 0.1 \ \text{(cm)} \tag{24.1}$$

where R_{50}, the beam quality specifier, is the depth at which the dose drops to 50% of the maximum central-axis dose, $D(d_{max})$ [10]. The reference depth is where the ion chamber's point of measurement must be placed when doing the calibration measurement. The point of measurement is the center of a cylindrical ion chamber, or on the inside of the front electrode of a plane-parallel ion chamber. If you are determining the beam quality by measuring ionization instead of dose, the beam quality parameter can be determined using

$$R_{50} = 1.029\,I_{50} - 0.06 \ \text{(cm)} \tag{24.2}$$

where I_{50} is the depth at which the ionization drops to 50% of the maximum central-axis ionization and

$2 \leq I_{50} \leq 10$ cm [11, 12]. The advantage of the new beam quality specifier is that d_{ref} (as defined above) is very close to R_{100} for lower energies and is monotonically increasing as a function of energy, which is not true of R_{100}. In the previous TG-21 calibration protocol, the measurement depth for electron beam calibration was at R_{100} or d_{max} [13], which can be quite variable at higher energies (depending on the design of the beamline).

When determining the beam quality using a cylindrical ionization chamber, the effective point of measurement must be used. For electron beams, the effective point of measurement is located upstream of the center of the ionization chamber by a distance of $0.5\,r_{cav}$, where r_{cav} is the radius of the ionization chamber [14, 15]. The software for computer-controlled water phantom systems allows users to specify the coordinate system relative to the effective point of measurement. However, if you do so, you must be careful to reset the coordinate system prior to calibration. Failing to do this would mean that the point of measurement would be deeper than d_{ref} and the calibration would be incorrect. Therefore, TG-51 recommends that measurement of the beam quality be done using center of ionization chamber coordinates. The entire curve can then be shifted upstream by $0.5\,r_{cav}$ to get the depth ionization curve. Equations (24.2) and (24.1) can then be used to determine R_{50} and d_{ref}.

The TG-51 protocol specifies that the calibration field size must be greater than or equal to 10×10 cm² for $R_{50} \leq 8.5$ cm, or 20×20 cm² for $R_{50} > 8.5$ cm. The source to surface distance (SSD) for the calibration may be any SSD from 90 to 110 cm. Because most typical accelerators use an applicator system with final collimation at 90–95 cm from the nominal source position and have a maximum energy of 20 MeV or less, the most commonly used calibration standard condition is a 10×10 cm² field at an SSD of 100 cm, although 15×15 cm² is also quite common. The measurement phantom for calibration must be liquid water and measure at least $30 \times 30 \times 30$ cm³. If the calibration beam is directed through a solid wall that is greater than 0.2-cm in thickness, all depths need to be scaled to water-equivalent depths.

Assuming that d_{ref} has been appropriately determined, the point of measurement of the ionization chamber is placed at d_{ref} and the ionization is measured for a set number of monitor units. The fully-corrected ionization, M, is calculated using

$$M = P_{ion}P_{TP}P_{elec}P_{pol}M_{raw} \tag{24.3}$$

where M_{raw} is the measured ionization, P_{ion} corrects for incomplete collection of ionization, P_{TP} corrects the reading to standard temperature (22°C) and pressure (1 atm), P_{elec} accounts for any innacuracies in the electrometer reading, and P_{pol} corrects the reading for polarity effects. The dose in water for an electron beam of specific quality, D_w^Q, is then given by the following equation from the TG-51 report [5].

$$D_w^Q = M \ P_{gr}^Q \ k'_{R_{50}} \ k_{ecal} \ N_{D,w}^{60\text{Co}} \qquad (24.4)$$

From left to right, M is the fully-corrected ionization reading, P_{gr}^Q is a gradient correction to account for the difference between the dose at d_{ref} and the dose at the ion chamber's effective point of measurement, $k'_{R_{50}}$ is a factor that converts the calibration factor for a standard electron beam to the calibration factor for the current electron beam quality, k_{ecal} is a factor that converts the calibration factor in a ^{60}Co beam to the calibration factor for a standard electron beam, and $N_{D,w}^{60\text{Co}}$ is the calibration factor (dose to water) in a ^{60}Co beam. The k_{ecal} factor depends on the ionization chamber chosen for the calibration and was introduced to prepare for a future where primary standards for electron beams may be available. The dose gradient at the reference depth depends on beamline details for electron beams and is not just a function of the beam quality; hence P_{gr}^Q must be explicitly included in the dose equation.

To summarize, the general process for calibrating an electron beam is to first determine the beam quality from I_{50} or R_{50}, then move the point of measurement of the ionization chamber to the d_{ref} depth and measure the ionization, then measure the ionization correction factors to get the fully-corrected ionization reading, and then move the ionization chamber deeper by 0.5 r_{cav} to determine the gradient correction factor. The dose to water at the reference depth, d_{ref}, can then be calculated using Equation (24.4). For the vast majority of institutions that prefer to calibrate a dose rate at the R_{100} depth, the measured dose to water should then be divided by the clinical percent depth dose at the reference depth to get the dose to water at R_{100}.

24.3.2 Reference Dosimetry Verification

The initial calibration is arguably the most important step in commissioning a device for radiation treatments and the electron modality should be handled in a similar manner to any other treatment modalities. As such,

good practice would dictate that a verification (second check) of this crucial step would be prudent and there are several ways to accomplish this task. In a larger department, there are likely multiple "calibrated" ionization chambers and electrometers, so this could be as simple as having a colleague repeat the setup and measurement using a different ionization chamber and electrometer. In a smaller department with just a single calibration system, you could have a second person repeat the measurement with the same equipment independently; however, if there is a problem with the dosimetry equipment, this process will not catch that problem. It is possible to hire a medical physicist specifically to verify the calibration using their own equipment, or a dosimetry service (e.g., IROC Houston Quality Assurance Center[1]) can be used to help verify the calibration [16, 17]. The latter is not as accurate but can catch gross errors in calibration. Depending on any other dosimeters that are available, it is also possible to cross-calibrate another dosimeter and then use that to verify the calibrations. An example of this might be to use an energy-independent dosimeter (e.g., p-type electron diode) and cross-calibrate this dosimeter for one beam energy and then check all of the other energies using that calibration factor. The cross-calibration could also be done on another machine, if available. In my experience, medical physicists often check the electron beam calibration using multiple methods.

24.3.3 Reference Dosimetry for Non-Standard Electron Modalities

For a normal linear accelerator, the calibration protocol, as described above is relatively straightforward and easy to apply. There are a couple of scenarios where it is not so obvious that the protocols can be applied; the first is intraoperative electron beam therapy (IOEBT or IOERT or IORT) and the second is total skin electron therapy (TSET or TSEI). For the former, the applicator or cone system is quite different from the standard linear accelerator and for the latter, there is usually no applicator and the distance to the treatment position is much greater than standard calibration protocols allow.

IORT can be carried out using a standard medical linear accelerator installed in the operating room [18], and to increase the safety of the procedure, the linear accelerator is often modified to disable photon (or x-ray) mode. The standard applicator system is modified to meet the needs of the IORT procedure, but standard electron applicators are not precluded for quality assurance purposes. Therefore, the calibration of each electron beam energy can be done using the standard calibration protocols as described above. Each IORT cone is then characterized by an output factor that relates the dose output with the IORT cone to the standard electron calibration geometry.

Because of the disadvantages of installing a linear accelerator in an operating room, a more common approach to IORT is to use a mobile linear accelerator. These devices have only been available for about the last 20 years and have their own challenges. The AAPM Report 92 (TG-72 [19]) provides a nice summary of the considerations involved in using such a system and Wootton et al. [20] describes commissioning and use of a newer model device. It recommends that the TG-51 calibration protocol be used for calibration of the beams with a 10-cm diameter circular applicator. These devices do not have normal electron applicators, but Monte Carlo calculations of the in-phantom water to air stopping power ratio for these devices shows that they are very close to those recommended in the calibration protocols, as might be expected [21]. One significant issue for calibrating these devices comes from their higher dose per pulse (relative to normal medical linear accelerators), which can cause the ion recombination correction, P_{ion}, to be more significant than normal medical linear accelerator electron beams. Depending on the ionization chamber chosen for the measurement, P_{ion} may be larger than 1.05, and AAPM Report 92 recommends switching to a different ionization chamber if this is the case [19]. Moretti et al. [22] described a method for calibrating a mobile linear accelerator using radiochromic film, which has little dependence on the dose per pulse. However, most physicists would likely be more comfortable using this technique as a verification of the in-water calibration according to current calibration protocols [23].

TSEI represents a different set of problems for reference dosimetry. For normal electron beams, the calibration condition is not very far removed from the irradiation condition. For TSEI, the applicator is removed, and the patient is treated very far away. Traditionally, TSEI calibrations are done at a location where the patient is treated [24]. Because of the lower energy of the beams, plane-parallel ionization chambers are strongly recommended. The approximately horizontal beam geometry implies that the beam has to be

incident on the wall of the water phantom. However, by the time the beam has exited the linear accelerator and made its way to the patient plane, the energy is very low and the reference depth for calibration will be within the wall unless you have a thin window in the phantom wall. Therefore, if calibration is done at the patient plane, solid phantoms are preferable. Since the current standard dose calibration protocols do not allow solid phantoms and very large distances, any dose measurement at the patient plane should really be considered an output verification and not a dose calibration.

It is possible to measure the dose rate using the TSEI mode at a standard SSD (e.g., 100 cm) in a water phantom, which would be compliant with the calibration protocols. The energy of the beam at 100-cm SSD will be higher than at the patient plane is likely to be high enough that a cylindrical ionization chamber would be allowed for the TG-51 calibration protocol. However, the higher dose rate used for the TSEI mode generally leads to more ion recombination, similar to the case for IORT. If P_{ion} is greater than 1.05, a different ionization chamber should be used to be compliant with the calibration protocols. AAPM Report 92 [19] gives an example where a cylindrical ionization chamber was not suitable at a dose rate of 10 Gy min^{-1} for IORT, but the P_{ion} for an advanced Markus (plane-parallel) ionization chamber was no greater than 1.03 and was therefore suitable for dose calibration. The dose rate for the TrueBeam™ (Varian Medical Systems, Palo Alto, CA) TSEI mode is 25 Gy min^{-1}. Since plane-parallel ionization chambers are also recommended for lower electron energies, using such a chamber would be recommended for the TSEI dose calibration.

To determine the relationship between the dose output at 100-cm SSD and at the treatment plane, it is not nearly as straightforward as it is for normal electron beams or IORT. The energy of the beam at the treatment plane for TSEI is much lower than at 100-cm SSD, and this change in energy needs to be considered. Using a dosimeter with less energy and dose rate dependence, such as radiochromic film or a semiconductor diode, can help. For example, one could calibrate at 100-cm SSD using a plane-parallel ionization chamber, then make a measurement of the output using another dosimeter in a Solid Water® (Gammex RMI, Middleton, WI) or equivalent phantom. The same phantom and dosimetry can then be put at the patient plane to determine the dose output at that point. It would also be prudent to have multiple ways to check the output. For example, Antolak et al. [25] used thermoluminescent dosimeters placed on the abdomen of TSEI patients and normalized the readings to the dose output of a standard 9-MeV electron beam (not the TSEI beam). While the spread of the readings was significant, which was indicative of the patient setup uncertainty, the average over all patients was within 1% of the desired dose.

The polarity effect is also significant when using smaller ionization chambers at larger distances. Das et al. [26] reported on measurements of bremsstrahlung dose for a TSEI beam and found that the length of cable in the beam significantly affected the results. Therefore, they recommended using dosimeters without cables when making measurements under those conditions. It is interesting to note that they did not mention anything about investigating the polarity dependence of the reading. When calibrating a TSEI beam using a small ionization chamber, Antolak and Hogstrom [24] found that correcting the raw readings for polarity was necessary. Anecdotally, the magnitude of the polarity effect does depend on the length of cable in the beam, but the polarity corrected reading is not very strongly dependent on the length of the cable in the beam and steps can be taken to minimize this (as recommended by Das et al. [26]). Therefore, I would not go as far as recommending that ionization chambers and other cabled dosimeters be avoided, but that cable lengths within the beams be minimized. A non-cabled dosimeter can be useful for a second check, however.

Assuming that the calibration is done at 100-cm SSD, the fact that the treatment condition is far removed from the calibration condition complicates the quality assurance of the procedure. The relative beam output and profile at the patient plane are strongly dependent on the energy of the beam [24]; therefore changes in beam output at the patient plane should not lead to immediate adjustment of the linear accelerator without further investigation.

24.3.4 Commissioning and QA Measurements

While calibration of the electron beams is arguably the most important aspect of electron beam dosimetry, commissioning of the electron beams is also very important. Commissioning the electron beams prepares them for clinical use, which in a modern clinic means measuring the data required for input into the treatment planning system and to validate the operation of the treatment planning system. The initial commissioning also establishes a baseline for the overall performance of the beam delivery system. In most clinics, the majority of the commissioning measurements are carried out in a scanning water phantom and AAPM Report 106 describes in detail the best practice for carrying out these measurements [2]. On an annual basis, the current performance level of the linear accelerator is checked using the same scanning water phantom. If required, beam parameters are adjusted to bring the performance back to the baseline. QA is also done on a daily and monthly basis, as prescribed by regulations and professionally-defined good practice, for example, AAPM Report 46 (TG-40 [27]). Daily and monthly QA does not generally require a large water phantom and methods for doing these dosimetric measurements will be described later.

24.3.5 Detector Choice

The two most common types of detectors used for commissioning and QA measurements are ionization chambers and semiconductor diodes. Ionization chambers are probably the most commonly used detectors, with cylindrical chambers being the most popular choice in North America due to AAPM calibration protocols allowing their use for energies down to 6 MeV [5]. In other parts of the world, well-guarded plane-parallel ionization chambers are preferred for electron beam measurements [7, 28]. For cylindrical ionization chambers, the user needs to be aware that the effective point of measurement is located at a point that is 0.5 r_{cav} upstream of the center of the ionization chamber, while the point of measurement in the calibration protocols is the center of the ionization chamber [3, 14, 29]. For plane-parallel ionization chambers, the effective point of measurement is on the inside of the front electrode, which is the same as the calibration point of measurement. For a typical cylindrical ionization chamber used for general purpose water phantom scanning, the active volume is approximately 0.1 cm^3, with a radius of approximately 3 mm and a length of approximately 6 mm. Plane-parallel ionization chambers used for water phantom scanning generally have a diameter of 5 mm or greater.

Semiconductor diodes, more specifically p-type, have several advantages for relative electron beam dosimetry. The sensitive volume of the diode is fairly small with a radius of 1–2.5 mm and a thickness of less than 0.1 mm. Despite the small volume, the nominal sensitivity (current per unit dose) is generally as large or larger than typical ionization chambers used for beam scanning. Similar to plane-parallel ionization chambers, the effective point of measurement is very well defined and is the position of the semiconductor die in the detector. The stopping power ratio (silicon to water) is not very energy dependent, and it is generally accepted that ionization and dose are interchangeable [3, 29, 30]. However, it is also considered good practice to routinely verify this by comparing to percent depth dose determined using ionization chambers. The response of diode detectors is reduced as a result of being irradiated (< 4%/kGy), so monitoring the response is warranted for extended measurement sessions. The temperature dependence of the response is approximately 0.3–0.4% K^{-1}, which is similar to ionization chambers.

Diamond detectors can also be considered for electron beam commissioning and QA. The Institute of Physics and Engineering in Medicine (IPEM) code of practice and the International Atomic Energy Agency (IAEA) calibration protocols specifically mention their use for directly measuring percent depth dose curves, similar to semiconductor diodes [29, 31]. However, AAPM has no specific guidance on the use of these detectors and the author has no practical experience with their use for electron beam dosimetry. As with semiconductor diode detectors, it is recommended that before a diamond detector be used for extensive dosimetric measurements, it is thoroughly commissioned, including comparisons with trusted ionization chamber measurements [32].

Despite the significant advantage of not having to convert the ionization signal to dose, semiconductor diode and diamond detectors are often not trusted for commissioning measurements. Ionization chambers are essentially mechanical devices with demonstrated long-term stability and very little variation from one device to another (of the same model). Because of this macroscopic sameness and the fact that ionization chambers are universally used for calibration, it is very easy to be confident in their use for commissioning measurements. Anecdotally, some physicists can cite instances where semiconductor diodes have led to bad measurements and they are therefore hesitant to trust them for measurements where ionization chambers can be used, even if they are not ideal. Ionization chambers can also occasionally fail (e.g., excessive leakage), but perhaps less often and physicists tend to still trust their use for most measurements.

The microscopic nature of the semiconductor and diode devices is also a potential issue. The measurement device is encapsulated and not visible to the naked eye. Examination with x-rays can be useful for confirming the location of the active volume but will not tell you if the semiconductor is p-type (preferred) or n-type. Over the years, there have been many different variations in construction of the devices. For example, there are photon-specific diodes with added filtration to provide better energy response when measuring x-ray beams. These devices are not suitable for electron beam measurements and it can be difficult to tell the difference between photon and electron diodes. Semiconductor diode detectors are also subject to radiation damage (p-type diodes are better in this regard) and if the diode is being used extensively, the perceived risk of the detector not performing as expected increases. In the author's opinion, many of these concerns are not warranted and it is very simple to ascertain that a particular diode is performing well prior to use. It would also be good practice to carefully track and label any semiconductor detectors in your possession to ensure that the appropriate detector is used.

24.3.6 Percent Depth Dose

When measuring percent depth dose in water, PDD_w, using a scanning water phantom, it is recommended to scan the detector starting at depth and moving toward the surface [2, 3]. If scanning is done in the reverse direction, surface tension effects can pull the water surface down with the detector for a few millimeters, leading to incorrect data at shallow depths. While it is easy to reduce surface tension effects using something like a few drops of liquid soap, it is still good practice to scan from depth toward the surface. As mentioned above, semiconductor diodes or diamond detectors can be used to measure percentage depth dose directly. However, it is also recommended that results are compared to ionization chamber measurements, so it is important to understand how to convert percent ionization to percent dose.

According to AAPM Report 99 [3], PDD_w is calculated from the percent depth ionization in water, PDI_w, using the following equation.

$$PDD_w(d) = PDI_w(d) \times \frac{\left(\bar{L}/\rho\right)_{air}^{w}(R_{50}, d) \cdot P_{fl}(E_d)}{\left(\bar{L}/\rho\right)_{air}^{w}(R_{50}, d_{max}) \cdot P_{fl}(E_{d_{max}})}$$

(24.5)

The two corrections are for the restricted stopping power ratio from water to air and the fluence correction. Prior to the implementation of the TG-51 calibration protocol, AAPM Report 32 (TG-25 [14]) recommended the use of the restricted stopping powers from the TG-21 calibration protocol [13], which were computed using monoenergetic electron beams. Burns et al. [10] improved upon this approach by calculating restricted stopping power ratios for realistic electron beams and parameterizing the restricted stopping power as a function of the depth of 50% dose, R_{50}, and the depth, d.

$$\left(\overline{L}/\rho\right)_{air}^{w}\left(R_{50},d\right)=$$

$$\frac{1.0752-0.50867\left(\ln R_{50}\right)+0.08867\left(\ln R_{50}\right)^2-0.08402\left(d/R_{50}\right)}{1-0.42806\left(\ln R_{50}\right)+0.064627\left(\ln R_{50}\right)^2+0.003085\left(\ln R_{50}\right)^3-0.1246\left(d/R_{50}\right)}$$

$$(24.6)$$

For the TG-51 calibration protocol, Equation (24.2) is used for calculating R_{50} from the measured value of I_{50} and Rogers [33] showed that using Equation (24.6) for the stopping power ratio produces dose estimates that are accurate at all depths to within 1% of the maximum dose on central axis. This equation is recommended by AAPM Report 99 [3] and is also consistent with recommendations in TRS398 [7]. The TG-51 calibration protocol also mentions that to be entirely consistent, the R_{50} value determined from the converted ionization curve should be compared with the R_{50} value previously determined using Equation (24.2). If they are significantly different, it is likely that an error was made along the way.

The second correction factor for conversion of ionization to dose is the fluence correction factor. In AAPM Report 99, the authors say to determine the replacement correction factor, P_{repl}, it can be divided into a gradient correction factor, P_{gr}, and a fluence correction factor, P_{fl}. In practice, P_{gr} is accounted for by using the effective point of measurement of the ionization chamber, which is why you don't really need to determine P_{repl}. For a plane-parallel ionization chamber, the effective point of measurement is the same as the point of measurement (inside surface of the front electrode), so P_{gr} is unity. For a cylindrical ionization chamber, commonly used in North America, the effective point of measurement is $0.5\,r_{cav}$ upstream of the center of the ionization chamber. There are two practical ways to handle this correction. The first method is to measure using the point of measurement (center of chamber) and then shift the

resulting curve upstream by $0.5\,r_{cav}$. The second method is to adjust the origin of the scanning coordinate system so that the center of the ionization chamber is at a depth of $0.5\,r_{cav}$ in the water phantom when the scanning depth coordinate reads zero. Either method is acceptable, but the author has a slight preference for shifting the curve after measurement. Modern scanning systems have the capability to store user preferences for each measurement device that is used, so scanning coordinates can be adjusted as simply as specifying a certain detector for the measurement. If you choose to have the system implicitly apply the gradient correction, then you risk having the point of measurement set incorrectly when calibrating the beam. No matter which method is chosen, having a standard process for doing this in a reproducible manner is most important.

The fluence correction factor accounts for the change in the primary electron fluence spectrum caused by the presence of the air cavity. For a well-guarded plane-parallel ionization chamber, the fluence correction factor is unity, which essentially means that the restricted stopping power ratio is the only correction needed to convert ionization to dose. For other plane-parallel ionization chambers, AAPM Report 48 (TG-39 [34]) should be used to get the fluence correction factor. For a cylindrical ionization chamber, the fluence correction factor is given in Table I of AAPM Report 99 [3]. Note that AAPM Report 99 and AAPM Report 32 [14] both indicate that the table is showing the replacement correction, even though it is really the fluence correction. For every ionization chamber diameter listed in the table, a second order polynomial fit reproduces the fluence correction factor to within 0.1% and the polynomial coefficients are listed in Table 24.1. Because P_{fl} is a slowly varying function of both the ionization chamber diameter and the mean energy at depth, $\overline{E}_d$, and the polynomial coefficients vary smoothly as a function of ionization chamber diameter, interpolation of the polynomial coefficients is almost the same as direct interpolation in the original table. However, the polynomial coefficients are easier to use in a spreadsheet or computer program, if needed (assuming that the correction is not already handled by the scanning software). The mean energy at depth is given by

$$\overline{E}_d = \overline{E}_0\left(1-\frac{d}{R_p}\right) \qquad (24.7)$$

TABLE 24.1   Polynomial Fit Coefficients for Cylindrical Ionization Chamber Fluence Correction Factor, P_{fl}

Chamber Diameter (mm)	A	B	C
3	0.9720	2.351	54.8
5	0.9547	3.841	89.0
6	0.9470	4.301	94.0
7	0.9385	4.957	105

Note: The cylindrical ionization chamber fluence correction factor is approximately given by $P_{fl} = A + B \times 10^{-3} \cdot \bar{E}_d + C \times 10^{-6} \cdot \bar{E}_d^2$, where $\bar{E}_d$ is the mean energy at depth d. The above coefficients reproduce Table I in AAPM Report 99 [3] to within 0.1% for all values in the table. Since P_{fl} is a slowly varying function of both energy and depth and the coefficients also vary slowly as a function of diameter, the coefficients can be interpolated to get the fluence correction factor for ionization chamber diameters not included in the table.

where $\bar{E}_0$ is the mean energy at the surface of the water phantom and R_p is the practical range of the electron beam [35]. The mean energy at the surface can be determined using the relationships given in the IPEMB code of practice [28] for R_{50} or I_{50} (in cm).

$$\bar{E}_0 = 0.656 + 2.059\,R_{50} + 0.022\,R_{50}^2 \qquad (24.8)$$

$$\bar{E}_0 = 0.818 + 1.935\,I_{50} + 0.040\,I_{50}^2 \qquad (24.9)$$

While it is recognized that there are more accurate data in the literature, differences between the simple approach described above and other approaches are clinically insignificant [3, 36].

Equation (24.5) implies that the depth d_{max} is known. However, if we are starting with an unknown beam, we will know the depth of maximum ionization, I_{100}, but the depth of maximum dose, d_{max} or R_{100}, is not necessarily the same. Therefore, what most users will really do is calculate the percent depth dose using the following equation.

$$PDD_w(d) = \frac{PDI_w(d) \cdot \left(\bar{L}/\rho\right)_{air}^w (R_{50}, d) \cdot P_{fl}(E_d)}{\max\left\{PDI_w(d) \cdot \left(\bar{L}/\rho\right)_{air}^w (R_{50}, d) \cdot P_{fl}(E_d)\right\}} \qquad (24.10)$$

In other words, the *PDI* (or ionization) values are multiplied by the restricted stopping power and fluence corrections, then the entire curve is normalized to the maximum of the corrected values. For most scanning systems, the computer analysis software will have these corrections built in. Once the correct operation of the software has been verified, converting ionization to dose is quite easy because the software can determine I_{50} and

hence R_{50} using Equation (24.2) and then all of the corrections can be made as described above.

Figure 24.1 shows how these correction factors work for a lower energy beam, 9 MeV, and a higher energy beam, 20 MeV. For the unshifted percent depth ionization curves, you'll notice that there is a slight upward curve near the surface. This is due to the fact that the ionization chamber is transitioning from being entirely submerged to only partially submerged. In this case, the radius of the ionization chamber is 3 mm, so the data for depths up to 3 mm is affected. Shifting the percent depth ionization curve upstream by 1.5 mm makes the curve look a little better, but the first 1.5 mm is still not reliable. The dose near the surface should have a steeper delta-ray buildup, but using just a cylindrical ionization chamber will not be able to resolve this feature. For the lower energy (9-MeV) beam, the stopping power and fluence correction factors have a fairly subtle effect, with the stopping power correction reducing the percent dose in the buildup region (relative to percent depth ionization) and the fluence correction factor slightly increasing the dose in the buildup region. For the higher energy (20-MeV) beam, you can see similar effects. However, because the change of energy is larger, the stopping power and fluence corrections have a greater effect, especially when it comes to the perceived therapeutic depth. The depth to 90% of maximum ionization is 5.5 cm, while the depth to 90% of maximum dose is approximately 6.0 cm. The depth of maximum dose is also about 1 cm deeper than the depth of maximum ionization. In addition, there is a larger difference between the curves with just the stopping power correction and the fully-corrected percent depth dose curve; hence the fluence correction factor is also more significant for the higher energy.

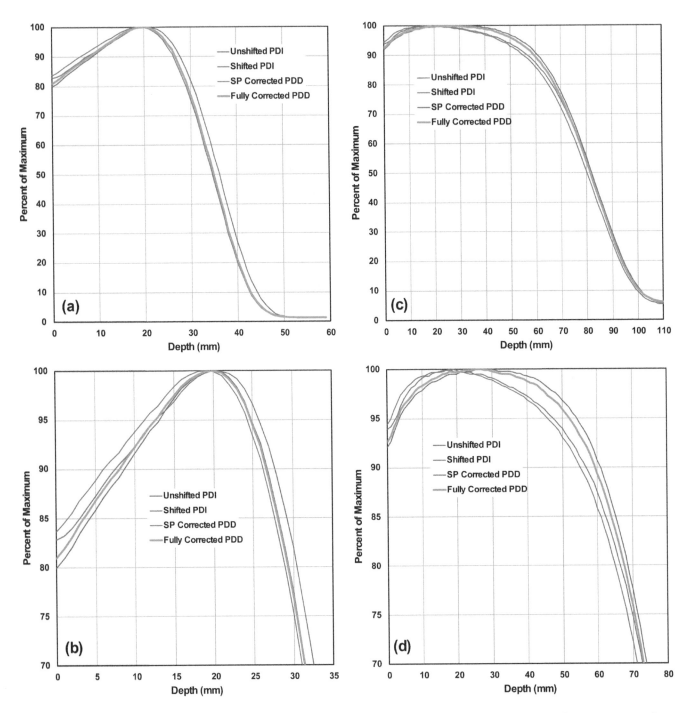

FIGURE 24.1   Percent depth ionization and percent depth dose for 15×15 cm² field size (a) 9-MeV, (b) 9-MeV zoomed in, (c) 20-MeV, and (d) 20-MeV zoomed in. The purple line is the percent depth ionization, the blue line is the same curve shifted upstream by 0.5r, the red curve includes the stopping power correction, and the thicker green curve is fully corrected for stopping power and the fluence correction factors.

24.3.7 Lateral Profiles

When measuring lateral profiles in a simple water phantom, the situation is quite a bit simpler than what was described above for percent depth dose. As long as the depth is constant and there are no added heterogeneities, the energy and fluence correction factor at each point on the profile are constant, so no corrections are needed as a function of off-axis position. The true profile at any depth is a convolution of a rectangular function (the beam aperture) and a dose-spread kernel that can be characterized by a standard deviation of lateral position. Because of the finite size of the detector, the

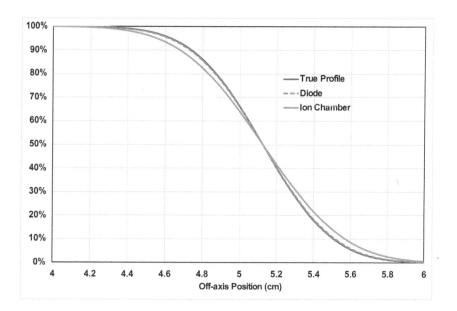

FIGURE 24.2 Calculated electron beam penumbra for a $10 \times 10 \ cm^2$ field size at a depth of 2.5 cm. The "True Profile" is a convolution of a rectangular function with a scattering kernel with a standard deviation (sigma) of 0.3 cm. The "Diode" profile is a moving average of the true profile with a width of 0.2 cm, which is a representative size for a semiconductor diode detector. The "Ion Chamber" profile is a moving average of the true profile with a width of 0.6 cm, which is a representative size for a cylindrical ionization chamber detector.

measured lateral profile is a convolution of a detector response function and the true profile and is therefore a smoothed version of the true profile. The dose-spread kernel is due to the angular variance of the electrons at the position of the electron aperture and the multiple Coulomb scattering in the water phantom [37]. Using the measured penumbra width, Antolak et al. [38] extracted the standard deviation of the scattering kernel for 12-MeV and 16-MeV beams. For both energies, the standard deviation is approximately 2 mm at the surface and 3–3.5 mm at the depth of maximum dose, increasing to a maximum of 11.5–14 mm at a depth of $0.8R_p$.

To illustrate the influence of detector size on the beam profile measurement, we will assume a uniform beam fluence and a dose-spread kernel standard deviation of 3 mm at a depth of 2.5 cm (approximate depth of maximum dose for 12 MeV electrons). The normalized dose profile is calculated using the first few terms of Equation (24.13) from Hogstrom et al. [37] and Figure 24.2 shows the penumbra of this profile for a $10 \times 10 \ cm^2$ field size. When we add a detector to the system, the measured profile is a convolution of the true profile and the detector response function. The effective scattering kernel standard deviation is the square root of the quadrature sum of the standard deviations of the true scattering kernel and the detector response

functions. If we assume that the response function of the detector is a rectangular function with the same width as the sensitive volume of the detector, w, the standard deviation of the response function is $w/2\sqrt{3}$. For a detector width of 2 mm, which is typical for a diode detector, the standard deviation is 0.58 mm. If the true scattering kernel standard deviation is 3.0 mm, the effective standard deviation is 3.06 and, as shown in Figure 24.2, the measured penumbra is almost identical to the true penumbra. If the detector width is increased to 6 mm (e.g., typical cylindrical ionization chamber), the detector response standard deviation is now 1.73 mm and the effective standard deviation is now 3.46 cm. In Figure 24.2, we see that the measured penumbra with the larger detector is visibly different from the true penumbra. Note that the detector profiles in Figure 24.2 were calculated using a moving average filter as an illustration of what the measured profiles might look like. At deeper depths, the detector size has less influence because the penumbra width increases, but it should be fairly clear that using a smaller detector (e.g., semiconductor diode) is desirable.

24.3.8 Output Factors

The output factor is the ratio of the dose per monitor unit for a give field size to the dose per monitor unit for a reference condition. The reference condition is usually the same field size and SSD that was used for the calibration of the linear accelerator (e.g., 10×10 cm^2 field size at 100-cm SSD). While the reference depth, d_{ref}, is the specified depth to use for calibrating the linear accelerator, it is much more common to use the clinical percent depth dose to nominally calibrate the linear accelerator (i.e., set 1 cGy per MU) at d_{max} or R_{100}. The AAPM Report 99 [3] defines the output factor, S_e using the following equation

$$ S_e\left(d_{max}(r), r, SSD\right) = \frac{\dot{D}\left(d_{max}(r), r, SSD\right)}{\dot{D}\left(d_{max}(r_0), r_0, SSD_0\right)} \quad (24.11) $$

where r is the field size for the output factor, r_0 is the reference field size, SSD is the SSD for the output factor measurement, SSD_0 is the reference SSD, and $\dot{D}$ is the dose rate (or dose per monitor unit). In this equation, the depth of maximum dose is determined for each irradiation condition. Having to determine the depth of maximum dose for each field size makes this measurement more complicated than a simple ratio of doses at a single depth. In a water phantom, the user could perform a depth dose scan (or convert ionization to dose) around d_{max} to determine its location and dose rate to compare to the dose rate at d_{max} for the reference field. However, for practical purposes, this process is generally more complicated than necessary.

The most common choice for output factor measurements is to use a nominal d_{max} depth for both measurements, with a reasonable choice being the depth of maximum dose of the reference field size. Once the nominal depth is determined, a simpler phantom (e.g., small water phantom with manual dosimeter positioning or solid phantom material) could be used to reproducibly place the effective measurement point of the detector at the desired depth. If an ionization chamber is used, doing the measurements at the same depth means that the stopping power and replacement corrections are also the same and the nominal output factor is simply a ratio of measured ionization values.

If the intent of the measurement is to check the dose output in the treatment planning system, using a nominal depth of maximum dose means that you can use the point dose tool to easily compare measurement and calculation. If it was desired to determine the output factor at the actual depths of maximum dose, clinical depth doses could be used to correct the measured values to

the depth of maximum dose. For the standard field size and SSD, this depth dose should be readily available. For the measured field size, the true depth dose may not be readily available (e.g., an irregular patient cutout), but there may be data for similar field sizes that may allow the user to closely approximate the true depth dose. If that is not good enough, it may be necessary to perform a depth dose scan.

If the output factors are needed as input data for the treatment planning system, the user needs to carefully check the physics manual to see what the treatment planning expects. The treatment planning system may only require output factors at a nominal depth of maximum dose to make data entry easier for the user since it would be easy for the computer algorithm to calculate the dose elsewhere.

If the user decides to use the actual depth of maximum dose, using a semiconductor diode or diamond detector that does not require depth-dependent correction factors would probably be more convenient. However, comparison with the treatment planning system becomes more complicated since you need to consider that the calculated depth dose may not be the same as the measured depth dose and how that can affect the interpretation of the results. Fortunately, this added complication is rarely necessary and should be avoided if possible.

The user might could also consider using the reference depth from the calibration, d_{ref}, instead of the nominal depth of maximum dose. The reference depth is usually close to d_{max}, but it can be deeper than d_{max} for higher energies. For semiconductor diodes, diamond detectors, and plane-parallel ionization chambers, where a gradient correction is not needed, placing the detector at d_{ref} is unambiguous. However, if a cylindrical ionization chamber is used, the user should ensure that the effective point of measurement of the ionization chamber is placed at d_{ref}. Since the reference calibration is done with the center of the ionization chamber at that depth, this could cause some confusion.

Given the above considerations, it is the author's recommendation that output factors be measured using a nominal depth of maximum dose, preferably with a semiconductor diode or diamond detector. If output factors at the true depth of maximum dose are necessary, clinical depth dose curves can generally be used to correct the measured value with sufficient accuracy for clinical use.

24.4 SOLID PHANTOM MEASUREMENTS

24.4.1 General Considerations

For any type of phantom measurement, there are two primary considerations for choosing an appropriate dosimeter and phantom. The first is whether the presence of the dosimeter significantly alters the radiation transport in the phantom material. For example, does placing a film between 2 slabs of phantom material alter how the electron beam traverses the material? The second consideration is how the energy response of the detector compares to the energy response of the phantom material. A good example of this for ionization chambers is the conversion of percent ionization to percent dose in an electron beam using depth-dependent stopping power and fluence correction factors. For that conversion, the assumption is that the beam is perpendicular to the water phantom surface so that the energy is primarily a function of depth. For any other geometry, the relatively simple formula for the stopping power correction in TG-51 may not be applicable. This is one reason why ionization chambers may not be the dosimeter of choice for a particular measurement. For example, semiconductor diode detectors were necessary to measure dose distributions behind heterogeneities such as rods and discs, used for evaluating electron beam dose algorithms [39, 40] and for the standard electron dose distribution data set measured by Boyd et al. [41].

24.4.2 Quality Assurance Measurements

For periodic quality assurance, solid phantom materials can be more convenient to use in place of setting up a water phantom. However, care must be taken to ensure that the chosen phantom material is appropriate for the measurement purpose. When AAPM Report 32 [14] was published, the most common solid phantom materials were plastics such as polystyrene and acrylic (PMMA) and the TG-21 calibration protocol allowed for reference dosimetry to be done in plastic phantoms [13]. When doing these calibrations, the absolute density of the plastic might not be the same as water, which meant that the depth of d_{max} might be different in the plastic as compared to d_{max} in water. Differences in scattering power meant that the actual dose at d_{max} would be different than simple density scaling would imply. The manufacturing of these plastic materials can also introduce additional uncertainties that complicate their use even more [3]. These phantoms can still be used for periodic quality assurance such as daily and monthly constancy checks, but some level of quality control should be exercised in their implementation.

Currently, there are several more advanced solid material water substitutes that are suitable for radiation dosimetry. Examples of such materials are Solid Water® (Gammex RMI, Middleton, WI, USA) and Virtual Water™ (Med-Cal, Verona, WI, USA), and there are several other similar materials on the market. They are generally sold in the form of solid slabs in a variety of thicknesses and thicker slabs can usually be ordered with machined cavities for insertion of standard ionization chambers. Their density and atomic composition are designed such that interactions with x-rays and electrons are much the same as in liquid water. Before putting any of these materials into routine use, they should be examining radiographically to verify that no unexpected voids exist in the material. One should also be careful to examine prior literature regarding the particular brand and model that is being used to ensure that it is suitable for the intended purpose. Some early formulations were known to be reasonably good substitutes in x-ray beams, but not so good for electron beams [42, 43].

24.4.3 Solid Phantom Film Measurements

When making measurements in solid phantoms using film, the phantom and film measurement system should fit together without any air gaps and the film edge should be flush with the edge of the phantom if the film is being irradiated edge-on [44, 45]. While Cherenkov emission is not a major energy loss mechanism for electron beams, the film may be sensitive to the resulting photons, which means that opaque phantom materials are preferred for electron beams. For radiographic films in paper jackets, the film may be left in the jacket, but it is recommended to put pinpricks in the jacket to release air inside the jacket and to allow for good compression of the film-phantom combination [45]

To assist with standard planar dosimetry measurements, Bova [46] designed a hinged film phantom using solid water slabs and pins for reproducible alignment. To allow for slight variations in film size from batch to batch, shims were available to get a better fit of the film within the phantom and to ensure coincidence of the edge of the film and the edge of the phantom. Unfortunately, this phantom is not commercially available to be the best of the author's knowledge. A different phantom-cassette design is commercially available

(Radiation Products Design, Albertville, MN 55301), but the author has no experience with the device.

If the film orientation is perpendicular (or at least not close to parallel) to the beam direction, the actual edge of the film relative to the edge of the phantom is not as important. However, the slabs used for the phantom must be flat, which is usually true for phantom slabs manufactured expressly for radiation dosimetry. However, off-the-shelf plastic slab materials will generally require machining to create a uniform-thickness slab with flat surfaces.

24.4.4 Other Solid Phantom Measurements

In addition to solid materials that mimic water, there are solid materials available that mimic muscle tissue, adipose, lung, bone, and other tissues. There are commercial phantoms available that have included heterogeneities, and these can be modified to allow for dosimeters to be placed inside the phantom. Hogstrom et al. [47] created such phantoms using patient CT data to help evaluate the accuracy of an electron dose calculation algorithm. In a later study, the same phantoms were used to evaluate customized bolus electron treatment plans [48]. The phantom had holes drilled out to accept small TLD capsules to measure the dose and because the capsules were relatively small, the dose perturbation of the dosimeters was reasonably small. As mentioned above, a key consideration for any type of phantom measurement for electron beams is minimizing the effect of the dosimeter on the dose distribution.

24.5 *IN VIVO* DOSIMETRY

There are many different kinds of dosimeters that are suitable for *in vivo* dosimetry for electron beams and there is a good chance that the reader already has access to at least one *in vivo* dosimetry system. For those that do not yet have access to such a system, Chapter 11 of this book [49] describes various systems that are available. For the most part, any *in vivo* dosimetry system that is suitable for use with x-ray beams can be used for measurements in electron beams, provided that you consider how the dosimeter may affect the irradiation itself. Some *in vivo* dosimeters may have inherent buildup that is useful for x-ray beams and it is generally not advised to use such dosimeters for electron beams.

When placing a dosimeter on the patient surface, it is important to remember that the dosimeter can perturb the electron beam. This perturbation may make it more difficult to get a good measurement and for larger dosimeters, it may affect the dose distribution inside the patient as well. Before doing any sort of *in vivo* dosimetry with electron beams, it may be useful to do some phantom measurements with the dosimetry system to ensure that you can obtain reasonable results under controlled conditions. To estimate the degree of dose perturbation inside the patient, it would be useful to place the dosimeter on a film phantom where the film is parallel to the beam direction and close to the phantom surface. This will allow you to make an informed decision regarding the utility of your particular system. The *in vivo* dosimeter should ideally be thin in the beam direction to minimize the perturbation. It should also attach firmly to the skin surface and it may be necessary to remove body hair so that air gaps between the dosimeter and patient do not affect the measurements [25].

24.6 SUMMARY

This chapter was limited to aspects of measuring dose distributions for therapeutic electron beams and there is much more that can be learned about their use in radiation therapy. The AAPM, IPEM, and IAEA documents mentioned in this chapter provide much more detail than what was provided here. Hogstrom and Almond [50] provide a reasonably comprehensive review of the physics of electron beam radiotherapy and AAPM Report 99 [3] supplements AAPM Report 32 [14] by providing several clinical examples of electron beam radiotherapy. Treatment planning aspects of electron beam radiotherapy have also been described elsewhere by the author [51].

NOTES

1. http://rpc.mdanderson.org/RPC/home.htm, accessed Oct 7, 2020.

REFERENCES

1. H. Huizenga and P. R. M. Storchi, "Numerical calculation of energy deposition by broad high-energy electron beams," *Physics in Medicine and Biology* 34(10), 1371–1396 (1989). https://doi.org/10.1088/0031-9155/34/10/003.

2. Indra J. Das, Chee-Wai W. Cheng, Ronald J. Watts, Anders Ahnesjo, John P. Gibbons, X. Allen Li, Jessica Lowenstein, Raj K. Mitra, William E. Simon, Timothy C. Zhu, "Accelerator beam data commissioning equipment and procedures: Report of the TG-106 of the Therapy Physics Committee of the AAPM," *Medical Physics* 35(9), 4186–4215 (2008). https://doi.org/10.1118/1.2969070.

3. Bruce J. Gerbi, John A. Antolak, F. Christopher Deibel, David S. Followill, Michael G. Herman, Patrick D. Higgins, M. Saiful Huq, et al., "Recommendations for clinical electron beam dosimetry: Supplement to the recommendations of Task Group 25," *Medical Physics* **36**(7), 3239–3279 (2009). https://doi.org/10.1118/1.3125820.

4. Bruce J. Gerbi, John A. Antolak, F. Christopher Deibel, David S. Followill, Michael G. Herman, P. D. Higgins, M. Saiful Huq, et al., "Erratum: 'Recommendations for clinical electron beam dosimetry: Supplement to the recommendations of Task Group 25' [Med. Phys. 36, 3239–3279 (2009)]," *Medical Physics* **38**(1), 548–548 (2010). https://doi.org/10.1118/1.3532916.

5. Peter R. Almond, Peter J. Biggs, Bert M. Coursey, William F. Hanson, M. Saiful Huq, Ravinder Nath, David W. O. Rogers, "AAPM's TG-51 protocol for clinical reference dosimetry of high-energy photon and electron beams," *Medical Physics* **26**(9), 1847–1870 (1999). https://doi.org/10.1118/1.598691.

6. Malcolm R. McEwen, Larry A. DeWerd, Geoffrey S. Ibbott, David S. Followill, David W. O. Rogers, Stephen Seltzer, Jan P. Seuntjens, "Addendum to the AAPM's TG-51 protocol for clinical reference dosimetry of high-energy photon beams," *Medical Physics* **41**(4), 041501 (2014). https://doi.org/10.1118/1.4866223.

7. IAEA, "Absorbed dose determination in external beam radiotherapy." *Technical Report Series* (IAEA, Vienna, Austria, 2000). https://doi.org/10.1097/00004032-200111000-00017.

8. R. C. Tailor and W. F. Hanson, "Calculated absorbed-dose ratios, TG51/TG21, for most widely used cylindrical and parallel-plate ion chambers over a range of photon and electron energies," *Medical Physics* **29**(7), 1464–1472 (2002). https://doi.org/10.1118/1.1487857.

9. R. C. Tailor, W. F. Hanson, G. S. Ibbott, "TG-51: Experience from 150 institutions, common errors, and helpful hints," *Journal of Applied Clinical Medical Physics* **4**(2), 102–111 (2003). https://doi.org/10.1120/jacmp.v4i2.2524.

10. D. T. Burns, George X. Ding, David W. O. Rogers, "R50 as a beam quality specifier for selecting stopping-power ratios and reference depths for electron dosimetry," *Medical Physics* **27**(7), 383–388 (1996). https://doi.org/10.1118/1.597893.

11. George X. Ding, David W. O. Rogers, Thomas R. Mackie, "Calculation of stopping-power ratios using realistic clinical electron beams," *Medical Physics* **22**(5), 489–501 (1995). https://doi.org/10.1118/1.597581.

12. M. Saiful Huq, Ning Yue, N. Suntharalingam, "Experimental determination of fluence correction factors at depths beyond dmax for a farmer type cylindrical ionization chamber in clinical electron beams," *Medical Physics* **24**(10), 1609–1613 (1997). https://doi.org/10.1118/1.597978.

13. AAPM, "A protocol for the determination of absorbed dose from high-energy photon and electron beams,"

Medical Physics **10**(6), 741–771 (1983). https://doi.org/10.1118/1.595446.

14. Faiz M. Khan, Karen P. Doppke, Kenneth R. Hogstrom, Gerald J. Kutcher, Ravinder Nath, Satish C. Prasad, James A. Purdy, Martin Rozenfeld, Barry L. Werner, "Clinical electron-beam dosimetry: Report of AAPM Radiation Therapy Committee Task Group No. 25," *Medical Physics* **18**(1), 73–109 (1991). https://doi.org/10.1118/1.596695.

15. IAEA, "Absorbed dose determination in photon and electron beams: An international code of practice," *Technical Report Series* (IAEA, Vienna, 1987), vol. 277.

16. J. Homnick, G. Ibbott, A. Springer, J. Aguirre, "TH-D-352-05: Optically stimulated luminescence (OSL) dosimeters can be used for remote dosimetry services," *Medical Physics* **35**(6 Part 27), 2994–2995 (2008). https://doi.org/10.1118/1.2962948.

17. J. Aguirre, P. Alvarez, D. Followill, G. Ibbott, C. Amador, A. Tailor, "SU-FF-T-306: Optically stimulated light dosimetry: commissioning of an optically stimulated luminescence (OSL) system for remote dosimetry audits, the radiological physics center experience," *Medical Physics* **36**(6 Part 13), 2591–2592 (2009). https://doi.org/10.1118/1.3181785.

18. Jatinder R. Palta, Peter J. Biggs, John D. Hazle, M. Saiful Huq, Robert A. Dahl, Timothy G. Ochran, Jerry Soen, Ralph R. Dobelbower, Edwin C. McCullough, "Intraoperative electron beam radiation therapy: Technique, dosimetry, and dose specification: Report of Task Force 48 of the Radiation Therapy Committee, American Association of Physicists in Medicine," *International Journal of Radiation Oncology, Biology, Physics* **33**(3), 725–746 (1995). https://doi.org/10.1016/0360-3016(95)00280-C.

19. A. Sam Beddar, Peter J. Biggs, Sha Chang, Gary A. Ezzell, Bruce A. Faddegon, Frank W. Hensley, Michael D. Mills, "Intraoperative radiation therapy using mobile electron linear accelerators: Report of AAPM Radiation Therapy Committee Task Group No. 72." *Medical Physics* **33**(5), 1476–89 (2006). https://doi.org/10.1118/1.2194447.

20. Landon S. Wootton, Juergen Meyer, Edward Kim, Mark Phillips, "Commissioning, clinical implementation, and performance of the Mobetron 2000 for intraoperative radiation therapy," *Journal of Applied Clinical Medical Physics* **18**(1), 230–242 (2017). https://doi.org/10.1002/acm2.12027.

21. Sergio Righi, Elio Karaj, Giuseppe Felici, Fabio Di Martino, "Dosimetric characteristics of electron beams produced by two mobile accelerators, Novac7 and Liac, for intraoperative radiation therapy through Monte Carlo simulation," *Journal of Applied Clinical Medical Physics* **14**(1), 6–18 (2013). https://doi.org/10.1120/jacmp.v14i1.3678.

22. E. Moretti, M. R. Malisan, A. Trianni, R. Padovani, "Use of radiochromic films to calibrate a high dose-per-pulse electron accelerator for intraoperative radiation therapy (IORT)," in *International Conference on Quality Assurance and New Techniques in Radiation Medicine* (IAEA, Vienna, Austria, 2006), pp. 488–489.

23. Evis Karaj, Sergio Righi, Fabio Di Martino, "Absolute dose measurements by means of a small cylindrical ionization chamber for very high dose per pulse high energy electron beams," *Medical Physics* **34**(3), 952–958 (2007). https://doi.org/10.1118/1.2436979.

24. John A. Antolak and Kenneth R. Hogstrom, "Multiple scattering theory for total skin electron beam design," *Medical Physics* **25**(6), 851–859 (1998). https://doi.org/10.1118/1.598295.

25. John A. Antolak, Jackson H. Cundiff, Chul S. Ha, "Utilization of thermoluminescent dosimetry in total skin electron beam radiotherapy of mycosis fungoides," *International Journal of Radiation Oncology, Biology, Physics* **40**(1), 101–108 (1998). https://doi.org/10.1016/S0360-3016(97)00585-3.

26. Indra J. Das, John F. Copeland, Harry S. Bushe, "Spatial distribution of bremsstrahlung in a dual electron beam used in total skin electron treatments: Errors due to ionization chamber cable irradiation," *Medical Physics* **21**(11), 1733–1738 (1994). https://doi.org/10.1118/1.597215.

27. Gerald J. Kutcher, Lawrence Coia, Michael T. Gillin, William F. Hanson, Steven Leibel, Robert J. Morton, Jatinder R. Palta, et al., "Comprehensive QA for radiation oncology: Report of AAPM Radiation Therapy Committee Task Group 40," *Medical Physics* **21**(4), 581–618 (1994). https://doi.org/10.1118/1.597316.

28. D. I. Thwaites, D. T. Burns, S. C. Klevenhagen, A. E. Nahum, W. G. Pitchford, "The IPEMB code of practice for electron dosimetry for radiotherapy beams of initial energy from 2 to 50 MeV based on an air kerma calibration," *Physics in Medicine and Biology* **41**(12), 2557–2603 (1996). https://doi.org/10.1088/0031-9155/41/12/001.

29. D. I. Thwaites, A. R. DuSautoy, T. Jordan, Malcolm R. McEwen, A. Nisbet, Alan E. Nahum, W. G. Pitchford, "The IPEM code of practice for electron dosimetry for radiotherapy beams of initial energy from 4 to 25 MeV based on an absorbed dose to water calibration," *Physics in Medicine and Biology* **48**(18), 2929–2970 (2003). https://doi.org/10.1088/0031-9155/48/18/301.

30. G. Rikner, "Characteristics of a P-Si detector in high energy electron fields." *Acta Radiologica: Oncology* **24**(1), 71–74 (1985). https://doi.org/10.3109/02841868509134368.

31. IAEA, "Review of data and methods recommended in the international code of practice for dosimetry IAEA Technical Reports Series No. 381. The use of plane parallel ionization chambers in high energy electron and photon beams." *IAEA-TECDOC* (IAEA, Vienna, 2000), vol. 1173.

32. Wolfram U. Laub and Richard Crilly, "Clinical radiation therapy measurements with a new commercial synthetic single crystal diamond detector," *Journal of Applied Clinical Medical Physics* **15**(6), 92–102 (2014). https://doi.org/10.1120/jacmp.v15i6.4890.

33. David W. O. Rogers, "Accuracy of the burns equation for stopping-power ratio as a function of depth and R50," *Medical Physics* **31**(11), 2961–2963 (2004). https://doi.org/10.1118/1.1803811.

34. Peter R. Almond, Frank H. Attix, Leroy J. Humphries, Hideo Kubo, Ravinder Nath, Steve Goetsch, David W. O. Rogers, "The calibration and use of plane-parallel ionization chambers for dosimetry of electron beams: An extension of the 1983 AAPM Protocol Report of AAPM Radiation Therapy Committee Task Group No. 39," *Medical Physics* **21**(8), 1251–1260 (1994). https://doi.org/10.1118/1.597359.

35. Dietrich Harder, "Einfluß Der Vielfachstreuung von Elektronen Auf Die Ionisation in Gasgefüllten Hohlräumen," *Biophysik* **5**(2), 157–164 (1968). https://doi.org/10.1007/BF01202901.

36. John A. Antolak, "Ionization to dose for electron beams: A comparison of three approaches (Abstract SU-FF-T-205)," *Medical Physics* **32**(6), 1997 (2005). https://doi.org/10.1118/1.1997933.

37. Kenneth R. Hogstrom, M. D. Mills, Peter R. Almond, "Electron beam dose calculations." *Physics in Medicine and Biology* **26**(3), 445–459 (1981).

38. John A. Antolak, Ernest Mah, John W. Scrimger, "Optimization of pencil beam widths for electron-beam dose calculations," *Medical Physics* **22**(4), 411–419 (1995). https://doi.org/10.1118/1.597606.

39. J. E. Cygler, Jerry J. Battista, John W. Scrimger, Ernest Mah, John A. Antolak, "Electron dose distributions in experimental phantoms: A comparison with 2D pencil beam calculations," *Physics in Medicine and Biology* **32**(9), 1073–1086 (1987). https://doi.org/10.1088/0031-9155/32/9/001.

40. Ernest Mah, John A. Antolak, John W. Scrimger, Jerry J. Battista, "Experimental evaluation of a 2D and 3D electron pencil beam algorithm," *Physics in Medicine and Biology* **34**(9), 1179–1194 (1989). https://doi.org/10.1088/0031-9155/34/9/004.

41. Robert A. Boyd, Kenneth R. Hogstrom, John A. Antolak, Almon S. Shiu, "A measured data set for evaluating electron-beam dose algorithms," *Medical Physics* **28**(6), 950–958 (2001). https://doi.org/10.1118/1.1374245.

42. D. I. Thwaites, "Measurements of ionisation in water, polystyrene and a 'solid water' phantom material for electron beams." *Physics in Medicine and Biology* **30**(1), 41–53 (1985). https://doi.org/10.1088/0031-9155/30/1/005.

43. Bruce Thomadsen, Chris Constantinou, Anthony Ho, "Evaluation of water-equivalent plastics as phantom material for electron-beam dosimetry," *Medical Physics* **22**(3), 291–296 (1995). https://doi.org/10.1118/1.597453.

44. J. Dutreix and A. Dutreix, "Film dosimetry of high-energy electrons." *Annals of the New York Academy of Sciences* **161**(1), 33–43 (1969). https://doi.org/10.1111/j.1749-6632.1969.tb34039.x.

45. Sujatha Pai, Indra J. Das, James F. Dempsey, Kwok L. Lam, Thomas J. Losasso, Arthur J. Olch, Jatinder R. Palta, Lawrence E. Reinstein, Dan Ritt, Ellen E. Wilcox, "TG-69: Radiographic film for megavoltage beam dosimetry," *Medical Physics* **34**(6), 2228–2258 (2007). https://doi.org/10.1118/1.2736779.

46. Francis J. Bova, "A film phantom for routine film dosimetry in the clinical environment." *Medical Dosimetry* **15**(2), 83–85 (1990). https://doi.org/10.1016/0958-3947(90)90040-O.

47. Kenneth R. Hogstrom, Michael D. Mills, Jeff A. Eyer, Jatinder R. Palta, David E. Mcllenberg, Raul T. Meoz, Robert S. Fields, "Dosimetric evaluation of a pencil-beam algorithm for electrons employing a two-dimensional heterogeneity correction," *International Journal of Radiation Oncology, Biology, Physics* **10**(4), 561–569 (1984). https://doi.org/10.1016/0360-3016(84)90036-1.

48. Robert L. Carver, Kenneth R. Hogstrom, Connel Chu, Robert S. Fields, Conrad P. Sprunger, "Accuracy of pencil-beam redefinition algorithm dose calculations in patient-like cylindrical phantoms for bolus electron conformal therapy," *Medical Physics* **40**(7), 71720 (2013). https://doi.org/10.1118/1.4811104.

49. D. Bollinger and A. Darafsheh, "Clinical considerations and dosimeters for in vivo dosimetry," in *Radiation Therapy Dosimetry: A Practical Handbook*, edited by A. Darafsheh (CRC Press, Boca Raton, FL, 2021), Chapter 11, pp. 151–171.

50. Kenneth R. Hogstrom and Peter R. Almond, "Review of electron beam therapy physics," *Physics in Medicine and Biology* **51**(13), R455–R489 (2006). https://doi.org/10.1088/0031-9155/51/13/R25.

51. John A. Antolak, "Electron beam treatment planning," in *Khan's Treatment Planning in Radiation Oncology*, edited by F. M. Khan, J. P. Gibbons, P. W. Sperduto (Wolters Kluwer, Philadelphia, PA, 2016), 4th ed., pp. 288–312.

CHAPTER 25

Proton Therapy Dosimetry

Michele M. Kim and Eric S. Diffenderfer

University of Pennsylvania
Philadelphia, Pennsylvania

CONTENTS

25.1 INTRODUCTION

Radiation dosimetry encompasses the measurement techniques and procedures that allow quantification and verification of the radiation dose delivered during radiation therapy. These techniques are applied throughout the processes of accepting an accelerator system, commissioning a treatment planning system, performing daily, monthly, annual, and patient treatment plan specific quality assurance (QA). Many aspects of radiation dosimetry are equivalent between different radiation sources and delivery modalities; however, a few aspects are particular to particle therapy with protons.

The fundamental advantage introduced by proton beam therapy over conventional x-ray and electron therapies is that the protons penetrate tissue and travel mostly with a straight trajectory before coming to rest and generating a high density of ionization in a small region. Exit dose beyond the Bragg peak is negligible, with a small contribution due to secondary particles and gamma rays produced in nuclear interactions. The highly localized deposition of proton dose enhances the normal tissue sparing capabilities of external beam therapy while maintaining a high dose to the tumor target and surrounding margin. However, the scattering and energy loss characteristics of protons lead to dosimetry and measurement techniques that differ from conventional electron radiation.

Calibration of proton beams requires measurement of absolute physical dose which typically differs from prescribed dose by a biological factor intended to account for the difference in biological effect [1, 2] such that the physical dose is multiplied by a relative biological effectiveness (RBE) factor. The product is called the RBE weighted dose and written using the International Commission on Radiation Units and Measurements suggested notation Gy(RBE) to indicate that the SI unit of Gy is modified by an RBE factor [3].

25.2 FUNDAMENTALS OF PROTON PHYSICS

Protons interact in matter with an atom or nucleus via several mechanisms. There are Coulombic interactions with atomic electrons, Coulombic interactions with the atomic nucleus, nuclear interactions with atomic nuclei, and Bremsstrahlung. Protons will lose kinetic energy continuously with multiple inelastic Coulombic interactions with the atomic electrons, hence depositing dose, as they penetrate matter. For the most part, protons travel in a straight line since the rest mass of protons are 1832 times that of an electron [4]. Protons and electrons having opposite charges causes attraction and the electrons can be released from the atoms. This results in ionization of atoms, and loose electrons go on to ionize further atoms in the vicinity of the atom from which it came. Protons on average lose relatively little energy in individual ionizations and do not get deflected as much. They will undergo hundreds of thousands of interactions per centimeter of material before eventually losing all of their energy and coming to rest.

Since protons are so much heavier than electrons, they are barely deflected by Coulomb interactions with atomic electrons. Passing through material, however, protons do experience a repulsive force when passing positively charged atomic nuclei. Nuclei are larger than electrons, resulting in deflections of protons at larger,

but still small, angles. Protons go through many of these interactions as it passes through matter, and statistically all of these interactions add up in a net angular and radial deviation, which has been called "multiple Coulomb scattering."

There are two main types of nuclear interactions as protons travel through a medium, elastic and nonelastic collisions. In elastic collisions between the proton and a nucleus, the nucleus is left intact while the proton loses a significant fraction of its energy and is deflected. In nonelastic collisions, the nucleus is broken apart as the proton loses a significant fraction of its energy and is deflected. Typically, the light fragment of the nucleus is knocked out with considerable speed as it leaves behind a heavy fragment that stays close to the place of interaction and is very ionizing.

Bremsstrahlung photons are generated when charged particles are passing in the field of an atomic nucleus. However, the likelihood of bremsstrahlung is roughly proportional to the inverse square of the particle mass, and therefore do not enact any clinical significance in proton therapy.

As protons travel through media, the dose deposited rises sharply near the end of their range, creating a Bragg peak. The shape of the Bragg peak is due to multiple effects. The first of which is the slow loss of energy that protons undergo due to Coulomb interactions with atomic electrons. The linear energy transfer (LET) or stopping power is the proton's linear rate of energy loss, measured in units of $MeV/g/cm^2$. This can be described by the Bethe–Block formula [5, 6]

$$\frac{S}{\rho} = -\frac{dE}{\rho dx}$$

$$= 4\pi N_A r_e^2 m_e c^2 \frac{Z}{A} \frac{z^2}{\beta^2} \left[\ln \frac{2m_e c^2 \gamma^2 \beta^2}{I} - \beta^2 - \frac{\delta}{2} - \frac{C}{Z} \right]$$

(25.1)

where S is the stopping power, ρ is the mass density of the absorbing material, $(-dE/dx)$ describes the energy loss rate, N_A is Avogadro's number, r_e and m_e are the electron radius and mass, respectively, z is the charge of the projectile, Z is the atomic number of the material, A is the atomic weight of the absorbing material, c is the speed of light, $\beta = v/c$ where v is the projectile velocity, $\gamma = (1 - \beta^2)^{(-1/2)}$, I is the mean excitation potential of the absorbing material, δ is the density correction from shielding of remote electrons by close electrons, and C is the shell correction term that is only important for low energies. Approximately, the energy loss is proportional to the inverse square of the proton's mean speed and $(Z/A)z^2$. The dose rises quickly as the protons slow down.

The Bragg peak is blurred due to two main effects. The first is called "range straggling," which are statistical fluctuations in the ionization process and causes a smearing out of the depth of penetration of the stopping protons. The magnitude of this is roughly 1% of the proton's range. The second effect is from the energy spread due to the specifics in producing high energy proton beams. There is never a pristine monoenergetic beam of protons, resulting in an energy spread of around 1% in magnitude.

Nuclear interactions of protons occur at a rate of around $1\%/g/cm^2$ until the last few millimeters of their end of range. Nuclear interactions reduce the number of primary protons in the beam. They also produce a halo of scattered primary protons and knocked-out secondary protons that travel long distances and create a tail to the lateral dose profile of a proton beam. Nuclear interactions also increase the RBE in the vicinity of interaction by creating heavily ionizing fragments with very high stopping power which then deposit dose. The nuclear interactions also create a halo of neutrons that mostly escape the patient without further interaction but are responsible for a small contribution to the dose inside and outside the primary radiation field. As protons travel in depth in a medium, the proton fluence decreases. Roughly, the 80% dose on the falling edge of the Bragg peak is very close to the same depth as the 50% fluence of the falling edge of the fluence distribution. The flat entrance plateau region of a proton depth dose curve is due to the combination between the rising energy deposition and the diminishing primary proton fluence.

Proton range is defined as the depth where half of the protons traveling in the medium have come to rest. It can be best described by the distal 80% point on the Bragg peak. Range is greater with higher energy proton beams. Lower energy proton beams tend to have narrower Bragg peaks since the range straggling effect is relative to its own proton range, namely ~1.5%, resulting in a smaller absolute broadening of the Bragg peak. The ratio of the Bragg peak to the plateau (or entrance) dose is higher for lower energy beams because of the narrower Bragg peak. The same energy is delivered in the Bragg peak in the last couple of g/cm^2 in a proton's

path, so a narrow peak is higher for the total energy in the peak to be constant. Proton range is often characterized by water-equivalent thickness (WET). The general equation is

$$t_w = t_m \frac{\rho_m}{\rho_w} \frac{\overline{S}_m}{\overline{S}_w} \qquad (25.2)$$

where t_w and t_m are thickness in water and material, ρ_w and ρ_m are the mass densities of water and material, and $\overline{S}_w$ and $\overline{S}_m$ are the mean proton mass stopping powers for water and material [7].

Monoenergetic proton beams will penetrate matter of a given density up to a depth which is defined by the beam energy. The relation between depth of penetration and energy is beneficial in the practical use of protons for radiation therapy.

To make clinical use of the very narrow proton Bragg peak, multiple monoenergetic Bragg peaks can be delivered to create a flat distal region of near-constant high dose, called the spread-out Bragg peak (SOBP). The monoenergetic peaks are not equally weighted, but the more proximal peaks have less weight and contribute the least to the dose delivered. While an SOBP does not deliver dose beyond the most distal high dose region, the entrance dose does increase due to the stacking of multiple Bragg peaks, and reduces the peak-to-plateau ratio that is seen in monoenergetic Bragg peaks.

Similar to photons, protons experience the inverse square effect. Compared to an initially parallel beam, proton fluence and dose at a point that is a distance r from the source will be reduced by a factor of $1/r^2$. This can be compensated by increasing the weights of the upstream Bragg peaks, but then will cause a much higher entrance dose. For this reason, proton beam source distances are at least 2 m from the isocenter so that the proximal tissue sparing can be maximized with the use of an SOBP.

Lateral dose distribution of a proton beam is due to four main effects that cause broadening of a narrow beam: multiple Coulomb scattering – near-Gaussian core, multiple Coulomb scattering – long tail, proton nuclear interactions, and neutron nuclear interactions. Multiple Coulomb scattering was worked on by Molière in 1947 [8]. This is the main cause of the spreading out of an initially infinitesimal pencil beam. The principal component is a nearly Gaussian distribution in the angle of deviation and in the lateral spread of a pencil beam. This lateral blurring is a function of depth within the beam. In addition to this Gaussian shape, there is a long tail that is due to large angle scattering in only one or a few collisions. This long tail portion of lateral dose spread has a relatively small amplitude and can be approximated by a second, broader Gaussian [9]. High energy protons that are scattered from the elastic and nonelastic nuclear collisions contribute to the tails of a proton pencil beam. As the depth increases, there is a larger halo of dose around the beam that can also be approximated by a Gaussian distribution. Dose can be underestimated if this halo dose is not considered while performing absolute dosimetry. Finally, there is a halo of neutrons that is generated but will usually escape the patient without further interaction. However, the neutrons can be responsible for low doses being deposited in and outside of the proton beam path. Since the lateral spread of a pencil beam is unavoidable, it is best to minimize the air gap between the patient surface and the final beam modifying material, whether that is a range shifter or an aperture.

25.3 DETECTORS

Many of the same detectors that are used in conventional x-ray radiation therapy are also useful in proton radiotherapy. Film, diodes, ionization chambers (ICs), scintillation, and array detectors are all widely used in proton therapy. There are many sources of information on commonly used ionizing radiation detectors including the textbooks by Attix [10] and Knoll [11]. However, there are certain characteristics of a detector that make it suitable for use in proton dosimetry. Detectors must be available for relative or absolute dosimetry with traceable calibration and exhibit response that is minimally dependent on beam energy. Namely the variation of detector response with proton stopping power ratio, or LET variation should be minimized for the particle energies relevant to proton therapy. For this reason, ICs constructed of tissue or air equivalent materials are typically used. Other detector types such as film, scintillators, solid state detectors, etc. can often exhibit a strong LET dependence which will alter their response compared to ICs as the beam LET changes in the Bragg peak.

25.3.1 Absolute Detectors

Absolute dosimetry involves measurement of the fundamental deposition of energy per unit mass of the active volume of the detector. There are a few technologies available for absolute measurement of dose which operate under one of three principles, ionometric measurement with an IC, chemical dosimetry such as Fricke dosimetry [12], and calorimetry. Additionally electron paramagnetic resonance (EPR) dosimetry with alanine [13, 14] and optically stimulated luminescence detectors (OSLDs) [15] are each becoming widely available for use in proton therapy. However, for the purpose of proton dosimetry the accepted international standards [3, 16, 17] for reference dosimetry use IC measurements with national standard traceable calibration certificates.

25.3.1.1 Ionization Chambers

ICs are the standard dose measurement device employed in the reference dosimetry protocols discussed in Section 25.6 and are generally recommended when access to a calorimeter or Faraday cup (FC) is not available. They can be constructed using tissue equivalent materials with a wide variety of shapes and sizes for convenience of use, and perform reliably with well-studied response variables, refer to Chapter 2 for detailed information on the construction and use of ICs for radiation dosimetry.

For protons, the largest uncertainty with measuring absorbed dose to water is with the beam quality correction factor, k_Q which depends on the ratios of stopping power and the energy required to produce an ion pair in air with the reference beam quality [17]. These values vary as a function of beam energy, LET, and the residual range of the proton beam. There are no reference proton beams available at calibration laboratories, so in most cases ^{60}Co is used as the primary reference quality. Stopping power calculations depend on the mean excitation energy value, I, for water and the chamber material (Section 25.2). Gomà et al. [18] used Monte Carlo techniques to calculate k_Q with I values from ICRU reports [19, 20] and more recent values from Andreo et al. [21] and Burns et al. [22]. They found agreement with tabulated values in IAEA TRS Report 398 within 2.3% and agreement with experimentally determined values within 1.1%.

Additionally, Vatnitsky et al. [23] used water calorimetry to compare k_Q for ICs in ^{60}Co reference proton beams. They found agreement with calculation within 1.2% for the entrance region of a monoenergetic 250 MeV proton beam and within 1.8% in the SOBP of a 155 MeV beam. Sorriaux et al. [24] used Monte Carlo methods to evaluated beam quality correction factor dependence on scanned proton beams compared to proton dose delivered as a broad beam. They found no additional uncertainty when using k_Q factors measured or computed for a broad beam with a scanned proton beam. Vatnitsky et al. [25] performed an international dosimetry intercomparison between 13 institutions and found that when using ICs with ^{60}Co calibration factors traceable to a standards laboratory, the variations between centers were within 3%.

Often it is preferable to use a parallel plate type IC for proton beam dosimetry over a cylindrical or Farmer type chamber. IAEA TRS Report 398 recommends the use of cylindrical chambers for reference dosimetry due to lower uncertainty of k_Q. They also recommend taking the central axis as the measurement point rather than the effective point of measurement at 0.75 times the inner radius. However Gomà et al. [26, 27] have since shown discrepancy between the tabulated k_Q values with experiment. The discrepancy can be traced to the recommendation to use the central axis as the measurement point with better agreement when the effective point is used. Bichsel [28] showed that a thin parallel plate detector filled with air is most appropriate to preserve the shape of the particle Bragg curve. Palmans et al. [29] found that the perturbation correction factor for plane parallel ICs was unity within experimental uncertainties. There are significant uncertainties associated with using cylindrical ICs in the sharp dose gradients around proton Bragg peaks. This leads to a large uncertainty in gradient correction factor and, with the finding of negligible contribution from perturbation corrections, therefore favors the use of a plane parallel chamber. However, in their Monte Carlo study Gomà et al. [18] deduced that there may be significant variation from unity of perturbation factors dependent on chamber type.

Large area parallel plate ICs have become available specifically to support proton depth profile measurement of pristine Bragg peaks for pencil beam scanning (PBS) delivery (See Section 25.4.2). Commissioning of PBS requires measurement of the unscanned integrated depth dose (IDD) profile collecting the full lateral profile of the beam which includes a small non-Gaussian contribution from secondary events known as the nuclear halo [9, 30]. Clasie et al. [31] modeled the nuclear halo contribution as a secondary Gaussian distribution that is nearly invariant with depth in phantom. Lin et al. [32] measured the nuclear halo dependency on proton nozzle design by measuring PBS field size effects on output. Collection efficiency can be increased by 3.5%, dependent on energy and measurement depth, for unscanned pencil beams by increasing the chamber diameter from 8.2 cm to 12 cm [33].

25.3.1.2 Calorimetry

Calorimetry is the gold standard for absolute dosimetry and the fundamental measurement technique for physical absorption of energy in matter and has been recommended for primary proton dosimetry standards [16, 34]. Based on measurement of heat energy deposited in a material when subjected to radiation, calorimetry is the most direct measurement of absorbed dose available and does not rely on detection of the products of radiation interaction such as ionization. Very small changes in temperature must be measured and often for this reason graphite calorimeters are used because compared to water it has a lower heat capacity but with similar radiological properties. Even so, there are a number of factors contributing to uncertainty in these measurements, which include the heat defect (deposited energy not resulting in a temperature change) and electronic stopping power ratio of water to the calorimeter material. The reader is advised to refer to Chapter 3 for additional details on the construction and use of calorimeters for absorbed dose measurement.

However, there are additional drawbacks when considering this dosimetry technology for use with proton beams and calibration services for direct calorimetry-based measurement of absorbed dose do not currently exist [16]. Corrections were calculated by Lourenço et al. using measurement and Monte Carlo techniques to correct for fluence differences between water and graphite at the same radiological depth [35]. The fluence corrections were found to be dependent on beam energy, measurement depth, and chamber geometry ranging from 0.99 for a 60 MeV beam at all depths, 0.99–1.04 for a 180 MeV beam, to 0.99–1.01 for a larger area detector when moving from the entrance to greater depths. Additionally, the heat defect is found to depend on the LET which varies with depth in ion beams [36–40].

Still there have been a number of studies performing absorbed dose measurements with calorimeters in proton beams and this work has gone a long way towards validating IC calibration and beam quality corrections. Medin et al. [38] used a sealed water calorimeter to measure k_Q for 10 MV photon and a 175 MeV proton fields. The comparison confirmed the photon beam quality values tabulated in TRS-398 but yielded a difference of

1.8% from the proton values. Medin [39] also measured absorbed dose in a high energy scanned proton beam with a water calorimeter and Farmer type ICs calibrated in a ^{60}Co beam to obtain k_Q factors for proton beams. The results agreed well with those tabulated in TRS-398 [17] and within 1.1% of prior measurements in a passively scattered beam. The discrepancy can be attributed to differences in ion recombination when using ICs in scanned proton beams. Sassowsky et al. [36] studied the use of water calorimetry in a scanned proton beam for evaluation of beam quality correction factors and the variation of heat effect due to LET as well as calorimeter response changes due to spatial-temporal variations in spot scanned proton beam delivery. Brede et al. [37] measured absorbed dose with a water calorimeter in a proton beam with uncertainty better than 1.8% and cross compared with IC measurements, themselves having uncertainty of 2.0%, for an agreement between calorimeter and ion chamber of 0.7%. Delacroix et al. [41] ran a comparison between calorimeter constructed of tissue equivalent material, various IC and FCs at four proton therapy centers and found an agreement of better than 1% between calorimeter and the other detectors. Vatnitsky et al. [23] developed a formalism for proton dosimetry with ICs based on absorbed dose to water calibration factors and beam quality correction factors. They used water calorimetry to measure $k_{Q\gamma}$ in a ^{60}Co beam and a proton beam to measure k_{Qp}. Sarfehnia et al. [40] used water calorimetry to measure absorbed dose to water in double scattered and scanned proton beams. The water calorimeter had been previously verified in photon and electron beams, and the overall uncertainty was 0.4% for double scattered and 0.6% for scanned proton beams, comparing very favorably with the overall uncertainty of 1.9% for the TRS-398 reference absorbed dose protocol.

25.3.1.3 Faraday Cup

FCs are designed to completely collect the total charge of a particle beam and can be used for indirect measurement of proton beam dosimetry. Verhey et al. [42] described an absorbed dose measurement method using a FC at the 160 MeV Harvard cyclotron and compared with measurements using a parallel plate IC, thimble type IC, and tissue equivalent calorimeter and found agreement at a range of depths within 2.6%. Vynckier et al. [43] developed a proton dosimetry protocol that recommends the use of a FC for calibrating an IC when a calorimeter is

not available. Cambria et al. [44] compared two different types of FC and an IC. They found good agreement between FC (better than 3.6%), but a larger difference between FC and IC up to 8.2%. Lorin et al. [45] used an FC to calibrate an IC and correct the deficiency of ICs due to ion recombination in the high ionization density created by scanned pulsed proton beams.

FCs will completely stop and collect the proton charge in a thick electrode leading to an accurate measure of the total charge on the FC. Secondary charged particles are created with recoil velocity sufficient to escape the FC, thereby affecting the charge measured in the device. Therefore, FCs are typically designed with a guard electrode to suppress escape electrons and ensure an accurate charge reading. Though Cascio and Gottschalk [46] showed that a simplified FC can be constructed to agree with a traditional FC within 1–5%, but without an electron suppression guard ring.

The elementary charge e is known with great precision and the total number of protons incident on the FC can be derived. Given the cross-sectional distribution of the beam, the total dose can be calculated from the energy and stopping power of the protons. A constant energy at the FC is assumed by Vynckier et al. [43] which provides the following formula for dose calculation from a FC,

$$D = N/a \times (S/\rho) \times (1.602 \times 10^{-10}) \qquad (25.3)$$

Here D is dose [Gy] to the medium, N is the number of protons, a is the area of the beam at the measurement position, (S/ρ) [MeV cm^2 g^{-1}] is mass stopping power in the medium for the protons at the known energy. Care must be taken in the application of this protocol to avoid contribution from low energy scattered protons which will deposit energy with a different stopping power. Demonstrating the sensitivity to measurement conditions, including the effects of scattered protons, Delacroix et al. [41] found deviations up to 17% between FC measurements at different proton centers. Grusell et al. [47] developed a method to use a FC to calibrate an IC in an uncollimated beam, thereby decreasing the contribution from scattered protons and the uncertainty of the stopping power of the protons in the gas filling the cavity of the IC.

25.3.2 Relative Detectors

Often the purpose of dosimetry measurements is to measure the dose relative to a reference measurement without regard for the absolute physical dose. This can take the form of point measurements for comparison with other measurements or 1D, 2D, and 3D relative dose distributions. In proton dosimetry and QA it is frequently useful to compare measured against benchmarked distributions, but it is impractical using a point dosimeter to measure with high spatial resolution. Therefore, when benchmarking constancy of spatial variation in delivered dose, it is more often useful to employ relative detectors to normalize out fluctuations between repeated measurements or to use high spatial resolution relative detectors for the comparisons.

25.3.2.1 Radiochromic Film
Film is very useful for relative proton dosimetry due to its convenience, high spatial resolution, availability in large dimensions that are easily custom fit to applications, and long-term storage capability. The use of self-developing radiochromic film has all but superseded silver emulsion film due to the clear advantage of not requiring the film developing equipment and chemicals. Recent formulations exhibit increased sensitivity, and removing the uncertainties introduced by the chemical developing process opens the possibility of absolute dosimetry with radiochromic films. Several groups have developed detailed handling, irradiation, scanning, and processing protocols for accurate film dosimetry [48–53].

The main issue faced with using radiochromic film for proton dosimetry is the quenching effect that increases with LET and leads to an under response in the Bragg peak [54–57]. MD-55 Gafchromic film (International Specialty Products (ISP), Wayne, New Jersey) is a radiochromic film that has been used for relative proton dosimetry of depth dose and lateral profiles for both modulated and unmodulated proton beams in the energy range 100–250 MeV [58]. The film exhibits good agreement with IC measurements, but under-responds in the distal region of the Bragg peak due to a quenching effect. Still they found agreement in the distal falloff depth (80–50% distance) was within 0.1–0.2 mm of IC measurements. Martisíková and Jäkel [59] found that Gafchromic EBT film response was comparable between Co-60 and protons. Another study by Mumot et al. [60] comparing MD-55 radiochromic film and a silicon diode found a 4% difference in a modulated SOBP and a 12% difference in an unmodulated Bragg peak, demonstrating the effect of LET variations on radiochromic film

response. Spielberger et al. [61] evaluated the response of silver emulsion film with different particle LET and found a clear suppression of response with higher LET. A linear dependence for radiochromic film quenching in proton radiation was found by Anderson et al. [62] who note that under response of >15% on the distal edge of the Bragg peak is possible and that LET calibration is necessary for accurate film calibration.

25.3.2.2 Thermoluminescent Detectors
Thermoluminescent detectors (TLDs) are discussed in Chapter 6, and they have had a limited role in proton dosimetry. TLDs have been studied for use in proton therapy and have shown good agreement with ICs and diamond solid state detectors [63–66]. While they can be made to small dimensions, they have a relatively large WET. However, the dose record for TLDs is temporally stable, and the primary use in proton therapy is for remote auditing of proton therapy sites by irradiation and shipment of TLDs embedded in standardized phantoms [67].

25.3.2.3 Optically Stimulated Luminescence Detectors
OSLDs are discussed in Chapter 7 and their operation is similar to TLDs, but light is used to stimulate luminescence instead of heat as in the case of TLDs. With their temporal stability and compact size, OSLDs are attractive for use in health physics settings and for surface measurements in proton therapy. OSLDs have been studied for use in proton therapy with minimal observed LET dependence for relatively low proton radiation [15, 68]. However, LET dependence has been observed for higher LET particles, and may be of interest for response due to secondary neutron radiation produced by the primary proton beam [69–71].

25.3.2.4 Solid State Diode Detectors
Solid state detectors in use for proton therapy include diamond, metal oxide field effect transistors (MOSFET), and doped semiconductor (p-type or n-type) detectors [42, 64, 72, 73]. These detectors can be constructed with very small proportions which lend them to high spatial resolution scanning measurements or small field dosimetry [73]. Additionally, diamond detectors have an active volume constructed with carbon and can be designed with near tissue equivalent composition. Fidanzio et al. [74] measured the LET dependence of a diamond detector in a 62 MeV proton beam. Grusell and Medin [75]

have observed an increase in sensitivity with LET of n-type silicon diodes after pre-irradiation. On the other hand, in this study the p-type diode did not exhibit LET correlated increased sensitivity. Onori et al. [76] measured significant variations in response from both silicon diode and diamond detector. They also observed a small dose rate effect with silicon diodes, while there was no observed effect from the diamond detector up to 5 Gy/min. Pacilio et al. [77] studied the response of a diamond detector and a p-type silicon diode in a 10–59 MeV proton beam. The diamond detector showed large quenching of response in the Bragg peak, where the p-type diode did not. However, the diode exhibited signs of radiation damage with a decrease in response after a high accumulated dose. The variable reports of LET dependence and susceptibility to radiation damage have probably limited the routine use of silicon diode detectors in proton dosimetry.

25.3.3 Multidimensional Detectors

Multidimensional detectors have seen a recent rise in use in proton therapy primarily due to a rapid increase in highly conformal and intensity modulated treatment techniques such as PBS and intensity-modulated proton therapy (IMPT). These techniques with the capability of generating high dose gradients with complex treatment volumes covered by multiple isocentric beam arrangements require enhanced dosimetry and QA.

25.3.3.1 Multielement Detectors

For ease of use and to potentially save considerable measurement time, several multichannel depth dose measurement devices have been developed. These devices typically consist of a stack of air-filled parallel plate ICs with intervening material designed to place the measurement depth of each IC at the equal geometrical and water equivalent depth. An early example developed at the GSI facilities in Darmstadt, Germany consisted of a multi-wire proportional chamber for position resolution which was coupled to a stack of large area parallel plate ICs for a 3D dose measurement system [78]. A multi-layer IC (MLIC) was developed along with a multi-pad IC (MPIC) to measure depth and lateral profiles of proton beams [79]. Bäumer et al. [33] evaluated a commercially available MLIC and note a small distortion of the Bragg peak when using an MLIC with large area parallel plate ICs. The authors note that the MLICs are likely not suitable for baseline commissioning data acquisition

but offer better than 0.4 mm range constancy measurement which can be used for QA. However, Takayanagi et al. [80] developed a dual ring MLIC and a correction technique that suppresses the difference between IDDs measured with the MLIC and in a water phantom.

Multilayer FCs have also been designed for measurements of proton fluence change with depth [81–84]. The dose can be derived from the fluence measurements in a multilayer FC through the relative stopping powers. However, the FC electrodes are typically constructed of copper which has a WET dependent on energy. Corrections must be made to obtain the range of a proton beam, but comparison against baseline measurements validated by a depth dose measurement with a water equivalent device can achieve good precision for range constancy measurements [84].

Multi-pad or arrays of ICs have been developed to combine good spatial resolution with the well-studied and reliable dose response of ICs [79]. Arjomandy et al. [85] characterized a commercial 2D array of ICs. The MatriXX detector (IBA Dosimetry, Schwarzenbruck, Germany) consists of 32 × 32 array of 4.5 mm diameter ICs with integrated multichannel electrometer and data acquisition system. The device gave excellent agreement with film and ion chamber measurements in water when used in a passively scattered proton beam [85, 86]. Lin et al. [87] developed a method to characterize the position and width of PBS pencil beams to submillimeter precision. Lin et al. [88] tested a newer version of the device adapted to the increased local proton dose rate that is inherent in actively scanned PBS systems. The small pencil beam size is delivered with a higher local intensity compared to passive delivery because of the increased delivery time scanning from spot to spot and layer to layer over the 3D treatment volume. The higher intensity is associated with increased ion recombination effects in the original MatriXX which is operated at a fixed voltage. The newer device, dubbed MatriXX PT, was designed with a smaller IC plate gap which minimizes the ion recombination and provides a significant improvement in absolute dose measurement in PBS beams.

25.3.3.2 Scintillating Detectors

Light emitting scintillation detectors can also be used for proton dosimetry to offer rapid measurement and high spatial resolution. The detectors consist of a scintillating screen plate coated with a scintillating material which is typically a gadolinium-based material ($Gd_2O_2S:Tb$).

The screen is coupled to a charge coupled device (CCD) camera in a light tight box [89]. The scintillation light output is dependent on ionization density and therefore LET, the scintillator is also known to experience quenching at high density regions such as in the Bragg peak [90]. The rapid turnaround time for image acquisition, high spatial resolution, and sensitivity make these detectors useful for 2D dosimetry at fixed depth [91]. A QA program for modulated PBS can be designed based around commercial versions of these detectors [92] and there have also been studies to use the detectors to characterize the nuclear halo effect that can alter PBS field size factors especially for small field sizes [9, 30].

Additionally, cone shaped scintillating screen detectors have been developed for isocentric QA and dosimetry. The devices consist of a CCD camera with a light tight coupling directed into a cone with the interior coated with a scintillating material. When a narrow pencil beam intersects with the cone, the light pattern that is produced in the interior scintillator is imaged by the CCD and used to reconstruct the pencil beam shape and path through the cone [93]. Cai et al. [94] developed a QA procedure for image guided proton therapy that employs a cone shaped detector that proved useful for verifying gantry angle, patient couch positioning and imager-radiation isocenter coincidence.

25.4 PROTON BEAM DELIVERY TECHNIQUES

There are several beam delivery techniques that are used in proton radiotherapy and it is important to identify the influence that each will have on the dosimetry techniques we employ. In general, it is necessary for particle therapy with projectiles much heavier than electrons (e.g. Protons, Helium, Carbon, etc.) to modulate the beam energy in order to spread the Bragg peak width over a usable area. Without sufficient energy modulation the Bragg peak is typically too narrow for effective treatment of a tumor larger than a few millimeters in diameter. This energy modulation is accomplished by either varying the beam energy transported into the treatment room in the case of a scanned modality, or in the case of a scattered modality by rapidly varying the energy in the treatment room through the use of a spinning modulator wheel.

25.4.1 Scattered Beams

Passive beam scattering is a technique that was first developed to cover the cross section and extent in depth of the planning target volume (PTV). Protons entering water or tissue in the same direction will start going in slightly different directions by a degree or so by a few centimeters of depth due to multiple Coulomb scattering. Scattering in a material increases as the proton slows down. Net scattering is roughly proportional to the square-root thickness of a thin sheet of material. The net energy loss is directly proportional to the thickness of the material. High Z (atomic number) materials scatter protons more strongly while low Z materials are more effective in slowing protons down. Combinations of high and low Z materials are used to control both energy loss and scattering.

The simplest passive scattering scheme utilizes a single scatterer such as lead to maximize scattering and minimize energy loss. However, this system is not very efficient, since the fraction of protons within the useful region of being within ±2.5% the desired dose is only around 5%, following the Gaussian spread of the scattered proton beam. The transverse dose distribution would also not be flat over this region. In addition, the protons would lose penetration energy in passing through the lead. This would also result in depth with a narrow Bragg peak with limited usefulness in providing coverage of larger targets. These limitations can be overcome by a double scattered system.

To obtain the depth-dose distribution that is useful for providing dose coverage of a target in the direction of the proton beam, a range modulator can be used to create an SOBP. The range modulator is typically a rotating object with steps that puts successively thicker layers of material into the beam path [95]. The thickness of each layer and the time the beam spends on each step are adjusted so that the resulting SOBP is flat. In addition to this device, a range shifter can be added in the beam path to provide coverage to more shallow targets so that the entire dose distribution is shifted to shallower depths. Range shifters are typically placed upstream (further from the patient) so that it can be smaller.

Double scattering systems (Figure 25.1) were developed to reduce energy loss and improve efficiency in proton delivery. This made the use of larger proton fields practical. A double scattering system is comprised of a uniform first scatterer that produces a Gaussian beam profile. The nonuniform second scatterer then modifies the Gaussian beam profile to produce a flat dose distribution at the target location [95]. One way to design the second scatterer is to have a flat portion blocked by a

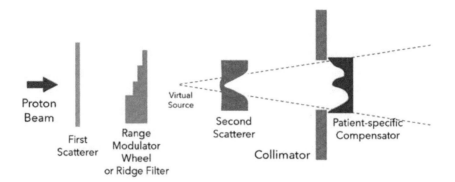

FIGURE 25.1 Double scattered beam delivery system schematic.

cylindrical plug or ring. The more preferred method has a contoured second scatterer so that the protons from the center of the profile are more strongly scattered. The contoured scatterer causes the center protons to lose more energy than the protons along the outer radius so an appropriately shaped energy compensator needs to be added. One drawback of double scattered systems is the sensitivity to beam steering. If the beam is off center by even a millimeter on the second scatterer, the dose distribution at the target will be tilted.

Double scattered systems require patient-specific hardware for conformation of the dose to the target volume. The first piece of hardware is an aperture, or a brass beam-stop with a hole that is shaped to the outer projection of the target in the beam's eye view. This can also be achieved with a multileaf collimator [96]. With the use of this hardware, there can be two areas of unwanted high dose proximal to the target towards the edges of the target (thus the edges of the range compensated portion) due to the constant range modulation of the field. This is not seen in scanned beams. The treatment planning system can be used to prepare files for automatic fabrication of patient-specific, field-specific hardware such as range compensators [97]. These can be designed to have different goals, such as guaranteeing target coverage with respect to alignment errors or patient internal organ motion [98].

Sharp dose distributions are the benefit of proton beams, but with poorly designed beam modifying devices, resulting dose distributions around the PTV boundary can be suboptimal. The final aperture closest to the patient results in sharp shadowing or dose fall off. The addition of any material after the aperture, such as a second scatterer, modulator, or range shifter, near the patient causes the beam to have less sharp dose fall off. The perturbing effect of any beam-modifying devices is

worse the further downstream it is located A single scatterer with upstream modulator produces the sharpest dose, and if a second scatterer is used to generate a large field, it should be as far upstream as possible. There is more scattering in the patient, therefore the dose gradient will be poor for deep targets no matter the nozzle design.

Large air gaps between the last beam modifying device and the patient will result in suboptimal lateral dose fall off due to scattering. In addition, large air gaps compromise the accuracy with which the range compensator deals with tissue inhomogeneities [99]. Distal dose falloff is determined by the Bragg peak. Distal fall off does not increase with depth in the patient; however, it does depend on range straggling and will thus increase with energy. A better distal fall off will be achieved if a beam of sufficient energy is used rather than a higher energy beam that has been degraded.

25.4.2 Scanned Beams

Recent proton systems have scanned beams. Proton charge enables magnetic scanning of narrow pencil beams. PBS involves a narrow beam of protons entering the patient at different locations by using magnetic fields to deflect the beam (Figure 25.2). PBS delivers dose inside the patient through sequential superposition of many physical pencil beams of small size. Usually there is no patient specific hardware or beam modifying devices involved with scanned beams.

Conformal 3D dose distribution is achieved by beam scanning. The width of an SOBP can be set along each pencil beam to the length of the target along that beam's axis. With this technique, the region of full prescription dose is confined to the target since the modulation width of the SOBP is varied along the target length. SOBP is varied by energy switching. The first layer of pencil beams delivered has the highest energy,

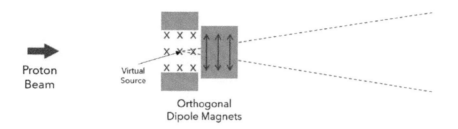

FIGURE 25.2   Pencil beam scanning system schematic.

covering the most distal edge of the target. Subsequent layers have lower energy with pencil beam weights that are optimized so that the addition of the entrance region of the earlier layers of pencil beams in addition to the Bragg peaks of lower energy layers add up to the prescription dose.

Scanned beams have multiple advantages, making them the preferred proton beam delivery system in the modern era. Nearly every proton in the delivery system enters the patient, so it is more efficient. This reduces activation of material near the patient since the protons are all deposited within the patient and not stopped by other materials. Therefore, the neutron dose is minimized except for those produced in the patient, which is an advantage for pediatric patients in reducing the induction of secondary cancers [99].

With the advent of PBS, intensity-modulated (IMPT) fields are achievable. Multiple inhomogeneous dose fields can be used to deliver a single conformal dose in the target but spare critical structures [99]. Another advantage of PBS is the simplification of the mechanical design and size of the gantry. Lateral and distal dose fall off can be improved with scanning systems.

A challenge in beam scanning is the sensitivity to organ motion during field delivery [100]. Beam spot scanning is usually applied only once, or a few times, over the whole target volume. Therefore, organ motion can quickly reduce the homogeneity of the dose in the target. Well-mobilized tumors, such as those in the head and neck, spinal cord, and pelvis, were treated first with beam scanning. Recently, motion management strategies and immobilization techniques are utilized in addition to techniques such as dose painting, where the field is split and delivered twice or more to ensure that the prescription is not completely dependent on one field. Careful selection of field angles and beam entry points within the patient can be incorporated to minimize the effects of organ motion as well.

25.5 MEASUREMENT TECHNIQUES

Many of the same techniques used for electron beam dosimetry can also be used for proton beam therapy. Differences arise from the great differential in mass between the two types of particle. The proton is approximately 2000 times heavier than the electron, and therefore is deflected by a much smaller amount from its path during interaction with the elementary and nuclear particles that make up the medium.

There are many ways to measure proton beams. Absolute dosimetry can be performed with calorimeters, FCs, and ICs, to name a few. Calorimeters are devices that measure changes in heat generated per unit mass of absorbing material. Corrections for calorimeters include a small estimate of the fraction of energy that goes into induction of chemical changes rather than heat.

FCs are measurement devices that are essentially a shielded, insulated block that is thick enough to stop the protons being measured [46]. They are used to measure proton fluence. The charge deposited in the block is proportional to the number of protons that have stopped in the block since the proton charge is very accurately known. With the knowledge of stopping power of the protons entering the FC and the proton fluence, the dose can be computed. Corrections using this device include any change in collected charge due to charged particles escaping the stopping material.

ICs are the most commonly used for absolute dosimetry. They are made with a pair of electrodes between which is a known amount of gas. As radiation passes the gas, it is ionized and the applied electric potential between the two electrodes will separate the positive ions and the electrons. The electrons are collected, and the total charge is measured. For proton pencil beams, ICs can be large enough to collect the whole beam to integrate the total dose. Small volume ICs often have cylindrical shapes with a central electrode wire or are made with two small parallel plates, which is useful for measurements of dose close to the surface.

Relative dosimetry is useful for characterizing the dose distribution both laterally and in depth. Furthermore, relative dosimetry can be used for beam line monitoring. Between the accelerator and the beam delivery system, monitors are used to measure the beam intensity and position along the beam line. Typically, these are large-area parallel-plate chambers. Machine output can also be measured with a large parallel-plate IC that is cross-calibrated against an absolute dosimeter. Diode arrays, CCD cameras, films, or scintillation detectors can also be used to measure dose distributions. Diode arrays are particularly useful for lateral dose distributions rather than in depth, where they exhibit LET dependencies.

25.6 DOSIMETRY PROTOCOLS

The International Atomic Energy Agency (IAEA) provides a series of technical reports (TRS) and TRS-398 provides a code of practice for reference dosimetry (beam calibration) for proton beams with energies in the range of 50–250 MeV. In order to do this calibration, a factor in terms of absorbed dose to water (N_{D,w,Q_0}) for a dosimeter is needed for a reference beam that has beam quality Q_0. For clinical proton beams, cylindrical and plane-parallel ICs are recommended for use as reference instruments for calibration. However, the dose to water uncertainty will be slightly higher for plane-parallel ICs than for cylindrical chambers due to the higher uncertainty for p_{wall} (the factor to correct the response of an IC for the non-medium equivalence of the chamber wall and any waterproofing material). Cylindrical chambers are however limited to proton beams with qualities at the reference depth $R_{res} \geq 0.5$ g/cm². Graphite wall chambers are preferred over plastic wall chambers due to their improved long-term stability and smaller chamber to chamber variations.

For cylindrical chambers, the reference point is taken to be on the central axis of the chamber at the center of the cavity volume.

Plane-parallel chambers can be used for dosimetry in proton beams but must be used for proton beams with qualities at the reference depth $R_{res} < 0.5$ g/cm². For these, the reference point is taken to be on the inner surface of the entrance window at the center. The cavity diameter of a plane-parallel ion chamber or the length of the cavity of a cylindrical chamber should not be larger than approximately half the reference field size. In addition, the outer

diameter of cylindrical chambers should not be larger than half the width of an SOBP. IAEA TRS-398 recommends using a plane-parallel chamber for relative dosimetry.

Beam quality specifiers for proton beams have been specified by the effective energy, which is defined as the energy of a monoenergetic proton beam having a range equal to the residual range R_{res} of the clinical proton beam. Residual range at a depth z is defined as

$$R_{res} = R_p - z \qquad (25.4)$$

where R_p is the practical range and is the depth at which the absorbed dose beyond the Bragg peak or SOBP falls to 10% of its maximum value. Both R_p an R_{res} have units of g/cm². Use of residual range was justified by the small energy dependence of water/air stopping-power ratios and since the effective energy is close to the maximum energy in the proton energy spectrum at the reference depth. Residual range is easily measurable, making it a good measure of beam quality. This will slightly underestimate the stopping power ratios in the middle of the SOBP but will not exceed 0.3%.

Beam quality should be measured in water with either the cylindrical or plane-parallel chamber reference point at the point of interest and with a clinical source-to-surface distance (SSD). The field size at the phantom surface should be 10×10 cm².

Absorbed dose to water at reference depth (z_{ref}) in a proton beam with quality Q and in the absence of the chamber is given by

$$D_{w,Q} = M_Q \, N_{D,w,Q_0} k_{Q,Q_0} \qquad (25.5)$$

where M_Q is the reading of the dosimeter with the reference point of the chamber positioned at z_{ref}. N_{D,w,Q_0} is the calibration factor in terms of absorbed dose to water for the dosimeter at the reference quality Q_0 and k_{Q,Q_0} is a chamber-specific factor that corrects for differences between the beam quality Q_0 and the actual quality being used Q. M_Q should be corrected for temperature-pressure, electrometer calibration, polarity effects, and ion recombination.

25.7 QUALITY ASSURANCE FOR PROTON THERAPY

QA aims to maximize the treatment effect while minimizing side effects of the treatment. The QA process is continually getting more complex as there are great advances in technology. QA criteria may be set by regulatory agencies at the national or state level. The report of the American Association of Physicists in Medicine's (AAPM) Task Group number 40, titled "Comprehensive QA for Radiation Oncology" is a resource that describes the physical aspects of QA for photon and electron beam therapy that can be applied to protons as well [101]. More recently, there has been Task Group report number 224, entitled "Comprehensive proton therapy machine quality assurance," detailing the specific machine QA requirements and methods for proton beam delivery systems [102]. Computerized treatment planning systems must undergo routine QA as well. IAEA has published a technical report on the topic of QA for treatment planning systems, and the AAPM Task Group report number 53 outlines QA for treatment planning systems [103, 104].

Routine QA can be divided by the frequency at which they should be performed. On a daily basis, proton beam output should be checked. Mechanical lasers, snout position alignment, beam delivery safety interlocks, patient monitoring systems, radiation monitors, and motion stops should be assessed. Monthly, the beam modulation system should be verified and checked. Any beam modifying devices should be checked as well as imaging devices. Annually, there should be extensive verification of the output under a wide variety of operating conditions, lateral profile flatness and symmetry, the virtual source location, MU linearity, dose per MU, and patient positioning devices should be checked for mechanical integrity and coincidence with the isocenter.

Patient-specific QA can be delivered using two dimensional ion chamber array detectors, such as the MatriXX PT (IBA Dosimetry, Germany) [105]. The planar dose distribution and treatment field profiles can be collected for each field. The measured isodose distributions (Figure 25.3(a)) can be compared with the planar dose calculated by the treatment planning system (Figure 25.3(b)). Profiles can be compared along the two directions at any point (Figure 25.3(c)). Gamma analysis provides information about the measured planar dose compared to the calculated dose (Figure 25.3(d)). Different depths can be used to analyze dose; however, a typical measurement is at the middle of the SOBP for each field. This can be achieved by using different thicknesses of water equivalent plastic on top of the ion chamber array.

Monthly QA procedures will often include measurement of system parameters that have a smaller chance of drifting over a short time scale [102]. Typically, these will include output constancy check with a small volume parallel plate IC at a benchmarked depth in water or plastic phantom. Additional dosimetry checks for PBS systems may include imaging and radiation isocenter alignment checks, spot positioning checks, and spot shape verification. A sample test pattern for PBS monthly QA is shown in Figure 25.4. The test pattern was measured in air with a 2D scintillation detector (Lynx, IBA Dosimetry) in panel (a) and compared to a baseline measurement in panel (b) through a gamma index evaluation [106].

25.8 UNCERTAINTIES

As in all external beam treatment modalities, dosimetry for proton beam therapy must take into account a number of uncertainties. Aside from the inherent dosimetry uncertainties, positioning uncertainties play a significant role, and in many cases have a larger risk in proton therapy due to changes in water equivalent path length. In most modern uses of proton therapy, the Bragg peak is used to take advantage of the superior dosimetric properties of protons relative to x-rays, electrons, or neutrons. As discussed in Section 25.2, the positional depth of the Bragg peak depends on both the beam energy and the stopping power of the material through which the proton beam passes. In the case of a shoot through technique, the variability of stopping power through the material and the uncertainty in its determination has much less impact dosimetrically.

This variability in path length is clear when visualizing a proton beam 'ray' passing through a complex heterogeneous material. When the material, or the patient as the case may be, shifts with respect to the ray a different path length to the terminal end of the Bragg peak will be realized.

Apart from the typical variability that is inherent in external beam therapy, the primary dosimetric uncertainty unique to proton therapy, as opposed to conventional treatment modalities (e.g. x-ray and electrons beams), is the uncertainty in the proton stopping power as the beam passes through a material.

The proton stopping power and subsequently the LET rises significantly at the maximum through distal edge of the Bragg peak. Mixing of this high LET in an SOBP which is created through the superposition of multiple

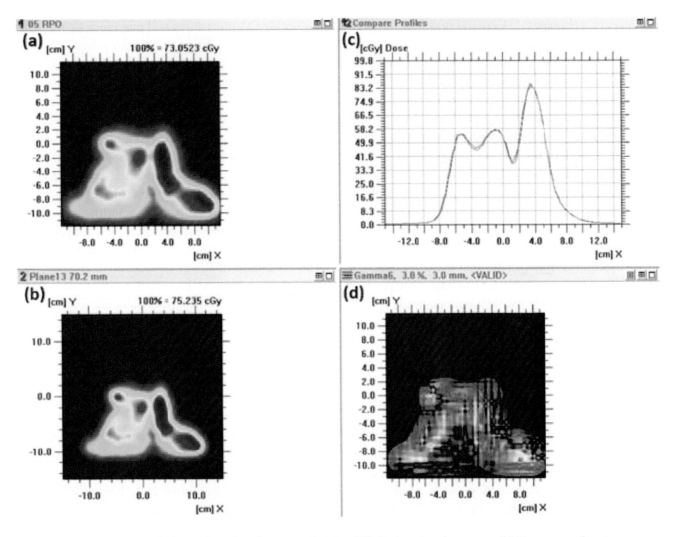

FIGURE 25.3   (a) Measured planar dose distribution with MatriXX 2D ion chamber array. (b) Treatment planning system calculated planar dose. (c) Profile comparison between measured and planning system calculated dose. (d) Gamma analysis with 3%/3 mm criterion.

energies causes an elevation of the average LET throughout the SOBP relative to the entrance region, Figure 25.5. The elevation is most pronounced at the distal edge due to the relative purity of low energy protons in this region, there are fewer contributions from high energy protons as compared to shallower depths. This increased LET, increases the biologically effective dose at the distal edge of the SOBP and contributes to increased uncertainty in the effective proton range [107, 108].

25.9 SUMMARY

Dosimetry for proton therapy makes use of many of the same techniques and equipment that are in widespread use with other modes of external beam radiotherapy. The significant issue to consider when designing a program for proton therapy is that the dosimetric properties

of protons depend strongly on the material where the protons deposit dose and the energy of the protons at the point where they deposit dose. This energy and material dependence comes out in the proton stopping powers and is also the primary reason why protons are an attractive alternative to high energy x-ray therapy. Protons stop in tissue and deposit an increased dose at the end of their range. This property can be exploited with the newest scanning and range modulation technologies to precisely sculpt proton dose around a tumor with the goal of sparing healthy tissues. A thorough understanding of the dosimetric properties of protons and the effects of stopping power variation will aid the physicist in choosing appropriate tools and methods to safely and effectively realize the potential of proton therapy.

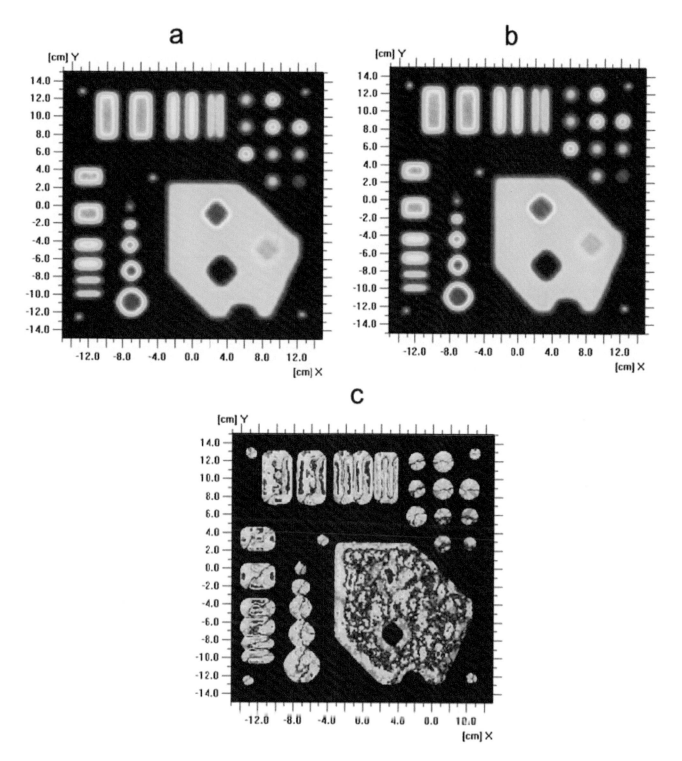

FIGURE 25.4 (a) Measured planar dose planar spot pattern in air with Lynx 2D scintillation detector. (b) Benchmarked 2D planar measurement. (c) Gamma analysis with 2%/2 mm criterion.

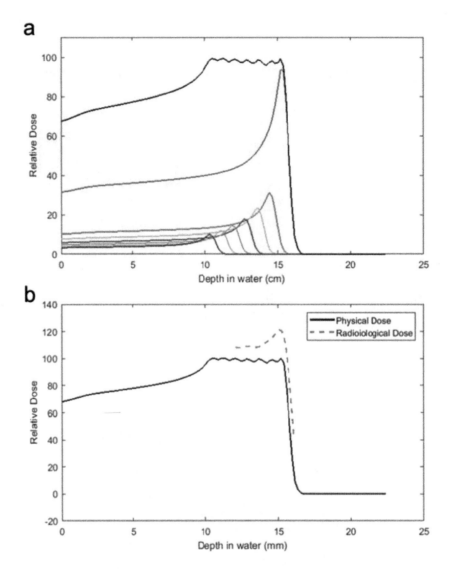

FIGURE 25.5 (a) Proton spread out Bragg peak (SOBP) generated through the superposition of multiple pristine Bragg peaks with weights chosen to produce a flat topped SOBP. (b) The high LET mixing at points in the distal portion of the SOBP lead to enhanced relative biological effective dose. (Adapted from [107] (Figure 9).)

REFERENCES

1. H. Paganetti, et al., "Relative biological effectiveness (RBE) values for proton beam therapy," *International Journal of Radiation Oncology, Biology, Physics* **53**(2), 407–421 (2002). Available at: http://www.ncbi.nlm.nih.gov/pubmed/12023146.

2. H. Paganetti, "Relative biological effectiveness (RBE) values for proton beam therapy. Variations as a function of biological endpoint, dose, and linear energy transfer," *Physics in Medicine and Biology* **59**(22), R419–R472 (2014). https://doi.org/10.1088/0031-9155/59/22/R419.

3. ICRU, *Report 59, Clinical Proton Dosimetry Part I: Beam Production, Beam Delivery and Measurement of Absorbed Dose, ICRU* (International Commission on Radiation Units and Measurements, Bethesda, MD, 1998). https://doi.org/10.1111/j.1440-1673.2001.0944a.pp.x.

4. W. D. Newhauser and R. Zhang, "The physics of proton therapy," *Physics in Medicine and Biology* **60**(8), R155–R209 (2015). https://doi.org/10.1088/0031-9155/60/8/R155.

5. H. Bethe, "Zur Theorie des Durchgangs schneller Korpuskularstrahlen durch Materie," *Annalen der Physik* **397**(3), 325–400 (1930). https://doi.org/10.1002/andp.19303970303.

6. F. Bloch, "Zur Bremsung rasch bewegter Teilchen beim Durchgang durch Materie," *Annalen der Physik* **408**(3), 285–320 (1933). https://doi.org/10.1002/andp.19334080303.

7. R. Zhang and W. D. Newhauser, "Calculation of water equivalent thickness of materials of arbitrary density, elemental composition and thickness in proton beam irradiation," *Physics in Medicine and Biology* **54**(6), 1383–1395 (2009). https://doi.org/10.1088/0031-9155/54/6/001.

8. B. Gottschalk, et al., "Multiple coulomb scattering of 160 MeV protons," *Nuclear Instruments and Methods in Physics Research Section B* **74**, 467–490 (1993).

9. E. Pedroni, et al., "Experimental characterization and physical modelling of the dose distribution of scanned proton pencil beams," *Physics in Medicine and Biology* **50**(3), 541–561 (2005). https://doi.org/10.1088/0031-9155/50/3/011.

10. F. H. Attix, *Introduction to Radiological Physics and Radiation Dosimetry* (Wiley, New York, 1986).

11. G. F. Knoll, *Radiation Detection and Measurement*, 3rd edn (Wiley, New York, 2000). Available at: http://he4.phys.tohoku.ac.jp/~booklet/seminar/files/Knoll_contents.pdf (Accessed: 14 January 2011).

12. H. Fricke and E. Hart, "Chemical dosimetry," in *Radiation Dosimetry: Instrumentation*, edited by F. Attix, W. Roesch, E. Tochilin (Academic Press, New York, 1966), pp. 167–239.

13. A. Ableitinger, et al., "Dosimetry auditing procedure with alanine dosimeters for light ion beam therapy," *Radiotherapy and Oncology* **108**(1), 99–106 (2013). https://doi.org/10.1016/j.radonc.2013.04.029.

14. A. Carlino, et al., "End-to-end tests using alanine dosimetry in scanned proton beams," *Physics in Medicine and Biology* **63**(5) (2018). https://doi.org/10.1088/1361-6560/aaac23.

15. J. R. Kerns, S. F. Kry, N. Sahoo, "Characteristics of optically stimulated luminescence dosimeters in the spread-out Bragg peak region of clinical proton beams," *Medical Physics* **39**(4), 1854–1863 (2012). https://doi.org/10.1118/1.3693055.

16. ICRU, "Report 78, prescribing, recording, and reporting proton-beam therapy," *Journal of the ICRU* **7**(2) (2007). https://doi.org/10.1093/jicru/ndn001.

17. IAEA, *Technical Report Series No. 398, Absorbed Dose Determination in External Beam Radiotherapy, IAEA Technical Report Series* (IAEA, Vienna, Austria, 2000). https://doi.org/10.1007/s00259-008-0767-4.

18. C. Gomà, P. Andreo, J. Sempau, "Monte Carlo calculation of beam quality correction factors in proton beams using detailed simulation of ionization chambers," *Physics in Medicine and Biology* **61**(6), 2389–2406 (2016). https://doi.org/10.1088/0031-9155/61/6/2389.

19. ICRU, "ICRU Report 37: Stopping powers for electrons and positrons," *Journal of the ICRU* (1984). https://doi.org/10.1093/jicru/ndm020.

20. ICRU, *Report 49, Stopping Powers and Ranges for Protons and Alpha Particles* (International Commission on Radiation Units and Measurements, Bethesda, MD, 1993).

21. P. Andreo, et al., "Consistency in reference radiotherapy dosimetry: Resolution of an apparent conundrum when ^{60}Co is the reference quality for charged-particle and photon beams," *Physics in Medicine and Biology* **58**(19), 6593–6621 (2013). https://doi.org/10.1088/0031-9155/58/19/6593.

22. D. T. Burns, et al., "Use of the BIPM calorimetric and ionometric standards in megavoltage photon beams to determine Wair and Ic," *Physics in Medicine and Biology* **59**(6), 1353–1365 (2014). https://doi.org/10.1088/0031-9155/59/6/1353.

23. S. M. Vatnitsky, J. V Siebers, D. W. Miller, "k_Q factors for ionization chamber dosimetry in clinical proton beams," *Medical Physics* **23**(1), 25–31 (1996).

24. J. Sorriaux, et al., "Consistency in quality correction factors for ionization chamber dosimetry in scanned proton beam therapy," *Medical Physics* **44**(9), 4919–4927 (2017). https://doi.org/10.1002/mp.12434.

25. S. Vatnitsky, et al., "Proton dosimetry intercomparison," *Radiotherapy and Oncology* **41**(2), 169–177 (1996). https://doi.org/10.1016/S0167-8140(96)01800-2.

26. C. Gomà, et al., "Proton beam monitor chamber calibration," *Physics in Medicine and Biology* **59**(17), 4961–4971 (2014). https://doi.org/10.1088/0031-9155/59/17/4961.

27. C. Gomà, et al., "Experimental validation of beam quality correction factors for proton beams," *Physics in Medicine and Biology* **60**(8), 3207–3216 (2015). https://doi.org/10.1088/0031-9155/60/8/3207.

28. H. Bichsel, "Calculated Bragg curves for ionization chambers of different shapes," *Medical Physics* **22**(11), 1721–1726 (1995). https://doi.org/10.1118/1.597535.

29. H. Palmans, et al., "Dosimetry using plane-parallel ionization chambers in a 75 MeV clinical proton beam," *Physics in Medicine and Biology* **47**(16), 2895–2905 (2002). https://doi.org/10.1088/0031-9155/47/16/305.

30. J. Harms, et al., "Nuclear halo measurements for accurate prediction of field size factor in a Varian ProBeam proton PBS system," *Journal of Applied Clinical Medical Physics* **21**(1), 197–204 (2020). https://doi.org/10.1002/acm2.12783.

31. B. Clasie, et al., "Golden beam data for proton pencil-beam scanning," *Physics in Medicine and Biology* **57**(5), 1147–1158 (2012). https://doi.org/10.1088/0031-9155/57/5/1147.

32. L. Lin, et al., "Experimental characterization of two-dimensional spot profiles for two proton pencil beam scanning nozzles," *Physics in Medicine and Biology* **59**(2), 493–504 (2014). https://doi.org/10.1088/0031-9155/59/2/493.

33. C. Bäumer, et al., "Evaluation of detectors for acquisition of pristine depth-dose curves in pencil beam scanning," *Journal of Applied Clinical Medical Physics* **16**(6), 151–163 (2015). https://doi.org/10.1120/jacmp.v16i6.5577.

34. J. T. Lyman, et al., *Protocol for Heavy Charged-Particle Therapy Beam Dosimetry: A report of Task Group 20 Radiation Therapy Committee American Association of Physicists in Medicine, New York* (1986). Available at: https://mail.aapm.org/pubs/reports/RPT_16.pdf.

35. A. Lourenço, et al., "Experimental and Monte Carlo studies of fluence corrections for graphite calorimetry in low- and high-energy clinical proton beams," *Medical Physics* **43**(7), 4122–4132 (2016). https://doi.org/10.1118/1.4951733.

36. M. Sassowsky and E. Pedroni, "On the feasibility of water calorimetry with scanned proton radiation,"

Physics in Medicine and Biology **50**(22), 5381–5400 (2005). https://doi.org/10.1088/0031-9155/50/22/011.

37. H. J. Brede, et al., "Absorbed dose to water determination with ionization chamber dosimetry and calorimetry in restricted neutron, photon, proton and heavy-ion radiation fields," *Physics in Medicine and Biology* **51**(15), 3667–3682 (2006). https://doi.org/10.1088/0031-9155/51/15/005.

38. J. Medin, et al., "Experimental determination of beam quality factors, k_Q, for two types of Farmer chamber in a 10 MV photon and a 175 MeV proton beam," *Physics in Medicine and Biology* **51**(6), 1503–1521 (2006). https://doi.org/10.1088/0031-9155/51/6/010.

39. J. Medin, "Implementation of water calorimetry in a 180 MeV scanned pulsed proton beam including an experimental determination of k_Q for a Farmer chamber," *Physics in Medicine and Biology* **55**(12), 3287–3298 (2010). https://doi.org/10.1088/0031-9155/55/12/002.

40. A. Sarfehnia, et al., "Direct absorbed dose to water determination based on water calorimetry in scanning proton beam delivery," *Medical Physics* **37**(7), 3541–3550 (2010). https://doi.org/10.1118/1.3427317.

41. S. Delacroix, et al., "Proton dosimetry comparison involving ionometry and calorimetry," *International Journal of Radiation Oncology, Biology, Physics* **37**(3), 711–718 (1997). https://doi.org/10.1016/S0360-3016(96)00536-6.

42. L. J. Verhey, et al., "The determination of absorbed dose in a proton beam for purposes of charged-particle radiation therapy," *Radiation Research* **79**(1), 34–54 (1979). https://doi.org/10.2307/3575020.

43. S. Vynckier, D. E. Bonnett, D. T. L. Jones, "Code of practice for clinical proton dosimetry," *Radiotherapy and Oncology* **20**(1), 53–63 (1991). https://doi.org/10.1016/0167-8140(91)90112-T.

44. R. Cambria, et al., "Proton beam dosimetry: A comparison between the Faraday cup and an ionization chamber," *Physics in Medicine and Biology* **42**(6), 1185–1196 (1997). https://doi.org/10.1088/0031-9155/42/6/014.

45. S. Lorin, et al., "Reference dosimetry in a scanned pulsed proton beam using ionisation chambers and a Faraday cup," *Physics in Medicine and Biology* **53**(13), 3519–3529 (2008). https://doi.org/10.1088/0031-9155/53/13/008.

46. E. W. Cascio and B. Gottschalk, "A simplified vacuumless Faraday cup for the experimental beamline at the Francis H. Burr Proton Therapy Center," *IEEE Radiation Effects Data Workshop* (2009), pp. 161–165. https://doi.org/10.1109/REDW.2009.5336294.

47. E. Grusell, et al., "Faraday cup dosimetry in a proton therapy beam without collimation," *Physics in Medicine and Biology* **40**(11), 1831–1840 (1995). https://doi.org/10.1088/0031-9155/40/11/005.

48. A. Niroomand-Rad, et al., "Radiochromic film dosimetry: Recommendations of AAPM Radiation Therapy Committee Task Group 55," *Medical Physics* **25**(11), 2093–2115 (1998). https://doi.org/10.1118/1.598407.

49. M. Fuss, et al., "Dosimetric characterization of GafChromic EBT film and its implication on film

dosimetry quality assurance," *Physics in Medicine and Biology* **52**(14), 4211–4225 (2007). https://doi.org/10.1088/0031-9155/52/14/013.

50. H. Bouchard, et al., "On the characterization and uncertainty analysis of radiochromic film dosimetry," *Medical Physics* **36**(6), 1931–1946 (2009). https://doi.org/10.1118/1.3121488.

51. A. Micke, D. F. Lewis, X. Yu, "Multichannel film dosimetry with nonuniformity correction," *Medical Physics* **38**(5), 2523–2534 (2011). https://doi.org/10.1118/1.3576105.

52. D. Lewis, et al., "An efficient protocol for radiochromic film dosimetry combining calibration and measurement in a single scan," *Medical Physics* **39**(10), 6339–6350 (2012). https://doi.org/10.1118/1.4754797.

53. G. Gambarini, et al., "Measurements of 2D distributions of absorbed dose in protontherapy with Gafchromic EBT3 films," *Applied Radiation and Isotopes* **104**, 192–196 (2015). https://doi.org/10.1016/j.apradiso.2015.06.036.

54. L. Zhao and I. J. Das "Gafchromic EBT film dosimetry in proton beams," *Physics in Medicine and Biology* **55**(10), N291–N301 (2010). https://doi.org/10.1088/0031-9155/55/10/N04.

55. B. Arjomandy, et al., "EBT2 film as a depth-dose measurement tool for radiotherapy beams over a wide range of energies and modalities," *Medical Physics* **39**(2), 912–921 (2012). https://doi.org/10.1118/1.3678989.

56. S. Reinhardt, et al., "Comparison of Gafchromic EBT2 and EBT3 films for clinical photon," *Medical Physics* **39**(August), 5257–5262 (2012).

57. B. R. Smith, et al., "LET response variability of Gafchromic™ EBT3 film from a ^{60}Co calibration in clinical proton beam qualities," *Medical Physics* **46**(6), 2716–2728 (2019). https://doi.org/10.1002/mp.13442.

58. S. M. Vatnitsky, "Radiochromic film dosimetry for clinical proton beams," *Applied Radiation and Isotopes* **48**(5), 643–651 (1997). https://doi.org/10.1016/S0969-8043(97)00342-4.

59. M. Martisíková and O. Jäkel, "Dosimetric properties of Gafchromic EBT films in monoenergetic medical ion beams," *Physics in Medicine and Biology* **55**(13), 3741–3751 (2010). https://doi.org/10.1088/0031-9155/55/13/011.

60. M. Mumot, et al., "The comparison of doses measured by radiochromic films and semiconductor detector in a 175 MeV proton beam," *Physica Medica* **25**(3), 105–110 (2009). https://doi.org/10.1016/j.ejmp.2008.06.001.

61. B. Spielberger, et al., "Experimental investigations of the response of films to heavy-ion irradiation," *Physics in Medicine and Biology* **46**(11), 2889–2897 (2001). https://doi.org/10.1088/0031-9155/46/11/309.

62. S. E. Anderson, et al., "A linear relationship for the LET-dependence of Gafchromic EBT3 film in spot-scanning proton therapy," *Physics in Medicine and Biology* **64**(5) (2019). https://doi.org/10.1088/1361-6560/ab0114.

63. C. Carlsson and G. Carlsson, "Proton dosimetry: Measurement of depth doses from 185-MeV protons by

means of thermoluminescent LiF," *Radiation Research* **42**(2), 207–219 (1970).

64. S. Vatnitsky, et al., "Application of solid state detectors for dosimetry of therapeutic proton beams," *Medical Physics* **22**(4), 469–473 (1995). https://doi.org/10.1118/1.597608.

65. M. G. Sabini, et al., "The use of thermoluminescent detectors for measurements of proton dose distribution," *Radiation Protection Dosimetry* **101**(1–4), 453–456 (2002). https://doi.org/10.1093/oxfordjournals.rpd.a006024.

66. J. R. Zullo, et al., "LiF TLD-100 as a dosimeter in high energy proton beam therapy – can it yield accurate results?," *Medical Dosimetry* **35**(1), 63–66 (2010). https://doi.org/10.1016/j.meddos.2009.03.001.

67. G. S. Ibbott, et al., "Challenges in credentialing institutions and participants in advanced technology multi-institutional clinical trials," *International Journal of Radiation Oncology, Biology, Physics* **71**(1 Suppl.), 71–75 (2008). https://doi.org/10.1016/j.ijrobp.2007.08.083.

68. C. S. Reft, "The energy dependence and dose response of a commercial optically stimulated luminescent detector for kilovoltage photon, megavoltage photon, and electron, proton, and carbon beams," *Medical Physics* **36**(5), 1690–1699 (2009). https://doi.org/10.1118/1.3097283.

69. G. O. Sawakuchi, et al., "Determination of average LET of therapeutic proton beams using Al_2O_3:C optically stimulated luminescence (OSL) detectors," *Physics in Medicine and Biology* **55**(17), 4963–4976 (2010). https://doi.org/10.1088/0031-9155/55/17/006.

70. D. A. Granville, N. Sahoo, G. O. Sawakuchi, "Calibration of the Al_2O_3:C optically stimulated luminescence (OSL) signal for linear energy transfer (LET) measurements in therapeutic proton beams," *Physics in Medicine and Biology* **59**(15), 4295–4310 (2014). https://doi.org/10.1088/0031-9155/59/15/4295.

71. E. G. Yukihara, et al., "Time-resolved optically stimulated luminescence of Al_2O_3:C for ion beam therapy dosimetry," *Physics in Medicine and Biology* **60**(17), 6613–6638 (2015). https://doi.org/10.1088/0031-9155/60/17/6613.

72. M. R. Raju, "The use of the miniature silicon diode as a radiation dosemeter," *Physics in Medicine and Biology* **11**(3), 371–376 (1966). https://doi.org/10.1088/0031-9155/11/3/302.

73. S. M. Vatnitsky, et al., "Dosimetry techniques for narrow proton beam radiosurgery," *Physics in Medicine and Biology* **44**(11), 2789–2801 (1999). https://doi.org/10.1088/0031-9155/44/11/308.

74. A. Fidanzio, et al., "A correction method for diamond detector signal dependence with proton energy," *Medical Physics* **29**(5), 669–675 (2002). https://doi.org/10.1118/1.1469634.

75. E. Grusell and J. Medin, "General characteristics of the use of silicon diode detectors for clinical dosimetry in proton beams," *Physics in Medicine & Biology* **45**, 2573–2582 (2000).

76. S. Onori, et al., "Dosimetric characterization of silicon and diamond detectors in low-energy proton beams," *Physics in Medicine and Biology* **45**(10), 3045–3058 (2000). https://doi.org/10.1088/0031-9155/45/10/320.

77. M. Pacilio, et al., "Characteristics of silicon and diamond detectors in a 60 MeV proton beam," *Physics in Medicine and Biology* **47**(8) (2002). https://doi.org/10.1088/0031-9155/47/8/403.

78. C. Brusasco, et al., "A dosimetry system for fast measurement of 3D depth-dose profiles in charged-particle tumor therapy with scanning techniques," *Nuclear Instruments and Methods in Physics Research, Section B: Beam Interactions with Materials and Atoms* **168**(4), 578–592 (2000). https://doi.org/10.1016/S0168-583X(00)00058-6.

79. D. Nichiporov, et al., "Multichannel detectors for profile measurements in clinical proton fields," *Medical Physics* **34**(7), 2683–2690 (2007). https://doi.org/10.1118/1.2746513.

80. T. Takayanagi, et al., "Dual ring multilayer ionization chamber and theory-based correction technique for scanning proton therapy," *Medical Physics* **43**(7), 4150–4162 (2016). https://doi.org/10.1118/1.4953633.

81. H. Paganetti and B. Gottschalk, "Test of GEANT3 and GEANT4 nuclear models for 160 MeV protons stopping in CH_2," *Medical Physics* **30**(7), 1926–1931 (2003). https://doi.org/10.1118/1.1586454.

82. S. Lin, et al., "More than 10 years experience of beam monitoring with the Gantry 1 spot scanning proton therapy facility at PSI," *Medical Physics* **36**(11), 5331–5340 (2009). https://doi.org/10.1118/1.3244034.

83. I. Rinaldi, et al., "An integral test of FLUKA nuclear models with 160 MeV proton beams in multi-layer Faraday cups," *Physics in Medicine and Biology* **56**(13), 4001–4011 (2011). https://doi.org/10.1088/0031-9155/56/13/016.

84. B. Tesfamicael, et al., "Technical Note: Use of commercial multilayer Faraday cup for offline daily beam range verification at the McLaren Proton Therapy Center," *Medical Physics* **46**(2), 1049–1053 (2019). https://doi.org/10.1002/mp.13348.

85. B. Arjomandy, et al., "Use of a two-dimensional ionization chamber array for proton therapy beam quality assurance," *Medical Physics* **35**(9), 3889–3894 (2008). https://doi.org/10.1118/1.2963990.

86. M. Varasteh Anvar, et al., "Quality assurance of carbon ion and proton beams: A feasibility study for using the 2D MatriXX detector," *Physica Medica* **32**(6), 831–837 (2016). https://doi.org/10.1016/j.ejmp.2016.05.058.

87. T. Lin, et al., "Design of a QA method to characterize sub-millimeter-sized PBS beam properties using a 2D ionization chamber array," *Physics in Medicine and Biology* **63**(10), 0–11 (2018). https://doi.org/10.1088/1361-6560/aabd89.

88. L. Lin, et al., "Use of a novel two-dimensional ionization chamber array for pencil beam scanning proton therapy beam quality assurance," *Journal of Applied Clinical Medical Physics* **16**(3), 270–276 (2015). https://doi.org/10.1120/jacmp.v16i3.5323.

89. S. N. Boon, et al., "Fast 2D phantom dosimetry for scanning proton beams," *Medical Physics* **25**(4), 464–475 (1998). https://doi.org/10.1118/1.598221.

90. S. N. Boon, et al., "Performance of a fluorescent screen and CCD camera as a two-dimensional dosimetry system for dynamic treatment techniques," *Medical Physics* **27**(10), 2198–2208 (2000). https://doi.org/10.1118/1.1289372.

91. S. Russo, et al., "Characterization of a commercial scintillation detector for 2-D dosimetry in scanned proton and carbon ion beams," *Physica Medica* **34**, 48–54 (2017). https://doi.org/10.1016/j.ejmp.2017.01.011.

92. S. Rana, et al., "Development and long-term stability of a comprehensive daily QA program for a modern pencil beam scanning (PBS) proton therapy delivery system," *Journal of Applied Clinical Medical Physics* **20**(4), 29–44 (2019). https://doi.org/10.1002/acm2.12556.

93. S. Rana and E. J. J. Samuel, "Feasibility study of utilizing XRV-124 scintillation detector for quality assurance of spot profile in pencil beam scanning proton therapy," *Physica Medica* **66**, 15–20 (2019). https://doi.org/10.1016/j.ejmp.2019.09.078.

94. W. Cai, et al., "Semi-automated IGRT QA using a cone-shaped scintillator screen detector for proton pencil beam scanning treatments," *Physics in Medicine and Biology* **64**(8), 0–10 (2019). https://doi.org/10.1088/1361-6560/ab056d.

95. A. M. Koehler, R. J. Schneider, J. M. Sisterson, "Range modulators for protons and heavy ions," *Nuclear Instruments and Methods* **131**(3), 437–440 (1975). https://doi.org/10.1016/0029-554X(75)90430-9.

96. E. S. Diffenderfer, et al., "Comparison of secondary neutron dose in proton therapy resulting from the use of a tungsten alloy MLC or a brass collimator system," *Medical Physics* **38**(11), 6248–6256 (2011). https://doi.org/10.1118/1.3656025.

97. M. S. Wagner, "Automated range compensation for proton therapy," *Medical Physics* 749–752 (1982). https://doi.org/10.1118/1.595123.

98. P. L. Petti, "New compensator design options for charged-particle radiotherapy," *Physics in Medicine and Biology* **42**, 1289–1300 (1997).

99. L. Hong, M. Goitein, M. Bucciolini, "Distal penetration of proton beams: The effects of air gaps between compensating bolus and patient Related content A pencil beam algorithm for proton dose calculations," *Physics in Medicine and Biology* **34**, 1309 (1989).

100. M. H. Phillips, et al., "Effects of respiratory motion on dose uniformity with a charged particle scanning method," *Physics in Medicine and Biology* **37**(1), 223 (1992).

101. G. J. Kutcher, et al., "Comprehensive QA for radiation oncology: Report of AAPM Radiation Therapy Committee Task Group 40," *Medical Physics* **21**(4), 581–618 (1994). https://doi.org/10.1118/1.597316.

102. B. Arjomandy, et al., "AAPM task group 224: Comprehensive proton therapy machine quality assurance," *Medical Physics* **46**(8), e678–e705 (2019). https://doi.org/10.1002/mp.13622.

103. B. Fraass, et al., "American association of physicists in medicine radiation therapy committee task group 53: Quality assurance for clinical radiotherapy treatment planning," *Medical Physics* **25**(10), 1773–1829 (1998). https://doi.org/10.1118/1.598373.

104. International Atomic Energy Agency, "Commissioning and quality assurance of computerized planning systems for radiation treatment of cancer," *Technical Reports Series 430* (2004).

105. B. Arjomandy, et al., "Verification of patient-specific dose distributions in proton therapy using a commercial two-dimensional ion chamber array," *Medical Physics* **37**(11), 5831–5837 (2010). https://doi.org/10.1118/1.3505011.

106. D. A. Low, et al., "A technique for the quantitative evaluation of dose distributions," *Medical Physics* **25**(5), 656 (1998). https://doi.org/10.1118/1.598248.

107. H. Paganetti, "Range uncertainties in proton therapy and the role of Monte Carlo simulations," *Physics in Medicine and Biology* **57**(11), R99–R117 (2012). https://doi.org/10.1088/0031-9155/57/11/R99.

108. A. Carabe, et al., "Range uncertainty in proton therapy due to variable biological effectiveness," *Physics in Medicine and Biology* **57**, 1159–1172 (2012). https://doi.org/10.1088/0031-9155/57/5/1159.

CHAPTER 26

Ion Range and Dose Monitoring with Positron Emission Tomography

Katia Parodi

Ludwig-Maximilians-Universität München
Garching b. München, Germany

CONTENTS

26.1 INTRODUCTION: *IN VIVO* DOSE VERIFICATION IN PARTICLE THERAPY

Light ion beams (i.e., protons and heavier ions up to charge $Z \approx 10$) offer the possibility of a highly precise and (particularly for $Z > 1$) biologically effective radiation therapy. This is especially due to their ability to concentrate the dose delivery and, for heavier ions, to reach also at the microscopic level an elevated ionization density for more effective tumor cell killing in a very localized region toward their end of range, so called Bragg peak [1]. In combination with recent technological advances such as pencil beam scanning delivery and intensity modulated treatment planning, all these advantages open new opportunities for dose escalation to the target and/or superior sparing of normal tissue and critical organs. This in turn promises improved clinical outcome for several tumor indications in comparison to conventional external beam radiotherapy. However, because of the increased physical selectivity, ion beam therapy is also more sensitive than conventional photon radiation to uncertainties in treatment planning and delivery, which may undermine the correct placement of the Bragg peak in the tumor, thus calling for methods of *in vivo* visualization of the beam range and, ideally, reconstruction of the applied dose. Such approaches should be able to provide *in vivo* confirmation of the successful tumor-conformal dose delivery. This would, on the one hand, improve treatment confidence and promote full clinical exploitation of the Bragg peak and the resulting dosimetric advantages of ion beams for safe dose escalation in clinical practice. On the other hand, *in vivo* monitoring should also provide a prompt feedback on any possible mismatch between the planned and actually delivered treatment in order to enable a treatment adaptation prior to the application of the next treatment fraction or, ideally, during the actual beam delivery for an almost real-time intervention (e.g., interruption of the delivery or, ultimately, on-the-fly treatment adaptation). This is deemed especially crucial for the emerging high-dose hypo-fractionated treatment regimens, where less or even no subsequent fractions are available for compensation of errors. Since the primary ions are to be stopped in the tumor, monitoring methods have to rely on penetrating secondary emissions induced by the ion irradiation, able to emerge from the patient for noninvasive detection. To this end, several monitoring methods are currently under investigation (see Chapters 27 and 28 of this book). This chapter focuses on Positron-Emission-Tomography (PET), which still represents the clinically most extensively investigated approach for a volumetric, *in vivo* and noninvasive verification of the actual treatment delivery and, in particular, of the ion beam range in the patient during or shortly after irradiation.

26.2 PET SIGNAL FORMATION

The unconventional application of the well-established PET nuclear medicine technique to ion therapy monitoring exploits the coincident detection of the annihilation photon pairs resulting from the β^+-decay of positron-emitting isotopes formed as a by-product of the therapeutic ion irradiation. This irradiation-induced β^+-activation provides a surrogate signal of the ion beam range in tissue, with a varying degree of correlation to the applied dose, due to the different PET signal formation depending on the primary ion species [2]. In fact, the mechanism of β^+-activation includes either target fragmentation only, or formation of both target and projectile positron-emitting fragments. Protons and light ions up to beryllium can only contribute to the β^+-activation of the target nuclei of the irradiated medium, since they either do not fragment at all (protons $Z = 1$), or cannot produce positron-emitting projectile fragments ($Z \leq 4$). Typical reaction products feature light isotopes such as ^{11}C, ^{15}O, ^{13}N, which exhibit half-lives $T_{1/2}$ of about 20, 2 and 10 min, respectively. Target fragmentation occurs all along the beam penetration path, as long as the beam energy is above the threshold for nuclear interaction, typically corresponding to about 1–4 mm residual ion range in tissue. In homogenous targets, this results in a characteristic track of β^+-activity with an approximately constant or slowly rising slope followed by a rather abrupt distal fall-off located few millimeters in front of the Bragg-peak (Figure 26.1, left, [2, 3]). For primary proton beams, the activation of the traversed tissue is mostly due to primary and secondary protons, while for heavier primary ions ($1 < Z \leq 4$), also longer ranging non-positron-emitting projectile fragments contribute a tail of β^+-emitting target fragments beyond the Bragg peak (Figure 26.1, middle, [2, 4]).

In addition to such a pedestal of β^+-activated target fragments, heavier ions ($Z \geq 5$) can also yield positron-emitting projectile fragments through the so called "auto-activation" mechanism [6]. Being the ion stopping process typically faster than the radioactive decay, these β^+-emitting projectile fragments mostly decay after having been slowed down to rest at their end of range. This offers specific advantages for beam range and dose verification, since the β^+-active isotopes of the primary ion beam are among the most abundant reaction products and accumulate shortly before the range of the primary stable ions [4], due to the nuclear reaction kinematics and the different stopping power. Hence, the auto-activation of heavy ion beams results in a marked activity maximum located shortly before the Bragg peak, superimposed onto the minor pedestal of β^+-active target fragments. This is shown in Figure 26.1, right, [2, 6] for a primary ^{12}C ion beam, where the activity maximum is originated from the major contribution of the long-lived ^{11}C and, to a lesser extent, of the short-lived ^{10}C ($T_{1/2} \approx 20$ s).

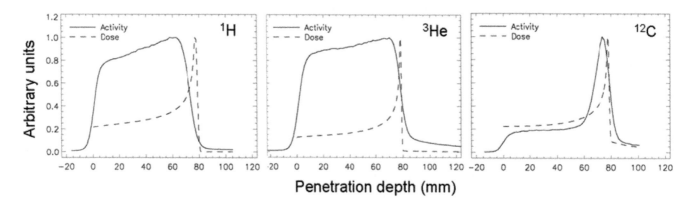

FIGURE 26.1 Calculated depth dose distribution (dashed line) and corresponding measured β^+-activity profile (solid line) for ^1H (left), ^3He (middle), and ^{12}C (right) irradiation of a homogeneous plastic target of polymethyl methacrylate, PMMA. (Taken from reference [2], adapted from refs. [3–5].)

26.3 PET SIGNAL IMAGING

After formation, β^+-active isotopes undergo β^+-decay at a random time depending on the isotope half-life. The emitted positron is slowed down in the surrounding medium, and most likely annihilates with an atomic electron (either as a free particle or in a bound state called positronium), after having (almost) entirely lost its kinetic energy, thus resulting in the (quasi) opposite emission of two annihilation gamma quanta of (approximately) 511 keV energy each. These energetic photons have a high probability to escape the patient and can be detected by surrounding detector pairs operated in coincidence, according to the standard nuclear medicine principles of PET imaging [7].

As for other monitoring techniques, the low amount of measurable events is among the major challenges for the unconventional application of PET imaging to the verification of ion treatment. In fact, the amount of theoretically available signal is limited by the reaction cross sections, resulting in irradiation-induced activity concentrations of approximately 0.2–10 kBq/Gy/cm³, depending on the ion type, time course of irradiation and anatomical site [8–10], for typical fraction doses of ~0.5–2 Gy [2]. In living tissue such activity does not only undergo radioactive decay, but is also subject to a complex hot chemistry and can be smeared out by diffusion processes or even dislocated far from the initial production place due to other physiological processes (e.g., perfusion). As a result of this so called biological "washout," the produced activity reduces more quickly not only due to the physical decay, but also to a so called biological decay, for which half-lives from 2 s up to 10,000 s have been reported, depending on ion species and tissue type [11–14]. Hence, all these considerations pose severe requirements to the efficient detection of the irradiation-induced PET signal, ideally at the treatment place.

However, different from diagnostic imaging, space restrictions from the patient couch, the beam direction and additional medical instrumentation constrain the geometry of the PET scanner to a typical dual-head design for integration at the irradiation place (so called in-beam or on-board). Such on-site installations are in principle the most appealing solution and the only possibility for an almost real-time monitoring. In fact, they enable detection of the formed activity directly during irradiation, thereby better preserving its correlation with the delivered dose by recording the major activity contribution from short-lived emitters (e.g., ^{15}O) and

minimizing the signal degradation from uncertainties in patient position and washout effects. However, they are technologically more demanding, requiring dedicated detector development and fast electronics synchronized with the beam delivery, ideally able to reject or at least suppress the undesired background of prompt radiation, e.g., prompt gamma (see Chapter 27 of this book) during actual beam extraction [15, 16]. Notable examples of such customized installations are the tomographic dual-head PET camera installed at the fixed beamline of the Gesellschaft für Schwerionenforschung (GSI Darmstadt, Germany [8]) and the planar dual-head camera integrated in the rotating gantry of the National Cancer Center (NCC Kashiwa, Japan [17]). These systems have been clinically used to image PET activity induced by actively scanned carbon ions during irradiation (only in the pauses of the pulsed delivery) and passively scattered protons immediately after end of irradiation, respectively. More recent latest-generation in-beam PET detectors just starting or close to start clinical operation include the planar dual-head camera installed at the fixed beamline of the Centro Nazionale di Terapia Oncologica (CNAO, Italy [18]) and the full-ring axially-shifted "openPET" design planned for integration in one treatment room of the National Institute for Radiological Sciences (NIRS, Japan [19]). In particular, the system at CNAO is being evaluated in a first clinical trial for proton and carbon ion beams, and offers a dedicated data acquisition and image reconstruction software for visualizing the build-up of activity (currently with best-quality events only recorded in the pauses of the pulsed delivery) during irradiation at ~10s time resolution [20] (Figure 26.2). However, due to the large development costs and challenges for data acquisition and reliable image reconstruction with unconventional detector geometries, such dedicated solutions are still restricted to few specialized institutions and not yet widely available.

Alternative clinically explored solutions relying on nuclear medicine instrumentation feature a movable neurological full-ring PET scanner in the treatment room (in-room, [21]), or fixed installations of full-ring PET [10, 22] and PET/CT (Compute Tomography) [13, 23] scanners in a separate room in walking distance of the treatment room (offline), for imaging from a couple of minutes up to ~20 min after irradiation, respectively. Here, major challenges entail accurate replication and fixation of the treatment position, for which an

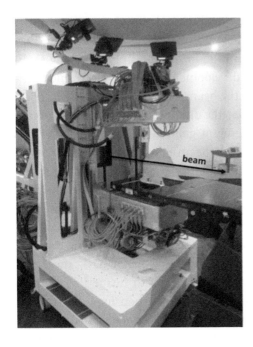

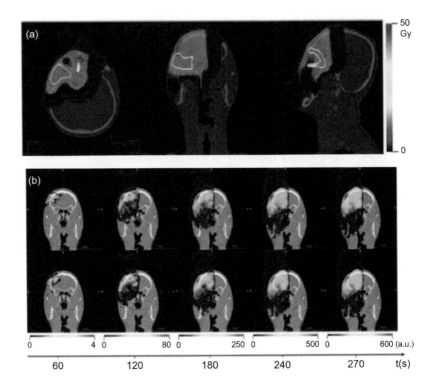

FIGURE 26.2 Left panel: The dedicated in-beam PET scanner in the horizontal beamline at CNAO, in measuring position. Right panel: Treatment plan (with dose colorwash superimposed onto the three orthogonal grayscale CT views) of a patient treated with scanned proton beams at CNAO (a), and corresponding PET activity (also in colorwash superimposed on the grayscale planning CT) measured with the dedicated in-beam PET scanner, reconstructed every 60 s and for the entire treatment time with additional 30 s after end of irradiation (from left to right) for two different treatment fractions (top and bottom row) (b). (Taken from [20].)

additional CT is beneficial to relate treatment and imaging anatomy (at the expense of minor additional dose), as well as loss of counting statistics due to physical and biological decay, requiring an according prolongation of the measuring time. The latter typically ranges from 5 min for in-room implementations, up to 30 min for offline solutions. Besides the reduction of counting statistics, an additional drawback of the time elapsed between activity formation and imaging is the already mentioned degradation of spatial correlation between the detectable PET signal and the original positron emitter production due to biological washout, which can directly impair the reliability of the PET signal as a surrogate of the delivered treatment dose. Moreover, imaging after completion of a given treatment fraction can only provide the PET signal produced by the integral activation of all the applied treatment fields. Unless special data processing techniques are employed to disentangle the inevitably overlapped signals, this limitation has so far restricted the range verification abilities of such a post-irradiation method to non-opposing beam portals for multi-field irradiation. Indeed, conventional geometries of full-ring PET scanners can offer improved tomographic imaging capabilities compared to customized in-beam geometries, but are challenged by the low counting statistics, well below the regimes of standard tracer imaging for which such scanners and reconstruction algorithms are designed [24].

According to the typical detector technologies (with pixelization of ~4–6 mm), along with the non-perfect co-linearity of the annihilation photons (~0.3° FWHM) and the finite positron range (few mm in tissue, depending on the endpoint energy of the considered isotope), PET imaging typically exhibits spatial resolutions of approximately 4–5 mm FWHM in the center of the scanner field-of-view [2]. However, such coarse resolution does not prevent the potential ability of mm-range verification accuracy, as experimentally demonstrated in controlled phantom studies with the GSI in-beam PET installation of moderate spatial resolution, as long as sufficient counting statistics is available [25]. Nevertheless, further improvements are expected by ongoing developments in PET imaging technologies both at the hardware (e.g., smaller crystal size, ultra-fast timing performance) and software (e.g., by incorporating the spatially dependent detector response in the image reconstruction process) level.

26.4 PET-BASED RANGE MONITORING

Owing to the described different β^+-activity formation processes and the basic physical difference between PET activation (due to nuclear interactions) and energy deposition (dominated by Coulomb inelastic interactions), there is a varying degree of correlation between the irradiation-induced PET signal and the beam range, typically defined as a certain percentage position of the dose fall-off. In particular, light ions limited to β^+-activation through target fragmentation only exhibit a weak spatial correlation between the depth-distributions of dose and positron emitters, whereby beam range information can be inferred from the position of the distal activity fall-off, typically few millimeters before the beam range (Figure 26.1, left, middle panel). For carbon ions, the correlation becomes more evident due to the peaked signal in both distributions (Figure 26.1, right). However, this improved correlation is somewhat outweighed by the lower amount of activity yield at the same range and delivered dose for carbon ions against protons, as experimentally shown by [26], where a factor of about 1/3 was observed for irradiation in a plastic (polymethyl methacrylate, PMMA) target. The reason for this lies in the fact that a much lower fluence of carbon ions is required to compensate for their higher energy loss rate (and enhanced biological effectiveness, when making the comparison at the same biological dose) with respect to protons. The correlation is also sensitive to the timing of the irradiation and imaging, which influences the amount of detected decays from the different produced isotopes, depending on their exact time of creation (based on the irradiation sequence), their half-life and the acquisition time window. In particular, in case of the peaked activity signal for heavy ions, the peak location can be changing over time depending on the contribution of different β^+-emitting projectile fragments, as experimentally observed for ^{16}O ion beams [27], while less relevant for primary ^{12}C ion beams where the peak is dominated by ^{11}C. For the lighter ions subject to β^+-activation through target fragmentation only, the fall-off position strongly depends on the tissue composition, determining the type of β^+-active target fragment, and

the energy dependence of the underlying cross section for positron emitter production.

Regardless of the level of correlation, range information can be obtained by comparing the activity distal-off position of two measurements from different treatment fractions [17], or between a measurement and an expectation, which can be obtained from an analytical [28–30] or stochastic (Monte Carlo) [23, 31, 32] calculation. The former approach can only provide a reproducibility check, without ensuring that the range inferred from the repeated PET measurements accurately matches the one intended at the treatment planning stage. Moreover, due to the above-mentioned considerations, it may be sensitive to the reproducibility of the timing of irradiation and data acquisition, which can also change between treatment fractions. On the contrary, the comparison of a measurement to a calculated prediction can enable an accuracy check of the intended delivery, provided that the PET computational engine shares similar physical beam models as the treatment planning system, particularly the same CT-range calibration curve used to transfer the well-characterized beam range in water to the patient tissue. However, in comparison to the dose calculation engine, more stringent requirements apply to the relevant nuclear models and/or used cross sections for positron emitter production, which play a crucial role in the PET formation process, while having only a negligible contribution on the dose calculation accuracy. Moreover, such a prediction of the irradiation-induced activity must also correctly model the patient-specific treatment plan, the fraction-specific time course of irradiation and PET data acquisition, as well as the detector-dependent image formation process [13, 31]. Here, a further challenge specific to the PET calculation engine is the correct representation of the patient anatomy also in terms of tissue composition, particularly with respect to the main elements (e.g., carbon, oxygen) contributing to the target activation, especially for the light ions subject only to this activation mechanism [33, 34]. Moreover, especially for in-room and offline approaches, the prediction model should also describe the above-mentioned washout processes, which are typically accounted for with different components (slow, medium, fast) of biological decay, with tissue- (and ideally also ion-) specific parameters [12–14, 23, 34]. Finally, both reproducibility and accuracy checks demand reliable methods to compare the two PET distributions in terms of their fall-off positions.

To this end, different methods have been proposed over the last decades, starting from simple point-like analysis of certain percentage levels of the distributions [13, 35], up to more sophisticated and automated approaches making use of the entire distal fall-off, e.g., with analysis of relative shifts in beam-eye-view or correlation coefficients in selected distal 3D regions of interest [36–39].

For the extensive pioneering clinical experience of actively scanned carbon ion and passively scattered proton irradiation monitored by the dedicated in-beam and on-board instrumentation at GSI and NCC, respectively, no detailed quantitative range analysis has been reported. However, several published results from these centers indicated the ability of on-site PET monitoring to detect deviations from the planned treatment (via accuracy checks at GSI, [8]) and inter-fractional changes (via both accuracy checks at GSI [3, 40] and reproducibility checks at NCC [17]). In particular, the availability of the in-beam PET monitoring at GSI triggered improvements of the CT-range calibration curve in the early phase of the pilot clinical project [41], due to systematic deviations observed between PET measurements and predictions in the first treated fractions [8]. PET-based detection of inter-fractional changes was mostly ascribed to varying patient anatomy or position, which introduced density modifications in the beam path resulting in the detected under- or over-range (Figure 26.3) [3, 17, 40]. In case of substantial anatomical changes, as later confirmed by an additional CT at NCC triggered by their regular PET-based reproducibility check, an adaptation of the treatment plan was performed [17]. Although it might be argued that such anatomical and positional changes should be nowadays detected more easily with in-room x-ray based volumetric image-guidance prior to treatment, the mentioned on-site PET-based approaches did not deliver any additional radiation exposure, as using the patient model based on the planning CT. Moreover, in-beam PET imaging could also detect transient modifications which might only occur during treatment, and therefore be missed by pre- or post-treatment imaging.

With the more recent in-room and offline PET(/CT) monitoring experiences for passively scattered as well as actively scanned proton and carbon ion irradiation at different facilities, quantitative range analysis showed that while reproducibility better than 1 mm could be observed (e.g., for glioma patients treated with both ion species [42]), accuracy was typically limited to within 1–5 mm for most of the examined cases [13, 39, 42–45]. However, this is mostly due to the remaining challenges of having an accurate prediction model, given the considerable uncertainties in the nuclear models/cross section data and patient composition for estimation of the physical activity production as well as limitations of the available washout models and related parameters, in addition to co-registration issues and statistical noise of the imaged weak and broad activity concentrations. Hence, improved results can be expected from the latest-generation dedicated in-beam PET scanners, for which initial promising phantom experiments at clinical doses suggested range retrieval capabilities within 1–2 mm [46].

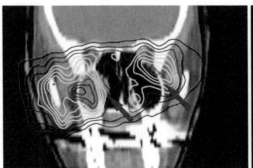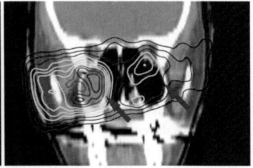

FIGURE 26.3 Example of discrepancy between measured (left) and predicted (right) activity (shown as colored isolines superimposed onto the grayscale planning CT), detected with the dedicated in-beam PET camera at GSI for a patient treated with a scanned carbon ion beam entering the patient from the right (left in the picture). Arrows indicate mismatches both in the distal maximum penetration depth of activity (right arrow), as well as within the nasal cavity, where a reduction of tissue has been suspected (left arrow). (Taken from [3].)

26.5 PET-BASED DOSE RECONSTRUCTION

The ultimate clinical exploitation of PET monitoring should ideally provide a solution to the challenging ill-posed inverse problem of retrieving the actual dose delivery from the measured activity, as already envisioned in one of the first pioneering studies on *in vivo* visualization of ion irradiation-induced activation [11]. A practical approximate solution to this very complex problem was first implemented in the pilot project at GSI, in order to provide the physicians with a rough estimation of the dosimetric consequences associated to the observed discrepancies between measured and expected in-beam PET signal [3, 40]. To this end, the proposed PET-guided dose quantification relied on a trial-and-error approach, which iteratively manipulated the planning CT (e.g., to mimic plausible anatomical changes, such as nasal cavity filling) or the treatment plan (e.g., change of isocenter or beam angle to imitate translational or rotational patient positional changes), performed a fast (~10–30 seconds) recalculation of the positron emitter distributions and, on demand, an analytical fast image reconstruction, until the trend observed in the experimental data could be reproduced [3]. In this case, the identified

most plausible modification served as the basis for performing a new dose calculation using the GSI treatment planning engine, to quantify the dosimetric deviation with respect to the original treatment plan and thus support physicians in making decisions whether to undertake additional investigations (e.g., repeated CT in case of a suspected anatomical change) prior to the delivery of the next treatment fraction (Figure 26.4, [3]). Although this method proved useful in some of the encountered cases at GSI, as reported in [3, 8, 40], it cannot provide an accurate dose quantification nor can be performed for each individual patient in the clinical routine.

More rigorous mathematical approaches were more recently proposed, which rely on analytical models of the correlation between the dose and the PET signal, and its inversion. More extensive investigations on this topic have been pursued for proton therapy, where the limitation to target fragments as well as the dominant activation mechanism through primary and secondary protons simplifies the relationship between PET and dose. In fact, both quantities directly depend on the proton energy distribution, weighted with the energy dependent mass stopping power in the case of

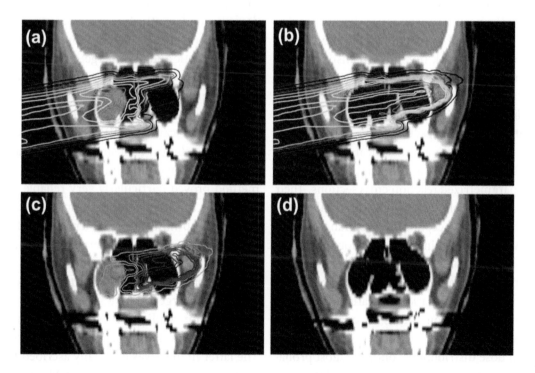

FIGURE 26.4   Planned dose distribution (colored isolines) calculated on the initial CT shown in grayscale (a), and its recalculation on the shown artificially manipulated CT based on the PET observations of figure [3] (b), both delivering a maximum physical dose of about 650 mGy. The difference of the two distributions results in minimum and maximum dose changes of −200 mGy and 495 mGy, respectively, fortunately located in uncritical anatomical regions (c). A new CT taken shortly afterwards (d) confirmed a very similar anatomical modification as conjectured (cf. CT shown in (b)). (Adapted from [3].)

the dose, or energy dependent cross section for the positron emitter production contributing to the resulting PET signal. Different groups proposed analytical formulations of this relationship either with so called positron emitters' species matrices [47] or filter functions [48, 49], specific for each given β^+-active isotope and calculated on the basis of analytical expressions or fits to Monte Carlo simulated data, respectively. Inversion of the relationship for dose reconstruction was based on deconvolution methods of the 1D–3D activity distribution as imaged with a PET scanner, applied to both *in silico* data for homogenous/heterogeneous phantoms and a clinical dataset, as well as offline PET measurements of passively scattered proton irradiation of an heterogeneous phantom and a clinical cranial indication [47, 49]. In the case of heavier ion beams like ^{12}C, the relationship becomes more complicated due to the presence of both projectile and target β^+-active fragments, including activity production through long ranging, non-positron-emitting fragmentation products, which especially contribute a distal tail beyond the activity and dose maximum. Here, the first approach proposed for dose quantification of ^{12}C and ^{16}O ions relied on the so called maximum likelihood estimation method, which was introduced in [50, 51] to estimate the position of the range (for monoenergetic beams) as well as of the distal and proximal edges of a spread-out Bragg peak (for more realistic clinical beams) given a certain distribution of measured positron annihilations. The most probable positions are determined by maximizing the likelihood between the measured data set and an expectation [50–53]. The latter is calculated starting from an analytical model of the spatial distribution of positron emitters produced in a target, using parametrized reaction cross sections. Their measurable annihilation events are then calculated analytically taking into account the individual isotope half-lives as well as the timing of irradiation and imaging. Finally, the detection process is modeled as the convolution with a position-dependent detector response, as obtained from Monte Carlo simulation. Since the positron emitter calculation entails all quantities also needed for determination of the shapes of the physical and biological dose distributions, once the most probable positions are determined, also the shapes of the most probable underlying dose distributions can be estimated [52, 53]. A final calibration step is then undertaken using measured data from the beam monitor system during the irradiation, as well as additional calculation

steps [53]. Hence, the method does not provide a direct absolute dose quantification, and heavily relies on the accuracy of several physical quantities (e.g., reaction cross sections) used for the analytical calculation in a given target. Moreover, despite the promising results, it has so far only been demonstrated with measured 1D data for monoenergetic (^{12}C, ^{14}N, ^{16}O, and ^{20}Ne) and passively modulated (^{12}C and ^{16}O) ion delivery to homogeneous targets, imaged for 500 s starting immediately after end of irradiation with an experimental dual-head in-beam positron camera [52]. Interestingly, the method has been also applied to monoenergetic and spread-out Bragg peak proton irradiation in [53, 54]. Very recently, another approach was proposed to predict dose distributions from PET images in carbon ion therapy [55]. It relies on a similar assumption as in [48], that the one-dimensional PET distribution can be described as a convolution of the depth dose distribution with a filter kernel. An evolutionary algorithm was then introduced to invert the relationship and predict the depth dose distribution from a measured PET distribution. Although the filter kernels could be obtained from a previously calculated library, for this work they were created on-the-fly, using Monte Carlo simulated predictions of the β^+-decay and depth dose distributions, and the same evolutionary algorithm. The approach was tested for monoenergetic and polyenergetic carbon ion irradiation of homogenous and heterogenous targets, simulated as well as imaged during (only in the pauses of the pulsed delivery) and up to at least 30 min after irradiation with the openPET scanner at NIRS. Moreover, a first attempt for *in silico* clinical data simulating a broad monoenergetic irradiation of a patient head was also reported. The method proved particularly very reliable for confirmation of the beam range, with an accuracy of 0.5 (simulation) and 1.0 mm (measurement) [55]. Moreover, very good agreement was also obtained for the dose reconstruction when using consistent simulated data in rather homogenous media, including the considered clinical case, while larger uncertainties and oscillations were observed for application to heterogeneous phantoms with sharp interfaces as well as all experimental data. The latter finding was especially ascribed to limitations of the physics models of the used Monte Carlo simulations, which failed to perfectly reproduce the measured activity distributions, thus resulting in inaccurate filter functions, which generated unavoidable inconsistencies in the deconvolution process. Moreover, similar to the

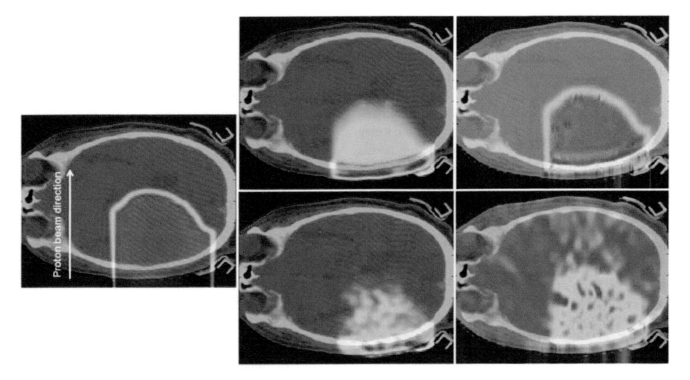

FIGURE 26.5 PET-based dose reconstruction for a head-and-neck patient treated with passively scattered proton delivery. The left panel shows the treatment planning dose in colorwash, superimposed onto the grayscale planning CT image. The top row shows the predicted activity distribution, calculated according to the filter formalism of [48] (left) and the reconstructed dose (right). The bottom row shows the PET signal measured offline (left) and the resulting dose reconstruction (right). (Reprinted with permission from [49]. ©Institute of Physics and Engineering in Medicine. Reproduced by permission of IOP Publishing. All rights reserved.)

work of [50–54], the method was so far only used in 1D, either applied to laterally integrated distributions, or by assembling data from 1D deconvolved depth dose distributions along each voxel line in beam direction for the 3D *in silico* clinical dataset (as also done in [49]). Hence, full extension of the method to 3D still needs to be accomplished.

In summary, although the different proposed methods generally show very encouraging results with dose agreement in some cases reaching accuracy within 2% [47] and mean relative errors (normalized to the maximum) below 1% [55], especially in the *in silico* studies, the so far only reported attempt to directly solve the dose quantification problem for offline PET data of a real clinical case showed considerable issues due to the very low counting statistics and biological washout effects (Figure 26.5, [49]). Hence, improvements of the PET image quality with online imaging workflows at next-generation in-beam detectors, along with a better understanding of remaining uncertainties in the knowledge of the reaction cross-sections, tissue stoichiometry and washout parameters will be crucial for enabling a more reliable direct PET-based dose reconstruction in the future.

26.6 CONCLUSION AND OUTLOOK

Despite the generally discussed shortcomings of PET monitoring as relying on an intrinsically "delayed" (depending on the isotope half-life) emission and being sensitive to biological washout, it still represents the most mature technique for 3D (and even 4D [56]) imaging with well-proven technologies for clinical implementation at relatively moderate efforts. As shown by the initial experience at different institutions with different ion species and beam delivery techniques, promising clinical results could be obtained even with suboptimal implementations, at least for favorable anatomical locations. Following these promising results, latest-generation in-beam PET scanners were recently developed, optimized for this specific application. Together with the current clinical trends of hypo-fractionation regimens, which call for increased fractional dose delivery, all these advances should help mitigate the major encountered drawbacks of low counting statistics and biological washout. Moreover, several activities are ongoing worldwide to improve the knowledge of cross section data (e.g., [57, 58]) and reliability of the models of the used computational algorithms for positron emitter production, typically relying (directly or indirectly) on Monte Carlo methods (e.g., [59, 60]). Also, different imaging techniques such as dual-energy CT and magnetic resonance imaging, which are receiving increasing attention in the context of ion beam therapy, might provide improved patient models in comparison to conventional CT for better tissue classification and more correct assignment of elemental composition, as suggested by the first promising studies of [33, 34] for the brain region. And while the results of the ongoing clinical trials, e.g., at CNAO, will allow better assessing the full potential of PET when using appropriate in-beam instrumentation specifically tailored to this application, there are also interesting developments in the nuclear medicine PET community which might considerably impact the capability of PET-based ion therapy monitoring in the future. In particular, ongoing developments toward next-generation detectors of ultrafast time-of-flight capabilities with coincidence time resolution below 10 ps [61] might open the perspectives of direct, event-by-event reconstruction of the activity produced during the irradiation, with minimal degradation of image quality in the case of limited angle geometries [62] and, with proper data acquisition, real-time treatment and range monitoring during actual "beam-on." Additional intriguing proposed approaches either rely on the possibility of exploiting irradiation-induced millisecond short-lived positron emitters like ^{12}N ($T_{1/2} = 11$ ms) for so called "beam-on PET" aiming to provide real-time feedback [63], or to resort to the initial idea of using β^+-radioactive beams for irradiation [64], now that facilities able to produce sufficiently high intensities for clinical irradiation are on the horizon [65]. Although limited to heavy ions (e.g., ^{11}C and ^{15}O), this method would offer a much more straightforward correlation between the PET activity maximum and the beam range [66], along with a considerable increase of counting statistics [67], thereby also simplifying the challenging task of dose reconstruction.

REFERENCES

1. M. Durante, R. Orecchia, J. S. Loeffler, "Charged-particle therapy in cancer: Clinical uses and future perspectives," *Nature Reviews Clinical Oncology* **14**, 483–495 (2017).

2. K. Parodi, "In vivo dose verification," in *The Physics of Proton Therapy*, edited by H. Paganetti (CRC Press), 1st edition.

3. K. Parodi, On the feasibility of dose quantification with in-beam PET data in radiotherapy with ^{12}C and proton beams, Ph.D. Thesis, Dresden University of Technology; 2004, in Forschungszentrum Rossendorf Wiss-Techn-Ber FZR-415 (2004).

4. F. Fiedler, P. Crespo, K. Parodi, M. Sellesk, W. Enghardt, "The feasibility of in-beam PET for therapeutic beams of ^{3}He," *IEEE Transactions on Nuclear Science* **53**, 2252–2259 (2006).

5. J. Pawelke, T. Bortfeld, F. Fiedler, T. Kluge, D. Möckel, K. Parodi, F. Pönisch, G. Shakirin, W. Enghardt, "Therapy monitoring with PET techniques," in *Proceedings of the Ion Beams in Biology and Medicine (IBIBAM) Meeting, Heidelberg, Germany, September 2007* (TÜV Media GmbH, Köln, 2007); ISBN 978-3-8249-1071-7, pp. 97–105.

6. C. A. Tobias, E. V. Benton, M. P. Capp, A. Chatterjee, M. R. Cruty, R. P. Henke, "Particle radiography and autoactivation," *International Journal of Radiation Oncology, Biology, Physics* **3**, 35–44 (1997).

7. D. W. Townsend, "Physical principles and technology of clinical PET imaging," *Annals of the Academy of Medicine of Singapore* **33**, 133–145 (2004).

8. W. Enghardt, P. Crespo, F. Fiedler, et al., "Charged hadron tumour therapy monitoring by means of PET," *Nuclear Instruments and Methods in Physics Research Section A* **525**, 284–288 (2004).

9. K. Parodi, H. Paganetti, E. Cascio, J. Flanz, A. Bonab, N. Alpert, K. Lohmann, T. Bortfeld, "PET/CT imaging for treatment verification after proton therapy – A study with plastic phantoms and metallic implants," *Medical Physics* **34**, 419–435 (2007).

10. S. Vynckier, S. Derreumaux, F. Richard, et al., "Is it possible to verify directly a proton-treatment plan using positron emission tomography?," *Radiotherapy & Oncology* **26**, 275–277 (1993).

11. G. W. Bennett, J. O. Archambeau, B. E. Archambeau, J. I. Meltzer, C. L. Wingate, "Visualization and transport of positron emission from proton activation in vivo," *Science* **200**, 1151–1153 (1978).

12. H. Mizuno, T. Tomitami, M. Kanazawa, et al., "Washout measurements of radioisotopes implanted by radioactive beams in the rabbit," *Physics in Medicine and Biology* **48**, 2269–2281 (2003).

13. K. Parodi, H. Paganetti, H. A. Shih, et al., "Patient study on in-vivo verification of beam delivery and range using PET/CT imaging after proton therapy," *International Journal of Radiation Oncology, Biology, Physics* **68**, 920–934 (2007).

14. C. Ammar, K. Frey, J. Bauer, C. Melzig, S. Chiblak, M. Hildebrandt, D. Unholtz, C. Kurz, S. Brons, J. Debus, A. Abdollahi, K. Parodi, "Comparing the biological washout of β+-activity induced in mice brain after 12C-ion and proton irradiation," *Physics in Medicine and Biology* **59**, 7229–7244 (2014).

15. K. Parodi, P. Crespo, H. Eickhoff, T. Haberer, J. Pawelke, D. Schardt, W. Enghardt, "Random coincidences during in-beam PET measurements at microbunched therapeutic ion beams," *Nuclear Instruments and Methods in Physics Research Section A* **545**, 446–458 (2005).

16. G. Sportelli, N. Belcari, N. Camarlinghi, G. A. Cirrone, G. Cuttone, S. Ferretti, A. Kraan, J. E. Ortuño, F. Romano, A. Santos, K. Straub, A. Tramontana, A. Del Guerra, V. Rosso, "First full-beam PET acquisitions in proton therapy with a modular dual-head dedicated system," *Physics in Medicine and Biology* **59**, 43–60 (2014).

17. T. Nishio, A. Miyatake, T. Ogino, et al., "The development and clinical use of a beam ON-LINE PET system mounted on a rotating gantry port in proton therapy," *International Journal of Radiation Oncology, Biology, Physics* **76**, 277–286 (2010).

18. M. G. Bisogni, A. Attili, G. Battistoni, N. Belcari, N. Camarlinghi, P. Cerello, S. Coli, A. Del Guerra, A. Ferrari, V. Ferrero, E. Fiorina, G. Giraudo, E. Kostara, M. Morrocchi, F. Pennazio, C. Peroni, M. A. Piliero, G. Pirrone, A. Rivetti, M. D. Rolo, V. Rosso, P. Sala, G. Sportelli, R. Wheadon, "INSIDE in-beam positron emission tomography system for particle range monitoring in hadrontherapy," *Journal of Medical Imaging* **4**, 011005 (2017).

19. H. Tashima, E. Yoshida, N. Inadama, F. Nishikido, Y. Nakajima, H. Wakizaka, T. Shinaji, M. Nitta, S. Kinouchi, M. Suga, H. Haneishi, T. Inaniwa, T. Yamaya, "Development of a small single-ring OpenPET prototype with a novel transformable architecture," *Physics in Medicine and Biology* **61**, 1795–809 (2016).

20. V. Ferrero, E. Fiorina, M. Morrocchi, F. Pennazio, G. Baroni, G. Battistoni, N. Belcari, N. Camarlinghi, M. Ciocca, A. Del Guerra, M. Donetti, S. Giordanengo, G. Giraudo, V. Patera, C. Peroni, A. Rivetti, M. D. D. R. Rolo, S. Rossi, V. Rosso, G. Sportelli, S. Tampellini, F. Valvo, R. Wheadon, P. Cerello, M. G. Bisogni, "Online proton therapy monitoring: Clinical test of a Silicon-photodetector-based in-beam PET," *Scientific Reports* **8**, 4100 (2018).

21. X. Zhu, S. España, J. Daartz, N. Liebsch, J. Ouyang, H. Paganetti, T. R. Bortfeld, G. El Fakhri, "Monitoring proton radiation therapy with in room PET imaging," *Physics in Medicine and Biology* **56**, 4041–4057 (2011)

22. Y. Hishikawa, K. Kagawa, M. Murakami, et al., "Usefulness of positron-emission tomographic images after proton therapy," *International Journal of Radiation Oncology, Biology, Physics* **53**, 1388–1391 (2002).

23. J. Bauer, D. Unholtz, F. Sommerer, et al., "Implementation and initial clinical experience of offline PET/CT-based verification of scanned carbon ion treatment," *Radiotherapy & Oncology* **107**, 218–226 (2013).

24. C. Kurz, J. Bauer, M. Conti, L. Guérin, L. Eriksson, K. Parodi, "Investigating the limits of PET/CT imaging at very low true count rates and high random fractions in ion-beam therapy monitoring," *Medical Physics* **42**, 3979–3991 (2015).

25. K. Parodi, F. Pönisch, W. Enghardt, "Experimental study on the feasibility of in-beam PET for accurate monitoring of proton therapy," *IEEE Transactions on Nuclear Science* **52**, 778–786 (2005).

26. K. Parodi, W. Enghardt, T. Haberer, "In-beam PET measurements of β+-radioactivity induced by proton beams," *Physics in Medicine and Biology* **47**, 21–36 (2002).

27. F. Sommerer, F. Cerutti, K. Parodi, A. Ferrari, W. Enghardt, H. Aiginger, "In-beam PET monitoring of mono-energetic (16)O and (12)C beams: Experiments and FLUKA simulations for homogeneous targets," *Physics in Medicine and Biology* **54**, 3979–3996 (2009).

28. A. Miyatake, T. Nishio, T. Ogino, "Development of activity pencil beam algorithm using measured distribution data of positron emitter nuclei generated by proton irradiation of targets containing (12)C, (16)O, and (40)Ca nuclei in preparation of clinical application," *Medical Physics* **38**, 5818–5829 (2011).

29. K. Frey, J. Bauer, D. Unholtz, et al., "TPSPET-A TPS-based approach for in vivo dose verification with PET in proton therapy," *Physics in Medicine and Biology* **59**, 1–21 (2014).

30. M. Priegnitz, F. Fiedler, D. Kunath, et al., "An experiment-based approach for predicting positron emitter distributions produced during therapeutic ion irradiation," *IEEE Transactions on Nuclear Science* **59**, 77–87 (2012).

31. F. Ponisch, K. Parodi, B. G. Hasch, W. Enghardt, "The modelling of positron emitter production and PET imaging during carbon ion therapy," *Physics in Medicine and Biology* **49**, 5217–5232 (2004).

32. K. Parodi, A. Ferrari, F. Sommerer, H. Paganetti, "Clinical CT-based calculations of dose and positron emitter distributions in proton therapy using the FLUKA Monte Carlo code," *Physics in Medicine and Biology* **52**, 3369–3387 (2007).

33. B. Berndt, G. Landry, F. Schwarz, T. Tessonnier, F. Kamp, G. Dedes, C. Thieke, M. Würl, C. Kurz, U. Ganswindt, F. Verhaegen, J. Debus, C. Belka, W. Sommer, M. Reiser, J. Bauer, K. Parodi, "Application of single- and dual-energy CT brain tissue segmentation to PET monitoring of proton therapy," *Physics in Medicine and Biology* **62**, 2427–2448 (2017).

34. J. Bauer, W. Chen, S. Nischwitz, J. Liebl, S. Rieken, T. Welzel, J. Debus, K. Parodi, "Improving the modelling of irradiation-induced brain activation for in vivo PET verification of proton therapy," *Radiotherapy & Oncology* **128**, 101–108 (2018).

35. C. H. Min, X. Zhu, K. Grogg, G. El Fakhri, B. Winey, H. Paganetti, "A recommendation on how to analyze in-room PET for in vivo proton range verification using a distal PET surface method," *Technology in Cancer Research & Treatment* **14**, 320–325 (2015).

36. A. Knopf, K. Parodi. T. Bortfeld, H. A. Shih, H. Paganetti, "Systematic analysis of biological and physical limitations of proton beam range verification with

offline PET/CT scans," *Physics in Medicine and Biology* **54**, 4477–4495 (2009).

37. S. Helmbrecht, A. Santiago, W. Enghardt, P. Kuess, F. Fiedler, "On the feasibility of automatic detection of range deviations from in-beam PET data," *Physics in Medicine and Biology* **57**, 1387–1397 (2012).

38. P. Kuess, S. Helmbrecht, F. Fiedler, W. Birkfellner, W. Enghardt, J. Hopfgartner, D. Georg, "Automated evaluation of setup errors in carbon ion therapy using PET: Feasibility study," *Medical Physics* **40**, 121718 (2013).

39. K. Frey, D. Unholtz, J. Bauer, J. Debus, C. H. Min, T. Bortfeld, H. Paganetti, K. Parodi, "Automation and uncertainty analysis of a method for in vivo range verification in particle therapy," *Physics in Medicine and Biology* **59**, 5903–5919 (2014).

40. W. Enghardt, K. Parodi, P. Crespo, F. Fiedler, J. Pawelke, F. Pönisch, "Dose quantification from in-beam positron emission tomography," *Radiotherapy & Oncology* **73**, S96–S98 (2004).

41. E. Rietzel, D. Schardt, T. Haberer, "Range accuracy in carbon ion treatment planning based on CT-calibration with real tissue samples," *Radiation Oncology* **23**, 2–14 (2007).

42. S. P. Nischwitz, J. Bauer, T. Welzel, H. Rief, O. Jäkel, T. Haberer, K. Frey, J. Debus, K. Parodi, S. E. Combs, S. Rieken, "Clinical implementation and range evaluation of in vivo PET dosimetry for particle irradiation in patients with primary glioma," *Radiotherapy & Oncology* **115**, 179–185 (2015).

43. A. C. Knopf, K. Parodi, H. Paganetti, T. Bortfeld, J. Daartz, M. Engelsman, N. Liebsch, H. Shih, "Accuracy of proton beam range verification using post-treatment positron emission tomography/computed tomography as function of treatment site," *International Journal of Radiation Oncology, Biology, Physics* **79**, 297–304 (2011).

44. C. H. Min, X. Zhu, BA. Winey, K. Grogg, M. Testa, G. El Fakhri, T. R. Bortfeld, H. Paganetti, H. A. Shih, "Clinical application of in-room positron emission tomography for in vivo treatment monitoring in proton radiation therapy," *International Journal of Radiation Oncology, Biology, Physics* **86**, 183–189 (2013).

45. J. Handrack, T. Tessonnier, W. Chen, J. Liebl, J. Debus, J. Bauer, K. Parodi, "Sensitivity of post treatment positron-emission-tomography/computed-tomography to detect inter-fractional range variations in scanned ion beam therapy," *Acta Oncologica* **56**, 1451–1458 (2017).

46. F. Pennazio, G. Battistoni, M. G. Bisogni, N. Camarlinghi, A. Ferrari, V. Ferrero, E. Fiorina, M. Morrocchi, P. Sala, G. Sportelli, R. Wheadon, P. Cerello, "Carbon ions beam therapy monitoring with the INSIDE in-beam PET," *Physics in Medicine and Biology* **63**, 145018 (2018).

47. E. Fourkal, J. Fan, I. Veltchev, "Absolute dose reconstruction in proton therapy using PET imaging modality: Feasibility study," *Physics in Medicine and Biology* **54**(11), N217–N228 (2009).

48. K. Parodi and T. Bortfeld, "A filtering approach based on Gaussian-powerlaw convolutions for local PET verification of proton radiotherapy," *Physics in Medicine and Biology* **51**, 1991–2009 (2006).

49. S. Remmele, J. Hesser, H. Paganetti, T. Bortfeld, "A deconvolution approach for PET-based dose reconstruction in proton radiotherapy," *Physics in Medicine and Biology* **56**, 7601–7619 (2011).

50. T. Inaniwa, T. Tomitani, T. Kohno, T. Kanai, "Quantitative comparison of suitability of various beams for range monitoring with induced beta+ activity in hadron therapy," *Physics in Medicine and Biology* **50**, 1131–1145 (2005).

51. T. Inaniwa, T. Kohno, T. Tomitani, "Simulation for position determination of distal and proximal edges for SOBP irradiation in hadron therapy by using the maximum likelihood estimation method," *Physics in Medicine and Biology* **50**, 5829–5845 (2005).

52. T. Inaniwa, T. Kohno, T. Tomitani, E. Urakabe, S. Sato, M. Kanazawa, T. Kanai, "Experimental determination of particle range and dose distribution in thick targets through fragmentation reactions of stable heavy ions," *Physics in Medicine and Biology* **51**, 4129–4146 (2006).

53. T. Inaniwa, T. Kohno, T. Tomitani, S. Sato, "Monitoring the irradiation field of ^{12}C and ^{16}O SOBP beams using positron emitters produced through projectile fragmentation reactions," *Physics in Medicine and Biology* **53**, 529–542 (2008).

54. T. Inaniwa, T. Kohno, F. Yamagata, T. Tomitani, S. Sato, M. Kanazawa, T. Kanai, E. Urakabe, "Maximum likelihood estimation of proton irradiated field and deposited dose distribution," *Medical Physics* **34**, 1684–1692 (2007).

55. T. Hofmann, M. Pinto, A. Mohammadi, M. Nitta, F. Nishikido, Y. Iwao, H. Tashima, E. Yoshida, A. Chacon, M. Safavi-Naeini, A. Rosenfeld, T. Yamaya, K. Parodi, "Dose reconstruction from PET images in carbon ion therapy: A deconvolution approach," *Physics in Medicine and Biology* **64**, 025011 (2019).

56. C. Kurz, J. Bauer, D. Unholtz, D. Richter, K. Herfarth, J. Debus, K. Parodi, "Initial clinical evaluation of PET-based ion beam therapy monitoring under consideration of organ motion," *Medical Physics* **43**, 975–978 (2016).

57. J. Bauer, D. Unholtz, C. Kurz, K. Parodi, "An experimental approach to improve the Monte Carlo modelling of offline PET/CT-imaging of positron emitters induced by scanned proton beams," *Physics in Medicine and Biology* **58**, 5193–213 (2013).

58. S. Salvador, J. Colin, D. Cussol, C. Divay, J.-M. Fontbonne, M. Labalme, "Cross section measurements for production of positron emitters for PET imaging in carbon therapy," *Physical Review C* **95**, 044607 (2017).

59. G. Battistoni, J. Bauer, TT. Boehlen, F. Cerutti, M. P. Chin, R. Dos Santos Augusto, A. Ferrari, P. G. Ortega, W. Kozłowska, G. Magro, A. Mairani, K. Parodi, P. R. Sala, P. Schoofs, T. Tessonnier, V. Vlachoudis, "The FLUKA code: An accurate simulation tool for particle therapy," *Frontiers in Oncology* **6**, 116 (2016).

60. A. Chacon, S. Guatelli, H. Rutherford, D. Bolst, A. Mohammadi, A. Ahmed, M. Nitta, F. Nishikido, Y.

Iwao, H. Tashima, E. Yoshida, G. Akamatsu, S. Takyu, A. Kitagawa, T. Hofmann, M. Pinto, D. R. Franklin, K. Parodi, T. Yamaya, A. Rosenfeld, M. Safavi-Naeini, "Comparative study of alternative Geant4 hadronic ion inelastic physics models for prediction of positron-emitting radionuclide production in carbon and oxygen ion therapy," *Physics in Medicine and Biology* **64**, 155014 (2019).

61. P. Lecoq, C. Morel, J. O. Prior, D. Visvikis, S. Gundacker, E. Auffray, P. Križan, R. Martinez Turtos, D. Thers, E. Charbon, J. Varela, C. de La Taille, A. Rivetti, D. Breton, J.-F. Pratte, J. Nuyts, S. Surti, S. Vandenberghe, P. Marsden, K. Parodi, J. M. Benlloch, M. Benoit, "Roadmap toward the 10 ps time-of-flight PET challenge," *Physics in Medicine and Biology* **65**, 21RM01 (2020).

62. P. Crespo, G. Shakirin, F. Fiedler, W. Enghardt, A. Wagner, "Direct time-of-flight for quantitative, real-time in-beam PET: A concept and feasibility study," *Physics in Medicine and Biology* **52**, 6795–6811 (2007).

63. H. J. T. Buitenhuis, F. Diblen, K. W. Brzezinski, S. Brandenburg, P. Dendooven, "Beam-on imaging of short-lived positron emitters during proton therapy," *Physics in Medicine and Biology* **62**, 4654–4672 (2017).

64. J. Llacer, A. Chatterjee, E. L. Alpen, et al., "Imaging by injection of accelerated radioactive particle beams," *IEEE Transactions on Medical Imaging* **3**, 80–90 (1984).

65. R. Augusto, T. Mendonca, F. Wenander, et al., "New developments of ¹¹C post-accelerated beams for hadron therapy and imaging," *Nuclear Instruments and Methods in Physics Research Section B* **376**, 374–337 (2016).

66. A. Mohammadi, H. Tashima, Y. Iwao, S. Takyu, G. Akamatsu, F. Nishikido, E. Yoshida, A. Kitagawa, K. Parodi, T. Yamaya, "Range verification of radioactive ion beams of 11C and 15O using in-beam PET imaging," *Physics in Medicine and Biology* **64**(14), 145014 (2019).

67. R. S. Augusto, A. Mohammadi, H. Tashima, E. Yoshida, T. Yamaya, A. Ferrari, K. Parodi, "Experimental validation of the FLUKA Monte Carlo code for dose and β^+-emitter predictions of radioactive ion beams," *Physics in Medicine and Biology* **63**, 215014 (2018).

CHAPTER 27

Prompt Gamma Detection for Proton Range Verification

Paulo Magalhaes Martins

German Cancer Research Center – DKFZ
Heidelberg, Germany
Instituto de Biofísica e Engenharia Biomédica – IBEB
Lisbon, Portugal

Riccardo Dal Bello and Joao Seco

German Cancer Research Center – DKFZ
University of Heidelberg
Heidelberg, Germany

CONTENTS

27.1 INTRODUCTION TO PROMPT GAMMA PHYSICS

27.1.1 Nuclear Reactions Leading to Prompt Gamma Generation

In charged particle radiotherapy, moderately relativistic proton and ion beams are delivered. The primary mechanism for energy loss is the collision with the atomic electrons. The direct interaction of the beam particles with the target nuclei has only a minor contribution in the stopping power. Still, the nuclear interactions are highly relevant for range verification methods.

Range verification is based on nuclear inelastic collisions. These are violent events where the projectile may change the target nucleus to a different isotope by knocking out light fragments or modifying its energy state. Because the produced nuclei are not at the ground level, they undergo a chain of de-excitation reactions. Secondary radiation is produced during these processes that can leave the patient and be measured with an external detector to retrieve the Bragg peak position.

Few general aspects of range verification based on inelastic nuclear interactions have to be highlighted. First, the secondary radiation is produced by interaction mechanisms that differ radically from the ones leading to the primary energy loss and the production of the Bragg peak. Therefore, the correlation of the deposited dose with the by-products of nuclear interactions exists but it is not direct and trivial. Second, the energies of the particles outgoing nuclear interactions can reach several MeV. This implies that the secondary radiation (e.g., neutrons, protons or gamma-rays) has a higher potential to leave the patient and be measured. Third, the time scale of nuclear interactions is the one of the strong force. The de-excitation processes that follow have characteristic times connected to each specific channel. These can range from several minutes (e.g., $T_{1/2} = 20.4$ min for the β^+ decay of $^{11}C_{\text{g.s.}}$) down to sub-nanoseconds (e.g., $T_{1/2} = 18.4$ ps for the γ decay of $^{16}O^*_{6.13\,\text{MeV}}$). In this chapter, we will focus on the fast processes that produce high energetic gamma radiation, i.e., *prompt gamma.*

Prompt gamma are one of the possible by-products of the inelastic nuclear reactions of the beam particles with the target nuclei. Figure 27.1 represents in a schematic way two possible inelastic collisions producing final states differing from the initial ones. Generally, after such a collision the beam particle is lost and does not contribute to the primary dose deposition in the Bragg peak. Typically, for a 16 cm deep Bragg peak, about 20% of the primary protons and about 50% of the primary ^{12}C ions undergo inelastic nuclear interactions before the end of the range [1]. The cross sections of such processes are energy dependent and

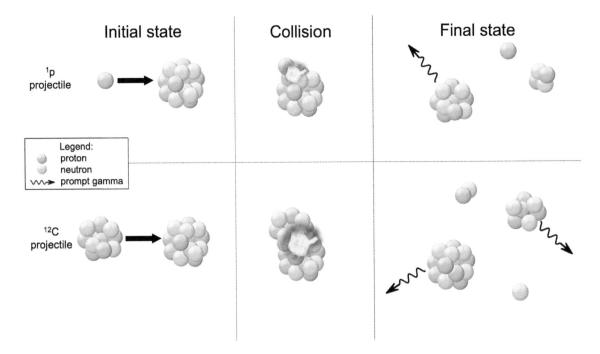

FIGURE 27.1 Schematic representation of nuclear interactions of the beam particles with the target nuclei. Top: proton–nucleus inelastic collision, specifically $^{16}O(p, p'\alpha)^{12}C^*$. Bottom: nucleus–nucleus inelastic collision, specifically $^{16}O(^{12}C, ^{10}B^*\,d\,n)^{15}O^*$.

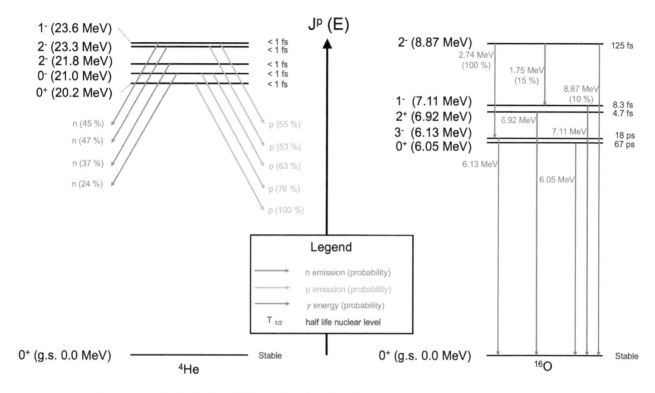

FIGURE 27.2 Nuclear energy levels for ^{4}He (left) and ^{16}O (right). The ground state and the first five states above it are shown. The principal de-excitation channels and their characteristic decay times are also reported. (Data from: National Nuclear Data Center, Brookhaven National Laboratory.)

they generally increase at lower collision energies. Therefore, the prompt gamma yield can be correlated with the residual energy of the projectile in the target, i.e., with the residual range.

Inelastic nuclear interactions produce excited nuclear states. Figure 27.2 shows the first five energy levels above the ground state for two light nuclei relevant for charged particle therapy. Similar properties are observed also for other nuclei such as p or ^{12}C. The principal de-excitation mechanisms are depicted as well. The first property to be noticed is that the energy separation for the nuclear levels is in the MeV range, which correlates directly with the energy of the secondary radiation produced in the de-excitation processes. A major difference is observed between the lightest nuclei and the heavier ones. Namely, the excited levels of p or ^{4}He nuclei do not relax by the emission of gamma radiation. The energy separation between the ground state and the first excited level is sufficient to open the channels for the emission of nucleons. A different situation is observed for ^{12}C or ^{16}O nuclei. Here, the energy separation between the ground state and the excited levels is smaller and the preferred de-excitation channel is the gamma isomeric transition. We observe that every nuclear level has a preferred

de-excitation channel. The ground state is reached either with a single transition or with a photon emission chain. The net result is that a series of gamma quanta are emitted, having well defined discrete energies. It is important to notice that every single nucleus has its unique energy levels. Therefore, the discrete energy spectrum of the de-excitation photons is a characteristic feature unique for every given isotope.

It is now possible to interpret the prompt gamma emission in the case of heavier projectiles, such as ^{12}C or ^{16}O. Here, both the target nucleus and the projectile can be excited and produce gamma radiation. As presented schematically in Figure 27.1, the gamma emission for the excited projectile happens while this nucleus is travelling away from the interaction zone in a relativistic moving frame. Therefore, the energy of the gamma quanta emitted by the projectile has to be corrected the relativistic Doppler factor. This does not apply for the prompt gamma emission from the target nucleus. Figure 27.2 also presents the characteristic times ($T_{1/2}$) that describe the time intervals in which the excited nuclear states exist after their production and before undergoing relaxation. These values are specific for every energy level and we can generally observe that $T_{1/2} = 1\,\text{ns}$.

Finally, we should also point out that there is a second mechanism for nuclear de-excitation replacing the single discrete emission with multiple photon emission. This presents a continuum spectrum. However, for the energies and the nuclei found in proton and ion therapy, the probability for the discrete emission is approximately an order of magnitude higher than the continuous one [2].

To summarize, the nuclear interactions of the particle beam with the target nuclei play a minor role in dose deposition but are important for range verification, in particular, the inelastic nuclear collisions that produce excited nuclear states. These can de-excite with the emission of gamma radiation that leaves the patient and can be measured with an external detector. The gamma radiation is emitted by light nuclei, such as ^{12}C or ^{16}O, but not by the lightest ones, such as p or ^{4}He. The prompt gamma energy spectrum presents discrete lines that are characteristic of the nuclei involved. These lines span up to approximately $E_\gamma = 10$ MeV with production times $T \ll 1$ ns.

27.1.2 Detecting Prompt Gamma and Sources of Background

The critical aspect for a successful application of Prompt Gamma Imaging (PGI) is an efficient detection of the secondary radiation that leaves the patient. The details depend on the specific PGI technique used and will be discussed in a following section. We will now focus on some general physics aspects that apply to all the systems.

Prompt gamma-rays have an energy spectrum that goes up to 10 MeV. Gamma quanta with energies greater than 1 MeV interact with materials preferentially through Compton scattering or pair production. The photoelectric effect, which is usually exploited for imaging at lower photon energies, plays here only a minor role. It is therefore likely for a single prompt gamma to undergo multiple interactions with partial energy deposition before being fully absorbed. This has an impact both on the transport of the gamma radiation from the nucleus to the detector as well as on the interactions within the detector.

Those interactions may happen within the patient, with the patient table and other equipment, or with the walls of the treatment room [3]. In such interactions, the gamma radiation is not absorbed but scattered with a different angle and its energy is reduced. This means that a significant number of events lose their original energy, time, and direction information before reaching the detector. Such *scattered events* are recorded but are not correlated with the Bragg peak position.

On the other hand, a photon that reaches the detector without previous interactions has still a significant probability of delivering just partial information. After a first interaction inside the detector, the scattered photon can leave it and deposit its remaining energy elsewhere. In such a case, only a part of its energy is recorded and therefore the original energy information is lost. Moreover, the fraction of energy deposited in the detector may be smaller than the one minimally measurable, which leads to a complete loss of the event.

Conversely, if there is a measurable energy in the detector, the correct time and direction information of the original prompt gamma are preserved. Therefore, these events can be treated either as background or as signal depending on whether the energy information is relevant for the specific PGI technique or not. We define such events as *partial energy deposition events*. Finally, if after the first interaction, also all the following ones happen inside the detector we observe a *full energy deposition event*. In this most desirable case, the system retrieves all the original energy, time, and direction information of the prompt gamma.

As mentioned before, prompt gammas are produced within nanoseconds following the interaction of the particle beam with the patient. We have seen how these particles are only one type among the multiple by-products of nuclear interactions. The prompt gamma emission channel competes with the production of other secondary particles on a similar time scale. Among the secondary particles, there are neutrons which are likely to reach the detectors. Such events generate a continuum energy spectrum which is hardly distinguishable from the one created by prompt gamma events. This spectrum contributes to the background that we observe in the measurements with the several PGI systems.

The neutrons can also interact with nuclei in the patient and produce further gamma radiation. One of the most important reaction is the hydrogen neutron capture ($_0^1\text{n} + _1^1\text{p} \rightarrow _1^2\text{d} + \gamma\,(2.22\,\text{MeV})$). The line associated to this reaction can be clearly identified in Figure 27.3 (left), since it is not correlated in time with the radio-frequency (RF) of the cyclotron. The mono energetic photon produced by this reaction is not directly correlated to the Bragg peak position and contributes to the background as well. We define all these processes as *neutron induced background*.

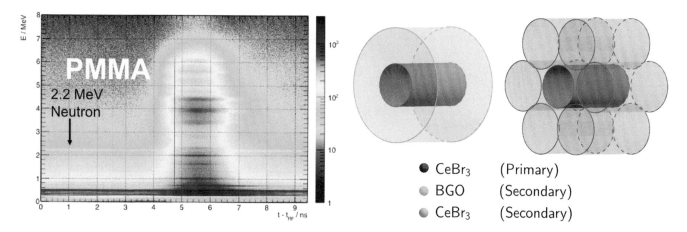

FIGURE 27.3   Left: prompt gamma lines and corresponding single- and double-escape peaks visible as time correlated horizontal stripes. The PMMA target was irradiated with a mono-energetic proton beam of 130 MeV in a setup without collimation. Right: schematic geometries of an anti-coincidence detector system: primary CeBr$_3$ with a BGO AC shielding (left); primary CeBr$_3$ with a CeBr$_3$ AC shielding (right). (Adapted from [30] and [24].)

For what concerns the charged nuclear fragments, only the most energetic ones will leave the patient and reach the detector system. However, before leaving the patient they may also undergo nucleus–nucleus interactions. Such events produce a further generation of gamma radiation, which is not directly correlated with the Bragg peak position. This phenomenon is more likely for ^{12}C ion beams, but not very relevant to proton treatments. We define this as *fragment induced background*.

27.1.3 Improving the Signal-to-Noise Ratio

The detailed knowledge of the physics processes regulating the production, transport and detection of the prompt gamma radiation can be exploited to differentiate the signal from the background. The implementation of the following techniques is specific to every PGI system. We will focus on the physics properties of the background. The identification of the differences between the signal and the background properties can highly improve the quality of the recorded signal.

An *anti-coincidence* detection configuration is depicted in Figure 27.3 (right). Such a system aims to suppress the events with partial energy deposition. The principle relies on the fact that the photon, after a partial energy deposition, leaves the primary detector carrying with it the residual energy. A secondary detector can be placed in the immediate proximity of the primary detector with the purpose of recording the escaping photon. This second energy deposition happens immediately after the first interaction in the primary detector, i.e., in coincidence. The event is discarded when the

PGI system identifies two events in coincidence. On the other hand, an energy deposition in the primary detector without any signal in the secondary detector can be interpreted as an event with full energy deposition and it should be recorded, i.e., in anti-coincidence.

The use of *time-of-flight* (TOF) is helpful to differentiate the prompt gamma from the scattered events, the neutrons and the fragment induced background [4]. Figure 27.4 (left) shows a typical TOF spectrum for a carbon beam. The spectrum is obtained by measuring the difference between the time given by the accelerator RF and the time signal given by the PGI system. One can identify a prominent peak, which is given by the prompt gamma radiation that travels without previous interactions towards the detector. Few nanoseconds after it, there is the delayed component caused by the scattered events. These photons originate together with the primary prompt gamma but undergo scattering before reaching the detector. Therefore, the total distance travelled is greater compared to the straight line from the point of origin in the patient to the PGI system. This increases the time interval needed to reach the detector and the scattered photons appear in the delayed component of the TOF spectrum. The constant and time independent background is given by the target activation and the neutron induced component. One further delayed component is observable in the case of the carbon ion irradiation, namely the fragment induced background. In this case, the TOF difference with respect to the prompt gamma peak is greater if compared to the scattered events. We can then select only the events in

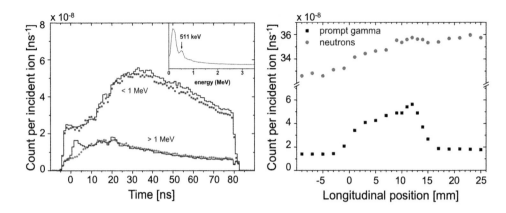

FIGURE 27.4 Left: TOF spectra measured for two target locations Z and for two selections on energy E deposited in a scintillator: $Z = 8$ mm (line), $Z = 26$ mm (dots), $E < 1$ MeV, and $E > 1$ MeV. Origin of time scale: ion impact on target. Right: detection rates ($E > 1$ MeV) as a function of longitudinal position of target, obtained for two different TOF ranges: $2 < T < 10$ ns (prompt γ-rays, squares) and $T > 10$ ns (neutrons, circles). (Adapted from [4].)

the TOF window in the prompt gamma peak region and suppress the background which is highly uncorrelated with the Bragg peak position.

The *energy selection* aims to simultaneously suppress a fraction of the scattered events and of the neutron induced component [5]. As discussed before, the prompt gamma-rays are emitted with a complex spectrum up to several MeV. If we take as an example the ^{16}O transitions presented in Figure 27.2, we observe that the direct emission from this nucleus happens mainly above $E_\gamma > 2.5$ MeV, while the neutron induced background is comprised mainly at lower energies. Moreover, we know that the scattered radiation usually reaches the detector after losing part of its energy. Taking into account all these effects, it is convenient to accept only the events above a certain energy threshold. The higher the energy threshold the better the background rejection. This however decreases the available statistics. Therefore, a compromise has to be reached and an optimization is needed for every specific PGI system. A further and more elegant approach for the energy selection consists in adopting an energy window only for certain specific transitions, e.g., the $E_\gamma = 6.13$ MeV emitted by the ^{16}O$^*_{6.13\ \text{MeV}}$ nucleus [6].

The applicability of each technique depends on the specific PGI system. Furthermore, multiple noise suppression techniques may also be adopted simultaneously.

27.2 IONS AND THEIR NUCLEAR REACTIONS

27.2.1 Monte Carlo Implementation

Conversely to the electromagnetic interactions, no rigorous model exists to describe the nuclear inelastic interactions, but several solutions have been proposed. A distinction should be made between the simulation of the prompt gamma production induced by proton irradiation or by heavier ions. In the former, we consider a nucleon–nucleus interaction that is generally treated by an *intra nuclear cascade (INC)*. In the latter, a more complex nucleus–nucleus interaction takes place and can be described by the *quantum molecular dynamics (QMD)* model. In both cases, after the interaction and the energy exchange, a pre-equilibrium and a de-excitation model describe the steps until reaching a stable state. The Monte Carlo codes most commonly used for the simulation of prompt gamma production are FLUKA [7], Geant4 [8], and MCNPX [9]. These three codes share the use of intra nuclear cascade models to simulate nuclear interactions. The specific implementation differs for each software and also within a given software by the exact choice of the parametrization. The following aspects are general and apply to all Monte Carlo software.

The INC is used to describe the interaction of the proton with a nucleus. It is the basis for the implementation of nuclear interactions in the modern Monte Carlo codes. The target nucleus is treated as an ensemble of quasi-free nucleons with an energy distribution approximated by a Fermi gas. The projectile is transported inside an effective time independent nuclear potential and undergoes multiple two-body interactions. The final state returns the single particles that escaped the nuclear potential and the modified target nucleus. The excited nucleus is then given as an input to the pre-equilibrium and de-excitation models to simulate the emission of prompt gamma radiation.

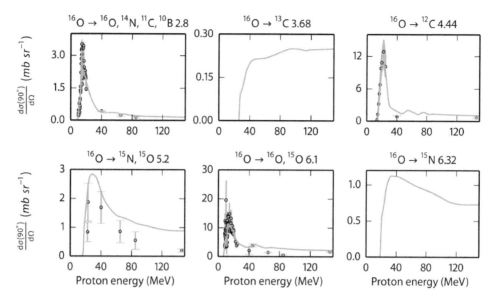

FIGURE 27.5   Differential cross sections of prompt gamma production induced by proton interactions with ^{16}O nuclei. The titles of the plots describe the nuclear channel and the corresponding energy of the gamma ray emitted in MeV. (Adapted from [15].)

The QMD is used to describe the nucleus–nucleus interaction. The collision cannot be modeled with a series of two-body interactions but rather considering all the target and the projectile nucleons as simultaneous participants of the interaction. Each nucleon is described by a Gaussian wave package and the total wave function is given by the direct product of all the nucleon wave functions. The equation of motion of the single nucleons are derived from the Hamiltonian of the system and the nuclear potential is updated at every step. This model has been proposed for the collision of heavy nuclei. However, several studies have optimized the parameters in the Geant4 implementation for the application with ^{12}C beams [2] and even protons [10].

27.2.2 Cross Sections

The implementation of the nuclear models in the Monte Carlo codes has to be benchmarked against experimental data. This is necessary to assess the reliability of the simulations, which are a fundamental component of the development of PGI systems. Multiple approaches can be used. Here we discuss the comparison between the integrated profiles and between the cross sections.

The integrated profiles include the data over the full prompt gamma energy spectrum above the threshold. Multiple transition lines are detected simultaneously at a given depth of the target. An example is shown in Figure 27.4 (right). The total yield is the result of both partial and total energy deposition events. The most modern Monte Carlo codes generally reproduce in a correct way the integrated experimental data and can be considered reliable for such applications.

The direct comparison of the cross sections for single de-excitation lines is a more fundamental approach. However, great discrepancies have been observed between different Monte Carlo codes [11]. Therefore, an experiment-based approach is preferred in such case. Figure 27.5 shows the cross sections for several reaction lines obtained during proton irradiation. The experimental data is used as input parameter for the modeling of the cross sections. A common feature can be observed for all the reaction lines. Namely, the maxima of the cross sections are located at low proton energies just before the drop to zero-values. The lower the proton energy, the shorter the residual range of the beam particle. Therefore, the maximum value of the prompt gamma yield in Figure 27.4 (right) has to be interpreted as a consequence of the location of the maxima at low energies in Figure 27.5.

27.3 PGI FOR PROTON THERAPY

27.3.1 Historical Development

The PGI as a technique for online verification of the ion range in particle therapy was first proposed in 2003 by Stichelbaut and Jongen at the PTCOG meeting [12]. We had to wait however until 2007 to have the first experimental results from a 38 MeV proton beam [13]. Until then, prompt gamma radiation emitted during particle therapy was always regarded as a noise source, mostly in the context of range verification through positron emission tomography [14]. The PET signal during the spill time was highly corrupted by the prompt-gamma background and measurements of PET data were only possible during the inter-spill time and after the treatment.

Since 2007, many groups emerged around the world trying to verify *in vivo* and online the range of the particle beams inside the patient. The leading groups were the ones in Boston [15], Belgium [16], Delft [17], Dresden [18], Korea [13], Lyon [4], Maryland [19], Milan [20], and Munich [21]. Each of these groups specialized in different approaches to solve the same problem: how to achieve by means of PGI a range uncertainty of 1 mm?

In recent years, the investment in this technology had a leap mostly due to the successful efforts of the Belgian and the Boston groups that rose the interest of major companies in developing their own prototypes. The knife-edge slit camera was first applied in the clinical setting in 2016 in Dresden [22]; while the prompt gamma spectroscopy (PGS) [15] was first tested for clinical use in patients in 2019 in Boston [23].

Meanwhile, the research has evolved to different clinical applications. While the former prototypes were mainly conceived for cyclotron-based facilities, a new device is being planned for synchrotron-based facilities [24]. Other competing devices are also heading towards clinical application by taking advantage of different features of the detection method, such as prompt-gamma timing [18] and the Compton camera [25].

Another possibility explored by prompt gamma spectroscopy is the determination of the elemental composition of irradiated tissues through characteristic prompt gamma rays emitted from ^{16}O and ^{12}C [19]. This technique could open new doors to the direct evaluation of the response of the tissues to particle therapy [26].

27.3.2 Cesium Iodide – CsI(Tl)

In 2006, C.-H. Min et al. proposed a multilayered collimator system to locate the distal dose falloff in proton therapy [5]. Such a system would capture neutrons and prevent unwanted gammas while measuring at right angles the prompt-gamma component resulting from the interaction of the proton beam with the patient tissue. One year later, a prototype composed by a collimator and a CsI(Tl) scintillator detector showed for the first time a clear correlation between the distribution of the prompt gammas and the location of the distal dose edge for a 38 MeV proton beam, thus confirming the previous simulations with MCNPX [5, 13]. A spatial resolution of 1–2 mm was already envisioned at that time. In 2012, a full-scale multislit collimation system based on CsI(Tl) was evaluated by Monte Carlo simulations for all therapeutic proton energies. This system requires however a two-dimensional distribution of the proton dose with 2D position sensitive gamma detectors [27] and lacks so far enough statistics when compared to the other available systems.

27.3.3 Knife-Edge Slit Camera

In 2012, Smeets et al. presented a pioneering prototype and methods for verification of the proton range through a slit camera [16]. The experimental work followed thorough simulations with MCNPX that defined the main parameters for a proton beam at 160 MeV, such as: collimator material and thickness; collimator slit width and angle; and scintillation material, thickness and segment width. A range estimation of 1–2 mm standard deviation has been achieved for a 160 MeV proton beam impinging on a homogeneous PMMA phantom, corresponding to doses in water at the Bragg peak of just 25 cGy.

This prototype has been acquired by IBA and evolved to a clinical prototype that was tested in humans for the first time at the OncoRay – Dresden in 2016 [22]. These two papers demonstrated great accuracy in the measurement of relative shifts between fractions. However, many improvements are being carried out to increase the absolute range verification capability of the camera [28]. Concurrently, further tests in humans took place at the University of Pennsylvania that showed a median precision in shift estimation after spot aggregation of 2.1 mm and a median inter-fraction variation of 1.8 mm [29]. Figure 27.6 shows the principle behind this system and the prototype itself mounted on a trolley, which is subject to the limitations of the camera positioning.

27.3.4 Prompt Gamma Spectroscopy

Also in 2012, Verburg and Seco proposed for the first time to use the energy spectrum of the prompt-gamma emission to retrieve an absolute range verification of the proton beam. As previously mentioned and shown in Figure 27.5, systematic simulation studies compared

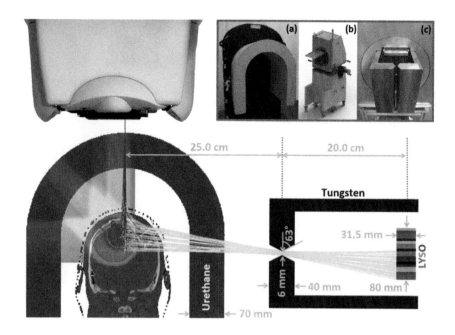

FIGURE 27.6   Schematic setup of the prompt gamma knife-edge slit camera system for imaging of the vertex field. Dark lines represent proton tracks, and light grey lines represent gamma rays selected by the collimator aperture. Upper right inset, from left to right: (a) the table-mounted U-shaped range shifter, (b) the camera trolley positioning system, and (c) the camera knife-edge slit collimator. (Adapted from [29].)

several Monte Carlo nuclear reaction models and available experimental data of discrete gamma line cross-sections [11]. The dedicated nuclear reaction codes, such as TALYS and EMPIRE, reproduced the experimental data more consistently than Geant4.

Further experimental work demonstrated the feasibility of measuring the discrete prompt-gamma rays for *in vivo* range verification of clinical proton beams [6]. The differential cross-sections were measured for 15 prompt gamma-ray lines from proton-nuclear interactions with ^{12}C and ^{16}O up to 150 MeV and an absolute range was determined with a standard deviation of 1.0–1.4 mm [15].

In 2018, a full-scale clinical prototype was presented [23]. Figure 27.7 shows the detection system, which consists of eight LaBr₃ scintillators and a collimator, mounted on a rotating frame. By making use of the previously experimentally determined cross-sections and a GPU-accelerated Monte Carlo simulation, an absolute proton range was determined for each spot of the distal energy layer with a mean statistical precision of 1.1 mm. Upcoming clinical studies are envisioned to assess the suitability of such system for range verification during patient treatments.

Concurrently, a recently formed group at the German Cancer Research Center – DKFZ demonstrated the use of CeBr₃ detectors for PGS at the proton facility in Dresden [30]. A semi-collimated setup was proposed for measuring the falloff region of the Bragg Peak. This configuration would be later adopted also by Hueso-González et al. [23].

27.3.5 Prompt Gamma Timing

A novel concept of prompt gamma-ray timing (PGT) has been proposed by Golnik et al. [18]. This is an uncollimated method that uses a gamma-ray detector of reasonable time resolution for measuring the finite transit time from entering the patient's body until stopping in the target volume. The underlying principle is the increasing transit time with the particle range that causes measurable effects in the PGT spectra that can be used for range verification.

This method has been demonstrated at a clinical proton facility and range differences of 5 mm in defined heterogeneous targets were identified with a single detector and 10^8 incident protons [31]. However, PGT spectra were observed to be smeared out by the bunch time spread and accelerator related proton bunch drifts against the RF of the accelerator have been detected [32]. A proton bunch monitor has been therefore developed to correct for potential bunch drifts and increase the robustness of the PGT method. Nevertheless, this technique is highly

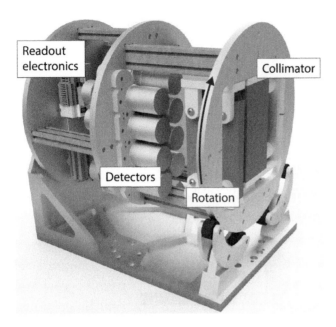 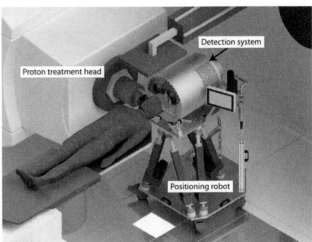

FIGURE 27.7 Left: 3D model of the clinical PGS prototype, which can rotate around its axis according to the beam incidence angle. The tungsten collimator is visible on the front plane, the eight scintillation detectors in the middle and the readout electronics on the back plate. Right: illustration of the gantry, proton treatment head, patient, top x-ray flat panel and the PGS system. The detector frame can rotate according to the gantry angle, and is mounted on a positioning robot, consisting of six actuators. The robot stands on a platform on wheels that is moved into the treatment room. (Adapted from [23].)

dependent on the number of gamma rays detected per incident proton. Large detector loads and a high acquisition throughput are therefore mandatory to draw statistically significant conclusions on range errors [33].

27.3.6 Compton Camera

The Compton camera (CC) belongs to the actively collimated systems and it requires position-sensitive gamma ray detectors with high resolution and efficiency, which are arranged in one scatterer, one absorber, or in several scatter planes. In contrast to the other PGI systems, it allows for two-dimensional or even three-dimensional images to be reconstructed as more gamma-rays and directions can be detected. However, the event coincidence needed between the different stages limits the overall efficiency. The major pitfalls are its highly demanding instrumentation in terms of spatial, time and energy resolution and the computationally intensive reconstruction algorithms [33]. Several groups around the world have been devoting significant efforts towards a clinical prototype for range verification [34–38] and some of them have abandoned it due to the technical complexity, electronics expense, low coincident efficiency, high detector load, radiation background, and the elevated percentage of random coincidences [33].

Draeger et al. are close to reach a clinical prototype [25]. Recent results reported prompt-gamma measurements with a small CC prototype placed at three different locations along the proton beam path. The data were combined to simulate measurements with a larger scale, clinical CC capable of imaging Bragg peak shifts for both hypo-fractionated and standard treatments. 3D images of the prompt gamma emission were produced and range shifts of 2 mm were detected for the delivery of a 2 Gy spot.

27.3.7 Oxygen Concentration

The spectroscopic capability of PGI is being further explored for measuring the elemental compositions of irradiated tissues. The nuclei emit a unique energy spectrum during proton irradiation [26]. It has been shown that the total emission from ^{16}O per unit dose is proportional to the concentration of oxygen within an irradiated volume of tissue [19]. PG emissions could then be used to accurately determine the concentration of key elements in tissue (O, C, N, Ca, K, etc.). Since the proton range highly depends on the density of the tissues, variations in density and concentration of the several elements could be tracked on a daily basis over the course of the treatment, thus providing information both on the tumor response and the Bragg peak position. Dedicated measurements have been made with HPGe detectors [19, 39, 40]. Verburg and Seco also reported the use of the PGS system to determine the relative content of ^{12}C and ^{16}O however with a large statistical uncertainty. In such a system, these concentrations are left as free parameters for the optimization process [15].

27.4 PGI FOR ION THERAPY

27.4.1 Time-of-Flight

Several groups reported the importance of TOF measurements for background reduction as the prompt component can be clearly differentiated from the delayed neutron-induced gamma component [6, 17, 18, 41]. This issue is even more relevant at synchrotron facilities where the microstructure of the beam is destroyed during the extraction of the beam. In cyclotron-based facilities, the TOF information is easily retrieved from the RF of the accelerator despite some challenges already mentioned concerning proton bunch drifts.

Other groups doing experiments on synchrotron-based facilities have used external detectors as primary trigger for TOF measurements [3, 42, 43]. The most commonly used is the plastic scintillator, e.g., EJ-200. These scintillators cope well with moderate intensities, provided they remain below 10^6 s^{-1}. This allows single particles to be identified and their arrival time to be correlated with the arrival time of the prompt gamma-rays on the secondary detector. In 2018, Martins et al. have shown at PTCOG a first small-scale prototype of a trigger system capable of tracking single particles at clinical intensities for all beam species accelerated at the Heidelberg Ion-Beam Therapy Center – HIT [44]. This single-particle trigger was able to detect single, double and triple hits within ion bunches. It also provided TOF information for prompt-gamma measurements with sub-ns resolution. Such prototype should not be compared to other detectors used for position tracking of the single particles, such as hodoscopes [45]. The major demands for a time-tracker in hadron therapy is its radiation hardness, single particle identification at clinical intensities and a very small thickness, since the minimum amount of material between the nozzle and the patient remains a primary concern. Krimmer et al. [46] and Biegun et al. [17] considered the TOF neutron rejection not to be effective for bunch widths larger than 5 ns (typically 80 ns) found

at synchrotrons. This has however been overcome with such novel single-particle trigger.

27.4.2 Linac-Based Experiments

Testa et al. were the first to perform prompt gamma experiments for heavier particles. By means of a TOF technique, Testa et al. measured the longitudinal profile of prompt gamma-rays emitted by 73 MeV/u ^{13}C irradiating a PMMA target [4, 47]. The experiment was performed at the GANIL facility (Caen, France) with a pulsed beam (beam pulses of ≈1 ns every 80 ns) with an intensity of 1 nA. The main detector, a cylindrical NaI(Tl), was placed at right angles to the beam direction behind a 20 cm thick lead collimator. The method used to reject the neutron component was similar to the one used in proton facilities. The gamma rays and the neutrons were discriminated by subtracting the given phase of the RF-signal from the accelerator to the detection time provided by the gamma ray detector. In Figure 27.4 (left), we showed a TOF spectra for two target locations (in the middle of the Bragg peak and 12 mm beyond the Bragg peak) and two energy regions. This allowed for the discrimination of the prompt gamma component, whose detection rate as a function of the longitudinal position Z was shown on the right.

Further experiments were performed at GANIL with a 95 MeV/u ^{12}C ion beam and compared with the results obtained at GSI with a 305 MeV/u ^{12}C ion beam [3]. An external plastic detector has been used at GSI to provide TOF information. A BaF$_2$ detector was chosen for gamma detection and a BC501 scintillator for neutron detection. The results show similar results at the two facilities both for the prompt gamma and the neutron components. The presented technique is applicable also in case of a continuous spill structure, provided the identification of the primary ions one by one is achieved.

27.4.3 Synchrotron-Based Experiments

Testa et al. also presented a systematic work with results based on the same TOF spectrum analysis, but considering the absolute prompt gamma yields and including the statistical and systematic uncertainties [48]. The experiments were performed at GANIL, GSI (Darmstadt, Germany), HIT (Heidelberg, Germany) and the WPE Proton Center (Essen, Germany). Carbon ions showed a prompt gamma production a factor five higher than protons. The values for the absolute prompt gamma yields for carbon ions accelerated at GANIL confirm the previous results from Agodi et al. [49]. The value obtained for 310 MeV/u carbon ions was $(79 \pm 2_{stat} \pm 23_{sys}) \times 10^{-6}$. Pinto et al. have also compared the experimental data with Geant4 using all applicable proton inelastic models. They observed that Geant4 overestimates the prompt-gamma emission yields by 40%. In order to improve the discrepancy observed, they used the QMD model with a wave packet width, $L = 1.3$ fm² [10].

Vanstalle et al. also proposed to benchmark Geant4 models for prompt gamma monitoring with carbon ions at GSI [42]. They also used external plastic scintillators for the trigger and BaF_2 scintillator for the detection of the prompt gamma rays. They measured the total yields and energy spectra of prompt gamma produced by the interaction of 220 MeV/u carbon ions with PMMA at 60° and 90° with respect to incoming beam. The reported values for both angles are similar and a prompt gamma yield of 1.06×10^{-2} was obtained at 90°. They observed a good match between the shapes of both simulated and experimental spectra, especially with INCL++ physics list. Mattei et al. reported on the prompt gamma yields measured during the irradiation of a PMMA target by ^{4}He, ^{12}C, and ^{16}O ion beams [43]. The experiments were performed with LYSO scintillators at 60° and 90° and a plastic start counter was used as well. The trigger rate was very low (below 6 kHz) being mostly dictated by the DAQ dead time. The results with carbon ions were compared with the four previous experiments [20, 42, 48, 49] and considerations were taken concerning the expected resolution of a slit camera for prompt gamma monitoring of particle therapy with heavier particles.

A novel approach inherited from the PGS system developed by Verburg and Seco is currently being developed at the DKFZ. First experiments concerning the performance of $CeBr_3$ scintillators to be used at HIT have been previously undertaken at OncoRay – Dresden [30]. A Monte Carlo optimization of the detector was carried out in order to maximize the signal-to-noise ratio after considering different detector geometries [24]. The presence of an active coincidence shield has been shown to increase the signal-to-noise ratio up to a factor of 3.5. Both papers investigated the capability of such detectors to detect low-energy lines down to 511 keV typically emitted by irradiated metals. This opens new perspectives for the detection of metal inserts in the beam path and unveils additional energy lines to be used for range

verification. The slightly lower energy resolution of these new $CeBr_3$ detectors, if compared to the crystals used by Verburg and Seco ($LaBr_3$) [15], is compensated by the inexistent intrinsic activity and lower cost. The major limitation of the Boston system is the cost of the $LaBr_3$ scintillating crystals [50].

27.5 PERSPECTIVES AND FUTURE CHALLENGES

Around 15 years have passed since PGI was first time proposed until the first application in men. The use of PET for monitoring charged hadron tumor therapy [51] has been shown to suffer from several effects, such as biological washout, lower falloff correlation, low motion and tissue reproducibility, thus yielding a minimum uncertainty of 6 mm in identifying the proton range [50, 52, 53]. PGI is today the most promising technique for real-time *in vivo* range verification. Either isolated or in combination with PET [54], PGI is expected to be installed at most of the proton therapy centers around the world. Only a center that can provide an imaging system for the position of the Bragg peak will be able to reduce the safety margins and combat the largest criticism to particle therapy: the range uncertainties [26, 50, 53].

The phase III clinical trials are essential to clarify the advantage of particle therapy over conventional photon therapy [55–57]. The PGI systems may play an important role in future clinical trials. In a first stage, such systems may gather as much data from daily treatments before drawing future necessary conclusions regarding its application in the clinical setting. The data acquisition should not conflict with the daily clinical workflow. Very early, one of the proposed uses was the delivery of small bunch of particles to the most distal spots or layers. This would demand a very short acquisition time with enough statistics and would serve as a guide for the full treatment.

The ultimate goal is to achieve an automatic feedback on the position of the Bragg peak and adjust the treatment accordingly. This demands very fast treatment planning routines, some of these already available in the research community [58–60]. However, the first studies are expected to track only the variations due to interfractional changes [61].

We presented the basic physical principles of PGI and the historical development of this technique. There are currently two prototypes [22, 23] ready for daily clinical use and that tackle the range uncertainty problem in different ways respectively: (1) fluence approach based on the knife-edge slit camera design by Smeets et al.; and (2) prompt-gamma spectroscopy approach based on the design by Verburg and Seco [15]. Both achieved the desired 1–2 mm accuracy and are able to perform both absolute and relative range verification. The challenge is now heading synchrotron facilities, where the beam structure is different and every ion beam species has its own features. Also here there are two different approaches to the same problem, either considering the absolute prompt gamma yields or the reaction channels associated to specific energy lines. This latter approach demands very precise experimental data from cross-sections, as a function of materials, ions species and energy. Moreover, in order to use the TOF information, an external trigger detector has to be placed between the nozzle and the patient. This is actually a problem shared, to a less extent, with the cyclotron-based facilities as proton bunch drifts along time have been reported [32].

Two other prototypes – Compton camera [25] and PGT [62] – are reaching the clinical stage, each of which standing on distinct proof-of-principles. The CC will enable 2D and 3D images of the Bragg peak and the PGT will be easier to translate to the clinics due to its uncollimated setup. Both have drawbacks, either the computation time and complicated instrumentation of the CC or the load and throughput demands of the PGT. Both the CC and the PGS system, due to its spectroscopic nature may be able to quantify the elemental compositions of irradiated tissues. This could be used to measure *in vivo* the tissue density and the ^{16}O and ^{12}C concentration of irradiated tumors.

The PGI is now a consolidated technique for range verification and we hope this will bring increased acceptance and applications to radiotherapy with protons, ^4He, ^{12}C, and ^{16}O.

REFERENCES

1. A. C. Kraan, "Range verification methods in particle therapy: Underlying physics and Monte Carlo modeling," *Frontiers in Oncology* 5 (2015).
2. G. Dedes, M. Pinto, D. Dauvergne, et al., "Assessment and improvements of Geant4 hadronic models in the context of prompt-gamma hadrontherapy monitoring," *Physics in Medicine and Biology* 59(7), 1747–1772 (2014).
3. M. Testa, M. Bajard, M. Chevallier, et al., "Real-time monitoring of the Bragg-peak position in ion therapy by means of single photon detection," *Radiation and Environmental Biophysics* 49(3), 337–343 (2010).
4. E. Testa, M. Bajard, M. Chevallier, et al., "Monitoring the Bragg peak location of 73 MeV/u carbon ions by means of prompt γ-ray measurements," *Applied Physics Letters* 93(9), 093506, 2008.
5. C. H. Min, C. H. Kim, M. Y. Youn, J. W. Kim, "Prompt gamma measurements for locating the dose falloff region in the proton therapy," *Applied Physics Letters* 89(18), 183517 (2006).
6. J. M. Verburg, K. Riley, T. Bortfeld, J. Seco, "Energy- and time-resolved detection of prompt gamma-rays for proton range verification," *Physics in Medicine and Biology* 58(20), L37–L49 (2013).
7. T. T. Böhlen, F. Cerutti, M. P. W. Chin, et al., "The FLUKA Code: Developments and challenges for high energy and medical applications," *Nuclear Data Sheets* 120, 211–214 (2014).
8. J. Allison, K. Amako, J. Apostolakis, et al., "Recent developments in Geant4," *Nuclear Instruments and Methods in Physics Research Section A* 835, 186–225 (2016).
9. C. J. Werner, J. S. Bull, C. J. Solomon, et al., "MCNP version 6.2 release notes," Technical report, Los Alamos National Lab. (LANL), Los Alamos, NM (United States) (2018).
10. M. Pinto, D. Dauvergne, N. Freud, J. Krimmer, J. M. Létang, E. Testa, "Assessment of Geant4 prompt-gamma emission yields in the context of proton therapy monitoring," *Frontiers in Oncology* 6, 10 (2016).
11. J. M. Verburg, H. A. Shih, J. Seco, "Simulation of prompt gamma-ray emission during proton radiotherapy," *Physics in Medicine and Biology* 57(17), 5459–5472 (2012).
12. F. Stichelbaut and Y. Jongen, "Verification of the proton beams position in the patient by the detection of prompt gamma-rays emission," in *39th Meeting of the Particle Therapy Co-Operative Group* (2003), p. 16.
13. C. H. Min, J. W. Kim, M. Y. Youn, C. H. Kim, "Determination of distal dose edge location by measuring right-angled prompt-gamma rays from a 38 MeV proton beam," *Nuclear Instruments and Methods in Physics Research Section A* 580(1), 562–565 (2007).
14. P. Crespo, T. Barthel, H. Frais-Kolbl, et al., "Suppression of random coincidences during in-beam PET measurements at ion beam radiotherapy facilities," *IEEE Transactions on Nuclear Science* 52(4), 980–987 (2005).
15. J. M. Verburg and J. Seco, "Proton range verification through prompt gamma-ray spectroscopy," *Physics in Medicine and Biology* 59(23), 7089–7106 (2014).
16. J. Smeets, F. Roellinghoff, D. Prieels, et al., "Prompt gamma imaging with a slit camera for real-time range control in proton therapy," *Physics in Medicine and Biology* 57(11), 3371–3405 (2012).
17. A. K. Biegun, E. Seravalli, P. Cambraia Lopes, et al., "Time-of-flight neutron rejection to improve prompt gamma imaging for proton range verification: A simulation study," *Physics in Medicine and Biology* 57(20), 6429–6444 (2012).
18. C. Golnik, F. Hueso-González, A. Müller, et al., "Range assessment in particle therapy based on prompt γ-ray

timing measurements," *Physics in Medicine and Biology* **59**(18), 5399–5422 (2014).

19. J. C. Polf, R. Panthi, D. S. Mackin, et al., "Measurement of characteristic prompt gamma rays emitted from oxygen and carbon in tissue-equivalent samples during proton beam irradiation," *Physics in Medicine and Biology* **58**(17), 5821–5831 (2013).

20. I. Mattei, G. Battistoni, F. Bini, et al., "Prompt-γ production of 220 MeV/u ^{12}c ions interacting with a PMMA target," *Journal of Instrumentation* **10**(10), P10034–P10034 (2015).

21. P. G. Thirolf, C. Lang, S. Aldawood, et al., "Development of a Compton camera for online range monitoring of Laser-accelerated proton beams via prompt-gamma detection," *EPJ Web of Conferences* **66**, 11036 (2014).

22. C. Richter, G. Pausch, S. Barczyk, et al., "First clinical application of a prompt gamma based in vivo proton range verification system," *Radiotherapy & Oncology* **118**(2), 232–237 (2016).

23. F. Hueso-González, M. Rabe, T. A Ruggieri, T. Bortfeld, J. M. Verburg, "A full-scale clinical prototype for proton range verification using prompt gamma-ray spectroscopy," *Physics in Medicine and Biology* **63**(18), 185019 (2018).

24. R. Dal Bello, P. Magalhaes Martins, J. Seco, "CeBr$_3$ scintillators for ^{4}He prompt gamma spectroscopy: Results from a Monte Carlo optimization study," *Medical Physics* **45**(4), 1622–1630 (2018).

25. E. Draeger, D. Mackin, S. Peterson, et al., "3D prompt gamma imaging for proton beam range verification," *Physics in Medicine and Biology* **63**(3), 035019 (2018).

26. K. Parodi and J. C. Polf, "In vivo range verification in particle therapy," *Medical Physics* **45**(11), e1036–e1050 (2018).

27. C. H. Min, H. R. Lee, C. H. Kim, S. B. Lee, "Development of array-type prompt gamma measurement system for in vivo range verification in proton therapy," *Medical Physics* **39**(4), 2100–2107 (2012).

28. J. Petzoldt, G. Janssens, L. Nenoff, C. Richter, J. Smeets, "Correction of geometrical effects of a knife-edge slit camera for prompt gamma-based range verification in proton therapy," *Instruments* **2**(4), 25 (2018).

29. Y. Xie, E. H. Bentefour, G. Janssens, et al., "Prompt gamma imaging for in vivo range verification of pencil beam scanning proton therapy," *International Journal of Radiation Oncology, Biology, Physics* **99**(1), 210–218 (2017).

30. P. Magalhaes Martins, R. Dal Bello, A. Rinscheid, et al., "Prompt gamma spectroscopy for range control with CeBr$_3$," *Current Directions in Biomedical Engineering* **3**(2), 113–117 (2017).

31. F. Hueso-González, W. Enghardt, F. Fiedler, et al., "First test of the prompt gamma ray timing method with heterogeneous targets at a clinical proton therapy facility," *Physics in Medicine and Biology* **60**(16), 6247–6272 (2015).

32. J. Petzoldt, K. E. Roemer, W. Enghardt, et al., "Characterization of the microbunch time structure of proton pencil beams at a clinical treatment facility," *Physics in Medicine and Biology* **61**(6), 2432–2456 (2016).

33. F. Hueso-González, F. Fiedler, C. Golnik, et al., "Compton camera and prompt gamma ray timing: Two methods for in vivo range assessment in proton therapy," *Frontiers in Oncology* **6**, 80 (2016).

34. S. W. Peterson, D. Robertson, J. Polf, "Optimizing a three-stage Compton camera for measuring prompt gamma rays emitted during proton radiotherapy," *Physics in Medicine and Biology* **55**(22), 6841–6856 (2010).

35. T. Kormoll, F. Fiedler, S. Schöne, J. Wüstemann, K. Zuber, W. Enghardt, "A Compton imager for in-vivo dosimetry of proton beams—a design study," *Nuclear Instruments and Methods in Physics Research Section A* **626–627**, 114–119 (2011).

36. S. Kurosawa, H. Kubo, K. Ueno, et al., "Prompt gamma detection for range verification in proton therapy," *Current Applied Physics* **12**(2), 364–368 (2012).

37. J. Krimmer, J. L. Ley, C. Abellan, et al., "Development of a Compton camera for medical applications based on silicon strip and scintillation detectors," *Nuclear Instruments and Methods in Physics Research Section A*, **787**, 98–101 (2015).

38. P. G. Thirolf, S. Aldawood, M. Böhmer, et al., "A Compton camera prototype for prompt gamma medical imaging," *EPJ Web of Conferences* **117**, 05005 (2016).

39. A. Schumann, J. Petzoldt, P. Dendooven, et al., "Simulation and experimental verification of prompt gamma-ray emissions during proton irradiation," *Physics in Medicine and Biology* **60**(10), 4197–4207 (2015).

40. L. Kelleter, A. Wrońska, J. Besuglow, et al., "Spectroscopic study of prompt-gamma emission for range verification in proton therapy," *Physica Medica* **34**, 7–17 (2017).

41. P. Cambraia Lopes, E. Clementel, P. Crespo, et al., "Time-resolved imaging of prompt-gamma rays for proton range verification using a knife-edge slit camera based on digital photon counters," *Physics in Medicine and Biology* **60**(15), 6063–6085 (2015).

42. M. Vanstalle, I. Mattei, A. Sarti, et al., "Benchmarking Geant4 hadronic models for prompt-γ monitoring in carbon ion therapy," *Medical Physics* **44**(8), 4276–4286 (2017).

43. I. Mattei, F. Bini, F. Collamati, et al., "Secondary radiation measurements for particle therapy applications: Prompt photons produced by ^{4}He, ^{12}C and ^{16}O ion beams in a PMMA target," *Physics in Medicine and Biology* **62**(4), 1438–1455 (2017).

44. P. Magalhaes Martins, R. Dal Bello, G. Hermann, et al., "Proceedings of the 57th annual meeting of the particle therapy cooperative group (PTCOG)," *International Journal of Particle Therapy* **5**(2), 58–229 (2018).

45. B. D. Leverington, M. Dziewiecki, L. Renner, R. Runze, "A prototype scintillating fibre beam profile monitor for ion therapy beams," *Journal of Instrumentation* **13**(05), P05030–P05030 (2018).

46. J. Krimmer, D. Dauvergne, J. M. Létang, É. Testa, "Prompt-gamma monitoring in hadrontherapy: A review," *Nuclear Instruments and Methods in Physics Research Section A* **878**, 58–73 (2018).

47. E. Testa, M. Bajard, M. Chevallier, et al., "Dose profile monitoring with carbon ions by means of prompt-gamma measurements," *Nuclear Instruments and Methods in Physics Research Section B* **267**(6), 993–996 (2009).

48. M. Pinto, M. Bajard, S. Brons, et al., "Absolute prompt-gamma yield measurements for ion beam therapy monitoring," *Physics in Medicine and Biology* **60**(2), 565–594 (2014).

49. C. Agodi, F. Bellini, G. A. P. Cirrone, et al., "Precise measurement of prompt photon emission from 80 MeV/u carbon ion beam irradiation," *Journal of Instrumentation* **7**(03), P03001–P03001 (2012).

50. J. Seco and M. F. Spadea, "Imaging in particle therapy: State of the art and future perspective," *Acta Oncologica* **54**(9), 1254–1258 (2015).

51. W. Enghardt, P. Crespo, F. Fiedler, et al., "Charged hadron tumour therapy monitoring by means of PET," *Nuclear Instruments and Methods in Physics Research Section A* **525**(1), 284–288 (2004).

52. M. Moteabbed, S. España, H. Paganetti, "Monte Carlo patient study on the comparison of prompt gamma and PET imaging for range verification in proton therapy," *Physics in Medicine and Biology* **56**(4), 1063–1082 (2011).

53. A. C. Knopf and A. Lomax, "In vivo proton range verification: A review," *Physics in Medicine and Biology* **58**(15), R131–R160 (2013).

54. K. Parodi, "On- and off-line monitoring of ion beam treatment," *Nuclear Instruments and Methods in Physics Research Section A* **809**, 113–119 (2016).

55. D. K. Ebner and T. Kamada, "The emerging role of carbon-ion radiotherapy," *Frontiers in Oncology* **6**, 140 (2016).

56. M. Durante, R. Orecchia, J. S. Loeffler, "Charged-particle therapy in cancer: Clinical uses and future perspectives," *Nature Reviews Clinical Oncology* **14**(8), 483–495 (2017).

57. M. Durante, "Proton beam therapy in Europe: More centres need more research," *British Journal of Cancer* **120**(21), 777–778 (2019).

58. J. M. Verburg, C. Grassberger, S. Dowdell, J. Schuemann, J. Seco, H. Paganetti, "Automated Monte Carlo simulation of proton therapy treatment plans," *Technology in Cancer Research & Treatment* **15**(6), NP35–NP46 (2016).

59. S. Mein, K. Choi, B. Kopp, et al., "Fast robust dose calculation on GPU for high-precision ^{1}H, ^{4}He, ^{12}C and ^{16}O ion therapy: The FRoG platform," *Scientific Reports* **8**(1) (2018).

60. L. Tian, G. Landry, G. Dedes, et al., "Toward a new treatment planning approach accounting for in vivo proton range verification," *Physics in Medicine and Biology* **63**(21), 215025 (2018).

61. S. Schmid, G. Landry, C. Thieke, et al., "Monte Carlo study on the sensitivity of prompt gamma imaging to proton range variations due to interfractional changes in prostate cancer patients," *Physics in Medicine and Biology* **60**(24), 9329–9347 (2015).

62. T. Werner, J. Berthold, W. Enghardt, et al., "Range verification in proton therapy by prompt gamma-ray timing (PGT), steps towards clinical implementation," in *IEEE Nuclear Science Symposium Conference Record 2017* (2017), pp. 1–5.

CHAPTER 28

Acoustic-Based Proton Range Verification

Kevin C. Jones

Rush University Medical Center
Chicago, Illinois

CONTENTS

28.1 INTRODUCTION

The spatial dose deposition provided by particle therapy is fundamentally different than that from photon therapy. Megavoltage photons exhibit low skin dose, a peak at 1–4 cm depth, and a monotonic decrease (inverse square and exponential) in dose thereafter. In contrast, heavy charged particles deposit a large fraction of their total dose in the last few centimeters of travel in the Bragg peak. Due to the absence of exit dose beyond the Bragg peak, the dose deposition provided by particle therapy is often considered superior to that of photons because structures immediately downstream from the target may be spared.

Although the lack of exit dose is advantageous, particle therapy irradiation is also associated with increased risks due to its strong dependence on depth [1, 2]. If the expected tumor depth does not match the actual depth, the dose deposited by a photon beam will change by < 1% per millimeter due to the low dose gradient beyond d_{max}. For particle beams, if the radiological depth is different than planned, the dose to the distal edge of the tumor may change from 100% to 0% (undershoot) or 0% to 100% for normal tissue downstream of the tumor edge (overshoot). Due to the steep gradient downstream of the Bragg peak, even a couple of millimeters of depth error may cause these drastic changes in delivered dose, and the biological repercussions are further exacerbated by the high relative biological effectiveness for low energy protons at the end of their range [3]. Although the energy of the beam exiting the treatment head and the range in water are known to high accuracy (≤600 μm) [4], expected and actual depths in a patient may differ for many reasons, including depth calculation uncertainties, setup errors, or patient anatomy changes [2]. One careful consideration predicts an uncertainty in proton range of up to 4.6% of the range +1.2 mm [1].

The uncertainty in delivered particle range and the associated risks have prevented using particle therapy to its full potential. Clinically, an uncertainty of up to 3.5% of the range +1–3 mm is assumed in planning for proton beams. Due to the possibility of overshooting, beam arrangements are chosen to avoid placing organs at risk directly downstream from the tumor [2]. In addition to considering the ideal, planned dose deposition, particle therapy plans are assessed for their robustness to possible range errors [5]. Robustness consideration increases calculation time and adds complexity to physician decision making.

Many researchers have focused on developing *in vivo* techniques for determining the particle dose deposition delivered to the patient. These considered techniques have largely been focused on the most popular form of particle therapy – proton therapy – and the field of study has been termed proton range verification. Different explored methods include positron emission tomography (PET) using proton-activated nuclei that decay with positron emission [6], prompt gamma detection using proton-activated nuclei that decay with gamma emission (imaging [7, 8], spectroscopy [9], and time-of-flight [10] techniques), and thermoacoustic-based techniques [11–26] which are the focus of this chapter.

In general, thermoacoustic techniques gather information about deposited energy by detecting the pressure waves emitted as the energy is converted to heat. When applied to particle range verification, the general thermoacoustic technique has been called ionoacoustics [15], RACT [16], and protoacoustics [13]. Because the majority of the literature is focused on determining the range of proton beams, the term protoacoustics will be used here. Protoacoustics is an attractive proton range verification technique because it is non-invasive, the signal may be collected *in vivo* in real-time, and the signal is linearly proportional to deposited dose (although a heat defect must be considered). In addition, the system is conceptually simple and is projected to be of low cost. The challenges facing clinical translation include low signal levels, potential errors due to heterogeneous sound speeds in tissue, limited acoustic transmission through air pockets and bone, and the difference in time-structure of proton pulses produced by most clinical accelerators compared to the ideal pulses for protoacoustic signal generation.

The goal of this chapter is to provide the reader with an understanding of the underlying physics of protoacoustics, the research that has been done in the field, and the challenges facing the technique.

28.2 UNDERLYING PHENOMENON

As the name suggests, thermoacoustic phenomena convert thermal energy into acoustic pressure waves. Fields of thermoacoustics differ based on the initial source of energy. For example visible light [27] (optoacoustics or photoacoustics), radio wave radiation [28], x-ray radiation [29], and particle beams [30] have all been used as excitation sources [31]. The underlying analytical equations that govern the emission of acoustic waves in response to these various forms of heat deposition are general to all of the aforementioned techniques.

In a lossless, homogeneous medium, the pressure p reaching a detector positioned at $\vec{r}$ at time t is given by two equations (Equations 28.1 and 28.2). For a delta-function, instantaneous heating pulse, the pressure is given by [18]:

$$p_\delta(\vec{r},t) = \frac{c\beta}{4\pi C_p} \frac{d}{dt} \left[\int_0^{2\pi} \int_0^{\pi} (tc)^2 \sin\theta d\theta d\phi \frac{E_{3D}(t=|\vec{r}-\vec{r}'|/c)}{tc} \right]$$

(28.1)

where c (m s^{-1}) is the sound speed, β is the thermal expansion coefficient (K^{-1}), and C_p is the specific heat capacity (J kg^{-1} K^{-1}). $E_{3D}(\vec{r}')$ is the three dimensional spatial energy deposition (J m^{-3}). The double integral in Equation (28.1) describes a spherical surface integral centered at the detector at $\vec{r}$, where the radial distance variable has been converted to the product of time and sound speed. The pressure wave reaching the detector originates on the spherical surface, and its arrival time is determined by its distance from the detector divided by the travel speed, $|\vec{r}-\vec{r}'|/c$. The magnitude of the pressure depends on how efficiently the deposited energy is converted into a pressure change at $\vec{r}'$. This conversion efficiency is described by the Grüneisen parameter, Γ, a dimensionless (Pa/(J/m^3) cancels out) material-specific parameter equivalent to the prefactor in Equation (28.1) (once c is brought out of the integral): $\Gamma = c^2\beta/C_p$. The pressure contributed by each point or voxel is further scaled by the inverse of the distance (tc in the denominator).

The integral in Equation (28.1) is always positive, and its maximum occurs at the time (or distance divided by c) that corresponds to the spherical shell that overlaps the most deposited energy E_{3D}. Each point is expanded in response to the heating (unless β is negative). Due to the time derivative, however, the detected pressure has positive and negative peaks. Conceptually, these positive and negative peaks are caused by gradients in the heated volume. The largest magnitude pressure peaks will be observed when there are large spatial gradients in deposited heat, E_{3D}.

If the heating excitation is deposited in a finite time (the heating pulse is not a delta-function), then the pressure detected is given by the convolution of $p_\delta(\vec{r}, t)$ with the temporal shape of the proton pulse $E_t(t)$ (s^{-1}):

$$p(\vec{r}, t) = p_\delta(\vec{r}, t) \otimes E_t(t). \qquad (28.2)$$

One can think of the excitation pulse E_t as the linear combination of many delta-function pulses, each of which initiates the emission of its own set of pressure waves. The detected pressure wave is the sum of all of these contributions. An interesting result of Equation (28.2) is that the magnitude of the detected pressure also depends on the gradients in the excitation pulse E_t along with the spatial gradients discussed above. Because p_δ has negative and positive peaks, convolution with a slowly varying E_t gives low intensity, broadened pressure waves.

These relationships between spatial and temporal gradients and the emitted pressure waves are elegantly encapsulated in the idea of stress confinement [27]. If the heat is deposited at a rate slower than the transit time across the heated volume, then spatial confinement is not achieved, pressure leaks out of the volume faster than heat is deposited, and the temporal characteristics of the heating pulse affect (generally broaden) the detected pressure waves. A simple rule-of-thumb equation states that stress confinement is achieved when the excitation pulse is shorter in time than L_p/c, where L_p is the characteristic linear dimension of the heated volume E_{3D}. Loss of stress confinement is usually detrimental, as the associated broadening results in loss of the desired dosimetric spatial information that would otherwise define the shape of the protoacoustic waves. Another consideration is that of thermal confinement, the idea that heat may diffuse out of the volume faster than E_t. Due to the slow rate of thermal diffusion compared to

L_p/c (the stress confinement threshold), thermal confinement is generally not a concern for protoacoustics.

If the medium is not homogeneous, c, β, and C_p will all vary in space, $\vec{r}'$. One can still conceivably apply Equation (28.1) with these varying material-specific properties brought into the integral, but the delay time between pressure emitted at $\vec{r}'$ and detected at $\vec{r}$ would depend on the speed of sound integrated along the line between these two points $t = |\vec{r} - \vec{r}'|/c_{avg}$. Equation (28.1) does not consider attenuation, scattering, or diffraction of acoustic waves. Therefore, propagation of thermoacoustic waves in heterogeneous media is typically simulated using more sophisticated techniques that solve the underlying wave equation.

28.3 PROTOACOUSTIC EMISSIONS

Proton dose depositions generate acoustic emissions [30]. The thermoacoustic origin of these emissions was proven in the seminal 1979 Sulak et al. paper [30], in which the authors demonstrate a linear relationship between acoustic amplitude and β/C_p. This relationship is shown by measuring proton-induced acoustic signals in different materials and in water at different temperatures. The most striking of these experimental results is the disappearance and inversion of acoustic amplitude as the water temperature is decreased below 4°C, the temperature at which water reaches its densest state and the expansion coefficient becomes negative. These results are shown in Figure 28.1. The discrepancy between the observed 6°C and expected 4°C inversion point is not resolved (although subsequent measurements [32] report 4°C). The authors also show that the acoustic emissions are not the result of microbubble formation – another potential underlying phenomenon – by demonstrating that the acoustic amplitude is independent of ambient pressure, nucleating ion content, and N_2 content.

In a homogeneous medium, protons deposit their dose in pencil beam kernels. A narrow pencil beam will generate a dose deposition with an approximately Gaussian radial profile – the width of which expands with depth – and an integrated depth profile that is fairly constant until the last couple centimeters of its range, when a large dose is deposited at the Bragg peak before abruptly stopping. Even if the protons are delivered in a broad field, one may think of this broad field as being composed of a linear combination of pencil beams, each of which is depositing its own pencil beam kernel.

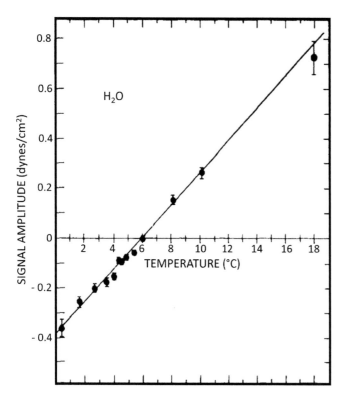

FIGURE 28.1 The protoacoustic amplitude measured in water as a function of initial temperature. An inversion point is observed at 6°. (Reprinted with permission from Elsevier: Elsevier, Nuclear Instruments and Methods, Experimental studies of the acoustic signature of proton beams traversing fluid media, L. Sulak, T. Armstrong, H. Baranger, M. Bregman, M. Levi, D. Mael, J. Strait. T. Bowen, A.E. Pifer, A. Polakos, H. Bradner, A. Parvulescu, W.V. Jones, J. Learned, 161, 203–217, 1979.)

Clinical proton beams have a depth penetration of up to 30 cm and radial full-width-half-max (FWHM) profiles of around 0.5–1.5 cm [33]. Clinical proton beams are commonly delivered as many pencil beams (pencil beam scanning – PBS) or as a broad beam (double scattering). For PBS delivery, the depth and horizontal/vertical positioning is controlled by changing the energy and rastering with magnetic fields. For double scattering delivery, the number of protons (current) is modulated as a function of time as a spinning wheel of varying thickness shifts the energy/range of the delivered protons to conform in depth to the far borders of the target. The double scattered beam aperture is controlled by collimating jaws and a field-specific compensator, which further conform the depth dose to the far edge of the tumor.

Conceptually, the proton dose deposition can be approximated by a cylinder (pre-Bragg peak plateau) capped by a high dose disc (the Bragg peak). Each of these shapes contributes to a macroscopic protoacoustic pressure wave. By placing a set of detectors in a line parallel to the proton dose deposition, the origin of the different pressure waves may be determined. Figure 28.2 shows the experimentally measured protoacoustic waves arriving at such a row of detectors [34]. In this figure, there are three sets of protoacoustic waves. The waves along A-E are due to the entrance window, and are not of particular interest to proton range verification. The waves along C-B-D are termed the γ waves, and those along A-B are the α waves [34]. Each wave is a bipolar pressure wave in which the compression (positive) peak precedes the rarefaction (negative) peak.

The γ wave is generated by the Bragg peak disc, and its arrival time depends on the distance between the Bragg peak and the detector, labeled as l. As a pseudo point source, the pressure falls as 1/R with increasing l [18, 30].

The α wave is generated by the cylindrical, pre-Bragg peak plateau, and its arrival time is related to the radial distance, labeled as s, between the proton beam axis and the detector. As a pseudo line source, the pressure falls as $1/\sqrt{R}$ when the detector is close to the beam axis (s is on the order of the range of the proton) [18]. When the detector is at the depth of the Bragg peak (for example, marked at B in Figure 28.2), the α wave overlaps with the γ wave ($s = l$). At detector depths deeper than the Bragg peak, all protoacoustic waves generated by the pre-Bragg peak cylinder contribute to the γ wave. Therefore, there is no distinct α wave at depths beyond the Bragg peak.

The hypothesis of protoacoustic proton range verification techniques is that the pressure waves carry information about the proton dose deposition. How this information is used distinguishes particular protoacoustic approaches. At one extreme, Radiation-Induced Acoustic Computed Tomography (RACT) collects the protoacoustic pressure traces from many (ideally all) detector positions surrounding the dose deposition, and reconstructs the 3D deposition though tomographic back-projection approaches [16]. On the other extreme, single or a sparse set of detectors are used to measure the Bragg peak and beam axis positions through time-of-flight calculations based on the arrival time of the different peaks [13, 15]. Specifically, $l \approx c\tau_\gamma$ and $s \approx c\tau_\alpha$, where τ_γ and τ_α are the arrival times of the protoacoustic γ and α waves, respectively.

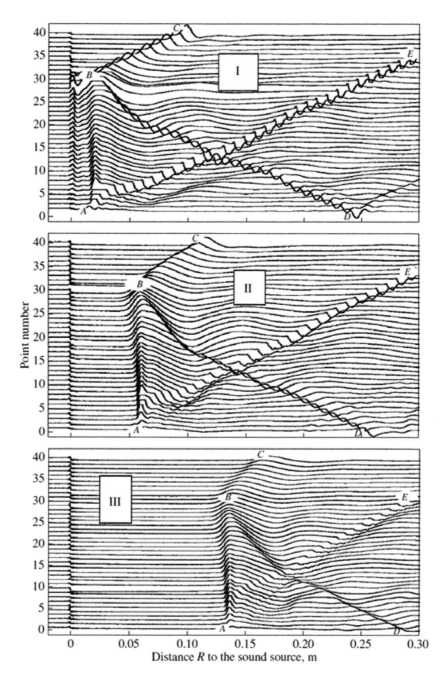

FIGURE 28.2 Protoacoustic emissions collected along a line parallel to the proton beam axis at offsets of 2.2 (I), 6.2 (II), and 13.6 (III) cm. (Reprinted with permission from Springer Nature: Springer Nature, Instruments and Experimental Techniques, Measuring the Ultrasonic Field Generated in Water upon the Deceleration of a Proton Beam, V. I. Albul, V. B. Bychkov, S. S. Vasil'ev, K.E. Gusev, V.S. Demidov, E.V. Demidova, N.K. Krasnov, A.F. Kurchanov, V.E. Luk'yashin, A. Yu. Sokolov, 47, 502–506, 2004.)

28.3.1 Pulse Width and Rise Time

As indicated in Equation (28.2) and described above with stress confinement, the temporal characteristics of the proton pulse affect the measured protoacoustic signal. If stress confinement is not achieved, the time derivative of the proton pulse is imprinted on the protoacoustic signal. Two important characteristics of the proton pulse's temporal shape are its pulse width, usually defined as the FWHM, and the rise time, usually defined by the time to rise from 10% to 90% or 20% to 80% of the maximum. Two of the commonly considered pulse shapes are rectangular and Gaussian.

For rectangular proton pulses, the rise and fall of the proton pulse generate a positive and negative protoacoustic peak, respectively, whose separation corresponds to the pulse width of the rectangular pulse [13, 15, 30]. Interestingly, if the rise and/or fall times are fast enough, the beginning and/or end of the proton pulse will satisfy stress confinement even if the pulse width is too long. For example, when the rectangular pulse width is increased the delay between positive and negative acoustic peaks also increases. Even as the pulse width is increased, the positive and negative features are unaffected, and retain their sharp features. In this case, the individual positive- and negative-associated protoacoustic peaks will themselves report directly on the spatial dose deposition. Although this might appear attractive – that quasi-stress confinement may be achieved by abruptly turning on and off a rectangular pulse even for a proton pulse with arbitrarily long pulse width – the dose deposited in the middle of the rectangular pulse generates no protoacoustic signal due to the equilibrium, constant heating. Given that protoacoustic measurements are primarily limited by deposited dose (described later), the dose-depositing – but protoacoustically non-contributing –

middle portion of a rectangular proton pulse make it non-ideal.

Gaussian proton pulses are easily modeled, and have been the focus of many simulations. Gaussian shapes have also been observed experimentally [17, 22, 25]. Unlike rectangular pulses, in which the central, flat portion does not contribute to the protoacoustic signal, Gaussian shapes are advantageous because the fall immediately follows the rise in proton amplitude. Studies investigating the ideal proton pulse have indicated that proton pulses with a shape that matches the spatially defined protoacoustic wave generated by a delta-function proton pulse, p_δ, are best suited to maximizing the protoacoustic signal-to-noise ratio (SNR) [18]. The reason for this is due to two effects. First, the efficiency – amplitude per deposited dose – of generating protoacoustic amplitude increases for shorter pulses (Figure 28.3a). Second, the SNR increases linearly with signal amplitude and with the square root of the number of averages. Although the efficiency effect suggests shorter pulses are better, optimal SNR is achieved by generating the highest possible amplitude per pulse when the cumulative dose limits the number of repeated irradiations (averages) that can be delivered (Figure 28.3b).

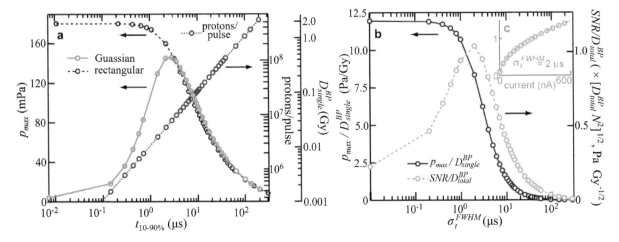

FIGURE 28.3 (a) The maximum simulated protoacoustic pressure (left axis) and protons per Gaussian pulse (right axis) is plotted as a function of 10–90% rise time (300 nA peak proton current, detector 5 cm distal to Bragg peak). (b) The maximum simulated protoacoustic signal per deposited dose (single pulse) is plotted on the left axis versus Gaussian proton pulse FWHM. The simulated SNR per total deposited dose is also plotted versus proton pulse length (b, right axis). Shorter pulses deposit less dose per pulse and, therefore, allow more averaging, but the absolute protoacoustic amplitude generated per pulse is lower than generated by longer pulses. As a result, for an equivalent total dose, the SNR is expected to peak at an intermediate pulse length. (c) As the proton pulse peak current is increased, the SNR per total deposited dose also increases. (Reprinted from K.C. Jones, C.M. Sehgal S. Avery, How proton pulse characteristics influence protoacoustic determination of proton-beam range: Simulation studies, Physics in Medicine and Biology 61, 2213–2242, 2016 © Institute of Physics and Engineering in Medicine. Reproduced by permission of IOP Publishing. All rights reserved.)

The highest amplitude is generated when the proton pulse matches the p_δ peaks because the convolution overlap is greatest. It follows that because p_δ depends on the spatial dimensions of the dose deposition, the ideal proton pulse will change depending on proton beam spot size and energy. For clinical beams, which have a spot size FWHM ~1 cm and penetrate ~1–30 cm, the ideal proton pulse for maximizing SNR is a Gaussian-like pulse with a FWHM ~ 5 μs.

28.3.2 Arrival Time

The protoacoustic pressure wave is bipolar, with positive and negative peaks. This bipolar structure complicates analysis. For time-of-flight based proton range verification, the arrival time of the protoacoustic wave must be identified [18, 24]. Intuitively, it is not clear which feature to identify as the arrival time (for example, positive and negative peaks are two possibilities). Simulations have indicated that the true arrival time for p_δ that gives $l = \tau_\gamma c$ is located somewhere between the positive and negative peak, but this "somewhere" depends on the measurement position relative to the Bragg peak [18]. Jones et al. proposed calibrating the system for a particular arrival time definition [18, 19]. For example, if one chooses to define the arrival time as the delay between the proton

pulse peak maximum and the initial protoacoustic positive peak maximum, then an additional correction, S_{inher}, is required to accurately convert the arrival time into distance to the Bragg peak: $l = \tau_\gamma c + S_{inher}$ [19]. This correction will depend on arrival time definition feature, proton pulse width and characteristics, and angular position of the detector relative to the Bragg peak.

28.3.3 Frequency

Both the spatial characteristics of the dose deposition and the temporal characteristics of the excitation proton pulse may shape the emitted protoacoustic waves. Therefore, the frequency of the protoacoustic spectrum – and the appropriate detectors – are affected by both the proton pulse and the proton spot size and energy [16, 30]. For clinically relevant proton beams, the p_δ has a frequency that peaks at ~50 kHz (Figure 28.4) [18]. Interestingly, the protoacoustic peaks emitted along the beam direction have higher frequency spectra than those emitted laterally because the Bragg peak is narrower in the depth direction than it is radially. Due to the expected <100 kHz frequency spectrum, the typical selected detector is the hydrophone, a water-proof transducer that is sensitive to lower frequency pressure waves than standard ultrasound imaging transducers.

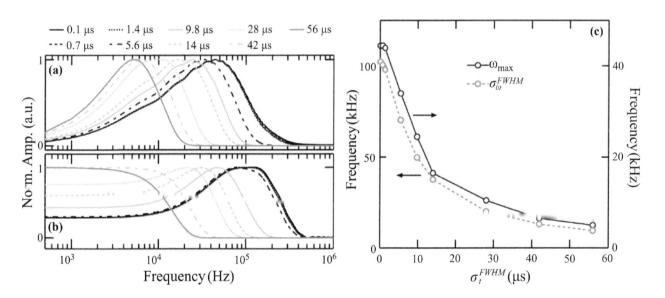

FIGURE 28.4   (a) Protoacoustic frequency spectra generated by Gaussian proton pulses simulated for detectors placed surrounding a 150 MeV dose deposition. (b) The protoacoustic frequency spectra simulated at a detector placed on the proton beam axis but deeper than the Bragg peak. (c) As the Gaussian proton pulse FWHM is increased, the frequency spectrum peaks at lower frequencies. (Reprinted from K.C. Jones, C.M. Sehgal S. Avery, How proton pulse characteristics influence protoacoustic determination of proton-beam range: Simulation studies, Physics in Medicine and Biology 61, 2213–2242, 2016 © Institute of Physics and Engineering in Medicine. Reproduced by permission of IOP Publishing. All rights reserved.)

28.4 CURRENT STATE OF PROTOACOUSTICS

Recent work in the field of protoacoustics has included both simulation and experiment. Simulations have been performed by numerically assessing the analytical equation (Equations 28.1 and 28.2) [16, 18, 19, 30, 35] and with the k-Wave MATLAB toolbox [13, 15, 23], a versatile program that allows inclusion of attenuation and heterogeneity by solving the wave equation with a psuedospectral model [36].

Experimental measurements require a detector, a triggered data collection instrument (oscilloscope), and a pulsed proton source. The detectors that have been used include hydrophones [11, 17, 22, 30, 37], ultrasound transducers for >1 MHz frequencies (low energy and short pulse proton beams) [15, 21, 26], strain gauges [38], and laser vibrometers [38]. For triggering data collection, direct electronic signals from the proton accelerator are often preferred [17], but scintillator-based prompt gamma detectors are versatile and do not rely on such access [22].

Although the above experimental components (detector and oscilloscope) are relatively simple, common, and/or inexpensive, the largest limitation on experimental studies has been (and will continue to be) access to appropriate proton beams. Limited access is partially due to the sparsity of proton centers, but it is further magnified by the disconnect between typical clinical proton beams and those required for efficient protoacoustic signal generation. For protoacoustic signal generation, the proton source must be pulsed and the pulses should have rise times of <20 μs. Ideally, the beams should be of clinical energies (up to 230 MeV), have instantaneous peak currents of 100s of nA, and have pulse widths ~5 μs. Generating proton pulses with <20 μs rise times has been challenging because the most common clinical proton accelerators – synchrotrons and isochronous cyclotrons – generate pulses with rise time of ~200 and 50 μs, respectively [39]. Therefore, researchers have utilized research accelerators with <1 μs short pulses [11, 15, 37], chopped the beam output [21], or they have modified clinical accelerators, such as modulating the output of an isochronous source [17]. Although these modifications are not (currently) appropriate for generating proton beams for treating patients, the newer clinical proton synchrocyclotron accelerators (e.g., IBA S2C2 [22] and Mevion S250 [25]) generate ideal proton pulses (4–5 pC in 10 μs for the IBA S2C2 [40]) for protoacoustic signal generation without modification.

28.4.1 Protoacoustic Image Formation

There are two general approaches to protoacoustic range verification: (1) using many detectors to generate an image and (2) using a few detectors to determine distances between Bragg peak and detector. Work toward the first (imaging) approach is described in this section, while the second (triangulation) is described in the next section.

Although there have been no experimental reports, simulations suggest that a cup-like array of 71 × 179 = 12709 detectors will provide sufficient protoacoustics signal-to-noise ratio to reconstruct a 3D image of a 1.6 cGy (at Bragg peak) clinical proton dose deposition with range verification accuracy of 1 mm [16] (homogeneous water medium, 1 μs rectangular pulse width, using typical noise for a hydrophone). Because the reconstruction relies on the number of projection viewpoints, many detectors are needed, and they must be positioned all around the dose distribution. If a higher proton dose is acceptable, the measurement can be collected with a smaller array of detectors that are rotated around the beam. For example, a single L-shaped line of 71 detectors may be rotated around the beam at 2° intervals during irradiation with 179 pulses in order to build up an equivalent dataset.

As shown in Figure 28.5, 2D protoacoustic dose deposition image reconstruction has been demonstrated experimentally using ultrasound transducer arrays to collect the high frequency protoacoustic signal generated by low energy (50 MeV), short pulse (1.5 μs) proton beams (>2000 Gy, 1.5 mm accuracy) [21]. This work is exciting because it produces 2D images of the Bragg peak using common ultrasound technology, but the differences in protoacoustic frequency generated by these low energy proton beams compared to clinical beams prevent use of these ultrasound transducers in proton therapy clinics.

28.4.2 Time-of-Flight Protoacoustic Range Verification

Instead of generating images based on signal collected from many detectors, the protoacoustic wave arrival time measured at one or a few detectors may be used to determine the Bragg peak position. This was first demonstrated experimentally by detecting the protoacoustic signal generated by low energy (20 MeV), short pulse (≤ 4.3 μs rectangular pulse with 3 ns rise time) proton beams (projected 1.6 Gy at the Bragg peak) with a single

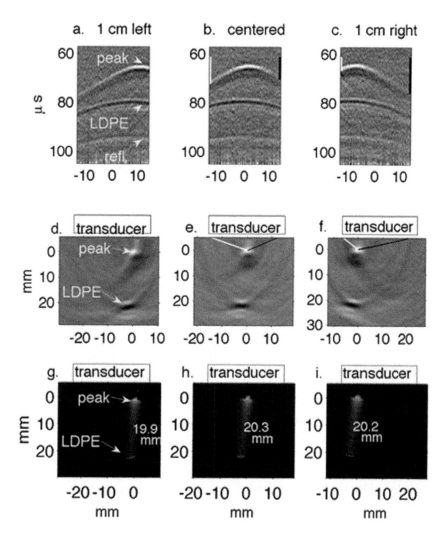

FIGURE 28.5 (a–c) The protoacoustic data generated by a 50 MeV proton beam was collected with an ultrasound transducer. The reconstructed Bragg peak (d–e) corresponds to the Monte Carlo calculated Bragg peak position (g–i). LDPE refers to the tank wall. The measurements were repeated in three configurations, in which the transducer is moved 1 cm left (a, d, g), centered (b, e, h), and 1 cm right (c, f, i) of the proton beam. (Reprinted from S.K. Patch, M.K. Covo, A. Jackson, Y.M. Qadadha, K.S. Campbell, R.A. Albright, P. Bloemhard, A.P. Donoghue, C.R. Siero, T.L. Gimpel, S.M. Small, B.F. Ninemire, M.B. Johnson, L. Phair, Thermoacoustic range verification using a clinical ultrasound array provides perfectly co-registered overlay of the Bragg peak onto an ultrasound image, Physics in Medicine and Biology 61, 5621–5638, 2016 © Institute of Physics and Engineering in Medicine. Reproduced by permission of IOP Publishing. All rights reserved.)

element 3.5 MHz ultrasound transducer [15], With the detector placed downstream of the Bragg peak, a range verification accuracy of ≤ 100 μm was demonstrated.

For proton beams of clinical energy and spot size measured in a water tank, Jones et al. reported a protoacoustic range verification standard deviation of 2.0 mm based on data collected at many detector positions surrounding the dose deposition (17 μs Gaussian-like proton pulse, up to 65 Gy) [19]. Extrapolation based on the detector noise in the treatment room led to a prediction of 2.2 mm range verification precision (standard deviation) with 2 Gy of deposited dose assuming a high

proton current of 860 nA (for the 17 μs proton pulse), as is shown in Figure 20.6. Although these measurements were performed at a clinical center with a clinical cyclotron, the proton pulsing was generated in an "off-label" procedure – currently unsuitable for treating patients – to achieve short pulses and high currents.

Characterization of range verification in water using protoacoustic signals generated by a clinical synchrocyclotron was reported by Lehrack et al. [22]. Synchrocyclotrons generate clinical proton pulses with FWHM of 1–10 μs, which is ideal for generating protoacoustic signals. For a detector placed at depth beyond

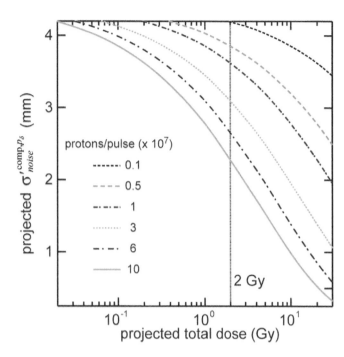

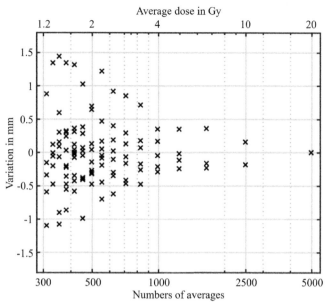

FIGURE 28.6 For the experimentally generated 17 μs excitation proton pulse, the projected precision (standard deviation) of measuring the protoacoustic time-of-flight arrival time is plotted versus total dose. If the peak proton current can be increased to ~900 nA (10 × 10⁷ protons/pulse), then a projected precision of 2.2 mm can be achieved with a dose of 2 Gy. (Reprinted with permission from John Wiley and Sons: John Wiley and Sons, Medical Physics, Acoustic time-of-flight for proton range verification in water, Kevin C. Jones, François Vander Stappen, Chandra M. Sehgal, Stephen Avery 43, 5213–5224, 2016.)

FIGURE 28.7 The proton beam range measured from the protoacoustic arrival time is plotted versus the number of averages and dose. The proton beam is generated by a synchrocyclotron source. (Reprinted from S. Lehrack, W. Assmann, D. Bertrand, S. Henrotin, J. Herault, V. Heymans, F. Vander Stappen, P.G. Thirolf, M. Vidal, J. Van de Walle, K. Parodi, Submillimeter ionoacoustic range determination for protons in water at a clinical synchrocyclotron, Physics in Medicine & Biology 62, L20–L30, 2017 © Institute of Physics and Engineering in Medicine. Reproduced by permission of IOP Publishing. All rights reserved.)

the Bragg peak, submillimeter accuracy (root mean square error of 0.8 mm) was achievable with <10 Gy of maximum deposited dose (2.5–3.7 μs FWHM proton pulses). Multiple beam energies and ranges were examined, and a 1 mm accuracy is achievable at <3 Gy of deposited dose, as is shown in Figure 28.7.

These initial water tank results investigating the relationship between deposited dose and accuracy are promising. The goal is to achieve 1 mm accuracy with doses < 2 Gy. Looking forward, moving to heterogeneous tissue is expected to decrease measured protoacoustic signal amplitude due to acoustic attenuation, but there is also higher expected initial pressure due to the higher Grüneisen parameter (Figure 28.1). Also, low-frequency acoustic noise from the patient (e.g., breathing, blood flow, digestion), which is not present for these studies, is expected to decrease the protoacoustic SNR. One method for increasing the SNR without additional dose, is to use multiple detectors.

28.4.3 Heterogeneity

Tissue heterogeneity presents challenges for both protoacoustic image reconstruction and time-of-flight range verification methods. Tissue heterogeneity affects both protoacoustic generation – through material-dependent Grüneisen parameters – and protoacoustic wave propagation – through material-dependent speed of sound and attenuation. As a result, the initial protoacoustic pressure distribution will be more complicated than the smooth pencil-beam dose distributions observed in homogeneous material. Furthermore, the macroscopic α and γ waves, which are the result of constructive interference from emissions all along the dose deposition, may be distorted due to the initial pressure distribution and/or by propagation through non-homogeneous material.

To characterize these effects, Jones et al. performed CT-based protoacoustic simulations of heterogeneous material [23]. Material-dependent properties were mapped based on Hounsfield unit values using parameters drawn from literature. Proton beams irradiated the liver and prostate, and the protoacoustic pressure waves reaching detectors on the patient's skin or in a transrectal probe were simulated. Under noiseless conditions, accuracies of ≤1.6 mm were reported (δ-function proton pulses) using multiple detectors to triangulate the Bragg peak position. These calculations were performed assuming the central axis of the proton beam was known, however. For a longer, 14 μs FWHM proton pulse, the accuracy suffered (≤5.8 mm). Attenuation in the liver case resulted in a loss in amplitude of a factor of two. Interestingly, time-of-flight distance calculations varied minimally (≤0.7 mm) when a generic water sound speed was used and compared to using a heterogeneity-corrected sound speed. Overall, the simulations showed that amplitudes and range verification accuracy are detrimentally affected by heterogeneity, but there are situations in which accurate range verification may be possible. For the considered cases, accuracy depended on detector placement.

Also through simulations, Patch et al. have demonstrated a method for accurately (≤ 2 mm) determining the Bragg peak position in heterogeneous material based on a two stage technique that combines beamforming and comparison to a database of dose distributions and their associated simulated protoacoustic emissions [25].

One proposed [15] and demonstrated [21, 26] method for neutralizing the effects of heterogeneous sound speeds is to overlay the protoacoustically identified Bragg peak on an ultrasound image. Because the heterogeneity will similarly affect the propagation of protoacoustic (lower frequency, one-way acoustic propagation) and ultrasound (higher frequency, two-way acoustic propagation) waves, any accumulated error will be shared by both modalities. Therefore, the Bragg peak position will be known relative to anatomy identified on the ultrasound image. This method has been demonstrated using low energy proton beams (50 and 15 MeV), with excellent results [21, 26].

28.4.4 Challenges

The largest challenges facing protoacoustic range verification methods are low signal-to-noise levels, errors induced by heterogeneity, and limited anatomical sites of application. Because acoustic waves are not transmitted through air pockets, possible sites for development include liver, pancreas, prostate, breast, and – if the low frequency protoacoustic waves can penetrate through skull – brain. On the technical side, standard synchrotron and isochronous cyclotron beam accelerators do not generate ideal proton pulses for protoacoustic signal generation. Newer synchrocyclotron proton sources, however, do produce and treat patients with proton pulses that are ideal for protoacoustic signal generation. If the protoacoustic range verification methods prove successful, protoacoustic detectors must be integrated into the clinical workflow (perhaps with additional ultrasound imaging), and the resulting range information must be acted on to either adjust the treatment in real-time or adapt for future fractions.

Current researchers in the field have not yet advanced to the stage of measuring protoacoustic signals in patients, an experiment performed in 1995 [12]. Unfortunately, the publication that resulted from this study is brief and the results are difficult to interpret. Looking forward, *in vivo* studies are necessary to characterize the protoacoustic detectability in the presence of biological acoustic noise. Verifying the accuracy of proton range verification methods *in vivo*, however, is complicated by the lack of a gold-standard technique against which they may be compared. Therefore, there are worthwhile, unanswered questions – particularly validation accuracy – that may be investigated through heterogeneous phantom or *ex vivo* studies.

28.5 CONCLUSIONS

Proton therapy has not realized its full potential due to range uncertainty. When dose is deposited by pulsed proton beams, thermoacoustic pressure emissions (protoacoustics) are generated. By collecting these emissions, the position of their origin – the Bragg peak – may be determined. Therefore, protoacoustics is a potential and promising proton range verification method. Challenges facing the technique include low acoustic signal and tissue heterogeneity.

REFERENCES

1. H. Paganetti, "Range uncertainties in proton therapy and the role of Monte Carlo simulations," *Physics in Medicine and Biology* **57**, R99–R117 (2012).
2. A.-C. Knopf and A Lomax, "*In vivo* proton range verification: A review," *Physics in Medicine and Biology* **58**, R131–R60 (2013).
3. H. Paganetti, "Relative biological effectiveness (RBE) values for proton beam therapy. Variations as a function of biological endpoint, dose, and linear energy transfer," *Physics in Medicine and Biology* **59**, R419–R472 (2014).

4. B. Clasie, N. Depauw, M. Fransen, C. Gomà, H. R. Panahandeh, J. Seco, J. B. Flanz, H. M. Kooy, "Golden beam data for proton pencil-beam scanning," *Physics in Medicine and Biology* **57**, 1147–1158 (2012).

5. J. Unkelbach and H. Paganetti, "Robust proton treatment planning: Physical and biological optimization," *Seminars in Radiation Oncology* **28**, 88–96 (2018).

6. A. Knopf, K. Parodi, T. Bortfeld, H. A. Shih, H. Paganetti, "Systematic analysis of biological and physical limitations of proton beam range verification with offline PET/CT scans," *Physics in Medicine and Biology* **54**, 4477–4495 (2009).

7. J. C. Polf, S. Avery, D. S. Mackin, S. Beddar, "Imaging of prompt gamma rays emitted during delivery of clinical proton beams with a Compton camera: Feasibility studies for range verification," *Physics in Medicine and Biology* **60**, 7085 (2015).

8. M. Priegnitz, S. Barczyk, L. Nenoff, C. Golnik, I. Keitz, T. Werner, S. Mein, J. Smeets, F. V. Stappen, G. Janssens, L. Hotoiu, F. Fiedler, D. Prieels, W. Enghardt, G. Pausch, C. Richter, "Towards clinical application: Prompt gamma imaging of passively scattered proton fields with a knife-edge slit camera," *Physics in Medicine and Biology* **61**, 7881 (2016).

9. J. M. Verburg and J. Seco, "Proton range verification through prompt gamma-ray spectroscopy," *Physics in Medicine and Biology* **59**, 7089 (2014).

10. C. Golnik, F. Hueso-González, A. Müller, P. Dendooven, W. Enghardt, F. Fiedler, T. Kormoll, K. Roemer, J. Petzoldt, A. Wagner, G. Pausch, "Range assessment in particle therapy based on prompt γ-ray timing measurements," *Physics in Medicine and Biology* **59**, 5399 (2014).

11. J. Tada, Y. Hayakawa, K. Hosono, T. Inada, "Time resolved properties of acoustic pulses generated in water and in soft tissue by pulsed proton beam irradiation – a possibility of doses distribution monitoring in proton radiation therapy" *Medical Physics* **18**, 1100–1104 (1991).

12. Y. Hayakawa, J. Tada, N. Arai, K. Hosono, M. Sato, T. Wagai, H. Tsuji, H. Tsuji, Acoustic pulse generated in a patient during treatment by pulsed proton radiation beam," *Radiation Oncology Investigations* **3**, 42–55 (1995).

13. K. C. Jones, A. Witztum, C. M. Sehgal, S. Avery, "Proton beam characterization by proton-induced acoustic emission: Simulation studies," *Physics in Medicine and Biology* **59**, 6549 (2014).

14. M. Ahmad, L. Xiang, S. Yousefi, L. Xing, "Theoretical detection threshold of the proton-acoustic range verification technique," *Medical Physics* **42**, 5735–5744 (2015).

15. W. Assmann, S. Kellnberger, S. Reinhardt, S. Lehrack, A. Edlich, P. G. Thirolf, M. Moser, G. Dollinger, M. Omar, V. Ntziachristos, K. Parodi, "Ionoacoustic characterization of the proton Bragg peak with submillimeter accuracy," *Medical Physics* **42**, 567–574 (2015).

16. F. Alsanea, V. Moskvin, K. M. Stantz, "Feasibility of RACT for 3D dose measurement and range verification in a water phantom," *Medical Physics* **42**, 937–946 (2015).

17. K. C. Jones, F. Vander Stappen, C. R. Bawiec, G. Janssens, P. A. Lewin, D. Prieels, T. D. Solberg, C. M. Sehgal, S. Avery, "Experimental observation of acoustic emissions generated by a pulsed proton beam from a hospital-based clinical cyclotron," *Medical Physics* **42**, 7090–7 (2015).

18. K. C. Jones, C. M. Sehgal, S. Avery, "How proton pulse characteristics influence protoacoustic determination of proton-beam range: Simulation studies," *Physics in Medicine and Biology* **61**, 2213–2242 (2016).

19. K. C. Jones, F. Vander Stappen, C. M. Sehgal, S. Avery, "Acoustic time-of-flight for proton range verification in water," *Medical Physics* **43**, 5213–5224 (2016).

20. S. Kellnberger, W. Assmann, S. Lehrack, S. Reinhardt, P. Thirolf, D. Queirós, G. Sergiadis, G. Dollinger, K. Parodi, V. Ntziachristos, "Ionoacoustic tomography of the proton Bragg peak in combination with ultrasound and optoacoustic imaging," *Scientific Reports* **6**, 29305 (2016).

21. S. K. Patch, M. K. Covo, A. Jackson, Y. M. Qadadha, K. S. Campbell, R. A. Albright, P. Bloemhard, A. P. Donoghue, C. R. Siero, T. L. Gimpel, S. M. Small, B. F. Ninemire, M. B. Johnson, L. Phair, "Thermoacoustic range verification using a clinical ultrasound array provides perfectly co-registered overlay of the Bragg peak onto an ultrasound image," *Physics in Medicine and Biology* **61**, 5621 (2016).

22. S. Lehrack, W. Assmann, D. Bertrand, S. Henrotin, J. Herault, V. Heymans, F. Vander Stappen, P. G. Thirolf, M. Vidal, J. Van de Walle, K. Parodi, "Submillimeter ionoacoustic range determination for protons in water at a clinical synchrocyclotron," *Physics in Medicine & Biology* **62**, L20 (2017).

23. K. C. Jones, W. Nie, J. C. H. Chu, J. V. Turian, A. Kassaee, C. M. Sehgal, S. Avery, "Acoustic-based proton range verification in heterogeneous tissue: Simulation studies" *Physics in Medicine & Biology* **63**, 025018 (2018).

24. W. Nie, K. C. Jones, S. Petro, A. Kassaee, C. M. Sehgal, S. Avery, "Proton range verification in homogeneous materials through acoustic measurements," *Physics in Medicine & Biology* **63**, 025036 (2018).

25. S. K. Patch, D. E. M. Hoff, T. B. Webb, L. G. Sobotka, T. Zhao, "Two-stage ionoacoustic range verification leveraging Monte Carlo and acoustic simulations to stably account for tissue inhomogeneity and accelerator-specific time structure – a simulation study," *Medical Physics* **45**, 783–793 (2018).

26. S. K. Patch, D. Santiago-Gonzalez, B. Mustapha, "Thermoacoustic range verification in the presence of acoustic heterogeneity and soundspeed errors – robustness relative to ultrasound image of underlying anatomy," *Medical Physics* **46**, 318–27 (2019).

27. M. Xu and L. V. Wang, "Photoacoustic imaging in biomedicine," *Review of Scientific Instruments* **77**, 041101 (2006).

28. S. K. Patch, D. Hull, W. A. See, G. W. Hanson, "Toward quantitative whole organ thermoacoustics with a clinical array plus one very low-frequency channel applied to prostate cancer imaging, *IEEE Transactions on Ultrasonics, Ferroelectrics, And Frequency Control* **63**, 245–255 (2016).

29. L. Xiang, B. Han, C. Carpenter, G. Pratx, Y. Kuang, L. Xing, "X-ray acoustic computed tomography with pulsed x-ray beam from a medical linear accelerator" *Medical Physics* **40**, 010701 (2013).

30. L. Sulak, T. Armstrong, H. Baranger, M. Bregman, M. Levi, D. Mael, J. Strait, T. Bowen, A. E. Pifer, P. A. Polakos, H. Bradner, A. Parvulescu, W. V. Jones, J. Learned, "Experimental studies of the acoustic signature of proton beams traversing fluid media," *Nuclear Instruments and Methods* **161**, 203–217 (1979).

31. S. Hickling, L. Xiang, K. C. Jones, K. Parodi, W. Assmann, S. Avery, M. Hobson, I. El Naqa, "Ionizing radiation-induced acoustics for radiotherapy and diagnostic radiology applications," *Medical Physics* **45**, e707–e721 (2018).

32. K. Graf, G. Anton, J. HÖSsl, A. Kappes, T. Karg, U. Katz, R. Lahmann, C. Naumann, K. Salomon, C. Stegmann, "Testing thermo-acoustic sound generation in water with proton and laser beams," *International Journal of Modern Physics A* **21**, 127–31 (2006).

33. L. Lin, C. G. Ainsley, T. D. Solberg, J. E. McDonough, "Experimental characterization of two-dimensional spot profiles for two proton pencil beam scanning nozzles" *Physics in Medicine and Biology* **59**, 493–504 (2013).

34. V. I. Albul, V. B. Bychkov, S. S. Vasil'ev, K. E. Gusev,V. S. Demidov, E. V. Demidova, N. K. Krasnov, A. F. Kurchanov, V. E. Luk'yashin, A. Y. Sokolov, "Measuring the Ultrasonic Field Generated in Water upon the Deceleration of a Proton Beam," *Instruments and Experimental Techniques* **47**, 502–506 (2004).

35. T. Terunuma, T. Sakae, Y. Hayakawa, A. Nohtomi, Y. Takada, K. Yasuoka, A. Maruhashi, "Waveform simulation based on 3D dose distribution for acoustic wave generated by proton beam irradiation," *Medical Physics* **34**, 3642–3648 (2007).

36. B. E. Treeby and B. T. Cox, "k-Wave: MATLAB toolbox for the simulation and reconstruction of photoacoustic wave fields," *Journal of Biomedical Optics* **15**, 021314 (2010).

37. V. I. Albul, V. B. Bychkov, K. E. Gusev, V. S. Demidov, E. V. Demidova, S. L. Konovalov, A. F. Kurchanov, V. E. Luk'yashin, V. I. Lyashuk, E. G. Novikov, A. A. Rostovtsev, A. Y. Sokolov, U. F. Feizkhanov, N. A. Khaldeeva, "Measurements of the parameters of the acoustic radiation accompanying the moderation of an intense proton beam in water," *Instruments and Experimental Techniques* **44**, 327–334 (2001).

38. W. Nie, K. C. Jones, A. Lieth, T. TeBeest, J. Foley, A. Kassaee, C. M. Sehgal, S. Avery, "TU-D-KDBRB1-3 comparison of acoustic detection methods for proton range verification," in *American Association of Physicists in Medicine 2018 Annual Meeting* (Nashville, TN, 2018).

39. "Prescribing, recording, and reporting proton-beam therapy (ICRU Report 78)," *Journal of the International Commission on Radiation Units and Measurements* **7**(2) (2007).

40. S. Henrotin, M. Abs, E. Forton, Y. Jongen, W. Kleeven, J. Van de Walle, P. Verbruggen, "Commissioning and testing of the first IBA S2C2," in *Proceedings of Cyclotrons 2016* (Zurich, Switzerland, 2016).

CHAPTER 29

Proton Radiography and Proton Computed Tomography

Xinyuan Chen and Tianyu Zhao

Washington University School of Medicine
St. Louis, Missouri

CONTENTS

29.1   INTRODUCTION

Proton therapy has been widely accepted and applied in cancer treatment because of its excellent 3D dose conformity. Compared with traditional photon radiation therapy, proton therapy is able to deposit most of its energy to a small targeted region without exiting dose due to the Bragg peak property of the protons. Therefore, radiation is delivered to tumor conformally while normal tissues beyond the target are spared by precise control of proton energy.

However, the superior accuracy in proton range can only be achieved with precise knowledge of the proton stopping power ratio (SPR) of different tissues inside the patient. Currently, the most reliable and common way of estimating proton SPR of a patient is stoichiometric calibration [1], which maps the electron density from a polyenergetic x-ray computed tomography (CT) image to the proton SPR. The accuracy of the estimated proton SPR from the CT image is degraded by a variety of factors, including acquisition parameters [2], imaging noise [3], and imaging artifacts [4], in addition to uncertainties associated with the stoichiometric calibration implicitly and explicitly, such as stoichiometric parameterization [2, 5], variation in human tissue composition and mean excitation energy [6, 7] in the calculation of theoretic proton SPR of human tissues. The overall uncertainty (1σ) in proton SPR for different tissue types has been estimated as 1.6% (soft tissue), 2.4% (bone), and 5.0% (lung) [8].

To overcome the intrinsic uncertainty in converting the CT Hounsfield Unit (HU) to proton SPR, it has been suggested that proton can be used directly as the imaging source to obtain *in vivo* integrated and pixelwise proton SPR, respectively, in proton radiography (pRG) and proton computed tomography (pCT). In principle, these imaging modalities can provide additional imaging for patient setup and reduce the uncertainty in proton range to less than 1%.

In this chapter, the basic physics of proton imaging, including proton interactions with matter and pRG/pCT reconstruction physics is introduced. Brief history of pRG/pCT development is reviewed. And finally, the most advanced technologies of pRG/pCT are introduced and discussed.

29.2 OVERVIEW OF PHYSICS OF PROTON IMAGING

29.2.1 Proton Interactions with Matter

Protons interact with matter mainly through Coulomb force between the electric field of the passing proton and the electric field of the orbital electrons of the material. This interaction can be described by the Bethe–Bloch equation [9, 10], where the energy loss rate of a proton is approximately proportional to inverse square of its velocity. Thus when a proton slows down in the material, its energy loss rate increases drastically, inducing a pronounced peak, named Bragg peak, toward the end of its range in the medium. The existence of Bragg peak restricts proton dose within a small region, assuring the dose conformity of proton therapy. Note that due to the statistical nature of proton interactions along the beam path, the Bethe–Bloch equation only characterizes the mean energy loss rate of a proton. The energy spectrum of protons widens as protons go deeper into the medium, thus range of individual proton varies. This phenomenon is called energy straggling [11].

Besides the Coulomb interaction with orbital electrons, protons may also interact with nuclei of the atoms through elastic Coulomb scattering, called "Multiple Coulomb Scattering" (MCS). Although this interaction has neglectable effect on the energy change of proton, it may change the direction of proton continuously, causing the so-called angular straggling. This is the major factor that limits the spatial resolution of pRG/pCT since protons don't travel in a straight path as primary photons do in the x-ray radiography and CT imaging.

In addition to these processes, a proton may also go through inelastic nuclear interactions at a rate of approximately 1% per cm in water at therapeutic energies. Such nuclear interactions remove the proton from the beam and reduce the primary fluence [12].

Increasing initial kinetic energy of the protons may improve spatial resolution of pRG/pCT via reduced MCS, and thus reduced angular straggling. However high initial kinetic energy may also lead to higher rate of inelastic nuclear interactions and more importantly, reduce energy contrast through the patient [13].

29.2.2 pRG/pCT Image Reconstruction

The primary mechanism of pRG/pCT contrast is the energy loss of each proton, unlike the fluence loss of photons in the x-ray radiography and x-ray CT. Therefore, the protons need to penetrate through the patient instead of being stopped inside patient within the target area as used in proton therapy. Accordingly, the initial kinetic energy of the protons in pRG/pCT is usually higher than that used in proton therapy. While typical initial kinetic energy for therapeutic applications ranges from 60 MeV (~3 cm range in water) to 230 MeV (~33 cm range in water), the typical initial kinetic energy for pRG/pCT is at or near the maximum energy of a medical accelerator, i.e., 230–250 MeV (~38 cm range in water) [14]. Although the 250 MeV proton beam doesn't have enough energy to completely pass through the hip region of a typical adult person, or the shoulder-to-shoulder distance in most male patients, it has sufficient range for scanning a human head and lung region of most people as long as the person's arms are raised out of the beam path. Therefore, many research efforts have been focused on developing proton imaging modalities that scan mainly the human head. Since proton SPR has little dependency on the initial kinetic energy of the protons used in pRG/pCT, the SPR acquired by proton transmission imaging is useful in the treatment planning for proton therapy.

Recall that the stopping power (S) is defined as the rate of energy loss per unit path length,

$$S = -\frac{dE}{dx} \qquad (29.1)$$

and the relative SPR is the ratio of stopping power (S) at a point of the material relative to that for water.

$$SPR = \frac{S_{material}}{S_{water}} \qquad (29.2)$$

To estimate relative proton SPR, a set of values of water equivalent path length (WEPL) is acquired by pRG/pCT. WEPL is defined as proton's path length in water that can produce the same amount of energy loss as that produced by proton passing through the imaging object. WEPL, written in Equation 29.3, is the line integral of SPR along a specific path length, which is analogous to ray projection in radiography.

$$WEPL\ (t,\theta)= \int\limits_{L_{t,\theta}} SPR(\vec{r})d\vec{r} \qquad (29.3)$$

where t and θ define the specific path that the proton goes through, $\vec{r}$ is the unit length that has the same direction as the passing proton.

In pRG, 2D images of mean WEPL can be used directly for the verification or correction of x-ray planning CT scans. In pCT, an additional step of reconstruction is needed to obtain SPR in a 3D volume from the WEPL measurements.

Based on different approaches to measure WEPL, pRG/pCT can be classified into two categories. One category is called proton integrating system, where WEPL is calibrated from the signal in a detector averaged over a large number of protons. The other category is called proton tracking system, where WEPL is calibrated from the residual energy or range of each proton. We will discuss them in the following sections.

29.3 HISTORY OF pRG/pCT DEVELOPMENT

First examples of pRG were demonstrated in the 1960s. In 1968, Koehler published the first image generated by pRG (Figure 29.1) using the proton beam at the Harvard Cyclotron (Cambridge, MA) [15]. The proton beam was scattered and broadened to approximately 2 cm in diameter by a lead scatterer, which reduced the beam energy from 160 MeV to 137 MeV. The imaging contrast was induced by a 100 μm thick, pennant-shape aluminum foil. A photographic film (Polaroid type-52 film) was placed close to the proton range to detect the sharp drop in proton fluence at this location. In spite of relative poor spatial resolution, the ability of energetic protons from an accelerator to produce high contrast image has been brought to people's attention. Many similar works on pRG were carried out afterwards. However they all used the sharp fall off of proton fluence to create contrast, thus the imaging objects were limited to have a small thickness [16, 17].

Later on in 1976, together with Cormack, Koehler published the first pCT design to detect small density

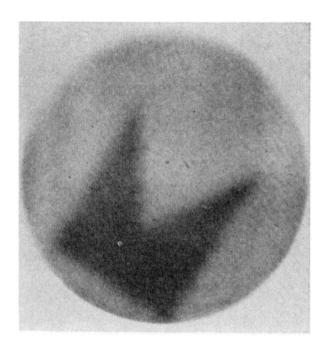

FIGURE 29.1 The first published proton radiograph from 1968 [15]. The contract was generated by a 100 μm thick, pennant-shape aluminum foil. (Reprinted from Koehler [15] with permission from The American Association for the Advancement of Science.)

difference (0.5%) [18]. The 158 MeV external beam of the Harvard cyclotron was used for that purpose. NaI scintillators coupled to photomultiplier tubes were used as detectors. The reconstruction was performed based on Abel's equation and the fact that the phantom was circularly symmetric. This system was considered as a proton integrating system, because all the protons within the proton beam were tracked as a whole. Although the reconstructed radial distribution of density difference was presented, no reconstructed image was included in the publication.

In 1978, a reconstructed image (Figure 29.2) using proton CT was first published by Hanson et al. at Los Alamos Laboratory (Los Alamos, NM) [19]. A 240 MeV proton beam was used as the imaging source. The advantage of low dose compared to x-ray CT was pointed out. The detector was composed of two parts: a hyperpure germanium (HPGe) detector and two layers of thick scintillation counters. The former was used to measure residual energy of each proton, the latter was used as a trigger of the HPGe detector so that the HPGe detector only recorded the events that happened within the central region, and thus increased the spatial resolution.

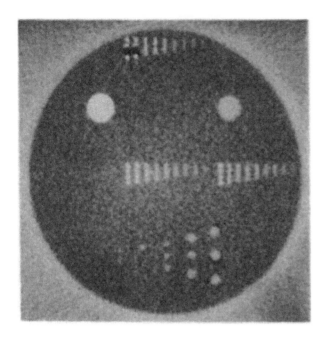

FIGURE 29.2 First image reconstructed from the Los Alamos proton CT scanner in 1978. ©1978 IEEE. (Reprinted from Hanson et al. [19] with permission from IEEE.)

The design concept of the detector has profound impact on the later on proton tracking systems.

In the following decade, Hanson's group took over the development of pCT by publishing a series of articles [20–22]. The HPGe detector was replaced by a stack of plastic scintillators in the later experiments. Human specimens (brain and heart) were scanned. A proton rate in excess of 10 kHz was obtained. The possibility of using curved projection path was also discussed. The Los Alamos work was a significant step forward, both conceptually and experimentally.

In 1987, Takada et al. [23] at Tsukuba were able to complete pCT scans in eight minutes using a multiple-pencil-beam scanning method in combination with moving slits, and a large magnetic spectrometer to measure the residual energy. In 1999, Zygmanski et al. at the Massachusetts General Hospital (MGH) [24] implemented a cone-beam CT system using scattered proton beam, and pointed out the advantages of obtaining stopping powers directly with pCT. The two systems are proton integrating systems, and assumed straight paths through the phantom.

29.4 CURRENT TECHNOLOGIES AND REMAINING CHALLENGES

29.4.1 Proton Integrating System

Proton integrating system is based on the assumption that signal measured in detector can be calibrated to

average proton WEPL of a number of protons through the imaging object. The structure of it is relatively simple compared to that of proton tracking systems. Usually, well confined proton pencil beams from a medical accelerator are used as the imaging source. A residual energy detector, such as a gadolinium oxysulfide scintillator coupled to an amorphous silicon matrix array is used to measure the averaged residual energy of the protons transmitted through the patient along a specific path. The reading of the detector will then be calibrated to the WEPL along the corresponding path. Finally, the proton SPR distribution within the patient can be calculated based on a set of values of WEPL.

Because of its compact structure, and its use of same proton beams with similar energies as the therapeutic proton beams, a proton integrating system has greater potential to be incorporated in proton therapy for treatment planning. However due to MCS, the spatial resolution of the proton integrating system is limited to a few millimeters, depending on the spot size and the thickness of the imaging object, and inferior to the proton tracking system, especially for the materials with high inhomogeneity, and at the edge region between different tissue types.

A recent work by Zhang et al. [25] showed pRG images (Figure 29.3) of an anthropomorphic head phantom, a range compensator and a frozen lamb's head. Time-resolved dose rate functions (DRFs) were measured by an x-ray amorphous silicon flat panel. Three methods of deriving SPR including root mean square (RMS) of DRFs only, intensity of DRFs only, and intensity-weighted RMS were implemented and compared. Interfaces between different materials were proved to be enhanced by incorporating the intensity information of DRFs. SPRs derived from both RMS only and intensity-weighted RMS were within ±1 for most of the Gammex phantom inserts, and with a mean absolute percentage error of 0.66% for all inserts.

Moreover, a Monte Carlo study has verified the feasibility of proton CT using multiple-layer ionization chamber (MLIC) as the detector [26]. The MLIC detector was able to record the residual energy of the proton beam at different depths. Thus the depth dose curve with a complete Bragg peak was plotted from the MLIC. The depth dose curve of the transmitted proton beam was then calibrated to WEPL. SPR distribution of the phantom was reconstructed with less than 1% deviation, and lower dose compared to x-ray CT.

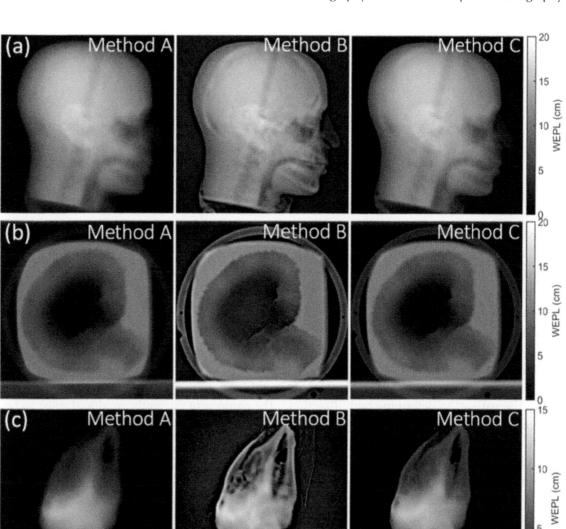

FIGURE 29.3 Proton radiography images derived by RMS of DRFs only (a), intensity of DRFs only (b), and intensity-weighted RMS (c) for an anthropomorphic head phantom, a range compensator and a frozen lamb's head by Zhang et al. [25]. (Reprinted from Zhang et al. [25] with permission from IOP.)

As demonstrated by the above-mentioned works, image quality of proton integrating systems has the potential to be improved in the future. The MCS effect may be corrected by the dose function recorded by the detector and more sophisticated statistical model of the proton beam.

29.4.2 Proton Tracking System

Proton tracking systems usually consist of two parts of detector: position sensitive detector (PSD) that tracks the direction and/or position of each proton, and residual energy range detector (RERD) to measure the residual energy of the protons exiting the imaging object. A schematic of an ideal proton-tracking pRG/pCT is shown in Figure 29.4. Based on the entrance and exiting positions, trajectory of individual proton inside the patient can be a straight line, or a most-likely path modeled statistically from estimated distribution of displacement and scattering angle [27], or a polynomial fitting path [28]. Studies [29] in homogeneous phantom showed that most-likely path is superior in imaging quality with fastest convergence and least residual error.

In 1999, an improved proton tracking system was developed by Pemler et al, where two scintillating fiber hodoscopes, one before and one after the patient, and a range telescope made of plastic scintillator tiles were

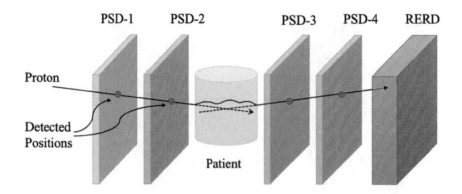

FIGURE 29.4 Schematic of the ideal proton-tracking pRG/pCT.

used to indicate both position and range of 10^6 protons/s [30]. The image was scanned in an area of 22.0×3.0 cm^2 with 1 MHz proton rates [30]. Between 2003 and 2013, a collaboration between Loma Linda University (LLU) and University of California Santa Cruz (UCSC), developed the first and the second generation of proton-tracking system using four silicon tracker planes (two before and two after the patient) to determine both direction and position of the incoming and outgoing proton. The proton rate was improved from 10 to 20 kHz to over 1 MHz, significantly reducing the acquisition time [31, 32]. By incorporating proton "most likely path" in an advanced iterative reconstruction method, the error of SPR was brought down to less than 1% [33].

A newly emerged group working on an Italian project for a PRonton IMAging (PRIMA) device proposed two generations of pCT system [34]. The systems were made of a silicon microstrip tracker and a YAG:Ce crystal calorimeter to capture single protons trajectory and residual energy, respectively. The first generation pCT system was constructed to have an active area of about 5×5 cm^2 and a data rate capability of 10 kHz. Two slices of tomographic reconstructions of the test phantom were displayed in Figure 29.5, with a sketch of the analyzed phantom on the left. An extended field of view (up to $\sim 5 \times 20$ cm^2) and an increased event rate capability up to 1 MHz was described for the second generation system.

A novel proton tracking system was proposed by the Proton Radiotherapy Verification and Dosimetry Applications (PRaVDA) consortium [35]. The pCT imaging system was based entirely on solid-state detector, making it possible to track multiple proton per readout cycle, which leads to a potential reduction in proton CT scan time. A 75-mm diameter PMMA sphere with substitute inserts was imaged by the fully solid-state

imaging system (Figure 29.6). Accuracy in SPR was measured to be $\leq 1.6\%$ for all the inserts.

Proton tracking systems require high-speed tracking of protons at the entrance and exit sides of the patient, with continuous recording of the residual energy. Although this approach records adequate information and is promising in theory for accurate pRG and pCT, the equipment currently employed must to be reduced in complexity and size if it is to be suitable in a clinical setting [36].

29.5 CONCLUSION

Range uncertainty, originating from the uncertainty in the mapping of electron density to proton stopping power, has been a unique challenge for proton therapy. The lack of capability to pin down proton SPR in the primary dataset for treatment planning, offsets the most

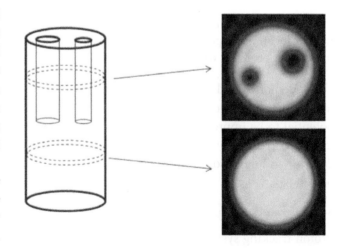

FIGURE 29.5 A sketch of the imaging phantom (left). Two slices of the tomographic reconstructions: one in the area with holes and one in the uniform region (right) [34]. (Reprinted from Scaringella et al. [34] with permission from IOP.)

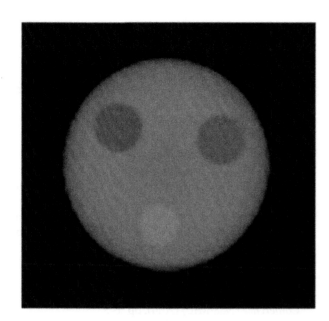

FIGURE 29.6 A slice of reconstructed spherical phantom containing three substitute inserts (top left: water equivalent, top right: adipose equivalent, bottom: average bone equivalent) [35].

anticipated benefit of precise delivery and sparing distal organs to a certain extent. Although the intrinsic uncertainty can be reduced with more sophisticated CT calibration or dual-energy CT scan, and its dosimetric impact could be mitigated with robust optimization, it never goes away as long as the primary dataset used for dose calculation is acquired with x-ray tomography. Alternatively, proton provides a promising way of imaging patients, providing either volumetric images with accurate relative proton SPRs, or projectional radiographic images that could be used for patient setup and range verification. The implementation of volumetric proton imaging becomes increasingly attractive and practical with mounting interests and maturing technologies. While multiple proton tracking systems show superior imaging quality and are closer to pre-clinical test, proton integrating systems provide integrated solutions for both volumetric imaging and projectional radiography that are compatible with most commercial proton therapy cyclotrons.

REFERENCES

1. U. Schneider, E. Pedroni, A. Lomax, "The calibration of CT Hounsfield units for radiotherapy treatment planning," *Physics in Medicine and Biology* **41**, 111 (1996).
2. S. España and H. Paganetti, "The impact of uncertainties in the CT conversion algorithm when predicting proton beam ranges in patients from dose and PET-activity distributions," *Physics in Medicine and Biology* **55**, 7557–7571 (2010).
3. A. V. Chvetsov and S. L. Paige, "The influence of CT image noise on proton range calculation in radiotherapy planning," *Physics in Medicine and Biology* **55** (2010).
4. K. M. Andersson, A. Ahnesjö, C. V. Dahlgren, "Evaluation of a metal artifact reduction algorithm in CT studies used for proton radiotherapy treatment planning," *Journal of Applied Clinical Medical Physics* **15**, 112–119 (2014).
5. B. Schaffner and E. Pedroni, "The precision of proton range calculations in proton radiotherapy treatment planning: Experimental verification of the relation between CT-HU and proton stopping power," *Physics in Medicine and Biology* **43**, 1579–1592 (1998).
6. A. Besemer, H. Paganetti, B. Bednarz, "WE-C-BRB-06: Clinical impact of uncertainties in the mean excitation energy of human tissues during proton therapy," *Medical Physics* **39**, 3944 (2012).
7. H. Paganetti, "Range uncertainties in proton therapy and the role of Monte Carlo simulations," *Physics in Medicine and Biology* **57**, R99 (2012).
8. M. Yang, X. Zhu, P. Parker, et al., "Comprehensive analysis of proton range uncertainties related to patient stopping-power-ratio estimation using the stoichiometric calibration," *Physics in Medicine and Biology* **57**, 4095 (2012).
9. H. Bethe, "Zur Theorie des Durchgangs schneller Korpuskularstrahlen durch Materie," *Annals of Physics* **397**, 325–400 (1930).
10. F. Bloch, "Zur Bremsung rasch bewegter Teilchen beim Durchgang durch Materie," *Annals of Physics* **408**, 285–320 (1933).
11. A. J. Lomax, "Charged particle therapy: The physics of interaction," *Cancer Journal* **15(4)**, 285–291 (2009).
12. J. F. Janni, "Energy loss, range, path length, time-of-flight, straggling, multiple scattering, and nuclear interaction probability: in two parts. Part 1. For 63 compounds Part 2. For elements $1 \leq Z \leq 92$," *Atomic Data and Nuclear Data Tables* **27**, 147–339 (1982).
13. J. Beringer, J. F. Arguin, R. M. Barnett, et al., "Review of particle physics," *Physical Review D* **86**, 010001 (2012).
14. M. Berger, J. Coursey, M. Zucker, *Stopping-Power and Range Tables for Electrons, Protons, and Helium Ions* (National Institute of Standards and Technology, 2000).
15. A. M. Koehler, "Proton radiography," *Science* **160**, 303–304 (1968).
16. D. R. Moffett, E. P. Colton, G. A. Concaildi, et al., "Initial test of a proton radiographic system," *IEEE Transactions on Nuclear Science* **22**, 1749–1751 (1975).
17. V. W. Steward and A. M. Koehler, "Proton radiography as a diagnostic tool," *Physics in Medicine and Biology* **18**, 591 (1973).
18. A. M. Cormack and A. M. Koehler, "Quantitative proton tomography: Preliminary experiments," *Physics in Medicine and Biology* **21**, 560–569 (1976).

19. K. M. Hanson, J. N. Bradbury, T. M. Cannon, et al., "The application of protons to computed tomography," *IEEE Transactions on Nuclear Science* **25**, 657–660 (1978).

20. K. M. Hanson, "Proton computed tomography," *IEEE Transactions on Nuclear Science* **26**, 1635–1640 (1979).

21. K. M. Hanson, J. N. Bradbury, T. M. Cannon, et al., "Computed tomography using proton energy loss," *Physics in Medicine and Biology* **26**, 965–983 (1981).

22. K. M. Hanson, J. N. Bradbury, R. A. Koeppe, et al., "Proton computed tomography of human specimens," *Physics in Medicine and Biology* **27**, 25–36 (1982).

23. Y. Takada, K. Kondo, T. Marume, K. Nagayoshi, I. Okada, K. Takikawa, "Proton computed tomography with a 250 MeV pulsed beam," *Nuclear Instruments and Methods in Physics Research Section A* **273**, 410–422 (1988).

24. P. Zygmanski, K. P. Gall, M. S. Z. Rabin, S. J. Rosenthal, "The measurement of proton stopping power using proton-cone-beam computed tomography," *Physics in Medicine and Biology* **45**, 511–528 (2000).

25. R. Zhang, K.-W. Jee, E. Cascio, G. C. Sharp, J. B. Flanz, H.-M. Lu, "Improvement of single detector proton radiography by incorporating intensity of time-resolved dose rate functions," *Physics in Medicine and Biology* **63**(1), 015030 (2017).

26. X. Chen, R. Liu, S. Zhou, et al., "A novel design of proton computed tomography detected by multiple-layer ionization chamber with strip chambers: A feasibility study with Monte Carlo simulation," *Medical Physics* **47**(2), 614–625 (2020).

27. D. C. Williams, "The most likely path of an energetic charged particle through a uniform medium," *Physics in Medicine and Biology* **49**, 2899–2911 (2004).

28. N. Krah, J. M. Létang, S. Rit, "Polynomial modelling of proton trajectories in homogeneous media for fast most likely path estimation and trajectory simulation," *Physics in Medicine and Biology* **64** (2019).

29. T. Li, Z. Liang, J. V. Singanallur, T. J. Satogata, D. C. Williams, R. W. Schulte, "Reconstruction for proton computed tomography by tracing proton trajectories: A Monte Carlo study," *Medical Physics* **33**, 699–706 (2006).

30. P. Pemler, J. Besserer, J. De Boer, "A detector system for proton radiography on the gantry of the Paul-Scherrer-Institute," *Nuclear Instruments and Methods in Physics Research* **432**, 483 (1999).

31. R. F. Hurley, R. W. Schulte, V. A. Bashkirov, "Water-equivalent path length calibration of a prototype proton CT scanner," *Medical Physics* **39**, 2438 (2012).

32. R. P. Johnson, V. Bashkirov, V. Giacometti, et al., "Results from a pre-clinical head scanner for proton CT," in *2014 IEEE Nuclear Science Symposium and Medical Imaging Conference (NSS/MIC)* (2014), pp. 1–5.

33. S. N. Penfold, R. W. Schulte, Y. Censor, A. B. Rosenfeld, "Total variation superiorization schemes in proton computed tomography image reconstruction," *Medical Physics* **37**, 5887–5895 (2010).

34. M. Scaringella, M. Bruzzi, M. Bucciolini, et al., "A proton Computed Tomography based medical imaging system," *Journal of Instrumentation* **9**(12), 12009 (2014).

35. M. Esposito, C. Waltham, J. T. Taylor, et al., "PRaVDA: The first solid-state system for proton computed tomography," *Physica Medica* **55**, 149–154 (2018).

36. G. Poludniowski, N. M. Allinson, P. M. Evans, "Proton radiography and tomography with application to proton therapy," *British Institute of Radiology* **88**, 20150134 (2015).

IV

Imaging Modalities

CHAPTER 30

Dosimetry of Imaging Modalities in Radiotherapy

George X. Ding

Vanderbilt University School of Medicine
Nashville, Tennessee

CONTENTS

30.1 INTRODUCTION

In modern radiation therapy, image-guided radiation therapy (IGRT) has been adopted as the standard of care to improve the geometric accuracy of patient positioning during radiotherapy [1–7] because it is capable of significantly reducing target positioning errors, hence enabling highly conformal treatments. There are many different imaging modalities used for image guidance in radiotherapy. The currently available technologies to perform IGRT can be divided, according to their operation, into planar systems, volumetric systems, and non-radiographic systems [8]. The dosimetry of imaging modalities used in radiotherapy involves ionizing radiation beams (such as x-rays) while non-radiographic systems use non-ionizing radiation (such as magnetic resonance imaging (MRI) systems, optical systems, ultrasound, etc). An overview of the state-of-the-art image-guidance techniques employed in radiation therapy can

be found in AAPM Medical Physics Monograph No 39 [9]. The following is a brief introduction of the imaging modalities and the dosimetry need for imaging devices.

30.1.1 Planar Imaging Modalities

Planar imaging modalities include electronic portal imaging devices (EPIDs) and stereoscopic x-ray imaging systems. These systems allow matching of planar, kilovoltage (kV) radiographs, fluoroscopy, or megavoltage (MV) images with digitally reconstructed radiographs (DRRs) from the planning CT [0].

Although the film was earlier commonly used as a verification method for field placement of treatment portal in radiation therapy [10], it has been gradually replaced by the electronic systems when EPID systems became available for IGRT both online and offline [11]. The developments driven by the need to image patient position before treatment delivery without using

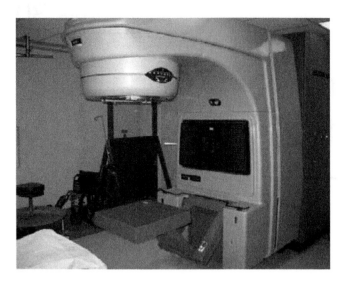

FIGURE 30.1 Conventional electronic portal imaging device (EPID). (Reproduced from AAPM TG-180 [1].)

radiographic film [12–20] led to the EPIDs. The EPID device is attached to the medical accelerator used for radiotherapy and is capable of obtaining a pair of orthogonal projected digital images instantly using radiotherapy megavoltage beams, as shown in Figure 30.1. The details on development and the technology of the EPID can be found in several review articles [21–23]. Currently medical accelerators made by Elekta and Varian for radiotherapy are equipped with these devices.

Room-mounted stereoscopic devices utilize angled x-rays tubes [24]. This type of device normally consists of a pair of x-ray tubes and image detectors mounted in the floor and on the ceiling. Examples of these devices are used by the CyberKnife (Accuray) system (Figure 30.2) and by the BrainLab ExacTrac X-ray (Brainlab) system (Figure 30.3).

30.1.2 Volumetric Systems

The need to visualize patient 3D volumetric images for accurate radiation delivery led to the clinical availability of MV-CBCT [25], MVCT [26–28], and kV-CBCT [29–32] in radiotherapy. Volumetric MV-CBCT images are reconstructed using projections acquired with the EPID and result in a greater dose than a pair of orthogonal MV portal images even with a low dose-rate therapy beam [33].

Current kV imaging devices are generally integrated into linear accelerators and are capable of acquiring both 2D radiographs and 3D volumetric kV-CBCT images [29–32]. Examples of kV-CBCT scanners integrated into

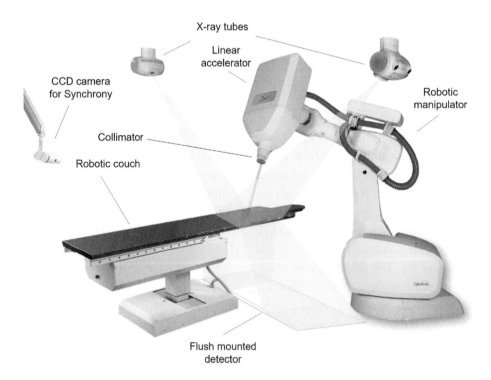

FIGURE 30.2 The CyberKnife™ system with its main components including IGRT systems. The stereotactic x-ray systems have ceiling mounted x-ray tubes and floor-mounted detectors. The x-ray systems are supplemented with a ceiling-mounted camera to determine the patient surface contour. (Reproduced from ICRU-91.)

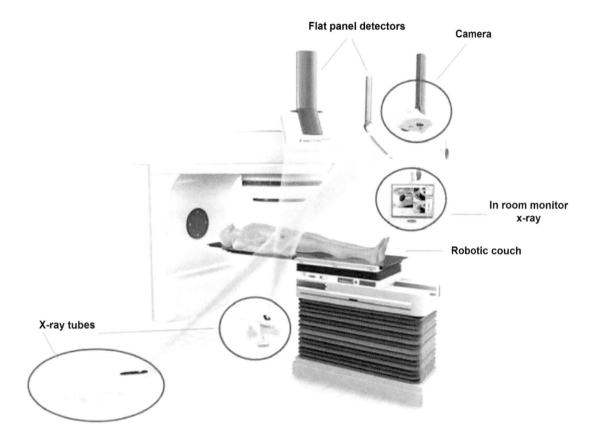

FIGURE 30.3 The ExacTrac™ system with floor-mounted x-ray tubes and ceiling-mounted flat panel detectors. (Reproduced from ICRU-91.)

a linac are Varian Medical Systems, Inc. (Palo Alto, CA) On Board Imaging (OBI) system (Figure 30.4) and an Elekta (Stockholm, Sweden) X-Ray Volume Imaging (XVI) system (Figure 30.4).

30.1.3 Need of Dosimetry for Imaging Dose

IGRT has now become the standard of care to improve the geometric accuracy of patient positioning during radiotherapy [1–7]. Unlike diagnostic imaging examinations, the patient may be imaged multiple times during any fraction in order to position the patient accurately during radiotherapy delivery. The additional radiation exposure to patients resulting from these x-ray imaging procedures may results in excessive dose to sensitive organs and potentially increases the chance of secondary cancers and, therefore, needs to be managed in order to minimize its risk [1, 2]. Based on considerations of clinical relevance, dose tolerances for critical organs and data available in the literature, AAPM TG-180 recommended 5% of the therapeutic target dose as the threshold beyond which the imaging dose should be accounted for as part of the total dose to radiotherapy patients [1].

While the commonly adopted radiation protection safety philosophy of As Low As Reasonably Achievable (ALARA) is still applicable to imaging dose, minimizing imaging doses, however, should not compromise image quality needed for target localization [1].

When there is reasonable expectation that the imaging dose will exceed 5% of the total prescribed dose, the imaging dose should be accounted for as part of the total dose to the patient [1]. The magnitude of imaging dose is dependent on many factors, including the frequency of imaging and the technique used. While the dose from a single imaging procedure is much lower than a fraction of therapeutic dose, the integral dose of repeated imaging procedures may not be negligible and should be accounted for [2].

In order to determine if the additional imaging doses exceed the recommended threshold, it is important to know the magnitude of additional imaging dose relative to the therapeutic dose. It is known that the imaging dose to the patient differs significantly depending on modalities and procedures used ranging from 0.1 cGy to 5.0 cGy for a single image acquisition, with the

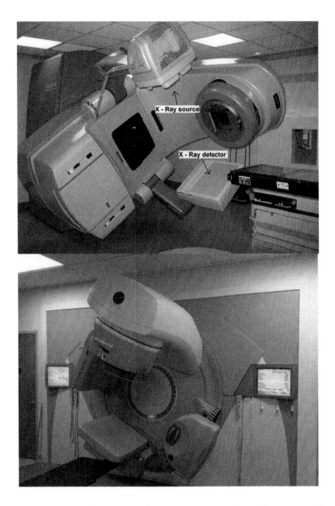

FIGURE 30.4 kV image devices integrated into linear accelerators: (a) Varian OBI system on a Varian Trilogy treatment unit ; (b) Elekta XVI on an Elekta Synergy treatment unit. (Reproduced from AAPM TG-180.)

exception of MV volumetric imaging [1]. Therefore, the knowledge of dosimetry for imaging dose is necessary. While the methods of determination of imaging dose resulting from the therapeutic megavoltage beams are generally available in radiotherapy clinics, the dosimeters for determining imaging dose resulting from kilovoltage beams are not generally available.

30.2 AVAILABLE METHODS FOR IMAGING DOSIMETRY

The commonly used imaging procedures in IGRT are kilovoltage radiographs and cone-beam computed tomography (kV-CBCT). kV imaging provides superior image contrast compared to MV imaging. The approaches used to determine kV imaging doses include experimental measurements [34–47] and calculations

using Monte Carlo methods [48–58]. The imaging doses resulting from MV beams can be calculated by using commercial treatment planning systems except for the beam that is only used for imaging, such as a the 2.5 MV beam in a Varian TrueBeam machine. However, for such a beam once the beam output is determined the imaging dose can be estimated with acceptable accuracy [59].

Although kV imaging is commonly used in the IGRT, it is relatively challenging to measure or to calculate kV imaging doses to the patient in a radiotherapy clinic. This is because measuring dose from kV beams requires specific detectors and expertise. AAPM TG-180 recommended two methods that can be used to estimate dose resulting from imaging procedures. One requires patient-specific dose calculations while the other makes use of tabulated values to estimate organ doses from a specific imaging procedure referred to as non-patient-specific imaging dose estimation [1].

30.2.1 Measurements of Output of Beams Used for Imaging

It needs to be emphasized that it is essential to determine the output of any beam that is used for imaging in order to facilitate AAPM TG-180 recommended methods to estimate the imaging dose to a patient. If the output of an imaging beam does not meet the manufacturer's specifications, the accuracy of estimated patient doses cannot be guaranteed by using AAPM TG-180 recommended methods.

When a therapy MV beam is used for imaging, its beam output is calibrated according to the dosimetry protocols [60, 61] and monitored on a daily basis. For an MV beam that is used for imaging only, its output can be measured according to the same dosimetry protocols [60, 61] based on measured values of percentage depth-dose (PDD) curve at depth of 10 cm [62].

The output for each image acquisition procedure, with specified protocol parameters, should be measured in air or in-phantom, according to the AAPM dosimetry protocols for kV [63] and MV [60] beams, to confirm that the measured dose is within the manufacturer-stated specifications at the time of acceptance of the imaging device. The exposure or absorbed dose from a kV beam can be measured by using a commercially available x-ray test devices, such as RaySafe X2, RaySafe Xi, which can measure kVp, half-value layer HVL and dose, etc. with a simple measurement setup, or by using a calibrated ionization chamber according to the dosimetry protocol [61, 63]. Although the methodology for measuring the HVL and dose with a calibrated ionization chamber is available from dosimetry protocols [61, 63], the calibration conditions recommended in these dosimetry protocols are often not applicable to imaging acquisition procedures, especially when the x-ray source is moving during the scan [1]. Although water is the most suitable medium for kV x-ray beam measurements, plastic phantom materials are more convenient in practice. By using an ionization chamber, a method [64] to determine the absorbed dose output resulting from a specific image acquisition procedure has been described based on a dosimetry protocol [61, 63].

30.2.2 Patient Imaging Dose Calculations

When the imaging beam is a therapeutic MV beam, the patient-specific imaging dose can be calculated based on the beam monitor units and field sizes used during image acquisition, unless the MV beam is an imaging-only one [62]. These MV dose calculations are patient CT based and can be included to therapeutic dose directly during treatment planning when it is necessary [65]. If the total number of imaging procedures is known, the repeated imaging dose can be summed and accounted for the total dose to the target and OARs. Miften et al. demonstrated this approach by showing optimized IMRT plans with and without MV-CBCT included in the process [65].

The dose calculation for kV beams is currently not available in a commercial treatment planning system, although a model-based algorithm has been shown to achieve an acceptable accuracy for kV energy beams [66]. Alaei et al. has shown that a commercial treatment planning system can be modified by characterizing kV beams to perform kV image dose calculations [67]. However, the modification of adding a kV beam in a commercial treatment planning system requires significant expertise and the dose calculation accuracy is limited with larger uncertainties to bones.

To meet the dose calculation accuracy demand, Monte Carlo techniques have been extensively used in radiotherapy in recent decades [68–70] and applied for kV imaging dose calculations [71]. Currently, patient organ doses resulting from kV imaging procedures reported in the literature are generally obtained from the Monte Carlo calculations [1, 57, 72]. This type of dose calculation requires the details of the imaged beam

acquisition parameters and therefore it is able to provide accurate personalized organ doses resulting from image guidance procedures from different patient size and image location [50, 72, 73].

Although Monte Carlo techniques are capable of providing accurate patient-specific organ dose calculations from different x-ray imaging procedures, the long computational times and expertise required for Monte Carlo simulations make it impractical in a clinical setting to calculate imaging doses in daily clinical routine practice. The success of Monte Carlo studies has provided detailed information about patient organ dose resulting from different image guidance procedures and led to a practical method of accounting for the additional dose to patients resulting from these imaging procedures, referred to as non-patient-specific imaging dose estimations [1].

Based on the studies reported in the literature [57, 58, 72, 73], the magnitude of patient organ doses from kV imaging is small relative to therapeutic dose. It has been shown that it is adequate to use simpler approaches that provide reasonable estimates of the imaging dose by using simple look-up tables to estimate organ doses from repeated imaging procedures. The approach is referred to as non-patient-specific imaging dose estimation [1] in which inter-patient variation and geometry dependence are small in most cases and the dose estimates could be provided in the form of simple look-up tables, which may be accurate enough to estimate the dose from repeated imaging procedures.

30.2.3 Patient Imaging Dose Estimations

The method of estimating organ doses from a specific imaging procedure, referred to as non-patient-specific imaging dose estimation, is recommended by AAPM TG-180 [1] in the absence of patient-specific dose calculations. Tables 30.1–30.3 are the examples of tabulated organ doses for specified image acquisition procedures associated with known mAs shown in Tables 30.4 and 30.5. For planar kV beams (Table 30.5), only beam entry doses are listed as the organ doses from planar beams are strongly dependent on the beam entry direction. Similar tabulated organ dose data for the Elekta kV imaging system are shown in Tables 30.6–30.8. These tabulated values are for specified mAs used for image acquisition are sufficient to estimate imaging dose to within 20% [72]. When users use a different mAs from listed in the image acquisition procedure, the organ dose can be scaled by

TABLE 30.1   Organ Doses for the Head & Neck and Brain Treatment Sites from Varian OBI v1.4 Using Standard Head kV-CBCT Scan

| | Standard Head, Brain | | | Standard Head, Head & Neck | |
Organ	D50 Range (cGy)	D10 Range (cGy)	Organ	D50 Range (cGy)	D10 Range (cGy)
Brain	0.21–0.33	0.27–0.40	Brain	0.15–0.22	0.16–0.23
Brainstem	0.19–0.30	0.22–0.32	Larynx	0.21–0.29	0.25–0.33
Chiasm	0.08–0.26	0.09–0.26	Oral Cavity	0.13–0.26	0.20–0.31
Eyes	0.03–0.31	0.04–0.35	Parotids	0.26–0.42	0.31–0.48
Optic Nerves	0.05–0.27	0.05–0.27	Spinal Cord	0.16–0.25	0.19–0.32
Pituitary	0.07–0.24	0.08–0.25	Thyroid	0.07–0.23	0.11–0.32
Spinal Cord	0.26–0.33	0.29–0.34	Esophagus	0.07–0.16	0.14–0.26
Skin	0.19–0.41	0.39–0.63	Skin	0.18–0.27	0.34–0.44
Bones	0.45–1.11	1.13–1.67	Bones	0.25–0.65	0.64–1.07

Note: D50 and D10 are minimum dose delivered to 50% and 10% of the organ volume, respectively.
Source: Reference [72] and kV-CBCT scan parameters for Varian OBI 1.4 shown in Table 30.4.

TABLE 30.2   Organ Doses for the Chest Treatment Site from Varian OBI v1.4 Using Low-dose Thorax kV-CBCT Scan

| | Low-Dose Thorax | |
Organ	D50 Range (cGy)	D10 Range (cGy)
Aorta	0.42–0.58	0.44–0.63
Lungs	0.30–0.61	0.43–0.72
Small Bowel	0.33–0.54	0.39–0.61
Esophagus	0.29–0.60	0.35–0.74
Kidney	0.43–0.54	0.49–0.59
Heart	0.31–0.55	0.41–0.63
Liver	0.31–0.51	0.38–0.61
Spinal Cord	0.32–0.57	0.35–0.78
Spleen	0.32–0.52	0.36–0.60
Stomach	0.28–0.57	0.31–0.62
Trachea	0.36–0.71	0.47–1.04
Skin	0.46–0.57	0.64–0.89
Bones	1.06–1.74	1.47–2.25

Source: Reference [72] and kV-CBCT scan parameters for Varian OBI 1.4 shown in Table 30.4.

TABLE 30.3   Organ Doses for the Pelvis Treatment Site from Varian OBI v1.4 Using Pelvis kV-CBCT Scan

| | Pelvis Scan, Prostate Isocenter | |
Organ	D50 Range (cGy)	D10 Range (cGy)
Bladder	1.36–2.20	1.72–2.69
Bowel	1.54–1.91	2.04–2.65
Femoral Heads	2.40–3.60	3.22–4.88
Prostate	1.19–1.79	1.33–1.89
Rectum	1.51–1.99	1.70–2.22
Skin	1.80–1.96	2.26–2.92
Bone	2.93–3.96	4.61–5.72

Source: Reference [72] and kV-CBCT scan parameters for Varian OBI 1.4 shown in Table 30.4.

TABLE 30.4   Parameters for kV-CBCT Specified Acquisition Techniques in Varian OBI 1.4

kV-CBCT	Name	Bow-Tie Filter	(kV)	(mAs)	Gantry Rotation (degrees)
OBI	Standard-dose Head	Full fan	100	145	200
OBI	Low-dose Head	Full fan	100	72	200
OBI	High-Quality Head	Full fan	100	720	200
OBI	Pelvis	Half fan	125	700	360
OBI	Pelvis Spot Light	Full fan	125	720	200
OBI	Low-Dose Thorax	Half fan	110	262	360
TrueBeam	Head	Full fan	100	147	200
TrueBeam	Pelvis	Half fan	125	1056	360
TrueBeam	Spotlight	Full fan	125	733	200
TrueBeam	Thorax	Half fan	125	264	360

Source: From reference [58].

TABLE 30.5 Parameters for kV Radiographs for Specified Acquisition Techniques in Varian OBI 1.4

Name	(kV)	(mAs)	Entry Dose
Head-AP	100	8	0.1 cGy
Head-Lat	70	5	0.05 cGy
Thorax-AP	75	5	0.1 cGy
Thorax-Lat	95	40	0.5 cGy
Pelvis-AP-Med	75	10	0.1 cGy
Pelvis-Lat-Med	105	80	1.6 cGy

Note: The clinical default OBI blades are set to $X_1 = X_2 = 13.3$ cm and $Y_1 = Y_2 = 10.3$ cm in all acquisition techniques. All six techniques were modeled with and without full-fan bow-tie filter.
Source: From Supplementary data in reference [58].

TABLE 30.6 Organ Doses for the Head & Neck Treatment Site from Elekta XVI kV-CBCT Scan Using S Cassettes, 100 kVp, 0.1 mAs/Acquisition, 360 Acquisitions, 345–190 Degree (IEC) Rotation

Head and Neck	
Organ	D50 Range (cGy)
Brainstem	0.06–0.08
Rt Eye	0.08–0.09
Lt Eye	0.13–0.13
Rt Parotid	0.05–0.06
Lt Parotid	0.16–0.17
Rt Cochlea	0.04–0.05
Lt Cochlea	0.09–0.12
Oral Cavity	0.09–0.11

Source: From reference [1].

TABLE 30.7 Organ Doses for the Pelvis Treatment Site from Elekta XVI kV-CBCT Scan Using M Cassette with Bowtie Filter, 120 kVp, 1.6 mAs/Acquisition, 650 Acquisitions, Full 360 Degree Rotation

Pelvis	
Organ	D50 Range (cGy)
Bladder	0.9–2.0
Rectum	1.1–1.9
Small Bowel	1.0–1.8

Source: Reference [1].

TABLE 30.8 Organ Doses for the Pelvis Treatment Site from Elekta XVI kV-CBCT Scan Using M Cassette Without Bowtie Filter, 120 kVp, 1.0 mAs/Acquisition, 650 Acquisitions, Full 360 Degree Rotation

Pelvis	
Organ	D50 Range (cGy)
Bladder	1.1–2.5
Rectum	1.3–2.4
Small Bowel	1.1–2.3

Source: Reference [1].

the ratio of mAs used and that list for the image acquisition. The tabulated organ doses can assist the clinician in: (1) determining if the imaging doses are expected to be close to the 5% threshold, (2) choosing a suitable IGRT protocol, and (3) accounting for the organ dose resulting from a specific image acquisition procedure over the course of treatment [1].

It is worth repeating here that it is essential to measure the output of any beam that is used for imaging in order to use this method to estimate the imaging dose to a patient. When the beam output of an imaging device exceeds the manufacturer's specifications, the imaging dose to the patient can be underestimated because the tabulated patient organ doses are based on the specific clinical default acquisition protocols.

30.3 IMAGING DOSE TO PATIENTS

Depending on the imaging modality and acquisition procedure, the dose resulting from single image guidance procedure varies significantly ranging from 0.1 cGy to 15 cGy [1]. In general organ doses from imaging decrease from MV-CBCT, to MVCT, to MV portal images, to kV-CBCT, to kV radiographs [1, 58].

Table 30.9 summarizes the range of imaging dose to soft tissues from different acquisition procedures. The listed range of imaging doses are based on reported data in the literature and are applicable to the default clinical image acquisition protocols set by manufacturers. If user's image acquisition parameters differ from the clinical default settings, the image doses can be obtained by scaling the listed dose with MV monitor unit for MV beams or mAs for kV beams [1]. For room-mounted

TABLE 30.9 Summary of the Range of Imaging Dose to Soft Tissues Resulting from a Single Imaging Procedure by Acquiring 3D or 2D Images for Patient Treatment Positioning [1]

Imaging Dose to Soft Tissues		
Acquisition Procedure	Low (cGy)	High (cGy)
MV-CBCT (Linac-based)	1.0	16.0
MVCT Lungs (TOMO unit)	0.8	2.5
A pair of portal images (6 MV)	2.0	5.0
A pair of portal images (2.5 MV)	0.5	2.0
kV-CBCT (Linac-based) head	0.1	0.5
kV-CBCT (Linac-based) chest	0.3	1.0
kV-CBCT (Linac-based) pelvis	1.0	3.0
A pair of radiographs (kV) head	0.01	0.2
A pair of radiographs (kV) chest	0.01	0.5
A pair of radiographs (kV) pelvis	0.01	2.0

Note: Organs are within the imaged region of interest.

stereoscopic devices equipped with angled x-rays tubes used by the CyberKnife (Accuray) or BrainLab ExacTrac X-ray (Brainlab) system, the dose from a pair of kV radiographs are generally less than that from linac-based kV systems due to extended distances between x-ray source and the patient.

30.4 SUMMARY

In this chapter, the imaging guidance doses to radiotherapy patients from commonly used imaging modalities and their dosimetry were briefly reviewed. Techniques and methods for imaging dose determinations were briefly described. Tabulated organ doses from commonly used image acquisition procedures were provided for a quick reference. The tabulated data can be used to estimate if the accumulated imaging dose from multiple imaging procedures exceeds the threshold of 5% of the therapeutic target dose beyond which imaging dose should be considered in the treatment planning process as recommended in AAPM TG-180 [1]. The provided data can be used by clinicians to make informed decisions in selecting an appropriate imaging procedure regarding the risk and benefits of the x-ray image guidance.

The goal of imaging guidance is to position the patient accurately so that the radiotherapy beams can be delivered to the target precisely. A variety of techniques to reduce the imaging dose are available. Although ALARA should always be the guiding principle in practice, balancing ALARA principle with the requirement for effective target localization requires that imaging dose be managed on the consideration of weighing risks and benefits to the patient [1].

REFERENCES

1. G. X. Ding, P. Alaei, B. Curran, R. Flynn, M. Gossman, T. R. Mackie, M. Miften, R. Morin, X. G. Xu, T. C. Zhu, "Image guidance doses delivered during radiotherapy: quantification, management, and reduction: Report of the AAPM Therapy Physics Committee Task Group 180," *Medical Physics* **45**(5), e84–e99 (2018).
2. M. J. Murphy, J. M. Balter, J. A. BenComo, I. Das, S. Jiang, C. M. Ma, G. Olivera, R. F. Rodebaugh, K. Ruchala, H. Shirato, F. F. Yin, "The management of imaging dose during image-guided radiotherapy: Report of the AAPM Task Group 75," *Medical Physics* **34**(10), 4041–4063 (2007).
3. D. M. Duggan, G. X. Ding, C. W. Coffey, 2nd, W. Kirby, D. E. Hallahan, A. Malcolm, B. Lu, "Deep-inspiration breath-hold kilovoltage cone-beam CT for setup of stereotactic body radiation therapy for lung tumors: Initial experience," *Lung Cancer* **56**(1), 77–88 (2007).
4. D. A. Jaffray and J. H. Siewerdsen, "Cone-beam computed tomography with a flat-panel imager: Initial performance characterization," *Medical Physics* **27**(6), 1311–1323 (2000).
5. D. Letourneau, J. W. Wong, M. Oldham, M. Gulam, L. Watt, D. A. Jaffray, J. H. Siewerdsen, A. A. Martinez, "Cone-beam-CT guided radiation therapy: Technical implementation," *Radiotherapy & Oncology* **75**(3), 279–286 (2005).
6. M. Oldham, D. Letourneau, L. Watt, G. Hugo, D. Yan, D. Lockman, L. H. Kim, P. Y. Chen, A. Martinez, J. W. Wong, "Cone-beam-CT guided radiation therapy: A model for on-line application," *Radiotherapy & Oncology* **75**(3), 271–278 (2005).
7. C. Thilmann, S. Nill, T. Tucking, A. Hoss, B. Hesse, L. Dietrich, R. Bendl, B. Rhein, P. Haring, C. Thieke, U. Oelfke, J. Debus, P. Huber, "Correction of patient positioning errors based on in-line cone beam CTs: Clinical implementation and first experiences," *Radiation Oncology* **1**, 16 (2006).
8. ICRU-91, "ICRU REPORT 91: Prescribing, recording, and reporting of stereotactic treatments with small photon beams," *Journal of the ICRU* **14**(2) (2014).
9. P. Alaei and G. X. Ding (eds.), "Image guidance in radiation therapy: Techniques, accuracy, and limitations," in *AAPM Medical Physics Monograph No 39* (Medical Physics Publishing, Madison, WI, 2018), Vol. **39**, ISBN 978-1936366-62-0 (Hardcopy), ISBN 978-1936366-63-7 (eBook).
10. V. Dyk (ed.), *The modern technology of radiation oncology: A compendium for medical physicists and radiation oncologists* (Medical Physics Publishing Corporation, Madison, WI, 2005), Vol. **2**, ISBN-13 978-1930524255, ISBN-10 1930524250.
11. A. L. Boyer, L. Antonuk, A. Fenster, M. Van Herk, H. Meertens, P. Munro, L. E. Reinstein, J. Wong, "A review of electronic portal imaging devices (EPIDs)," *Medical Physics* **19**(1), 1–16 (1992). https://doi.org/10.1118/1.596878
12. M. Strandqvist and B. Rosengren, "Television-controlled pendulum therapy," *The British Journal of Radiology* **31**(369), 513–514 (1958). https://doi.org/10.1259/0007-1285-31-369-513
13. H. Wallman and N. Stålberg, "A television-Röntgen system for pendulum therapy," *The British Journal of Radiology* **31**(370), 576–577 (1958). https://doi.org/10.1259/0007-1285-31-370-576
14. J. R. Andrews, R. W. Swain, P. Rubin, "Continuous visual monitoring of 2 mev. roentgen therapy," *The American Journal of Roentgenology, Radium Therapy, and Nuclear Medicine* **79**(1), 74–78 (1958).
15. S. Benner, B. Rosengren, H. Wallman, O. Netteland, "Television monitoring of a 30 MV x-ray beam," *Journal of Scientific Instruments* **7**(1), 29–34 (1962). https://doi.org/10.1088/0031-9155/7/1/302

16. N. A. Baily, R. A. Horn, T. D. Kampp, "Fluoroscopic visualization of megavoltage therapeutic x ray beams," *International Journal of Radiation Oncology, Biology, Physics* **6**(7), 935–939 (1980). https://doi.org/10.1016/0360-3016(80)90341-7

17. H. Meertens, M. Van Herk, J. Weeda, "A liquid ionisation detector for digital radiography of therapeutic megavoltage photon beams," *Plasma Sources Science and Technology* **30**(4), 313–321 (1985). https://doi.org/10.1088/0031-9155/30/4/004

18. J. Leong, "Use of digital fluoroscopy as an online verification device in radiation therapy," *Physics in Medicine and Biology* **31**(9), 985–992 (1986). https://doi.org/10.1088/0031-9155/31/9/004

19. M. van Herk and H. Meertens, "A matrix ionisation chamber imaging device for on-line patient setup verification during radiotherapy," *Radiotherapy and Oncology* **11**(4), 369–378 (1988). https://doi.org/10.1016/0167-8140(88)90208-3

20. A. Ezz, P. Munro, A. T. Porter, J. Battista, D. A. Jaffray, A. Fenster, S. Osborne, "Daily monitoring and correction of radiation field placement using a video-based portal imaging system: A pilot study," *International Journal of Radiation Oncology, Biology, Physics* **22**(1), 159–165 (1992). https://doi.org/10.1016/0360-3016(92)90995-T

21. A. L. Boyer, and H. Meertens, "A review of electronic portal imaging devices (EPIDs)," *Medical Physics* **19**(1), 1–16 (1992). https://doi.org/10.1118/1.596878

22. P. Munro, "Portal imaging technology: past, present, and future," *Seminars in Radiation Oncology* **5**(2), 115–133 (1995). https://doi.org/10.1016/S1053-4296(95)80005-0

23. L. E. Antonuk, "Electronic portal imaging devices: A review and historical perspective of contemporary technologies and research," *Physics in Medicine and Biology* **47**(6), R31–R65 (2002). https://doi.org/10.1088/0031-9155/47/6/201

24. M. Murphy, M. Balter, J. BenComo, I. Das, S. Jiang, C. M. Ma, G. Olivera, R. Rodebaugh, K. Ruchala, H. F. Yin, "The management of imaging dose during image-guided radiotherapy: Report of the AAPM Task Group 75," *Medical Physics* **34**, 4041–4063 (2007).

25. E. C. Ford, J. Chang, K. Mueller, K. Sidhu, D. Todor, G. Mageras, E. Yorke, C. C. Ling, H. Amols, "Cone-beam CT with megavoltage beams and an amorphous silicon electronic portal imaging device: Potential for verification of radiotherapy of lung cancer," *Medical Physics* **29**(12), 2913–2924 (2002). https://doi.org/10.1118/1.1517614

26. T. R. Mackie, J. Balog, K. Ruchala, D. Shepard, S. Aldridge, E. Fitchard, P. Reckwerdt, G. Olivera, T. McNutt, M. Mehta, "Tomotherapy," *Seminars in Radiation Oncology* **9**(1), 108–117 (1999).

27. T. R. Mackie, T. Holmes, S. Swerdloff, P. Reckwerdt, J. O. Deasy, J. Yang, B. Paliwal, T. Kinsella, "Tomotherapy: A new concept for the delivery of dynamic conformal radiotherapy," *Medical Physics* **20**(6), 1709–1719 (1993).

28. T. R. Mackie, J. Kapatoes, K. Ruchala, W. Lu, C. Wu, G. Olivera, L. Forrest, W. Tome, J. Welsh, R. Jeraj, P. Harari, P. Reckwerdt, B. Paliwal, M. Ritter, H. Keller, J. Fowler, M. Mehta, "Image guidance for precise conformal radiotherapy," *International Journal of Radiation Oncology, Biology, Physics* **56**(1), 89–105 (2003).

29. D. A. Jaffray, D. G. Drake, M. Moreau, A. A. Martinez, J. W. Wong, "A radiographic and tomographic imaging system integrated into a medical linear accelerator for localization of bone and soft-tissue targets," *International Journal of Radiation Oncology, Biology, Physics* **45**(3), 773–789 (1999).

30. A. S. Shiu, E. L. Chang, J. S. Ye, M. Lii, L. D. Rhines, E. Mendel, J. Weinberg, S. Singh, M. H. Maor, R. Mohan, J. D. Cox, "Near simultaneous computed tomography image-guided stereotactic spinal radiotherapy: An emerging paradigm for achieving true stereotaxy," *International Journal of Radiation Oncology, Biology, Physics* **57**(3), 605–613 (2003).

31. M. Uematsu, T. Fukui, A. Shioda, H. Tokumitsu, K. Takai, T. Kojima, Y. Asai, S. Kusano, "A dual computed tomography linear accelerator unit for stereotactic radiation therapy: A new approach without cranially fixated stereotactic frames," *International Journal of Radiation Oncology, Biology, Physics* **35**(3), 587–592 (1996).

32. K. M. Yenice, D. M. Lovelock, M. A. Hunt, W. R. Lutz, N. Fournier-Bidoz, C. H. Hua, J. Yamada, M. Bilsky, H. Lee, K. Pfaff, S. V. Spirou, H. I. Amols, "CT image-guided intensity-modulated therapy for paraspinal tumors using stereotactic immobilization," *International Journal of Radiation Oncology, Biology, Physics* **55**(3), 583–593 (2003).

33. J. Pouliot, A. Bani-Hashemi, J. Chen, M. Svatos, F. Ghelmansarai, M. Mitschke, M. Aubin, P. Xia, O. Morin, K. Bucci, M. Roach, 3rd, P. Hernandez, Z. Zheng, D. Hristov, L. Verhey, "Low-dose megavoltage cone-beam CT for radiation therapy," *International Journal of Radiation Oncology, Biology, Physics* **61**(2), 552–560 (2005).

34. J. R. Perks, J. Lehmann, A. M. Chen, C. C. Yang, R. L. Stern, J. A. Purdy, "Comparison of peripheral dose from image-guided radiation therapy (IGRT) using kV cone beam CT to intensity-modulated radiation therapy (IMRT)," *Radiotherapy & Oncology* **89**(3), 304–310 (2008).

35. M. Stock, A. Palm, A. Altendorfer, E. Steiner, D. Georg, "IGRT induced dose burden for a variety of imaging protocols at two different anatomical sites," *Radiotherapy & Oncology* **102**(3), 355–363 (2011). https://doi.org/10.1016/8140(11)00607-4 [pii] 10.1016/j.radonc.2011.10.005

36. D. E. Hyer and D. E. Hintenlang, "Estimation of organ doses from kilovoltage cone-beam CT imaging used during radiotherapy patient position verification," *Medical Physics* **37**(9), 4620–4626 (2010).

37. A. Palm, E. Nilsson, L. Herrnsdorf, "Absorbed dose and dose rate using the Varian OBI 1.3 and 1.4 CBCT system," *Journal of Applied Clinical Medical Physics* **11**(1), 3085 (2010).

38. S. Kim, S. Yoo, F. F. Yin, E. Samei, T. Yoshizumi, "Kilovoltage cone-beam CT: Comparative dose and

image quality evaluations in partial and full-angle scan protocols," *Medical Physics* **37**(7), 3648–3659 (2010).

39. M. W. Kan, L. H. Leung, W. Wong, N. Lam, "Radiation dose from cone beam computed tomography for image-guided radiation therapy," *International Journal of Radiation Oncology, Biology, Physics* **70**(1), 272–279 (2008). https://doi.org/S0360-3016(07)04081-3 [pii] 10.1016/j.ijrobp.2007.08.062

40. A. Shah, E. Aird, J. Shekhdar, "Contribution to normal tissue dose from concomitant radiation for two common kV-CBCT systems and one MVCT system used in radiotherapy," *Radiotherapy & Oncology* **105**(1), 139–144 (2012). https://doi.org/S0167-8140(12)00221-6 [pii] 10.1016/j.radonc.2012.04.017

41. N. Wen, H. Guan, R. Hammoud, D. Pradhan, T. Nurushev, S. Li, B. Movsas, "Dose delivered from Varian's CBCT to patients receiving IMRT for prostate cancer," *Physics in Medicine and Biology* **52**(8), 2267–2276 (2007).

42. C. Walter, J. Boda-Heggemann, H. Wertz, I. Loeb, A. Rahn, F. Lohr, F. Wenz, "Phantom and in-vivo measurements of dose exposure by image-guided radiotherapy (IGRT): MV portal images vs. kV portal images vs. cone-beam CT," *Radiotherapy & Oncology* **85**(3), 418–423 (2007). https://doi.org/S0167-8140(07)00527-0 [pii] 10.1016/j.radonc.2007.10.014

43. S. C. Jeng, C. L. Tsai, W. T. Chan, C. J. Tung, J. K. Wu, J. C. Cheng, "Mathematical estimation and in vivo dose measurement for cone-beam computed tomography on prostate cancer patients," *Radiotherapy & Oncology* **92**(1), 57–61 (2009).

44. G. X. Ding and A. W. Malcolm, "An optically stimulated luminescence dosimeter for measuring patient exposure from imaging guidance procedures," *Physics in Medicine and Biology* **58**(17), 5885–5897 (2013). https://doi.org/10.1088/0031-9155/58/17/5885

45. R. M. Al-Senan and M. R. Hatab, "Characteristics of an OSLD in the diagnostic energy range," *Medical Physics* **38**(7), 4396–4405 (2011).

46. V. Schembri and B. J. Heijmen, "Optically stimulated luminescence (OSL) of carbon-doped aluminum oxide (Al_2O_3:C) for film dosimetry in radiotherapy," *Medical Physics* **34**(6), 2113–2118 (2007).

47. N. Tomic, S. Devic, F. DeBlois, J. Seuntjens, "Comment on "Reference radiochromic film dosimetry in kilovoltage photon beams during CBCT image acquisition" [Medical Physics 37, 1083–1092 (2010)]," *Medical Physics* **37**(6), 3008 (2010).

48. J. C. Chow, M. K. Leung, M. K. Islam, B. D. Norrlinger, D. A. Jaffray, "Evaluation of the effect of patient dose from cone beam computed tomography on prostate IMRT using Monte Carlo simulation," *Medical Physics* **35**(1), 52–60 (2008).

49. G. X. Ding, D. M. Duggan, C. W. Coffey, "Accurate patient dosimetry of kilovoltage cone-beam CT in radiation therapy," *Medical Physics* **35**(3), 1135–1144 (2008).

50. G. X. Ding and C. W. Coffey, "Radiation dose from kilovoltage cone beam computed tomography in an image-guided radiotherapy procedure," *International Journal of Radiation Oncology, Biology, Physics* **73**(2), 610–617 (2009).

51. G. X. Ding, P. Munro, J. Pawlowski, A. Malcolm, C. W. Coffey, "Reducing radiation exposure to patients from kV-CBCT imaging," *Radiotherapy & Oncology* **97**(3), 585–592 (2010). https://doi.org/S0167-8140(10)00468-8 [pii] 10.1016/j.radonc.2010.08.005

52. Y. Zhang, Y. Yan, R. Nath, S. Bao, J. Deng, "Personalized assessment of kV cone beam computed tomography doses in image-guided radiotherapy of pediatric cancer patients," *International Journal of Radiation Oncology, Biology, Physics* **83**(5), 1649–1654 (2012). https://doi.org/S0360-3016(11)03480-8 [pii] 10.1016/j.ijrobp.2011.10.072

53. Y. Zhang, Y. Yan, R. Nath, S. Bao, J. Deng, "Personalized estimation of dose to red bone marrow and the associated leukaemia risk attributable to pelvic kilo-voltage cone beam computed tomography scans in image-guided radiotherapy," *Physics in Medicine and Biology* **57**(14), 4599–4612 (2012). https://doi.org/10.1088/0031-9155/57/14/4599

54. B. R. Walters, G. X. Ding, R. Kramer, I. Kawrakow, "Skeletal dosimetry in cone beam computed tomography," *Medical Physics* **36**(7), 2915–2922 (2009).

55. E. Spezi, P. Downes, E. Radu, R. Jarvis, "Monte Carlo simulation of an x-ray volume imaging cone beam CT unit," *Medical Physics* **36**(1), 127–136 (2009).

56. P. Downes, R. Jarvis, E. Radu, I. Kawrakow, E. Spezi, "Monte Carlo simulation and patient dosimetry for a kilovoltage cone-beam CT unit," *Medical Physics* **36**(9), 4156–4167 (2009).

57. E. Spezi, P. Downes, R. Jarvis, E. Radu, J. Staffurth, "Patient-specific three-dimensional concomitant dose from cone beam computed tomography exposure in image-guided radiotherapy," *International Journal of Radiation Oncology, Biology, Physics* **83**(1), 419–426 (2011). https://doi.org/S0360-3016(11)02860-4 [pii] 10.1016/j.ijrobp.2011.06.1972

58. G. X. Ding, and P. Munro, "Radiation exposure to patients from image guidance procedures and techniques to reduce the imaging dose," *Radiotherapy & Oncology* **108**(1), 91–98 (2013). https://doi.org/10.1016/j.radonc.2013.05.034

59. G. Ding and P. Munro, "The imaging dose to patients from a 2.5 MV imaging beam." *International Journal of Radiation Oncology Biology Physics* **99**(2), E654 (2017). https://www.redjournal.org/article/S0360-3016(17)33233-9/pdf.

60. P. R. Almond, P. J. Biggs, B. M. Coursey, W. F. Hanson, M. S. Huq, R. Nath, D. W.O. Rogers, "AAPM's TG-51 protocol for clinical reference dosimetry of high-energy photon and electron beams," *Medical Physics* **26**(9), 1847–1870 (1999).

61. P. Andreo, D. T. Burns, K. Hohlfeld, M. S. Huq, T. Kanai, F. Laitano, V.G. Smyth, S. Vynckier, "Absorbed dose determination in external beam radiotherapy: An international code of practice for dosimetry based on standards for absorbed dose to water," in *Technical Report Series, IAEA* (International Atomic Energy Agency, Vienna, 2000), Vol. 398.

62. G. X. Ding and P. Munro, "Characteristics of 2.5MV beam and imaging dose to patients," *Radiotherapy & Oncology* **125**(3), 541–547 (2017). https://doi.org/10.1016/j.radonc.2017.09.023

63. C. M. Ma, C. W. Coffey, L. A. DeWerd, C. Liu, R. Nath, S. M. Seltzer, J. P. Seuntjens, "AAPM protocol for 40-300 kV x-ray beam dosimetry in radiotherapy and radiobiology," *Medical Physics* **28**(6), 868–893 (2001).

64. G. X. Ding and C. W. Coffey, "Beam characteristics and radiation output of a kilovoltage cone-beam CT," *Physics in Medicine and Biology* **55**(17), 5231–5248 (2010). https://doi.org/S0031-9155(10)51181-3 [pii] 10.1088/0031-9155/55/17/022

65. M. Miften, O. Gayou, B. Reitz, R Fuhrer, "IMRT planning and delivery incorporating daily dose from megavoltage cone-beam computed tomography imaging," *Medical Physics* **34**(10), 3760–3767 (2007).

66. J. M. Pawlowski and G. X. Ding, "An algorithm for kilovoltage x-ray dose calculations with applications in kV-CBCT scans and 2D planar projected radiographs," *Physics in Medicine and Biology* **59**(8), 2041–2058 (2014). https://doi.org/10.1088/0031-9155/59/8/2041

67. P. Alaei, G. X. Ding, H. Guan, "Inclusion of the dose from kilovoltage cone beam CT in the radiation therapy treatment plans," *Medical Physics* **37**(1), 244–248 (2010).

68. F. Verhaegen and J. Seuntjens, "Monte Carlo modelling of external radiotherapy photon beams," *Physics in Medicine and Biology* **48**(21), R107–R164 (2003).

69. D. W. Rogers, "Fifty years of Monte Carlo simulations for medical physics," *Physics in Medicine and Biology* **51**(13), R287–R301 (2006).

70. J. Seco and F. Verhaegen (eds.), *Monte Carlo Techniques in Radiation Therapy* (CRC Press, 2016), ISBN 9781138199903.

71. E. Spezi, P. Alaei, G. X. Ding, "Calculations of the imaging dose with emphasis on kilovoltage imaging," in *Image Guidance in Radiation Therapy: Techniques, Accuracy, and Limitations, AAPM Monogram No.39*, edited by P. Alaei and G. X. Ding (Medical Physics Publishing, Madison, WI, 2018), ISBN 978-1936366-62-0 (Hardcopy) ISBN 978-1936366-63-7 (eBook).

72. A. P. Nelson and G. X. Ding, "An alternative approach to account for patient organ doses from imaging guidance procedures," *Radiotherapy & Oncology* **112**(1), 112–118 (2014). https://doi.org/S0167-8140(14)00265-5 [pii] 10.1016/j.radonc.2014.05.019

73. P. Alaei, E. Spezi, M. Reynolds, "Dose calculation and treatment plan optimization including imaging dose from kilovoltage cone beam computed tomography," *Acta Oncologica* **53**(6), 839–844 (2014). https://doi.org/10.3109/0284186X.2013.875626

Index

Note: Locators in *italics* represent figures and **bold** indicate tables in the text.

Printed and bound by CPI Group (UK) Ltd, Croydon, CR0 4YY

24/10/2024

01778292-0020